AI-Enabled Electronic Circuit and System Design

Ali Iranmanesh • Hossein Sayadi
Editors

AI-Enabled Electronic Circuit and System Design

From Ideation to Utilization

Springer

SVPTI

Editors
Ali Iranmanesh
Electrical Engineering and Computer Science
Silicon Valley Polytechnic Institute
Los Altos Hills, CA, USA

Hossein Sayadi
Computer Engineering and Computer Science Department
California State University, Long Beach (CSULB)
Long Beach, CA, USA

ISBN 978-3-031-71435-1 ISBN 978-3-031-71436-8 (eBook)
https://doi.org/10.1007/978-3-031-71436-8

This Springer imprint is published by the registered company Springer Nature Switzerland AG
The registered company address is: Gewerbestrasse 11, 6330 Cham, Switzerland

Preface

Artificial intelligence (AI) has increasingly become a vital tool in the design and development of electronic circuits and systems. This emergence is driven by the need for more efficient, accurate, and innovative design processes to keep pace with the growing complexity and performance demands of modern electronic devices.

The importance of AI in electronic design is multifaceted, offering significant benefits across various aspects of the design process. AI significantly reduces the time required for the design cycle of electronic circuits and systems by automating routine and complex tasks, enabling designers to achieve faster turnarounds, and accelerating the time-to-market for new products. AI algorithms excel in optimization problems, allowing for the creation of designs that maximize performance while minimizing cost and power consumption, which is crucial in fields like consumer electronics, automotive, and telecommunications. As electronic systems become more complex, AI provides the tools necessary to manage and leverage this complexity, facilitating innovative design approaches that might be impractical or impossible with conventional methods and leading to the development of cutting-edge technologies. AI also enhances the accuracy and reliability of electronic designs by predicting potential issues and optimizing design parameters, reducing the likelihood of errors that can lead to costly rework or failures in the field. Its ability to learn and adapt makes AI highly scalable, capable of handling a wide range of design challenges from simple circuits to highly integrated systems-on-chip (SoCs), making it a versatile tool in the designer's arsenal. Furthermore, by streamlining the design process and reducing the need for extensive manual intervention, AI helps lower development costs, benefiting startups and smaller companies that may not have extensive resources. In conclusion, the integration of AI in the design of electronic circuits and systems represents a significant leap forward in the field of electronic engineering. It enhances efficiency and innovation while addressing the growing demands for performance, complexity management, and cost reduction. As AI technology continues to evolve, its role in electronic design is expected to expand, further transforming the landscape of modern electronics.

This book intends to provide a comprehensive exploration of the innovative intersection between artificial intelligence and electronic design. This volume, consisting of 16 chapters authored by experts in the field, highlights the transformative impact of AI on various aspects of electronic circuit and system design, from initial concept to practical application.

Chapter 1 delves into the use of AI in automating circuit design and modeling. It discusses both device-level and logic/circuit-level design considerations, providing an overview of state-of-the-art techniques while examining the challenges and opportunities that AI/ML methodologies bring to circuit modeling.

In Chap. 2, AI technologies and methods are presented to enhance modern electronic design processes. A knowledge-based method for design capture at a high level of abstraction during the early design phase is introduced, supported by an ontology-defined knowledge base that facilitates a consistent description of electrical systems and their components.

Chapter 3 introduces an AI-enabled memory design methodology that leverages AI techniques to extract valuable data insights for hardware designers. This methodology, applied to video and deep learning applications, has demonstrated significant power savings and reduced hardware implementation costs compared to the state-of-the-art.

Chapter 4 proposes a method to combine RTL generation with early timing estimations to accelerate the digital design flow. An AI-enabled Static Timing Analysis (STA) flow is introduced, which involves predicting pin-to-pin delays and slews using machine learning models trained on RTL Intermediate Representation (IR).

Chapter 5 explores the use of graph learning for signoff interconnect delay estimation. This chapter details how global message-passing graph representation learning on RC graphs can lead to ultra-fast net delay estimation, improving accuracy and runtime efficiency across various technology nodes.

Chapter 6 discusses the fundamentals of AI-enabled placement algorithms and the application of reinforcement learning to placement optimization problems in 2D and 3D VLSI systems. This chapter outlines a framework for designing and characterizing these algorithms, identifying practical needs and challenges that require innovative solutions.

Chapter 7 provides an in-depth look at AI-powered solutions within the FPGA CAD design flow. It covers high-level synthesis, logic synthesis, placement, routing, and timing optimization, comparing AI-based methods with traditional approaches to highlight the advantages of integrating AI in these critical steps.

Chapter 8 offers a detailed exploration of 3-D packaging techniques and technologies, emphasizing the role of AI in enhancing design efficiency, precision, and reliability throughout the various stages of 3-D integration, from concept generation to testing.

Chapter 9 addresses AI methods in hardware security contexts, including side-channel analysis, reverse engineering of circuits, and analysis of physical unclonable functions (PUFs). It concludes with a summary and outlook on the use of AI in hardware security.

Chapter 10 explores the critical area of zero-day malware detection in AI-enabled cybersecurity. This emerging field is crucial for maintaining the overall security landscape, emphasizing the significance and impact of AI techniques in detecting previously unknown threats.

Chapter 11 focuses on AI attacks on power, electromagnetic (EM), and direct memory access (DMA), excluding architecture-based side-channel attacks. This chapter highlights the potential vulnerabilities and countermeasures in these areas.

Chapter 12 discusses how reinforcement learning can be leveraged to strengthen real-time defenses against sophisticated cyber threats. The convergence of hardware-assisted security and adaptive learning offers a dynamic shield against emerging threats, ushering in a new era of proactive cybersecurity strategies.

Chapter 13 surveys recent approaches to integrating machine learning techniques into conventional hardware design flows with a focus on security objectives. It discusses the benefits, limitations, and potential future research directions in ML-enabled hardware security.

Chapter 14 provides an in-depth exploration of specialized hardware accelerators designed to enhance AI applications. It covers the transition from traditional computing systems to advanced AI-specific hardware, addressing the demands of AI algorithms and the inefficiencies of conventional architectures.

Chapter 15 introduces associative learning in a realistic context using a cognitive-inspired robotic system and neuromorphic hardware. This chapter explores the potential of associative memory for addressing data security challenges in deep learning.

Finally, Chap. 16 discusses various optimization techniques aimed at reducing computational power and memory requirements of deep neural network models during the inference phase. It highlights the Posit number system as a method for minimizing these requirements in neural network implementations.

In summary, this book, with its diverse range of topics and expert contributions, offers readers a comprehensive insight into the revolutionary impact of AI on electronic circuit and system design, spanning from ideation to practical application. We hope you find it both informative and engaging.

Los Altos Hills, CA, USA — Ali Iranmanesh
Long Beach, CA, USA — Hossein Sayadi

Special thanks to following chapter reviewers:

Prof. Kyle Juretus – Drexel University
Dr. Hechen Wang, Research Scientist – Intel Labs
Dr. Zhixing Li – Synopsys
Prof. Shih-Hsu Huang – Chung Yuan Christian University
Prof. Rakesh Mahto – Cal State Fullerton
Dr. Suyash Ranjan – Qualcomm
Prof. Abdulrahman Mahmoud – Harvard University
Dr. Abishai Daniel – Micron
Prof. Zhong Guan – UC Santa Barbara
Dr. Ankur Guha Roy – Broadcom
Prof. Richard Yarnell – University of Central Florida
Dr. Vaibhav Verma – Qualcomm
Dr. Aathreya S Bhat – NVIDIA Corporation
Prof. Mehran Mozaffari Kermani – University of South Florida
Dr. Avani Dave – Intel Corp.
Prof. Xinfei Guo. Shanghai – Jiao Tong University
Dr. Gopal Raut – CDAC Bangalore
Dr. Anand Iyer – Microsoft
Prof. Yuanqing Cheng – Beihang University
Prof. Na Gong – University of South Alabama
Dr. Kang Jun Bai – Air Force Research Laboratory
Dr. Arijit Banerjee – Advanced Micro Devices
Prof. Zhangying He – California State University, Long Beach

Contents

1 AI-Assisted Circuit Design and Modeling 1
Deepthi Amuru and Zia Abbas

2 Linking System and Circuit Design by AI Techniques 41
Frank Wawrzik, Johannes Koch, Khushnood Adil Rafique, Sören Kwasigroch, Moritz Herzog, and Christoph Grimm

3 AI-Enabled Efficient Memory Design for Data-Intensive Applications 83
Na Gong, Jinhui Wang, Wei Jin, Hritom Das, and Ali Haidous

4 AI-Enabled Static Timing Analysis at Early Stages of the Digital Design Flow 109
Daniela Sánchez Lopera and Wolfgang Ecker

5 Harnessing Graph Learning for Efficient Timing Signoff 155
Xinfei Guo, Linyu Zhu, and Yichen Cai

6 AI-Enabled Placement for 2D and 3D ICs 189
Abdullah Mansoor, Malgorzata Chrzanowska-Jeske, and Marcin Jeske

7 Enhancing FPGA CAD Flow with AI-Powered Solutions 225
Chunxiao Lin and Yang Yi

8 AI-Enabled 3D Integration 257
Vandana Kumari and Manoj Kumar Majumder

9 AI-Enabled Hardware Security 309
Emanuele Strieder, Johanna Baehr, and Matthias Hiller

10 On AI-Enabled Cybersecurity: Zero-Day Malware Detection 343
Hossein Sayadi and Zhangying He

11 Machine Learning Applications and Attacks Using Side Channel Analysis 387
Chaitanya Bhure, Geraldine Shirley Nicholas, and Fareena Saqib

12 Adapt and Defend: Reinforcement Learning for Hardware-Assisted Security 427
Hossein Sayadi and Zhangying He

13 Machine Learning-Enhanced Analysis and Design for Trustworthy Integrated Circuits 467
Kaixin Yang, Yinghua Hu, Dake Chen, Chunxiao Lin, Yang Yi, Peter A. Beerel, and Pierluigi Nuzzo

14 Hardware Accelerators for Artificial Intelligence 497
S. M. Mojahidul Ahsan, Tamzidul Hoque, Md Sakib Hasan, Mrittika Chowdhury, and Anurag Dhungel

15 Implementation of Associative Learning Using Cognitive-Inspired Robotic System 537
Tianze Liu, Noah Zins, Yan Zhang, and Hongyu An

16 Enabling Memory-Augmented Neural Networks for Efficient Edge Applications 565
Elias Khatami Zenozian, Mehdi Kamal, Ali Afzali-Kusha, and Massoud Pedram

Index 607

Chapter 1
AI-Assisted Circuit Design and Modeling

Deepthi Amuru and Zia Abbas

1.1 Introduction

Over the past 50 years, the Integrated Circuit (IC) industry has consistently evolved to accommodate the needs of portable electronic devices. We have witnessed the development of intricate and sophisticated electronic devices and systems through relentless innovation and the application of advanced IC design and manufacturing techniques. The semiconductor sector continues to push its limits in various directions, with upcoming technologies like 2.5D and 3D-IC on the horizon [1], challenging the scalability and reliability of IC design flow. The pursuit of smaller, smarter, faster, and more energy-efficient electronics remains insatiable, presenting an impending challenge of dealing with a steep increase in IC design complexity over the years. The advent of edge computing is extending its reach across diverse domains such as Internet of Things (IoT), vehicular networks, smart grid applications, distributed robotics, and augmented reality. These applications demand solutions that offer low latency and highly efficient Power, Performance, and Area (PPA) specifications while maintaining sustainable environmental goals.

Electronic Design Automation (EDA) has been immensely successful in managing the exponential growth in IC capabilities from the past 60 years, which now involve billions of transistors. EDA has completely transformed the way electronic engineers approach the design and manufacture of integrated circuits. However, accurately modeling circuit behavior and assessing the performance of advanced designs has become progressively challenging for EDA tools as transistor sizes approach atomic levels. Skilled designers at various levels of abstraction lend their expertise to explore the design landscape, create efficient schematic designs,

D. Amuru (✉) · Z. Abbas
Center for VLSI and Embedded Systems Technology (CVEST), International Institute of Information Technology, Hyderabad, Telangana, India
e-mail: deepthi.amuru@research.iiit.ac.in; zia.abbas@iiit.ac.in

A. Iranmanesh, H. Sayadi (eds.), *AI-Enabled Electronic Circuit and System Design*,
https://doi.org/10.1007/978-3-031-71436-8_1

produce layouts that satisfy multiple PPA constraints, enhance performance, integrate appropriate testing measures, conduct reliability analysis, and incorporate essential feedback mechanisms to ensure precise chip functionality aligned with application specifications. Despite these endeavors, many processes are iterative, labor-intensive, time-consuming, and heavily reliant on manual intervention and do not guarantee a fully optimized solution within market timelines. Furthermore, locating human expertise to address specific problems has become increasingly difficult, directly impacting chip yield. The emergence of technologies like edge computing is ushering in new challenges for EDA. AI (Artificial Intelligence) assistance offers a reliable solution to address these hurdles.

This chapter provides a concise overview of the traditional Application-Specific Integrated Circuit (ASIC) design flow in Sect. 1.2, while Sect. 1.3 presents a brief introduction to artificial intelligence and machine learning techniques. Section 1.4 continues with the exploration of AI-assisted device design and modeling, followed by AI-assisted circuit design in Sect. 1.5. Section 1.6 addresses the challenges and opportunities in this domain. The chapter concludes with Sect. 1.7, followed by future trends discussed in Sect. 1.8.

1.2 Traditional IC Design, Modeling, and Their Challenges

The ASIC design that strives to create specialized solutions for a variety of applications has a complex, repetitive, and time-consuming process flow; its complexity linearly increases with the design size. Full custom ASICs require significant design time and effort; however, they provide the highest level of customization and optimization. Semi-custom ASICs balance customization and design complexity, making them popular for many applications. They require less design effort compared to full-custom ASICs but may not achieve the same level of optimization.

The major steps involved in the front-end design of a typical semi-custom are presented in Fig. 1.1. The process requires an EDA tool set at each step. EDA refers to a suite of algorithms and software tools for creating electronic systems, including ICs and Printed Circuit Boards (PCBs). A brief of ASIC design flow focused on the front-end design steps is presented below:

Design Specifications The initial and crucial phase in IC design is the creation of the chip design specifications. This document serves as the foundational blueprint for the entire design process, providing designers with a clear framework. It ensures that the final IC fulfills the intended functionality and meets various performance criteria, including speed, power, and area constraints. Additionally, the specification covers aspects such as power budget, clocking requirements, reliability, yield, chosen technology, security, IP considerations, project timelines, and milestones. This comprehensive document plays a vital role in managing expectations and promoting effective communication among the diverse stakeholders involved in the design process.

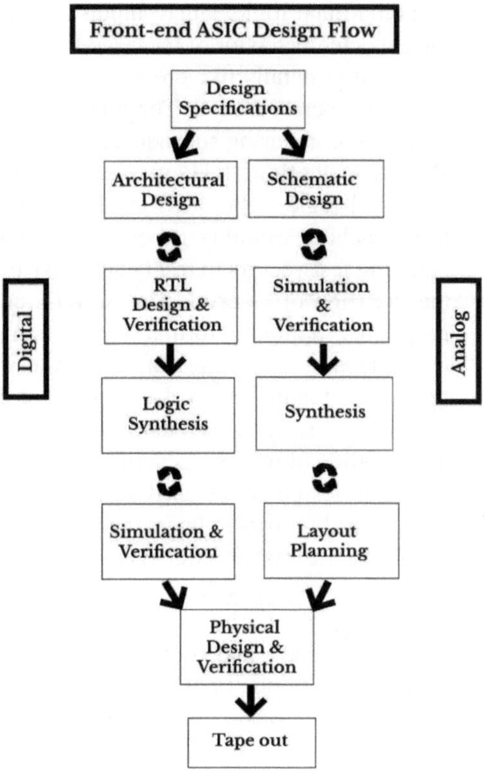

Fig. 1.1 A typical ASIC design flow, focusing on front-end design

Digital Design Flow

- **Architectural Design:** The architectural design phase in digital IC designing involves creating a high-level abstract representation of the IC based on chip specifications. It encompasses the identification of major functional blocks and their interactions, along with the creation of behavioral descriptions using hardware description languages like VHDL or Verilog. Additionally, this phase includes exploring technology options and evaluating various architectural choices to meet specified performance constraints. In brief, the architectural design phase sets the groundwork for the Register-Transfer-Level (RTL) design phase and plays a pivotal role in determining block-level specifications and interactions. A well-defined architecture provides a clear direction for the subsequent stages of the design process, ensuring that the final IC aligns with its intended functionality and performance criteria.
- **RTL Design:** RTL design focuses on describing the digital logic of a circuit with a primary emphasis on data movement and operations between registers. It acts as a bridge between architectural design and the physical implementation of the circuit. Within RTL descriptions, vital elements include clocking signals (clock and reset) and control signals that govern data flow. This design approach encompasses both sequential logic, where state information is preserved, and

combinational logic, where inputs solely determine outputs. RTL code outlines the expected behavior of the circuit rather than specifying its structure. Additionally, timing details like clock periods and setup/hold times are incorporated to ensure proper operation. The RTL design is then synthesized into gates and laid out on silicon during subsequent stages of the design flow.

- **RTL Verification:** RTL verification ensures the correctness and functionality of digital logic at the RTL level. A typical verification environment comprises test benches, stimulus generators, and monitors. RTL designs are tested by applying test vectors to test benches written in VHDL or Verilog. It also involves the insertion of assertions in the RTL code to check for specific conditions or properties, applying formal verification techniques, corner-case analysis, and stress testing to ensure the correctness and robustness of the design in all operating conditions. A key aspect of RTL verification involves performing regression testing in each iteration to confirm the design's functionality following any modifications. RTL verification is crucial for identifying and resolving potential design issues early in the development phase, thereby minimizing the risk of costly errors in later stages.
- **Logic Synthesis:** Logic synthesis entails transforming a high-level design description into a netlist comprising gates and flip-flops tailored for a specific technology. Synthesis tools aim to optimize by reducing gate count, enhancing power efficiency, and boosting overall performance. This process encompasses tasks like mapping, optimization, analysis, and verification, all aimed at guaranteeing that the resultant circuit fulfills specified criteria in performance, power consumption, and area utilization. The synthesized netlist serves as the basis for subsequent stages in physical design and the manufacturing process.
- **Digital Logic Simulation and Verification:** After synthesis, the design undergoes comprehensive verification encompassing logic, timing, and power analysis. Logic verification, employing techniques like simulation and formal verification, ensures the design's continued functionality. Timing analysis is crucial to confirm adherence to specified timing constraints, involving the estimation of signal propagation delays. Additionally, power consumption is a vital concern, with logic synthesis tools estimating usage and potentially implementing optimizations to reduce energy consumption. Simulation and logic verification are pivotal steps that follow synthesis to guarantee the design's robustness and accuracy.

Analog Design Flow

- **Schematic Design:** Schematic design is the fundamental step for designing an analog IC. It involves various tasks such as selecting electronic components, exploring architectural possibilities, determining circuit topology, and setting up transistor biasing to guarantee operation within designated regions while accounting for stability. This process is often iterative, with engineers conducting a sequence of refinements to the design by modifying component values, topologies, and other parameters based on simulation outcomes and design objectives.

- **Analog Circuit Simulation and Verification:** The schematic design undergoes various simulations, including DC analysis (for biasing points), AC analysis (for frequency response), transient analysis (for time-domain behavior), and specific analyses if required. These simulations verify that the circuit adheres to defined performance standards and behaves as intended across diverse operating conditions. Based on simulation outcomes, the schematic design is iteratively refined to guarantee the circuit meets its performance specifications.
- **Physical Design and Verification:** IC physical design involves the detailed layout and interconnection of components within the chip, focusing on enhancing performance and power efficiency. It is a critical backend step in digital and analog IC design processes. In digital IC physical design, key stages encompass floorplanning, placement, routing, clock tree synthesis, and achieving timing closure. On the other hand, in analog IC physical design, the primary focus is on layout design. To ensure compliance with manufacturing and functional requirements, verification steps such as functional, timing, power, and physical verification are executed. They include verification of the physical design against schematic capture, design rules, electrical characteristics, and reliability.
- **Tape Out:** Once the design passes all verification checks, it is prepared for fabrication by generating the final files, which are sent to the semiconductor foundry.

The major design stages in custom IC design flow are presented in Fig. 1.2. During the schematic capture phase, pre-layout circuit simulations analyze the basic operation of analog blocks with initial behavioral models. A detailed circuit design and analysis using a testbench setup and HDL simulations is performed to verify performance specifications. Design modifications are done through corner analysis and worst-case performance verification to ensure the optimized design variables meet performance parameters.

Subsequently, a layout is generated from the schematic. Design reuse or Engineering Change Orders (ECOs) allow for modifications without a complete redesign. Physical verification, including DRC (Design Ruel Check) and LVS

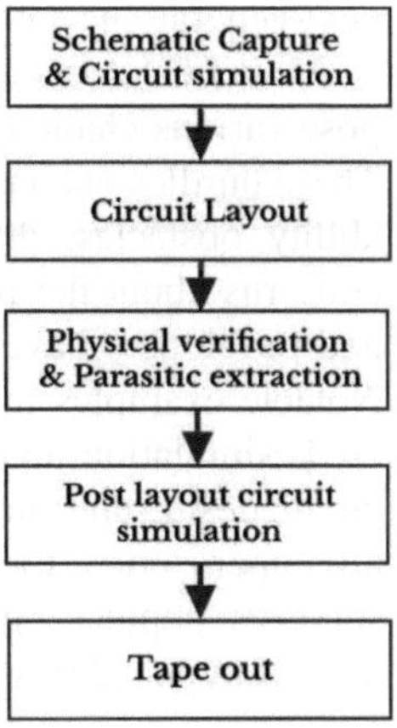

Fig. 1.2 Custom IC design flow (recreated based on [2])

(Layout versus Schematic) checks, ensures compliance with design rules [2]. Parasitic extraction then identifies parasitic resistance and capacitance from metal interconnects, and post-layout simulations evaluate their impact on circuit performance. This netlist is also used for IR drop and electromigration analysis.

Post-layout simulations are compared with pre-layout simulations to ensure the analog block meets performance specifications, accounting for physical effects. Once confirmed, the design is ready for tape-out, where the GDSII file, an industry-standard database format, is generated and sent to the foundry for chip fabrication.

EDA tools are pivotal throughout the custom or semi-custom IC design process, from concept inception to chip fabrication. The effectiveness and achievement of each design stage rely on the proficiency of the EDA tools and the skill of the designers involved. These tools and resources empower designers to create, verify, optimize, and validate their designs. A range of well-regarded EDA tools is available in academic and industrial settings, with notable offerings from leading companies such as Cadence, Synopsys, and Siemens (formerly Mentor Graphics). These encompass Cadence Innovus, Synopsys VCS, Siemens ModelSim, and QuestaSim (Siemens Questa Advanced Simulator) for digital implementations. In high-level synthesis (HLS), platforms like Cadence Genus [3], Siemens Oasys AMS [4], and Catapult HLS [5] empower designers to leverage standard C++ and SystemC to articulate functional intent, thus enabling a shift to a more efficient abstraction level for both ASIC and FPGA design and validation. Xilinx Vivado suite [6] is another popular platform for FPGA-based (Field Programmable Gate Array) digital implementations, from RTL design to Physical implementation.

For analog mixed-signal (AMS) and RF (radio frequency) designs, tools such as Cadence Virtuoso, Spectre [7], Synopsys Design Compiler, IC Compiler [8], and Siemens Tanner serve from design entry through simulation, layout, physical verification, and sign-off. Specific tools like Synopsys' PrimeTime [9], PrimePower, StarRC, Silvaco's SmartSpice [10], and Aldec's Riviera Pro [11] offer essential functionalities for processes like timing analysis, power evaluation, and other critical analyses. Furthermore, Silvaco TCAD and Victory are a widely favored platform for process/device-level simulations, instrumental in developing and advancing next-generation semiconductor devices [12].

Most EDA tools are bound by licenses and tied to specific platforms, which can pose various challenges for designers and organizations involved in ASIC design. These hurdles encompass limited accessibility, potential compatibility issues, scalability obstacles, difficulties in securing vendor support, and, in some cases, concerns about the potential obsolescence of the tools. Consequently, researchers and developers have contributed to the development of open-source alternatives. Notable examples include Icarus Verilog [13], GHDL [14], and MyHDL [15] for HDL simulation and synthesis and NGSPICE [16], and eSim [17] for electronic circuit design and simulation, supporting both digital and analog domains. Additionally, open-source tools like Qflow [18], Graywolf [19], VTR, and OpenRoad [20] serve as popular options for synthesis and placement tasks, while Magic [21] and KLayout [22] are relied upon as open-source layout viewer and editor tools.

Although these open-source solutions offer robust capabilities for ASIC design, it is important to note that they may have varying maturity levels, documentation, and support compared to their commercial counterparts.

The relentless advancement and refinement of EDA tools are a steadfast response to the escalating complexity of modern chip designs. This trajectory extends into cutting-edge domains like edge computing, where the demand for highly specialized, compact, and efficient integrated circuits is rapidly growing. Yet, within the current EDA landscape, many processes at each stage entail substantial manual intervention. These routines are resource-intensive, requiring significant time and effort, and often fall short in leveraging insights and knowledge gleaned from prior implementations. This inefficiency underscores a critical need for more streamlined and intelligent methodologies in IC circuit design.

Integrating AI and Machine Learning (ML) into the EDA framework emerges as a transformative opportunity. By infusing AI/ML techniques into the IC circuit design process, designers can unlock the potential for unparalleled levels of optimization and automation. This fusion holds promise for revolutionizing design space exploration, enabling the identification of optimal configurations and solutions with unprecedented efficiency. This not only facilitates the identification of optimal design solutions but also encourages innovation and the pursuit of novel design paradigms. Furthermore, by automating repetitive and labor-intensive tasks, AI/ML can significantly expedite the design process, freeing up valuable time and resources for more strategic and creative endeavors.

Machine learning applied to circuit design has not been as extensively investigated as its use in logic synthesis and physical design, leaving ample opportunity for further exploration and innovation in this area. This chapter delves into the application of AI in circuit design, examining the methods employed in the literature, potential uses, challenges, and future trends.

1.3 Brief on Artificial Intelligence and Machine Learning Techniques

Artificial intelligence involves creating computing systems that can solve problems, learn from experience, and make decisions like humans. ML and Deep Learning (DL) are key parts of AI. ML enables machines to autonomously learn from past data on their own, without being directly programmed. DL, a specialized area of ML, improves these learning abilities even more.

ML encompasses the processes of learning and self-correction in response to the introduction of new data. It demonstrates adeptness in handling structured and semi-structured data. On the other hand, AI is capable of managing structured, semi-structured, and unstructured data. AI/ML algorithms can be categorized into three principal types: supervised, unsupervised, and reinforcement learning (RL).

1.3.1 Supervised Learning

Supervised learning, a fundamental approach in ML, involves training an algorithm to recognize patterns in a labeled dataset, where the correct examples are provided, and then using that learned knowledge to predict labels for new, unseen data. Supervised learning encompasses both regression and classification techniques. Regression is a statistical approach that focuses on understanding the relationship between continuous dependent variables and independent variables [23, 24]. It enables the creation of predictive models that estimate outcomes using new input data. On the other hand, classification within supervised learning involves the analysis of data to derive models that describe significant data classes [25].

Classification and regression involve learning function(s) $Y = f(X)$ that aim to predict and map independent variables Y to dependent variables X. This learning function predicts the corresponding class label y for a given input(s) X. Supervised learning finds its application in image recognition, medical diagnostics, weather forecasting, and so on. It also proves valuable in VLSI circuit analysis and modeling. The developed predictive models map the I/O relationship of a circuit and act as circuit prototypes running several orders faster than the traditional simulators. For example, linear and polynomial regressions were employed in [26] to build predictive models that estimate leakage power and total power of standard digital cells against process variations in 16-nm CMOS technology. The models exhibited a very close match with SPICE simulations (reported an error $\leq 1\%$). A similar approach was presented with a gradient boosting algorithm in [27] for the estimation of leakage power and propagation delay in digital cells, reporting a speedup of 10,000× against the SPICE environment. A technology-independent circuit sizing for fundamental analog circuits in 0.18-μm technology using artificial neural networks (ANNs) was proposed in [28]. It reported 90% accuracy against traditional modeling.

Supervised learning relies heavily on having a large amount of accurately labeled training data, which can be particularly challenging in specialized fields such as VLSI due to factors like cost, resource constraints, and privacy concerns. However, strategic choices of algorithms and techniques for transferring knowledge from trained models to other relevant models can alleviate the demand for extensive training data.

1.3.2 Unsupervised Learning

Unsupervised learning utilizes unlabeled datasets to discover hidden patterns. In this paradigm, algorithms autonomously discern relationships and groupings without the presence of explicit predefined output labels. Clustering, dimensionality reduction, and associated mining are the techniques in unsupervised learning for revealing intrinsic structures and dependencies, which include exploratory data analysis and anomaly detection [29]. Such an analysis proves pivotal in the applications where generation of labeled datasets poses a challenge, such as the VLSI domain.

Examples of unsupervised learning in VLSI design encompass hotspot detection through feature extraction and pattern matching [30] and feature extraction of circuit leakage and switching power from the circuit topology [31]. In [31], feature extraction is achieved by encoding the circuit's state for leakage power estimation and by encoding circuit transitions for switching energy estimation.

1.3.3 Reinforcement Learning

Reinforcement learning is an AI computational approach, where an agent learns to sequence decisions through interactions with an environment. Unlike supervised learning, RL operates in scenarios where explicit guidance is limited, relying on a system of rewards and penalties to steer the learning process. The agent explores its surroundings, performs actions, and receives feedback in the form of rewards or penalties, learning through trial and error or within an optimization framework. It adapts its strategy to maximize cumulative rewards over time [32]. Advanced RL algorithms, including deep RL, harness neural networks to manage intricate and high-dimensional state spaces, enabling machines to address tasks ranging from game playing to robotics [33]. AutoCkt is a deep reinforcement learning framework designed for analog circuit design [34]. It is trained on a sparse subsample of the design space, which helps to reduce convergence time when deploying to meet new design specifications. GCN-RL Circuit Designer [35] leverages graph convolutional neural networks and RL to transfer knowledge across various technology nodes and circuit topologies. Placement optimization [36] and iEDA [37], an open-source physical design toolkit, are other examples of reinforcement learning.

The selection of the learning paradigm for circuit modeling depends on the specific task and the nature of available data. However, among various approaches, supervised learning has been dominant, particularly in tasks such as circuit characterization and performance predictions.

1.4 AI-Assisted Device Design and Modeling

Moore's law has been the driving force propelling the swift evolution of IC technology for over 50 years. To counteract the deceleration of Moore's law, the industry is venturing into the realm of novel materials like FinFETs, tunnel FETs (TFETs), nanowire FETs (NWFETs) [38], nanosheet FETs (NSFETs) [39], negative capacitance FETs [40], and other breakthrough technologies such as heterogeneous, 3D integration. These innovative transistors could outperform conventional silicon transistors in both speed and energy efficiency. As device dimensions shrink to the nanoscale, the intricacy of interacting physical mechanisms and the influence of random dopant fluctuations (RDF) and statistical effects substantially rise, making the application of conventional deterministic models challenging. Device variability emerges as a significant hurdle in advanced nanometer technologies, denoting the scenario where devices of the same type exhibit diverse electrical characteristics due to random fluctuations in the manufacturing process. This variability complicates the design of circuits that can withstand these deviations. Therefore, it is imperative to conduct thorough transistor modeling, IC simulations, and performance evaluations before actual device fabrication in an effort to save time and cost. A fast and efficient compact device model facilitates effective design and modeling of transistors.

1.4.1 Device Characterization and Modeling

In the ever-evolving landscape of semiconductor technology, the creation and analysis of cutting-edge transistor devices, such as NSFETs, demand a meticulous examination of their I-V (current-voltage) and C-V (capacitance-voltage) characteristics before the fabrication of IC. This scrutiny is imperative to ensure the resulting semiconductor devices' reliability, performance, and operational effectiveness. The I-V characteristics are crucial benchmarks, offering insights into the transistor's behavior under varying operational conditions and voltages, revealing details about threshold voltage (V_{th}), saturation, and off-state leakage (I_{off}). Simultaneously, C-V characteristics provide essential information about the transistor's capacitance and charge storage capabilities, influencing signal propagation and speed. A thorough analysis of these characteristics is pivotal in refining transistor design, allowing engineers to tailor device parameters to meet specific performance criteria. By scrutinizing I-V and C-V behaviors, potential issues such as excessive power consumption, leakage currents, or nonideal switching behavior can be identified and addressed during the design phase, mitigating the risk of costly post-manufacturing modifications. Furthermore, as nanoscale devices introduce complex interplay of quantum effects and material properties, studying I-V and C-V characteristics becomes increasingly vital. This proactive approach is indispensable for achieving the desired functionality, reliability, and performance in semiconductor devices, ultimately contributing to the success of IC manufacturing and the advancement of state-of-the-art technologies.

Over time, various physics-based and empirical models have been utilized for device modeling of transistors. Designers employ these models to predict the electrical characteristics of transistors (I-V, C-V, etc.) and other devices under different operating conditions. Physics-based models, such as Berkley Short Channel IGFET Model (BSIM) and BSIM-CMG [41], and empirical models, like LUT (Look-Up Table)-based models, are accurate and generalizable but involve solving complex and nonlinear mathematical equations. They rely on extensive SPICE (Simulation Program with Integrated Circuit Emphasis) simulations and may encounter convergence issues for larger circuits. These models are widely incorporated in Process Design Kits (PDKs) for IC design. However, as transistors scale down to nanometers, nonidealities become more mathematically intricate, posing challenges in finding closed-form solutions for device models. Additionally, creating accurate physics-based models for each rapidly evolving next-generation device or technology is complex and time-consuming and demands a high level of expertise.

An emerging and promising approach is AI-assisted device modeling, which involves machine learning and deep learning. This approach has the potential to greatly enhance the precision and effectiveness of device modeling for advanced nanometer technologies while reducing computational requirements. AI-assisted models emphasize data-oriented modeling, removing the necessity for an in-depth study of the device's physical characteristics and allowing for the rapid creation of accurate surrogate device models. These models can predict the electrical characteristics of devices under various design and process parameter variations with high accuracy and efficiency, even for complex devices and operating conditions. This aids in achieving a fast turnaround time (TAT) and aims to produce reliable models that substantially improve computational efficiency. This enables rapid adoption of AI/ML in novel devices, allowing for easy and quick evaluation of transistor transfer characteristics and estimating any underlying impacts before establishing the underlying physics.

1.4.1.1 Methodology

The primary objective of automated modeling at this stage is to estimate the complex input-output relations under the influence of bias voltages and numerous geometrical and process parameters within varied operating conditions at very high computational speeds. These impacts peak as the transistor dimensions reach near-atomic sizes, quantum effects become increasingly significant and altogether impose a nonlinear impact on the device characteristics, making it quite complex to model them through traditional physical equations. EDA fraternity successfully developed predictable device performance models, including these effects, using AI/ML techniques. The methodology for transistor characterization and device modeling is discussed below.

To create a compact AI/ML model for studying the characteristics of a Field-Effect Transistor (FET) under modeling, the initial task involves identifying the specifications to be modeled and the key inputs that influence them. The FET could be an existing device like CMOS or FinFETs or a novel transistor design being explored for suitability in sub-nano technologies such as 3 nm. Subsequently, the generation of data samples for training the AI/ML model becomes crucial, wherein the influence of various device parameters (inputs) on the device characteristics (outputs or specifications) is comprehensively captured. Depending on the specific characteristics being monitored, the data samples are produced through DC, AC, and transient analysis. In most of the implementations in the literature, the drain current, I_{DS} being the predominant nonlinear element in a transistor, is modeled by varying the bias voltages. Temperature effects, variational analysis due to geometric parameters, and process effects can be added to the modeling for studying the combined impact on the drain current (Eq. 1.1). Another advantage of ML modeling is a quick derivation of small-signal model parameters from I-V (transistor current versus voltage) characteristics as the first derivatives of drain current. An accurate AI-assisted compact model should accurately predict the transistor specifications such as I-V or C-V characteristics and be efficient in small-signal analysis.

$$I_{DS} = f(V_{GS}, V_{DS}, V_{th}, T, G, P) \tag{1.1}$$

where V_{GS} is the gate to source voltage, V_{DS} is the drain to source voltage, V_{th} is the threshold voltage, T represents temperature, and G and P represent a set of l and k number of geometric parameters $\{g_1, g_2, ..., g_l\}$ and process parameters $\{p_1, p_2, ..., p_k\}$, respectively. For existing transistor designs with available PDKs or model files, simulators like SPICE can serve as platforms for data generation. However, novel transistor designs utilize Sentaurus to generate electrical, thermal, and optical characteristics. The training sample size for the AI/ML model depends on several factors, like the captured characteristics, the number of considered device parameters, their impact on specifications, the model type, and the optimization algorithm employed. In many cases, it relies heavily on the expertise and intuition of the designer and ML engineer, considering their skills and experience.

For example, for building statistically aware power estimation models using polynomial regression, [26] claimed the requirement of 50,000 training samples for achieving an accuracy of $\leq 1\%$ with SPICE models, whereas PVT-aware leakage and delay estimation models suiting 16nm CMOS and FinFET technologies are built over 15,000 training samples using a gradient boosting algorithm in [27]. They also showcase an accuracy $\leq 1\%$ against SPICE simulations. Therefore, the training sample requirements depend on several factors such as the use case, application, expertise of the designer in capturing the relevant data, application of suitable ML algorithm, and the skill of the ML engineer in tuning the model. The data generation may also involve iterations based on the model's accuracy achieved after training.

In another case study [42], the average electric field (AEF) of FeFET is modeled using DNNs. The obtained AEF values are clustered, and the cluster centroid is chosen as a representative AEF value. This approach is used because direct value prediction is challenging due to the uneven distribution of AEF values. The training samples are simulated on TCAD (Technology Computer-Aided Design) with a drain-source bias of 50 mV and a V_{GS} range of ± 4 V in 50 mV steps. A total of 14,495 simulations were employed for training. This setup achieved excellent generalization to unseen polarization configurations, with an error of approximately 1% compared to precise TCAD simulations. The TCAD simulator is replaced with this trained DNN model for the switching dynamics model. However, when the same DNN model was trained with a larger voltage step of 100 mV and evaluated on polarization configurations and gate voltages outside the training data, the error increased to approximately 3%. This highlights the importance of capturing valid training data covering all design specification margins. Although the trained model can generalize predictions to inputs outside the training data range, this may lead to increased prediction errors.

Additional details and a typical setup for data generation are described in Sect. 1.4.1.2 and Fig. 1.4. Once the training data is collected, the subsequent step is to select a suitable AI/ML algorithm(s) that can efficiently map the relationships between transistor characteristics and parameters. The ML engineer makes this choice based on their expertise and understanding of the data intricacies in close collaboration with digital/analog designers. Continuous feedback from both communities is crucial for attaining an accurate surrogate model. At this stage, the primary goal of AI-assisted modeling is to develop a reliable, compact model that efficiently characterizes the device, is generalizable to various electrical, physical, and design parameters, and is computationally several orders faster than traditional simulators, contributing to TAT. Meeting these requirements may necessitate multiple iterations of the modeling process until a satisfactory level of accuracy is achieved with minimal training time.

The diagram in Fig. 1.3 illustrates a standard workflow for device modeling. The initial step involves generating a dataset through TCAD device simulations to effectively capture the I-V and C-V characteristics with appropriate input

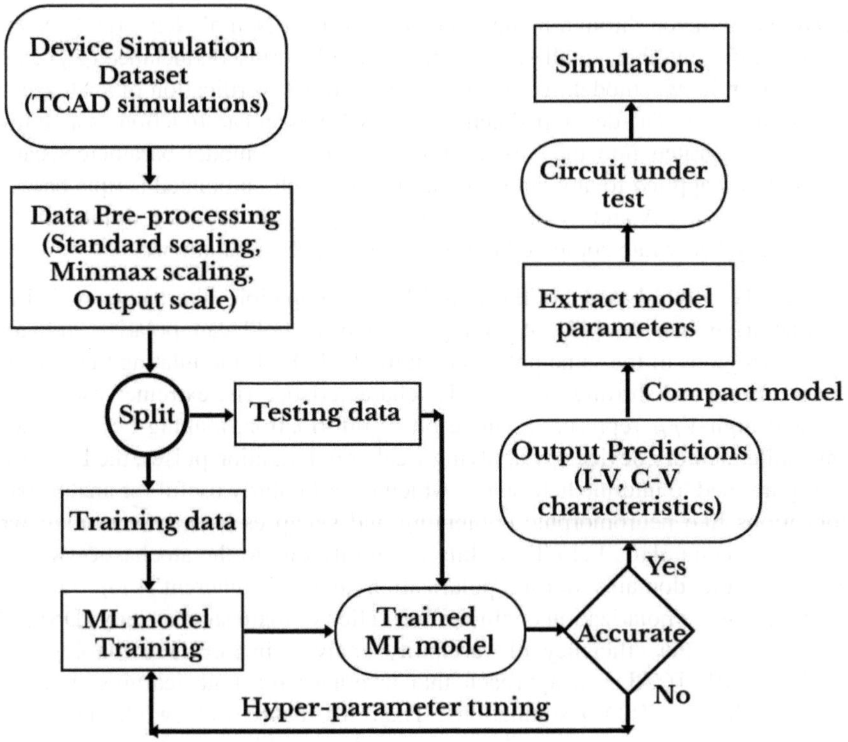

Fig. 1.3 The workflow of the AI-assisted transistor modeling

excitations. This dataset typically exhibits continuous distributions in both input and output parameters, making regression algorithms of supervised learning the suitable choice for modeling. Subsequently, the normalized dataset is partitioned into training and testing data, often following split percentages such as 90-10 or 80-20. An AI/ML model undergoes training on specified training data, continuing until it meets a stopping criterion, which could be a specified number of iterations or a set accuracy threshold (often Mean Square Error (MSE) in regression scenarios). The selection of hyperparameters is crucial for achieving accurate modeling and swift convergence, influenced by the data distribution and the ML engineer's experience. In certain scenarios, optimization algorithms like the Genetic Algorithm (GA) or Ant Bee Colony (ABC) algorithm, as mentioned in [43], can be employed to identify optimal tuning parameters for the model. Widely used Python open-source libraries like Scikit [44], TensorFlow, and Keras [45] are commonly applied for implementing ML/DL algorithms.

The trained model's performance is evaluated using testing data samples to verify the desired level of accuracy. If accuracy constraints are not met, the ML model undergoes tuning and retraining with new hyperparameter values until the required accuracy is achieved. The resulting ML model, with satisfactory accuracy, can predict output specifications such as drain current, subthreshold swing, capacitances, etc., based on the simulations in the dataset. In cases where the I_{DS} is predicted through models such as Artificial Neural Networks (ANNs) is infinitely differentiable, small-signal parameters like transconductance (g_m) can be derived from it. ML modeling also helps with a quick verification of such second-order parameters. The developed generalized ML model can function as a compact model if integrated into the SPICE flow. The trained model parameters can be extracted and applied to any circuit under test through automated scripts based on Verilog or Verilog-A and Python [46], facilitating highly accurate simulations at a significantly faster rate compared to traditional SPICE simulations.

Case Study HfO2-based FeFET is a CMOS-compatible alternative with lower manufacturing barriers [47]. Applying an electric field can polarize individual domains (regions of the same polarity) within the FeFET, modulating the threshold voltage (V_{TH}) and shifting the $I_D - V_G$ characteristic. The extreme cases of low-V_{TH} and high-V_{TH} represent a single bit of information, making FeFET a dense nonvolatile memory device. By applying weaker polarization pulses, the FeFET can be programmed to intermediate states, which is particularly useful for analog-based applications like neuromorphic computing and synapses [48], where small write pulses can accumulate. FeFET exhibits variability due to the stochastic switching process of the domains during polarization and the inherent V_{TH} variability depending on the polarization configuration called domain stochasticity (DS) [49].

To enhance the efficiency of variability analysis in FeFET technology, [42] introduced ML-TCAD, an approach that leverages machine learning models to replace traditional TCAD simulations. Their ML-based framework significantly accelerated device-to-device and cycle-to-cycle variability analysis, achieving a speedup of 13,600×. In their study, a FeFET device was constructed by integrating a

10-nm thick ferroelectric (FE) layer in place of the high-K layer. To ensure realistic ferroelectric behavior, the Preisach model [50] in TCAD was calibrated using measurements from a metal–FE–metal capacitor. The simulation data included I-V characteristics and average electric field (AEF) values, generated by sweeping the percentage of positively polarized (P_{FE+}) domains from 5 to 95% in 5% increments. The polarization configuration, represented by a 10×10 domain grid, served as the input for computing AEF and drain current. Two ML models were developed: The first one is a deep neural network (ResNet-18 [51]) to predict AEF, followed by a clustering method to find a representative AEF value, and the second one is a regression model for predicting drain current. Both models demonstrated high accuracy, with a weighted mean absolute percentage error of less than 1%.

1.4.1.2 Training Data Generation

To develop a concise AI-assisted transistor model, a significant amount of data samples is required to capture the nonlinear characteristics of the modeled device effectively. Generating the necessary training samples involves DC and transient SPICE simulations using standard BSIM models. Alternatively, background data from device design and simulation can be collected through waveforms and log files to extract essential training information. Utilizing accumulated simulation data for modeling reduces the time and effort needed for data generation. The number of training samples needed for ML or DL model development depends on the model's type and complexity. Previous works often employ a trial-and-error approach, with a correlated training data generation scheme providing continuous accuracy feedback. A feedback loop with simulation CAD tool(s) helps achieve the optimal training data size. State-of-the-art works used SPICE simulations through TCAD implementations for data generation; Sentaurus was the most popular for novel transistor designs. Few were based on the standard datasheets such as BSIM and IBM.

A comprehensive transistor characterization data generation setup is depicted in Fig. 1.4. The I_{DS} is collected for various bias voltages, such as V_{GS} and V_{DS}, depending on the transistor's operational region. It is crucial to capture the strong dependence of device characteristics on changing temperatures, varied manufacturing processes, and design parameters based on the targeted application. These dependencies can be appropriately included in the sweep data. The simulation setup needs to be tailored based on the modeling requirements. For instance, if the objective is to monitor I_{DS} alone (operating point calculation), then DC analysis will suffice. For frequency and small-signal analysis, estimating second-order effects such as g_m and output conductance (g_{DS}), AC analysis can be used and transient analysis for any time calculations. In specific cases, V_{th} can be varied directly or indirectly. Designers determine the typical values of bias voltages, process, and design parameters (mostly related to standard library models such as BSIM or IDRS reports) based on the type of transistor being characterized and the target application.

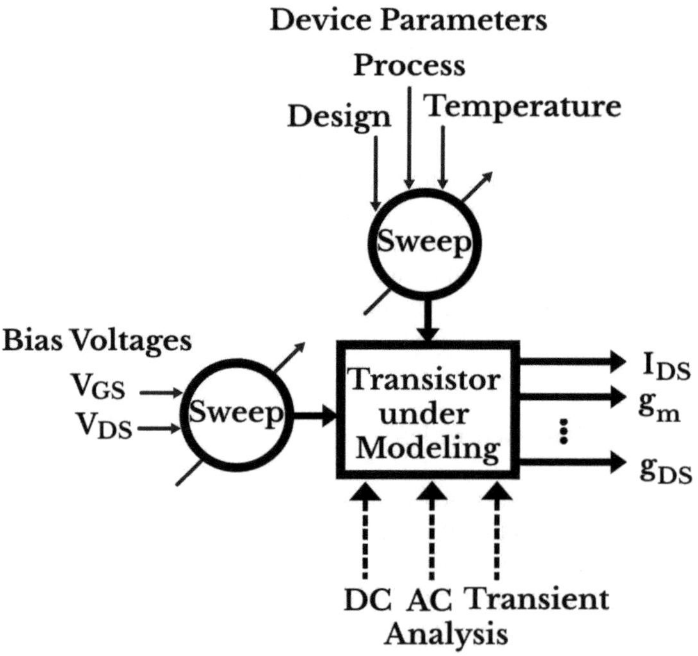

Fig. 1.4 Simulation setup for transistor characterization

1.4.1.3 State-of-the-Art Works

The EDA community has been actively engaged in the development of automated procedures for variational analysis of transistors and devices since the 1980s [52]. Response Surface Modeling (RSM), a statistical and mathematical methodology, has been widely explored in VLSI circuit design for many years [53, 54]. It provides a systematic approach to understanding and predicting how interactions between various factors affect the performance and reliability of ICs. RSM helps in efficiently selecting experimental points (training samples) to characterize the response surface, often using techniques like Design of Experiments (DoE) [55]. This method allows for creating accurate models with minimal experimental effort, reducing computational costs and time compared to full-scale simulations. By analyzing these response surfaces, designers can optimize speed, power consumption, and reliability to achieve desired performance metrics and make informed trade-offs between competing objectives.

The incorporation of AI in the modeling of devices and transistors has made substantial contributions over time, addressing challenges in device characterization, process variation modeling, compact modeling, model parameter extraction and optimization, and reliability modeling, encompassing factors like aging, stress, electromigration, and temperature effects to estimate the long-term performance of devices. Recent years have witnessed the prevalence of AI-assisted device models, primarily based on ANNs and Polynomial Regression (PR) [56, 46, 43, 57, 58, 59, 60]. These techniques offer a unique opportunity to unveil the underlying relationships between device designs and specifications. Notably, these approaches exhibit significant computational efficiency, being several orders of magnitude faster than traditional CAD simulators while maintaining accuracy. A comprehensive overview of key contributions and their modeling specifics are detailed in Table 1.1. Approximation of early voltage and MOSFET characteristics based on PR is presented in [56]. The efficiency of ANNs in modeling I-V and C-V characteristics of N-type MOSFETs applied to complex circuits—XOR gate, ring oscillators, and 6T Static Random-Access Memory (SRAM) resulted in an average error of less than 1% [46]. The proposal of employing different layers of NNs for predicting Silicon Carbide (SiC) MOSFET characteristics in various operating regions is discussed in [43]. Autoencoders (AEs) are utilized for extracting latent features from the dataset, followed by modeling using third-order PR, resulting in accurate I-V and C-V characteristics modeling with as few as 50 training samples [59]. Furthermore, the derivation of g_m by the first derivative and Drain-Induced Barrier Lowering (DIBL) through a suitable dataset is explored. To derive a small-signal model, [57] proposes combining an ANN I-V model with Taylor series expansion, with verification against implementations of CMOS inverters and cascode current mirrors. In [60], the high/low threshold voltage (HV_{th}/LV_{th}), on-state current (I_{on}), and subthreshold slope (SS) are extracted from raw polarization map data for modeling purposes. The resulting model predictions are 10^6 times faster than traditional TCAD simulations and achieve over 98% accuracy.

Previous works have reported two modeling approaches: global and local or single models [46]. Global fitting models cover multiple devices in a single model across various operating regions, offering the advantage of evaluating circuit performances based on different structural parameters with minimal TCAD usage. However, they require complex ML models (such as ANN networks) and take a longer time to converge. Local fitting models are considered a good option for monitoring single devices at faster simulation speeds with high accuracy. The most influential geometry and process parameters, in addition to bias voltages, were identified as gate length (L_g), sheet thickness and width (T_{sheet} and W_{sheet}, respectively), spacer length (L_{sp}), and oxide thickness (T_{ox}) for the device under modeling. While Adam optimizer and tanh activation function are commonly used in ANN modeling in most works, [58] has identified that for accurate results, the $I_{DS} - V_{DS}$ relationship needs to be modeled using the Sigmoid activation function and $I_{DS} - V_{TG}$ through the tanh function as two separate ANN models. Further, to alleviate challenges associated with DL approaches, such as asymmetric I-V modeling, nonzero current at $V_{DS} = 0$, and overfitting, hybrid analytical approaches coupled with AI-assisted models have been proposed as intriguing prospects for modeling next-generation MOSFETs [59]. The typical modeling workflow of such a setup is depicted in Fig. 1.5.

Table 1.1 Summary of the state-of-the-art AI-assisted models for device modeling and characterization

Reference	Device under modeling	Technology node (nm)	Parameter sweep	Output specifications	AI/ML model
[56]	MOSFET	180	$V_{DS}, V_{GS}, V_{th}, \lambda$	I_{DS}	PR
[46]	NSFET	3	$V_{DS}, V_{GS}, L_g, T_{\text{sheet}}, W_{\text{sheet}}, L_{sp}, T_{ox}$	I - V, C - V	ANN—two hidden layers for I-V (20 × 15 and 15 × 10) and one hidden layer for C-V (10 × 5 and 5 × 5 neurons) for global and local models, respectively
[43]	SiC MOSFETs	SCT3080KL [61]	V_{DS}, V_{GS}, temperature	I_{DS}	ANN—two hidden layers (10 × 10 neurons)
[57]	MOSFET	35	V_{DS}, V_{GS}	I_{DS}	ANN—15 and 11 hidden layers for *n* and *p* transistors
[58]	Thin TFET	Not specified	V_{DS}, V_{TG} (top gate voltage)	I_{DS}	ANN (two hidden layers (7 × 7)) + Physics based model
[59]	GAAFET	1.5	V_{DS}, V_{GS}	I_{DS}, g_m, g_{DS}	ANN—three hidden layers (6 × 6 × 6)
[62]	3D n-type FinFET	L_g 15 – 25 nm	V_{DS}, V_{GS}	I - V, C - V	AE (five layers—40 × 3 × 3 × 3 × 40) + PR (third order)
[60]	FeFET	28	Polarization map	I_{on}, HV_{th}, LV_{th},SS	Fully connected NN—6 hidden layers, 49 inputs, 4 outputs (49×–256 × 128 × 96 × 64 × 48 × 24 × 4)

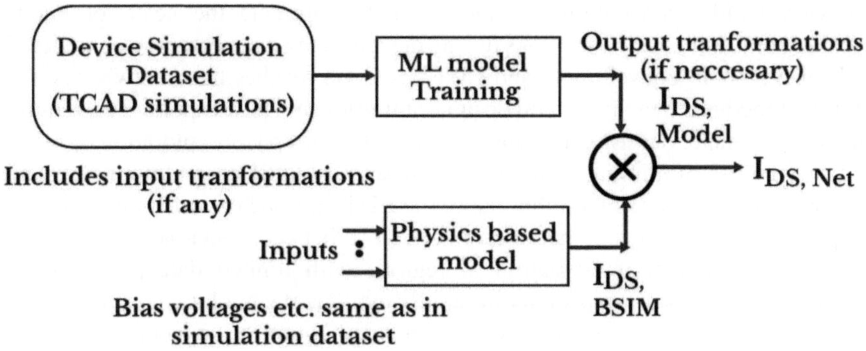

Fig. 1.5 The workflow of the AI-assisted transistor modeling along with physics model (recreated from [59])

1.4.2 Device Sizing

Choosing appropriate device sizing is a crucial aspect of analog and digital circuit design. Device dimensions (transistors and passive elements) determine the circuit performance, power consumption, and overall functionality. This process requires a careful trade-off analysis that considers the application's specific requirements and constraints, along with the details of the underlying manufacturing process. Illustratively, in analog circuitry such as amplifiers, the sizing of transistors becomes a determinant of gain and bandwidth. Larger transistors offer higher gain but often at the expense of reduced bandwidth, highlighting the delicate balance required in the sizing decision. Moreover, the sizing of devices significantly impacts the linearity and noise characteristics of analog circuits, influencing the overall signal integrity. In digital design, device sizing extends its influence to the gate's ability to drive multiple loads, a parameter known as fan-out. Appropriate sizing ensures that the gate can effectively drive subsequent gates without compromising performance integrity. This precise sizing enhances the design's robustness, protecting it from variations in input levels and uncertainties in the manufacturing process. The sizing decision becomes a strategic maneuver to optimize the trade-off, ensuring the circuit meets its operational goals while maintaining efficiency. The choice of transistor sizing is not arbitrary but is intricately tied to the specific process technology employed and the broader constraints encapsulated in the PPA framework.

The traditional approach to circuit design involves repetitive sizing and simulation to meet specific specifications, a process that relies on complex mathematical modeling linking device geometries and biasing currents to transistor small-signal models. However, this method, even with skilled designers, can be time-consuming and involves numerous iterations, especially as circuit complexity increases. The EDA community has introduced SPICE-generated LUTs to meet the time-to-market demands [63]. These LUTs contain small-signal parameters of transistors across various voltage sweeps, offering a one-time effort for capturing SPICE model behavior. Unlike conventional methods, LUTs eliminate the need for iterative tweaking and approximation issues during device sizing and performance trade-offs.

However, both conventional and LUT-based approaches have drawbacks. Conventional methods heavily rely on designer intuition and past experiences, requiring manual parameter adjustments and limiting design space exploration. This can increase the risk of missing optimal solutions and is inefficient and time-consuming for complex circuits. On the other hand, the LUT approach is static and dependent on precomputed data and involves a trade-off between accuracy and memory efficiency. Inefficient interpolation in regions with limited data points reduces flexibility and accuracy, especially in rapidly evolving technology nodes.

Researchers are exploring the use of automated optimization techniques and ML to navigate the vast design spaces in analog circuit design, aiming to identify optimal configurations with minimal simulation effort.

1.4.2.1 Problem Statement

The problem of circuit or device sizing can be described as an optimization task focused on minimization as

$$F_i(X|P_j) < C_i, j = 1, 2, \ldots, k \tag{1.2}$$

where X is the sizing vector with N design parameters $X = \{X_1, X_2, \ldots X_n\}$ including transistors and passive elements, P is the set of k predefined PVT conditions, $\{p_1, p_2, \ldots, p_k\}$ suiting application needs, and $C = \{C_1, C_2, \ldots C_m\}$ is the design constraint set with M constraints. Here, $X \in D$, where D_s is the design space with $D_s = \{D_1, D_s, \ldots D_n\}$ for each parameter. The circuit sizing can be represented as a design space exploration challenge, where traditional or automated methods must navigate a vast array of design parameters to identify a feasible solution that meets all constraints within a limited time frame and simulation budget.

1.4.2.2 State-of-the-Art Works

Various automated design space exploration methodologies suggested in the literature can be broadly categorized as follows:

- **Optimization Algorithms:** Designers employ advanced optimization techniques to identify effective solutions in the pursuit of meeting the design objectives and tight time-to-market deadlines [64]. Circuit sizing can be automated in CAD tools by leveraging optimization techniques such as convex optimization, Bayesian optimization (BO), and evolutionary algorithms (EAs) [65, 66]. They are effective in systematically navigating the design space and determining optimal device sizes for analog circuits. By automating the iterative process, these algorithms minimize the reliance on manual adjustments. The outlined

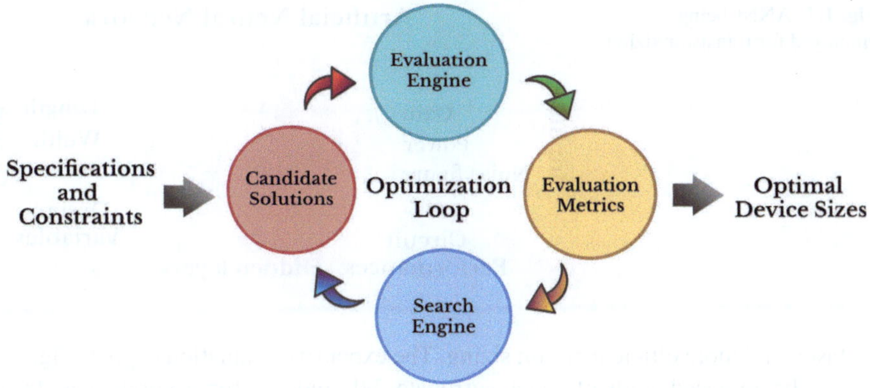

Fig. 1.6 Workflow of optimization algorithms in analog circuit sizing (recreated from [67])

workflow for addressing the circuit sizing problem using optimization algorithms is depicted in Fig. 1.6. In this process, an optimization algorithm systematically adjusts sizing values in iterative cycles to enhance yield while adhering to specified constraints. The efficacy of the candidate solutions generated by the optimization engine is assessed using an evaluation engine, such as the Cadence Spectre simulator, against designer-defined parameterized test benches [67]. The refinement of candidate solutions continues iteratively until optimal transistor sizing is achieved.

While optimization algorithms offer benefits, they also come with specific limitations. For instance, the Genetic Algorithm (GA) is prone to premature convergence, leading the population to converge to suboptimal solutions before reaching the global optimum [68]. Simulated annealing is vulnerable to getting trapped in local minima, whereas particle swarm optimization (PSO) experiences a slow convergence rate [69]. The BO technique involves creating a probabilistic surrogate model for the objective function. Following the initial sampling, this model is progressively refined using new data that optimizes the acquisition function [70, 71]. Although widely adopted, it should be acknowledged that this approach comes with extended processing time [69]. On the other hand, Differential Evolution (DE), although robust, is computationally intensive and sensitive to parameter settings [72]. The crucial task of algorithm selection and parameter tuning underscores the need for meticulous consideration in optimization endeavors. Later, methods based on RSM that produce concise analytical models across a selected design space hypercube were proposed [73]. These models prove valuable for designers, establishing a well-suited collection of sizing-focused design models. Such models can then be integrated into an optimization-based sizing tool.

- **Machine Learning Models:** The combination of ML and optimization algorithms for circuit sizing represents a powerful strategy in electronic design. This synergy enables designers to balance exploration and exploitation, leading to

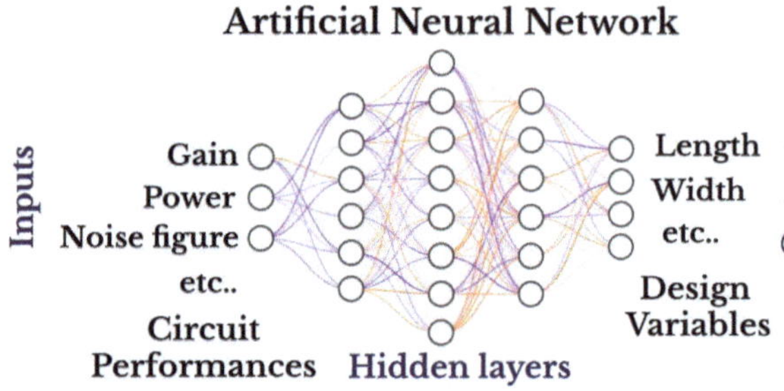

Fig. 1.7 ANNs being employed for transistor sizing

faster and more efficient circuit sizing. The expensive evaluation engine (Fig. 1.6) can be replaced with efficient surrogate ML models for accurate and faster predictions. Regression models within ML can be trained using simulation data to forecast the performance of analog circuits across different device sizes. This empowers designers to adopt a data-driven approach for decision-making regarding optimal sizing, eliminating the need for extensive manual simulations.

NNs are the most popular AI modeling approach among other methods to capture the intricate interactions within analog circuits [74]. This proves especially valuable for managing complex relationships between devices in intricate circuits, a scenario where manual approaches may encounter challenges. Training NNs typically involves a large simulated dataset that includes different configurations of design variables. These variables could include parameters like the length and width of transistors in a circuit designed to measure its performance figures. The trained NN, utilizing the simulation dataset, becomes equipped with the knowledge to predict optimal design variables for a given circuit's specifications (Fig. 1.7). This dataset may also encompass other design parameters, including capacitive load, bias components, and process variations, ensuring accurate capture of variational impacts. An example of an ANN-based technology-independent circuit sizing model, as presented in [28], demonstrated up to 90% accuracy in estimating transistor sizing in a 0.18 μm technology while meeting multiple circuit specifications. Notably, the ANN was trained using simulations from various technologies, including 1.5 μm, 0.5 μm, 0.35 μm, and 0.25 μm, with the 0.18 μm technology being unknown to it. The methodologies introduced by İslamoğlu et al. [64], Wolfe and Vemuri [75], Garitselov et al. [76], Lourenço et al. [77], Wang et al. [78], Harsha and Harish [79] involve the utilization of ANNs for analog circuit sizing. Optimization algorithms run on these methodologies instead of traditional simulators, aiming to deliver design solutions that are both fast and accurate. For a more comprehensive exploration of the application and effectiveness of ANNs in automating the sizing of analog designs, [80] provides detailed insights. For reducing the data dependency of ML models and enhancing the circuit yield, cluster-based approaches can be employed as proposed by Canelas et al. [81]. Table 1.2 outlines recent distinctive contributions in AI-assisted transistor sizing.

Table 1.2 Recent contributions to AI-assisted transistor sizing

Reference	AI-assisted Method	Methodology and contribution
[72]	GP + BI	Multi-objective optimization using GDE3 and Bayesian Inference (MODEBI) with Gaussian Process (GP) results in a fast and robust complex hyperspace exploration that involves design variables and PVT corners with many constraints
[82]	GNN + BO	Graph Neural Network (GNN) model, an extension of ParaGraph [83], integrates parasitic resistance and coupling capacitance into its framework, leveraging graph embeddings from ParaGraph. ParaGraph focuses on predicting net parasitics and device parameters using circuit schematic graphs. GNN application for parasitic prediction during SPICE simulations and surrogate modeling circumvents the need for iterative layout generation, useful when automated layout tools are limited. The experiments on 16-nm FinFET-based three-stage receiver amplifier and transmitter line-driver show that the GNN model with self-attention graph embedding readout (GNN-attn) provides a faster convergence rate (up to three times) compared to GNN-BLR (GNN-attn with the last layer replaced with Bayesian Linear Regression) and NN-BLR (NN with the last layer as BLR). BO is also involved in updating design parameters in each iteration
[67]	VAE + DNN + BO	The integrated device can be described using a set of latent variables obtained from the trained VAE (Variational Autoencoder) model, effectively replacing the discrete-valued parameterizations that are typically associated with individual devices. This integration of VAE facilitates an automated analog circuit sizing approach using BO in conjunction with a combination of VAE and an FCNN. The successful application of this methodology extends to mixed-variable optimization scenarios
[84]	ANN + DE	Addressing the possible failure of ML models in modeling for complete and stringent circuit specifications, a global optimization framework that involves infill sampling criterion and DE, based on which an ANN model is constructed in every iteration over the best candidate solutions till the stopping criterion is met. The framework reduces the simulation requirement to hundreds and is 20 times faster than DE
[85]	DNN + EA	An evolutionary algorithm designed to adapt to variations in analog circuit sizing, employing a DNN prediction model across various technology nodes, achieved a performance improvement of up to 79x faster than conventional EA optimization methods. Further, a force-directed model with the estimated distance (force) of the current design to the passed and failed designs of the previous flow is added as a fitness function to EA. This aids in guiding the optimization process toward better yield, reducing the required MC simulations
[86]	ResNN + GA	A multi-objective optimization framework based on Residual Neural Networks (ResNNs) demonstrates significant reductions in delay and leakage of CMOS 22-nm high-K digital cells, achieving up to 36.7% and 18.8% improvements, respectively
[87]	DNN + EA + Simulator	A framework for optimization using DNN assistance combines global and local optimization engines: evolutionary algorithms and DNN-based methods. This approach yields superior design solutions with minimal simulations, focusing on various opamp designs

- **Reinforcement Learning:** RL models can adjust to alterations in design requirements or technology nodes. Through continuous learning and optimization based on feedback, these models can dynamically adapt and refine device sizing, showcasing greater flexibility in response to evolving design specifications. An ML-based optimization scheme, AutoCkt, utilizes deep RL to determine the post-layout circuit parameters based on a specified target specification [34]. It gains extensive insight into the entire design space using a sparse subsampling technique. The findings indicate that AutoCkt achieves convergence and fulfills all desired specifications significantly quicker than other methods: It is, on average, 40 times faster than a conventional genetic algorithm and 9.6 times faster than the leading current approach when accounting for layout parasitic considerations. RobustAnalog [88], a circuit design framework via multitask RL for optimization under variational space, reduces the optimization time to 14×–30× compared to EA, BO, and Deep Deterministic Policy Gradient (DDPG) [89] with smaller simulation cost. RobustAnalog achieved a 100% success rate in accomplishing all the corner tasks of a strongARM latch, two-stage operational transimpedance amplifier (OTA) and folded OTA sizing meeting multiple design constraints. A brief of the methodology follows:
  - The training set is clustered using the K-means algorithm based on the corners.
  - RL agent produces actions and sends them to each environment (corner-specific subsets), followed by circuit simulation and rewards assignment.
  - If the subset successfully completes all tasks during the agent evaluation, the sizing solution is then tested across the entire test stack.
  - The loop of training subset generation, adaptive pruning, and actor-critic interface iterates till a complete solution is obtained.

Parasitic resistances and capacitances introduced during the layout phase substantially impact the circuit performances, causing signal delays, decreased signal integrity, and increased power consumption. This underscores the critical necessity of accounting for layout-induced parasitics in the sizing of analog circuits. In addressing this, automated device sizing approaches cognizant of layout parasitics have been introduced, as outlined in [90, 91]. These methods guarantee that design parameters are optimized by analyzing post-layout circuit performance, underscoring the critical role of accounting for parasitic elements throughout the design phase.

1.5 AI-Assisted Circuit Design and Modeling

Simulation and modeling play a crucial role in assessing and validating the functionality of AMS and digital circuits before their production and deployment, especially in nano and sub-nano technologies. Detecting differences in functional and electrical performance at the outset of the design phase can enhance IC yield, a metric significantly dependent on simulation tool capabilities. This analysis involves extensive Monte Carlo (MC) simulations using conventional tools like HSPICE and Prime Time, which gauge circuit performance by incorporating intricate mathematical representations of electronic components and their interactions within an IC.

Accurate modeling requires an examination and prediction of variations in circuit performance arising from manufacturing discrepancies, noise, voltage fluctuations, temperature variations, etc. This involves considering parameters such as transistor length, width, doping concentrations, interconnect characteristics, variations in operating conditions, and more. Statistical models are employed to account for these variations, ensuring robust circuit design and optimizing yield. This modeling is vital for addressing manufacturing uncertainties before the physical design phase, ensuring reliable IC performance across different production runs. Despite providing precise results, traditional simulations for statistical modeling entail extensive computational time [92].

The complexity of modern ICs, densely packed with numerous components in a limited space, poses challenges in accurately modeling and simulating their behavior, especially considering the growing manufacturing variations and quantum effects. The need for faster development cycles calls for swift and accurate models, sometimes necessitating a trade-off with exhaustive simulations. To address this, the EDA community has proposed the development of surrogate models. These models approximate circuit performance based on statistical parameters, bypassing the need for prolonged simulation runs with traditional tools and significantly reducing computational overhead.

In this scenario, researchers have explored different AI-driven methods to prototype circuit performance and evaluate its ability to withstand various design and process variations. The primary objective is to optimize the estimation of performance reliably, aiming for accelerated speeds and enhanced efficiency.

1.5.1 Methodology

AI/ML surrogate modeling is widely employed in predicting circuit performances, considering design parameters and manufacturing process variations, encompassing fluctuations in operating voltages and temperatures. This objective is pervasive across digital and AMS applications. In digital circuit modeling, the focus lies on assessing key performance metrics such as propagation delay and power, being the crucial factors. Some instances also involve modeling design area [93].

Many approaches utilize logic cell characterization, concentrating on logic gates as leaf cells. The performance estimations of these logic gates can then be leveraged for complex circuit predictions, offering advantages in resource reutilization and faster simulation speeds. In contrast, circuit-specific modeling approaches develop AI-assisted surrogate models tailored to particular circuits and their unique statistical variations. However, these models may not be easily repurposed for other applications.

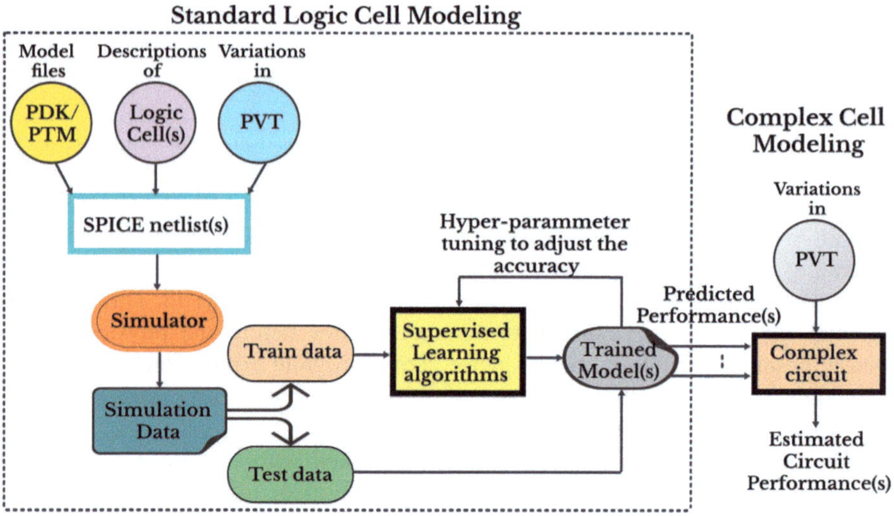

Fig. 1.8 ML-based logic-level characterization and extended complex cell estimations

The fundamental stages involved in constructing AI/ML surrogate models, based on logic cells for performance prediction, utilizing regression techniques, are delineated in Fig. 1.8. To initiate this process, SPICE netlists are generated in combination with the PDK/PTM files specific to the target technology node, encompassing logic cell descriptions and distributions of PVT variations tailored to the application. Subsequently, simulation data is produced via MC simulations, encompassing a sweep across diverse variations.

This simulation data is then partitioned into training and testing datasets. Supervised learning, involving either regression or classification, is applied to the training dataset. The resultant model is scrutinized for accuracy against the test dataset, and fine-tuning of model parameters is executed to enhance accuracy. These trained ML models are invoked by complex circuits based on their structure, enabling them to predict performance changes in response to applied PVT variations. It is noteworthy that the standard cell modeling process necessitates a one-time effort for each library cell and each performance metric of interest within a selected technology node. Importantly, these models can be reused across a diverse array of complex cell structures. Power and delay are the most prominent performance metrics of interest at nano and sub-nano nodes.

Alternatively, in AMS designs employing corner-based analysis, classification techniques come into play, forecasting the corner associated with circuit performance under various input conditions.

1.5.2 State-of-the-Art Works

Over the past two decades, the literature has witnessed the emergence of various methodologies dedicated to analyzing variability in the performance of digital circuits. The main goal of these methods is to accurately assess how PVT variability affects circuit performance. One noteworthy contribution introduced a method for estimating statistical parametric yield, to assess the overall parametric yield of MOS circuits [52, 98]. Table 1.3 presents ML models proposed in the literature for standard cell characterization and their performance estimation. In Table 1.3, MPR denotes Multiple Polynomial Regression, involving the modeling of more than two inputs [26], and LGBM denotes Light Gradient Boosting Machine, utilized as a gradient boosting methodology to enhance the performance of weak learning models [27].

In most of these studies, the physical variations examined include transistor characteristics like oxide thickness, threshold voltage, effective channel dimensions (with gate length, fin height, and thickness representing equivalent FinFET parameters). Additionally, other device parameters taken into account include body doping and source/drain doping concentrations.

At the circuit level, employing NN-based power estimation through input/output (I/O), cell number, and memory activation counters offers significant cost and complexity reductions, particularly during high-speed runtime [105, 106]. The application of ANNs for voltage and temperature-aware statistical analysis of leakage power achieved a notable enhancement, accelerating runtime speeds by a remarkable factor of 100. Additionally, the errors in the mean and standard deviation of MC statistical leakage estimations were kept below 1% and 2%, respectively. The characterization of transistor stacks in CMOS gates based on Support Vector Machines (SVMs) increased the runtime by 17× for power estimation [107]. The delay estimation model proposed in [92] demonstrates a 5% error compared to HSPICE without requiring any pre-simulation steps.

Case Study To expedite transistor sizing optimization, [86] developed a framework that integrates trained Residual Neural Network (ResNN) models with optimization algorithms for power-delay optimization of digital standard cells. Characterization data was generated from SPICE netlists for 22-nm High-K CMOS technology using PTM models, as illustrated in Fig. 1.8, by varying Process, Voltage, and Temperature (PVT) conditions. Ten process parameters, including physical and electrical equivalents of oxide thickness, nominal gate oxide thickness, source/drain junction depth, and channel doping concentration, were varied by $\pm 3\sigma$ from their mean. Temperature variations ranged from $-55\,^{\circ}$C to $-125\,^{\circ}$C, and voltage variations were $\pm 0\%$ from the nominal value. The process parameters also accounted for random variations in transistor length (22–33 nm) and width (44–440 nm) to assess the impact of transistor sizing on leakage power and propagation delay. The effect of load on delay was modeled as load capacitance and incorporated into delay estimations. Leakage and delay under these variations were estimated through SPICE MC simulations using DC and transient analysis, respectively.

Table 1.3 State-of-the-art AI-assisted models for statistical logic cell characterization and performance estimation

References	Modeled performance	AI model	Key take-away
[94, 95, 96]	Delay, Power	RSM	One or two orders of magnitude speedup at the expense of 2–7% decrease in accuracy
[97]	Delay, power	RSM	Optimized gate delay cells tolerant to variations at 16-nm technology were effectively implemented in mod-2 counter and 15-stage ring oscillator to minimize the maximum delay encountered
[52, 98]	Parametric yield	RSM	Two orders of accuracy improvement on SRAM over traditional approach
[99]	Leakage	RSM	Optimization strategies for power under delay constraints, incorporating temperature effects on Short gate and asymmetric gate work function (SG/ASG) n/p-FinFET at 22 nm technology, achieve a 50% power reduction at 373K, with an associated 9.9% increase in area
[100]	Delay, slew	BI	A combination of sparse sampling and priors of timing model coefficients from old libraries results in building new library in target technology with 15× reduced simulations
[101]	Leakage	PR	5× faster simulation at the expense of 1% accuracy loss
[102]	Delay	PR	The Burr distribution (a three-parameter probability distribution function) yields results of 3.62%, 3.49%, and 2.78% on ISCAS benchmark circuits using 90 nm, 45 nm, and 22 nm technologies, respectively, which are higher compared to those obtained with two-parameter distributions
[93]	Power, delay, energy	PR	95% accurate models from 180 to 7 nm nodes. However, process and temperature variations were not considered during modeling
[26]	Leakage and total power	MPR	Accurate modeling with <1% error over 16-nm CMOS logic cells and complex circuits compared to SPICE
[103]	Leakage power	ANN	Accurate FinFET logic cell models with <1% error over SPICE
[27]	Leakage power and delay	LGBM	CMOS and FinFET standard logic cell modeling resulting in <1% error compared to SPICE simulations and reutilization of logic cell models for complex circuit modeling
[104]	Power	ET, PR	Extra tree (ET) regressors work better than PR while modeling Resistive Load Inverter (RLI) circuits with 61% of accuracy

They developed a ResNN network composed of a sequence of residual layers—specifically, stacked linear and ReLU layers. Each ResNN model, tailored for a standard digital logic cell, is trained using characterization data for approximately 1000 epochs, achieving an R^2 score of around 0.99, indicating about 1% error compared to SPICE simulations. Further, genetic algorithms were applied to these trained ResNN models to determine the optimal transistor sizing in a couple of seconds, resulting in up to 36.74% reduction in leakage and 18.75% reduction in delay compared to the SPICE simulator.

For further exploration of ML-based power estimation techniques and timing error prediction at the Register Transfer Logic (RTL) level, refer to [108, 109, 110]. [111].

We observed that most of the works mention a mean error against MC SPICE simulations. It is essential to thoroughly evaluate the developed model at worst-case corners/conditions to ensure that the errors do not reach a level that could lead to chip design failures. Therefore, the worst-case observed error should also be employed as a metric in circuit design ML applications.

1.5.3 Reliability Verification

Reliability verification is critical in IC design, ensuring robustness against electrical and physical rules to prevent premature failures or degradation over time. This process is vital for guaranteeing the safe and effective operation of ICs under elevated temperatures over extended periods, meeting specified performance criteria. There has been a significant increase in reliability issues in nanometer devices over time. They include:

Electrical and Physical Checks Verification of design against electrical and physical rules to prevent electromigration, timing violations, and thermal issues.

Aging and Degradation Analysis Evaluation of aging effects such as NBTI (Negative Bias Temperature Instability) and HCI (Hot Carrier Injection) using SPICE models to predict performance degradation over the operational lifetime [112].

Temperature and Stress Testing Testing under varied temperature and stress conditions verifies the IC's reliable operation within its designated operating parameters.

Design Robustness Assessing the robustness of the design against environmental variations, manufacturing variations, and external disturbances.

Traditionally, reliability analysis, which includes timing, power integrity checks, and simulations, relies on specialized tools like Cadence, often requiring considerable time depending on circuit complexity. In recent years, ML has gained traction in IC reliability verification [113]. ML techniques analyze aging effects such as NBTI and HCI, predicting their long-term impact on IC performance [114, 115]. They excel in uncovering complex relationships and patterns in reliability data, offering insights beyond traditional methods. ML models trained on historical fault data classify potential reliability-affecting anomalies. Moreover, ML optimizes testing strategies by identifying critical tests to maximize coverage while minimizing time and costs [116]. Ultimately, ML enhances quality assurance by detecting defects early in manufacturing, thereby improving yield and reducing reliability issues in final IC products.

Ye et al. [117] introduced an SVM-based regression model for predicting voltage droop during dynamic timing analysis. Voltage droop, caused by high-switching currents from logic gates, becomes more problematic at higher clock frequencies and smaller technology nodes. The study demonstrated that the SVM predictor imposes minimal overhead, particularly advantageous for large circuits. In [118], a Graphical Convolutional Network (GCN) classifier was developed to enhance the testability of circuit designs through observation point insertion. The approach involved transforming a circuit netlist into a graph representation using GCNs, where each graph node incorporated testability-related attributes obtained from the SCOAP tool. Through training with historical data, the GCNs were optimized to determine optimal locations for observation point insertion, treated as a binary classification task. Specifically, the decision for each gate in the design focused on whether to add an observation point on its output port. Evaluation on industrial designs demonstrated that the GCN-based approach outperformed existing commercial tools in enhancing testability.

To address the computational speed limitations of traditional FEM (Finite Element Method), approaches like COMSOL [119], [120] introduced a GAN-based (Generative Adversarial Networks) method for electrostatic analysis in modeling time-dependent dielectric breakdown (TDDB) aging. This innovative approach utilizes GANs to synthesize VLSI layout images containing both vertical and horizontal routing tiles. These images help in learning the electrical potential distribution, incorporating essential boundary conditions such as VDD and GND necessary for calculating the electric field. The researchers opted for Wasserstein GAN (WGAN) [121] due to its improved stability during training compared to conventional GANs and CGANs, which often encounter the vanishing gradient problem. The proposed method achieves a remarkable 213× acceleration over COMSOL while maintaining approximately 97% accuracy, demonstrating significant computational efficiency gains with minimal compromise in accuracy. Thakur et al. [122] and Amuru et al. [111] provide a review of ML techniques in circuit testing and reliability analysis.

1.6 Challenges and Opportunities

Harnessing the potential of AI/ML surrogate modeling presents a compelling avenue for expediting design processes and enhancing the performance estimation and optimization of ICs. The incorporation of AI-assisted models yields several benefits:

- **Understanding Complex Patterns and Inter-relationships:** AI/ML models excel in discerning intricate patterns and relationships within statistical datasets. Their ability to comprehend complex structures enables a deeper understanding of the nuanced factors influencing circuit performance.
- **Decoupling Modeling Complexity from IC Complexity:** Unlike traditional methods, the complexity of AI/ML models does not escalate at the same pace as that of the complex circuit itself. This decoupling ensures that the modeling process remains manageable and efficient even as IC designs become increasingly intricate.
- **Eliminating Manual and Rule-Based Approaches:** The integration of AI/ML in circuit design and modeling obviates the need for manual and rule-based approaches and complex mathematical representations, leading to a significant acceleration of the analysis phase. Automation through intelligent models streamlines processes, saving time, and resources.
- **Enhancing Accuracy in Performance Estimations:** AI-assisted models consistently outperform traditional methods by providing more accurate performance estimations. The inherent ability to learn from data patterns contributes to higher precision in predicting IC behavior under diverse conditions.
- **Adaptability to Circuit Design Variations:** AI/ML models exhibit a remarkable capacity to adapt to variations and changes in circuit designs. This adaptability results in more flexible solutions that can accommodate a spectrum of scenarios, fostering versatility in the face of evolving design requirements.

However, comprehending the complexities involved in developing automated AI/ML procedures for IC modeling is paramount. Foremost among thc challenges is the necessity for an extensive volume of training data, a demand imposed by the ML algorithms. Acquiring and curating datasets for VLSI processes is particularly difficult and time-consuming without automating the data collection process across the stages. To tackle the problem of data scarcity, it is crucial to establish an organized approach for automatic data collection across the IC design stages.

Addressing the data scarcity is further facilitated by implementing a benchmarking system and making statistical analysis datasets publicly available through platforms such as IEEE dataport [123]. This approach ensures a significant acceleration and enhances the trustworthiness of surrogate model development.

Another critical concern lies in the potential generation of biased outcomes by ML models if the bias-variance trade-off is not thoroughly examined before modeling. Therefore, it is essential to gather and use diverse data that covers all aspects of the design, using suitable data analysis methods. Mitigating challenges in generalizing AI models, while steering clear of overfitting and under-fitting, remains a recurrent challenge in training ML algorithms. Recommended compensatory techniques include cross-validation, regularization, and ensemble methods [24]. However, circuit designers face a challenge in interpreting ML models that prototype circuit behavior. This issue can be resolved by creating a codesign environment that encourages close collaboration between designers and ML engineers.

Automated procedures are often prone to potential attacks. Before deploying these models, it is crucial to assess and implement appropriate countermeasures against security vulnerabilities. Furthermore, the efficacy of ML models is highly contingent on extensive data, and enhancing trust in the model's efficiency requires a dedicated testing scheme that includes typical corner cases from circuit designers, covering multiple design vulnerabilities.

1.7 Conclusion

In conclusion, design tools incorporating automated optimization techniques featuring AI surrogate models mark a departure from heavy dependence on expert designers, offering valuable support to less experienced engineers. These tools harness insights from historical data, utilizing learned patterns to facilitate well-informed decision-making. The adoption of AI-assisted surrogate models at both the device and circuit levels has aggregated considerable attention from researchers in recent years. These models offer swift and reliable predictions of device characteristics and complex input/output and inter-parameter relationships for both existing and emerging technologies under research, obviating the need for composite mathematical representations. ML models enable quick assessment of real-time performance during design, bypassing the requirement for lengthy simulations. The simulation speed can be significantly increased, chip yield can be improved, and turnaround time and design costs can be reduced. Additionally, they offer concise and accurate predictions of circuit behavior, reducing dependence on extensive lookup tables. This proves particularly advantageous in designs constrained by memory limitations. Furthermore, implementing a hierarchical modeling approach, involving transistor or logic-level standard cell characterization with adaptive feedback, underscores the essential attributes of scalability and universality in surrogate models, enabling the modeling of a wide array of complex circuits. The synergistic fusion of AI and circuit modeling not only enhances the efficiency of the design process but also signifies a paradigmatic shift in overcoming the time limitations inherent in traditional simulation methods. The capacity for swift and precise predictions establishes AI-assisted surrogate models as a foundational element in propelling the evolution of electronic design.

1.8 Future Trends

Addressing the concerns in applying AI to VLSI circuit design and modeling (as mentioned in Sect. 1.6), current research has geared up toward artificial data generation techniques, transfer learning techniques, and deep reinforcement methodologies. Most ML algorithms require a substantial volume of data for accurate modeling, which entails considerable effort in generating a large amount of simulation data for training in the IC domain. One alternative to reduce the manual effort and simulation cost is to initiate a systematical procedure to collect data during the circuit design and testing process in the form of log data, etc., which can be processed and used for training AI/ML algorithms. Stable artificial data generative schemes such as GANs and variational encoders generate realistic and high-quality synthetic data resembling a given data distribution [124]. Their ability to perform suitable interpolation in latent space can be useful in adding data points to the limited training data, aiding in improving the performance of AI/ML models.

An emerging and rapidly advancing field in recent literature is transfer learning [125, 126]. This approach is pivotal for leveraging knowledge gained in modeling one application to train another closely related application, utilizing insights obtained from prior modeling efforts. The implementation of transfer learning is instrumental in scenarios where enhanced model performance is sought while mitigating the need for extensive task-specific data. Another noteworthy technique is meta-learning [127], which involves training a model on a diverse set of tasks, enabling it to rapidly adapt and learn new tasks with minimal data. Meta-learning has found applications in domains such as image processing and autonomous systems and is anticipated to extend its relevance to the circuit domain shortly. Both transfer learning and meta-learning significantly contribute to the efficiency, adaptability, and generalization capabilities of machine learning models. These techniques play vital roles in overcoming challenges associated with data scarcity, task diversity, and dynamic environments.

The integration of an efficient recommendation system into EDA tools has the potential to significantly accelerate the circuit design process, empowering designers by expediting design procedures and facilitating quicker, more informed decision-making. These recommendation systems leverage cutting-edge technologies, such as reinforcement learning and Graph Neural Networks (GNNs), to provide optimal circuit parameters. By harnessing reinforcement learning, the systems learn from past design experiences and dynamically adapt their recommendations enhancing the circuit design process. Furthermore, integrating GNNs improves the recommendation system's capability to grasp and model intricate relationships within the design domain. The combined use of reinforcement learning and GNNs streamlines design procedures and facilitates extensive exploration of the design space, aiming to identify optimal circuit parameters aligned with performance objectives and design constraints. This results in a more thorough and effective search process, minimizing the time required for designers to make well-informed decisions. Cao [31], Wang et al. [35] underscore the successful application of reinforcement learning and GNNs in circuit design, emphasizing the potential of incorporating such recommendation systems within EDA tools for improved design efficiency and optimization.

The integration of diverse ML modeling techniques, combining reinforcement learning and transfer learning strategies, holds the potential to revolutionize circuit design and modeling. This comprehensive approach is poised to usher in an era of intelligent automation, paving the way for sophisticated and automated paradigms in the field of EDA in the near future.

References

1. Kawano, M. (2021). Technology trends in 2.5d/3d packaging and heterogeneous integration. In *2021 5th IEEE Electron Devices Technology & Manufacturing Conference (EDTM)* (pp. 1–3).
2. Custom IC Design (2024). https://community.cadence.com/cadence_blogs_8/b/cic/posts/custom-ic-design-flow-methodology-introduction-737044421. Accessed: Jun 15, 2024.
3. Cadence Genus (2023) https://www.cadence.com/en_US/home/tools/digital-design-and-signoff/synthesis/genus-synthesis-solution.html. Accessed: November 24, 2023.
4. Oasys-rtl (2023). https://eda.sw.siemens.com/en-US/ic/oasys-rtl/. Accessed: November 24, 2023.
5. Siemens $catapult_{HLS}$ (2023). https://resources.sw.siemens.com/en-US/fact-sheet-catapult-high-level-synthesis-and-verification. Accessed: November 24, 2023.
6. Deschamps, J.-P., Sutter, G. D., & Cantó, E. (2012). *Guide to FPGA implementation of arithmetic functions* (Vol. 149). Springer Science & Business Media.
7. Brunvand, E. (2009). *Digital VLSI chip design with cadence and synopsys CAD tools* (1st ed.). Addison-Wesley.
8. Deshpande, N., & Sowmya, K.B. (2020). A review on ASIC flow employing EDA tools by synopsys. *SSRG International Journal of VLSI & Signal Processing (SSRG-IJVSP), 7*(1), 15–19.
9. PrimeTime (2023). https://www.synopsys.com/implementation-and-signoff/signoff/primetime.html. Accessed: November 10, 2023.
10. SmartSpice (2023). https://silvaco.com/examples/smartspice/index.html. Accessed: November 10, 2023.
11. Riviera-PRO (2023) https://www.aldec.com/en/products/functional_verification/riviera-pro. Accessed: November 10, 2023.
12. Blaise, P. (2019). Ab initio simulation of advanced materials and devices: Current challenges. In *2019 IEEE International Electron Devices Meeting (IEDM)*, (pp. 24.3.1–24.3.4).
13. Iverilog (2023). https://iverilog.fandom.com/wiki/User_Guide. Accessed: November 14, 2023.
14. ghdl (2023). https://readthedocs.org/projects/gritbub-ghdl/downloads/pdf/latest/. Accessed: November 10, 2023.
15. myhdl (2023). https://docs.myhdl.org/en/stable/manual/index.html. Accessed: september 01, 2023.
16. NGSPICE (2024). https://ngspice.sourceforge.io/. Accessed: Jun 10, 2024.
17. eSim (2024). https://esim.fossee.in/. Accessed: Jun 10, 2024.
18. qflow (2023). http://opencircuitdesign.com/qflow/. Accessed: July 09, 2023.
19. Graywolf (2024). https://github.com/rubund/graywolf. Accessed: Jun 10, 2024.
20. Openroad (2023). https://openroad.readthedocs.io/en/latest/main/README.html. Accessed: July 21, 2023.
21. Magic (2023). http://opencircuitdesign.com/magic/. Accessed: May 14, 2023.
22. Klayout (2023). https://www.klayout.de/doc-qt5/manual/lvs_overview.html. Accessed: Augest 09, 2023.
23. Kotsiantis, S. B., Zaharakis, I., & Pintelas, P. (2007). Supervised machine learning: A review of classification techniques. *Emerging Artificial Intelligence Applications in Computer Engineering, 160*(1), 3–24.
24. Goodfellow, I., Bengio, Y., & Courville, A. (2016). *Deep learning*. MIT Press. http://www.deeplearningbook.org
25. Han, J., Kamber, M., & Pei, J. (2012). 8 - classification: Basic concepts. In J. Han, M. Kamber, & J. Pei (Eds.), *Data mining (Third Edition)*. The Morgan Kaufmann Series in Data Management Systems (3rd ed., pp. 327–391). Morgan Kaufmann.
26. Amuru, D., Zahra, A., & Abbas, Z. (2019). Statistical variation aware leakage and total power estimation of 16 nm vlsi digital circuits based on regression models. In A. Sengupta, S. Dasgupta, V. Singh, R. Sharma, & S. K. Vishvakarma (Eds.), *VLSI design and test* (pp. 565–578). Springer Singapore.

27. Amuru, D., Ahmed, M. S., & Abbas, Z. (2020). An efficient gradient boosting approach for PVT aware estimation of leakage power and propagation delay in cmos/finfet digital cells. In *2020 IEEE International Symposium on Circuits and Systems (ISCAS)* (pp. 1–5).
28. Kahraman, N., & Yildirim, T. (2008). Technology independent circuit sizing for fundamental analog circuits using artificial neural networks. In *2008 Ph. D. Research in Microelectronics and Electronics* (pp. 1–4). IEEE.
29. McKinney, W. (2010). Data structures for statistical computing in python. In S. van der Walt & J. Millman (Eds.), *Proceedings of the 9th Python in Science Conference* (pp. 56 – 61).
30. Zhuo, C., Yu, B., & Gao, D. (2017). Accelerating chip design with machine learning: From pre-silicon to post-silicon. In *2017 30th IEEE International System-on-Chip Conference (SOCC)* (pp. 227–232).
31. Cao, L. (2002). Circuit power estimation using pattern recognition techniques. In *Proceedings of the 2002 IEEE/ACM International Conference on Computer-Aided Design* (pp. 412–417).
32. Sutton, R. S., & Barto, A. G. (2018). *Reinforcement learning: An introduction*. A Bradford Book.
33. Jeerige, A., Bein, D., & Verma, A. (2019). Comparison of deep reinforcement learning approaches for intelligent game playing. In *2019 IEEE 9th Annual Computing and Communication Workshop and Conference (CCWC)* (pp. 0366–0371).
34. Settaluri, K., Haj-Ali, A., Huang, Q., Hakhamaneshi, K., & Nikolic, B. (2020). Autockt: Deep reinforcement learning of analog circuit designs. In *Proceedings of the 23rd Conference on Design, Automation and Test in Europe*, DATE '20 (pp. 490–495). EDA Consortium.
35. Wang, H., Wang, K., Yang, J., Shen, L., Sun, N., Lee, H.-S., & Han, S. (2020). Gcn-rl circuit designer: Transferable transistor sizing with graph neural networks and reinforcement learning. In *2020 57th ACM/IEEE Design Automation Conference (DAC)* (pp. 1–6).
36. Agnesina, A., Chang, K., & Lim, S. K. (2020). Vlsi placement parameter optimization using deep reinforcement learning. In *2020 IEEE/ACM International Conference On Computer Aided Design (ICCAD)* (pp. 1–9).
37. Li, X., Huang, Z., Tao, S., Huang, Z., Zhuang, C., Wang, H., Li, Y., Qiu, Y., Luo, G., Li, H., Shen, H., Chen, M., Bu, D., Zhu, W., Cai, Y., Xiong, X., Jiang, Y., Heng, Y., Zhang, P., Yu, B., Xie, B., & Bao, Y. (2024). ieda: An open-source infrastructure of eda. In *Proceedings of the 29th Asia and South Pacific Design Automation Conference*, ASPDAC '24 (pp. 77–82). IEEE Press.
38. Song, T. (2019). Opportunities and challenges in designing and utilizing vertical nanowire FET (V-NWFET) standard cells for beyond 5 nm. *IEEE Transactions on Nanotechnology, 18*, 240–251.
39. Kim, T., Jeong, J., Woo, S., Yang, J., Kim, H., Nam, A., Lee, C., Seo, J., Kim, M., Ryu, S., Oh, Y., & Song, T. (2023). Ns3k: A 3-nm nanosheet fet standard cell library development and its impact. *IEEE Transactions on Very Large Scale Integration (VLSI) Systems, 31*(2), 163–176.
40. Amrouch, H., van Santen, V. M., Pahwa, G., Chauhan, Y., & Henkel, J. (2020). Ncfet to rescue technology scaling: Opportunities and challenges. In *2020 25th Asia and South Pacific Design Automation Conference (ASP-DAC)* (pp. 637–644).
41. Chauhan, Y. S., Venugopalan, S., Karim, M. A., Khandelwal, S., Paydavosi, N., Thakur, P., Niknejad, A. M., & Hu, C. C. (2012). Bsim: Industry standard compact MOSFET models. In *2012 Proceedings of the European Solid-State Device Research Conference (ESSDERC)* (pp. 46–49).
42. Thomann, S., Novkin, R., Li, J., Hu, Y., Xiong, J., & Amrouch, H. (2024). ML-TCAD: Accelerating FeFET reliability analysis using machine learning. *IEEE Transactions on Electron Devices, 71*(1), 213–222.
43. Liu, Y., Zhang, W., Zhu, Z., Dong, X., & Deng, W. (2021). DC model for SiC MOSFETs using artificial neural network optimized by artificial bee colony algorithm. *AIP Advances, 11*(11), 115219.

44. Pedregosa, F., Varoquaux, G., Gramfort, A., Michel, V., Thirion, B., Grisel, O., Blondel, M., Prettenhofer, P., Weiss, R., Dubourg, V., Vanderplas, J., Passos, A., Cournapeau, D., Brucher, M., Perrot, M., & Duchesnay, E. (2011). Scikit-learn: Machine learning in Python. *Journal of Machine Learning Research, 12*, 2825–2830.
45. Abadi, M., Barham, P., Chen, J., Chen, Z., Davis, A., Dean, J., Devin, M., Ghemawat, S., Irving, G., Isard, M., Kudlur, M., Levenberg, J., Monga, R., Moore, S., Murray, D. G., Steiner, B., Tucker, P., Vasudevan, V., Warden, P., Wicke, M., Yu, Y., & Zheng, X. (2016). Tensorflow: A system for large-scale machine learning. In *Proceedings of the 12th USENIX Conference on Operating Systems Design and Implementation*, OSDI'16 (pp. 265–283). USENIX Association.
46. Woo, S., Jeong, H., Choi, J., Cho, H., Kong, J.-T., & Kim, S. (2022). Machine-learning-based compact modeling for sub-3-nm-node emerging transistors. *Electronics, 11*(17), 2761.
47. Böscke, T. S., Müller, J., Bräuhaus, D., Schröder, U., & Böttger, U. (2011). Ferroelectricity in hafnium oxide: CMOS compatible ferroelectric field effect transistors. In *2011 International Electron Devices Meeting* (pp. 24.5.1–24.5.4).
48. Mulaosmanovic, H., Ocker, J., Müller, S., Noack, M., Müller, J., Polakowski, P., Mikolajick, T., & Slesazeck, S. (2017). Novel ferroelectric fet based synapse for neuromorphic systems. In *2017 Symposium on VLSI Technology* (pp. T176–T177).
49. Chatterjee, S., Thomann, S., Ni, K., Chauhan, Y. S., & Amrouch, H. (2022). Comprehensive variability analysis in dual-port fefet for reliable multi-level-cell storage. *IEEE Transactions on Electron Devices, 69*(9), 5316–5323.
50. Jerry, M., Chen, P.-Y., Zhang, J., Sharma, P., Ni, K., Yu, S., & Datta, S. (2017). Ferroelectric fet analog synapse for acceleration of deep neural network training. In *2017 IEEE International Electron Devices Meeting (IEDM)* (pp. 6.2.1–6.2.4).
51. He, K., Zhang, X., Ren, S., & Sun, J. (2016). Deep residual learning for image recognition. In *2016 IEEE Conference on Computer Vision and Pattern Recognition (CVPR)* (pp. 770–778).
52. Cox, P., Yang, P., Mahant-Shetti, S.S., & Chatterjee, P. (1985). Statistical modeling for efficient parametric yield estimation of mos vlsi circuits. *IEEE Transactions on Electron Devices, 32*(2), 471–478.
53. Alvarez, A. R., Abdi, B .L., Young, D. L., Weed, H. D., Teplik, J., & Herald, E. R. (1988). Application of statistical design and response surface methods to computer-aided vlsi device design. *IEEE Transactions on Computer-Aided Design of Integrated Circuits and Systems, 7*(2), 272–288.
54. Young, D. L., Teplik, J., Weed, H. D., Tracht, N. T., & Alvarez, A. R. (1991). Application of statistical design and response surface methods to computer-aided VLSI device design II. desirability functions and taguchi methods. *IEEE Transactions on Computer-Aided Design of Integrated Circuits and Systems, 10*(1), 103–115.
55. Lamidi, S., Olalere, R., Yekinni, A., & Adesina, K. (2024). Design of experiments (DOE): Applications and benefits in quality control and assurance. In S. Z. Qamar & N. Al-Hinai (Eds.), *Quality control and quality assurance* (Chap. 2). IntechOpen.
56. Khan, M. A. H., Rahman, A. S.M. Z., Muntasir, T., Acharjee, U. K., & Layek, M. A. (2011). Multiple polynomial regression for modeling a mosfet in saturation to validate the early voltage. In *2011 IEEE Symposium on Industrial Electronics and Applications* (pp. 261–266). IEEE.
57. Hammouda, H. B., Mhiri, M., Gafsi, Z., & Besbes, K. (2008). Neural-based models of semiconductor devices for spice simulator. *American Journal of Applied Sciences, 5*(4), 385–391.
58. Li, M., İrsoy, O., Cardie, C., & Xing, H. G. (2016). Physics-inspired neural networks for efficient device compact modeling. *IEEE Journal on Exploratory Solid-State Computational Devices and Circuits, 2*, 44–49.
59. Kao, M.-Y., Kam, H., & Hu, C. (2022). Deep-learning-assisted physics-driven mosfet current-voltage modeling. *IEEE Electron Device Letters, 43*(6), 974–977.
60. Choe, G., Ravindran, P. V., Lu, A., Hur, J., Lederer, M., Reck, A., Lombardo, S., Afroze, N., Kacher, J., Khan, A. I., & Yu, S. (2022). Machine learning assisted statistical variation

analysis of ferroelectric transistors: From experimental metrology to predictive modeling. In *2022 IEEE Symposium on VLSI Technology and Circuits (VLSI Technology and Circuits)* (pp. 336–337).
61. ROHM-sct3080kl (2023). https://www.rohm.com/products/sic-power-devices/sic-mosfet/sct3080kl-product. Accessed: Augest 09, 2023.
62. Mehta, K., & Wong, H.-Y. (2021). Prediction of finfet current-voltage and capacitance-voltage curves using machine learning with autoencoder. *IEEE Electron Device Letters, 42*(2), 136–139.
63. Jespers, P. G. A., & Murmann, B. (2017). *Introduction* (pp. xii–xiv). Cambridge University Press.
64. İslamoğlu, G., Çakici, T. O., Afacan, E., & Dündar, G. (2019). Artificial neural network assisted analog ic sizing tool. In *2019 16th International Conference on Synthesis, Modeling, Analysis and Simulation Methods and Applications to Circuit Design (SMACD)* (pp. 9–12).
65. Bäck, T. (1996). *Evolutionary algorithms in theory and practice: Evolution strategies, evolutionary programming, genetic algorithms*. Oxford University Press.
66. Wong, Y. C., Syafeeza, A. R., & Hamid, N. A. (2015). A transistor sizing tool for optimization of analog CMOS circuits: Tsop. *International Journal of Engineering and Technology, 7*(1), 140–146.
67. Touloupas, K., & Sotiriadis, P. P. (2022). Mixed-variable bayesian optimization for analog circuit sizing through device representation learning. *Electronics, 11*(19), 3127.
68. Lourenço, N., & Horta, N. (2012). GENOM-POF: Multi-objective evolutionary synthesis of analog ICs with corners validation. In *Proceedings of the 14th Annual Conference on Genetic and Evolutionary Computation*, GECCO '12 (pp. 1119–1126). Association for Computing Machinery.
69. Fayazi, M., Colter, Z., Afshari, E., & Dreslinski, R. (2021). Applications of artificial intelligence on the modeling and optimization for analog and mixed-signal circuits: A review. *IEEE Transactions on Circuits and Systems I: Regular Papers, 68*(6), 2418–2431.
70. Lyu, W., Xue, P., Yang, F., Yan, C., Hong, Z., Zeng, X., & Zhou, D. (2018). An efficient bayesian optimization approach for automated optimization of analog circuits. *IEEE Transactions on Circuits and Systems I: Regular Papers, 65*(6), 1954–1967.
71. Touloupas, K., & Sotiriadis, P. P. (2022). Locomobo: A local constrained multiobjective bayesian optimization for analog circuit sizing. *IEEE Transactions on Computer-Aided Design of Integrated Circuits and Systems, 41*(9), 2780–2793.
72. Vişan, C., Pascu, O., Stănescu, M., Şandru, E.-D., Diaconu, C., Buzo, A., Pelz, G., & Cucu, H. (2022). Automated circuit sizing with multi-objective optimization based on differential evolution and bayesian inference. *Knowledge-Based Systems, 258*, 109987.
73. Daems, W., Gielen, G., & Sansen, W. (2003). Simulation-based generation of posynomial performance models for the sizing of analog integrated circuits. *IEEE Transactions on Computer-Aided Design of Integrated Circuits and Systems, 22*(5), 517–534.
74. Afacan, E., Lourenço, N., Martins, R., & Dündar, G. (2021). Review: Machine learning techniques in analog/RF integrated circuit design, synthesis, layout, and test. *Integration, 77*, 113–130.
75. Wolfe, G., & Vemuri, R. (2006). Extraction and use of neural network models in automated synthesis of operational amplifiers. *IEEE Transactions on Computer-Aided Design of Integrated Circuits and Systems, 22*(2), 198–212.
76. Garitselov, O., Mohanty, S. P., & Kougianos, E. (2012). Fast-accurate non-polynomial metamodeling for nano-CMOS PLL design optimization. In *Proceedings of the 2012 25th International Conference on VLSI Design*, VLSID '12 (pp. 316–321). IEEE Computer Society.
77. Lourenço, N., Rosa, J., Martins, R., Aidos, H., Canelas, A., Póvoa, R., & Horta, N. (2018). On the exploration of promising analog IC designs via artificial neural networks. In *2018 15th International Conference on Synthesis, Modeling, Analysis and Simulation Methods and Applications to Circuit Design (SMACD)* (pp. 133–136).

78. Wang, Z., Luo, X., & Gong, Z. (2018). Application of deep learning in analog circuit sizing. In *Proceedings of the 2018 2nd International Conference on Computer Science and Artificial Intelligence*, CSAI '18 (pp. 571–575). Association for Computing Machinery.
79. Harsha, M. V., & Harish, B. P. (2020). Artificial neural network model for design optimization of 2-stage op-amp. In *2020 24th International Symposium on VLSI Design and Test (VDAT)* (pp. 1–5).
80. Rosa, J. P. S., Guerra, D. J. D., Horta, N. C. G., Martins, R. M. F., & Lourenço, N. C. C. (2020). *Using ANNs to size analog integrated circuits* (pp. 45–66). Springer International.
81. Canelas, A., Martins, R., Póvoa, R., Lourenço, N., & Horta, N. (2017). Efficient yield optimization method using a variable k-means algorithm for analog IC sizing. In *Design, Automation & Test in Europe Conference & Exhibition (DATE), 2017* (pp. 1201–1206).
82. Liu, M., Turner, W. J., Kokai, G. F., Khailany, B., Pan, D. Z., & Ren, H. (2021). Parasitic-aware analog circuit sizing with graph neural networks and bayesian optimization. In *2021 Design, Automation & Test in Europe Conference & Exhibition (DATE)* (pp. 1372–1377).
83. Ren, H., Kokai, G. F., Turner, W. J., & Ku, T.-S. (2020). Paragraph: layout parasitics and device parameter prediction using graph neural networks. In *Proceedings of the 57th ACM/EDAC/IEEE Design Automation Conference*, DAC '20. IEEE Press.
84. Budak, A. F., Gandara, M., Shi, W., Pan, D. Z., Sun, N., & Liu, B. (2022). An efficient analog circuit sizing method based on machine learning assisted global optimization. *IEEE Transactions on Computer-Aided Design of Integrated Circuits and Systems, 41*(5), 1209–1221.
85. Song, L.-Y., Kuo, T.-C., Wang, M.-H., Liu, C.-N. J., & Huang, J.-D. (2022). Fast variation-aware circuit sizing approach for analog design with ml-assisted evolutionary algorithm. In *2022 27th Asia and South Pacific Design Automation Conference (ASP-DAC)* (pp. 80–85).
86. Agarwal, K., Jain, A., Amuru, D., & Abbas, Z. (2022). Fast and efficient resnn and genetic optimization for PVT aware performance enhancement in digital circuits. In *2022 International Symposium on VLSI Design, Automation and Test (VLSI-DAT)* (pp. 1–4).
87. Lberni, A., Marktani, M. A., Ahaitouf, A., & Ahaitouf, A. (2024). Analog circuit sizing based on evolutionary algorithms and deep learning. *Expert Systems with Applications, 237*, 121480.
88. Shi, W., Wang, H., Gu, J., Liu, M., Pan, D. Z., Han, S., & Sun, N. (2022). Robustanalog: Fast variation-aware analog circuit design via multi-task RL. In *2022 ACM/IEEE 4th Workshop on Machine Learning for CAD (MLCAD)* (pp. 35–41).
89. Lillicrap, T. P., Hunt, J. J., Pritzel, A., Heess, N., Erez, T., Tassa, Y., Silver, D., & Wierstra, D. (2019). *Continuous control with deep reinforcement learning*.
90. Budak, A. F., Zhu, K., Chen, H., Poddar, S., Zhao, L., Jia, Y., & Pan, D. Z. (2023). Joint optimization of sizing and layout for ams designs: Challenges and opportunities. In *Proceedings of the 2023 International Symposium on Physical Design*, ISPD '23 (pp. 84–92). Association for Computing Machinery.
91. Lourenço, N., Martins, R., Canelas, A., Póvoa, R., & Horta, N. (2016). AIDA: Layout-aware analog circuit-level sizing with in-loop layout generation. *Integration, 55*, 316–329.
92. Kim, D.-W., & Choi, T.-Y. (2000). Delay time estimation model for large digital CMOS circuits. *VLSI Design, 11*(2), 161–173.
93. Stillmaker, A., & Baas, B. (2017). Scaling equations for the accurate prediction of CMOS device performance from 180 nm to 7 nm. *Integration, 58*, 74–81.
94. Mutlu, A. A., & Rahman, M. (2005). Statistical methods for the estimation of process variation effects on circuit operation. *IEEE Transactions on Electronics Packaging Manufacturing, 28*(4), 364–375.
95. Brusamarello, L., Wirth, G., Roussel, P., & Miranda, M. (2011). Fast and accurate statistical characterization of standard cell libraries. *Microelectronics Reliability, 51*, 2341–2350.
96. Miranda, M., Roussel, P., Brusamarello, L., & Wirth, G. (2011). Statistical characterization of standard cells using design of experiments with response surface modeling. In *2011 48th ACM/EDAC/IEEE Design Automation Conference (DAC)* (pp. 77–82).

97. Basu, S., Thakore, P., & Vemuri, R. (2007). Process variation tolerant standard cell library development using reduced dimension statistical modeling and optimization techniques. In *8th International Symposium on Quality Electronic Design (ISQED'07)* (pp. 814–820).
98. Miranda, M., Zuber, P., Dobrovolný, P., & Roussel, P. (2011). Variability aware modeling for yield enhancement of SRAM and logic. In *2011 Design, Automation Test in Europe* (pp. 1–6).
99. Chaudhuri, S., & Jha, N. K. (2014). FinFET logic circuit optimization with different FinFET styles: Lower power possible at higher supply voltage. In *2014 27th International Conference on VLSI Design and 2014 13th International Conference on Embedded Systems* (pp. 476–482).
100. Yu, L., Saxena, S., Hess, C., Elfadel, I. A. M., Antoniadis, D., & Boning, D. (2015). Statistical library characterization using belief propagation across multiple technology nodes. In *2015 Design, Automation & Test in Europe Conference & Exhibition (DATE)* (pp. 1383–1388). IEEE.
101. Cheng, L., Gupta, P., & He, L. (2009). Efficient additive statistical leakage estimation. *IEEE Transactions on Computer-Aided Design of Integrated Circuits and Systems, 28*(11), 1777–1781.
102. Moshrefi, A., Aghababa, H., & Shoaei, O. (2018). Statistical estimation of delay in nano-scale CMOS circuits using burr distribution. *Microelectronics Journal, 79*, 30–37.
103. Gourishetty, S., Mandadapu, H., Zahra, A., & Abbas, Z. (2019). A highly accurate machine learning approach to modelling PVT variation aware leakage power in FinFET digital circuits. In *2019 IEEE Asia Pacific Conference on Circuits and Systems (APCCAS)* (pp. 61–64).
104. Bhavesh, M. D., Anilkumar, N. A., Patel, M. I., Gajjar, R., & Panchal, D. (2022). Power consumption prediction of digital circuits using machine learning. In *2022 2nd International Conference on Artificial Intelligence and Signal Processing (AISP)* (pp. 1–6).
105. Hou, L., Zheng, L., & Wu, W. (2006). Neural network based VLSI power estimation. In *2006 8th International Conference on Solid-State and Integrated Circuit Technology Proceedings* (pp. 1919–1921).
106. Stockman, M., Awad, M., Khanna, R., Le, C., David, H., Gorbatov, E., & Hanebutte, U. (2010). A novel approach to memory power estimation using machine learning. In *2010 International Conference on Energy Aware Computing* (pp. 1–3). IEEE.
107. Garg, L., & Sahula, V. (2012). Variability aware support vector machine based macromodels for statistical estimation of subthreshold leakage power. In *2012 International Conference on Synthesis, Modeling, Analysis and Simulation Methods and Applications to Circuit Design (SMACD)* (pp. 253–256).
108. Zhou, Y., Ren, H., Zhang, Y., Keller, B., Khailany, B., & Zhang, Z. (2019). Primal: Power inference using machine learning. In *2019 56th ACM/IEEE Design Automation Conference (DAC)* (pp. 1–6).
109. Zhang, Y., Ren, H., & Khailany, B. (2020). Grannite: Graph neural network inference for transferable power estimation. In *2020 57th ACM/IEEE Design Automation Conference (DAC)* (pp. 1–6).
110. Jiao, X., Rahimi, A., Narayanaswamy, B., Fatemi, H., de Gyvez, J. P., & Gupta, R. K. (2015). Supervised learning based model for predicting variability-induced timing errors. In *2015 IEEE 13th International New Circuits and Systems Conference (NEWCAS)* (pp. 1–4).
111. Amuru, D., Zahra, A., Vudumula, H. V., Cherupally, P. K., Gurram, S. R., Ahmad, A., & Abbas, Z. (2023). AI/ML algorithms and applications in VLSI design and technology. *Integration, 93*, 102048.
112. Alam, M., Kang, K., Paul, B. C., & Roy, K. (2007). Reliability- and process-variation aware design of VLSI circuits. In *2007 14th International Symposium on the Physical and Failure Analysis of Integrated Circuits* (pp. 17–25).
113. Chen, T., Sun, Q., & Yu, B. (2021). Machine learning in nanometer ams design-for-reliability : (Invited paper). In *2021 IEEE 14th International Conference on ASIC (ASICON)* (pp. 1–4).
114. Landeros Muñoz, N., Valero, A., Tejero, R. G., & Zoni, D. (2022). Gated-CNN: Combating NBTI and HCI aging effects in on-chip activation memories of convolutional neural network accelerators. *Journal of Systems Architecture, 128*, 102553.

115. Chen, Y.-G., Lin, I.-C., & Wei, Y.-C. (2021). A novel NBTI-aware chip remaining lifetime prediction framework using machine learning. In *2021 22nd International Symposium on Quality Electronic Design (ISQED)* (pp. 476–481).
116. Gaber, L., Hussein, A. I., & Moness, M. (2021). Fault detection based on deep learning for digital VLSI circuits. *Procedia Computer Science, 194*, 122–131. 18th International Learning & Technology Conference 2021.
117. Ye, F., Firouzi, F., Yang, Y., Chakrabarty, K., & Tahoori, M. B. (2016). On-chip droop-induced circuit delay prediction based on support-vector machines. *IEEE Transactions on Computer-Aided Design of Integrated Circuits and Systems, 35*(4), 665–678.
118. Ma, Y., Ren, H., Khailany, B., Sikka, H., Luo, L., Natarajan, K., & Yu, B. (2019). High performance graph convolutionai networks with applications in testability analysis. In *2019 56th ACM/IEEE Design Automation Conference (DAC)* (pp. 1–6).
119. COSMOL (2024). https://www.comsol.com/comsol-multiphysics. Accessed: Jun 09, 2024.
120. Lamichhane, S., Peng, S., Jin, W., & Tan, S. X.-D. (2021). Fast electrostatic analysis for VLSI aging based on generative learning. In *2021 ACM/IEEE 3rd Workshop on Machine Learning for CAD (MLCAD)* (pp. 1–6).
121. Arjovsky, M., Chintala, S., & Bottou, L. (2017). *Wasserstein gan.*
122. Thakur, G., Jain, S., & Sohal, H. (2022). Current issues and emerging techniques for VLSI testing: A review. *Measurement: Sensors, 24*, 100497.
123. https://ieee-dataport.org/, 201X.
124. Srivastava, P., Kumar, P., & Abbas, Z. (2023). Qualitative data augmentation for performance prediction in VLSI circuits. *arXiv preprint arXiv:2302.07566.*
125. Liu, J., Hassanpourghadi, M., Zhang, Q., Su, S., & Chen, M. S.-W. (2020). Transfer learning with bayesian optimization-aided sampling for efficient AMS circuit modeling. In *2020 IEEE/ACM International Conference On Computer Aided Design (ICCAD)* (pp. 1–9).
126. Wu, Z., & Savidis, I. (2022). Transfer learning for reuse of analog circuit sizing models across technology nodes. In *2022 IEEE International Symposium on Circuits and Systems (ISCAS)* (pp. 1033–1037).
127. https://builtin.com/machine-learning/meta-learning/, 2023.

Chapter 2
Linking System and Circuit Design by AI Techniques

Frank Wawrzik, Johannes Koch, Khushnood Adil Rafique, Sören Kwasigroch, Moritz Herzog, and Christoph Grimm

2.1 Introduction

Current electronic and system design methods comprise a flow from eliciting requirements to creating system models and architecture simulations to then testing and implementing the models and requirements in hardware (and software). All of these stages are served by their predominant popular tools. For example, requirements are gathered and structured in Doors. System models are often created in SysML [1], with tools like Enterprise Architect, PTC Integrity Modeler, Cameo, MagicDraw, Papyrus, or others. At this stage requirements, an overall overview, and testable simulation models can be well represented and linked. System engineers and integrators work at this level.

At architecture level, SystemC [2, 3] is used for HW/SW Systems. It offers an abstract way to simulate electronic systems and their environment. Often we find combinations of simulators and hybrid approach with, for example, co-simulating mechanical systems in Modelica. At register transfer level, VHDL and other languages (respectively, tools) are used. To support the integration and exchange of models, in particular, IP-XACT is used that has a focus on interfaces and lower levels of abstraction.

In this chapter, we describe the complementary use of AI techniques that complement the existing tools and processes. The objective is to link the different levels of abstraction and even the context at application layer, e.g., sensors in the Internet of Things. For this purpose, we use an ontology-defined knowledge base. The ontology comprises basic classes inspired by the standard ISO26262 that include not only functions, parts, components, and systems but also sensors,

F. Wawrzik (✉) · J. Koch · K. A. Rafique · S. Kwasigroch · M. Herzog · C. Grimm
Chair Design of Cyber-Physical Systems, Gottlieb-Daimler-Straße, Kaiserslautern, Germany
e-mail: wawrzik@cs.uni-kl.de; johannes.koch@cs.uni-kl.de; khushnood.rafique@cs.uni-kl.de; soeren.kwasigroch@cs.uni-kl.de; m_herzog15@cs.uni-kl.de; grimm@cs.uni-kl.de

A. Iranmanesh, H. Sayadi (eds.), *AI-Enabled Electronic Circuit and System Design*,
https://doi.org/10.1007/978-3-031-71436-8_2

actuators, microcontrollers, processors, and their properties and that link their functions to other parts and properties. By the end of Sect. 2.4, we will demonstrate how our basic semantic model is a very complete, systematic, and well-founded approach to describing physical designs and in a more comprehensive context the wider domain of electronics in general. This has many advantages like finally a model upon which every process and interaction, EDA tool, manufacturing process, and digital twin can be based to name but one. The open world assumption and integratability of ontologies and knowledge graphs allows us to extend the knowledge and build upon the already existing data.

In this book chapter, we elaborate on research areas related to electronic design. We exemplify our research throughout this book chapter with various examples. These include, but are not limited to, the following:

- Operational Amplifier in a tire pressure metering system (TPMS) in Sect. 2.2)
- Delta-Sigma D/A converter (in Sect. 2.3)
- Hardware Accelerator with Neural Net (in Sect. 2.4)
- General Electronics Hardware (e.g., processors, microcontrollers, sensors, converter, etc.) (in Sect. 2.4)
- Automotive Electrical System (in Sect. 2.7)

One research area is the new advancements in AI technologies and their integration into the design process. Figure 2.1 gives a high-level overview. In this book chapter, we take the reader, engineer, academic, researcher, and student by the hand and walk him through all the phases of the integrated process of knowledge-based electronic design capture. As can be seen in the figure, the design, requirements, and specification are captured in SysMD and SysMLv2, respectively, whereas SysMD is based on a subset of the SysMLv2 metamodel; both work interchangeably. If desired, the user can also select the SysMLv2 design language and directly create the model in SysMLv2. Section 2.2 will explain this phase in detail. A digital twin, which runs on a backend, is the virtual representation of our physical design,

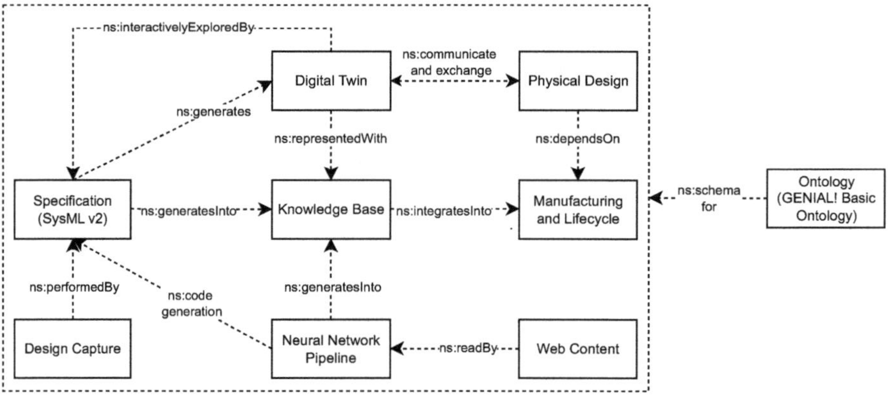

Fig. 2.1 Design capture and concept generation with SysMLv2

described by the ontology and explained in Sect. 2.3. Section 2.4 will then show our ontology reference model in detail. Further, the content of our knowledge base is described. The next two sections describe how we use various machine learning methods to build both the knowledge base in Sect. 2.5 and code generation for our system design in Sect. 2.6. Section 2.7 finally shows another application of our system design framework and tool on modern design architectures of the automotive electrical system (here frequently referred to as "boardnet"). This section also shows an example of the application of ontology-based reasoning to augment the design process. Section 2.8 concludes, summarizes the work, and contextualizes its importance. In the remainder of this section, we give an introduction of SysMLv2 and our SysMD (System Markdown) language and tool.

2.1.1 SysMLv2 and SysMD

This section introduces SysMLv2, an evolution of the widely adopted Systems Modeling Language, and SysMD, which builds on SysMLv2 and provides means for improved documentation, and linking expressions and constraints to a (symbolic/numeric) solver to check consistency of models.

2.1.1.1 SysMLv2

As products and systems grow in complexity across diverse domains, the Systems Modeling Language (SysML) has emerged as a widespread standard for systems engineering. However, the complexity of SysML itself and the expertise required for its proper use presents challenges. SysMLv2 addresses the limitations of its predecessor by introducing advancements that cater to the evolving needs of the industry. Unlike its predecessors, SysMLv2 addresses the importance of human readability and formal analysis capabilities. It goes beyond traditional modeling languages like UML and SysML, offering a more intuitive syntax and enhanced analytical features. SysMLv2 provides the following key features:

1. **Textual representation:** SysMLv2 introduces a textual representation format that complements diagrams.
2. **Standardized REST API:** A standardized REST API via which models and data can be exchanged is given as part of the standard.
3. **Metamodel and semantic libraries:** A set of base classes is defined on top of which profiles can be developed.

2.1.1.2 SysMD: Bridging the Gap in Systems Modeling

While SysMLv2 addresses the challenges posed by complex modeling languages, there is another aspect that needs consideration—bridging the gap for domain experts who lack modeling expertise. SysMD (System Modeling with Documentation) is a novel framework that makes systems modeling more inclusive, accessible, and adaptable. SysMD has the following key features:

1. **Integration of documentation and models:** One of the SysMD's standout features is its seamless integration of modeling with documentation. By combining Markdown documents with models, it enhances the maintainability of knowledge bases. This integration allows for the inclusion of formatted text, tables, and references to external documents.
2. **Inclusive approach for various stakeholders:** SysMD recognizes that systems modeling is not exclusive to modeling experts. Its syntax is designed to be accessible to individuals without prior modeling expertise, making it easier to use and understand for various stakeholders, including domain and management experts.
3. **Continuous consistency checking:** To support formal analysis, SysMD incorporates continuous consistency checking. This feature employs constraint propagation techniques, enabling swift evaluations of design element dependencies. This advantage plays out in agile development processes where quick assessments and feedback are required.

In this chapter, we decided to display our models in SysMD, because for the average reader the models are easy to comprehend. At the same time, again, we want to emphasize the interoperability to and equivalence of SysMLv2, for those who want to reuse our models and adapt the presented framework. SysMD operates as a specialized subset within the broader context of SysMLv2. SysMLv2, being a comprehensive evolution of systems modeling languages, provides a robust framework for expressing complex relationships and structures. Within this framework, SysMD emerges as a tailored solution, designed to enhance accessibility and ease of use. The interoperability between SysML v2 and SysMD ensures that models and the results of constraint analysis achieved through SysMD can be effortlessly integrated into the broader systems modeling ecosystem facilitated by SysMLv2. This symbiotic relationship not only acknowledges the nuanced needs of various stakeholders but also emphasizes the cohesive integration of specialized frameworks within the overarching language, facilitating a smooth and unified approach to systems modeling.

2.2 Design Capture with SysMLv2

Electronic systems face a variety of challenges and tasks. Among them are how to interface optimal with sensors and implement voltage amplification of samples, analog to digital conversion, and the data transmission to name a few. These mostly integrated systems, covering both analog and digital circuits, are often also named analog-mixed signal systems (AMSs). However, for new chip designs, the reusability and the automation for the analog components are not as high as for

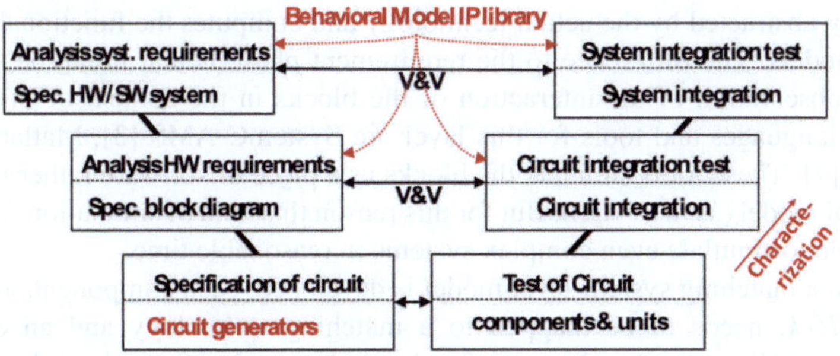

Fig. 2.2 V-Model for electrical system design [4]

the digital designs. This is stated by Prautsch et al. in [4]. Therefore, the design time for analog models is still one of the major bottlenecks [5], even with well-structured design processes. The design for new integrated chips (ICs) or other electrical systems are often following the V-Model [4]. As Fig. 2.2 shows, there is a top-down and bottom-up phase. First, requirements are specified, and then models of the system level and the block level are created in their respective diagrams. After the block models are created, the circuit designer starts creating models of the concrete implementation. More details to these abstraction layers will be described in Sect. 2.2.1.

SysMLv2 is a system engineering language standardized by the OMG [1]. It can be used both on the requirements analysis phase and in the verification and validation phase (*V&V*). Each level of the verification and validation phase, which is displayed on the right branch, is always adjacent to one of the levels of the abstraction level for modeling on the left branch. Each derived circuit will always be tested against its specification on its own first, before an integrated test takes place. The models are combined together as block models and tested against this block-level specification. This part of the V-model is also called the bottom-up phase [4].

The propagation of performances from the circuit models to higher abstraction layers should reduce computing resources. On higher abstraction layers, the simulation speed is higher, which is desired for integration tests. To address this issue, we propose our own *hierarchical modeling and verification* method, which will be illustrated in the following subsection. It will be demonstrated on an example system-level model of a tire pressure meter system (TPMS) described shortly in Sect. 2.2.2.

2.2.1 Abstraction Layers in the AMS Design

The AMS design process is roughly separated in three abstraction layers. They are introduced in Fig. 2.2 in their appropriate design phase. The *system-level* model contains the complete intended design and constitutes known ideal block models. It is mostly abstracted by the actual technology and computes the functionality of the block models. The difference to the requirement phase is that it can be simulated for the observation of the interaction of the blocks in the context of this system. Typical languages and tools for this layer are SystemC AMS [3], Matlab [6], and QEMO [7]. These tools simulate the blocks as a piece of software rather than as an electrical model (as in SPICE). But for this reason this kind of simulation framework is capable to simulate even complex systems in reasonable time.

When a matching system-level model is developed, each component, also called *IP* or *block*, needs to be mapped to a matching technology and an electronic component. For some blocks predefined solutions exist. Sometimes however we need to adapt properties like bit accuracy, sizing, and others. Because of these increased details of the model, the time for exhaustive simulation of one IP becomes a challenging task. Here models from hardware description languages are simulated with high accuracy like SPICE or can be synthesized on programmable logic like FPGA via VHDL AMS or Verilog AMS. The difficulty in the circuit-level abstraction layer is that various constraints (for instance, the range of voltage or temperature) need to be satisfied for the theoretical circuit models to be valid. This constraint often needs expert knowledge and complicate automation from block level to circuit level [8]. Some research approaches in this kind of model need to be mentioned to compare state of the art: PyCells are parameterized and scalable models [5] and generators for resistors or capacitors. They provide placement of these passive components also in a scalable manner [9].

2.2.2 Example Tire Pressure Meter System (TPMS)

This demonstrator model was created in the BMBF project AnastASICA. It gives insight how the circuit performance with dependencies to circuit parameters can be made visible in a system context. Its structure is depicted in Fig. 2.3. It consists of the analog component *PressureSensor*, two *negating operational amplifiers (OpAmps)*, and a 10-bit ADC. Lastly, a digital *DetectionUnit* gives a status about the development of the pressure in the tire. In this example only the amplifier model receives performance values from the circuit-level model. The remaining models are kept idealized such that the simulation time is low. This TPMS is also specified in SysMLv2 which helped discussions with partners in the project early on.

2.2.3 System Modeling and SysMLv2

In the design process it is common to implement prototypes for checking if the functionality of intended system fits to the specification and get fast feedback. It is also a good way to present the first understanding of the specification to the

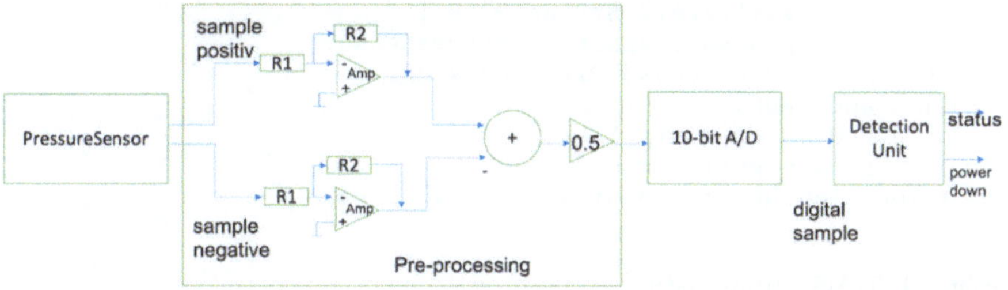

Fig. 2.3 Block diagram of the tire pressure meter system

stakeholders. For this purpose, it can be beneficial in the AMS design to present the results of a system model simulation that demonstrate the (unclear) specified functionalities. Therefore, the first model will be created in the top-down phase of the process.

The SysMLv2 language processed in the SysMD notebook was used as requirement tracking and documentation tool in the AnastASICA project for the TPMS model. It should help the system model designer to keep an overview of the specification and get a scale over the main quantities. However this model is only intended for static requirements and constraint checking. Dynamic behavior is simulated in the SystemC AMS system model, respectively.

The system model of the TPMS resulted from the following SysMLv2 specification:

```
part def ReceiverUnit: Component; // ...
part def SlowLossDetection: Function;  // ...
part def SuddenLossDetection: Function;  // ...
part def PressureSensor: Component {
    attribute samplerate: Quantity [kHz] = 1.0 [kHz];
    attribute vout: Voltage [mV] = 100.0 [mV];
}
part def PreProcessing: Component {
    attribute samplerateIn: Real [kHz] = PressureSensor::
    samplerate;
    attribute amplification: Real(1 .. 1000);
    attribute voutPreProcess: Real(0 .. 1) [V] = PressureSensor::
    vout * amplification;
}

part def ADC: Component {
    vinADC: Real (0 .. 1) = PreProcessing::voutPreProcess;
    samplerateIn : Real [kHz] = PreProcessing::samplerate;
    samplerateOut: Real [Hz] = 100.0 [Hz];
    decimation: Real(2..100) = samplerateIn/samplerateOut;
}

part def Microcontroller: Processor;
part def DetectionSW: Software;

part def AnastASICA::SensorUnit {
```

```
    attribute asilLevel: String = "B";
    attribute pressureSensor: PressureSensor;
    attribute preProcess: PreProcessing;
    part adc: ADC;
    part cpu: Microcontroller;
    part sw: DetectionSW;
    allocation hwsw: Executes from sw to cpu;
}
```

Listing 2.1 TPMS systems model

We omitted here the requirements for the *Microcontroller* and *DetectionSW* so that we can focus on one block model and its refinements later on. The above model defines the decomposition of the system model in its block model components. For the specification of block model attributes, SysMLv2 requests a type (e.g., *DetectionSW* is of the type *sw (software)*). Additionally, it offers to assign a unit (e.g., *kHz* or *V*).

The *PreProcessing* model amplifies the value of the *PressureSensor* to the range the *ADC* is defined which is $[0.0, 1.0]V$. In this current model SysMLv2, therefore, the constraint net evaluating dependencies would constrain the amplification from *[1 .. 10]*.

Such requirement model gives the system model designer and the circuit model designer the dimension of its model and helps to communicate with common understanding in mind. In the AnastASICA project, this TPMS system model was then written with SystemC AMS. The focus of the model is the preprocessing function by a high-level negating operational amplifier. There are the output_range_min, output_range_max, and the offset error that were implemented in the SystemC AMS model. Each of these performances is characterized by 69 design parameters and their respective deviations and propagated to the OpAmp model.

2.2.4 Refinement of Requirement Model

While exploring a design, sometimes missing parts in the specification can be found. During a design process, sometimes missing information occurs, e.g., due to incomplete or inconsistent specification. This could occur due to other expert knowledge that is considered or, when furother parts with nonideal properties are added to the system model. Also a change request, which is added the specification, will trigger to reevaluate the overall feasibility. These changes need to be added to the specification and the first models to not lose track of the current design goal. In the *TPMS* model the hierarchical approach to shift circuit performances to the blocks of the system model was realized. Here the input of the circuit designer is a list of nonideal properties its model could be characterized for the sensitivity to design parameters. For the verification and validation purposes, the system designer makes a decision which performance should be considered on the system context. The method of the hierarchical approach followed the proposal of Zivkovic et al. [10]. The main idea is to approximate circuit-level performances as affine kernels stored in affine forms the quantities of affine arithmetic [11]. This has the benefit that the save inclusion of each reachable value can be ensured as well that linear sensitivities can be stored by default. The method to compute the affine form for a performance can be summarized as follows:

1. Select for the performance x the $n \in \mathrm{N}$ parameter $p_1, ..., p_n$ and its possible range $[\underline{p_i}, \overline{p_i}]$ for which the influence should be tracked.
2. Determine a meaningful working point x_0 in the parameter space of the n parameters where the circuit performance should be characterized.
3. Characterize the deviations as the first-order partial derivation $x_j = \frac{\delta x}{\delta p_j} \restriction_{x=x_0}$ in the working point x_0.
4. Normalize the sensitivity (*normalization called* x_j) such that the parameter can be modeled as $p_i \in [-1, 1]$.
5. Via optimization determine maximum $\overline{x}$ and minimum $\underline{x}$ of x, such that missing constants $\overline{p_{n+1}} = \overline{x} - (x_0 + \sum_{i=1}^{n} |x_i|)$ to the upper bound and $\underline{p_{n+1}} = (x_0 - \sum_{i=1}^{n} |x_i|) - \underline{x}$ can be determined.
6. Define the affine form for the performance x as $\tilde{x} = x_0 + \sum_{j=1}^{n} x_j * p_j + [\underline{p_{n+1}}, \overline{p_{n+1}}]$.

In the project AnastASICA affine forms were used to propagate a certain performance of a circuit to a block-level SystemC AMS model. Before the start of the simulation, an assignment for the n parameters $p_i \in [-1, 1]$ will be selected and linked to the affine form to get a floating point value for the performance. However, each parameter gets only one unique value for all performances. Therefore the block model is a parameterized model that simulates a certain range of implementations. The OpAmp model gets the *output_range* and the *offset* error. Therefore, the stable output of this model is calculated as

$$output = \max(\min(\text{output\_range\_max}, \text{amplification*input}), \text{output\_range\_min}) + \text{offset}.$$

And also such refinement can be added to the requirement model. Then the *component preprocessing* looks as follows:

```
[..]
part PreProcessing {
    attribute output_range_min: Real(-2 .. 0)[V]
    attribute output_range_max: Real(1 .. 2.5)[V]
    attribute offset : Real(0 .. 0.7)[V]
    attribute samplerateIn: Real [kHz] = PressureSensor::
    samplerate,
    attribute amplification: Real(1 .. 1000),
    attribute vouttmp: Real (0 .. 1) V = min(output_range_max,
    PressureSensor::vout * amplification)
```

```
    attribute voutPreProcess: Real(0 .. 1) [V] = max(
    output_range_min,vouttmp)+ offset
    [..]
}
```

Listing 2.2 Refinement of TPMS systems model

With such refinement you can add further constraints from the circuit models to the *PreProcessing* specification such that a tighter *amplification* and *voutPreProcess* will be observed. Therefore the TPMS model with this OpAmp circuit with the given *output_range_min*, *output_range_max*, and *offset* will still be to the specification.

2.2.5 Summary and Conclusion

In this section utilization of SysMLv2 for electrical design capture was explained. Typical AMS design phases and the abstraction layers for different phases were described. SysMLv2 will help model designers to communicate over the same requirements and provide an option to store the documentation near the system model specification. During the bottom-up verification phase, the specification model can also be used to check the specification regarding its consistency when affine kernel of performance of the chosen circuit model is inserted in the SysMLv2 model. However, for these bottom-up applications, the tool support for characterization and optimization of a circuit model has to be available.

2.3 Digital Twin as a Connector Between the Engineer and the System Model

Within the topic of AI, a physical design that is connected to the knowledge base and the systems model yields the integration of the development and operation (DevOps) cycle for our approach. Artificial intelligence within DevOps can be a useful tool for the engineer. This engineer can observe results about the models, and with the connection to the knowledge base and artificial intelligence, the results of the modeled system can be predicted.

Defining a digital twin is tedious since there are multiple definitions of a digital twin. In this section, the following definition is used.

Definition 1 A digital twin is an exact representation of the structure of an object in reality. The object in reality is called physical twin. The physical twin has to be connected with the digital twin.

The digital twin can be automatically generated from the systems model allowing for the registering of the elements within the digital twin. Comparing with state of the art, multiple works were published considering AutomationML (Schröder et al.

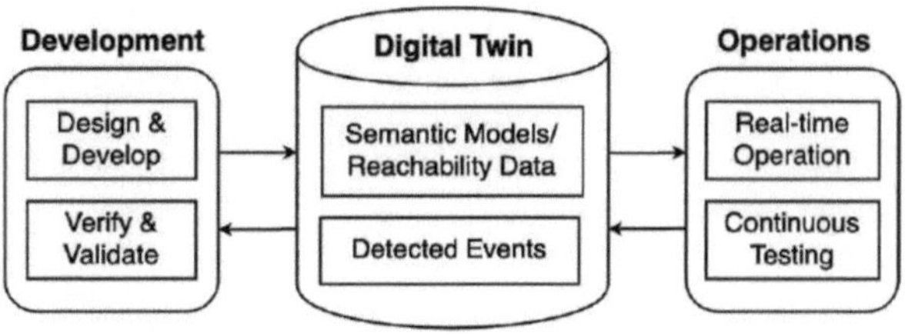

Fig. 2.4 Digital twin and physical twin operation and interaction [12]

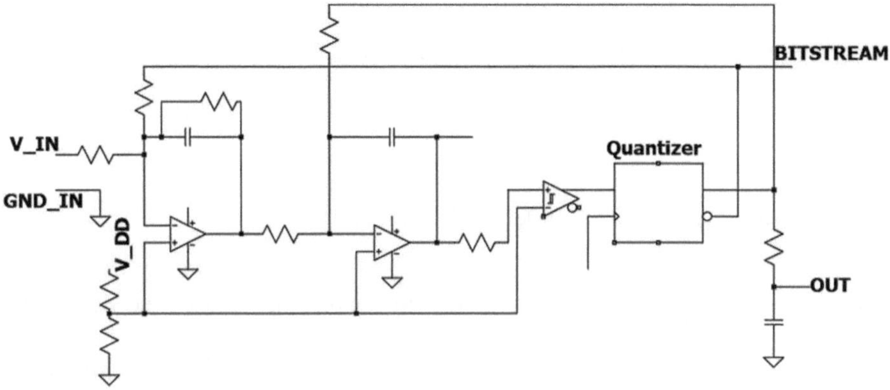

Fig. 2.5 Used schematic of the $\Delta\Sigma$ converter according to [14, 15]

[13]) and SysMD. This enables the creation of the public interface for the digital twin, which is used for the connection between the digital twin and the physical twin. A generation of the interface between the digital twin and the knowledge base is automatically constructed. With the automatically generated interface the data can be mapped to the given property of the physical twin and also the digital twin, allowing for a traceback from each signal. An AI-based anomaly detection and predictive maintenance is enabled by the approach. Figure 2.4 gives a high-level overview.

Within Schröder et al. [13] AutomationML is translated into an XML format. From this XML format, the user interface and the digital twin are created. But these are not interconnected with a knowledge base, and the connection between the digital twin and the physical twin is not clear.

$\Sigma\Delta$ is a general concept for the control of systems [14, 15]. This can be used as an analog to digital converter. Hereby our implementation is similar to Van De Plassche [14]. The resulting circuit is displayed in Fig. 2.5.

The system model of the $\Sigma\Delta$ converter is displayed in Fig. 2.6.

In the following, the described models are used to explain the generated systems and describe the findings resulting from the models. We show a short introduction of the SysMD model for the digital twin and how the digital twin interacts with the knowledge base. Also we enable the possibility for an engineer to view the results. If the systems can be controlled, a digital twin is able to control the system. The

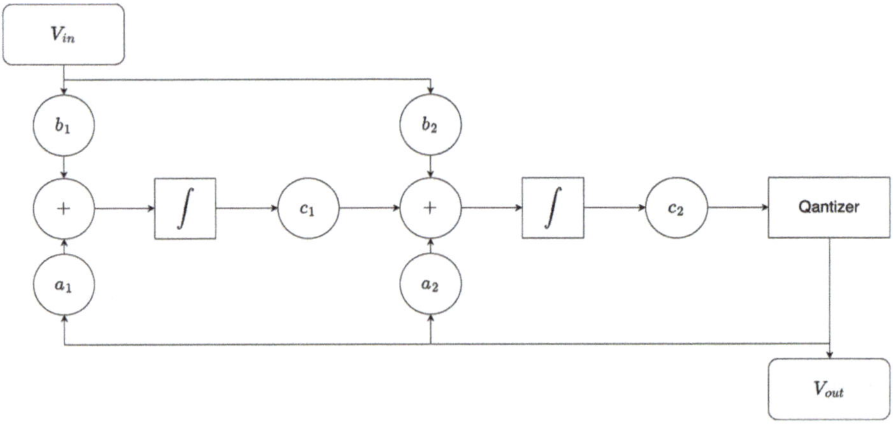

Fig. 2.6 System model of the $\Delta\Sigma$ model according to [14, 15]

engineer is also able to check the values and rework the values if needed. In the following the SysMD model for the $\Delta\Sigma$ is described, and the features that need to be added according to our work related to our dependency description language [16] are described.

2.3.1 Description SysMD Model

Within this chapter the definition of a communication within the component graph and the internal structure of the digital twin is not discussed. First, the system model is parsed. Therefore the public interface has to be marked with recognizable keywords allowing for the automatic parsing. If this is done, the model looks like the model within Listing 2.3.

```
import SI; import ISM;
part DeltaSigma {
    part a [1..2]: Factor;
    part b [1..2]: Factor;
    part c [1..2]: Factor;
    part adder [1..2]: Summing;
    part integrator [1..2]: Integrator;
    part quant:Quantizer;
    attribute inputVoltage: Voltage measurable input;
    attribute outputVoltage: Voltage measurable output;
    connect inputVoltage to a::inputVoltage;
    connect outputVoltage to b::inputVoltage;
    [...]
    connect a1::outputVoltage to adder1::inputVoltage1;
    connect c1::outputVoltage to adder2::inputVoltage3;
```

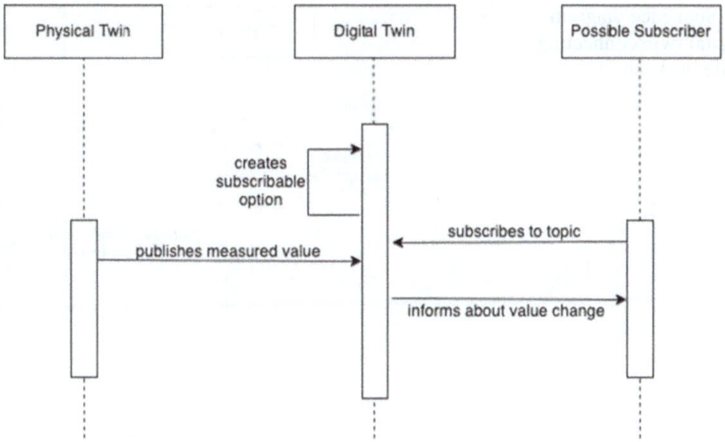

Fig. 2.7 Sequence diagram of the digital twin connections and the possible activities

```
    connect integrator1::inputVoltage to c2::inputVoltage;
    connect outputVoltage to quant::inputVoltage;
}
```

Listing 2.3 ΔΣ systems model

The code of the SysMD parser shown within Listing 2.3 is parsed and should allow for the creation of the System Model shown in Fig. 2.6. This allows the generation of the Digital Twin from the SysMD code.

Within the automatic generation process, the public interface of the digital twin is also generated. This structures the communication between the physical twin and the digital twin. For the digital twin, a test application has been implemented. This application utilizes the MQTT protocol. MQTT was chosen since it allows for a quick and easy setup of the interfaces. Also, the MQTT protocol is known for handling huge datasets. It has to be kept in mind that the underlying protocol of MQTT defines that there is a real-time communication possible, which is needed for the control with a digital twin. The communication is implemented in the following way (Fig. 2.7). The implementation creates automatically the topics of the digital twin. After the creation of the topics, the physical twin can subscribe to the created topics and push its values. The implementation with MQTT allows the use of clients, which allow for the observance of the data. With the same clients, a digital and physical twin can be controlled. An example of this is the display and control in a control room.

Another factor is the connection to the AGILA backend. The AGILA backend is the backend for our tool. The SysMD notebook is the frontend application. This connection to the backend is used to download the systems models. The digital twin utilizes the AGILA backend to store the data within a database. Currently we use the Neo4j graph database. The connection between the AGILA backend and the digital twin is realized with a REST API. The API is modeled according to the SysMLv2

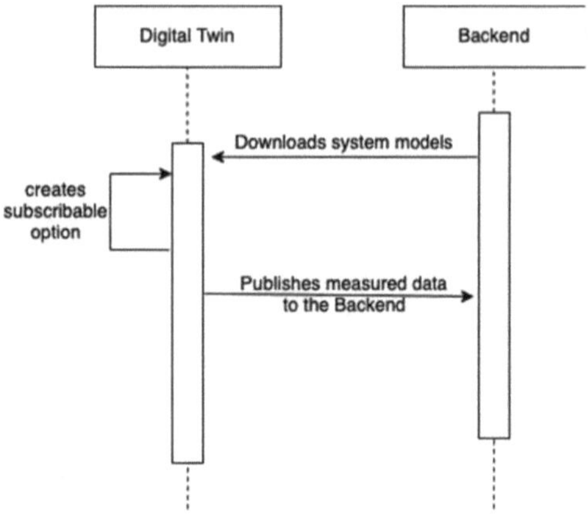

Fig. 2.8 Sequence diagram of the digital twin connecting to the agile backend

REST API. To this data also the structure of the system model is added. This allows later on to review the data. This is a crucial step to store the data in a data lake. With Fig. 2.8 the connection between the AGILA backend and the digital twin application is displayed as a sequence diagram.

Currently, it is established that the data can be viewed live via the MQTT interface and also later on with the data contained within the database. This allows the engineers to check the data always and at any time.

This connection allows the storage of the data in the knowledge base. If the data from the physical twin and the digital twin is within the knowledge base, the data is processed further.

2.3.2 Further Processing of the Data

Currently, there are two ways to access the data. One is the access to the live data. An example is a human machine interface (HMI) in a control center. The second option is the data stored in the knowledge base. The data in the knowledge base can be reviewed later on. But it is to consider that a physical twin generates a lot of data, even in the given example of the $\Delta\Sigma$ converter. With that approach, the already present structure, within the SysMD model, is preserved. Automation of the processing of the data is allowed. The processing of the data can be filtering, anomaly detection, or algorithmic analysis.

The filtering step has to be done for the denoise or simple removal of unnecessary parts of the signal. Especially important is the filtering step for noisy data measured by any sensor. The filtering of noisy data needs to be done before the data is stored or viewed by the engineer.

Considering anomaly detection, the digital twin is used to check for anomalies. These can be anomalies that might be present if the system is failing or if the system is running in an overload situation. This allows to predict the maintenance dates.

Hereby the digital twin denoises the data, checks for anomalies, and generates a prediction report. This prediction report is sent out to the engineers and the operator. To allow for this step, the reasoning for AI needs to be trained beforehand. This needs to be carefully monitored, while training. Another possibility for anomaly detection is if the behavior model is known, like the use case of the $\Delta\Sigma$ modulator. Without an AI anomaly detection can be implemented in that case. Hereby, the data can be calculated, if the input matches the expected output. This allows to predict if the signal is correct or anomalies are found within the signal.

To allow for an easier understanding of the data to the engineer, in combination with the system design, the digital twin's data needs to be connected with the systems model. Implementation of this is done in the SysMD Notebook in the form of the display of the generated data right beside the model and interactively within the model. This allows the engineer to make changes quicker and produce a more stable output of the system later on. Adding to that goal of a digital twin is one of the preferable tools for an engineer.

Since the data is also stored in the knowledge base, a review of the data can be done later on as well. This allows to check if a defect is present in part right at the beginning of the engineering process, allowing for quicker reactions to faulty design decisions and thus reducing the number of failures in the field.

2.3.3 Summary and Conclusion

In this section, we described how to work with a digital twin for an engineer is laid out, and how the digital twin is connected to a knowledge base. A very important step is the filtering step. This is due to the fact that most algorithms require filtered data. Also, sensors mostly have noise in the data that is transmitted to the digital twin. After the data is filtered, algorithms like anomaly detection and predictive maintenance become feasible. Especially in the field of AI anomaly detection is possible even without a model present. This makes AI-based predictive maintenance more affordable to implement. With a digital twin also the display of the data can be changed, especially in the development process. This is because the digital twin is connected to the model, due to the generation process. This allows an implementation, where the data can be interactively viewed within the model. The storage of the data within the knowledge base allows for the usage of the data later on, for analysis and the usage of the data at a later point in time.

2.4 Knowledge Base

This section is structured into three parts with a preface. In the preface we will motivate the work and explain the background of the field. First, in part one, we will introduce the background of our design principles in knowledge building.

By this, we will show what our semantic structure looks like, meaning what are the basic constructs and axioms of our knowledge base and how we describe basic use cases. Often ontology engineers do not have the complete picture of the potential vision of the semantic web in mind and thus focus on building domain ontologies, which often do not get reused. Here we provide some background information and examples that improved our approach. After the background, in the second part, we will introduce the ontology. This was work that originated from a PhD thesis [17] in order to support describing the automotive supply chain and automating communication processes. One of the results was a machine-processable and understandable ontology ".owl" file to describe the domain. It is used to describe all of our computational models. The ontology comprises basic classes that include not only parts, components, systems, sensors, actuators, microcontrollers, processors, and their properties but also their functions and dependencies to other parts and properties. By the end of the second part, we will demonstrate how our basic semantic model is a very complete, systematic, and well-founded approach to describing physical designs and in a more comprehensive context the wider domain of electronics in general. This has many advantages like finally a model upon which every process and interaction, EDA tool, manufacturing process, and digital twin can be based to name but one. The third part of this subchapter describes in more detail the content of our knowledge base. Whereas the first and second parts provide the basics, the third part outlines the knowledge content. We continuously add to the knowledge base in machine- and human-acquired ways and supervise its content. We give examples of machine-acquired content, parts, and properties. We highlight how this approach fits into the circuit design world and helps to integrate a variety of processes and applications.

2.4.1 Motivation

Knowledge bases are not new and originated as part of expert systems in the early 1960s and 1980s (commercial). At this time it was possible to describe simple structures like isA relationships, instances, and some relationships. However, only until the emergence of the semantic web and specifically the ontology web language (OWL), knowledge bases gained widespread use and adoption. It needed a deep philosophical and methodological approach and a well-founded language and logic in order to describe information that the computer was able to understand and process. An ontology is a specification and formalization of a domain of discourse [18]. This definition is a way of saying that with classes, instances, relationships, and data values, any domain, its entities, and how the entities are related can be well described. In this way the ontology can be understood as a domain description. By using definitions and axioms, the domain gets specified, and thus the user and/or machine can use this information to correctly make use of the domain.

Currently, state of the art in the semantic web is that an expert, the ontology engineer, creates and engineers a domain ontology that is then used in the project for addressing general interoperability issues. Ontologies are built for specific domains and published at various domain-specific conferences, including module suites with accompanying knowledge bases. Wikidata [19] and DBpedia [20] are examples of the large knowledge bases currently online on the web. Overall the research field deals with everything related to ontology and knowledge engineering comprising evaluation of ontologies, design patterns, semantic matching, knowledge graph construction with machine learning, reasoning, market of semantic technologies, top-level ontologies, semantic search, standardization, graph databases, higher order languages, and more.

Many ontology models can be found in the domain of IoT. So the initial idea was to achieve interoperability for the devices in smart homes and other IoT infrastructures. Now the trends and applications spread too many domains describing laws, sports, innovation, finance, and so on. Also the applications are now becoming far wider; one new application is, for example, in the semiconductor supply chain to improve exchange and coordination. Another trend is to reason over the knowledge and improve reasoners.

Now that we see more and more ontologies being developed the research trend is shifting toward constructing larger knowledge graphs or knowledge bases with automated means. Often natural language processing, different machine learning techniques, and now more recently also ChatGPT are used. This research area is called knowledge graph construction.

2.4.2 How to Design Well-Founded Knowledge Bases

2.4.2.1 Ontology Design Process

After acquiring the knowledge and understanding of best practices and ontology engineering development methods, the ontology can be constructed. For this we chose an ontology language. There is a variety of languages with different expressivity and purposes, but the most common one and best standardized is the Ontology Web Language (OWL). Designing any ontology most commonly comprises the following steps:

1. Competency Questions
2. Collecting Domain Entities and Lists of Definitions
3. Ontology Development
4. Ontology Validation
5. Ontology Publication
6. Maintenance

At the beginning of each design process of an ontology, the requirements have to be elicited. We address the question of how to represent the domain in form of competency questions. Each question should provide an answer for what and how to represent a certain kind of knowledge. For example, when we want to have sensors in the knowledge base, it needs to be discussed and determined what we understand as a sensor. What sensors should be represented? Is the class sensor abstract enough to represent all the desired components? Is the list of entities complete to address the issue that should be solved by the ontology? Second, we create a list of all entities and their respective definitions. This process should always include not only the ontology expert but also everyone involved with building the applications and the end user using the final product and system. If this stage has been completed, the ontology engineer starts implementing the ontology in OWL. Classes, object properties (relationships), instances, definitions, and metadata are created. Further axioms are then to be added for some models, if necessary or desired for the applications. Often the ontology is a domain, reference, or core ontology with a lightweight axiomatization, called lightweight. In the next step we validate the ontology, either by human evaluation or by a tool like OOPS! [21]. OOPS! checks for various best practices and implementation issues—for example, if labels are applied, metadata is missing, or if the ontology is available on the web. After evaluation and validation, the ontology then gets published with both documentation and the OWL file. Often permanent identifiers are a good choice. In this way the ontology can be referred to with a consistent namespace and is not subject to retrieval errors in case of changing servers. Finally, the ontology should in general be maintained. It is a living artifact which may change, and the domain users may discover new or additional usages and may require changes to the ontology domain model and versions.

2.4.2.2 Top-Level Ontologies

Top-level ontologies are now more frequently part of current ontology module suits and publications in the literature. They comprise entities that are common to all domains and describe the most basic constituents of reality. Here, many top-level ontologies distinguish between continuants and occurrents. Continuants are entities that continue or persist in time, like objects, whereas occurrents are entities that happen in time, like, for example, events or processes. The advantages of top-level ontologies are that they improve conceptualizing a domain and separate concerns in order to create domain models more easily. Interoperability of domain models and reusability are two decisive advantages of top-level ontologies. For our ontology model we reused the Basic Formal Ontology (BFO). Figure 2.9 illustrates the taxonomic structure and classes of BFO. For a more detailed explanation of all classes, we refer to [22, 17].

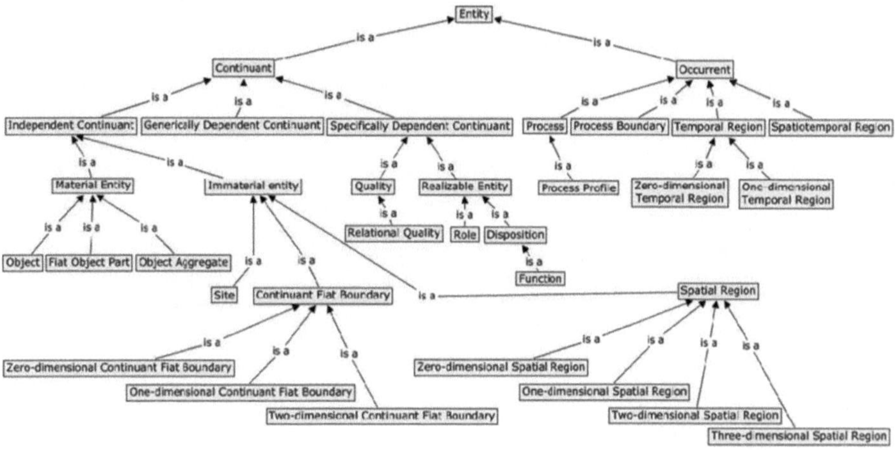

Fig. 2.9 BFO 2.0 all classes [23]

2.4.3 An Ontology for Electronic Design: The GENIAL! Basic Ontology (GBO)

2.4.3.1 Overview

GBO [24] was designed to facilitate to communication along the supply and value chain in automotive. Here the problem is that when different participants of the value chain communicate, there are often miscommunications and not a seamless integration between views and data. The ontology was able to describe the domain of electronics in order for simulators and tools to work together. At the same time, automotive OEM, Tier 1 (supplier) and Tier 2 (semiconductor manufacturer), was able to make use of the domain for their exchange processes. After continued evaluation it was recognized that GBO can serve as a representation for the entire domain of electronics. The ISO 26262 standard is a standard in safety and automotive to ensure quality control and functioning of electronic components. It was combined with the top-level BFO and axiomatized and refined to yield the ontology domain model. The taxonomic structure and a few axioms are illustrated in Fig. 2.10. The following are GBO's main classes and definitions (they are a selection taken from [17]):

- *Context*—the circumstances that form the setting for an event, statement, or idea and in terms of which it can be fully understood, e.g., Market, legal frame conditions
- *Element*—continuant that is (according to ISO 26262) a system (3.163), components (3.21) (hardware or software), hardware parts (3.71), or software units (3.159)

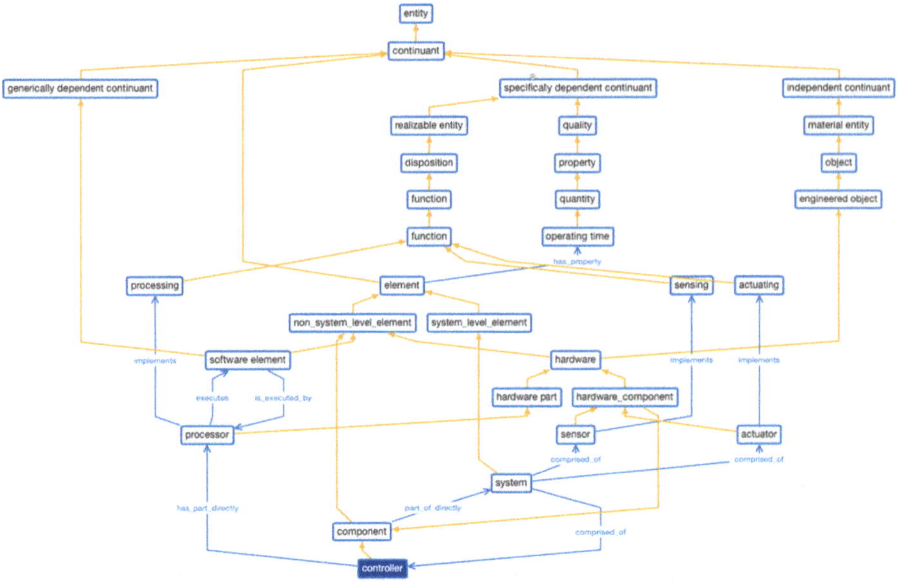

Fig. 2.10 Explicit definition of a GBO controller with its superclasses and relationships [24]

- *Component*—according to ISO26262, nonsystem-level element (3.41) that is logically or technically separable and is comprised of more than one hardware part (3.71) or one or more software units (3.159)
- *Measure*—a bfo:quality that is the amount of quantities
- *Property*—a quality or characteristic of an element, hardware, software, or function. A property has a Boolean, integer, or real variable through the measure class. Specifically properties cannot be quantified and quantified. But quantified properties are quantities.
- *Unit*—a quality that is any standard used for comparison in measurements
- *Function*—a bfo:function that an element (e.g., system, component, hardware, or software) implements
- *Dependency*—a gbo:property that is a Boolean or an arithmetic function/equation on properties, used to be calculated with GENIAL constraint propagation, as string in LaTex syntax

2.4.3.2 Reasoning and Axiomatization

Reasoning refers to the process of making deductions and conclusion. In the case of knowledge representation, there are two major types: ontology-based reasoning and rule-based reasoning. Rule-based reasoning is if-then conditions that can be applied over the ontology model, for example, in the Semantic Web Rule Language (SWRL). In the case of ontology-based reasoning, we here refer to conclusions

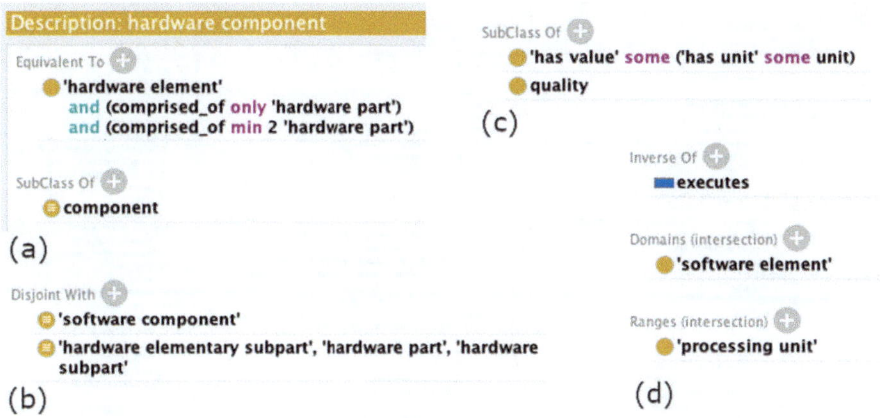

Fig. 2.11 Small collection of GBO axioms [24]

drawn from axioms based on OWL 2 DL. Based on the definition of each axiom in the ontology model, the resulting inference can be displayed. Axiomatizations are important for computerized support. Models without many axioms can comprise domain models and can be used to represent the domain. However, they are not very useful for classification tasks, accurate semantics, additional automatic construction of implicit knowledge, and consistency checking. These are additional benefits and opportunities that not only comprise better domain models but are practically useful as well. For this reason the ontology (GBO) described in the next section was built for consistency checking. It comprises the following features:

- Existential restrictions
- Universal restrictions
- Cardinality restrictions
- Domain/range axioms
- Covering axioms
- Transitive object properties with sub-properties
- Inverse and symmetric properties
- Disjoint and equivalent classes

Figure 2.11 shows a selection of some types of axioms to explain the reasoning functions. In (a) we see the formal definition of hardware component, which is a subclass of component. Only hardware parts are parts of hardware components. And there need to be at least two hardware parts as parts of hardware components that it classifies as a proper hardware component. In (b) we see the disjointness axioms of hardware component. A hardware component is neither another type of hardware than the others, nor is it a software component. In (c) a quantity is formally defined. It is a class that has values, which again need to have a unit associated with the values. In (d) software (software element) and a processing unit are related to each other with domain and range axioms. In this way the relationships execute and

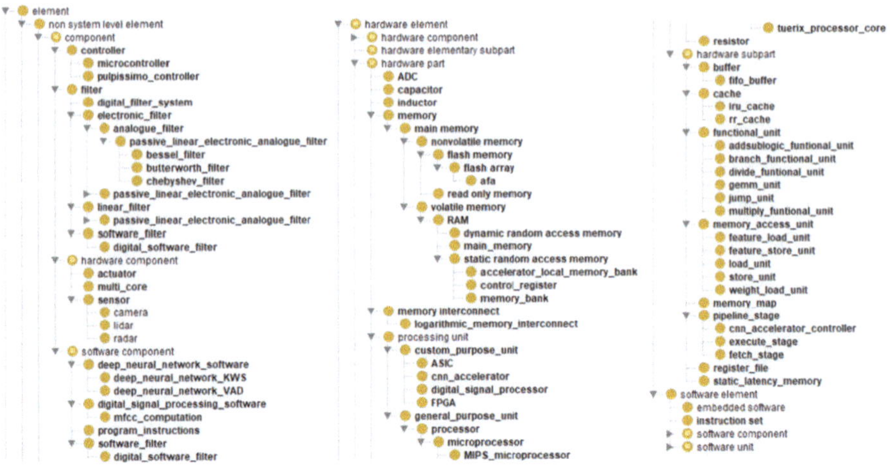

Fig. 2.12 Taxonomy of hardware elements [17]

executed by are supposed to be used between those types of classes; or in other words, every class that executes a software will be inferred to be a processor (main type of processing unit), because only processors execute software. In the same way, regarding every instance of software that is executed by some generic instance, that generic instance becomes a type of processing unit.

2.4.4 Electronic Knowledge Base of Systems, Hardware, and Software

After an ontology is created, the process can begin of populating that ontology domain model with content in the form of instances, triples, and their respective classes. This can be used for feedback to adjust the domain description and also for the building of the domain. In this subsection, we give an illustration of some of the content of our current knowledge base. Figure 2.12 shows basic components of a use case of a neural net hardware accelerator showing mostly hardware parts the systems consists of. We can see filter, neural network software, memory taxonomy, general and custom purpose processors, and lower level units like caches, functional units pipeline stages, and so on. All these models have been constructed with their relationships (i.e., has part) and were validated by the reasoner to ensure the consistency of the structure and vocabulary. As can be intuited from the figure, this yields high-quality representations for knowledge base design.

The use case model was manually constructed by a knowledge engineer. Several discussions followed with hardware design experts to improve the model. Whereas manually constructed domain knowledge bases are very time-consuming and error-prone, the area of automatic knowledge base construction currently offers promising

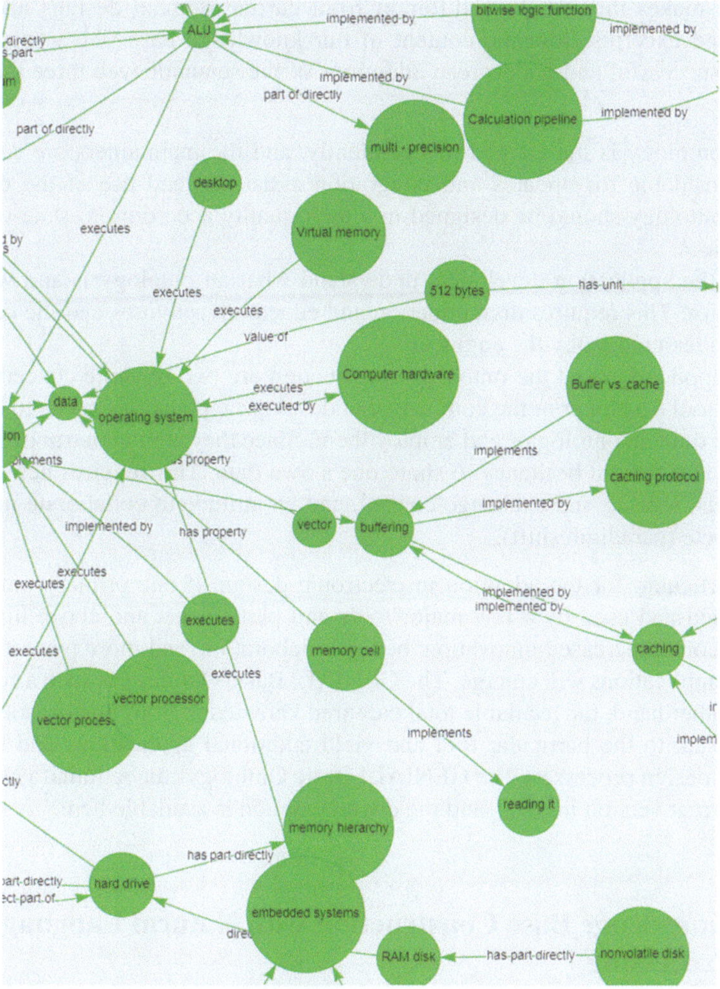

Fig. 2.13 Connected electronic knowledge graph

potential. Machine learning methods have matured to a degree that they become feasible for knowledge base construction. Figure 2.13 gives an outline of a graph that we constructed with machine learning methods to further illustrate our current content. Section 2.5 discusses our construction framework in more detail.

2.4.5 Summary and Conclusion

In this section we motivated background and current ontology engineering and introduced the GENIAL! Basic Ontology for electronic design. We elaborated on what makes this ontology different from current domain designs and finally showcased excerpts from the content of our knowledge base. We want to stress that for successful and widespread adoptions of the semantic web three factors are decisive:

1. The ontology is maintained independently, and the maintainers are supportive and available for updates and integration assistance and use of the ontology. The ontology should be designed in a high quality according to state-of-the-art practices.
2. That the application developers understand what an ontology is and what it is used for. This requires dissemination and education activities and the capability for understanding by the engineer.
3. The application and the ontology construction are two separate concerns. Each EDA tool developer or the company that develops applications needs to be aware of the domain ontologies and employ them. Since they are in control of the code, there is a general hesitancy to share one's own data. This requires new business models, sharing and exchange control, and an attitude to collaborate instead of compete (paradigm shift).

In particular, for the adoption in electronic design, if one primary ontology is established and used by a few major tools and participants and above hindrances are overcome, increased innovation, better collaboration, and more powerful cross-domain applications will emerge. The GENIAL! Basic Ontology is such a candidate. On the other hand, the available total captured knowledge from diverse sources can be fed back to the particular tool and yield additional applications and uses and improve design processes. The GENIAL! Basic Ontology can be found via this link in its current version in owl,[1] and the documentation is available here.[2]

2.5 Knowledge Base Construction with Natural Language Processing

2.5.1 Introduction

In the neural network pipeline, we explore AI and NLP applications like Knowledge Base Construction (KBC) through Named Entity Recognition (NER) and code generation with transformer neural networks. KBC, a vital NLP facet, leverages NER to make textual information machine-processable, reducing manual data processing efforts. Our approach employs a bidirectional long short-term memory (bi-LSTM) network to construct knowledge bases aligned with ISO26262 defi-

[1] https://github.com/wawrzik/GENIALOntologies/blob/master/OntologyModuleSuite/\GENIAL!BasicOntology/GENIALOntBFO.owl.

[2] https://wawrzik.github.io/GENIALOntologies/index-en.html.

nitions from microelectronics text. Training involved a tailored dataset, ensuring domain-specific fit. The model, validated on electronic text documents, establishes relationships between entities, forming graph structures responsive to user queries. Unlike prior segments, we emphasize data acquisition methodologies and neural network training. Ongoing work involves code generation using AI, NLP, and transformer networks, alongside a specialized hybrid Markdown syntax generator.

2.5.2 Background

Within the GENIAL! initiative in Germany, our focus is on the GENIAL! Basic Ontology (GBO), which serves as a shared lexicon for microelectronic systems. We have developed a robust ontology and tools to enhance communication across the value chain. The ontology ensures consistent term utilization, boosting data quality, as detailed in [25]. Despite reasoning challenges, leveraging an efficient reasoner like Konclude [26] proves effective, paving the way for reasoning with larger datasets. We contribute insights on the construction of knowledge graphs within the microelectronics domain, integrating the subtle distinctions of our ontology. Our novel approach lets ontology axioms define machine classification, offering a unique capability to generate precise knowledge automatically.

2.5.3 Motivation and Approach

Diverging from conventional Named Entity Recognition (NER) approaches, our methodology investigates the nuanced distinctions the Long Short-Term Memory (LSTM) model can make in defining classes. Unlike focusing on broad entities like places or dates, we aim for subtler distinctions, such as discerning between *hardware part* and *hardware subpart*. Figure 2.14 illustrates the process of providing text as input, specific to the domain of microelectronics, and establishing relationships between identified classes.

Our approach constructs knowledge bases with input from domain experts, incorporating both textual information and executable, machine-processable components through the SysMD [27] notebook. This Markdown-based [28] integration

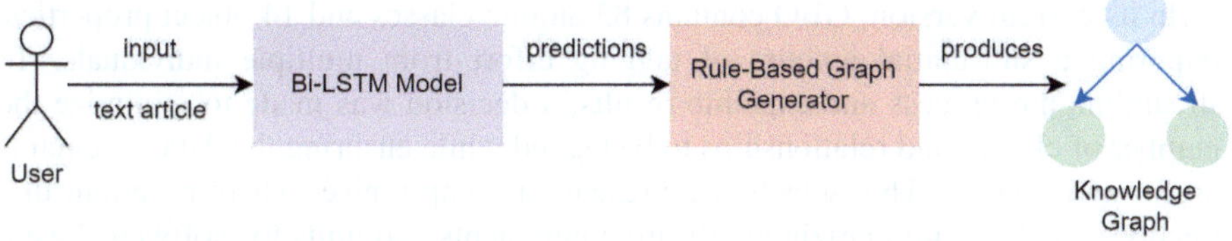

Fig. 2.14 Knowledge graph construction process with Bi-LSTM

Table 2.1 Tag counts

Labels	Number of samples
Hardware part (hwp)	1426
Component (comp)	319
Hardware component (hwc)	902
Hardware subpart (hwsp)	362
Software (sw)	364
System (sys)	583
Measure (mea)	400
Unit (unit)	520
Function (func)	597
Quantity (qt)	1770

enables the inclusion of various media types and offers functionalities like "Analyze" and "Recommend." The amalgamation of parts of vehicles and electrical components combines human-defined and computer-defined knowledge, facilitating contributions from nonexperts through a feedback loop, ultimately saving time. The bidirectional translation between the SysMD model and the knowledge graph establishes a comprehensive understanding of the electrical systems and related models, allowing for efficient refinement and classification.

2.5.3.1 Dataset

We engaged the expertise of two assistants in the data labeling process, dedicating approximately 20 hours to training them in ontology vocabulary. The training included an introductory session and detailed instructions on the initial set of tags. To ensure the quality of their work, continuous supervision was provided, offering corrections and addressing any queries that arose during the training process. The tagging activities carried out for 20 hours per week over 3 months yielded the tag counts depicted in Table 2.1.

In constructing our dataset, we adopted the IOB (Inside, Outside, Beginning) tagging scheme proposed by Ramshaw et al. [29]. This tagging format classifies sentences by labeling each token as "B" if it marks the start of a named entity, "I" if it lies within a named entity, and "O" if it does not belong to any named entity. Punctuation marks and spaces were treated similarly, tagged as "O." A total of 2307 sentences were annotated using this labeling scheme. The model was trained on approximately 1845 sentences (80% of the entire dataset).

In its current version, GBO contains 83 atomic classes and 14 object properties, requiring a substantial amount of tagging effort from multiple individuals. To streamline the process and expedite results, a decision was made to minimize the number of classes and relationships to be tagged while ensuring the dataset remains useful and precise. This selection represents a compromise, wherein certain distinctions, such as summarizing software components and units to "software," were made to simplify the classification. Some classes, including "hardware elementary subpart," were omitted due to limited available articles for tag coverage.

2.5.3.2 Bi-LSTM

In crafting our model, we leveraged the capabilities of a bidirectional Long Short-Term Memory (bi-LSTM) network, a neural architecture adept at processing input in two directions [30]. This design intertwines hidden layers in opposite directions, converging at the output layer, ensuring assimilation of information from both the future and the past. The choice of the bi-LSTM network is rooted in the established effectiveness of LSTM networks, a subtype of Recurrent Neural Networks (RNNs), particularly in addressing sequence prediction challenges [30]. Renowned for their success in word-level predictions like NER, RNNs, including LSTMs, can capture historical information, enhancing the network's ability to predict current outputs [31, 32].

In evaluating NER tasks, relying solely on model accuracy can be misleading due to the prevalence of the "O" tag or "null tag." The number of examples labeled as "null tag" often surpasses those from any other class. The model tends to correctly recognize a significant portion of "no tag" samples, contributing to inflated accuracy. Therefore, for our model evaluation, we employed the F1-score [33], along with other metrics such as precision and recall on the test data (see Table 2.2).

2.5.3.3 Relationship Establishment

After training and validating the model, we established relationships between the identified elements. Our methodology for constructing knowledge bases involved a two-step process. Firstly, we employed the NER strategy to recognize pertinent elements in the provided text. Subsequently, we created a graph using these identified elements, where elements served as nodes, and edges were formed based on the context.

Table 2.2 Classification report [34]

	Precision	Recall	F1-Score
comp	0.65	0.63	0.64
func	0.35	0.38	0.36
hwc	0.66	0.59	0.62
hwp	0.65	0.64	0.64
hwsp	0.56	0.52	0.54
mea	0.68	0.88	0.77
qt	0.73	0.67	0.70
sw	0.51	0.58	0.54
sys	0.46	0.53	0.49
unit	0.75	0.81	0.78

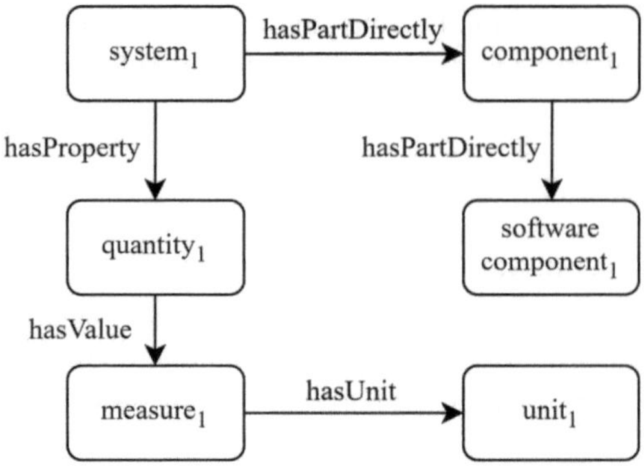

Fig. 2.15 Knowledge graph generated from NER predictions (adapted from [34])

Table 2.3 Predictions

Tokens	Predictions
Visual	B-hwp
Processing	I-hwp
Unit	I-hwp
,	O
Speed	B-qt
,	O
Real	B-func
–	I-func
Time	I-func

A rule-based algorithm was used to establish relationships between different entities with nested if-else loops. "List Pred" is selected here for demonstration purposes. Figure 2.15 visually depicts the graph created from "List Pred": {system$_1$, quantity$_1$, measure$_1$, unit$_1$, component$_1$, sw_component$_1$, function$_1$, quantity$_2$, measure$_2$, hw_component$_1$, system$_2$}.

2.5.4 *Results*

In evaluating NER tasks, we addressed accuracy challenges posed by the prevalent "no tag" issue by adopting the F1-score alongside precision, recall, and support metrics. Precision varied across classes, with "unit" leading at 75%, while recall for "measure" was the highest at 88%. The F1-score, considering both precision and recall, proved essential, especially in imbalanced datasets, visual processing unit

Examining underclassification and overclassification impacts, increasing class diversity improved the overall F1-score. Evaluating the model on a Wikipedia article on "Graphics Processing Unit" showcased accurate predictions, illustrated in Table 2.3. Extracted information was visualized through knowledge graphs, as shown in Fig. 2.15 from one of our first successful experimentations, which demonstrates entity relationships.

2.5.5 *Conclusion*

We applied Named Entity Recognition (NER) to construct knowledge bases following ISO26262 definitions within the GENIAL! Basic Ontology. While demonstrated in the microelectronics domain, this approach is versatile, applicable to various tasks with labeled data. The results are satisfactory, providing knowledge engineers with informative graphs. Future work involves expanding the dataset for improved model performance, refining graph generation using transformer neural networks, and enriching results with additional ontology tags. Diversifying expert articles and integrating OWL reasoners and human feedback are part of our considerations.

2.6 SysMD Code Generation with Transformer Neural Network

2.6.1 *Introduction*

Transformative progress in text-based learning has been achieved through transformers, particularly large language models (LLMs) like ChatGPT (by OpenAI). These models, trained on extensive datasets using transformer encoders, excel in translating natural language into code snippets. Despite the proficiency of models like ChatGPT and Bard [35] in code generation, there is a growing focus on high-level code generation.

Initiatives like CodeBERT [36] leverage pretrained transformers for code articulation from natural language. Models such as CodeT5 [37] and CodeBERT, utilizing transformer architectures like T5, have gained recognition for their focus on code understanding and generation. Our model is tailored for SysMD [27] code generation, offering simplicity for professionals less familiar with modeling languages.

Inspired by the transformer encoder and decoder model from "Attention is All You Need" [38], our architectural framework enhances this foundation. Rigorous testing demonstrated commendable performance, akin to GPT-3 [39], with compiler-compatible code generation. This success marks a promising advancement in transformer-based language models for SysMD code generation.

2.6.2 *Background*

SysMD [27], a modeling tool and language based on SysMLv2 [1], aims to democratize systems modeling by making it accessible to domain experts with minimal modeling background. Using intuitive, near-natural language statements and seamless integration with Markdown files, SysMD combines code and documentation within a single document. The language focuses on building maintainable knowledge bases by incorporating Markdown documents with model comments, enhancing comprehensiveness and accessibility. SysMD's user-friendly syntax and semantics, even for non-modelers, simplify the modeling of relationships, constraints, and needs, along with continuous consistency checking. Complying with the SysMLv2 standard, SysMD supports various modeling languages, providing an inclusive approach to systems modeling. Overall, SysMD promotes knowledge sharing and collaboration by making the creation of connected, well-documented models easier through Markdown and a recommender system.

2.6.3 Methodology

The development of the transformer model was executed using PyTorch Lightning [40], a high-level wrapper that expedited development cycles and maintained project structure. Tokenization, a fundamental aspect in natural language processing, involved the evaluation of various tokenizers with SpaCy [41] emerging as the optimal choice. The Torchtext vocabulary library [42] was employed in conjunction with the tokenizer, efficiently managing words based on frequency and aiding in data representation. The project data, structured in JSON format, featured natural language text and SysMD code, initially validated using Python code from the CodeSearchNet dataset [43].

To enhance the model's robustness, data augmentation strategies were implemented, primarily involving the replacement of function and variable names with numerical representations. This not only expanded the training dataset but also introduced regularization, bolstering the model's ability to avoid overfitting. The tuning of model hyperparameters played a crucial role, with adjustments made to key parameters such as learning rate, embedding dimension, dropout, optimizer, and loss function. Careful tuning of these parameters, including the choice of the Adam optimizer [44] over Stochastic Gradient Descent (SGD) [45] for superior performance with the transformer architecture, contributed to the optimization of the model's learning efficacy and overall performance.

2.6.4 Results and Evaluation

The Bilingual Evaluation Understudy (BLEU) [46] score is a method for assessing machine language translation models, utilizing a mathematical formula to compare translated and original texts based on n-gram sequences. Ranging between 0 and 1, the BLEU score indicates the degree of similarity between translations and originals, with higher scores signifying greater resemblance. Studies have identified a correlation between BLEU scores and human evaluation, validating its effectiveness in approximating human judgment of translation quality. However, BLEU has limitations, being a corpus-based metric that may overlook individual sentence nuances and struggle with semantic and grammatical aspects. With our first prototype built on a small dataset of approximately 11463 SysMD training code samples and their corresponding natural language texts with 1433 validation data samples, our results were quite adequate with a BLEU score of 0.38. The score was acceptable given the small dataset available for training. Listing 2 demonstrates a sample output of the prototype for a natural language piece of text input.

```
NL Input: The cooling system of a car is responsible for
    regulating the temperature of the engine. It is made up of
    several components such as the radiator, water pump, &
    thermostat.

SysMD Output:
Global hasA Package CoolingSystem.
CoolingSystem defines
Cooling isA Component.
CoolingSystem hasA Package CoolingComponents.
CoolingSystem::CoolingComponents imports AirIntakeSystem.
CoolingSystem::CoolingComponents defines
Radiator isA Component;
WaterPump isA Component;
Thermostat isA Component.
```

Listing 2.4 SysMD code generated from natural language

2.6.5 Conclusion

This section presents a SysMD code generation method using transformer models. Inspired by "Attention is All You Need," our model, akin to GPT-3, showcases acceptable performance. SysMD, based on SysMLv2, offers a user-friendly approach to systems modeling, combining natural language statements and Markdown integration. The methodology involves transformer development with PyTorch Lightning, optimal tokenization using SpaCy, and data augmentation for robustness. Despite constraints, BLEU score evaluation yielded satisfactory results, marking a promising step forward in transformer-based language models for SysMD code generation. This research opens avenues for democratizing systems modeling and enhancing knowledge sharing. Further refinement and exploration, especially with larger datasets, hold the potential for significant advancements in SysMD code generation and automatic generation of other modeling languages.

2.7 Automotive Electrical System (Boardnet) Application

2.7.1 Introduction to Automotive Boardnet Design

Within the automotive sector, the electrical system responsible for supplying electricity, facilitating communication, and facilitating data interchange among the numerous electronic components of a vehicle is known as the boardnet. The need for smart, connected cars is growing, and as automotive technology develops, boardnet design plays a more important role. In order to achieve the highest levels of vehicle performance, safety, and efficiency, electronic system integration must be done effectively. An overview of the difficulties and significance of boardnet design in contemporary automobiles is given in this section. There were only rather simple lighting and ignition components in early car electrical systems. But the boardnet has become notably more complex as car technology has improved. These days, they comprise advanced driver assistance systems (ADAS), sensors, actuators, power distribution, electronic control units (ECUs), or entertainment systems. The vehicle electrical system functions as the vehicle's central nervous system, enabling interaction between all of its various components and ensuring smooth functioning.

2.7.1.1 Evolution of Boardnet Architecture

The rising complexity of vehicle systems and the need for greater degrees of automation and communication have led to a development in boardnet architecture. The electrical systems in early cars were crude, with manual controls and few features. The automobile industry has seen a shift over the years due to developments in electronics and computer technologies, ushering in the era of intelligent vehicles. Wiring harnesses became longer and heavier as more electronic components were added, which resulted in financial and spatial limitations. Multiplexed wiring solutions were created in response to this problem in order to minimize the amount of cables needed. Boardnet architecture was further impacted by the need for quicker communication protocols and larger data rates as cars moved toward electrified propulsion and ADAS.

2.7.1.2 Requirements and Challenges

Modern cars must adhere to strict regulations and face difficult boardnet design issues. Among the main obstacles that designers and engineers must overcome are the growing quantity of electrical parts, varying voltage levels, specifications for data transfer, electromagnetic compatibility, and functional safety. Striking an acceptable compromise between the weight, expense, and complexity of the wire harness and the growing number of electronic components is one of the main issues. The need for power and data transmission rises with the integration of new car technologies like sophisticated driver assistance systems and electric powertrains, placing more strain on the boardnet. It is crucial to ensure functional safety and dependability, particularly when it comes to autonomous driving. To guarantee ongoing operation in the event of a failure, redundant communication pathways, fail-operational capabilities, and fault tolerance must all be included into the boardnet. Additionally, as cars become more connected and digitalized, the boardnet is vulnerable to cybersecurity attacks. To shield drivers and passengers from possible dangers, it is essential to prevent illegal access and ensure data integrity.

2.7.2 *Automotive Electrical System Architectures*

2.7.2.1 Traditional Boardnet Architectures

Traditional boardnet systems depended on many control units and discrete wiring harnesses for different purposes prior to the development of modern technology. With many control units handling different functions like air conditioning, entertainment, ADAS, and engine management, the car would have a large number of ECUs dispersed all over it. Although this architecture was functional for a long time, it had a few drawbacks. The number of ECUs rose, which resulted in higher prices, weight, and complexity. Furthermore, the vast wire harness and the absence of connectivity between ECUs presented difficulties for the additional integration of new features.

2.7.2.2 Disruptive Paradigm Shift

New mobility and use ideas are causing a disruptive paradigm change in the vehicle sector. Boardnet design needs have changed significantly with the advent of autonomous and linked driving and the drive toward electrification. Higher technological skills from the boardnet are required for automated driving. It must respond instantly to control actuators and manage the growing amount of data produced by sensors and cameras. Simultaneously, in order to facilitate vehicle-to-everything (V2X) connection, quicker and more dependable communication protocols are required due to connectivity needs.

2.7.2.3 Intelligent Zonal and Centralized Boardnet Concepts

The idea of intelligent zonal and centralized boardnet architectures arises to solve the shortcomings of conventional systems and fulfill automated driving needs. By combining functions and lowering the quantity of ECUs, these novel ideas hope to create a system that is more streamlined and effective. A zonal architecture incorporates several ECUs into bigger components called zones, each of which performs a distinct functionality. These zones work together, which reduces transmission costs and improves data flow and power distribution management. Centralized architectures, on the other hand, focus on crucial operations into a few numbers of highly effective processing units. Centralized designs provide for more complex integration of information, effective information interchange, and better adaptability for subsequent enhancements by centralizing computing capacity.

2.7.3 AI-Enabled Semantic Modeling for Enhanced Integration

In order to build and integrate current boardnet systems, artificial intelligence (AI) and semantic modeling are essential. Algorithms powered by artificial intelligence (AI) have the ability to arrange and connect electronic components in the best possible way, which can simplify wiring and increase energy efficiency and data flow. With the use of semantic modeling, a more thorough and effective boardnet design is made possible by a higher level knowledge of the relationships between components. Designers can find the best configurations and make data-driven decisions by using AI to examine massive information and simulate various situations. Furthermore, the integration of digital twins allows for real-time simulation and validation of boardnet designs. Digital twins provide a virtual representation of the physical system, enabling engineers to test and optimize the boardnet's performance before physical implementation. This approach significantly reduces development time and costs while ensuring greater accuracy and reliability.

2.7.3.1 Advantages of AI-Driven Boardnet Design

AI-driven boardnet design has many benefits over conventional techniques. First, it makes design more methodical and effective, which cuts down on expenses and development time. Large-scale data analysis and sophisticated scenario simulation are capabilities of AI algorithms, which can result in optimal designs that satisfy cost, safety, and performance requirements. Second, a comprehensive approach to boardnet design that takes into account several elements at once is made possible by the combination of AI and semantic modeling. This leads to enhanced compatibility with upcoming technologies and greater coordination between various vehicle systems. Third, functional safety and dependability are improved by AI-driven boardnet design. By recommending duplicate paths or fail-operational procedures and identifying possible failure sites, AI algorithms may strengthen the boardnet. Lastly, quicker response to evolving consumer preferences and regulatory needs is made possible by AI-based design processes. Before putting various configurations into actual automobiles, engineers may test them and determine if they comply with rules by employing digital twins.

2.7.3.2 Future Prospects and Challenges

As AI continues to evolve, the future prospects for boardnet design are promising. Advanced AI algorithms, including machine learning and deep learning techniques, will further improve the efficiency and accuracy of boardnet optimization. Predictive AI models will enable proactive maintenance and real-time adaptability of the boardnet to changing conditions. One of the key challenges for AI-driven boardnet design is ensuring data security and privacy. As vehicles become more connected and collect vast amounts of data, safeguarding sensitive information from cyber threats is crucial. Another challenge is standardization. As the automotive industry moves toward centralized architectures and shared computing resources, establishing industry-wide standards for communication protocols, data formats, and interfaces becomes essential. Moreover, AI-driven boardnet design requires skilled personnel with expertise in AI algorithms, data analytics, and semantic modeling. Bridging the gap between traditional automotive engineering and AI will be vital to leverage the full potential of AI in boardnet design.

2.7.3.3 Semantic Model of the Boardnet and Reasoning Application

We built a comprehensive model of various aspects of boardnet design, which includes topologies, variants, types and libraries of components and parts, their interconnections, specifications, constraints, requirements, and properties. This work was published in recent works. In [47], for example, we further describe the application and comparison of SysMD and OWL. Both modeling languages have different approaches that need to be consolidated. We provide reasoning implementations for enhancement of the construction process. Figure 2.16 shows a small excerpt of a classification of relevant "components" contained in the boardnet model. The classifications are an assignment to GBO superclasses, which are explained in Sect. 2.4. The correctness of the model has been confirmed by the reasoner, which does not show any errors. Domain controllers are disjoint to zonal controllers, but central controllers are currently not disjoint with the other types. Additionally, axioms are present that define that zonal architectures do not contain domain controllers, but only zonal controllers and central controllers are allowed. This is additionally a way to implement additional control checks, which make a model sound and do not allow the engineer on the SysMD side to make logical or model-related other mistakes. Figure 2.17 further illustrates the instantiations of the model. It is also an excerpt, and it has to be noted that even as a small knowledge base, the amount and complexity of the axiom definitions already comprises a significant strain on computing resources. What can be observed in the figure are the many connections of wires and the different instances of the components and parts. The instances comprise a structural hierarchy of has_parts relations which comprise the automotive and its subsystems like the boardnet and all its hierarchical parts and their interconnections. The inferred properties or relationships are highlighted

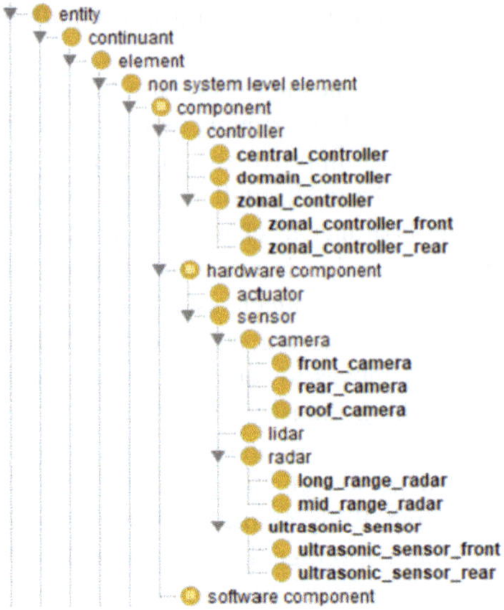

Fig. 2.16 Domain excerpt of automotive electrical system architecture components classification within GBO

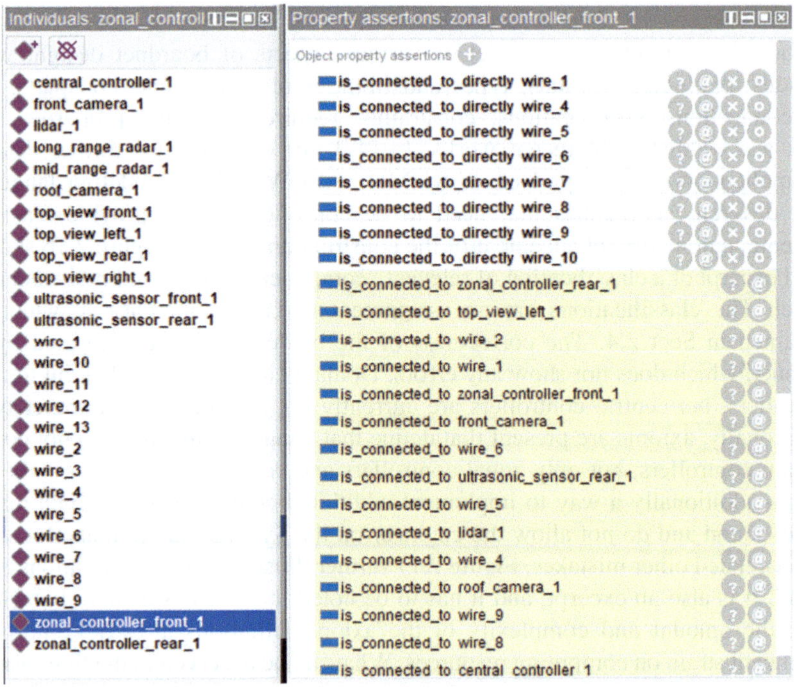

Fig. 2.17 Reasoning with instances of connected components

in yellow. What makes the model useful and more intelligent compared to common system models are the following axiomatizations:

- Subclass of "is_connected_to_directly" to "is_connected_to" object property
- Transitivity of "is_connected_to" and intransitivity of "is_connected_to_directly"
- Property chain of "has_part" with "is_connected_to"

This allows to (1) make explicit all connections from and to all components and (2) construct automatically all the missing has_parts relationships for the other components: In other words—and as an example—if the zonal controller is part of a zonal architecture and the ultrasonic sensor is connected to that zonal controller, the ultrasonic sensor is now also a part of that specific zonal architecture type.

2.7.4 Preliminary Conclusions

A crucial component of contemporary vehicle development is the boardnet design, particularly when it comes to autonomous and linked operation. OEMs can reinvent vehicle electronic system architectures and produce cars that are safer, more adaptive, and effective by incorporating AI-driven semantic modeling and intelligent architectures. It is anticipated that the future of the automobile industry will be greatly impacted by the incorporation of AI technology into boardnet design, which will spur innovation and transform car electronics. To fully utilize AI in boardnet design and other areas, more study and cooperation in research and industry will be necessary as this subject develops. Vehicle electronic system design will get increasingly complex as AI research continues to progress, guaranteeing that cars in the years to come will be at the cutting edge of automotive technological advancements.

2.8 Conclusion

In the pursuit of revolutionizing systems modeling and design capture, this comprehensive exploration unfolds a narrative that spans the spectrum of AI-enabled technologies and methodologies. At its core lies the strategic integration of SysMLv2 and its specialized subset, SysMD, offering a nuanced approach to modeling that prioritizes accessibility and interoperability.

The decision to adopt SysMD as the display medium stems from its inherent readability, making complex models comprehensible for a broad audience. However, it is crucial to emphasize that SysMD operates not in isolation but as an integral part of the broader SysMLv2 framework. This strategic coupling ensures that the models and analyses conducted in SysMD seamlessly integrate into the broader systems modeling ecosystem.

Design capture, a pivotal phase in the development of complex systems, benefits significantly from the utilization of SysMD. Its role in facilitating effective communication among model designers, coupled with its applicability in verification phases, positions SysMD as a valuable tool in the engineer's toolkit.

The emergence of digital twins as connectors between engineers and system models amplifies the transformative potential of AI. Through the filtering of data and the application of advanced algorithms, digital twins not only enhance the reliability of data but also enable applications like anomaly detection and predictive maintenance, making AI-based solutions more accessible and affordable.

Knowledge bases, underpinned by ontologies like the GENIAL! Basic Ontology, are identified as the backbone of informed decision-making in electronic design. The challenges of adoption, including the need for maintenance, education, and collaborative paradigms, are acknowledged. The potential of a shared ontology to foster innovation and collaboration in electronic design becomes evident.

The application of Named Entity Recognition (NER) to construct knowledge bases demonstrates versatility, particularly in the microelectronics domain. This approach, although showcasing satisfactory results, opens avenues for future improvements, including dataset expansion, refined graph generation, and the integration of advanced machine learning methods.

The unveiling of a SysMD code generation method, leveraging transformer models, marks a significant stride toward democratizing systems modeling. The user-friendly approach of SysMD, combined with advancements in transformer-based language models, holds promise for automatic code generation across various modeling languages.

Finally, the exploration of AI-enabled semantic modeling in boardnet design for modern automotive development underscores the transformative potential of AI. The integration of AI technologies in boardnet design has far-reaching implications for vehicle electronics, promising efficiency, safety, and adaptability.

In summation, this collective endeavor transcends individual components of the overall approach, offering a holistic perspective on the symbiotic relationship between AI and systems modeling. From design capture to digital twins, knowledge bases, code generation, and boardnet integration, the narrative converges on a future where AI serves as the catalyst for a more efficient, interconnected, and innovative approach to systems modeling and design capture. The collaborative synergy between AI and SysMLv2 sets the stage for continued advancements, inviting researchers and practitioners to explore the transformative possibilities in diverse domains.

Acknowledgments This work has been supported by the GENIAL! project with funding from the BMBF under grant agreement no. 16ES0865K. This work has been supported by the AnastASICA project with funding from the BMBF under grant agreement no. 16ES0991. This work has also been funded by the Federal Ministry of Education and Research (Germany), research project KI4BoardNET under grant no. 16ME0782.

References

1. Bajaj, M., Friedenthal, S., & Seidewitz, E. (2022). Systems modeling language (SysML v2) support for digital engineering. *Insight, 25*, 19–24.
2. Panda, P. R. (2001). SystemC: a modeling platform supporting multiple design abstractions. In *Proceedings of the 14th international symposium on Systems synthesis* (pp. 75–80).
3. Vachoux, A., Grimm, C., & Einwich, K. (2003). SystemC-AMS requirements, design objectives and rationale. In *2003 Design, Automation and Test in Europe Conference and Exhibition* (pp. 388–393).
4. Prautsch, B., et al. (2020). AnastASICA–towards structured and automated analog/mixed-signal IC design for automotive electronics. In *ANALOG 2020; 17th ITG/GMM- Symposium* (pp. 1–6).
5. Crossley, J., et al. (2013). BAG: A designer-oriented integrated framework for the development of AMS circuit generators. In *2013 IEEE/ACM International Conference on Computer-Aided Design (ICCAD)* (pp. 74–81).
6. Pietruszka, W. D. (2012). *MATLAB® und Simulink® in der Ingenieurpraxis*. Berlin: Springer.
7. Bellard, F. (2005). QEMU, a fast and portable dynamic translator. In *USENIX annual technical conference, FREENIX Track* (Vol. 41, p. 46).
8. Scheible, J., & Lienig, J. (2015). Automation of analog IC layout: Challenges and solutions. In *Proceedings of the 2015 Symposium on International Symposium on Physical Design* (pp. 33–40).
9. Prautsch, B., Wittmannm, R., Eichler, U., Hatnik, U., & Lienig, J. (2021). Generators, Templates, and Code Generation for Flexible Automation of Array-Style Layouts. In *SMACD/PRIME 2021; International Conference on SMACD and 16th Conference on PRIME* (pp. 1–4)
10. Zivkovic, C., Grimm, C., Olbrich, M., Scharf, O., & Barke, E. (2018). Hierarchical verification of AMS systems with affine arithmetic decision diagrams. *IEEE Transactions on Computer-Aided Design of Integrated Circuits and Systems, 38*, 1785–1798.
11. Stolfi, J., & de Figueiredo, L. H. (2003). An introduction to affine arithmetic. *Trends in Computational and Applied Mathematics, 4*, 297–312.
12. Saratha, S. D. C., Grimm, C., & Wawrzik, F. (2021). A Digital Twin with runtime- verification for industrial development-operation integration. In *2021 IEEE International Conference on Engineering, Technology and Innovation (ICE/ITMC)* (pp. 1–9).
13. Schroeder, G. N., Steinmetz, C., Pereira, C. E., & Espindola, D. B. (2016). Digital Twin Data Modeling with AutomationML and a Communication Methodology for Data Exchange. In *IFAC-PapersOnLine* (Vol. 49(30), pp. 12–17). ISSN: 24058963. Amsterdam: Elsevier B.V.
14. Van De Plassche, R. (1978). A sigma-delta modulator as an A/D converter. en. *IEEE Transactions on Circuits and Systems, 25*, 510–514. ISSN: 0098-4094. http://ieeexplore.ieee.org/document/1084506/ (2023).
15. Adhikari, S., & Grimm, C. (2023). Modeling switched capacitor sigma delta modulator nonidealities in SystemC-AMS en. In *2010 Forum on Specification & Design Languages (FDL 2010)* (IET, Southampton, UK, 2010) (pp. 216–221). ISBN: 978-2-9530504-3-1. https://digital-library.theiet.org/content/conferences/10.1049/ic.2010.0155
16. Grimm, C., et al. (2021). APPEL—AGILA ProPErty and Dependency Description Language. In *MBMV 2021; 24th Workshop* (VDE, online, Mar. 2021). ISBN: 978-3-8007-5500-4. https://ieeexplore.ieee.org/abstract/document/9399724/metrics#metrics (2023).
17. Wawrzik, F. (2022). *Knowledge Representation in Engineering 4.0* (Vol. 152, (2)). doctoralthesis. Kaiserslautern: Technische Universität. http://nbn-resolving.de/urn:nbn:de:hbz:386-kluedo-68887.
18. Gruber, T. R. (1995). Toward principles for the design of ontologies used for knowledge sharing. In *International Journal of Human-Computer Studies* (Vol. 43, pp. 907–928).
19. Vrandečič, D., & Krötzsch, M. (2014). Wikidata: A Free Collaborative Knowledgebase. *Commun. ACM, 57*, 78–85. ISSN: 0001-0782. https://doi.org/10.1145/2629489

20. Lehmann, J., et al. (2015). DBpedia—A large-scale, multilingual knowledge base extracted from Wikipedia. *Semantic Web, 6*, 167–195. http://dblp.uni-trier.de/db/journals/semweb/semweb6.html#LehmannIJJKMHMK15
21. Poveda-Villalón, M., Gómez-Pérez, A., & Suárez-Figueroa, M. C. (2014). OOPS! (OntOlogy Pitfall Scanner!): An On-Line Tool for Ontology Evaluation. *International Journal on Semantic Web and Information Systems, 10*, 7–34. ISSN: 1552-6283. https://doi.org/10.4018/ijswis.2014040102
22. Arp, R., Smith, B., & Spear, A. D. (2015). *Building Ontologies with Basic Formal Ontology*. ISBN: 0262527812. New York: The MIT Press.
23. Abhishek, K. (2016). An ontology based decision support for tuberculosis management and control in India. *International Journal of Engineering and Technology, 8*, 2860–2877.
24. Wawrzik, F., & Lober, A. (2021). A Reasoner-Challenging Ontology from the Microelectronics Domain. In *ISWC 2021 The 20th International Semantic Web Conference- Semantic Reasoning Evaluation Challenge (SemREC)*.
25. Wawrzik, F., & Lober, A. (2021). A Reasoner-Challenging Ontology from the Microelectronics Domain. In G. Singh, R. Mutharaju, & P. Kapanipathi (Eds.), *Proceedings of the Semantic Reasoning Evaluation Challenge (SemREC 2021) co-located with the 20th International Semantic Web Conference (ISWC 2021), Virtual Event, October 27th, 2021* (Vol. 3123, pp. 1–12). CEUR-WS.org. http://ceur-ws.org/Vol-3123/paper1.pdf
26. Steigmiller, A., Liebig, T., & Glimm, B. (2014). Konclude: System description. *Web Semantics: Science, Services and Agents on the World Wide Web, 27–28*, 78–85.
27. Dalecke, Š., Rafique, K. A., Ratzke, A., Grimm, C., & Koch, J. (2022). SysMD: Towards "Inclusive" Systems Engineering. In *2022 IEEE 5th International Conference on Industrial Cyber-Physical Systems (ICPS)* (pp. 1–6).
28. Gruber, J. (2012). Markdown: Syntax. http://daringfireball.net/projects/mark-down/syntax. Retrieved on June **24,** 640.
29. Ramshaw, L. A., & Marcus, M. P. (1999). *Natural language processing using very large corpora* (pp. 157–176). Berlin: Springer.
30. Hammerton, J. (2003). Named entity recognition with long short-term memory. In *Proceedings of the seventh conference on Natural language learning at HLT-NAACL 2003* (pp. 172–175).
31. Medsker, L. R., & Jain, L. (2001). Recurrent neural networks. *Design and Applications, 5*, 64–67.
32. Masoud, M., Pereira, B., McCrae, J., & Buitelaar, P. (2021). Automatic Construction of Knowledge Graphs from Text and Structured Data: A Preliminary Literature Review. In D. Gromann, et al. (Eds.), *3rd Conference on Language, Data and Knowledge (LDK 2021)* (Vol. 93, pp. 19:1–19:9). Dagstuhl, Germany: Schloss Dagstuhl—Leibniz-Zentrum für Informatik. ISBN: 978-3-95977-199-3. https://drops.dagstuhl.de/opus/volltexte/2021/14555
33. Yacouby, R., & Axman, D. (2020). Probabilistic Extension of Precision, Recall, and F1 Score for More Thorough Evaluation of Classification Models. In *Proceedings of the First Workshop on Evaluation and Comparison of NLP Systems* (pp. 79–91). New York: Association for Computational Linguistics. https://aclanthology.org/2020.eval4nlp-1.9
34. Wawrzik, F., Rafique, K. A., Rahman, F., & Grimm, C. (2023). Ontology learning applications of knowledge base construction for microelectronic systems information. *Information, 14*, 176.
35. Aydin, Ö. (2023). Google Bard generated literature review: metaverse. *Journal of AI, 7*, 1–14.
36. Feng, Z., et al. (2020). CodeBERT: A pre-trained model for programming and natural languages. arXiv preprint arXiv:2002.08155.
37. Wang, Y., Wang, W., Joty, S., & Hoi, S. C. (2021). Codet5: Identifier-aware unified pre-trained encoder-decoder models for code understanding and generation. arXiv preprint arXiv:2109.00859.
38. Vaswani, A., et al. (2017). Attention is all you need. In *Advances in neural information processing systems* (Vol. 30).
39. Dale, R. (2021). GPT-3: What's it good for? *Natural Language Engineering, 27*, 113–118.
40. Falcon, W. A. (2019). PyTorch lightning. *GitHub, 3*.

41. Honnibal, M., Montani, I., Van Landeghem, S., & Boyd, A. (2021). SpaCy 2: Natural language understanding with Bloom embeddings, convolutional neural networks and incremental parsing. To appear.
42. Peters, M. E., et al. (2018). TorchText: Data loading and preprocessing for NLP. arXiv preprint arXiv:1805.08303. https://arxiv.org/abs/1805.08303
43. Husain, H., Wu, H.-H., Gazit, T., Allamanis, M., & Brockschmidt, M. (2019). Code-SearchNet Challenge: Evaluating the State of Semantic Code Search. arXiv preprint arXiv:1909.09436. https://arxiv.org/abs/1909.09436.
44. Kingma, D. P. & Ba, J. (2014). Adam: A method for stochastic optimization. arXiv preprint arXiv:1412.6980. https://arxiv.org/abs/1412.6980.
45. Bottou, L., & Bousquet, O. (2010). Large-Scale Machine Learning with Stochastic Gradient Descent. In *Proceedings of COMPSTAT'2010.* http://leon.bottou.org/papers/bottou-bousquet-2010
46. Papineni, K., Roukos, S., Ward, T., & Zhu, W.-J. (2002). BLEU: a method for automatic evaluation of machine translation. In *Proceedings of the 40th annual meeting of the Association for Computational Linguistics (ACL)* (pp. 311–318). https://www.aclweb.org/anthology/P02-1040.pdf
47. Wawrzik, F., Koch, J., Post, S., & Grimm, C. (2023). AI-Enabled semantic modeling for enhanced boardnet integration in automotive design. In W. Karwowski, T. Ahram (Eds.) *Artificial Intelligence, Social Computing and Wearable Technologies* (AHFE Open Access, vol 113). New York: AHFE International. https://doi.org/10.54941/ahfe1004200

Chapter 3
AI-Enabled Efficient Memory Design for Data-Intensive Applications

Na Gong, Jinhui Wang, Wei Jin, Hritom Das, and Ali Haidous

3.1 Introduction

"*The memory is the computer*" [1], particularly for today's data-intensive applications, such as mobile videos and deep learning systems. Mobile video streaming on YouTube, Vimeo, and Netflix has increased quickly. According to a recent survey, approximately 70% of the digital video content audience in the United States was reported to watch videos on their smartphones [2]. However, with a mobile video decoding chip, embedded memories occupy over 65% of the core area of a video decoder chip [3, 4] and contribute to over 30% of the system's power consumption [5]. This use-case is only expected to grow for the next-generation video formats, H.265/HEVC and H.266/VVC, which has 2x–3x greater memory demands when compared to H.264. Similarly, efficient memory design is of paramount importance to support deep learning systems. In particular, the machine learning model training and inference process needs continuous model updating (e.g., iterative synaptic weight update) and requires intensive memory access. On-chip SRAM occupies 56% of the silicon area and contributes to over 60% power consumption of the entire system [6]. Accordingly, the development of effective power-efficient memory techniques is highly desired for mobile video and deep learning applications due

N. Gong (✉) · J. Wang · A. Haidous
Department of Electrical and Computer Engineering, University of South Alabama, Mobile, AL, USA
e-mail: nagong@southalabama.edu

W. Jin
Department of Computer Science and Engineering, University of North Texas, Denton, TX, USA

H. Das
Department of Electrical and Computer Engineering, Oklahoma State University, Stillwater, OK, USA

A. Iranmanesh, H. Sayadi (eds.), *AI-Enabled Electronic Circuit and System Design*,
https://doi.org/10.1007/978-3-031-71436-8_3

to the concerns of increasing energy costs of mobile and Internet of Things (IoT) edge devices and the dominant energy contribution of embedded memory.

Traditionally, mobile hardware designers, including memory designers, concentrate on hardware level, energy-efficient design techniques in order to accommodate the large amount of data; this process usually comes with significant implementation overhead (e.g., silicon area, delay) to solve the memory-failure problem [3, 7]. An important aspect, but usually ignored by memory designers, is that the data itself can be transformed into useful knowledge for hardware design. Recently, it has been observed that modern data-intensive applications, including videos and deep learning, have three common data characteristics [8]: (i) redundant inputs, (ii) multiple acceptable outputs, and (iii) statistical computations. Those intrinsic characteristics provide substantial opportunities for data relationship discovery and pattern identification. In this chapter, we will provide an overview of how to introduce advanced artificial intelligence (AI) techniques to discover data relationships and data knowledge hidden in large datasets, which will be then incorporated to develop memory techniques. The presented AI-enabled memory design methodology is shown in Fig. 3.1, which represents a new hardware design paradigm which switches from pure hardware-centric design to AI-enabled data-aware hardware design. Specifically, we have developed three different AI-assisted memory design methodologies for power efficiency optimization: region-of-interest (ROI)-aware low-power memory design, data pattern-based low-power memory design, and data pattern-based self-recovery memory design.

The organization of this chapter is as follows. Section 3.2 presents the background. A review of related low-power memories is provided in Sect. 3.3. In Sect. 3.4, we present our developed AI-enabled memory design technique to further enhance the power efficiency of memories. Section 3.5 presents a research roadmap that contains open research problems and potential solutions. Finally, we conclude this chapter in Sect. 3.6.

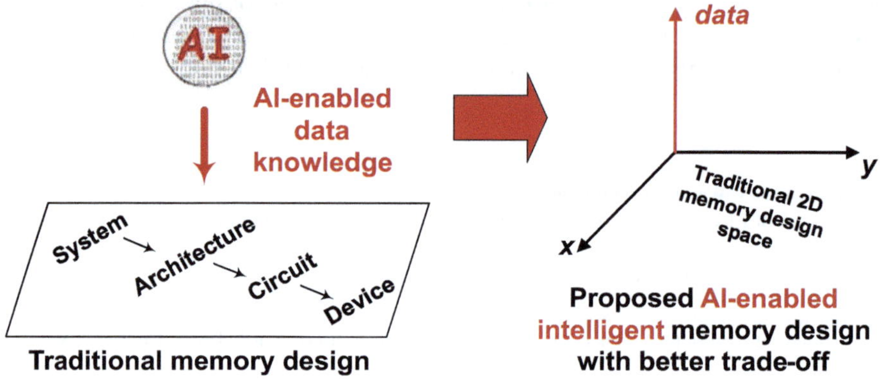

Fig. 3.1 AI-enabled memory design process

3.2 Background

In our analysis, we focus on memory design in two data-intensive applications, one is video steaming and the other one is deep learning. This section will present the background and the significance of memories for those two applications. Also, as traditional CMOS technologies are facing potential technology limits, emerging memory technologies are becoming more and more important to support those data-intensive applications, and therefore this section will also introduce emerging memory technologies for data-intensive applications.

3.2.1 Memory in Video Streaming

The development in mobile video streaming and processing has been significantly dependent on the performance of video memory. In particular, to optimize data transmission, video compression technologies have played a crucial role, ensured smoother streaming experiences while conserved bandwidth. However, the process of compressing and decompressing videos places substantial demands on the resources of mobile devices, particularly their energy consumption. This energy-intensive behavior poses challenges for low battery life and poor device performance. This is prompting ongoing efforts in optimizing video processing algorithms, memory, and peripheral hardware to strike a balance between quality, efficiency, and power consumption. To enhance compression efficiency, different codecs, such as H.265/HEVC (high-efficiency video coding) and AV1, have been developed. Due to that, it reduces the amount of data required for streaming without compromising quality. In addition, dedicated video processing units (VPUs) and neural processing units (NPUs) help to improve energy efficiency and performance of the video processing.

Figure 3.2 illustrates a conventional H.264 video decoding and display process. After entropy decoding, inverse quantization (IQ), and inverse transformation (IT), the motion compensator utilizes the previous reconstructed frames, which are stored in the reference frame memory, along with the transmitted motion vectors, to construct new frames. Hence, the frames are decoded, and the mobile-display controller periodically sends them from the frame buffer to the display panel. Multiple memories are utilized during this process. Thus, memory management is crucial in mobile video decoding systems, as these memories not only store decoded video frames but also significantly impact power consumption and overall system cost. As a result, efficient memory techniques are extremely important to prolong battery life and enhance the user experience by reducing power consumption without sacrificing video quality.

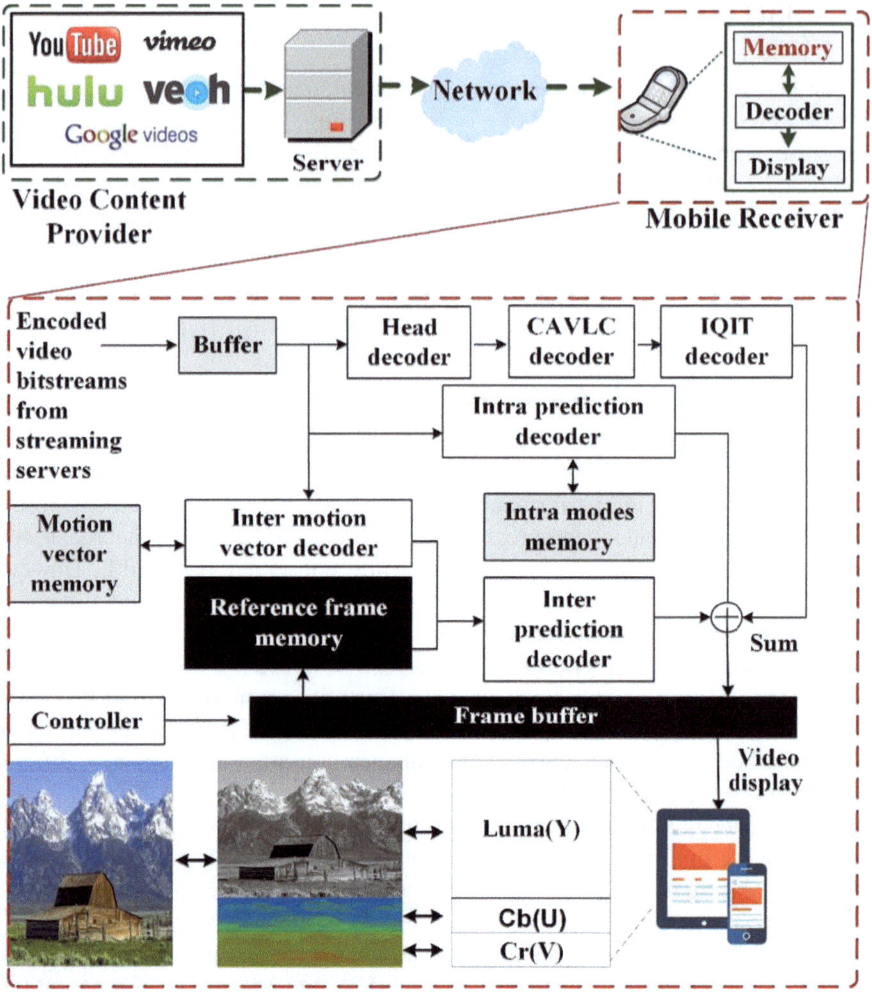

Fig. 3.2 Memory and video steaming

3.2.2 *Memory in Deep Learning*

As another popular data-intensive application, deep learning is transforming many modern artificial intelligence (AI) systems. However, such deep learning-enabled benefit does come with its cost, such as the associated demanding resources, particularly memories. This is due to the essential operation of deep learning systems. Specifically, the operation of deep neural networks is based on many connected, biological pathways that are present within animals, such as the brain or visual cortex. During the training process, the majority of memory usage by neural networks comes from the weights that are repeatedly updated throughout

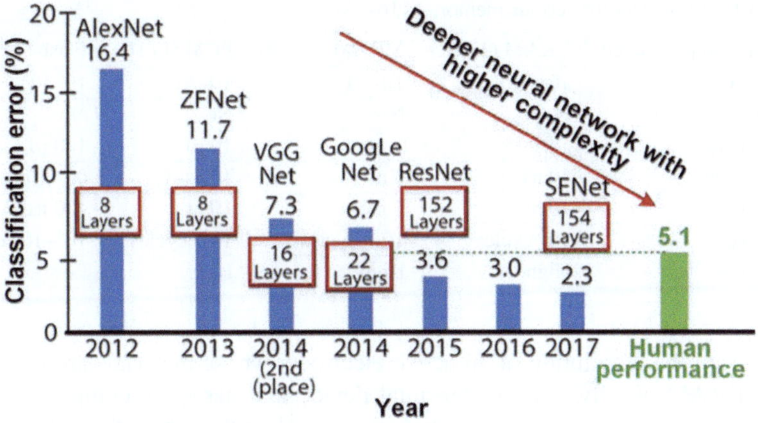

Fig. 3.3 The classification performance and size of deep learning from ImageNet Large Scale Visual Recognition Challenge (ILSVRC). (Adapted from [7])

the training process. In particular, from the classification performance perspective, in order to enhance the predication accuracy, neural networks continue to grow and they become deeper and wider. As illustrated in Fig. 3.3, many advanced neural networks have over 100 layers with millions to hundreds of millions of parameters which require significant storage resource [9, 10]. Thus, power-efficient memory techniques are extremely important for deep learning systems.

3.2.3 Emerging Memory Technologies

Recently, intensive research efforts have been made to develop emerging memory technologies, particularly for data-intensive applications. Most of these new technologies are nonvolatile memories, emerging memories as listed in Table 3.1, such as Resistive Random Access Memory (RRAM) [11, 12], Phase Change Memory (PCM) [13], and Magnetic Random Access Memory (MRAM) [14], have received more attention from both academia and industry. Such emerging memories are inherently nonvolatile and showing great potential for power efficiency. They utilize resistance/state change to represent bit information in memory. Sometimes even one cell, such as RRAM, has more than 100 states (levels) to replace more than 7-bit traditional memory cells (SRAM or DRAM) [15], further improving their power efficiency. In this section, typical emerging memories will be briefly introduced.

RRAM is to use a layer of conductive filament (formation and disruption of filament) to change the resistance of an insulator layer which is sandwiched between two metal electrodes [19, 20], so it is also named filamentary memory. There are two categories for RRAMs: one is oxide-based RRAM, which generates the conductive filament from the formation of oxygen vacancies in the resistive layer; the other

Table 3.1 Parameters for various memories [16–18]

Comparison parameters	RRAM (1T1R)	MRAM (1T1R)	PCM (1T1R)	Flash (1T)
Size (F^2)	10–30	10–30	10–30	10–30
Volatility	No	No	No	No
Write energy	1 pJ	1 pJ	10 pJ	100 pJ
Write speed	10 ns	5 ns	10 ns	0.1–1 ms
Read speed	10 ns	5 ns	10 ns	100 ns
Endurance	10^7 cycles	10^{15} cycles	10^{12} cycles	10^4–10^6 cycles
Scalability	High	High	High	Low

results from the dissolution of an active electrode, whose materials are usually Ag or Cu. RRAM usually is a two-terminal device that has great compatibility with CMOS technology, and its manufacture steps are less than that of a standard Flash memory. It also has lower programming voltage (<3 V) than that of a Flash memory (20 V), and result in 1 pJ write energy as listed in Table 3.2. Furthermore, RRAM is a high speed memory and its typical writing time is 10 ns [21], only one-tenth of a Flash memory.

PCM basically uses the mechanism as electrical conductivity of amorphous chalcogenide alloy films. The current-induced Joule heating generates the phase transition. $Ge_2Sb_2Te_5$ *(GST)* or *GeTe* chalcogenide are two popular materials. PCM has low voltage operation (<3 V) and high write/read speed (10 ns). A relatively high endurance, about 10^{12} cycles, has been verified and demonstrated. PCM does not require any initialization step (e.g., RRAM requires a formation process). The device characteristics can be determined by material and manufacture modification (doping, etc.) [22, 23].

MRAM employs the magnetism instead of the electrical charges to store bits and provide nonvolatility [19]. There are two categories of MRAM. First one is the toggle MRAM, which has a 1M1T structure (one magnetic tunnel junction (MTJ) + one transistor) to form a much high-density structure. Another one is to use a spin-torque transfer property, where the spin of electrons polarizes the current to build the desired magnetic state of the free layer for programming or writing information to an MRAM. As compared with other nonvolatility memory, MRAM has the fast switching speed with 5 ns reading or writing times and very good endurance (up to 10^{15} cycles) [18].

The impact of those emerging memory technologies on low-power memory design and the proposed AI-enabled memory design will be presented in the following sections.

3.3 State of the Art

This section presents state-of-the art of memory designs for videos and deep learning systems. Section 3.3.1 presents general-purpose memories and application-

Table 3.2 AI-enabled low-power memory design methodology

Categories	Utilized AI techniques	AI-enabled knowledge for hardware design	Target applications	Major advantages	Limitations
AI-enabled content-adaptive video memory	Decision tree	Number of truncated bits for various videos	Videos	Low hardware implementation overhead	Limited prediction accuracy
AI-enabled content-adaptive memory	Machine learning facial detection model	Facial detection	Videos	Save compute resources and power from video encoding	Can only detect faces; a more complex ROI algorithm can identify additional regions
AI-enabled self-recovery memory	Association rule mining	Data association and relationship	Videos and deep learning	Discovered data association relationship	Can mine frequent patterns; other data mining techniques are needed for additional data knowledge discovery

specific memory designs using current CMOS technologies (e.g., more-than-6T bitcells, assist techniques, error-correction-code (ECC)). Section 3.3.2 further discusses low-power memory designs using emerging memory technologies.

3.3.1 Low-Power CMOS Memory Design

3.3.1.1 CMOS General-Purpose Memory Design

Low power is always a key criterion for embedded memory design. There are different techniques to reduce the power consumption of embedded memories, such as voltage scaling, device scaling, and body biasing. Among them, voltage scaling is particularly effective to reduce the memories' power consumption due to the strong correlation of dynamic and leakage power consumption on supply voltage. However, Voltage-scaled SRAMs are prone to functionality failures. Accordingly, different design techniques have been developed, which mainly fall into the following three aspects to provide an optimized power solution for embedded memories.

More-than-6T bitcells 6T SRAM bitcell is the conventional memory design solution for embedded memory with a weak read stability and a higher voltage operation compared to more-than-6T bitcells. As a result, to gain low-voltage operation, most of the existing designs adopt more-than-6T bitcells such as 8T [24], 9T [25], 10T [7], and 12T [26]. The main weakness of memories with more-than-6T bitcells is the significant silicon area overhead. Also, those bitcells usually suffer from additional design issues, such as 8T's write half-select disturb [24], which limits the energy gains.

Complex assist schemes Complex assist schemes, such as boosted wordline [27], negative bitline [28], and dual-rail supply [29], are developed to lower the memory's operation voltage. There is always a trade-off between performance and design cost. In addition, power optimization is often achieved with significant design complexity with a design area overhead. At the same time, a power penalty is observed due to the voltage regulations or boosting circuits.

Error-correction techniques As memory failures occur at low voltages, in order to correct the caused bit flips, error-correction techniques have been implemented, with extra peripheral and area overhead. Various error-correction techniques have been developed for memories such as error-correction codes [30]. However, the improvements in embedded memory power efficiency are often achieved with significant silicon area overhead. For examples, about 75% and 36% silicon area overhead are considered by using conventional hamming code-74 (ECC74) and hamming code-1511 (ECC1511), respectively [31].

Overall, a general weakness of the existing general-purpose memory solutions is their high implementation or design cost. This high design overhead might be acceptable for general-purpose systems. However, they are not sufficient to satisfy the special storage needs for data-intensive applications such as video data processing and deep learning.

3.3.1.2 Application-Specific Memory Design

Application-specific properties are analyzed to reduce the implementation cost of memory design and manufacture. Those techniques can consider the context of the target applications, thereby further optimizing the design for the specific target applications' requirements.

Hybrid bitcells The performance of different applications such as video and deep learning are dependent on the memory's architecture. The first set of application-specific techniques attempts to utilize contribution of different data bits and design memories with hybrid bitcells for a specific application. Several design efforts have investigated application resilience of videos or deep learning to approximations with "good enough" or "acceptable" output and additional power savings. Chang et al. [32] presented a hybrid (6T + 8T) SRAM to achieve quality-power optimization. Specifically, reliable 8T cells, with a lower failure probability, were used to store the most significant bits (MSBs) which have a larger impact on the output quality for video applications as compared to least significant bits (LSBs). Alternatively, area-efficient 6T cells were adopted to store LSBs to minimize the silicon area overhead. In [33], a heterogeneous sizing scheme was presented to design hybrid memory for video applications, thereby reducing the failure probability of 6T bitcells for MSBs. Moreover, MSBs allow larger bitcells compared to LSBs to harden the MSBs. As another design technique, a split data-aware 10T and 8T (SDA-10T-8T) memory is presented by our research group for mobile video applications. With hybrid 10T and 8T bitcells, the developed memory enables reliable operation at 0.36 V under process variation and aging effects [34].

Adaptive ECC Approximate memory design is always preferable to optimize power consumption with acceptable performance. Recently, there were several attempts to enable application-specific approximate memory with adaptive ECC. ECC can compensate for the performance with area and power overhead. For example, Frustaci et al. present an approximate memory design with selective ECC (15,11) which can switch between no ECC and ECC1511 based on an application's quality target [35]. Here, memory's LSBs are utilized to store the ECC parity bits, and it was able to protect 2.75 MSBs per byte using one LSB per byte. To further reduce the implementation cost and optimize the memory performance, in our recent work [31], we developed an approximate SRAM with three power-quality trade-off levels, hamming code-74 (ECC74), hamming code-1511 (ECC1511), and no ECC. In the developed memory, its ECC74 mode protects two MSBs per byte using 1.5 LSBs per byte, and its ECC1511 mode protects 5.5 MSBs per byte using two LSBs per byte. However, this memory only works when the memory failure rate is very low, or when the memory is running at its nominal supply voltage. In addition, another approximate memory design technique is developed in [36], which combines selective MSB hardening with ECC74 to provide various design options for trading off between power and quality. Our proposed design was achieving a 45.84% decrease in power delay product (PDP) compared to the memory in [31].

Data-aware memory designs As a main contributor to the overall power consumption of a memory system, dynamic power is mainly dependent on the switching activities of the memory's bitlines. Researchers have also studied data characteristics, such as switching activities, for memory design. For example, Sinangil et al. [5] designed a low bitline switching SRAM for the most significant bits (MSBs) in the video data. Due to that, it reduces the readout's power consumption of the memory. Another data-aware memory was developed to enable runtime adaptation based on data statistics [37]. The developed memory utilized 10T memory bitcells and associated conditional pre-charge circuitry to adapt the statistical dependencies present in the data bits. However, inadequate exploration of data characteristics is the main weakness of the application-specific memory designs. Usually, the video data is humongous, and it is not easy for hardware designers to observe the inherent video data behaviors directly from the large volume of data. Accordingly, hardware designers require a deeper understanding and systematic information of inherent behavior information, including input, output, and data patterns. Unfortunately, a simple and conventional exploration of data characteristics is not enough to design an optimized memory design with power efficiency. As a result, we introduce AI-based approaches to discover data knowledge for hardware design process, which will be detailed in Sect. 3.4. In addition, AI-based approach can perform a holistic search for the best performing memory solution to enhance the performance of video processing and deep learning applications.

3.3.2 Low-Power Emerging Memory Design

Similarly, low-power design techniques using emerging memory technologies are very important. Although both reading and writing are routine operations in emerging memories, the writing power is much more than that of the reading. This is because writing time is a major contributor to overall system running time, and the write current is much higher than the read or standby current [38] in emerging memories. Therefore, as for low-power emerging memory design, researchers usually focus on writing power reduction. In the following three paragraphs, three writing power reduction technologies are introduced, respectively, for RRAM, MRAM, and PCM.

A number of pulses compression (NPC) method is implemented for RRAM [39, 40] to reduce writing power. In an RRAM-based neuromorphic system, the conventional method to change the conductance of an RRAM for weight update will follow the value of weight change that is calculated by the ANN algorithm, and the periphery circuit will generate corresponding signals to enable the pulse generator for producing different numbers of positive/negative pulses and tuning the conductance of an RRAM. With this NPC method, the number of pulses can be reduced to exactly one and keeps the original width of writing pulses, which greatly reduce writing power and latency. The overhead as slow learning process only happens in several epochs at the beginning, and has no significant impact for the entire weight update. When this NPC method is applied for RRAM-based classification scenario, five ANN algorithms are utilized to demonstrate that it consumes less power than those without the novel method. The power saving is from 7.7% to 26.9% [39].

MRAM always has a high writing power. Usually during MRAM writing, even if the value in MRAM does not change, there still is a significant write current flowing through it [41] and results in much power consumption. A low-power technology using aforementioned write activation circuit is proposed to avoid unnecessary writing power when the value remains unchanged [41] in MRAM. Experimental results from Leon2 processor indicates, with the proposed technology, the system averagely saves 70% of the total writing power. In addition, the area overhead can be negligible, only 0.68% [41].

In PCM, a Flip-N-Write method is used to reduce write power [42, 43]. It is an architectural technology with a more efficient read-modify-write operation to replace a traditional PCM writing operation. In a writing cycle, a tag bit is used to identify whether the bits in a line are flipped. When the line is read, the tag bit is used to determine whether the bits in the line should be flipped [43]. As a result, the reduction in bit updates with Flip-N-Write as compared with the conventional write scheme is 56% (32-bit word) to 63% (2-bit word) [42].

Based on the existing low-power memory designs using CMOS and emerging technologies, a new dimension—data knowledge, which is enabled by AI technologies, will be added to achieve better trade-off during the design process.

3.4 AI-Enabled Low-Power Memory Design Methodology

In our research, we have utilized AI techniques to assist with low-power memory design for videos and deep learning applications. As shown in Fig. 3.4, those AI-enabled techniques can be divided into three different categories: (i) AI-enabled content-adaptive memory for video applications, (ii) region-of-interest (ROI)-aware memory for video applications, and (iii) AI-enabled self-recovery memory for video and deep learning applications. In those memory designs, one general feature is that the introduction of AI techniques to low-power memory design process through the interaction of both hardware and software views enables a new dimension for power savings. Table 3.2 presents a big picture of the developed techniques in those three categories. It can be seen that different AI techniques can help designers obtain important information for memory design, such as the region-of-interest (ROI) in videos. Additionally, the suitable AI techniques, such as advanced data mining techniques particularly suited to mobile video data applications, can be used

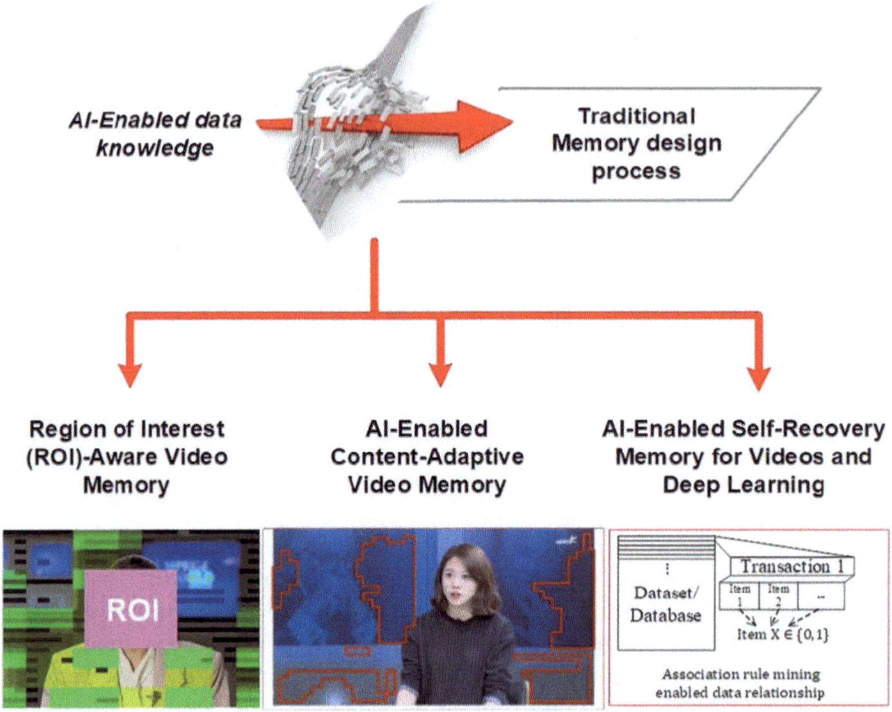

Fig. 3.4 AI-enabled low-power memory design methodology

to discover meaningful data patterns, thereby achieving intelligent memory with high power efficiency. AI techniques can also help designers to connect different design parameters, such as the relationship between the number of truncated bits and video content in the designed bit truncation memory [44]. The details of each design technique will be detailed in the following subsections.

3.4.1 AI-Enabled Content-Adaptive Video Memory

Our team has successfully developed an innovative video memory system that leverages AI technology to intelligently adapt to various types of content, resulting in significant power savings [44]. Our research has yielded two crucial discoveries: Firstly, our study revealed that videos with similar peak signal-to-noise ratio (PSNR) values may exhibit vastly different visual quality. Secondly, we observed a clear relationship between banding distortion and plain macroblocks (MBs). Specifically, videos with a high concentration of plain MBs tend to exhibit lower visual quality. This discovered knowledge enables us to intelligently adjust video quality, ultimately enhancing hardware's energy efficiency.

In order to account for the MB information, we derived a single, averaged percentage for the entire sample by calculating the variance of pixel luminance values for a specific MB [45]:

$$
\begin{aligned}
V_{\mathrm{MB}} &= \sum_{i=0}^{15} \sum_{i=0}^{15} \left(P\left(i, j\right) - \rho_{\mathrm{MB}}\right)^2 \gg 8 \\
\mathrm{MB} &= \begin{cases} \text{Plain} & \text{if}\,(V_{\mathrm{MB}} \le \mathrm{Th}_{\mathrm{Low}}) \\ \text{Textured} & \text{Else} \end{cases}
\end{aligned}
\tag{3.1}
$$

where ρ_{MB} and V_{MB} are the average luminance and variance of luminance values in a given MB, respectively. The value we used for $\mathrm{Th}_{\mathrm{Low}}$ was 1.25, as determined in [46] through regression analysis.

Figure 3.5 presents two video samples that have similar peak signal-to-noise ratio (PSNR) values but different percentages of plain macroblocks (MBs) with two least significant bits (LSBs) truncated. The distribution of plain MBs and the resulting banding distortion effect are illustrated in the figure. It is worth noting that there is a clear relationship between the banding distortion and the presence of plain MBs; videos with higher amounts of plain MBs, especially when densely distributed, generally exhibit decreased visual quality. Based on this observation, we have developed a content-adaptive model to predict the number of truncated LSBs for different videos. To streamline the computational process, we focus on the average plain MB percentage per video frame and specifically analyze low-motion videos with a stationary camera or featuring a reporter.

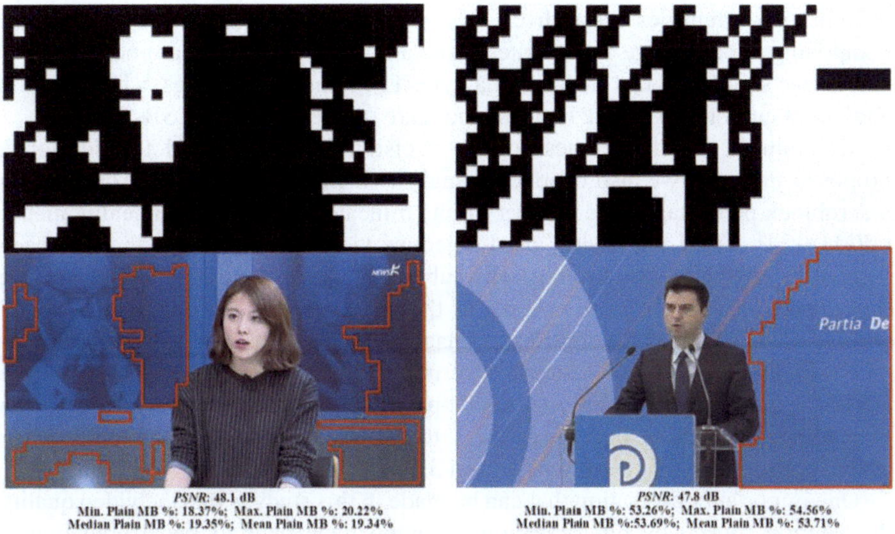

Fig. 3.5 Plain MBs visualization and video output comparison of two videos with varying plain MB % (with 2 LSBs truncated). White: plain MBs [44]

One of the key technical challenges we encountered was determining the optimal number of least significant bits (LSBs) to truncate for different videos [44]. We conducted subjective video testing to address this and used the data collected to develop an AI model. We conducted two sets of subjective video studies to gather viewers' feedback. In each study, participants were asked to watch multiple versions of the same video. They viewed both the original video and a truncated version. Then, they rated the quality on a scale from 1 to 5 based on their perception (5 for imperceptible, 4 for perceptible but not annoying, 3 for slightly annoying, 2 for annoying, and 1 for very annoying). A target average score of 4.0 or higher was set as the threshold for acceptable video quality. The first study involved 10 participants, and the second involved 13 participants. Each participant viewed 7 videos from a pool of 34 sample videos in the first study and 9 videos in the second study. We categorized the video samples into different regions by using the average scores for various amounts of truncated LSBs. This categorization helped us develop models that establish a connection between the average plain MB percentage and the number of LSBs that can be truncated. Our analysis revealed an inverse relationship between the plain MB percentage and the acceptable number of truncated LSBs. Leveraging this relationship and the subjective data collected from participants, we developed a decision tree model, as illustrated in Fig. 7a. Following the tree from top to bottom based on the plain MB percentages, we can determine the number of truncated LSBs suitable for different videos.

Based on the developed decision model, we designed a low-power viewer-aware dynamic bit-truncation memory, which is shown in Fig. 7b. This memory consists of 512 words with 64 bits, totaling 32 kb 6T SRAM bit cells. In order to implement viewer-aware bit truncation for the least significant bits (LSBs), two distinct bitline conditioning circuitries are utilized. The regular bitline conditioning circuitries, comprising a pre-charge unit, write driver and sense amplifier, are connected to the 4 most significant bits (MSBs) in a byte. The remaining bitlines, which contain additional circuitry to enable bit truncation, are utilized for the 4 LSBs in a byte.

To evaluate the effectiveness of the decision tree model and the developed proposed memory, we also conducted analysis including the comparison of plain macroblock percentages, the number of bits truncated, and the video quality metric (VQM) [34]. VQM is a widely used objective video quality metric that has been demonstrated to correlate strongly with subjective viewer ratings. The decision tree model performed well for almost all of the videos, with only one video (tagged wF6lvdXXwc4) out of 20 considered unacceptable by most participants. In terms of power efficiency, input patterns covering all data switching possibilities were tested for memory. Compared to normal operation, the average power consumption of reading and writing operations for 1 to 4 LSB truncations can enable power savings of 13.54%, 20.10%, 26.83%, and 33.31%, respectively.

One important observation that can be made in this study is that the video quality degradation caused by bit truncation is region dependent; if the degradation is present in the more important regions in a video frame (i.e., regions of interest (ROI)), the video quality loss is much more noticeable. For example, for those different videos, the lowest acceptance rate is 73% for the video tagged 2AQ6rhVhwRc, which featured banding close to the viewer's focus point, depicting a kitten playing with a string in the video. This conclusion motived us to introduce ROI awareness, which is enabled by AI techniques, to the memory design process. The developed ROI-aware memory further enhanced power efficiency, which will be discussed in the next section.

3.4.2 ROI-Aware Low-Power Bit Truncation Memory Design

We have utilized AI-enabled ROI awareness in designing video memory for additional power savings. Specifically, this technology has been employed in the design process for bit truncation memory. As discussed earlier, bit truncation is a commonly used low-power design technique for memories, but it can considerably impact the output quality of applications such as videos. Our studies have focused on the performance of the human visual system (HVS) and have revealed that viewers tend to concentrate more on specific areas of a video, referred to as regions of interest (ROIs). For example, in video conferencing applications, viewers typically pay more attention to the face regions than other areas (Fig. 3.6a). Similarly, viewers are most focused on facial regions in consecutive frames in video surveillance. Therefore, the impact of bit truncation on ROIs can significantly diminish the visual quality. Consequently, if truncation-induced banding distortion appears in ROIs, it can have a negative effect on a viewer's experience.

The ROI-aware video storage technique involves both video encoding and decoding processes, as illustrated in Fig. 3.6b. In traditional mobile video streaming, the mobile device initially requests a video for display from the cloud. The streaming servers then encode the requested video and transmit the encoded bitstream to the mobile device for decoding and display. Multiple memories are required during this process to store the intermediate and final results of the frame data, including the reference macroblock, frame memory, and display memory. These memories are accessed frequently and significantly impact the system's overall cost and power consumption.

The proposed technique optimizes power consumption by extracting ROIs in the cloud server and transmitting the truncation region data with the encoded bitstream to the mobile device. This approach reduces the mobile device's computational overhead. The mobile device's hardware video decoder receives the truncation region data and makes memory bit-truncation decisions to achieve more significant power savings with minimal perceived quality loss. The truncation decision logic of the mobile device hardware can be further optimized to improve power consumption, offering the option of either no truncation or 3-bit truncation for the truncation regions.

Our memory design is AI-enabled and ROI-aware, and it has been thoroughly verified using a frame buffer [47]. The detailed frame buffer circuit is shown in Fig. 3.7, which has M words, each consisting of N bits. To assess the functionality and measure the average power consumption of this circuitry, we created a 128-

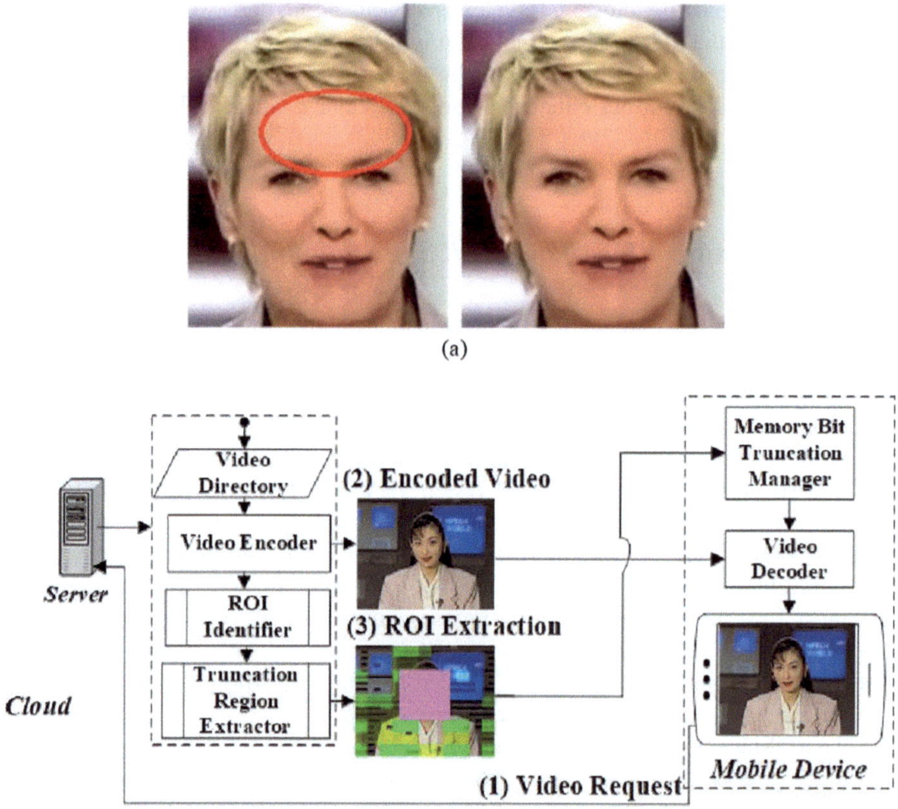

Fig. 3.6 (**a**) Importance of ROI in the video storage process. (**b**) AI-enabled ROI-aware video streaming system. (Adapted from [48])

word by 24-bits memory array. In this design, the input and output pins are referred to as data [23:0] and out [23:0], respectively. The bits 23-16 are designated as the Y byte, bits 15-8 as the U byte, and bits 7-0 as the V byte. As shown in Fig. 3.7, our memory implementation includes a driver and a sense amplifier for writing and reading data. We have also incorporated enabled bit truncation, controlled by T_Y, T_U, and T_V signals. When all these control signals are de-asserted (set to logic “0”), the frame buffer operates as a traditional memory device, with the sense amplifier operating using a supply voltage (VDD) and a pre-charge signal phi2b. However, when the T_Y signal is asserted (set to logic “1”), the peripheral circuitry generates two signals: y! (the inverted value of T_Y) and y_pre! (the inverted value of the pre-charge enables signal). These signals control the sense amplifier for the 3 least significant bits (LSBs) of the Y byte, enabling truncation. During this process, the VDD for this sense amplifier remains grounded, and the pre-charge signal is reactivated, resulting in reduced power consumption compared to normal operation. The 3-LSBs are generated as “100” during the read back

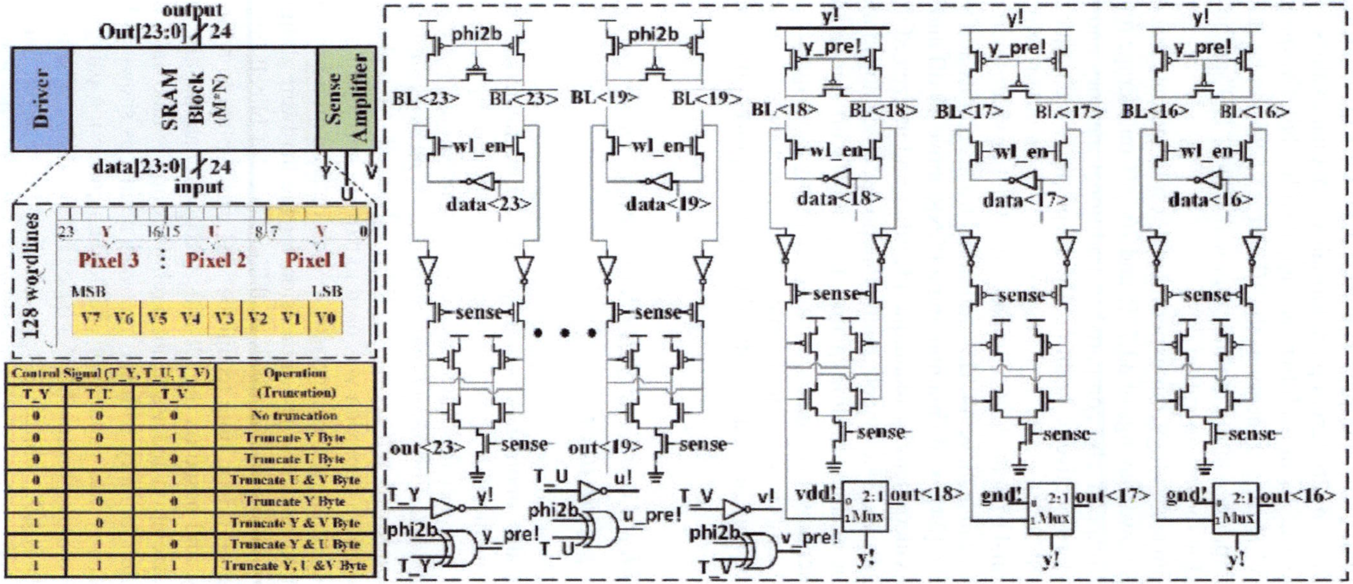

Fig. 3.7 AI-enabled ROI-aware memory design. (Adapted from [48])

operation using three 2:1 multiplexers instead of regular data output. When bit truncation is asserted, these multiplexers select "100" through control signals y! u! or v! Otherwise, these multiplexers pass normal readout data values. Additionally, the VDD of all the 3 LSBs of each byte is controlled by the corresponding control signals y! u! and v! During truncation, the VDD for LSBs can be powered off to save power consumption, and the multiplexers will select "100" as the output data, thereby achieving low-power operation.

Based on a 45 nm technology, our experimental results indicate that the proposed technique achieves power savings of 83.79% and 76.56% on average for videos with and without ROI, respectively, compared to traditional memory and state-of-the-art technology [34], respectively. Additionally, the proposed technique can enhance video quality (WPSNR values) by an average of 20.17% for videos with ROI and by 26.46% for additional truncated bits compared to [34]. For videos without ROI, the proposed technique can achieve 44.61% additional truncated bits and 8.26% power savings compared to [34], while maintaining a PSNR above 40 dB and 0.95 SSIM.

Our developed video memory technique identifies and utilizes ROI regions and therefore intelligently balances video quality and power efficiency in mobile device video storage. This enables runtime quality adaptation with significantly reduced pixel bits and further power savings compared to existing techniques. The developed ROI-aware video memory is distinct from existing viewer-aware or data-dependent schemes and can be simultaneously utilized to optimize power efficiency.

3.4.3 AI-enabled Low-Cost Self-Recovery Memory for Videos and Deep Learning

In our study, we also applied AI techniques and investigated data patterns to enable low-power memory design for videos and deep learning. During the design process, we aim to answer this question: *instead of avoiding/correcting the faults with significant hardware overhead, can we embrace memory faults by enabling AI-enabled data-informed self-recovery?* With advanced AI techniques, a two-dimensional data-pattern approach was used to explore horizontal data-association and vertical data-correlation characteristics. Such data relationship discovery and pattern identification enable a new dimension for the hardware design space and bring self-recovery ability to memories in the presence of bitcell failures. Specifically, the association rule mining technique [48] was applied to a bit-level, video data pattern study, and based on the identified frequent data patterns; utilizing the discovered data correlation and association relationship in video data, we presented a data-enabled video memory with self-recovery ability, as shown in Fig. 13 [49]. The presented memory was capable of enable data correction using the identified patterns, based on a multiplexers (MUX)-based bitline design scheme. Each bitline was connected to a MUX, which was controlled by the received fault positions. If a fault was indicated, self-recovery is enabled by selecting the data pattern. With

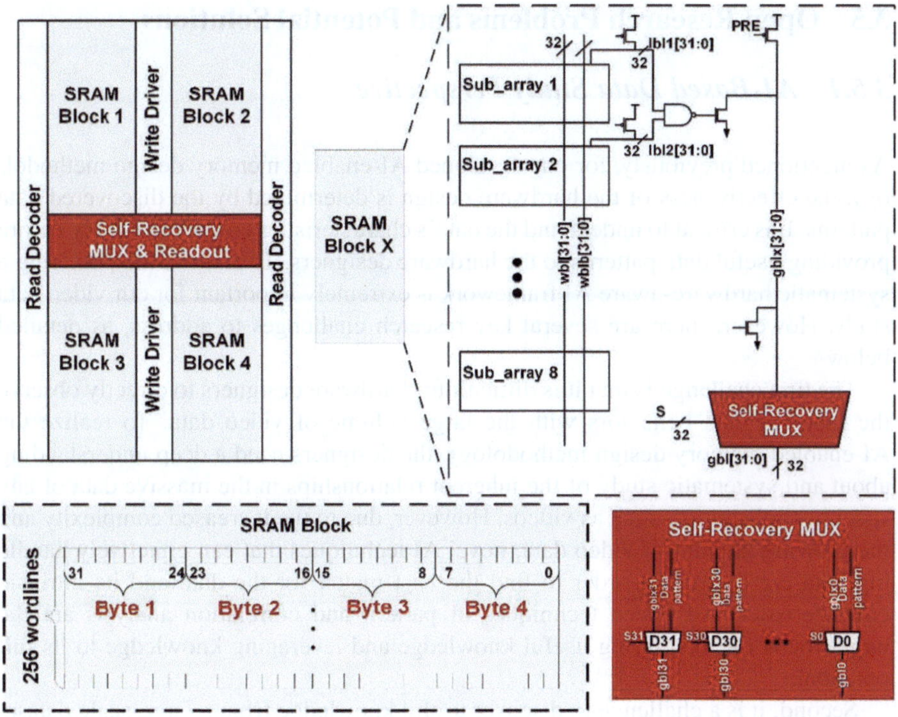

Fig. 3.8 AI-enabled self-recovery memory. (Adapted from [49])

a data-pattern-enabled self-recovery ability, the developed memory exhibits a low implementation cost (7.94%) and reliable operation at the near-threshold voltage (0.5 V) to support good video quality.

The developed AI-enabled memory design technique has also been applied to deep learning, as shown in Fig. 3.8. We conducted an off-line data-mining-assisted, synaptic data-correlation/association relationship study and, based on it, designed a memory that can store synaptic weights efficiently with a data self-recovery ability in the presence of memory failures [50]. A key conclusion was that within the first eight MSBs of weight values, multiple MSBs have strong association to the most significant bit of exponent values. Therefore, the most significant bit of exponent values was stored in robust memory bitcells to achieve self-recovery of other relevant bits based on data association/correlation. Our results demonstrated that using data-mining discovered data characteristics, as compared to traditional memory, the AI-enabled memory enables 45.6% and 83.2% reduction in active power and leakage power savings, respectively, with low implementation cost (3.17%) and less than 1% degradation in classification accuracy.

3.5 Open Research Problems and Potential Solutions

3.5.1 *AI-Based Data Study Perspective*

As mentioned previously, for our developed AI-enabled memory design methodology, the effectiveness of the hardware design is determined by the discovered data patterns. It is critical to understand the data's characteristics comprehensively before providing useful data patterns to the hardware designers. To achieve this, building a systematic hardware-aware AI framework is extremely important for our video data study. However, there are several key research challenges to address, as detailed below.

The first challenge is that it is difficult for hardware designers to directly observe the inherent data behaviors with the large volume of video data. To realize the AI-enabled memory design methodology, the designers need a deep understanding about and systematic study of the inherent relationships in the massive data of any specific applications, such as videos. However, due to the increased complexity and the growing amount of video data, novel AI techniques that can effectively handle big data are desired in order to find the best match for the data and its intrinsic characteristics. Advanced techniques of pattern and correlation analysis are the target focus for identifying useful knowledge and leveraging knowledge to its full potential.

Second, it is a challenge to discover useful knowledge from a large-scale dataset with a mixed set of data categories. For example, in a big video dataset, different video data categories may display various behaviors and patterns. One promising research direction is to develop a two-layer, systematic AI framework with (i) a top layer that focuses on general data patterns and relationship analysis in a large-scale setting, with the identified knowledge used for design-time hardware design, and (ii) a bottom layer that focuses on context-specific data patterns and relationship discovery, with a relatively smaller scale that is limited to individual video categories and/or similar video categories, with the discovered knowledge mainly used for runtime hardware adaption.

Third, even data patterns can be identified, it might be still challenging to apply them to the hardware design process. For example, in our video data study, the set of frequent patterns derived from traditional association rule mining methods, as we observed, has several issues. First, the obtained frequent patterns are too huge for subsequent interpretation by the designer to determine their usefulness for hardware design. When applied to larger-scale video data, we are expecting more data patterns will be generated. This will create exciting pattern-discovery opportunities but will also make it much more challenging for designers to discover useful patterns. Second, many smaller and shorter sub-patterns are generated, but they are not very meaningful for hardware designers. This is mainly due to the a priori principle [51] used in the association mining algorithms: if an itemset is frequent, then all of its subsets must also be frequent. In other words, a large pattern contains an exponential number of smaller, frequent sub-patterns. To address this challenge, compact representations of Frequent Itemset are desired. Two such representations, in the form of maximal and closed frequent item sets, are worth exploring, based on which a more powerful hardware-driven AI framework could be built to help designers fully utilize the most representative data patterns discovered during the design process.

3.5.2 Hardware Design Perspective

From the hardware design process, there are also several challenges. First, implementation cost is a major design consideration for the developed AI-enabled memory design methodology. As mentioned earlier, in our previous video data study, general data insight that is discovered in large-scale video data is not dependent on a specific video format, category, or other video context; therefore, the data knowledge can be introduced to the design-time memories in order to benefit all videos without dynamic memory adaption mechanisms. However, for context-specific data patterns in smaller video datasets, hardware adaption is required to handle different video data. Similarly, in the ROI-aware video system design process, the ROI identification brings additional implementation cost and power consumption in the video encoding process. Enabling hardware adaption with a low implementation cost is a major challenge for the designers.

In addition, our current studies on AI-enabled memory designs have been focused on SRAM design based on traditional CMOS technology. As CMOS technology is approaching its physical limit, it is desirable to explore EMs to enlarge design space. However, when the developed AI-enabled memory design methodology is applied for EMs, the heterogeneous integration between EMs and peripheral CMOS circuitry is always a challenge, due to the following reasons. First, variability in RRAM [52, 53], MRAM, and PCM, such as nonlinearity [54], device-to-device variations [55], cycle-to-cycle variations [56, 57], and on/off ratio variations, is costly [58]. If each EM cell performs much different and its characteristics vary in time, programming for EMs to a desired state becomes a personalized endeavor and it is an impossible work for mass production as it consumes overlarge time and power for supporting CMOS circuitry. Second, scalability is another concern [59, 60]. CMOS circuitry is developed very fast, even beyond 5 nm with quite small supply voltage. However, most of the emerging memories have big size, sometimes even use micrometer technology, and need a high voltage for programming. In addition, while emerging memories are integrated with CMOS circuitry, how to fabricate the integrated system together, such as enabling adaptable power supply for emerging memories and CMOS circuitry, is another critical challenge.

3.5.3 Application Perspective

Our current study focuses on videos and deep learning. Even for those two applications, challenges still exist at the application level. For mobile video data, the characteristics highly depend on the context of the video frames, including bit-rate and resolution can be controlled at the application level. For example, video content may be coded with different bit-rates, the application then selects the required one depending on decoder capability. Therefore, the diversity of mobile videos becomes another challenge for a good understanding of mobile video data characteristics. Also, since one potential useful mobile video characteristic stems from the fact that application functionality is defined on a continuous scale of output quality, video quality is another important aspect of video delivery. However, there is no unique metric to measure video output quality. Therefore, there exists a choice between multiple metrics to measure video output quality. Similarly, for deep learning applications, the data storage strongly depends on a specific deep learning model and its dataset. How to extract general data patterns for numerous deep learning models remains a core question.

As a result, to explore the full potential of AI-enabled low-power memory for different applications, multidisciplinary collaboration is highly needed between AI scientists, hardware designers, and application experts. We face the challenge of comprehensively studying the data characteristics, considering various application-level contexts and quality metrics, thereby moving AI-enabled low-power memory design from its initial stages of exploration to broader adoption.

3.6 Conclusion

This chapter presents the unique advantages of AI-enabled efficient memory design methodology and a roadmap for future directions. With advanced AI techniques introduced to the ASIC design process, it effectively transforms the data into useful knowledge and actionable information for efficient ASIC design and develop novel memories that exploit the discovered data relationship for substantial energy efficiency. We anticipate that this interdisciplinary research direction will invite broad exploration in memory and other ASIC design, AI community, and their intersections.

Acknowledgments The work was supported by the National Science Foundation under grants CCF-1514780, CCF-1855706, CNS-2211215, and OIA-2218046.

References

1. Faraboschi, P. (2014). Keynote at the HiPEAC 10th anniversary.
2. Mobile video in the United States - Statistics & Facts. [Online]. Available: https://www.statista.com/topics/2725/mobile-video-in-the-united-states/#topicOverview.
3. Wang, J.-S., Chang, P.-Y., Tang, T.-S., Chen, J.-W., & Guo, J.-I. (2011). Design of sub-threshold SRAMs for energy-efficient quality-scalable video applications. *IEEE Journal on Emerging and Selected Topics in Circuits and Systems, 1*(2), 183–192. https://doi.org/10.1109/JETCAS.2011.2158345

4. Liu, T., & Lee, C. (2006). Memory-hierarchy-based power reduction for H. 264/AVC video decoder. In *2006 International symposium on VLSI design, automation and test* (pp. 1–4). IEEE. https://doi.org/10.1109/VDAT.2006.258171
5. Sinangil, M. E., & Chandrakasan, A. P. (2014). Application-specific SRAM design using output prediction to reduce bit-line switching activity and statistically gated sense amplifiers for up to 1.9$\times$ lower energy/access. *IEEE Journal of Solid-State Circuits, 49*(1), 107–117. https://doi.org/10.1109/JSSC.2013.2280310
6. Chen, T., et al. (2014). DianNao: A small-footprint high-throughput accelerator for ubiquitous machine-learning. *ACM SIGARCH Computer Architecture News, 42*(1), 269–284. https://doi.org/10.1145/2654822.2541967
7. Noguchi, H., et al. (2007). A 10T non-precharge two-port SRAM for 74% power reduction in video processing. In *IEEE computer society annual symposium on VLSI (ISVLSI '07)* (pp. 107–112). IEEE. https://doi.org/10.1109/ISVLSI.2007.2
8. Venkataramani, S., Chakradhar, S. T., Roy, K., & Raghunathan, A. (2015, June). Approximate computing and the quest for computing efficiency. In *Proceedings of the 52nd annual design automation conference* (pp. 1–6). ACM. https://doi.org/10.1145/2744769.2751163
9. Wang, J., Zhang, J., Bao, W., Zhu, X., Cao, B., & Yu, P. S. (2018, July). Not just privacy: Improving performance of private deep learning in mobile cloud. In *Proceedings of the 24th ACM SIGKDD international conference on knowledge discovery & data mining* (pp. 2407–2416). ACM. https://doi.org/10.1145/3219819.3220106
10. Image classification on ImageNet. [Online]. Available: https://paperswithcode.com/sota/image-classification-on-imagenet
11. Luo, Q., et al. (2017, Dec). 8-Layers 3D vertical RRAM with excellent scalability towards storage class memory applications. In *2017 IEEE international electron devices meeting (IEDM)* (p. 2.7.1–2.7.4). IEEE. https://doi.org/10.1109/IEDM.2017.8268315
12. Fu, J., Liao, Z., & Wang, J. (2019). Memristor-based neuromorphic hardware improvement for privacy-preserving ANN. *IEEE Transactions on Very Large Scale Integration (VLSI) Systems, 27*(12), 2745–2754. https://doi.org/10.1109/TVLSI.2019.2923722
13. Kim, T., & Lee, S. (2020). Evolution of phase-change memory for the storage-class memory and beyond. *IEEE Transactions on Electron Devices, 67*(4), 1394–1406. https://doi.org/10.1109/TED.2020.2964640
14. Smullen, C. W., Mohan, V., Nigam, A., Gurumurthi, S., & Stan, M. R. (2011, Feb). Relaxing non-volatility for fast and energy-efficient STT-RAM caches. In *2011 IEEE 17th international symposium on high performance computer architecture* (pp. 50–61). IEEE. https://doi.org/10.1109/HPCA.2011.5749716
15. Fu, J., Liao, Z., & Wang, J. (2022). Level scaling and pulse regulating to mitigate the impact of the cycle-to-cycle variation in Memristor-based edge AI system. *IEEE Transactions on Electron Devices, 69*(4), 1752–1762. https://doi.org/10.1109/TED.2022.3146801
16. Oboril, F., Bishnoi, R., Ebrahimi, M., & Tahoori, M. B. (2015). Evaluation of hybrid memory technologies using SOT-MRAM for on-chip cache hierarchy. *IEEE Transactions on Computer-Aided Design of Integrated Circuits and Systems, 34*(3), 367–380. https://doi.org/10.1109/TCAD.2015.2391254
17. Salahuddin, S., Ni, K., & Datta, S. (2018). The era of hyper-scaling in electronics. *Nature Electronics, 1*(8), 442–450. https://doi.org/10.1038/s41928-018-0117-x
18. Gebregiorgis, A., et al. (2022). A survey on memory-centric computer architectures. *ACM Journal on Emerging Technologies in Computing Systems, 18*(4), 1–50. https://doi.org/10.1145/3544974
19. Molas, G., & Nowak, E. (2021). Advances in emerging memory technologies: From data storage to artificial intelligence. *Applied Sciences, 11*(23), 11254. https://doi.org/10.3390/app112311254
20. Fu, J., Liao, Z., Gong, N., & Wang, J. (2019). Mitigating nonlinear effect of Memristive synaptic device for neuromorphic computing. *IEEE Journal on Emerging and Selected Topics in Circuits and Systems, 9*(2), 377–387. https://doi.org/10.1109/JETCAS.2019.2910749

21. Wang, C., et al. (2017). Ultrafast RESET analysis of HfO $_x$ -based RRAM by sub-nanosecond pulses. *Advanced Electronic Materials, 3*(12), 1700263. https://doi.org/10.1002/aelm.201700263
22. Cheng, H. Y., et al. (2012, Dec). A thermally robust phase change memory by engineering the Ge/N concentration in (Ge, N)xSbyTe z phase change material. In *2012 international electron devices meeting* (p. 31.1.1–31.1.4). IEEE. https://doi.org/10.1109/IEDM.2012.6479141
23. Navarro, G., et al. (2018). Phase-change memory: Performance, roles and challenges. In *2018 IEEE international memory workshop (IMW)* (pp. 1–4). IEEE. https://doi.org/10.1109/IMW.2018.8388845
24. Kim, T.-H., Liu, J., & Kim, C. H. (2009). A voltage scalable 0.26 V, 64 kb 8T SRAM with V$_{\min}$ lowering techniques and deep sleep mode. *IEEE Journal of Solid-State Circuits, 44*(6), 1785–1795. https://doi.org/10.1109/JSSC.2009.2020201
25. Chang, M.-F., Chang, S.-W., Chou, P.-W., & Wu, W.-C. (2011). A 130 mV SRAM with expanded write and read margins for subthreshold applications. *IEEE Journal of Solid-State Circuits, 46*(2), 520–529. https://doi.org/10.1109/JSSC.2010.2091321
26. Qureshi, M. K., & Chishti, Z. (2013, June). Operating SECDED-based caches at ultra-low voltage with FLAIR. In *2013 43rd annual IEEE/IFIP international conference on dependable systems and networks (DSN)* (pp. 1–11). IEEE. https://doi.org/10.1109/DSN.2013.6575314
27. Hirabayashi, O., et al. (2009, Feb). A process-variation-tolerant dual-power-supply SRAM with 0.179μm^2 cell in 40nm CMOS using level-programmable wordline driver. In *2009 IEEE international solid-state circuits conference - digest of technical papers* (pp. 458–459,459a). IEEE. https://doi.org/10.1109/ISSCC.2009.4977506
28. Wang, D. P., et al. (2007). A 45nm dual-port SRAM with write and read capability enhancement at low voltage. In *2007 IEEE international SOC conference* (pp. 211–214). IEEE. https://doi.org/10.1109/SOCC.2007.4545460
29. Kushida, K., et al. (2013, Nov). A 27% active and 85% standby power reduction in dual-power-supply SRAM using BL power calculator and digitally controllable retention circuit. In *2013 IEEE Asian solid-state circuits conference (A-SSCC)* (pp. 25–28). IEEE. https://doi.org/10.1109/ASSCC.2013.6690973
30. Hamming, R. W. (1950). Error detecting and error correcting codes. *Bell System Technical Journal, 29*(2), 147–160. https://doi.org/10.1002/j.1538-7305.1950.tb00463.x
31. Das, H., Haidous, A. A., Smith, S. C., & Gong, N. (2021). Flexible low-cost power-efficient video memory with ECC-adaptation. *IEEE Transactions on Very Large Scale Integration (VLSI) Systems, 29*(10), 1693–1706. https://doi.org/10.1109/TVLSI.2021.3098533
32. Chang, I. J., Mohapatra, D., & Roy, K. (2011). A priority-based 6T/8T hybrid SRAM architecture for aggressive voltage scaling in video applications. *IEEE Transactions on Circuits and Systems for Video Technology, 21*(2), 101–112. https://doi.org/10.1109/TCSVT.2011.2105550
33. Kwon, J., Chang, I. J., Lee, I., Park, H., & Park, J. (2012). Heterogeneous SRAM cell sizing for low-power H.264 applications. *IEEE Transactions on Circuits and Systems I: Regular Papers, 59*(10), 2275–2284. https://doi.org/10.1109/TCSI.2012.2185335
34. Gong, N., Jiang, S., Challapalli, A., Fernandes, S., & Sridhar, R. (2012). Ultra-low voltage split-data-aware embedded SRAM for mobile video applications. *IEEE Transactions on Circuits and Systems II: Express Briefs, 59*(12), 883–887. https://doi.org/10.1109/TCSII.2012.2231018
35. Frustaci, F., Khayatzadeh, M., Blaauw, D., Sylvester, D., & Alioto, M. (2015). SRAM for error-tolerant applications with dynamic energy-quality management in 28 nm CMOS. *IEEE Journal of Solid-State Circuits, 50*(5), 1310–1323. https://doi.org/10.1109/JSSC.2015.2408332
36. Das, H., Haidous, A. A., Smith, S. C., & Gong, N. (2023). Approximate memory for low-power video applications. *IEEE Access, 11*, 57735–57744. https://doi.org/10.1109/ACCESS.2023.3283409
37. Duan, C., Gotterba, A. J., Sinangil, M. E., & Chandrakasan, A. P. (2017). Energy-efficient reconfigurable SRAM: Reducing read power through data statistics. *IEEE Journal of Solid-State Circuits, 52*(10), 2703–2711. https://doi.org/10.1109/JSSC.2017.2731814

38. Bennett, D. MRAM optimizes system energy consumption, *Electronic Design*. Accessed 30 January 2024. [Online]. Available: https://www.electronicdesign.com/technologies/embedded/digital-ics/memory/article/21801309/mram-optimizes-system-energy-consumption
39. Liao, Z., Fu, J., & Wang, J. (2021, Aug). Ameliorate performance of Memristor-based ANNs in edge computing. *IEEE Transactions on Computers, 70*(8), 1299–1310. https://doi.org/10.1109/TC.2021.3081985
40. Liao, Z., Fu, J., Ding, C., & Wang, J. (2022, March). Pulse truncation enabled high performance and low energy Memristor-based accelerator. In *SoutheastCon 2022* (pp. 473–478). IEEE. https://doi.org/10.1109/SoutheastCon48659.2022.9764063
41. Bishnoi, R., Oboril, F., Ebrahimi, M., & Tahoori, M. B. (2014, March). Avoiding unnecessary write operations in STT-MRAM for low power implementation. In *Fifteenth international symposium on quality electronic design* (pp. 548–553). IEEE. https://doi.org/10.1109/ISQED.2014.6783375
42. Cho, S., & Lee, H. (2009, Dec). Flip-N-Write: A simple deterministic technique to improve PRAM write performance, energy and endurance. In *2009 42nd annual IEEE/ACM international symposium on microarchitecture (MICRO)* (pp. 347–357) Accessed 31 January 2024. [Online]. Available: https://ieeexplore.ieee.org/document/5375405
43. Liu, H.-K., et al. (2021). A survey of non-volatile main memory technologies: State-of-the-arts, practices, and future directions. *Journal of Computer Science and Technology, 36*(1), 4–32. https://doi.org/10.1007/s11390-020-0780-z
44. Edstrom, J., et al. (2019). Content-adaptive memory for viewer-aware energy-quality scalable mobile video systems. *IEEE Access, 7*, 47479–47493. https://doi.org/10.1109/ACCESS.2019.2908997
45. Shafique, M., et al. (2017). Application-guided power-efficient fault tolerance for H.264 context adaptive variable length coding. *IEEE Transactions on Computers, 66*(4), 560–574. https://doi.org/10.1109/TC.2016.2616313
46. Shafique, M., Molkenthin, B., & Henkel, J. (2010, March). An HVS-based adaptive computational complexity reduction scheme for H.264/AVC video encoder using prognostic early mode exclusion. In *2010 design, automation & test in Europe conference & exhibition (DATE 2010)* (pp. 1713–1718). IEEE. https://doi.org/10.1109/DATE.2010.5457091
47. Haidous, A., Oswald, W., Das, H., & Gong, N. (2022). Content-adaptable ROI-aware video storage for power-quality scalable mobile streaming. *IEEE Access, 10*, 26830–26848. https://doi.org/10.1109/ACCESS.2022.3156274
48. Agrawal, R., Imieliński, T., & Swami, A. (1993). Mining association rules between sets of items in large databases. *ACM SIGMOD Record, 22*(2), 207–216. https://doi.org/10.1145/170036.170072
49. Edstrom, J., Chen, D., Gong, Y., Wang, J., & Gong, N. (2019). Data-pattern enabled self-recovery low-power storage system for big video data. *IEEE Transactions on Big Data, 5*(1), 95–105. https://doi.org/10.1109/TBDATA.2017.2750699
50. Edstrom, J., Gong, Y., Chen, D., Wang, J., & Gong, N. (2017). Data-driven intelligent efficient synaptic storage for deep learning. *IEEE Transactions on Circuits and Systems II: Express Briefs, 64*(12), 1412–1416. https://doi.org/10.1109/TCSII.2017.2767900
51. Wang, J., Han, J., & Pei, J. (2003, Aug). CLOSET+: Searching for the best strategies for mining frequent closed itemsets. In *Proceedings of the ninth ACM SIGKDD international conference on knowledge discovery and data mining* (pp. 236–245). ACM. https://doi.org/10.1145/956750.956779
52. Oli-Uz-Zaman, M., Khan, S. A., Oswald, W., Liao, Z., & Wang, J. (2022). Stuck-at-fault immunity enhancement of Memristor-based edge AI systems. *IEEE Journal on Emerging and Selected Topics in Circuits and Systems, 12*(4), 922–933. https://doi.org/10.1109/JETCAS.2022.3207687
53. Oli-Uz-Zaman, M., et al. (2022, May). Reliability improvement in RRAM-based DNN for edge computing. In *2022 IEEE international symposium on circuits and systems (ISCAS)* (pp. 581–585). IEEE. https://doi.org/10.1109/ISCAS48785.2022.9937260

54. Khan, S. A., Oli-Uz-Zaman, M., & Wang, J. (2023). PAWN: Programmed analog weights for non-linearity optimization in Memristor-based neuromorphic computing system. *IEEE Journal on Emerging and Selected Topics in Circuits and Systems, 13*(1), 436–444. https://doi.org/10.1109/JETCAS.2023.3235658
55. Fu, J., Liao, Z., Liu, J., Smith, S. C., & Wang, J. (2021). Memristor-based variation-enabled differentially private learning systems for edge computing in IoT. *IEEE Internet of Things Journal, 8*(12), 9672–9682. https://doi.org/10.1109/JIOT.2020.3023623
56. Fu, J., Liao, Z., & Wang, J. (2020, Sept). Cycle-to-cycle variation enabled energy efficient privacy preserving technology in ANN. In *2020 IEEE 33rd international system-on-chip conference (SOCC)* (pp. 66–71). IEEE. https://doi.org/10.1109/SOCC49529.2020.9524794
57. Fu, J., Liao, Z., & Wang, J. (2023, Oct). Cycle-to-cycle variation suppression in ReRAM-based AI accelerators. In *2023 IEEE physical assurance and inspection of electronics (PAINE)* (pp. 1–6). IEEE. https://doi.org/10.1109/PAINE58317.2023.10317995
58. Adam, G. C., Khiat, A., & Prodromakis, T. (2018). Challenges hindering memristive neuromorphic hardware from going mainstream. *Nature Communications, 9*(1). https://doi.org/10.1038/s41467-018-07565-4
59. Oli-Uz-Zaman, M., et al. (2022). Mapping transformation enabled high-performance and low-energy Memristor-based DNNs. *Journal of Low Power Electronics and Applications, 12*(1). https://doi.org/10.3390/jlpea12010010
60. Oli-Uz-Zaman, M., Khan, S. A., Oswald, W., Liao, Z., & Wang, J. (2023, Jun). Reconfigurable mapping algorithm based stuck-at-fault mitigation in neuromorphic computing systems. In *Proceedings of the Great Lakes symposium on VLSI 2023, in GLSVLSI '23* (pp. 261–266). Association for Computing Machinery. https://doi.org/10.1145/3583781.3590208

Chapter 4
AI-Enabled Static Timing Analysis at Early Stages of the Digital Design Flow

Daniela Sánchez Lopera and Wolfgang Ecker

4.1 Introduction

In theory, the digital design flow follows a top-down approach. However, in practice, it becomes an iterative process. It starts with the initial specifications, which are converted into register-transfer level (RTL) blocks using hardware description languages (HDLs). Logic synthesis transforms these RTL blocks into corresponding gates from a target technology. Afterward, the gate-level netlist is translated into a physical layout. Timing sign-off checks ensure the final layout meets the timing constraints. If timing violations occur, the design flow iterates back to implement corrective measures at previous stages, as shown in the flow of Fig. 4.1a. Even though electronic design automation (EDA) has created a compendium of tools automating the design flow, this is still an iterative and lengthy process.

Given the importance of the RTL design phase, RTL generation frameworks, such as Chisel [1], PyVerilog [2], and MetaRTL [3], are employed to accelerate this stage. RTL generation frameworks generate different CPU cores, systems on chips (SoCs), and intellectual property (IP) blocks while re-using existing models, parameters, and generators. Replacing the manual RTL design with the automated generation, as shown in the flow of Fig. 4.1b, reduces concept, design, and verification efforts. Nevertheless, the RTL micro-architectures generated by RTL generators do not consider the design's quality of results (QoRs) after physical synthesis. Instead, they rely on the engineer's expertise in coding the generator to implement efficient micro-architectures regarding power, performance, and area (PPA).

One of the most important aspects of a digital design is timing closure, as the design's correct functionality should be guaranteed at the target clock frequency.

D. Sánchez Lopera (✉) · W. Ecker
Technical University of Munich and Infineon Technologies AG, Munich, Germany
e-mail: daniela.sanchez-lopera@tum.de; wolfgang.ecker@infineon.com

A. Iranmanesh, H. Sayadi (eds.), *AI-Enabled Electronic Circuit and System Design*,
https://doi.org/10.1007/978-3-031-71436-8_4

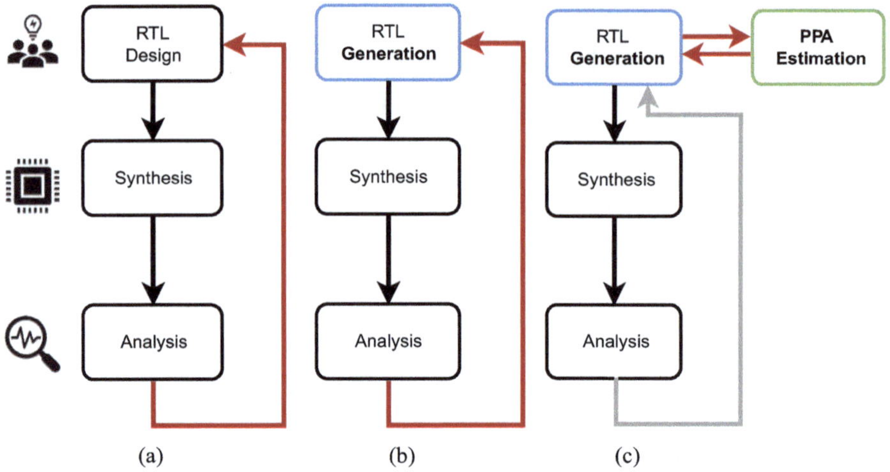

Fig. 4.1 Digital design flows. (**a**) Conventional flow. (**b**) Automated flow. (**c**) Optimized flow

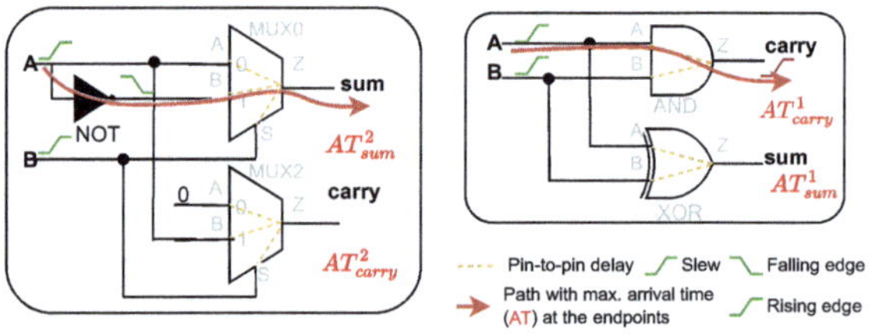

Fig. 4.2 Different RTL micro-architectures and timing results for a half-adder [6]

Static timing analysis (STA) uses the synthesized netlist, a target technology library, and timing constraints. Even though early timing analysis could be done along the digital design flow, only after routing does the timing analysis consider the timing behavior of gate cells, wires, and parasites. Thus, timing analysis at the latest stages of the design flow is more certain. However, it implies executing all stages of the physical synthesis flow, such as floor planning, placement, clock tree synthesis, and routing. This flow involves different sub-tools and subteams, resulting in very late timing feedback of an RTL design.

Addressing timing sign-off violations at RTL is crucial to accelerate the design flow [4] mainly because the RTL micro-architecture and HDL coding styles impact the QoRs obtained from the synthesized netlist [5]. For instance, Fig. 4.2 shows two micro-architectures for the same half-adder, resulting in different arrival times (ATs) at the design's outputs and, thus, different performance based on a target clock frequency. The best RTL micro-architecture in the conventional digital design flow is selected based on previous experiences and expert knowledge. Similarly, RTL generators rely on engineers' expertise in building timing-efficient generators. Thus, RTL generation frameworks require early information about the timing behavior of the generated designs. Acquiring this information is time-consuming, as synthesis and static timing analysis (STA) runs are required. Accurately back-annotating timing to RTL is also challenging, as the boundaries of the RTL modules may not be preserved after synthesis.

This chapter proposes the combination of RTL generation with early timing estimations to accelerate the digital design flow as shown in the flow in Fig. 4.1c. Recent approaches use AI to estimate the QoRs of a design at different stages, particularly at the high-level and logic synthesis [7], as well as in physical synthesis [8]. Many applications utilize ML models to predict timing-related metrics, including arrival times (ATs) [9] and slack [10]. Inspired by the timing analysis flow, which builds a timing graph and propagates delays through all nodes, Graph Neural Networks (GNNs) [11] are being used [12, 9, 10]. Nevertheless, they take a pre-routing gate-level netlist as input and lose the mapping to the RTL micro-architecture. They estimate timing-related metrics for exploring the design space but at lower levels of abstraction than RTL. Sengupta et al. [5] map the HDL code to an abstract syntax tree (AST) and estimate post-placement power and total negative slack (TNS). They compare different RTL variants regarding their trade-off between TNS and power. However, they fall short in identifying specific critical blocks that can be optimized, thus hindering micro-architecture search.

We introduce the AI-enabled STA flow depicted in Fig. 4.3. This flow comprises two main steps: predicting pin-to-pin delays and slews and using those predictions to conduct the conventional block-based STA approach proposed in [13]. This block-based approach represents each combinatorial gate on a design as a block and propagates and accumulates the input delays and the component delays block by block instead of along a timing path. Our proposed flow takes as input an RTL IR, which is an AST representation of the design's ports, constants, connections, and primitive components, such as logic gates and comparators. Machine learning (ML) models are trained to predict pin-to-pin delays and slews of primitive components. These primitive components are the building blocks of any design, allowing us to apply our flow to any evaluation circuit. The predicted values are the inputs to the block-based STA approach. ATs, RATs, and slacks are computed for all design components. This early timing feedback of an RTL IR enables the comparison of

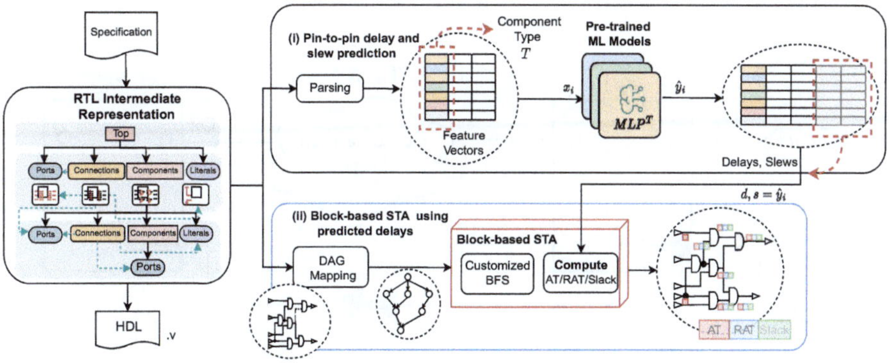

Fig. 4.3 Proposed AI-enabled STA flow [14]

different micro-architectures without running RTL generation, synthesis, or STA tools.

The rest of this chapter is structured as follows: Sect. 4.2 provides a background on STA. Section 4.3 summarizes seminal papers estimating timing metrics with and without ML techniques. Section 4.4 introduces the data generation flow from specifications to gate-level netlists. Generated data is used to train different ML models for delay and slew estimation, as described in Sect. 4.5. Finally, estimated delay and slew values are leveraged in Sect. 4.6 to perform the block-based STA computation flow. Moreover, Sect. 4.7 presents how such early estimation could ease manual or automated search for optimal micro-architectures using RTL generation tools. Finally, Sect. 4.8 concludes this chapter and suggests future research directions.

4.2 Static Timing Analysis

In the digital design flow, timing analysis determines whether a design works at the required clock frequency. This analysis can be dynamic, using simulation over different input vectors, or *static*, where no input vectors are required to simulate the complete circuit. **Static timing analysis (STA)** propagates gate and wire delays and slews from the primary inputs to the design endpoints (input to sequential elements or primary outputs), covering all timing paths. STA is faster and more efficient than dynamic analysis but cannot detect false timing paths.

The accumulated delay at the designs' endpoints is the **arrival time (AT)** for that given path. In synchronous circuits like the one in Fig. 4.4, the AT represents the time the data arrives at the input of the receiving sequential element. It can also be defined as the time required for the input to travel through the data path represented as a dashed green line. The **required arrival time (RAT)** is when the data must be at a design's endpoint to ensure the correct functionality of the whole design. The RAT is also defined as the time the clock signal travels through the clock path, highlighted as a dashed red line in Fig. 4.4. Thus, the RAT is determined based on the clock period and timing constraints.

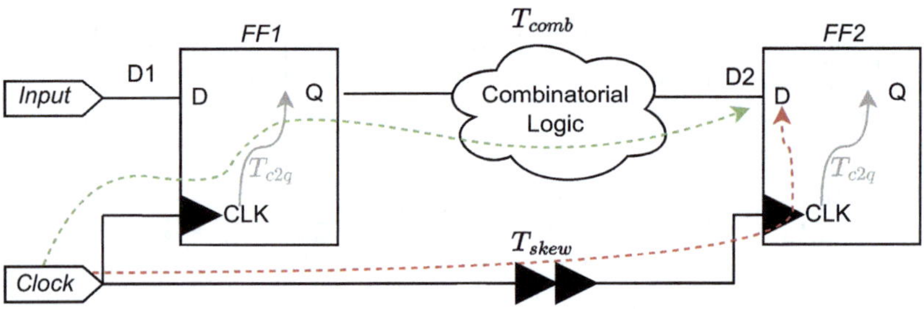

Fig. 4.4 Basic synchronous circuit

The difference between the RAT and AT is called **slack**. This is a severity measure: A positive slack means that the cells in the path could be replaced by cells with higher delay (and potentially, lower area) without causing timing violations, a negative slack means that the signal through the path is late, and the cells in it should be optimized or replaced by faster ones (with potentially higher area). The path with the worst slack is called the critical path of the circuit [13]. STA tools find critical paths of the design by calculating whether the slack is met or violated for all paths under a target clock period T_{clk}.

Two checks can be conducted to ensure a synchronous design works at the target clock frequency: **setup** and **hold**. These checks confirm whether the data input is stable for some time before and after the clock edge. This is to avoid incorrect data being captured or latched through the data path. The setup check determines if the synchronous input comes to the sequential element and is stable by a time T_{setup} before capturing the edge of the clock. The maximum setup constraints are described in Eq. 4.1.

$$T_{cq} + T_{logic} + T_{setup} \leq T_{clk} + T_{skew}, \tag{4.1}$$

where T_{c2q} is the time from the clock signal to the output of the sequential element or flip-flop *FF1*, and the T_{setup} of the flip-flop *FF2*. Hold checks determine if the input is stable at time T_{hold} after the edge of the clock:

$$T_{c2q} + T_{comb} \geq T_{hold} + T_{skew}. \tag{4.2}$$

STA tools require a gate-level netlist and a technology library describing the target CMOS technology. ATs are calculated by accumulating the delays of the gates in the combinatorial path or cloud of Fig. 4.4. The **delays** through each gate are known as propagation delays. They are the time required for a signal to go from an input to an output pin. The pin-to-pin delays are affected by many factors, such as the functionality of the gate, fan-outs, fan-ins, and slews or input transition times. The **fan-out** is the capacitive load at the output. The **slew** is a signal's transition time. Slews propagate along the timing path, i.e., the slew of the output pin of a gate is the input slew of the following connected components in the timing graph [15].

Technology library files [16] describe wire and gate delays and slews as nonlinear delay models (NLDM). These models result from multiple simulations of the gate under different operating conditions regarding process, voltage, and temperature (PVT). The technology library files use the lookup table (LUT) method for function approximation. Specifically, the gate delay is defined as a LUT with two variables as indexes: input slew and fan-out load. During timing analysis, the STA tool uses interpolation or extrapolation to calculate delays and output slews for input values that are not specified or outside the defined indexes. Moreover, gate and wire delay models are scaled according to the current PVT conditions, which physically affect the cell behavior. The selected PVT values are considered fixed; thus, STA is deterministic [17].

STA is an efficient method for detecting timing violations for setup and holding checks without requiring input vectors. However, running STA tools for different RTL variants consumes time and computation resources. STA requires obtaining a gate-level netlist for the RTL design, i.e., running (at least logic) synthesis tools. The STA tool traverses all paths in the netlist. Per each gate in each path, the tool performs inter- or extrapolation of delay and slew models to get timing information of every pin-to-pin connection. Thus, a faster method is required to approximate the timing behavior of a digital design. It is important to clarify that this chapter focuses on combinatorial designs and does not target sequential designs, multi-cycle operations, multiple clocks, nor clock analysis.

4.3 Related Work

State-of-the-art research has recognized the need for a faster method to obtain timing metrics for digital designs rather than running synthesis or STA tools. Some studies focus on estimating component delays, while others estimate maximum ATs or slacks directly. Moreover, some works use RTL designs as input, while others utilize gate-level netlists after placement or routing. Specific works target FPGA hardware platforms instead of ASIC designs. This section summarizes these related works, focusing on the methods used for estimating timing-related metrics. Accordingly, the related works are classified into non-ML-based and ML-based approaches, as illustrated in Fig. 4.5.

4.3.1 Non-ML-Based Approaches

Estimating the delay of a component can be interpreted as an approximation of a complex function. Different established methods can address such approximation tasks, including analytical, macro-modeling, and polynomial-based approaches [18].

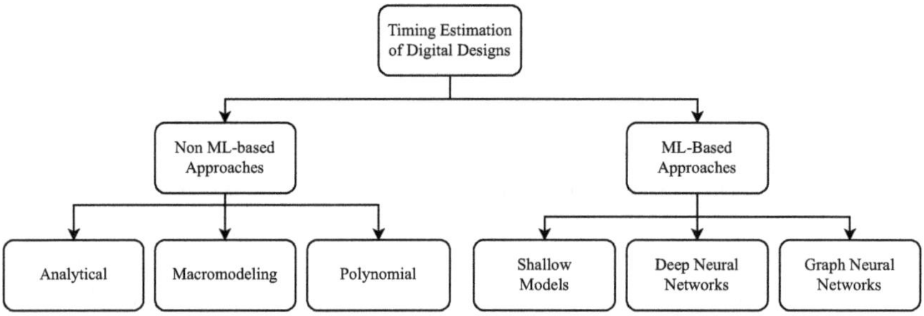

Fig. 4.5 Classification of related works to timing estimation of digital designs with and without ML models

4.3.1.1 Analytical

Analytical methods find an exact formula to describe the function being approximated. In this case, they relate a gate's delay to other design features based on the underlying physical and mathematical models.

Wallace et al. [19] estimate the delay of a combinatorial cloud before its mapping to a gate-level netlist. They use a logarithmic function to model the delay through each component based on their complexity and fan-out. The component type and fan-out values are obtained from the designs before logic synthesis. These values are then scaled using constant coefficients or technology-dependent factors calculated using the delay of NOT and NAND gates, fan-out limit, and standard load.

Qiu et al. [20] estimate post-routing delays given the placement of a circuit. Scaling factors like net degree, route length, and net delay are extracted from sample designs. Those factors are terms of the estimation functions, which can be quadratic or linear equations. The proposed *delay sampling* method computes the delay as a quadratic function of the pin-to-pin placement length. The quadratic function is used for new designs with pre-calculated scaling factors and the current wire's placement length.

4.3.1.2 Macro-Modeling

Macro-modeling builds an abstract model and decomposes a complex function into smaller blocks that can be approximated more easily. Typically, macro-modeling is employed to estimate design metrics based on high-level circuit descriptions. In this method, the analyzed design metric is connected to a specific set of parameters and coefficients, forming a template. These parameters are computed by using sample designs. For instance, Macii et al. [21] estimate the delay at the behavioral level. Technology-independent gate-level netlists are mapped to binary decision diagrams. For delay estimation, the model relates various types of the complexity of a Boolean function to its gate-level implementation, such as size, the number of inputs, and density.

4.3.1.3 Polynomial

Polynomial models use equations to approximate functions. They are regression-based, where the delay is the dependent variable, and design parameters are the independent variables. *QUEST* [22] is a tool that uses *best-fit* polynomial estimation for area and delay from RTL descriptions. QUEST transforms the RTL code into a technology-independent form, resembling a logic synthesis tool. But it only targets three logic gates: AND, OR, and NOT. The initial parameter extraction for the best-fit polynomial uses a few test designs. QUEST reaches lower estimation errors when testing with a hundred designs and four technology libraries.

4.3.2 ML-Based Approaches

ML's success in solving NP-complex problems, classification, and regression tasks in various fields has caught the attention of the EDA community [23, 24]. ML has been applied across different stages of the design flow for analog and digital circuits on different hardware platforms and processes. Four main areas of action have been recognized: predicting optimal configurations, approximating design metrics without running EDA tools, optimizing PPA by exploring the design space, and optimizing designs through reinforcement learning (RL) [8].

This book showcases ML applications in the EDA field. ML models have estimated design metrics such as congestion, wire length, and power [8]. This chapter focuses on ML models used for approximating the timing of digital designs. Classical methods listed in Sect. 4.3.1 are constrained to the type of components, computationally expensive for larger designs, or do not consider signal transition times. In contrast, ML-based models learn patterns in the data without explicit knowledge of the mathematical or physical model. ML-based models are data-driven, i.e., they are trained on large datasets and learn relationships between data features and ground-truth labels. In addition to being faster during inference, ML-based approximation functions are scalable and effective for complex functions when data is available.

Figure 4.6 shows the training and inference flows of ML models used for design metric prediction. In the training flow, circuits are passed through EDA tools to perform certain stages of the design flow until a target stage. This is called the *input stage*. The representation of the circuit at this stage is used for feature extraction. A further stage is executed to collect the ground-truth labels or objectives. This is called the target prediction stage. In the inference flow, features are extracted from the same input stage representations, and objectives are unnecessary.

Shallow and deep ML methods have been applied to estimate timing metrics. Usually, training circuits are mapped to Euclidean data, such as vectors or matrices, without considering that EDA objects are represented naturally as graphs. This Euclidean mapping loses valuable information about the circuit's connectivity. The

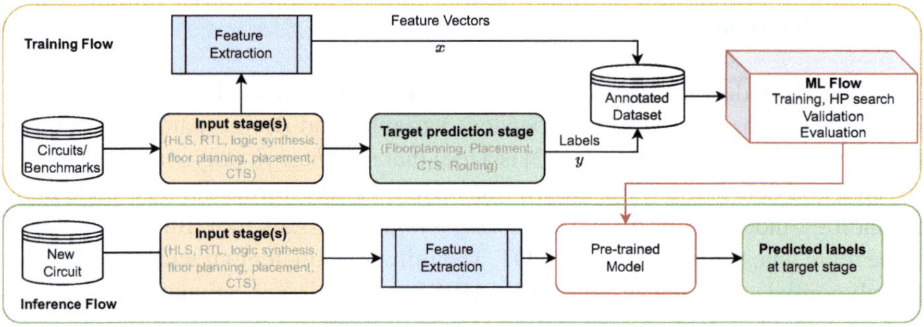

Fig. 4.6 Training and inference flows of ML models predicting digital design metrics

GNNs [11] framework is an opportunity to solve EDA tasks directly using graph structures, e.g., for intermediate RTL, gate-level, or transistor netlists [25]. Thus, we classify the ML-based applications to timing estimation based on the type of NNs used for Euclidean or graph-structured data.

4.3.2.1 Shallow Methods and Deep Neural Networks

ML models using Euclidean structures estimate delays or ATs of circuit components at the early stages of the design flow, such as at HLS [26], logic synthesis [27], and placement [28, 29].

The Pyramid framework [26] estimates the optimal post-route target clock frequency and utilization of a design at HLS. Pyramid models timing as a regression problem, where the input features come directly from the HLS reports, and the objectives to predict are the maximum clock frequency, throughput, and area. They create C-to-FPGA metrics for benchmark circuits and train ML models such as RFs, support vector machines (SVM), and NNs.

From logic synthesis, Ramesh et al. [27] use NNs to predict the post-route ATs of a circuit after performing STA and statistical STA. The NN is trained with results from commercial tools for ASIC synthesis and timing analysis. To that end, open-source gate-level netlists are used, and their features include the number of inputs and outputs, gate count, the number of nets, dynamic power, and gates in the critical path.

Martin et al. [28] propose a net delay estimator for timing-driven FPGA placement. They use regression models such as MLPs, gradient boost regression (GBR), and RFs. The latter gives better results. All models are trained with more than 300 benchmarks based on a 20-nm technology. The training dataset is built by extracting features for each benchmark, where the most relevant are the number of pins, the half-perimeter wire length of the longest path, and average congestion.

Finally, Barboza et al. [29] predict post-route net delays, slews, and endpoints slack of netlists after placement. The models are trained using open-source benchmarks and a technology library [16]. The selected features include driver and sink capacitance, the distance between the target and sink node, maximum driver input slew, and context sink locations. They explore three ML-based regression models: Lasso, RFs, and NNs. An NN is trained to reduce an asymmetric mean square error (MSE) loss to add a penalty for underestimations.

4.3.2.2 Graph Neural Networks

GNNs are widely used in the digital design flow [30]. Figure 4.7 shows the end-to-end flow when solving an EDA task using GNNs. EDA objects are used as graphs, i.e., features and ground-truth labels symbolized with flags, from EDA tools, are annotated to a graph structure. These are depicted in Fig. 4.7 using a feature matrix **X** and flags per node. A GNN extracts node, edge, or graph embeddings, which

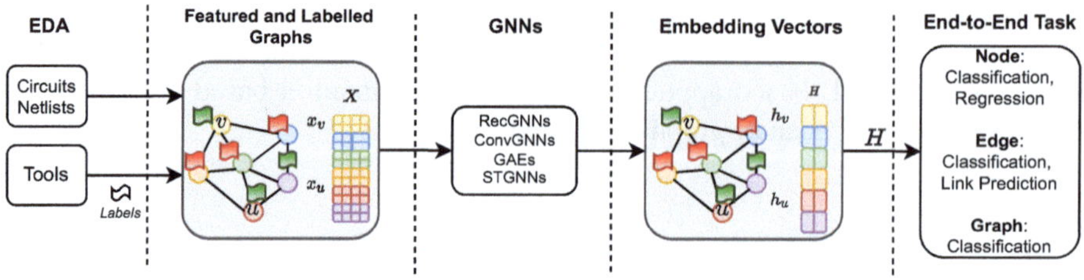

Fig. 4.7 End-to-end flow when using GNNs [30]

are represented as an embedding matrix **H**. This matrix is the input to Euclidean ML methods that perform classification or regression at the edge, node, or graph level. GNNs are used to estimate timing-related metrics, including post-routing ATs [31] and total negative slack (TNS) [32, 5]. Inspired by conventional STA, some approaches estimate component delays and slews and then propagate them through the timing paths [9, 33].

Shrestha et al. [31] highlight the benefit of an early timing profile to avoid corrective circuit changes later in the design flow. They estimate the post-routing AT of a design's timing paths. To consider prediction results in three different scenarios, they take the input netlists after three stages: floor planning, placement, and CTS. They use open-source benchmarks and a 65-nm technology to generate over a thousand physical layouts, from which timing paths are extracted and used for training. Results show that the prediction improves the closer the input stage is to the objective stage.

Lu et al. [32] propose an end-to-end framework to predict post-route TNS and detect failed runs earlier in the flow. They use GNNs to encode the netlist after detailed and global placement and CTS. The netlists from three sequential stages of the design flow are mapped to featured graphs and encoded by a global GNN. The node embeddings are then used as inputs to an MLP predicting the TNS per stage. As their final goal is to predict whether a design will meet the timing sign-off constraints, they go further and map the physical synthesis flow to a sequential flow. Thus, the outputs of the global GNN are considered a time series and are processed using an LSTM architecture.

Ye et al. [9] present a GNN for path-based timing analysis. They propose a three-step flow: GNN encoding, gate slew and delay prediction, and critical-path AT calculation. Their GNN encoding handles multidimensional edge features. The resulting encoding vectors are input to an MLP model, estimating the path-based cell slews and delays. The AT is the cumulative addition of the predicted cell delays. Similarly, Guo et al. [33] present an end-to-end timing prediction engine. They map pre-routing netlists to DAGs and estimate global post-routing timing metrics such as AT and slacks at the design's endpoints. Their model targets two tasks: the cell delay and slew calculation and the AT and slack propagation. The complete model is jointly trained by accumulating the losses of each task.

Approaches in [31, 32, 33], and [9] build input graphs from pre-route gate-level netlists. However, the design's quality after physical synthesis depends on the quality of the initial RTL [4]. Motivated by this, Sengupta et al. [5] map the Verilog RTL to an AST and use ML (shallow, deep, and GNN-based) models to predict the dynamic-power-performance Pareto front of an RTL. They compare different RTL variants and HDL coding styles. Their approach reaches higher prediction accuracy and is faster than running synthesis, power, and timing analysis tools.

4.3.3 Summary

Similar to non-ML-based approaches, this chapter recognizes the need for a timing estimation at the early stages of the design flow. Classical approaches are limited to a subset of component types or sample designs. Even though they are mostly technology independent, they do not take advantage of the abundant data provided by EDA tools and are not data-driven. In contrast, ML approaches generate technology-specific training datasets. This chapter creates different datasets and leverages NNs, GNNs, and RFs to estimate component delays and slews. Using estimated delays, block-based STA is conducted to obtain ATs, RATs, and slacks for each component in an RTL IR.

Our chapter is comparable to [28], as the delay estimation is mapped to a regression problem. However, Martin et al. target FPGA platforms and extract ground-truth labels after placement. Most ML approaches take input features from stages where the RTL design is already determined, such as placement and CTS. Even though their predictions have higher quality and are closer to sign-off results, they use gate-level netlists as inputs, losing the mapping to the RTL. Similar to our work, Sengupta et al. [5] recognize the importance of the RTL design, its micro-architecture, and coding styles. They compare the TNS of different micro-architectures but cannot pinpoint critical components in the design. This chapter borrows inspiration from approaches in [9, 33] leveraging component delay and slew estimation to estimate path ATs and slacks.

4.4 Training Data Collection Flow

The abundance of available data enables ML models to estimate the timing of digital designs. The *divide and conquer* strategy in the digital design flow has led to a compendium of tools and subteams working standalone. The consequences of this strategy are twofold: Iterations or corrective measures along the design flow are costly, and each tool generates a large amount of data involving designs, netlists, layouts, and reports. The open-source toolchain OpenROAD [34] has been developed to integrate standalone tools into a single physical synthesis toolchain and take advantage of the generated data by providing a data collection and storage

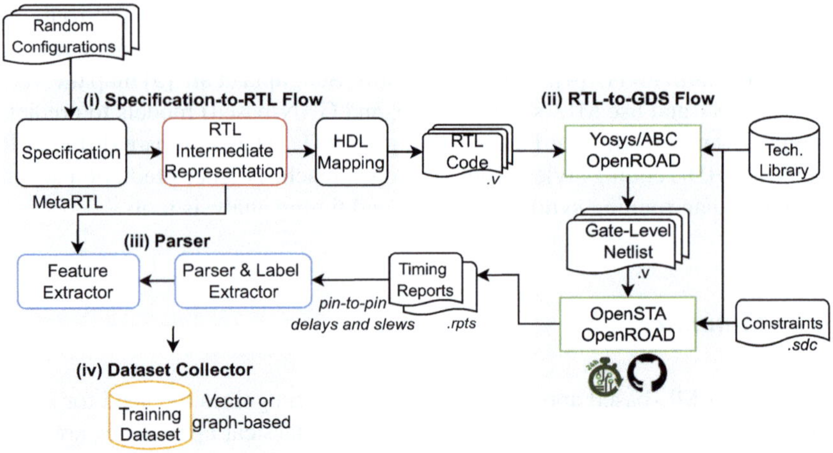

Fig. 4.8 Data collection flow [7]

infrastructure called METRICS 2.0 [35]. The latter enables more ML applications to predict tools and flow outcomes. Recognizing the importance of stages prior to physical synthesis, OpenROAD has connected its toolchain to Yosys [36], ABC [37], and Chisel [1], an RTL generation tool [38]. Drawing inspiration, we built our in-house data collection flow as shown in Fig. 4.8.

The data collection process has four components: RTL generation, synthesis, STA, reports parsing, and collecting. At a high level, we generate random configurations, determining the nature of our training circuits. An in-house RTL generation tool generates an RTL IR and a Verilog-based RTL for each configuration. The generated design is synthesized using open-source tools. The gate-level netlist is then used for STA. The generated reports are preprocessed, and relevant information, such as pin-to-pin delays or endpoints' AT, are extracted. Different datasets are generated to cover diverse component types, technologies, and data structures. Therefore, the parser and dataset collectors are explained later in the chapter. This section focuses on the RTL generation and synthesis flows.

4.4.1 An Automated Specification-to-RTL Flow

Thanks to the advancement of programming languages, RTL generation frameworks are being developed [1, 2, 3]. They provide highly customizable circuit generators while taking advantage of new programming paradigms. Chisel [1] uses Scala to describe designs, compiles them using the flexible-intermediate representation of RTL (FIRRTL), and generates the output Verilog code. PyVerilog [2] provides Python modules for Verilog generation and RTL design analysis. MetaRTL [3] applies the model-driven architecture (MDA) principle [39] to RTL generation using Python. MetaRTL uses templates, models, and metamodels to construct customizable generators that, from informal specification, produce HDL-based RTL.

This chapter employs the in-house generation framework MetaRTL, as its benefits are twofold: Python is the programming language used for hardware generation and ML applications. Second, formal properties are generated, and it is possible to verify MetaRTL designs following a four-eyes principle [40]. The formal verification capabilities of our in-house framework motivate a complete flow from specification to layout even further. However, formal verification is beyond the scope of this chapter.

4.4.1.1 MetaRTL Structure

Following an MDA, MetaRTL leverages metamodels and templates to connect and generate three different models of a design, as shown in Fig. 4.9: the initial specifications or model of things (MoT), the design micro-architecture or model of design (MoD), and finally, the mapping to an HDL or model of view (MoV). These models conform to their respective metamodels called MetaSpec, MetaRTL, and MetaView, respectively.

This chapter uses the intermediate model and the HDL design to extract features and ground-truth labels, respectively. The intermediate model or RTL IR represents the micro-architecture of the design as an AST of a design's submodules, components, ports, constants, and connections. Figure 4.10 shows the tree structure of an RTL IR, with three levels of hierarchy below the top root. Connections link the current hierarchy's ports, constants, and components and interconnect the different hierarchy levels. Features to train ML models are selected from the RTL IRs for many reasons. Firstly, the IR is a technology and platform-independent representation, making our methods easy to generalize to other hardware platforms, HDLs, and CMOS technologies. Moreover, the RTL IR comprises descriptive, behavioral, sequential, and primitive or basic components. Particularly, the primitive components are the fundamental logic functions of digital designs, including reduced, logical, and bitwise logic components like LAND, RAND, and BAND,

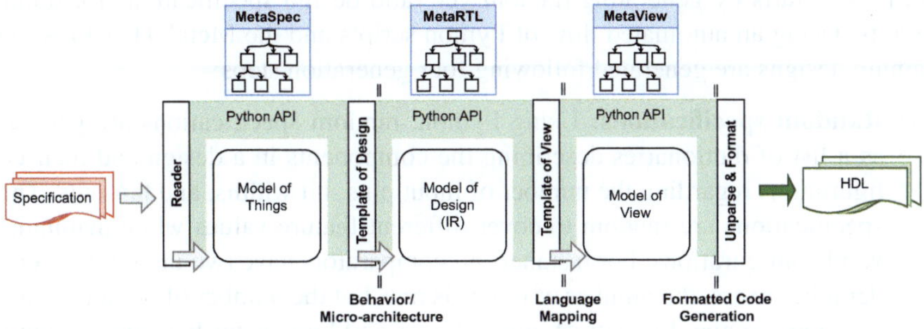

Fig. 4.9 MetaRTL building blocks [3]

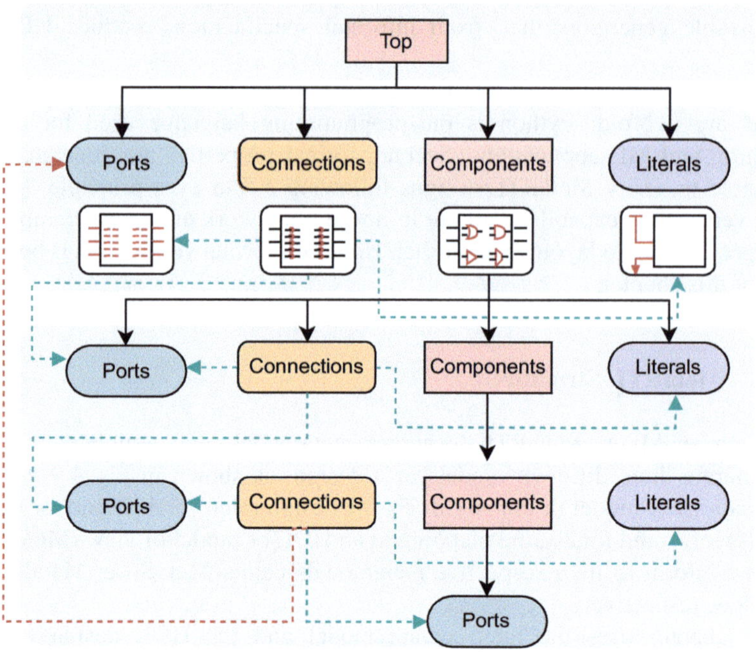

Fig. 4.10 RTL IR structure used in MetaRTL [3]

arithmetic operators like addition and multiplication, multiplexers and demultiplexers, and comparators like greater than (GT) and less than (LT). Estimating the timing behavior of basic components makes the proposed approach generalizable to any hardware design, from simple half-adders to complete SoCs.

4.4.1.2 Generating Training Circuits

This chapter predicts the pin-to-pin delays and slews of basic components, such as logic gates, arithmetic, multiplexers, and comparators. The data collection flow in Fig. 4.8 starts by generating random yet valid design specifications for training circuits. Using an automated flow of Python scripts and the MetaRTL API, several training designs are generated following four generation steps:

(i) **Random specifications:** Using Python, random specifications are generated as a list of dictionaries describing the components in a design and their configurations regarding the number of input pins, bit widths, and fan-outs. These specifications are random to cover different feature values while maintaining valid configurations. For instance, all comparators have two input pins. For the demultiplexers, the number of inputs is one, but the number of outputs changes in a wider range. In contrast, multiplexers and logic gates have one output but several inputs. Figure 4.11 shows some RTL IRs of the constructed dataset

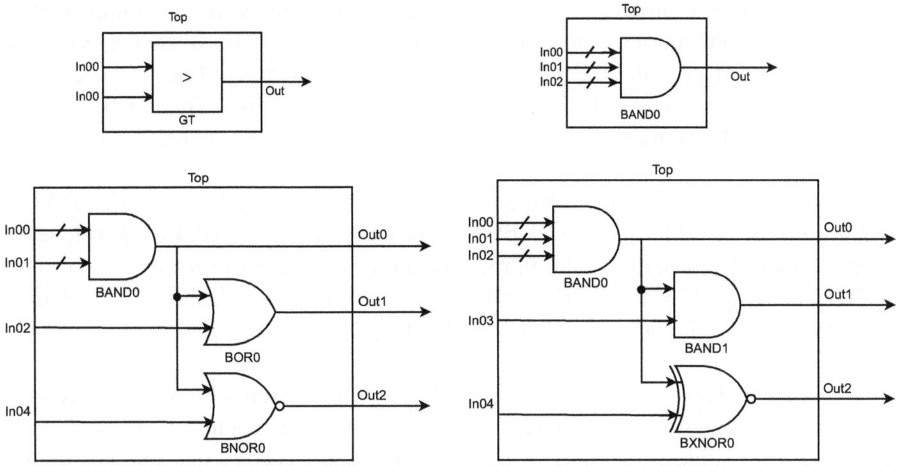

Fig. 4.11 Example of RTL IRs in the training dataset

using one single GT comparator, a single bitwise AND with three inputs, or three components with different inputs and bit widths.

(ii) **Templates of design (ToDs):** Dictionaries containing design specifications are read using Python to build proper ToDs. The ToD is a Python program that invokes MetaRTL libraries to instantiate ports, connections, submodules, and primitive components. The ToD instantiates the components and configurations from the random specification using MetaRTL objects. One ToD is generated per each randomly generated specification.

(iii) **Models of design (MoDs) or RTL IRs**: MetaRTL automatically converts a ToD into an MoD. This conversion step is executed per each generated ToD, resulting in a tree-like representation saved as an XML file.

(iv) **Verilog-based HDL**: MetaRTL automatically converts the MoD into an HDL view. In this way, a Verilog file is generated per each random specification.

4.4.2 *RTL-to-Gate-Level Flow*

Figure 4.8 shows that the generated Verilog is the input to the synthesis flow, utilizing predefined tool parameters, clock, and timing constraints.

4.4.2.1 Logic Synthesis

Yosys and ABC are standalone tools connected to the OpenROAD toolchain to perform logic synthesis. Logic synthesis can be performed hierarchically or flattened. The latter flattens the design into one big module, facilitating cross-boundary optimizations. However, this process can result in the renaming of RTL components, connections, and ports, making the mapping between the RTL and gate-level netlist complex. This chapter uses hierarchical synthesis to keep RTL boundaries and component names in the gate-level netlist. This facilitates the mapping of RTL components in timing reports and the back-annotation of delays, slews, and ATs.

The inputs to the logic synthesis tool are the Verilog-based RTL and the technology library file [16] available in a process design kit (PDK). The latter is a set of data files describing the fabrication process of a chip for the design tools. A PDK, developed by a foundry, contains technology details, design rules, standard cell libraries, and layout information [41]. This chapter uses a proprietary 40-nm PDK.

Our data collection flow does not go deeper into design stages like placement or routing. As suggested in related works, the closer the input design stage is to the prediction stages, the better results can be achieved. As this chapter targets the RTL IR as input, timing reports after logic synthesis are a good choice. Our flow could be extended by running the open-source tools from the OpenROAD toolchain. However, this will cause an increase in runtime for generating training data. Even though timing analysis after logic synthesis does not consider wire or parasitic effects, it provides a good intuition about the initial timing behavior of the netlist.

4.4.2.2 Timing Analysis

OpenSTA [42] is a gate-level static timing analyzer that can be used to verify the timing of designs and is incorporated into the OpenROAD toolchain. OpenSTA requires the gate-level netlist, the technology library, and predefined timing constraints in Synopsys® design constraints (SDC) format as inputs. In the technology library, each cell's timing behavior or NLDMs are described for a specific corner case regarding PVT values. The timing constraints define the clock configuration, such as skew and period, but also the delay, slew, and load of primary ports.

Table 4.1 lists key STA parameters for the 40-nm technology. Different datasets are created by varying the slews and delays of primary ports in different ranges. The specific values used will be mentioned in each section. Moreover, we explore different flows to perform synthesis and timing analysis simultaneously with one or *N* different timing constraints, as shown in Fig. 4.12. Two flows are proposed: one for a few different configurations and another for *N* different configurations. The latter flow is more efficient as it uses the same netlist for all STA runs. Each run can be parallel with its separated timing constraints and generate separated timing reports. This speeds up the generation of timing reports for the training circuits.

Using Python, TCL scripts are automatically generated to configure OpenSTA and generate specific timing reports. This automated creation of TCL scripts allows to generate OpenSTA commands for different pin names, path groups, and types without manual intervention. These scripts specify the pins through the path, type,

Table 4.1 STA configuration parameters when using the 40-nm technology [18]

Setup	Name	Value	Unit
Ports	Input delay	Varies*	ns
	Output delay	Varies*	ns
	Input load	0.0075	fF
	Output load	0.015	fF
	Input slew	Varies*	ns $[rise, fall]$
Clock	Period	6.6	ns
	Latency	[0.1, 1.0]	ns $[min, max]$
	Transition	[0.1, 0.2]	ns $[rise, fall]$
	Skew	[0.6, 0.15]	ns $[setup, hold]$
Operating conditions	Voltage	1.08	V
	Temperature	−40	$°C$
	Process	1.30	–

* Different selected ranges per dataset.

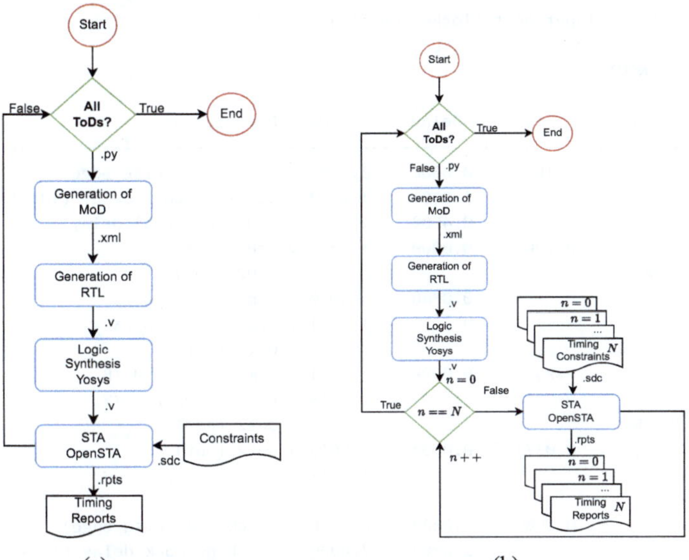

Fig. 4.12 Flows running one or parallel STA setups per the same netlist. (**a**) One STA configuration. (**b**) N parallel STA configurations

group of the path, and timing arc. OpenSTA is configured to analyze all four types of timing paths, i.e., rising and falling for the minimum and maximum edges. Relevant timing information is extracted from generated reports and serves as ground-truth labels for our ML models. Figure 4.13 shows an RTL IR and one of the generated timing reports. A timing report comprises four main parts: header, data path, RAT,

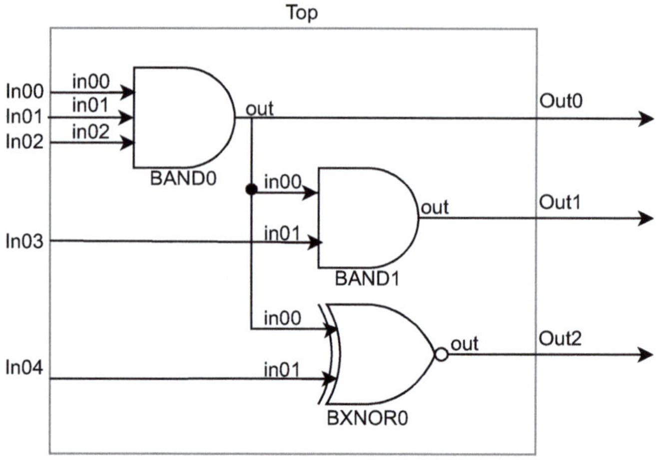

(a)

```
Startpoint: In02 (input port clocked by clk)
Endpoint: Out2 (output port clocked by clk)
Path Group: INOUT
Path Type: max
Fanout      Cap      Slew     Delay      Time   Description
--------------------------------------------------------------------------------
                   0.1000    0.0000    0.0000   clock clk (rise edge)
                             1.0000    1.0000   clock network delay (ideal)
                             0.4000    1.4000 v input external delay
                   0.2000    0.0000    1.4000 v In02 (in)
     1   0.0024                                 In02 (net)
                   0.2000    0.0000    1.4000 v comp_BAND0/_0_/C
                   0.0416    0.1771    1.5771 v comp_BAND0/_0_/X
         0.0070                                 Out0 (net)
                   0.0416    0.0000    1.5771 v comp_BXNOR0/_0_/B
                   0.0371    0.0847    1.6618 v comp_BXNOR0/_0_/Y
     1   0.0000                                 Out2 (net)
                   0.0371    0.0000    1.6618 v Out2 (out)
                                       1.6618   data arrival time

                   0.1000    6.0000    6.0000   clock clk (rise edge)
                             1.0000    7.0000   clock network delay (ideal)
                            -0.6000    6.4000   clock uncertainty
                             0.0000    6.4000   clock reconvergence pessimism
                            -0.8000    5.6000   output external delay
                                       5.6000   data required time
--------------------------------------------------------------------------------
                                       5.6000   data required time
                                      -1.6618   data arrival time
--------------------------------------------------------------------------------
                                       3.9382   slack (MET)
```

(b)

Fig. 4.13 Example of an RTL IR and one of its generated timing reports. (**a**) Sample RTL IR. (**b**) One generated timing report

and slack calculation. The header specifies the path's start, endpoint, type, and group. The data path delay shows the pin-to-pin delays for each gate in the path. One RTL IR component is sometimes mapped to multiple gates in the netlist. In that case, the gate delays are accumulated to represent the delay of that component at the RTL IR. The data arrival time aggregates all individual cell delays along the path. Finally, the RAT and slacks are calculated.

The flow in Fig. 4.8 is compatible with any RTL generation framework with an AST representation as RTL IR. The AST representation allows us to describe a design in terms of modules and components, which timing behavior we want to estimate. Commercial synthesis tools could also be used to collect ground-truth labels for the synthesis flow. This work employs open-source tools, as they provide more transparency in the design flow and allow thousands of runs without incurring high license costs. Open-source tools enable multiple ML applications and research in the field, even within industrial design flows [43, 44].

4.5 Pin-to-Pin Component Delay and Slew Estimation

The first step to estimate timing for an entire RTL design involves analyzing individual components, i.e., determining how long a signal travels from the input to the output pin of each component. At the gate level, this analysis is straightforward because technology libraries provide timing models for each gate. At RTL, the challenges for the estimation are twofold. First, the Verilog-based RTL is an abstract representation of data flow across registers and logical operations, i.e., no concrete gates are implemented. To overcome this, RTL IRs are used to map the RTL design to a structural description using components and modules. Second, mapping components in RTL or RTL IR to gates in the target technology is not a one-to-one or unique process, as logic optimization also impacts how the mapping is performed.

4.5.1 Problem Statement

Given an RTL IR describing a combinatorial logic, estimations of the delays and slews over each pin-to-pin connection in the design are required. Pin-to-pin values are needed to detect critical pins inside individual components. For instance, in the schematic view of Fig. 4.14, delay and slew values must be predicted for all the yellow highlighted pin-to-pin connections.

Timing estimation of RTL IRs is challenging, as it implicitly approximates the technology mapping and optimizations done by the logic synthesis tool and the timing models described in the technology library. Moreover, the timing models are also influenced by PVT conditions, gate type, and pin-specific features such as load and transition times. This chapter proposes ML models to estimate pin-to-pin delays and slews of primitive components in an RTL IR. By doing this, the execution

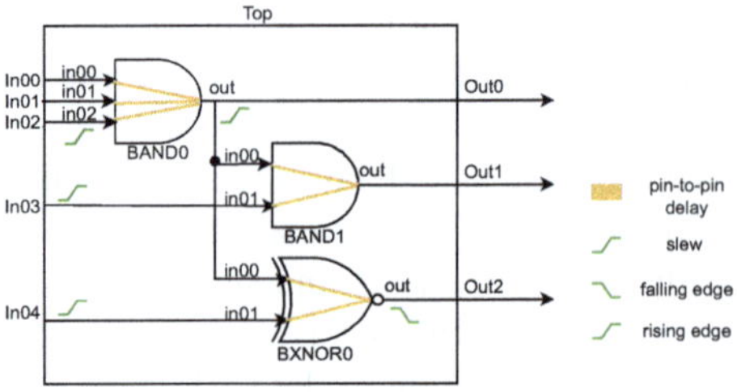

Fig. 4.14 RTL IR of a combinatorial circuit with pin-to-pin delays to estimate

of RTL generation, logic synthesis, and STA is replaced with an ML-based flow. As highlighted in Fig. 4.6, the main benefit of ML models is in the inference flow, where early estimations are obtained without running EDA tools.

The delay and slew estimation is defined as a regression task. ML regression can model complex and nonlinear patterns with less risk of overfitting. It is also more scalable to large datasets and high-dimensional spaces, and lastly, it generalizes better than simple curve-fitting techniques. Regression models are trained in supervised learning to learn the relationship between data features and ground-truth labels. The input to a regression model is a feature vector $\mathbf{v_i}$ where i is a pin-to-pin connection of a component within a design. The vector $\mathbf{v}$ encodes information on the selected features describing the pin-to-pin connection. For instance, the RTL IR in Fig. 4.14 with seven pin-to-pin connections highlighted in yellow provides seven dataset samples. The model's output $\mathbf{y_i}$ is a vector representing the regression objectives for each pin-to-pin connection i. This chapter uses delays d and output slews s^{OUT} as objectives and considers rising r and falling f edges, i.e., $\mathbf{y_i} = [s_f^{OUT}, s_r^{OUT}, d_f, d_r]$, where each objective is determined by its maximum and minimum values. Thus, the model's output is a vector of eight elements.

There are several steps required to solve a regression task with ML models. First, features are selected and extracted from the training data. Datasets containing features and objectives are built and split for training, validation, and testing. Second, ML model architectures that suit the training data structure are selected. Afterward, the models' hyperparameters are tuned, and the models are trained. Finally, trained models are evaluated over unseen data. The following subsections provide more details of those steps for our regression task.

4.5.2 Dataset Collection and Feature Selection

After generating RTL IRs of training circuits, datasets are built using meaningful features and structures that match the nature of the task. Drawing inspiration from STA tools and the timing NLDMs, we choose features for each pin-to-pin connection to give information about the type of component, fan-outs, bit widths, the number of pins, and current pin position. These features are extracted from the RTL IR, i.e., the input to our models is at a higher abstraction level than

Table 4.2 Component types covered in tabular datasets DS-I and DS-II [18]

Category	Type	No. Circuits	No. Samples
Logical[1,2]	AND, NAND, OR, NOR, XOR, XNOR, NOT	200.9K	2564K
Arithmetic	CMULT, CDIV, CMOD, CREM, CPLUS, CMINUS, HWPLUS, HWMINUS, HWMUL	68.6K	665K
Multiplexing	MUX[2], DEMUX	1.0K	78K
Comparators	GT, GTEQ, LT, LTEQ, EQ, NEQ, ISPOS	5.5K	126K

[1] Three fashions: logical, reduced, and bitwise.
[2] Component types considered in DS-I.

the gate-level netlist. STA tools propagate the slew along the timing graph and consider it input to delay models. Thus, we evaluate the model's performance with and without considering slew, treating it as a feature and the output slew as an objective. The inference is done sequentially to ensure proper consideration of the slew propagation through the timing graph [45]. In total, three different datasets are generated [45, 18, 46].

4.5.2.1 Tabular Datasets

In a tabular dataset, each row represents a pin-to-pin connection of a logic component in an RTL IR. The columns are the selected features and objectives. We generate two tabular datasets that cover different component types in an RTL IR: a basic dataset *(DS-I)* and a generalized dataset *(DS-II)*. For the datasets DS-I and DS-II, we generate timing reports with two and four variants of slews at the primary inputs, respectively. Table 4.2 displays the count of circuits and samples per component category for both tabular datasets.

Dataset I (DS-I)

The training circuits used in this dataset contain twenty types of RTL IR components, including multiplexers, inverters, and various types of logic gates used bitwise, logical, and reduced. Each training design is generated randomly, with the number of inputs ranging from one to twelve and the bit widths ranging from one to thirty-two. The number of components per design varies from zero to six to account for different fan-outs, and single-component designs have no connected components to the output pin. Overall, DS-I comprises 42.9K training samples.

Dataset II (DS-II)

DS-I is enlarged to encompass additional component types, such as demultiplexers, arithmetic operators, and comparators. This expanded dataset, DS-II, includes 37 primitive component types listed in Table 4.2. The input bit widths range from one to thirty-two for up to thirty-two inputs. Different component types have different configurations. For instance, the GT comparator has two inputs and one output, while logic gates have several inputs. Thus, the number of inputs for *GT* is fixed to two, while for logic gates, it is randomly chosen from the $[1, 32]$ range. This results in an imbalanced dataset concerning the component type, where the number of circuits using logical gates is $36.5\times$ larger than the number of circuits for comparators. In total, DS-II consists of 3.4M training samples.

Features of Tabular Datasets

Table 4.3 lists the selected features x and objectives y. These features aim to represent the pin connection with the available information from the RTL IR. Inspired on the NLDMs, the slew values are the only features used that are directly related to the pin delays and extracted from the timing reports. The component type x_0 is used to reflect the gate's complexity and its implementation patterns. x_1 identifies the pin-to-pin connection from the current pin to the output among all possible considered x_2, highlighting the signal's path through the component. Pin-specific features are considered, such as the bit widths x_3, x_4, x_9, and x_{10} and the number of connected components x_5, x_6. The number of outputs x_8 correlates with the component implementation, especially for multiple output components as the demultiplexer [18]. For both tabular datasets, the categorical feature x_0 is one-hot encoded. x_0 is mapped to an $N \times T$ sparse matrix, where N is the number of samples and T the number of component types supported, i.e., 20 for DS-I, and 37 for DS-II.

Table 4.3 Features and objectives of tabular datasets DS-I and DS-II [18]

Features and Objectives	Symbol	Size	Type	Description
Component type	x_0	T	`bin`	One of T[1] possible types
Input number	x_1	1	`int`	Current input pin considered
No. inputs	x_2	1	`int`	Random number $\in [1, B]$[2]
Bit width input	x_3	1	`int`	Random number $\in [1, 32]$
Bit width output	x_4	1	`int`	Random number $\in [1, 32]$
Connections to input	x_5	1	`int`	Connected components $\in [0, 6]$
Connections to output	x_6	1	`int`	Connected components $\in [0, 6]$
Input slew	x_7	4	`float`	Transition times for current pin
No. outputs[3]	x_8	1	`int`	Random number $\in [1, 32]$
Sum bit width inputs[3]	x_9	1	`int`	Sum over all inputs bit width
Avg. bit width inputs[3]	x_{10}	1	`float`	Average over all inputs bit width
Output slew	y_1	4	`float`	Transition times for output pin
Delay	y_2	4	`float`	Delay from current to output pin

[1] If the training set is DS-I, then $T = 20$. Otherwise, $T = 37$.
[2] If the training set is DS-I, then $B = 12$. Otherwise, $B = 32$.
[3] Extra features considered for DS-II.

Additionally, we map the bit widths per input to a zero-padded vector, resulting in an $N \times B$ matrix, where $B = 12$ and $B = 32$ for DS-I and DS-II, respectively. A *MinMaxScaler* normalizes the features and four objectives in the [0, 1] range.

4.5.2.2 Graph Dataset (DS-III)

The training circuits used for DS-III contain one to six basic components, including only logic operators. The number of inputs to a component varies from one to eight. Similarly, each input's bit widths vary between one, two, four, and eight bits. The input slews of the primary input ports vary to cover four corner points. For both rising and falling edges, the four $[min, max]$ tuples considered are [0, 0], [0.5, 0, 5], [1.0, 1.0], and [0.0, 1.0]. Running STA for each generated training design and all four input transition times produces a final dataset length of 56336 graphs [46]. In the graph-based dataset, instead of mapping pin-to-pin connections to rows of a table, the RTL IR is mapped to a featured DAG following the pin node convention, i.e., pins are the nodes, and their connections are the edges, as illustrated in Fig. 4.15.

Features of Graph Dataset

After selecting the graph's topology, numerical features are annotated to nodes and edges. These features resemble those used in tabular datasets, but some are unnecessary when considering the entire DAG as a training sample. Table 4.4 shows the node features on top, followed by the input slew edge feature and the edge objectives. Each edge label considers rising and falling edges and minimum and maximum corners. Node and edge features and labels are normalized to be [0, 1]. The node feature indicating the component type is one-hot encoded, resulting in a vector of size 1×23. Notice that the number of RTL-supported components is 19. The remaining four types indicate primary port pins, start-end nodes, or unknown components, which might come during the evaluation of unseen designs. Hence, the node features are one-dimensional vectors of 28 elements, while the edge feature vectors have a 1×4 dimension. The edge label vectors are of size 1×8 [46].

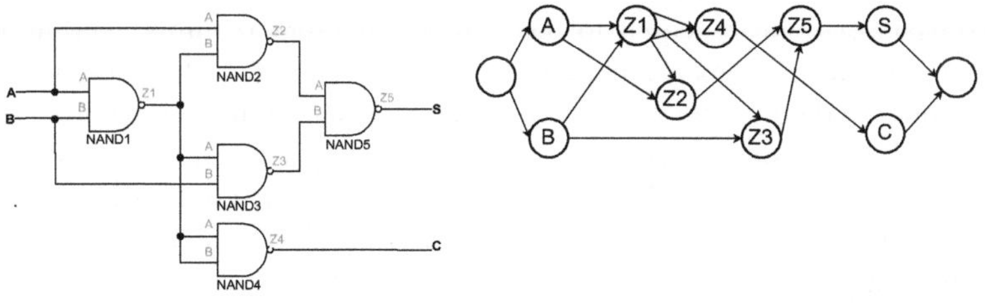

Fig. 4.15 Example of RTL IR and its mapping to a DAG

Table 4.4 Features and objectives for nodes and edges of graph dataset DS-III [46]

Feature and objectives	Symbol	Size	Type	Description
RTL node type	x_0	23	`int`	One-hot encoded node type
No. inputs	x_1	1	`int`	Number of input connections to the node
No. outputs	x_2	1	`int`	Number of output connections of the node
Bit width	x_3	1	`int`	Bit width of the node
Depth level	x_4	1	`int`	Level of the node in the DAG
Edge inverter	x_5	1	`int`	Whether the component inverts the timing arcs
Input slew	x_6	4	`float`	Input transition time from the input pin
Output slew	y_0	4	`float`	Output transition time of the output pin
Delay	y_1	4	`float`	Delay through the gate from input to output pin

4.5.3 Machine Learning Models

The second step in the ML flow is model selection and training. Vector- and graph-based models are employed to cope with the data structure of the three training datasets [45, 18, 46]. Tabular datasets are processed with multilayer perceptron (MLP) networks, where linear input layers process the features of each row in the dataset. Graph-based datasets are preprocessed using GNNs [11]. GNNs encode the graph's topological and feature information into lower- or higher-dimensional vectors. The resulting vectors are inputs to an MLP trained for a regression task at the edge level. More details about the selected models are given in the following [45, 18, 46].

4.5.3.1 MLP-Based Models

Two architectures, a simple *MLP* [47] and tabular *TabMLP* [48], are compared for the tabular datasets.

Multilayer Perceptron (MLP) Network

Figure 4.16a illustrates the layers of an MLP: input, hidden, and output. The input layer consists of neurons equal to the number of input features. The number of hidden layers and neurons are hyperparameters to be tuned. The output layer has neurons equal to the number of objectives. The layers are fully connected, meaning each element in the subsequent layer results from applying the activation function to the weighted sum of all values in the current layer. The component type feature x_0, indicating the gate's functionality, is a categorical variable. Categorical labels can be encoded differently to find an appropriate numerical representation. One-hot encoding is commonly used to generate sparse binary vectors of size N, where N is the total number of categories. The MLP in Fig. 4.16a uses a one-hot encoding

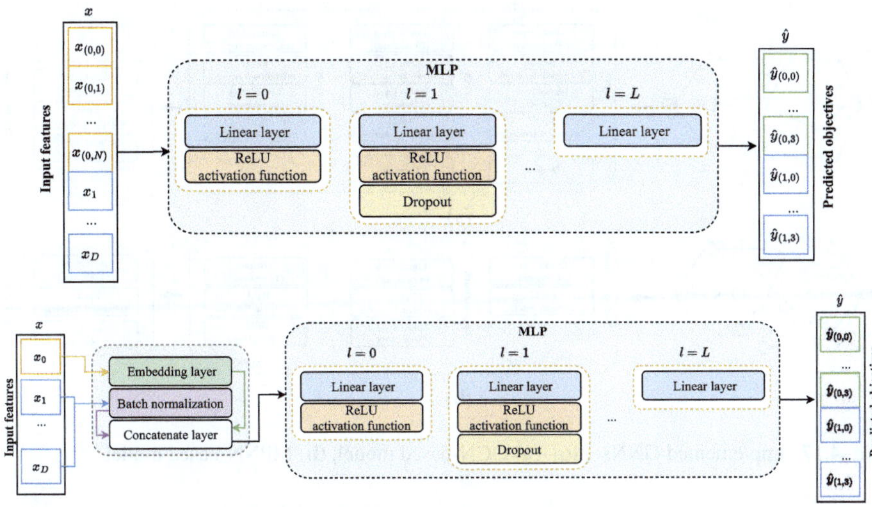

Fig. 4.16 Implemented MLP and TabMLP models [45, 18]

format for x_0, transforming this feature into a vector of size N. Each hidden layer uses a ReLU [49] activation function with a dropout layer to avoid overfitting. Training the MLP model with DS-I, the numbers of input neurons are 41 and 43 with and without input slew features, respectively. For DS-II, input slew features are always considered. As the supported component types and the maximum number of inputs increase, the input layer of the MLP model has 80 neurons [18].

Tabular Multilayer Perceptron (TabMLP) Network

The TabMLP architecture illustrated in Fig. 4.16b uses an embedding layer to encode the categorical feature x_0 into an N-dimensional vector. Embedding layers get the numerical value per category as inputs and, during training, learn the optimal N-dimensional encoding vector for each category. Unlike one-hot encoding, this approach reduces dimensionality to less than the number of categories and utilizes nonbinary vectors, thereby decreasing sparsity [50]. Figure 4.16b shows that the resulting embedding vector of x_0 is concatenated with the rest of the features passed by a batch normalization layer. Finally, the concatenated vector is passed through an MLP model. For DS-I, N is equal to 10. The total numbers of numerical input features are 21 and 23, with and without considering the slew, respectively. The number of neurons in the output layer is four, including the prediction of output slew for both edges. For DS-II, N is 19, and the total number of features is 43 [18].

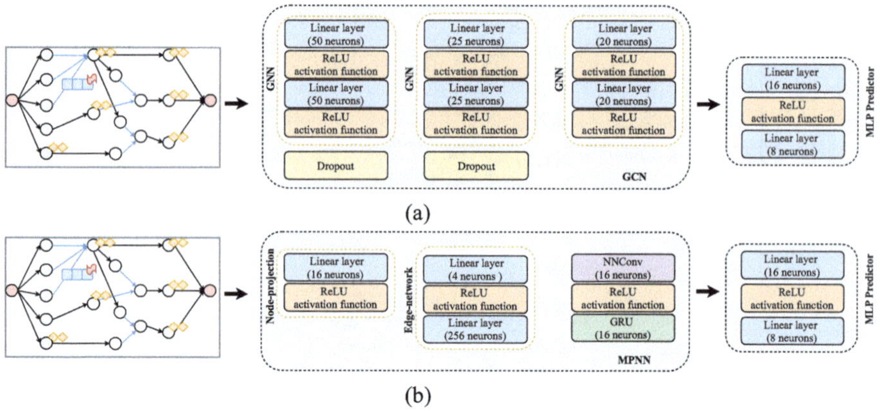

Fig. 4.17 Implemented GNNs [46]. (**a**) GCN-based model. (**b**) MPNN-based model

4.5.3.2 GNN-Based Models

The inputs to the MLP and TabMLP models are feature vectors of each pin-to-pin connection in a design. This mapping of EDA objects to the Euclidean space has been done to cope with ML techniques. However, it overlooks that EDA objects are intuitively graphs. GNNs are introduced by Gori et al. [11] as an NN framework that directly operates on graphs. GNNs learn the graph embedding vectors, denoted as $\mathbf{h}_u$ $\forall u \in V$, that encode the neighborhood information of each node u. GNNs consider both topological and feature information. They employ a message-passing strategy, where node embeddings or *messages* are transmitted through the neighborhood, and nonlinear differentiable functions are applied to update the node embeddings [51].

This chapter implements an extended *Graph Convolution Network [52] (GCN)* and a *Message-Passing Neural Network (MPNN) [53]* illustrated in Fig. 4.17 [46]. The input to the models is a graph representing the RTL IR as shown in Fig. 4.15, where nodes and edges are annotated with numerical features. These GNN models obtain node embeddings while considering node and edge features represented as x_n and x_e, respectively. As shown in Fig. 4.7, GNN-based models require a GNN to extract graph embeddings and an MLP predictor for classification or regression at the node, edge, or graph level. The following explains the GNN architectures and the MLP predictor used.

Graph Convolutional Network (GCN)

Mathematically, the propagation along the L layers of a GCN follows Eq. 4.3:

$$\mathbf{H}^{(l+1)} = \sigma\left(\tilde{\mathbf{D}}^{-\frac{1}{2}}\tilde{\mathbf{A}}\tilde{\mathbf{D}}^{-\frac{1}{2}}\mathbf{H}^{(l)}\mathbf{W}^{(l)}\right), \tag{4.3}$$

where $\mathbf{H}^{(l)}$ is the stacked matrix of vector embeddings $\mathbf{h_n}$ per n nodes in layer l. In $l = 0$, the embeddings of the nodes are their feature vectors, and $\mathbf{H} = \mathbf{X}$. Thus, $\mathbf{H} \in \mathbb{R}^{n \times d}$. $\sigma(\cdot)$ is a nonlinear activation function. GCNs incorporate self-connections to each node of the initial graph by adding the identity matrix of order I_n to $\mathbf{A}$, i.e., $\tilde{\mathbf{A}} = \mathbf{A} + \mathbf{I_n}$. The new $\tilde{\mathbf{A}}$ is multiplied side by side with its diagonal degree matrix $\tilde{\mathbf{D}}$. The $\mathbf{W}^{(l)}$ matrix contains the learnable parameters of each l layer [46].

GCNs do not support message passing with multidimensional edge features. To tackle this, an extended GCN architecture is implemented, as shown in Fig. 4.17a, and involves three L GNN and two dropout layers. Each GNN layer is composed of two linear layers. The first linear layer concatenates the edge embeddings $\mathbf{h_e}$ and applies a ReLU activation function to build a message, which is then aggregated to the neighboring embedding vectors. The second linear layer concatenates the message with the current node's embedding vector $\mathbf{h_n}$. The resulting embeddings are transformed using a ReLU function to obtain the final updated node embedding.

Message-Passing Neural Network (MPNN)

In contrast to GCNs, MPNNs consider edge features $\mathbf{x^e}$. MPNN builds a message $\mathbf{m}_u$, using an MLP $M(\cdot)$, that travels to the node u from its neighbors $v \in N(u)$ and considers edge features $\mathbf{x}^e_{(u,v)}$, as shown in Eq. 4.4. All messages arriving at the node u are aggregated. The message m is passed to an MLP $F(\cdot)$ to update the node embeddings following Eq. 4.5.

$$m_u^{l+1} = \sum_{v \in N(u)} M(h_u^l, h_v^l, x_{u,v}^e). \tag{4.4}$$

$$h_u^{l+1} = F(h_u^l, m_u^{l+1}). \tag{4.5}$$

Figure 4.17b shows the implemented MPNN. The node features undergo an initial projection by passing through a linear layer and a ReLU function. Subsequently, the message-passing strategy is performed twice. During each iteration, the edge features and their hidden embedding vectors are transformed by an edge network consisting of two linear layers and a ReLU function. The resulting edge embeddings are inputs to an *NNConv*, a graph convolution layer that multiplies them with the node embeddings of neighboring nodes. The product is added to the current node embeddings. Lastly, a gated recurrent unit (GRU) [54] combines the current node embeddings with the node features.

MLP Predictor

The MPNN and extended GCN node embeddings are fed into an MLP predictor. Specifically, the MLP takes a concatenated vector of the embedding vectors of the source nodes $\mathbf{h}_{src}$ and destination nodes $\mathbf{h}_{dst}$ of the edges. Thus, the input dimension of the MLP is $1 \times 2D$, where D is the size of the output node embedding vector. The output layer of the MLP has eight neurons, representing rising and falling edges' maximum and minimum output slews and delays, listed in Table 4.4.

The GCN- and MPNN-based models are trained with a stochastic gradient descent optimizer. Barboza et al. [55] suggest that the loss function for predicting slews and delays should penalize underestimations, i.e., the cases when the predicted $\hat{y}$ value is lower than the truth y value. This asymmetric MSE loss function is formulated in Eq. 4.6, where the first term of the sum corresponds to the conventional MSE loss, and N is the total number of samples:

$$Asymm.\ MSE = \frac{1}{N}\sum_{i=1}^{N}(y_i - \hat{y}_i)^2 + \frac{1}{N}\sum_{i=1}^{N}(|y_i - \hat{y}_i| + y_i - \hat{y}_i)^2. \tag{4.6}$$

4.5.4 Evaluation Circuits

We employ unseen hardware adder designs to evaluate the trained models [18, 46].

The simplest adder is the Ripple Carry Adder (RCA), which uses N full adders, and the previous stage's carry-out bit becomes the current stage's carry-in. At any particular time, only one full adder instance is used [56]. A more complex architectural version is the *parallel prefix* adder, as it splits the computation into three stages: preprocessing, carry propagation, and postprocessing [57]. Examples of parallel prefix adders that optimize further for area or performance are Sklansky Adder (SKA) [58], Brent-Kung Adder (BKA) [59], and Kogge-Stone Adder (KSA) [60].

Using MetaRTL, different adder designs and micro-architectures are implemented for 4, 8, 16, 32, 64-bit with one carry input and carry output bit. Their RTL IRs comprise logic components such as bitwise XOR, AND, and OR. The maximum number of inputs per component is two. To evaluate the GNN-based models, each design is mapped to a featured DAG. Figure 4.18 shows averaged details of the adder DAGs across all architectures with and without vectorization (V) and for each

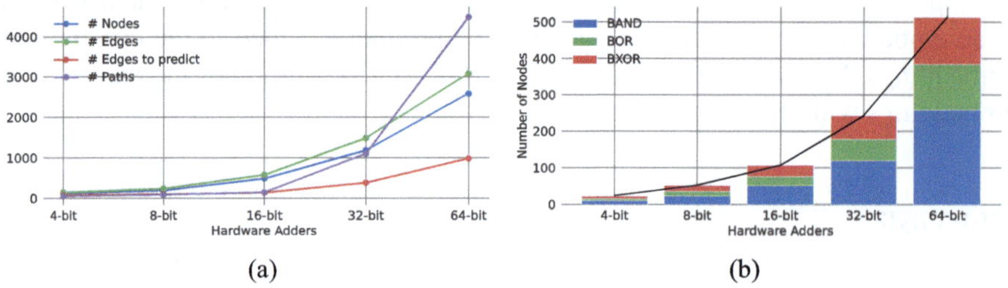

Fig. 4.18 Averaged graph information over all adders for each bit width [46]. (**a**) Number of nodes and edges. (**b**) Number of nodes of each logic type

bit width. The number of nodes, edges, edges to predict, and paths increases as the adder's bit width increases.

4.5.5 Experimental Results

We trained different models using all the mentioned ML architectures and collected datasets. After running the inference flow for hardware adders' designs, we analyze the results for each training dataset and architecture. The features of adder designs are mapped to tabular or graph datasets. Slews and delays of pin-to-pin connections are predicted and compared to pin-to-pin results from timing reports when using the same synthesis and STA configuration. For DS-I, experiments are conducted with and without considering slew as a feature [45, 18]. We employ the same architectures for DS-II but always consider slews [18]. Finally, DS-III uses graphs as inputs and GNN-based models [46].

4.5.5.1 Inference Using MLPs and Dataset I

This experiment evaluates the performance of the MLP and TabMLP models trained with DS-I. Specifically, it examines the impact of simultaneously learning and propagating slew values. Only 4-bit hardware adders are used in the test dataset.

Table 4.5 shows the average R^2 values for estimating fall and rise delays. The MLP model lacking slew information exhibits low estimation accuracy. To increase the accuracy, we incorporate the input slew as a feature. This result confirms the intuition that the pin-to-pin delay depends on the input slew, as explicitly seen in the timing models of technology libraries. However, the input slew is known only for the primary ports. The predicted output slew must be propagated and used as input slew for subsequent pin connections. Thus, the inference flow for a complete circuit must be done sequentially, pin by pin. Even though sequential inference is slower than parallel inference, where all pin-to-pin connections are considered simultaneously, it aligns with the conventional timing analysis flow. The final columns of Table 4.5 support this strategy by displaying high R^2 values. The same behavior is observed with TabMLP. Based on the results for DS-I, the following experiments always learn and propagate slew values. One limitation of propagating the predicted values is the simultaneous propagation of the prediction error along the timing paths. To avoid larger errors at the path endpoints and a significant reduction in the prediction

Table 4.5 R^2 of MLP and TabMLP with and without slew propagation [18]

Slew	No			Yes			Propagated		
Test Circuit	BKA	KSA	SKA	BKA	KSA	SKA	BKA	KSA	SKA
MLP	0.31	0.36	0.37	0.95	0.92	0.95	0.94	0.92	0.94
TabMLP	0.31	0.30	0.37	0.84	0.87	0.91	0.84	0.88	0.90

Table 4.6 Results of models trained with DS-II for different adders [18]

Test	R^2				MAE [ns]				MSE [ns^2]			
circuit	MLP	TabMLP	DT	RF	MLP	TabMLP	DT	RF	MLP	TabMLP	DT	RF
BKA	0.66	0.75	0.51	0.55	8.06	7.30	8.98	8.76	0.98	0.73	1.46	1.23
KSA	0.68	0.77	0.52	0.53	7.70	7.11	8.54	8.81	0.97	0.72	1.40	1.24
SKA	0.68	0.81	0.53	0.53	8.03	6.32	9.34	9.71	1.01	1.71	1.66	1.46
Average	0.67	**0.77**	0.52	0.54	7.94	**6.91**	8.96	9.09	0.99	**0.68**	1.51	1.31

accuracy, we aim to reduce the error per each component, and therefore, we explore different ML architectures, and later in this chapter, we train one ML model per each component type.

Inference is run for both models, and results are compared to timing reports generated with different timing constraints for the primary ports. The MLP model achieves, on average, 6.76 ns MAE, 0.96 ns^2 MSE, and R^2 equal to 0.87. The experiments are repeated using the TabMLP model. This achieves, on average, 9.49 ns MAE, 1.39 ns^2 MSE, and R^2 equal to 0.79.

4.5.5.2 Inference Using MLPs and Dataset II

In contrast to DS-I, DS-II considers more RTL IR primitives, i.e., one single model is trained to predict the timing behavior of pins within different components in a design. The results show that the TabMLP using embedding layers generalizes to more RTL primitives than the MLP model using one-hot encoding. Additionally, inference is run for three different timing constraints to evaluate the stability of the models across different slews at primary inputs. Using DS-II, shallow methods such as random forests and decision trees are trained [18].

Table 4.6 shows the prediction results when running inference over the 4-bit adders with shallow and deep ML models. The deep models outperform the shallow regressors in all evaluation metrics and circuits.

4.5.5.3 Inference Using GNNs and Dataset III

The MPNN and the GCN-based models predict the slews and delays for edges connecting a node's input and output pin in the DAG. Delay values for the edges connecting primary ports or input pins are not considered. Inference is executed for the eight implemented adders and all bit widths. Figure 4.19 shows the obtained R^2 coefficient for the GCN-based model. Across all adders and bit widths, the average R^2 coefficient is 0.78. The best results are achieved for 4-bit and 8-bit bit widths across all adder types. However, adders with higher bit widths, such as 16-, 32-, and 64-bit, include pins with a bit width larger than 8-bit, and these cases have not yet been covered in the training data. The R^2 scores with MPNN are worse than those

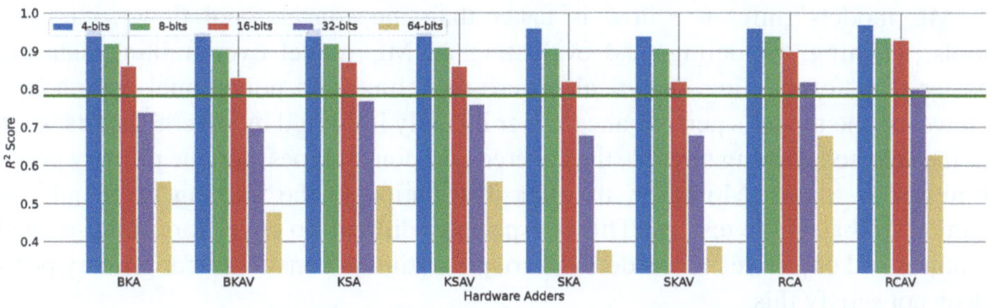

Fig. 4.19 Average R^2 of models trained with DS-III for adders using GCN [46]

Table 4.7 Performance of MLP, TabMLP, GCN, and MPNN models when trained with different datasets and evaluated over hardware adders [45, 18, 46]

Test dataset	Training dataset	Model	R^2	MAE [ns]	MSE [ns^2]
4-bit	DS-I	MLP	0.87	6.76	0.96
		TabMLP	0.79	9.49	1.39
	DS-II	MLP	0.67	7.94	0.99
		TabMLP	0.77	6.91	0.68
	DS-III	MPNN	−0.83	20.67	7.04
		GCN	**0.96**	**2.64**	**0.16**
all	DS-II	MLP	0.59	7.00	1.00
		TabMLP	0.73	7.00	1.00
	DS-III	MPNN	−1.00	22.89	7.81
		GCN	**0.78**	**0.16**	**2.76**

in Fig. 4.19, indicating inadequate performance on larger or unseen graphs. This confirms the known weakness of MPNNs in distinguishing between two isomorphic graphs, where topologically identical graphs are embedded to the same point [61].

4.5.6 Summary

Table 4.7 summarizes the experiment results, revealing the complexity of the pin-to-pin delay estimation task. According to the table, the GCN-based model outperforms all other methods in terms of performance across 4-bit and tested adders. Taking inspiration from technology libraries helps to consider proper features representing the nature of the task better, as shown when adding slew as input and objective features. However, despite these improvements, the R^2 coefficient remains lower than 0.80, and we must note that this model is solely trained for logic components. As real designs comprise many component types, it is essential to have a generalizable approach. Each component type exposes unique functionalities, pin configurations, bit widths, and time behaviors. Thus, considering all component types to train one model adds noise to the training data and worsens the prediction quality.

ML models' inference flow is faster than the conventional flow with EDA tools. Training, validating, and evaluating an ML model over training data are easier than formulating analytical, macro-modeling, and polynomial approaches. However, the model's performance is significantly impacted by several factors, such as the selected training circuits, the covered component types, and the primary port's timing constraints. Moreover, the data distribution in the training set should be similar to that of the test set. This is especially difficult to cover for timing-related features and objectives. Considering a couple of timing constraints for primary ports does not satisfy this.

In summary, the models and experiments in this section show the complexity of the delay estimation task and motivate the steps conducted in the next section. First, more strict configurations for training circuits are used to cover configurations needed rather than all possible ones. Second, an ML model per each component type is trained to focus on the features and timing behavior of one component type at a time. Moreover, more different timing constraints are covered for the training circuits regarding the delay and input slews of primary ports.

4.6 AI-Enabled Static Timing Analysis

Figure 4.3 shows the main steps of the proposed AI-enabled STA flow: pin-to-pin delay prediction and STA computation flow. Rather than relying on fixed pin-to-pin delay values or interpolating and extrapolating data from a technology library, we fetch delays from pretrained models. Building upon the findings from Sect. 4.5, the pin-to-pin delay estimation is performed by training T different MLPs, where T is the total number of component types. The propagation of the slew and the signal ATs through the timing paths is block-based until the design's endpoints are reached. RATs are backpropagated, and slacks are calculated per each pin in the design. This section describes how ML-predicted delays are leveraged to perform block-based STA without running synthesis or STA tools [14].

4.6.1 Problem Statement

Given an RTL IR describing a combinatorial logic, early estimations of timing design metrics such as ATs, RATs, and slacks are needed to compare different RTL variants and pinpoint critical blocks in the design. Pin-to-pin delay and slew estimations are required to calculate global design metrics such as ATs, RATs, and slacks. For instance, given the RTL IR in Fig. 4.20, delay and slew values should be predicted for every pin-to-pin connection highlighted in yellow, and ATs, RATs, and slacks should be annotated to every pin.

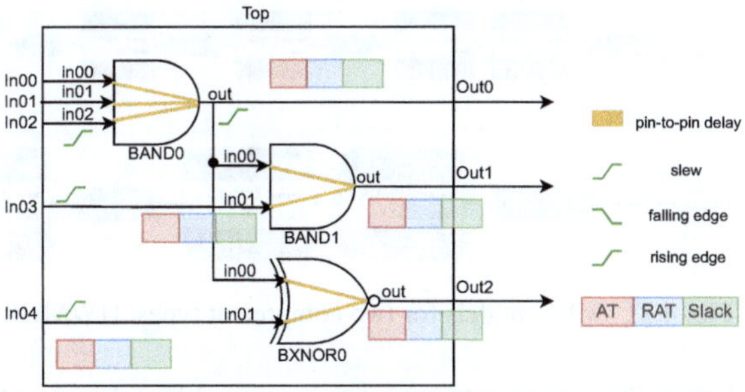

Fig. 4.20 RTL IR of a combinatorial circuit with pin-to-pin delays and ATs, RATs, and slacks to estimate

4.6.2 Consolidation of Pin-to-Pin Delay Models

Like Sect. 4.5, MLPs predict pin-to-pin delays and slews of primitive components within an RTL IR. Three significant differences are implemented in the final AI-enabled STA flow [14]:

4.6.2.1 Selection of Training Circuits

We generate circuit specifications for each component type by randomly choosing valid configurations for the number of input pins, bit widths, and fan-outs. These configurations are based on those used in existing hardware modules like hardware adders, multipliers, RISC-V, and SoC designs. Compared to the datasets generated in Sect. 4.5, the configurations are selected carefully, so the ones employed in real designs are covered during training data. This avoids the generation of valid but not necessary configurations. Following the same data collection flow in Sect. 4.4, 30.6K designs are automatically generated. These are less than the circuits generated for DS-I, DS-II, and DS-III.

4.6.2.2 Varying Timing Constraints for Training Data

For each netlist, STA is run in parallel with different input slews for the primary inputs. This is beneficial to increase the coverage of the ground-truth labels, as described in Sect. 4.4. Instead of selecting just a few corner cases, this section follows the parallel flow in Fig. 4.12b. STA is run 81 times in parallel for the same netlist. The remaining timing constraints in Table 4.1 are constant during training and evaluation.

4.6.2.3 One Model per Component Type

One MLP model is trained for each component type T. This decision is based on the recognition that different component types possess distinct features and timing behaviors to predict, necessitating separate training. For the example, the number of

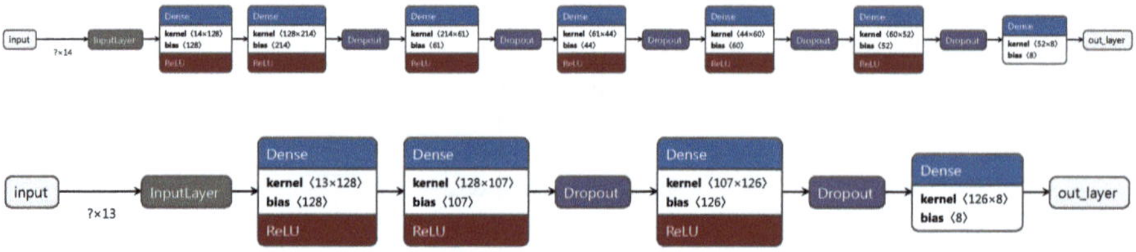

Fig. 4.21 MLP models for two component types: HWMUL (top) and GT (bottom)

inputs in a multiplexer can vary, making the number of inputs a relevant feature. In contrast, demultiplexers have only one input, rendering this feature redundant. Hence, certain features are excluded from the feature vectors $\mathbf{x}$ based on the component type.

A total of 26 MLPS are trained with different numbers of hidden layers (between two and four). The first layer has neurons equal to the number of features for the respective component type. The output layer always has eight neurons, where the first four neurons predict the slew and the last four the delay. Hyperparameter search is conducted to find the optimal number of layers, neurons per hidden layer, the optimizer, loss function, and dropout and learning rates. Examples of the selected architectures are shown in Fig. 4.21, generated by Netron [62], for the component types HWMUL and GT. The different MLPs, denoted by MLP-T, are trained with respective datasets DT-T during 50 epochs.

4.6.3 Static Timing Analysis Flow

The ML models estimating pin-to-pin delay and slew values are incorporated into a block-based STA computation flow proposed in [13]. The advantage of the block-based compared to path-based STA lies in its complexity being determined by the total number of components instead of the number of paths. Moreover, a block-based approach calculates global design metrics such as RATs and slacks per each pin of the design.

The inputs to the block-based STA flow are graphs. Thus, RTL IRs are mapped to a DAG structure following the pin node convention, i.e., pins are the nodes, and their connections are the edges, as illustrated in Fig. 4.15. A customized breadth-first search (BFS) is implemented to visit a node after all its predecessors have already visited it and annotated their corresponding AT values. After this preliminary step, the three core stages of the STA flow are executed. First, the forward pass or AT propagation takes place. In this step, the delay and transition times of the primary inputs get propagated throughout the timing graph, adding the ML-predicted delays per edge and propagating the maximum slews and AT. Second, the backward pass or RAT computation occurs. The RATs at the design endpoints are calculated based on the given output delays and clock configuration. These RATs are then backpropagated through the graph. Finally, the slack per node is calculated as the difference between the AT and RAT. All three steps consider maximum and minimum delays for rising and falling edges and inversions of the timing arcs caused by components such as the inverter gate. The following sections give a more detailed overview of each step.

4.6.3.1 Graph Traversal

The timing analysis's first step is finding the order to traverse the graph. The standard BFS algorithm traverses a graph in a breadth ward motion, i.e., starting from the top node, it explores all the vertices at the current depth before moving on to the vertices at the next depth level. Doing so guarantees that it visits all the nodes that are reachable from the starting node. This search is unsuitable for the block-based timing analysis, as one node should be visited if all *predecessors* p are visited before [63].

The inputs to the customized BFS are the list of nodes in the DAG and primary inputs. The output is a list called Q that contains the traversal order of the nodes. The algorithm starts by determining the maximum length of all paths from the top node to the end node in the graph. A queue q_l for each level of the graph is initialized. The primary inputs are appended to the first-level queue. The BFS iterates through each level from the first to the maximum level of the graph. The queue of each level q_l is iterated until it is empty. In each iteration, the first node of q_l is extracted. The algorithm checks whether its predecessors are already in the visited list Q. In that case, the current node is added to Q if not already added. Finally, the node's successors are added to the queue of their respective levels. The algorithm continues this process until all levels have been processed.

4.6.3.2 Forward Pass: AT Propagation

Given the delays, D of primary inputs, and the traversal order Q defined by the customized BFS, the ATs for all nodes in the design are calculated. For each node n in Q, the maximum slew from its predecessors to the current node is propagated to obtain the input slew s^{in}. If the edge between a predecessor and the current node is a wire, the delay is set to zero. If the pin-to-pin connection connects pins of the same component type, T, then the ML inference flow is executed. To that end, the pin-to-pin connection features are collected from the flow depicted in Fig. 4.8. The consolidated ML models use tabular datasets, where the features are columns, and each pin-to-pin connection is a row sample of the dataset. In contrast to Sect. 4.4, consolidated models are pretrained per each component type T. The pin features are used as input to the pretrained MLP for the type T. This model predicts the delay D and output slew s^{out} for that connection, as Sect. 4.4 explains. After all predecessors of a node are visited, the node's AT can be calculated using Eq. 4.7.

$$AT(n) = max_{\forall p \in pred(n)}[AT(p) + D(p, n)], \forall n \in Q \quad (4.7)$$

The flow is conducted for the four corner cases, rising and falling, for maximum and minimum delays. Some components, such as NOT, XOR, and NAND, invert their timing arcs. This information is listed in the cell description from the technology liberty file as timing sense. In these cases, the ATs for rising and falling edges are computed by inverting the edges for the delays, i.e., using falling and rising delays, respectively.

4.6.3.3 Backward Pass: RAT Propagation and Slack Calculation

The next step is to calculate the RAT of all nodes in the graph. To that end, the RATs at the primary outputs are calculated and defined based on the given output external delay and the clock configuration, as described in Sect. 4.2. In this chapter, the clock frequency is given by the design requirements. Afterward, the traversal order Q found using the customized BFS is reversed Q^R. Given a list of primary outputs with their respective RATs and the delay matrix D for all pin-to-pin connections found in the forward pass calculations, the backward pass flow propagates the RAT to all nodes n in Q^R. For each node n, the backward pass calculates the RAT for each successor. This is done by subtracting the delay between the current node and its successor from the RAT value of the successor node. The minimum RAT value among all successors is then assigned to the current node. In general, for all successors $s \in successors(n)$, Eq. 4.8 is computed. Lastly, the current node is marked as a visited node, and the next node is iterated.

$$RAT(n) = min_{\forall s \in succ(n)}[RAT(s) - D(n,s)], \forall n \in Q^R \tag{4.8}$$

The final step of the block-based STA flow is subtracting the calculated RATs and ATs for every node in the graph. The subtraction is called slack and is calculated in Eq. 4.9.

$$Slack_{(n)} = RAT_{(n)} - AT_{(n)}, \forall n \in Q \tag{4.9}$$

4.6.4 *Experimental Results*

Our AI-enabled block-based STA accurately computes ATs, RATs, and slacks for each component in an RTL IR. We compare the calculated ATs with the maximum ATs obtained at the design endpoints from OpenSTA reports across all paths and inputs. The efficiency of our flow is assessed using industrial designs, including encoders and hardware adders, which consist of 158 primary outputs. Table 4.8

Table 4.8 Computed ATs vs. ATs from timing reports after logic synthesis

Test dataset	Training dataset	Model	R^2	MAPE [%]	ME [ns]
Industrial	DS-T	MLP-T	0.92	9.56	3.66

compares the error metrics between the computed ATs and the STA results obtained after logic synthesis. Averaging over all timing edges, the computed ATs have a high correlation coefficient R^2 of 0.92. This high correlation can be attributed to the consolidated T models for pin-to-pin delay and slew estimation and their training using timing reports obtained from hierarchical netlists. The prediction achieves an MAPE of 9.56% and an ME of 3.66 ns. These errors arise from certain industrial designs comprising unsupported component types, i.e., types for which no trained MLP can predict the output delay or slew. These unsupported types are considered ideal wires, impacting the overall calculation and elevating the error metrics. Future work should extend our flow to consider other primitive component types, such as LUTs and encoders [14].

4.6.5 Summary

This section summarizes the consolidated timing estimation engine. The AI-enabled STA consists of two major steps: the pin-to-pin delay and slew estimation, which takes inspiration from the work presented in Sect. 4.4, and the conventional block-based STA flow [13]. Predicted delays and slews are predicted and propagated along the timing paths of a circuit. Moreover, the block-based approach allows calculating global design metrics per pin, such as ATs, RATs, and slacks. The calculated slack is valuable feedback: If the slack is negative, it guides the engineer in coding the RTL micro-architecture through the critical blocks. A positive and higher slack gives the designer more flexibility to add functionality or optimize the area utilization. Nodes with the same worst negative slack compose the critical path, i.e., the worst path violating timing constraints.

The proposed AI-enabled block-based STA reaches a high R^2 coefficient over unseen and complex test designs, and it is a main contribution to the state of the art. Ye et al. [9] and Guo et al. [10] also estimate delay, slews, and ATs but follow different approaches. Ye et al. [9] calculate the ATs accumulating the predicted cell delays per path. In contrast, this chapter follows the block-based approach for two reasons. First, the complexity does not scale exponentially with the number of components. Second, it provides meaningful timing metrics per pin and not only at the design's endpoints. On the other hand, Guo et al. [10] use ML models to predict ATs and slacks directly. This implies generating additional training data for the AT prediction and constrains the solution space for the training data distribution. Works in [10] and [9] use pre-routing netlists as inputs, losing details of the RTL micro-architecture and enabling design exploration at later stages of the design flow. In contrast, our proposed flow takes the RTL IR as input and enables optimization at the early stages of the design.

In conclusion, having a block-based STA operating at the RTL IR gives valuable insights into the timing behavior of the final RTL without running RTL generation, synthesis, and STA tools. This facilitates the search for optimal micro-architectures, as it is easy and fast to get an early timing estimation of a design after slight changes or corrective measures. Without AI, modifications to the design will cause reiteration of the flow and rerun RTL generation, synthesis, and analysis.

4.7 Potential Application

This section outlines how timing estimations of a design can ease and automate the micro-architecture search, especially inside an RTL generation framework. For instance, given the different micro-architectures of a half-adder in Fig. 4.2 and the same input timing constraints, different ATs, RATs, and slacks are expected at the sum and carry output ports.

4.7.1 Micro-Architectural Transformations

Despite advances in logic synthesis tools, the quality of the generated gate-level netlist still depends on the quality of the input RTL [5, 4]. Logic synthesis tools perform logic optimization and technology mapping. However, the logic optimization step can be more effective with an optimized RTL. Therefore, the micro-architecture search becomes a critical yet easy step inside an RTL generation framework.

Micro-architecture search can be further automated within an RTL generation framework through design *transformations*. The latter are generators transforming components of an RTL IR while maintaining the same functionality. Transformations add specific components, such as a debugger JTAG module [64], to a design while preserving its overall functionality. The benefit of these transformations is that they reduce design efforts, as a single transformation can be applied to any input RTL IR without requiring manual modifications for updating ports, structures, or connections [64]. Thus, these transformations facilitate the automated and efficient modification of small micro-architecture details within an RTL IR.

4.7.1.1 Multiplexer Trees

Multiplexers are one interesting component in digital designs and are found in many data paths, varying in the number of input ports and bit widths. An n-to-1 multiplexer has n data inputs, $m = \lceil log_2(n) \rceil$ selection bits, and one output. The multiplexer selects a data input given the selection bits and forwards it to the output. Using HDL, multiplexers are represented as *case* structures, where the number of inputs corresponds to the number of possible values the output can take. The RTL IR represents multiplexers as basic primitives with n inputs.

During logic synthesis, multiplexers are mapped to a set of logic gates or multiplexers available in the target technology, commonly 2-to-1 or 3-to-1 multiplexers.

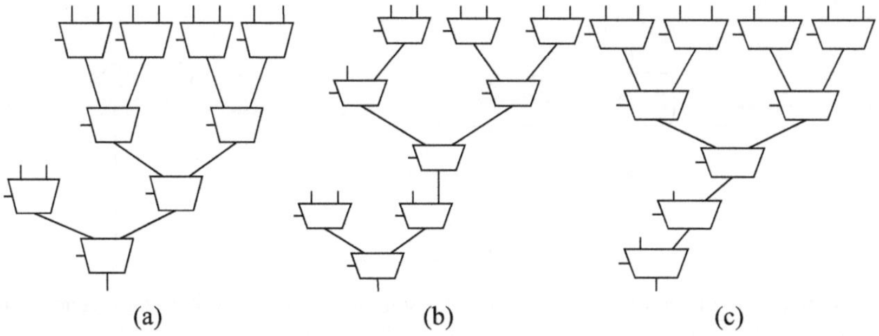

Fig. 4.22 Different decompositions of a 10-to-1 multiplexer tree

Thus, for multiplexers at RTL with a higher number of inputs, a multiplexer tree is constructed and restructured considering the CMOS multiplexers available and the AT of the inputs. If the technology has only 2-to-1 multiplexers, the synthesis tool decomposes big multiplexers into a tree of 2-to-1 multiplexers. However, little effort is spent finding the best tree structure, i.e., the best tree depth and order of inputs according to the ATs, so the latest inputs come later in the tree. The latter has motivated many works to optimize multiplexer trees for power [65], performance [66], or area [67].

The higher the number of inputs of a multiplexer at RTL, the higher the options for restructuring the multiplexer tree. A balanced tree and the cascade decompositions are the edge cases. For an n-to-1 multiplexer, the number of possible decompositions can be viewed as the number of structurally different binary trees with $n-1$ nodes. For a binary tree with n nodes, the number of different structures is given by the Catalan number [68] C_n in Eq. 4.10. Consequently, the number of possible decompositions for an n-to-1 multiplexer is C_{n-1}.

$$C_n = \frac{(2n)!}{(n+1)!n!} \tag{4.10}$$

Figure 4.22 depicts different possible decompositions for a 10-to-1 multiplexer. In cases when the depth of the multiplexer tree is higher than m, an encoder is needed to preserve the multiplexing logic based on the original and new selection bits. Given this large number of possible decompositions, which rises exponentially with n, it is clear that choosing the optimal decomposition is a nontrivial task.

4.7.2 *Micro-Architecture Search*

The optimal multiplexer tree decomposition depends on the structure of the binary tree, the order of the inputs, and their ATs. Therefore, the micro-architecture search involves two main components: transformations to explore different decomposition

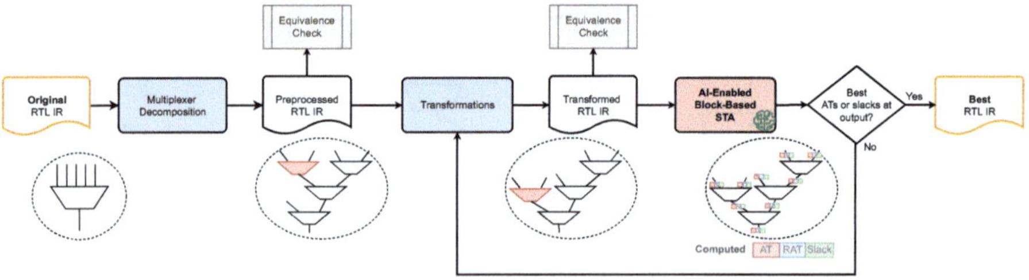

Fig. 4.23 Micro-architecture search flow at RTL using *transformations* of an RTL generation tool and the proposed AI-enabled block-based STA computation flow

trees and timing information to guide the search. It is possible to transform the multiplexer trees using an RTL generation framework and automated transformations to RTL IRs. Timing information from STA tools requires a gate-level netlist and, thus, an HDL-based RTL. This conventional flow is costly regarding runtime, as every tree decomposition in the RTL IR would need to be converted into its HDL description, synthesized, and analyzed. Our proposed AI-enabled STA flow takes RTL IRs as inputs and quickly computes timing metrics per each multiplexer in the transformed tree.

Figure 4.23 shows the basic workflow for a micro-architecture search for the best multiplexer tree. First, a multiplexer component with N inputs is found in the RTL IR. This is decomposed as a balanced tree to use 2-to-1 multiplexer components available in the CMOS technology. The transformed RTL IR is the input to the AI-enabled STA, providing all relevant timing metrics per each pin in the tree. In particular, the ATs at inputs and outputs are decisive for the search. The former helps to find better tree structures, moving the latest input signals to the bottom of the tree. The ATs at the outputs guide the search for the best decomposition. Trees with lower ATs or better slacks at the outputs are saved as best trees. These steps are repeated until the best tree is found. As the original RTL IR is transformed multiple times in the flow, equivalence checks are required to guarantee the original and transformed designs have the same functionality.

The flow in Fig. 4.23 can be performed manually or automatically using optimization algorithms. The latter vary in complexity from random searches to more sophisticated methods such as genetic algorithms (GA) or reinforcement learning (RL). Optimization algorithms aim to find the most optimal solution to a problem by maximizing or minimizing an objective function. In the case of the multiplexer trees, the objective function is linked to the calculated ATs or slacks at the output of the tree, both before and after the transformations. Optimization algorithms also require automatic exploration of different solutions in the search space, i.e., automatically choosing the transformation to apply next. In RL, this is referred to as the action space, while in a GA, new solutions are obtained by mutation or a combination of previous solutions.

Multiplexer trees can be transformed by enumerating all pins, swapping a pair of inputs and multiplexers, or relocating subtrees accordingly. As suggested by the

Catalan number in 4.10, the higher the number of inputs, the higher the number of possible tree structures. Additionally, the different ATs at the inputs of the tree influence each solution. Experimental results have shown that random search is feasible for up to seven inputs. For a higher number of inputs, more sophisticated optimization algorithms should be employed. Future work will replace manual or random search with algorithms like RL. In RL, problems are solved by iteratively taking decisions or actions and observing their effect on the environment to guide the next decision to maximize a reward function. RL has already demonstrated its efficacy in optimizing EDA tasks such as micro-architecture selection of parallel prefix adders [69] and logic synthesis [70]. Consequently, RL can be a suitable solution for the automated micro-architecture search when the actions correspond to the transformations inside an RTL generation framework and the reward function is calculated with the proposed AI-enabled STA flow.

4.8 Summary and Conclusions

This book chapter illustrates how AI enables STA at the early stages of the RTL design without running RTL generation, synthesis, and timing analysis tools. To that end, an automated data collection flow is set using an in-house RTL generation and open-source synthesis tools, going from design specifications to gate-level netlists in an automated manner. Datasets with fixed parameters and configurations are generated to train ML models. From generated RTL IRs, features are extracted in tabular and graph structures so that MLP, Tabular MLP- and GNN-based models are trained to predict the pin-to-pin delays and slews. Different training circuits, dataset structures, and model architectures are compared. Specially, deep and graph-based ML models are compared to evaluate the benefit of representing the input designs and their features as tables or graph structures. The usage of graphs allows embedding the designs using GNNs, which is an ML framework being leveraged for multiple EDA tasks. The power of GNNs relies on the encoding of the input graph using its node and edges features but also its topology information. ML predictions are leveraged to compute the well-known block-based STA.

A consolidated AI-enabled STA flow is presented using predicted delays to compute block-based timing metrics per pin. Our flow has several advantages. Early timing estimations can be obtained for any design at its early stages. Given an RTL IR composed of primitive operators, the proposed flow trains an ML model per type to predict delay and transition times through that component. ATs, RATs, and slacks are calculated for any RTL IR. Second, our flow running inference of ML models is faster than running different tools for RTL generation, (at least logic) synthesis, and STA. Like all ML techniques, they can be easily trained to target new component types and technologies. The presented data generation flow allows automatic training and evaluation circuit generation. Moreover, using parallelism, the time and energy to generate new training data and train ML models can be reduced to 1 week. Future research could focus on the transfer learning capabilities of the models specially across same component types but different technology nodes. Finally, computed STA design metrics can guide engineers coding RTL generators to find better micro-architectures.

Our current approach creates one ML model for each component type and technology node. Moreover, it does not consider wire delays and assumes no noise and timing constraints as fixed values. Even though this simplified approach lacks generalization across different technology nodes, wire models, and corner cases, the main purpose of this work is to estimate the timing of a design at the early stage of the RTL generation. In this scenario, the hardware engineer has a target technology node, a fixed corner case, and defined timing constraints. Instead of exploring different technologies and constraints, the engineer aims to compare different RTL architectures. In such cases, our ML-based approach offers a practical and adaptable solution, where a data generation flow automatically and quickly generates training data and new ML models shall be trained. However, this lack of generalization reflects the limitations of the AI-enabled timing estimation. One single ML model cannot cover different technology nodes or learn from the data across same component types but different technologies.

According to Ren et al. [71], ML models still face challenges in solving complex tasks of the digital design flow unless the problem is limited to a smaller subset. The solution space of ML models is limited by the training data, which may be biased or subject to distribution shifts. Even with a data collection flow generating various RTL variants and collecting their timing results, it remains impossible to train an ML model that covers all possible primitive components in an RTL IR, feature ranges, technology nodes, target clock frequencies, design rules, and timing constraints. Our ML-based timing estimation has achieved satisfactory performance when the problem is constrained to a specific set of timing constraints, component types, and a technology node.

Future research in ML for timing analysis should focus on using ML alongside classical optimization and heuristic algorithms, as suggested in [71]. This integration should be computationally efficient while enhancing the interpretability of ML-based predictions. These two aspects would increase the acceptance of AI solutions across hardware engineers in industry and academia.

Competing Interests The work described herein is partly funded by the German Federal Ministry for Economic Affairs and Energy (BMWi) as part of the research project ProgressivKI (19A21006C). This study was partly funded by Infineon Technologies AG. The content in this chapter is part of ongoing Ph.D. research work at the Technical University of Munich.

References

1. Bachrach, J., Vo, H., Richards, B., Lee, Y., Waterman, A. et al. (2012). Chisel: Constructing hardware in a Scala embedded language. In *Design Automation Conference (DAC)* (pp. 1216–1225).

2. Takamaeda-Yamazaki, S. (2015). PyVerilog: A Python-based hardware design processing toolkit for Verilog HDL. In *Applied Reconfigurable Computing* (pp. 451–460). Berlin: Springer.
3. Schreiner, J., Findenigy, R., & Ecker, W. (2016) Design centric modeling of digital hardware. In *International High-Level Design Validation and Test Workshop (HLDVT)* (pp. 46–52).
4. Synopsys, Synopsys unveils RTL architect to accelerate design closure. https://www.prnewswire.com/news-releases/synopsys-unveils-rtl-architect-to-accelerate-design-closure-301023175.html
5. Sengupta, P., Tyagi, A., Chen, Y., & Hu, J. (2022) How good is your Verilog RTL code? A quick answer from machine learning. In *International Conference on Computer-Aided Design (ICCAD)* (pp. 1–9).
6. Sánchez Lopera, D., Subedi, I., & Ecker, W. (2023) Using graph neural networks for timing estimations of RTL intermediate representations. In *Workshop on Machine Learning for CAD (MLCAD)* (pp. 1–6).
7. Alcorta, E. S., Gerstlauer, A., Deng, C., Sun, Q., Zhang, Z., Xu, C., Wills, L. W., Sánchez Lopera, D., Ecker, W., Garg, S., & Hu, J. (2023). Special session: Machine learning for embedded system design. In *International Conference on Hardware/Software Codesign and System Synthesis (CODES+ISSS)* (pp. 28–37).
8. Huang, G., Hu, J., He, Y., Liu, J., Ma, M., et al. (2021). *Machine Learning for Electronic Design Automation: A Survey* (vol. 26). New York: ACM.
9. Ye, Y., Chen, T., Gao, Y., Yan, H., et al. (2023). Graph-learning-driven path-based timing analysis results predictor from graph-based timing analysis. In *Asia and South Pacific Design Automation Conference (ASP-DAC)* (pp. 547–552).
10. Guo, Z., Liu, M., Gu, J., Zhang, S., Pan, D. Z., & Lin, Y. (2022). A timing engine inspired graph neural network model for pre-routing slack prediction. In *Design Automation Conference (DAC)* (pp. 1207–1212). New York: ACM.
11. Gori, M., Monfardini, G., & Scarselli, F. (2005). A new model for learning in graph domains. In *International Joint Conference on Neural Networks* (vol. 2, pp. 729–734).
12. Lu, Y.-C.,, Nath, S., Khandelwal, V., & Lim, S. K. (2021). Doomed run prediction in physical design by exploiting sequential flow and graph learning. In *International Conference on Computer Aided Design (ICCAD)* (pp. 1–9).
13. Hitchcock, R. B., Smith, G. L., & Cheng, D. D. (1982) Timing analysis of computer hardware. *IBM Journal of Research and Development, 26*(1), 100–105.
14. Sánchez Lopera, D., Kunzelmann, R., Kaja, E., & Ecker, W. (2024). Fake Timer: An engine for accurate timing estimation in register transfer level designs. In *2024 25th International Symposium on Quality Electronic Design (ISQED)* (pp. 1–8). New York: IEEE, just accepted.
15. IEEE standard for integrated circuit (IC) delay and power calculation system (2000). *IEEE Std 1481-1999.*
16. *Liberty user guides and reference manual suite* (2013).
17. *Synopsys timing constraints and optimization user guide* (2010)
18. Sánchez Lopera, D., Servadei, L., Prebeck, S., & Ecker, W. (2022) Early RTL delay prediction using neural networks. *Microprocessors and Microsystems (MICPRO), 94.* https://www.sciencedirect.com/science/article/pii/S0141933122002010
19. Wallace, D. E., & Chandrasekhar, M. S. (1990). High-level Delay Estimation for Technology-independent Logic Equations. In *International Conference on Computer-Aided Design (ICCAD)*. New York: IEEE.
20. Qiu, J., Reda, S., & Hassoun, S. (2010). Fast, accurate a priori routing delay estimation. In *International Workshop on System Level Interconnect Prediction (SLIP)* (pp. 77–82). New York: ACM.
21. Macii, A., Macii, E., Odasso, G., Poncino, M., & Scarsi, R. (1999). Regression-based macromodeling for delay estimation of behavioral components. In *Great Lakes Symposium on VLSI (GLS)* (pp. 188–191). New York: IEEE Computer Society.
22. Srinivasan, A., Huber, G. D., & LaPotin, D. P. (1998). *Accurate Area and Delay Estimation from RTL Descriptions* (Vol. 6(1), pp. 168–172). New York: IEEE.

23. Krizhevsky, A., Sutskever, I., & Hinton, G. E. (2012). Imagenet classification with deep convolutional neural networks. In *Advances in Neural Information Processing Systems (NIPS)* (vol. 25).
24. Yang, K. K., Wu, Z., & Arnold, F. H. (2019). Machine-learning-guided directed evolution for protein engineering. *Nature Methods, 16*(8), 687–694.
25. Ren, H., Nath, S., Zhang, Y., Chen, H., & Liu, M. (2022). Why are graph neural networks effective for EDA problems? In *International Conference On Computer-Aided Design (ICCAD)* (pp. 1–8).
26. Makrani, H.M., Farahmand, F., Sayadi, H., Bondi, S., et al. (2019). Pyramid: Machine learning framework to estimate the optimal timing and resource usage of a high-level synthesis design. In *International Conference on Field Programmable Logic and Applications (FPL)* (pp. 397–403).
27. Ramesh, S., & Jayaparvathy, R. (2019). Artificial neural network model for arrival time computation in gate level circuits. In *Automatika: časopis za automatiku, mjerenje, elektroniku, računarstvo i komunikacije* (Vol. 60(3), pp. 360–367).
28. Martin, T., Grewal, G., & Areibi, S. (2021). A machine learning approach to predict timing delays during FPGA placement. In *International Parallel and Distributed Processing Symposium Workshops* (pp. 124–127). New York: IEEE.
29. Barboza, E. C., Shukla, N., Chen, Y., & Hu, J. (2019). Machine learning-based pre-routing timing prediction with reduced pessimism. In *Design Automation Conference (DAC)* (pp. 1–6).
30. Sánchez Lopera, D., Servadei, L., Kiprit, G. N., Wille, R., & Ecker, W. (2023) A comprehensive survey on electronic design automation and graph neural networks: Theory and applications. *Transactions Design Automation Electronic Systems (TODAES), 28*(2), 2. https://doi.org/10.1145/3543853
31. Shrestha, P., Phatharodom, S., & Savidis, I. (2022). Graph representation learning for gate arrival time prediction. In *Workshop on Machine Learning for CAD (MLCAD)* (pp. 127–133).
32. Lu, Y.-C., Nath, S., Khandelwal, V., & Lim, S. K. (2021). Doomed run prediction in physical design by exploiting sequential flow and graph learning. In *International Conference On Computer-Aided Design (ICCAD)* (pp. 1–9).
33. Guo, Z., Liu, M., Gu, J., Zhang, S., Pan, D. Z., & Lin, Y. (2022). A timing engine inspired graph neural network model for pre-routing slack prediction. In *Design Automation Conference (DAC)* (pp. 1207–1212).
34. Ajayi, T., Chhabria, V. A., Foga, M.ça, Hashemi, S., Hosny, A., Kahng, A. B., et al. (2019). INVITED: Toward an open-source digital flow: First learnings from the OpenROAD project. In *Design Automation Conference (DAC)* (pp. 1–4).
35. Hashemi, S., Ho, C.-T., Kahng, A. B., Liu, H.-Y., & Reda, S. (2018). METRICS 2.0: A machine-learning based optimization system for IC design. In *Workshop on Open-Source EDA Technology*. Computer Science (pp. 1–4).
36. Wolf, C., Glaser, J., & Kepler, J. (2013). Yosys-A free Verilog synthesis suite.
37. Brayton, R., & Mishchenko, A. (2010). ABC: An academic industrial-strength verification tool. In *Computer Aided Verification* (pp. 24–40).
38. Chen, J., Jiang, I. H.-R., Jung, J., Kahng, A. B., Kim, S., Kravets, V. N., Li, Y.-L., Varadarajan, R., & Woo, M. (2021). DATC RDF-2021: Design flow and beyond ICCAD special session paper. In *International Conference On Computer Aided Design (ICCAD)* (pp. 1–6).
39. Truyen, F. (2006). The fast guide to model driven architecture. *Cephas Consulting Corp.*
40. Devarajegowda, K., Ecker, W., & Kunz, W. (2019). How to keep 4-eyes principle in a design and property generation flow. In *Methods and Description Languages for Modelling and Verification of Circuits and Systems (MBMV)* (pp. 1–6).
41. A guide to advanced process design kits. https://semiengineering.com/a-guide-to-advanced-process-design-kits/
42. OpenSTA. https://github.com/The-OpenROAD-Project/OpenSTA
43. Sánchez Lopera, D., Kashyap, P. et al. (2021). Using open-source EDA tools in an industrial design flow. In *Design and Verification Conference and Exhibition Europe (DVCON)* (pp. 1–8).

44. Lück, C., Sánchez Lopera, D., Wenzek, S., & Ecker, W. (2022). Industrial experience with open-source EDA tools. In *Workshop on Machine Learning for CAD (MLCAD)* (pp. 143–143)
45. Sánchez Lopera, D., Servadei, L., Kasi, V. P., Prebeck, S., & Ecker, W. (2021). RTL delay prediction using neural networks. In *Nordic Circuits and Systems Conference (NorCAS)* (pp. 1–7).
46. Sánchez Lopera, D., & Ecker, W. (2022). Applying GNNs to timing estimation at RTL. In *International Conference on Computer-Aided Design (ICCAD)*.
47. Pal, S. K., & Mitra, S. (1992). *Multilayer Perceptron, Fuzzy Sets, Classification* (Vol. 3(5), pp. 683–697). New York: IEEE.
48. Pytorch-Widedeep (2021). https://github.com/jrzaurin/pytorch-widedeep
49. Agarap, A. F. (2018). Deep learning using rectified linear units (RelU). arXiv preprint arXiv:1803.08375.
50. Huang, X., Khetan, A., Cvitkovic, M., & Karnin, Z. (2020). TabTransformer: Tabular data modeling using contextual embeddings. ArXiv. https://arxiv.org/abs/2012.06678
51. Wu, Z., Pan, S., Chen, F., Long, G., Zhang, C., & Yu, P. S. (2021). A comprehensive survey on graph neural networks. *Transactions on Neural Networks and Learning Systems, 32*(1), pp. 4–24.
52. Welling, M., & Kipf, T. N. (2016). Semi-supervised classification with graph convolutional networks. In *International Conference on Learning Representations (ICLR)*.
53. Gilmer, J., Schoenholz, S. S., Riley, P. F., Vinyals, O., & Dahl, G. E. (2017). Neural message passing for quantum chemistry. In *International Conference on Machine Learning (PMLR)* (pp. 1263–1272).
54. Cho, K., Van Merriënboer, B., Bahdanau, D., & Bengio, Y. (2014). On the properties of neural machine translation: Encoder-decoder approaches. arXiv preprint arXiv:1409.1259.
55. Barboza, E. C., Shukla, N., Chen, Y., & Hu, J. (2019). Machine learning-based pre-routing timing prediction with reduced pessimism. In *Design Automation Conference (DAC)* (pp. 1–6)
56. Ghaderi, A., & Frounchi, J. (2016). *Power and Speed Analysis of CMOS-based Multipliers using VEDIC Techniques*, vol. 11 (pp. 62–68).
57. Rani, C. S., & Ramesh, C. (2014). Design and implementation of high performance parallel prefix adders (pp. 5899–5903).
58. Sklansky, J. (1960). Conditional-sum addition logic. *IRE Transactions on Electronic Computers, 2*, 226–231.
59. Brent & Kung (1982). A regular layout for parallel adders. *Transactions on Computers, 100*(3), 260–264.
60. Kogge, P. M., & Stone, H. S. (1973). A parallel algorithm for the efficient solution of a general class of recurrence equations. *Transactions on Computers, 100*(8), 786–793.
61. Balcilar, M., Héroux, P., Gauzere, B., Vasseur, P., Adam, S., & Honeine, P. (2021). Breaking the limits of message passing graph neural networks. In *International Conference on Machine Learning (PMLR)* (pp. 599–608).
62. Roeder, L. (2023). lutzroeder/netron. https://github.com/lutzroeder/netron
63. Li, B. (2010). Hierarchical statistical static timing analysis considering process variations, Ph.D. dissertation, Technische Universität München.
64. Han, Z., Wang, D., Rutsch, G., Li, B., Prebeck, S. S., Sánchez Lopera, D., et al. (2021) Aspect-oriented design automation with model transformation. In *International Conference on Very Large Scale Integration (VLSI-SoC)* (pp. 1–6). New York: IEEE.
65. Narayanan, U., Leong, H. W., Chung, K.-s., & Liu, C. L. (1997). Low power multiplexer decomposition. In *International Symposium on Low Power Electronics and Design (ISLPED)* (pp. 269–274). New York: ACM.
66. Thakur, S., Wong, D., & Krishnamoorthy, S. (1996). Delay minimal decomposition of multiplexers in technology mapping. In *Design Automation Conference (DAC)* (pp. 254–257). New York: ACM. http://ieeexplore.ieee.org/document/545582/
67. James, R., Mythili, P., Anilkumar, C., & Vijayakumari, C. (2013). A modified technique for the optimal design of combinational digital circuits with multiplexers using genetic algorithm. In *International Conference on Computational Intelligence and Information Technology (CIIT)* (pp. 225–229). New York: Institution of Engineering and Technology. https://digital-library.theiet.org/content/conferences/10.1049/cp.2013.2595

68. Stanley, R. P. (2015). *Catalan Numbers* Cambridge: Cambridge University Press.
69. Roy, R., Raiman, J., Kant, N., Elkin, I., Kirby, R., Siu, M., Oberman, S., Godil, S., & Catanzaro, B. (2021). PrefixRL: Optimization of parallel prefix circuits using deep reinforcement learning. In *Design Automation Conference (DAC)* (pp. 853–858). New York: IEEE.
70. Hosny, A., Hashemi, S., Shalan, M., & Reda, S. (2020). DRiLLS: Deep reinforcement learning for logic synthesis. In *Asia and South Pacific Design Automation Conference (ASP-DAC)* (pp. 581–586). New York: IEEE. https://ieeexplore.ieee.org/document/9045559/
71. Ren, H., et al. (2023). Machine learning and algorithms: Let us team up for EDA. *IEEE Design & Test, 40*(1), 70–76.

Chapter 5
Harnessing Graph Learning for Efficient Timing Signoff

Xinfei Guo, Linyu Zhu, and Yichen Cai

5.1 Introduction

In the contemporary landscape of expanding chip sizes and escalating complexity, design engineers grapple with substantial challenges in adhering to aggressive tape-out schedules while striving to deliver high-quality designs within the constraints of limited computing resources. Among the critical signoff checks, achieving timing closure poses a particular challenge, given the rising operating speeds inherent in modern integrated circuits (ICs). To meet the demands of timing closure, the conventional design flow necessitates multiple iterations of timing checks. The introduction of additional Engineering Change Orders (ECOs) further adds complexity to the process, potentially resulting in schedule shifts [1, 2]. It is widely recognized that advanced technology nodes introduce a multitude of design corners and scenarios, significantly complicating the task of timing estimation [3].

In the initial stages of the design process, timing estimation often relies on simplified delay models that permit marginal errors. Precision in timing analysis is not mandatory at this point. As a design approaches the signoff stage, obtaining accurate path delay information becomes crucial for discovering any remaining timing violations. This process typically begins with extracting the Standard Parasitic Exchange Format (SPEF) netlist from a routed design. Accurate path

X. Guo (✉)
University of Michigan – Shanghai Jiao Tong University Joint Institute, Shanghai Jiao Tong University, Shanghai, China

State Key Laboratory of Integrated Chips and Systems (SKLICS), Shanghai, P.R. China
e-mail: xinfei.guo@sjtu.edu.cn

L. Zhu · Y. Cai
University of Michigan – Shanghai Jiao Tong University Joint Institute, Shanghai Jiao Tong University, Shanghai, China
e-mail: linyuzhu@sjtu.edu.cn; cai_yichen@sjtu.edu.cn

A. Iranmanesh, H. Sayadi (eds.), *AI-Enabled Electronic Circuit and System Design*,
https://doi.org/10.1007/978-3-031-71436-8_5

delay can then be obtained through SPICE simulations or static timing analysis (STA) engines with Signal Integrity (SI) mode enabled. However, these methods are extremely resource-intensive and time-consuming, prompting the need for faster and more accurate timing estimation techniques suitable for signoff.

Path delay composes of both cell and interconnect delays. Cell delay is usually derived directly from cell libraries (.lib) using lookup tables based on wire capacitance and input slew. In contrast, interconnect delay calculation is more complex, often involving intricate modeling and "mysterious" formulas, even when detailed RC (resistance and capacitance) information is available from the SPEF file [4, 5]. Traditional approaches, such as Elmore delay models [6] and more sophisticated methods like D2M [7], only yield approximate results. These methods can deviate significantly from the true values, making them unreliable for signoff purposes. Furthermore, in advanced technology nodes, complex netlists introduce significant coupling effects and irregular RC networks, complicating the estimation process even further.

In interconnect delay estimation during signoff, a key trade-off has traditionally been balancing accuracy and runtime efficiency. However, current methods tend to fall at opposite ends of this spectrum. On the one side, highly accurate techniques can be very slow, while faster approaches often compromise on precision. The emergence of machine learning (ML) has revolutionized this landscape, offering promising new avenues for the development of Electronic Design Automation (EDA) techniques [8, 9]. Particularly, the recent deployment of ML techniques into timing estimation has offered a promising middle ground on the trade-off curve [4, 10, 11, 12]. In [11], an approach employing ML was suggested for pre-routing timing prediction with the aim of mitigating pessimism. Another study in a similar category was presented in [13], focusing on timing prediction during the placement stage using ML models. These strategies primarily aimed to bridge the gaps between pre- and post-routing analyses, with the caveat that the chosen features did not incorporate wire topology information, a crucial factor for precise signoff timing estimation. Han et al. [14] introduced a tool using a big-data approach to address the timing correlation problem across different stages using a hierarchy of ML models. Kahng et al. [12] proposed gathering a set of timing models, including Elmore delay and D2M, as features for timing estimation with ML models. Meanwhile, another innovative work by Yang et al. [1] suggested the application of an ML model for fast timing prediction, followed by the rapid generation of ECO to fix remaining timing violations. Additionally, [15] promoted the use of Artificial Neural Networks (ANNs) to correlate initial timing estimates with golden SPICE simulations, enhancing the accuracy of the effective capacitance (C_{eff}) estimation model. In an alternative approach, [4] introduced a novel ML-based wire timing estimator designed for both tree and non-tree net structures. This method extracted topological features directly from RC networks and converted non-tree nets into tree structures. The estimation was performed using the gradient boosting ML model, XGBoost [16]. Similarly, [17] presented a technique that started with a non-Signal Integrity (SI) timing report and estimated additional delay caused by coupling by considering total resistance and capacitance. While these methods showed accurate signoff timing estimation in specific scenarios, they had notable limitations. For example, [16] required extensive feature extraction, involving over ten features and a loop-breaking procedure, which made it cumbersome and impractical for rapid signoff processes. The work [17], on the other hand, depended on external timing estimators and did not fully incorporate the wire topology information introduced during detailed routing. To address these challenges, we propose leveraging graph learning techniques. This approach transforms the net structure into a graph and predicts the RC net delay from output pins to input pins without the need for exhaustive feature extraction or complex preprocessing. By utilizing global message passing graph representation learning directly on RC graphs, our method streamlines the estimation process, offering a more efficient and practical solution for signoff caliber timing estimation.

Graph learning featured by Graph Neural Networks (GNNs) has been explored within the EDA community in recent years, as highlighted in a comprehensive survey by Lopera et al. [18]. In the realm of signoff power estimations, [19] proposed the use of GNNs. In timing prediction, [20] mapped RTL designs to directed graphs with multidimensional node and edge features as inputs for GNN models. Similarly, at the preplacement stage, [10] employed a customized graph attention network to predict net length and timing. Additionally, [21] proposed a GNN-based engine for pre-routing slack prediction. In the studies conducted by Ye et al. [22, 23], a rapid and precise wire timing estimator was introduced. This was achieved through a novel graph learning architecture capable of generating representations for wire paths by aggregating local structural information and global relationships across entire RC nets. Despite the studies showcasing the efficacy of GNNs in predicting timing across various design stages with commendable accuracy, many of them still necessitated a substantial set of features, leading to additional runtime consumption for feature extraction. Furthermore, it is noteworthy that these graph learning-based timing prediction models do not explicitly tackle the challenges associated with cross-corner timing prediction. Comprehensive solutions that efficiently handle feature requirements and explicitly account for cross-corner timing prediction are essential for advancing the robustness and applicability of such models. In this chapter, we begin by introducing a customized graph learning model designed to be highly sensitive to interconnect topology. As shown in Fig. 5.1, this model leverages parasitic information from the extracted SPEF, streamlining the typically tedious feature extraction process and enabling rapid net delay estimation during the signoff phase. Additionally, incorporating preprocessed graph features can further enhance estimation accuracy, with only a slight increase in runtime. To address the imperative of achieving fast and accurate timing signoff across diverse process corners, we then present a comprehensive learning-based framework. This framework seamlessly integrates learning-based reference corner selection and graph learning-based interconnect timing prediction into the broader timing signoff steps and ECO processes. Notably, it only necessitates information about a single known corner while adeptly predicting all unknown corners. Through evaluations conducted on two widely used industry processes, our proposed framework outper-

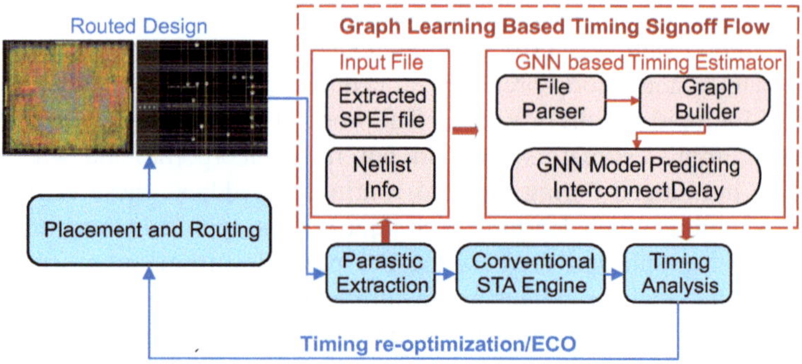

Fig. 5.1 The framework of the conventional timing signoff flow and the proposed graph learning-based timing signoff flow

forms alternative ML models and existing strategies, demonstrating a noteworthy accuracy enhancement and runtime reduction.

The remainder of this chapter is structured as follows: Sect. 5.2 discusses the details of the proposed graph learning-based interconnect delay estimator, specifically designed for single-corner timing prediction. Additionally, in Sect. 5.3, learning-based cross-corner timing prediction framework is introduced. The chapter concludes with a summary and final remarks in Sect. 5.4.

5.2 Graph Learning-Based Model for Fast and Accurate Interconnect Delay Prediction at Single Corner

5.2.1 Interconnects in Advanced Process Nodes

In a routed design, a net comprises wires connecting cell pins, typically originating from one output pin and terminating at multiple input pins. The routing process often divides a net into tree-like segments based on available resources, and each segment is laid out using different metal layers connected by vias. In advanced technologies like FinFETs or gate-all-around (GAA) FETs, intricate layout designs lead to a surge in connections among different interconnects. The increasing complexity of parasitic models and significant coupling effects in advanced technology nodes call for innovative parasitic extraction (PEX) methods for interconnect modeling. These methods must accurately capture both parasitic and layout-dependent effects. This complexity often leads to intricate net topologies, including deep trees, loops, and non-tree segments, each displaying highly variable behaviors. Figure 5.2 illustrates the diverse range of net behaviors and RC values observed from an SPEF file extracted from a large-scale (multimillion gate) industrial design in a sub-10-nm technology node. Table 5.1 summarizes the maximum, minimum, and mean values of four key metrics, highlighting the extensive variability across nets and

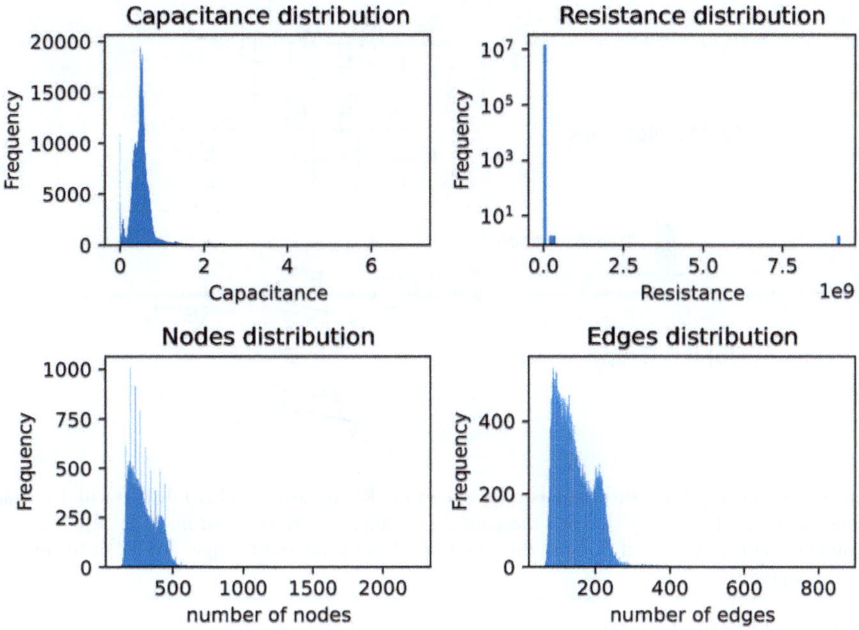

Fig. 5.2 Distributions of parasitic and net topology in a large-scale industrial design with sub-10-nm technology node (node and edge count are obtained based on graph representation of the net shown in Sect. 5.2.2)

Table 5.1 Statistics from an SPEF file in a large-scale industrial design with sub-10-nm technology node

	Max	Min	Mean
Resistance (Ω)	9.35G	0.1 m	1.37k
Capacitance (fF)	7.09	~ 0	0.5
Node count (per net)	860	63	294
Edge count (per net)	2244	126	146

the broad spectrum of RC values. Given this diversity, accurate and efficient interconnect delay estimation necessitates a novel approach to representing complex RC networks, moving beyond traditional closed-form approximations or analytical methods.

5.2.2 *Transformation of RC Networks into Graph Representation*

As shown in Fig. 5.3, the transformation of an RC network into a graph involves assigning each node V to represent a capacitance and each edge E to represent a resistance between two nodes. This process results in the creation of an RC graph denoted as $G = (V, E)$. The features associated with edges and nodes

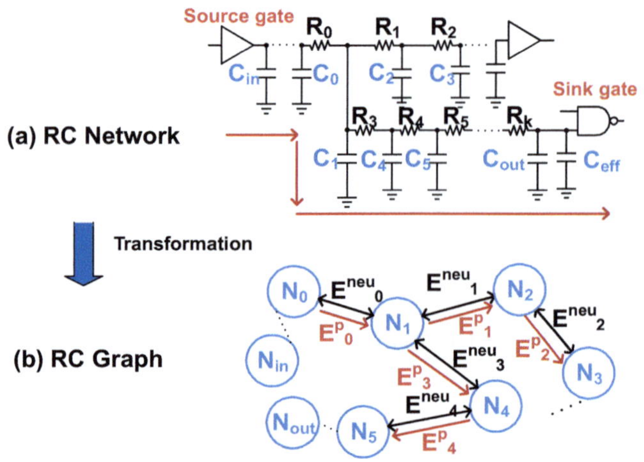

Fig. 5.3 Graph representation of a net. (**a**) shows the RC network, and (**b**) displays its RC graph representation. Here, C_{eff} denotes the gate capacitance of the sink, while N_{out} represents the combined value of C_{out} and C_{eff}. In this graph, E_i^{neu} is the neutrality edge, and E_i^p is the positive edge (adapted from [5])

Table 5.2 List of features for the customized GNN model

Feature type	Name	Size (type)
Node feature	Output node or non-output	1 (Boolean)
	Labeled or non-labeled	1 (Boolean)
	Effective capacitance	1 (float)
	Node load	1 (float)
Edge feature	Resistance (positive)	1 (float)
	Resistance (negative)	1 (float)
	Resistance (neutral)	1 (float)

are detailed in Table 5.2. The node driving fanout gates for a net is termed the "output node," and the gate capacitance from the fanout cells is defined as the "node load" for the output node. Output nodes that possess golden timing data are labeled for training and testing purposes. In a heterogeneous graph, connections between nodes involve different edge types, contrasting with the single edge type found in homogeneous graphs. Recent studies have shown that heterogeneous graphs can generate meaningful vector representations for each node, which are beneficial for various downstream applications. Additionally, leveraging a heterogeneous graph allows for the implementation of different GNN layers tailored to various edge types [24]. In our RC network representation, we utilize a heterogeneous graph that includes three edge types: positive edge, negative edge, and neutral edge. The positive edge directs from a source node to an output node, while the neutrality edge is undirected. Conversely, the negative edge runs in the opposite direction to the positive edge. Notably, all three edge types correspond to the same resistance value.

5.2.3 Customized GNN Model for Interconnect Delay Prediction

In this section, we explore a customized graph learning model[1] designed for efficient and accurate signoff timing prediction. Recognizing that the regression performance tends to degrade with an increasing number of GNN layers due to the emergence of "over-smoothing," we address this issue by eliminating all supervised layers in the anterior layers. Instead, we reserve multilayer perceptrons (MLPs) for the final layers to avoid an excessive number of supervised weight parameters that could exacerbate over-smoothing. Another challenge arises from the general computational expense associated with GNN layers. To mitigate this, we carefully control the graph traversal order and the scope of message propagation at each iteration, ensuring a balanced trade-off between model depth and runtime efficiency. In this section, we delve into the details of the proposed model, outlining the strategies employed to address the issues of over-smoothing and computational efficiency.

In our experiments, we explore three primary types of prediction tasks on graphs as outlined by Wang et al. [25]: graph-, edge-, and node-level tasks. A graph-level task involves predicting a single property for the entire graph. Edge-level tasks focus on predicting properties or the presence of edges, while node-level tasks explore properties for each node within the graph. In our scenario of predicting delays, we have observed that the node-level regression task demonstrates the most promising performance for predicting RC graph and node delays.

The Elmore delay model [6] serves as a well-known method for estimating interconnect delay from a source node to a leaf node. It achieves this by summing the resistance along the path to the ith node, multiplied by the capacitance at the branch's end. Drawing inspiration from this model, we determine that the output node must encapsulate comprehensive graph information for effective node regression tasks. This necessity arises because delay calculations require RC values across entire branches, including those extending to other nodes. While larger RC values do not necessarily translate to significant delay components unless they contribute substantially to the delay calculation as a product [4], the overall RC product holds more significance than individual values. Therefore, in a GNN model, the receptive field—the span of influence—should include more hops than the number of node levels between the source and output nodes. This approach ensures effective capture of these critical relationships within the graph structure.

Figure 5.4 illustrates the overall structure of our specialized graph learning model designed for estimating delays based on RC networks at a single corner. Detailed explanations of the model notations can be found in Table 5.3. The model consists of $2N_g + 2$ GNN layers and 5 MLP layers. Each GNN layer performs a message passing function across four types: negative passing, neutral passing,

[1] An earlier version of this model was introduced in our previous work [5].

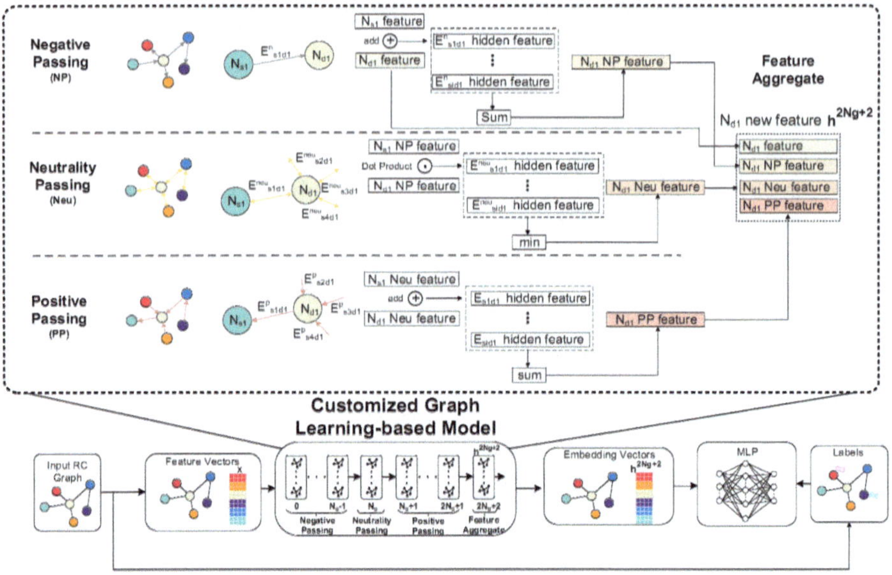

Fig. 5.4 The customized GNN model for predicting interconnect delay at single corner (modified based on [5])

Table 5.3 Customized GNN model notations

Notation	Description
$E^{n,neu,p}$	Edge for Negative, Neutrality, and Positive edge types
h^{2N_g+2}	Node embedding in layer $2N_g + 2$
N_d	Destination node
N_g	Count of non-output nodes
N_s	Source node
Negative passing	Message passing function for edge type Negative
Neutrality passing	Message passing function for edge type Neutrality
Positive passing	Message passing function for edge type Positive

positive passing, and feature aggregation. In negative passing, node messages are updated through addition and reduced via summation, propagating along negative edge types. Neutral passing updates node messages using dot product operations and reduces them through minimum operations along neutrality edges. Positive passing updates node messages through addition and reduces them via summation along positive edges. The feature aggregation layer concatenates node embedding features, which are then processed through a five-layer fully connected neural network. The node embeddings produced by the final layer are used to predict the net delay from inputs to outputs. To manage computational efficiency, message passing is selectively applied to specific parts of the graph during each iteration.

5.2.4 Augmentation with Additional Feature

In the context of final signoff, the accuracy of delay estimation holds paramount importance. Previous research has demonstrated that incorporating additional features in machine learning models can enhance accuracy. Drawing inspiration from this, we augment our model by introducing pre-calculated Elmore delay as an additional feature. Elmore delay is a key feature also employed by state-of-the-art machine learning-based interconnect delay estimators [4]. While the calculation of Elmore delay is not straightforward for complex net structures involving loops, a loop-breaking algorithm is typically applied [4], introducing additional runtime for preprocessing. Despite the preprocessing runtime, leveraging Elmore delay simplifies the introduced graph learning model. Since the Elmore model already captures global graph information, our model is streamlined to only include eight GNN layers with a receptive field limited to the 8-hop neighborhood, embedding local information exclusively. This reduction in complexity maintains the accuracy of the model while enhancing computational efficiency.

5.2.5 Evaluation Results

5.2.5.1 Experimental Setup

Our models are implemented using PyTorch and DGL [25], a graph learning framework. The experiments were conducted on a Linux machine equipped with a 3.7GHz AMD 5900X CPU with 128 GB RAM. The training and test dataset is constructed from an extracted SPEF file of a routed multimillion gate high-speed industrial design in a sub-10-nm technology node.[2] Labels are obtained from the internal signoff timing analysis tool, resulting in a total of 104.65k labeled output nodes. The dataset is divided into training and testing data using an 8:2 ratio with a random split. To ensure fair comparisons, we also trained the dataset using the XGBoost model [16], a gradient boosting machine learning model widely used in state-of-the-art signoff interconnect delay estimators [4]. Four sets of experiments were conducted:

1. **RC-GNN**: the proposed graph learning GNN model without Elmore delay as an additional feature
2. **RC-GNN (acc)**: an accuracy-enhanced RC-GNN model with additional features as described in Sect. 5.2.4
3. **XGBoost**: the XGBoost model with the same feature set as in the first experiment
4. **XGBoost (acc)**: the XGBoost model with the same feature set as in the second experiment

[2] In adherence to confidentiality protocols, the design details cannot be revealed.

Accuracy and runtime metrics for these four models are collected and compared to evaluate their performance in interconnect delay estimation.

5.2.5.2 Accuracy Comparison

In Fig. 5.5, the correlation between predicted delay values (Y axis) and golden values (X axis) is shown for all four experiments. The detailed comparisons are also summarized in Table 5.4. Results consistently show that RC-GNN outperforms XGBoost across various metrics, achieving a notable R^2 correlation of 0.994 with a reduced feature set. This superiority is evident in the correlation plots, where most data points align closely with the ideal correlation line for RC-GNN. Although RC-GNN exhibits slightly higher Max Error, this is primarily due to outlier nets with exceptionally high resistance that require specialized handling. Further analysis reveals that the mean error within the top 20% errors in RC-GNN

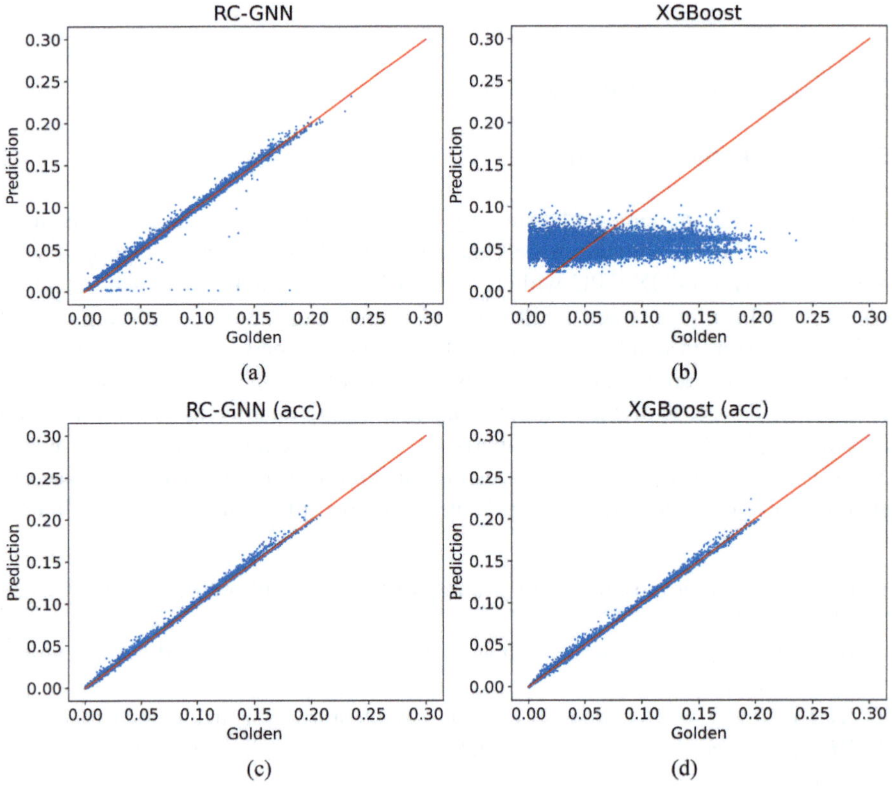

Fig. 5.5 Predicted interconnect delay based on various models versus golden values from signoff tools (the unit of the delay is in *ps*) [5]. (**a**) RC-GNN model with reduced feature set. (**b**) XGBoost model (baseline) with reduced feature set. (**c**) RC-GNN model with complete features. (**d**) XGBoost model (baseline) with complete features

Table 5.4 Comparison of our graph learning models with the recently proposed machine learning-based approach (XGBoost) for interconnect delay prediction in signoff (adopted from [5])

	RC-GNN	RC-GNN (acc)	XGBoost [4]	XGBoost (acc) [4]
Total data size	105K	104.651K	105K	105K
Train data size	84K	84K	84K	84K
Test data size	21K	21K	21K	21K
Mean absolute error	**1.55 ps**	**0.93 ps**	33.85 ps	0.96 ps
R^2	**0.99**	1.00	0.02	1.00
Mean squared error	**1.1e–05**	2.6e–6	1.8e–3	2.6e–6
Explained variance score	**0.99**	1.00	0.02	1.00
Median absolute error	**1.05 ps**	0.60 ps	28.21 ps	**0.59 ps**
Max Error	180.0 ps	**20.7 ps**	**175.6 ps**	27.6 ps
Top max 20% mean error	**4.0 ps**	2.8 ps	74.1 ps	2.8 ps

remains significantly lower than that of XGBoost, maintaining accuracy within a few *ps*.

Introducing elmore delay as an additional feature enhances RC-GNN, demonstrating comparable or improved accuracy across all metrics. Particularly, significant reductions in Max Error are observed while maintaining high model fit. Overall, RC-GNN (acc) emerges as the most effective model among the four. Both graph learning-based models consistently exhibit superior accuracy compared to the baseline XGBoost approach.

5.2.5.3 Error Analysis

To gain a detailed insight into the distribution of errors within our proposed graph learning-based model RC-GNN, we analyze the distributions of absolute error and error percentages depicted in Fig. 5.6. Red lines in the figure indicate mean errors among the largest 20% delay values, serving as reference points. The analysis reveals an uneven distribution of estimation errors across the dataset. Longer nets with larger delay values exhibit fewer errors, addressing significant influence on timing critical paths, which are crucial for closing timing in signoff. Conversely, shorter nets show higher absolute error ratios, often featuring larger errors but are less critical in the timing closure process. These observations boost confidence in the model's predictive capabilities.

5.2.5.4 Runtime Comparison

In addition to accuracy, the efficiency measured by runtime is crucial for evaluating a timing estimation model. Runtime consists of two components for each model: The first component involves building the graph and preprocessing the features, while the second component pertains to the actual inference time. Figure 5.7 provides the

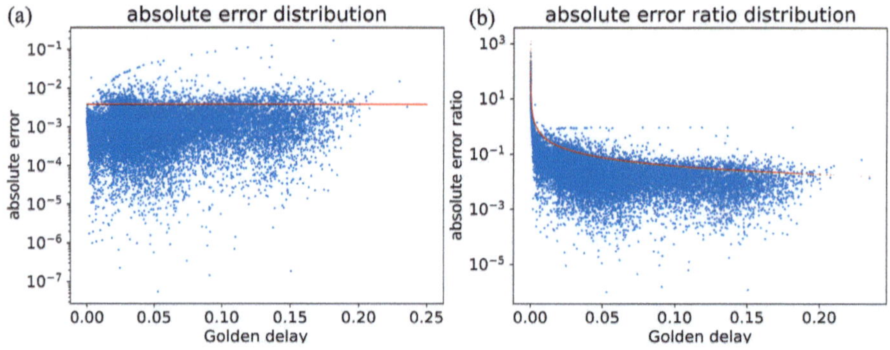

Fig. 5.6 The error distribution of the proposed graph learning-based model, with red lines indicating the mean error for the top 20% of delay values (adopted from [5]). (**a**) The distribution of absolute errors (the deviation of the estimated value from the reference or golden value). (**b**) The distribution of absolute error percentages (the ratio of the absolute error to the reference or golden value)

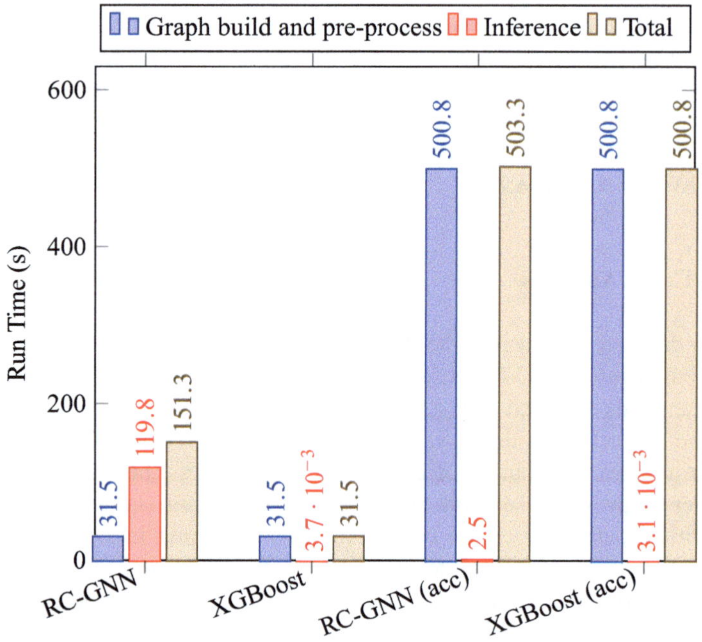

Fig. 5.7 Runtime breakdown of the proposed graph learning-based interconnection timing prediction models and the baseline models

runtime breakdown for all four experiments. Using the same feature set as inputs, RC-GNN has slower inference compared to the XGBoost counterpart because our GNN model is deeper and involves global message passing. It is important to note

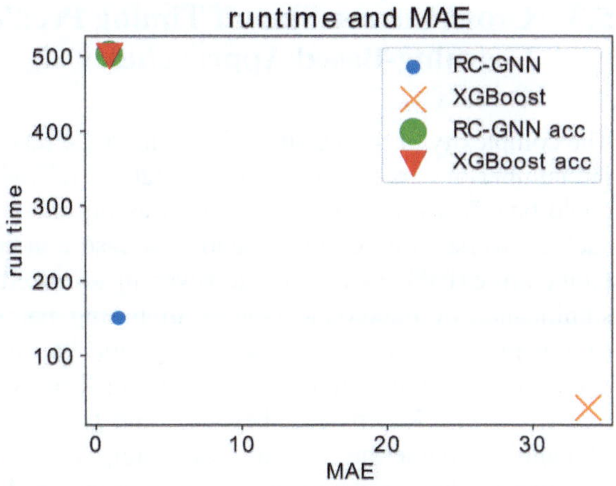

Fig. 5.8 Estimation accuracy (MAE) and total runtime (in s) for various models discussed in this chapter (adopted from [5])

that all steps were executed using CPUs only for now, and further acceleration of graph learning models can be achieved by leveraging GPUs.

While the XGBoost leads to the lowest runtime, its accuracy is significantly poorer, as indicated in Table 5.4. To comprehensively evaluate the efficiency and accuracy of the proposed models, we present a plot of Mean Absolute Error (MAE) against runtime in Fig. 5.8. Compared to the XGBoost (acc) model, the RC-GNN model achieves nearly identical accuracy levels but operates approximately four times faster. Although the RC-GNN (acc) model offers slightly improved accuracy, it involves additional runtime overhead due to feature extraction and processing. In conclusion, for most signoff scenarios, a lightweight version of the graph learning model is preferable, effectively balancing accuracy and efficiency.

5.2.6 Summary

Up to this point, we have explored a customized approach based on graph learning for efficient and accurate prediction of signoff interconnect delays at a single RC corner. By leveraging a transformed RC graph and a tailored graph learning model, we emphasize the suitability of graph learning for this task, achieving an enhanced balance between accuracy and runtime. Evaluating the proposed model using an industrial design in an advanced node, we demonstrate its superiority over the recently proposed XGBoost-based interconnect delay estimator in runtime while maintaining comparable accuracy levels. Importantly, this model eliminates the need for time-consuming feature extraction and feature processing steps. As technology nodes progress, the proliferation of process corners poses a significant challenge in timing prediction during signoff. In the next section, we will discuss how a graph learning-based model can predict cross-corner interconnect timing with limited information from a single corner.

5.3 Cross-Corner Signoff Timing Prediction with Learning-Based Approaches

The complexity of timing signoff is primarily a result of the exponential increase in the number of corners and scenarios [26, 27]. Traditionally, the operational space could be effectively constrained by assessing timing at a select few design points, such as worst-case conditions and best-case conditions for process, voltage, and temperature (PVT) variations. However, in advanced technology nodes, the growing significance of interconnect delay in timing paths, coupled with the shrinking routing pitch and the impact of coupling capacitance, has elevated the role of process variations in metallization on overall timing. The RC extraction corners now include worst C, worst RC, best C, best RC, and typical, exacerbating critical challenge of handling numerous corners. Checking timing for over hundreds of corners or scenarios has become time-consuming, given the limitations of computational resources. Furthermore, when timing failures occur in specific corners or scenarios, designers must address these cross-corner or cross-mode violations through labor-intensive ECO iterations before the tape-out phase. Each iteration typically involves a distinct extraction run for each RC corner and a separate analysis run for each corner/mode combination in STA. In complex designs, the proliferation of hundreds or thousands of ECOs to address various design issues leads to numerous iterations for extraction and timing analysis. Building upon the promising results showcased in Sect. 5.2, in this section, we delve further into the feasibility of employing a graph learning-based model[3] for predicting signoff timing across process corners, with a particular focus on RC corners.

5.3.1 Prior Cross-Corner Signoff Timing Prediction Approaches

Several studies have focused on cross-corner delay estimation, employing observed RC corners to predict unknown corners and enhance signoff efficiency. In [29], a method was introduced for swift and precise hold timing analysis across all corners. The objective was to determine the worst-case hold slack at all corners using a limited set of full STA runs. Similarly, [30] proposed a branch-and-bound method to pinpoint the worst-case process corner for chip signoff timing analysis. Likewise, [31] expedited the derivation of worst-case timing results compared to a conventional multi-corner STA by estimating the worst-case timing across various process corners. This allowed for evaluating the upper bound of critical path timing through a single-pass analysis instead of running STA for all corners. However, in advanced technology nodes, finding a strictly tight upper

[3] An earlier version of this model was introduced in our work [28].

bound that avoids overdesign can pose challenges. In [32], a data-driven approach based on multivariate linear regression was explored to predict timing analysis at unobserved corners using results from observed corners. The approach employed a backward stepwise selection strategy to determine which corners to observe and which to predict. The underlying idea was that timing results for all corners exist within a low-dimensional ambient space and can be reconstructed using results from observed corners, capitalizing on correlations across all corners. Nevertheless, this approach demanded multiple observed corners for relatively good coverage, rendering it time-consuming. The selection of observed corners necessitated careful consideration, with no guarantee of universality for all designs, prompting reselection for new designs, as the chosen policy impacts accuracy. In [33], under a similar assumption, the authors introduced nonlinear regression models for STA prediction with improved accuracy. They proposed using a statistical analysis tool to select key STA results as model inputs, thereby saving additional STA runs. While maintaining high predictive accuracy, this approach proved less practical for signing off numerous RC corners. In [34], a novel nonlinear machine learning-based timing prediction model was introduced, considering all iterative STA optimizations in a unique space and leveraging their close correlations. An interesting observation from this work was that path delays at different corners are strongly correlated, with correlations dependent on the topology and structure of the path. However, it is worth noting that they only included coarse-grained path features and did not delve into net structures or RC corners.

In summary, previous research on cross-corner timing prediction often necessitated multiple corners to be designated as "known" or "observed" corners to estimate timing for the remaining "unknown" corners. This approach involved running signoff tools multiple times to generate reference data for the known corners, leading to additional runtime overhead. Moreover, previous efforts predominantly concentrated on enhancing the accuracy and efficiency of the models themselves, with less emphasis on addressing the challenges associated with integrating these models into a standard STA flow. In this section, we present a learning-based framework for cross-corner timing signoff that adopts a "one-for-all" approach, utilizing only one known corner to predict interconnect delays across the rest. The primary features of this framework are listed below:

1. The introduced framework encompasses two distinct interconnect delay estimation approaches: One prioritizes precision, while the other focuses on delivering rapid results. This dual approach enables flexible accuracy and efficiency tradeoffs, catering to different design scenarios.
2. To adjust to diverse designs, the framework uses a novel learning-based method to identify the most informative known corner, even with limited data. This optimizes delay prediction for each unique design.
3. Prioritizing smooth integration with signoff and timing ECO workflows, the framework ensures a streamlined and efficient process. Its effectiveness across two industry-standard technology nodes confirms not only superior accuracy and speed but also seamless integration, outperforming existing methods.

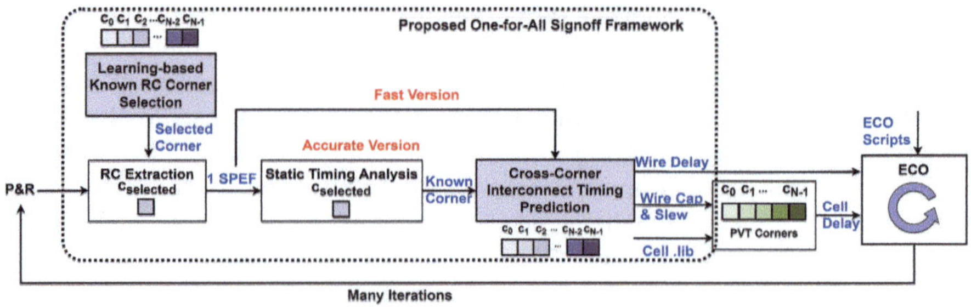

Fig. 5.9 Overview of the learning-based cross-corner signoff timing prediction framework (adopted from [28])

5.3.2 Overview of Learning-Based Cross-Corner Signoff Timing Framework

Figure 5.9 illustrates the proposed learning-based framework, which comprises two main components highlighted in purple. These components are a learning-based RC corner selection strategy and a cross-corner interconnect timing prediction model. The RC corner selection strategy intelligently identifies and selects the most suitable known corner for a specific design, enabling targeted analysis. On the other hand, the interconnect timing prediction model accurately predicts timing based solely on the selected corner.

To perform precise timing analysis, the process begins by extracting an RC network and generating an SPEF file from the identified corner (denoted as "Accurate Version"). Subsequently, STA is performed to determine the timing details specific to the selected corner. Alternatively, a fast version of the process (denoted as "Fast Version") bypasses STA and directly utilizes the SPEF file to quickly approximate timing data. A significant advantage of this approach compared to existing signoff methodologies is the elimination of laborious parasitic extraction and timing analysis for all RC corners. These tasks typically demand substantial runtime and computing resources. By avoiding these resource-intensive steps, the proposed framework accelerates the time-consuming ECO iteration cycle.

5.3.3 Learning-Based Known RC Corner Selection

Previous approaches, such as [32], have typically employed a greedy-based search policy to traverse all corners and select known corners, which are then fixed for all designs. However, this fixed selection may not be optimal for different designs, necessitating the reselection of known corners. To overcome this limitation, our proposed approach introduces a learning-based known corner prediction strategy based on a Convolutional Neural Network (CNN) classifier, as depicted in Fig. 5.10.

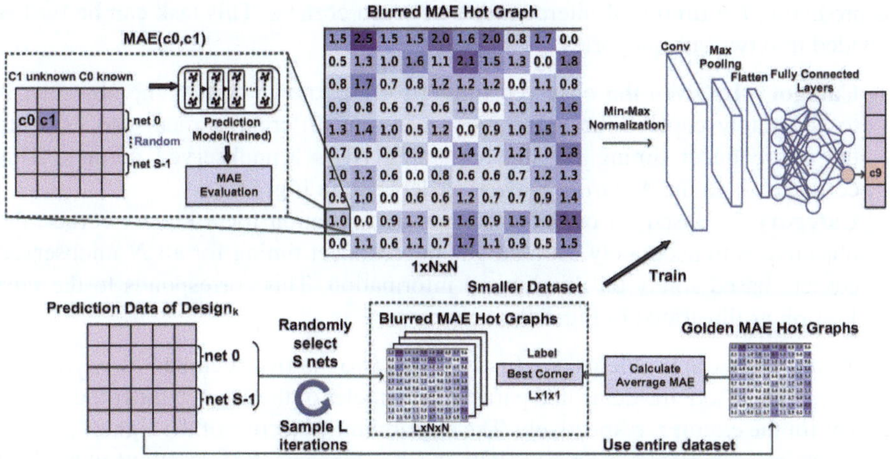

Fig. 5.10 Learning-based known corner prediction strategy that employs a CNN classifier

For a new design, a small set of S nets is randomly sampled to undergo the actual signoff process. Typically, S ranges from 100 to 500 in our experiments. These nets are evaluated by calculating the Mean Absolute Error (MAE) between the predicted cross-corner timings and the golden timing values. Specifically, the MAE($c0$, $c1$) represents the MAE for the predicted timing value of the unknown corner $c1$ using the known corner $c0$. Since the S nets can be considered as a small batch from a larger dataset, the MAE varies across different samples, resulting in a $1 \times N \times N$ "blurred" MAE hot graph denoted as $MAE(c_i, c_j)_{0\leq i<N}^{0\leq j<N}$. This serves as one of the input features for the CNN model. The CNN model consists of convolution layers, pooling layers, and fully connected layers. It takes the MAE hot graph as input to classify the best corner for the given design. In terms of training the CNN model, the best predictive corner is determined for each design from the entire prediction dataset and saved as the labels. This process is repeated for each design for a total of L times, resulting in $L \times N \times N$ labeled graphs. By formulating the corner selection process as a classification problem, we introduce an adaptive solution for selecting the best-known corner for new designs and process nodes. This approach allows for the optimal selection of a known corner based on the specific characteristics of each design, enhancing the accuracy and efficiency of the overall framework.

5.3.4 Graph Learning-Based Cross-Corner Interconnect Timing Prediction

In this section, we firstly assume that the sole known corner is given based on the prior discussed selection methodology, and then the primary challenge remaining is predicting the timing of interconnects between corners. This task can be further divided into two subcategories:

- **Category 1**: Given the observed timing for the known corner, the objective is to accurately capture the correlations between different corners and estimate the interconnect timing for the remaining $N - 1$ unobserved corners. This corresponds to the *Accurate Version* as depicted in Fig. 5.9.
- **Category 2**: Based on only the parasitic information for a known corner, the objective is to accurately estimate the interconnect timing for all N unobserved corners based solely on this limited information. This corresponds to the *Fast Version* as illustrated in Fig. 5.9.

To resolve these problems, we firstly define two matrices denoted as $\mathbf{y}_{i,j}(z)$ and $\mathbf{p}_{i,j}(z)$, which store the delay and parasitic parameter of the jth net under the ith RC corner for the circuit z, respectively. The overarching objective of the signoff process is to obtain a complete matrix $\mathbf{Y}(z)$ with the assistance of $\mathbf{P}(z)$. Previous work [32] made a hypothesis that many columns of matrix $\mathbf{Y}(z)$ are linearly dependent. Linear regression model is thus employed:

$$f_w(\mathbf{Y}(z)^{i\in\mathcal{I}}) = w^\top \mathbf{Y}(z)^{i\in\hat{\mathcal{I}}}$$

where $\mathcal{I} = \{observed\ l\ corners\}$ and $\hat{\mathcal{I}} = \{unobserved\ N - l\ corners\}$. Given data $\mathcal{D}_M = \{\mathbf{Y}(z)^{i\in\mathcal{I}}, \mathbf{Y}(z)^{i\in\hat{\mathcal{I}}}\}$, where $\mathbf{Y}(z)^{i\in\mathcal{I}} \in \mathbb{R}^{M\times l}$, and $\mathbf{Y}(z)^{i\in\hat{\mathcal{I}}} \in \mathbb{R}^{M\times(N-l)}$, the model is then trained:

$$w_t \leftarrow w_{t-1} - \mathbf{Y}(z)^{i\in\mathcal{I}^\top}\left(\mathbf{Y}(z)^{i\in\mathcal{I}} w - \mathbf{Y}(z)^{i\in\hat{\mathcal{I}}}\right)$$

In their case, the input data dimension is $M \times l$, necessitating signoff timing data from l observed corners. However, there is a desire to minimize the number of observed corners in order to save runtime. The selection of these observed corners is crucial for achieving accurate predictions. The selection of these l corners is crucial for prediction accuracy. Furthermore, the assumption that the selection of these corners remains constant across designs overlooks the possibility that new designs may introduce new RC typologies. $\mathbf{P}(z)$ encompasses unique interconnect topology features and is indispensable for timing prediction. The proposed cross-corner interconnect timing prediction strategy is illustrated in Fig. 5.11. Firstly, we introduce a Graph Mapping (GM) function which maps the original Euclidean matrix $\mathbf{P}(z)$ to graph matrix $\mathbf{P}^g(z)$.

$$\mathcal{GM}(\cdot) : \mathbf{P}(z) \to \mathbf{P}^g(z)$$

The mapping and dataset generation process is detailed in Algorithm 1. This mapping distills complex wire topology into higher dimensional feature space. Each node $v \in \mathcal{V}_{i,j}$ is paired with their node attributes $a(v)$.

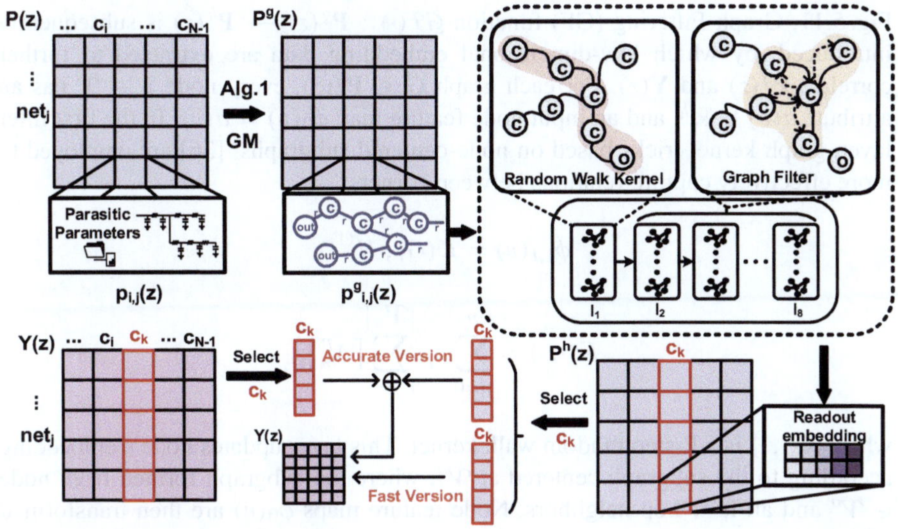

Fig. 5.11 The proposed cross-corner interconnect timing prediction strategy (accurate version and fast version) (adopted from [28])

Algorithm 1: Graph mapping

Input: Matrix $\mathbf{P}(z)$;
Output: Matrix $\mathbf{P}^g(z)$;
for $\mathbf{p}_{i,j}$ *in* $\mathbf{P}(z)$ **do**
  Assign capacitance as c-node set $\mathcal{V}^c_{i,j}$ from $\mathbf{p}_{i,j}$;
  Assign output nodes as o-node set $\mathcal{V}^o_{i,j}$;
  Assign resistance as edge set $\mathcal{E}_{i,j}$ from $\mathbf{p}_{i,j}$;
  $G(\mathcal{V}_{i,j}, \mathcal{E}_{i,j})$ with adjacency matrix $\mathcal{A}_{i,j} \in \mathbb{R}^{|\mathcal{V}_{i,j}| \times |\mathcal{V}_{i,j}|}$;
  Capacitance value, Elmore delay $\sum_k R_{ks} C_k \rightarrow$ node attribute $a(v)$;
  while $|\mathcal{V}_{i,j}| < |\mathcal{E}_{i,j}| + 1$ **do**
    $G(\mathcal{V}_{i,j}, \mathcal{E}_{i,j}) \xrightarrow{\textit{breaking loop } [4]} \hat{G}(\hat{\mathcal{V}}_{i,j}, \hat{\mathcal{E}}_{i,j})$
  end while
  $\mathbf{p}^g_{i,j} = \hat{G}(\hat{\mathcal{V}}_{i,j}, \hat{\mathcal{E}}_{i,j})$;
end for
return Matrix $\mathbf{P}^g(z)$

For each graph $G \in \mathbf{P}^g(z)$, G is considered as heterogeneous graph, where nodes do not uniformly influence each other. Output nodes $\mathcal{V}^o$ are designated as o-nodes, and all capacitance is assigned to c-nodes. Due to the homophily assumption of GNNs [35], heterophilous graphs with nodes having different labels tend to be weakly learned in common GNNs, such as Graph Convolutional Networks (GCNs). By bypassing GNN layers and incorporating the concept of convolution kernel filters from Convolutional Neural Networks, graph filters without trainable weights are used to extract local structural information around each node. As shown in

Fig. 5.11, Graph Filtering (GF) function $\mathcal{GF}(\cdot) : \mathbf{P}^g(z) \rightarrow \mathbf{P}^h(z)$ is subsequently introduced, by which low-dimensional embedding data are extracted to further correlate $\mathbf{P}^g(z)$ and $\mathbf{Y}(z)$. For each graph $G \in \mathbf{P}^g(z)$, every node $v \in \mathcal{V}$ has an attribute $a(v) \in \mathbb{R}^{d_0}$ and an input node feature map $\phi_0(v) = a(v)$. In the first filter layer, graph kernel tricks based on node-centered subgraphs [36] are employed to more effectively capture relations between corners:

$$\phi_{1,i}(v) = K(G_v, H_i^{(1)})$$

$$K(\cdot,\cdot) = \sum_{p=0}^{P} \lambda_p \sum_{i,j=1}^{|\mathcal{V}_\times|} [A_\times^p]_{ij}$$

where $K(\cdot,\cdot)$ is a P-step random walk kernel. This layer updates node's embedding according to the subgraph centered at $\mathcal{V}^o$, where the subgraph formed from node $\in \mathcal{V}^o$ and all its 1-hop neighbors. Node feature maps $\phi_0(v)$ are then transformed into $\phi_1(v)$ in a way that aggregates the local information of neighbors to the central node $\mathcal{V}^o$. Message passing-based graph filters are then employed in the subsequent layers, whose node feature map is defined as

$$\phi_l(v) = combine\Big(\phi_{l-1}(v), aggregate\Big(\phi_{l-1}(u)\Big)\Big), u \in Neighborhood(v)$$

Then graph-level readout layer is deployed to generate the embedding for the entire graph. Representations from all layers are concatenated:

$$p_{i,j}^h = concat\big(\sum_{v \in \mathcal{V}^o} \phi_l(v) \big| l = 0, 1, \ldots, L\big)$$

In the fast version strategy, a regression model is used to learn the correlation between $\mathbf{P}{c_k}^h(z)$ (the information for a known corner) and the columns of $\mathbf{Y}^h(z)$ (interconnect timing for all corners). This regression model helps estimate the interconnect timing for all N unobserved corners based solely on the limited information available for the known corner. In the accurate version strategy, the correlation is learned between the concatenated information $\mathbf{P}{c_k}^h(z)$ and $\mathbf{Y}_{c_k}(z)$ (interconnect timing for the known corner c_k) and the columns of $\mathbf{Y}^h(z)$. This concatenated information provides more comprehensive input for the regression model, allowing for a more accurate estimation of the interconnect timing for the remaining $N-1$ unobserved corners. By leveraging regression models and learning the correlations between different variables, both strategies aim to improve the accuracy of interconnect timing estimation by effectively capturing the relationships between known corners and the overall interconnect timing.

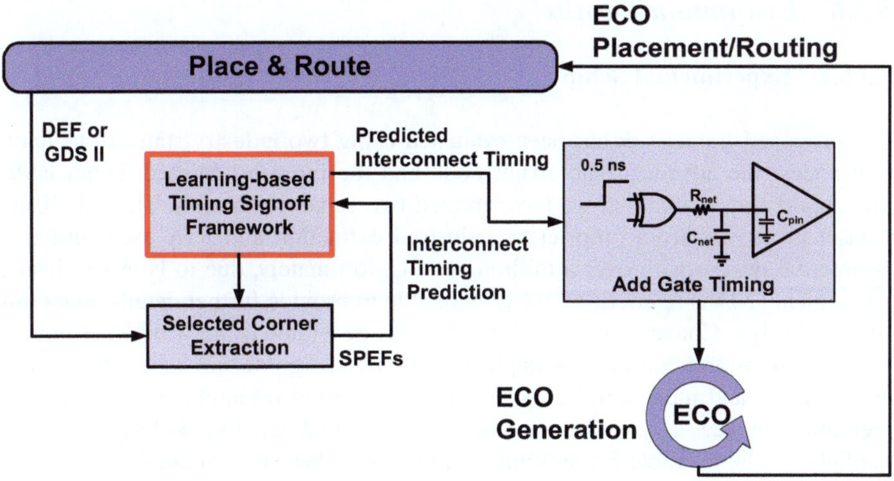

Fig. 5.12 The integration of the proposed learning-based timing signoff steps with ECO process, including ECO initiation and ECO iterations

5.3.5 Integration with ECO Flow

In contrast to previous approaches that overlooked the integration and compatibility of their frameworks with the current design flow, we have proactively addressed this concern in our approach. We have ensured a smooth incorporation of the proposed framework into a conventional ECO flow, leveraging existing commercial implementation tools. The flow is shown in Fig. 5.12. To achieve this integration, our approach includes a carefully designed interface that allows for seamless integration with state-of-the-art signoff tools. This interface enables the cross-corner interconnect timing information to be seamlessly incorporated into the overall design flow. Once the cross-corner interconnect timing information is integrated, the framework gathers gate timing information and generates comprehensive timing reports for all corners. These timing reports serve as the foundation for employing user-defined or automatically generated ECO scripts. These ECO scripts are executed using implementation tools to facilitate incremental place and route, a crucial step in addressing timing issues. Typically, the ECO process involves multiple iterations, ranging from ten to over hundreds, before achieving the final timing closure. The number of iterations can vary depending on the complexity of the design. Throughout these iterations, the comprehensive approach of our framework ensures that it seamlessly aligns with the existing design flow and workflow. By considering the integration and compatibility aspects with the current design flow, our approach provides a practical and efficient solution for addressing timing issues, allowing for smoother and more effective implementation of engineering changes while minimizing disruptions to the overall design process.

5.3.6 *Evaluation Results*

5.3.6.1 Experimental Setup

Our proposed framework has been evaluated using two industry-standard technology nodes: the advanced sub-10 nm node and the more established 28 nm node. To support this evaluation, we have created two separate datasets. The sub-10 nm dataset is derived from proprietary industrial extraction and STA tools, utilizing an internal design with over a million nets. Unfortunately, due to Non-Disclosure Agreement (NDA) restrictions, we are not able to provide further details about this specific design. Conversely, the 28 nm dataset is generated using widely used open-source designs and commercial implementation and signoff tool suites. In order to ensure a fair and meaningful comparison with existing research, such as the ones mentioned in [22, 23], we have included the identical benchmark designs that are publicly available. Table 5.5 presents a summary of these benchmarks.

The framework is implemented using PyTorch and the Deep Graph Library (DGL) [25], a popular toolkit for graph-based learning. The prediction model is both trained and evaluated on a Linux system featuring a 3.7 GHz AMD 5900X CPU and 128 GB of RAM. The datasets are divided into training and testing sets for each design. Evaluation is performed using Mean Absolute Error (MAE) against signoff tools for prediction accuracy. In both technology nodes, the number of RC corners is 10, denoted as $[c_0, \ldots, c_9]$. Runtime metrics are also recorded for each process.

The selection of baseline models for the comparison of cross-corner prediction accuracy was conducted with careful consideration. The study by Kahng et al. [32] is chosen to represent the state of the art in cross-corner signoff interconnect delay prediction, utilizing a linear regression (LR) model. Moreover, [4] introduces one of the most recent learning-based models for signoff interconnect delay prediction in a single corner, employing the powerful XGBoost model. In addition to reproducing these two prior works, our framework incorporates other regression or machine learning models, including Support Vector Machine (SVM), Random Forest (RF), and Multi-Layer Perceptron (MLP) models. By including these diverse models, we aim to provide a comprehensive comparison that evaluates the effectiveness of our introduced framework in cross-corner interconnect timing predictions.

Table 5.5 Benchmark statistics

Process	Benchmark	#Cells	#Nets	#FFs
28 nm	B19	13,863	15,394	1980
	PCI_BRIDGE	8518	9269	3337
	DES_PERT	1506	1634	192
	AES-128	5363	5432	935
	SALSA	11,292	12,749	1674
	OPENGFX	8220	8534	1584
sub-10 nm	In-house design (confidential)	Over 1 million nets		

5.3.6.2 Interconnect Delay Prediction Accuracy in Sub-10 nm

In Fig. 5.13, we present a heatmap visualization of the MAE for predicted interconnect delay in the sub-10 nm node, considering one known corner and predicting for all possible corners. The X-coordinate (c_x) represents the selected known corner (c_0–c_9), while the Y-coordinates (c_y) denote the predicted corners. Figure 5.13a through f displays the results when using signoff data from the known corner, while Fig. 5.13h illustrates the fast version, which predicts results based solely on the SPEF of the known corner. Additionally, we compare the proposed accurate version against the previous best results across all models, as shown in Fig. 5.13g. In this comparison, "0×" indicates unchanged accuracy, while "n×" indicates that the accurate version is n times more accurate than the previous best result under the same condition. The comparison demonstrates that our proposed accurate model outperforms all previous approaches, achieving a maximum of 7× accuracy improvement. For the proposed fast version, the MAE remains close to the previous best values in most cases, indicating its ability to swiftly predict interconnect delay with acceptable accuracy levels. The cross-corner interconnect delay prediction results in the sub-10 nm node are presented in Table 5.6, which includes the mean, minimum, and MAE values when using c_i as the known corner. Notably, the accurate version (acc) outperforms the other two strategies, while the fast version (fa) achieves an acceptable level of accuracy compared to the previous best (pb) version.

Prior methodologies, as demonstrated in [32], typically require more than one known corner. To assess the effectiveness of our approach, we conducted experiments to determine the minimum number of known corners required among all previous approaches to achieve a similar MAE level as our proposed single-corner prediction strategy using c_x as the known corner. The results are summarized in Table 5.7, where each MAE value is averaged among all predicted values in the unknown corners. The findings reveal that an average of five corners is needed for the previous best methods to reach a similar MAE level achieved by our proposed accurate version using only one corner. This implies a significant reduction in both runtime and computational resources required for collecting reference data while achieving improved accuracy.

5.3.6.3 Interconnect Delay Prediction Accuracy in 28 nm

In a similar evaluation methodology to the sub-10nm node, we assessed our proposed framework in the 28 nm node and present the results in Table 5.8. The table includes the average MAE values across all unobserved corners, as well as the runtime for each benchmark design. The findings demonstrate a consistent trend compared to the sub-10-nm node evaluation. The accurate version outperforms previous best strategies by orders of magnitude, with only a marginal increase in

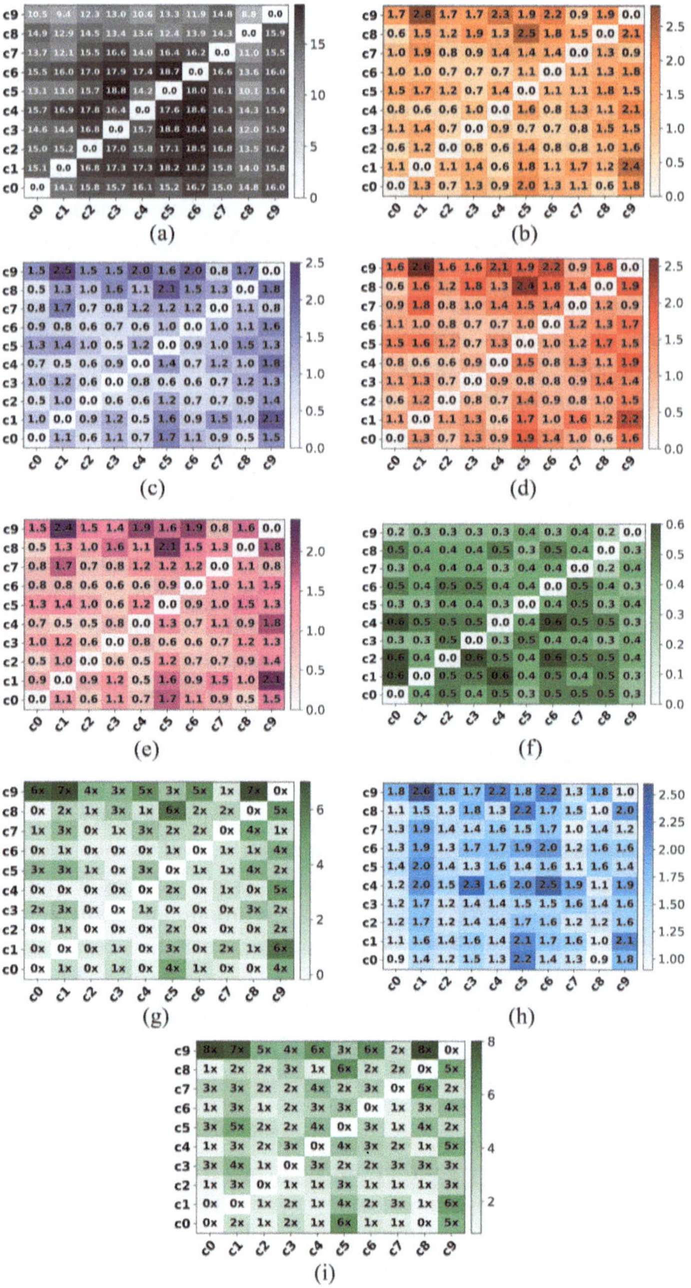

Fig. 5.13 MAE (in ps) heatmap for cross-corner interconnect delay prediction results using various regression and learning-based approaches in the sub-10-nm technology node, with one known corner (adopted from [28]). (**a**) Support Vector Machines (SVM). (**b**) Random Forest (RF). (**c**) Multilayer Perceptron (MLP). (**d**) Linear regression (DATE '19 [32]). (**e**) XGBoost (DAC '20 [4]). (**f**) The proposed accurate version (denoted as "acc"). (**g**) Accurate Version over Previous Best (denoted as "pb") across (**a**)–(**e**). (**h**) The Proposed Fast Version (denoted as "fa"). (**i**) Accurate Version over Fast Version (acc over fa)

Table 5.6 A summary of the predicted MAE (ps) (mean/min/max) of the unobserved corners when using c_i as the observed corner in the sub-10 nm node (adopted from [28])

Benchmark	Corner	0	1	2	3	4	5	6	7	8	9
Industrial design	bp	1.5/0.8/2.4	1.2/0.5/2.1	1.0/0.7/1.7	0.8/0.6/1.5	1.0/0.5/1.5	0.8/0.5/1.8	0.8/0.6/1.3	0.8/0.5/1.4	1.1/0.5/2.1	0.9/0.5/1.7
	fa	1.8/1.3/2.6	1.5/1.1/2.2	1.4/1.2/1.9	1.6/1.3/2.0	1.5/1.3/2.0	1.8/1.2/2.5	1.5/1.4/1.7	1.4/1.4/1.7	1.6/1.1/2.1	1.4/1.2/2.2
	acc	**0.3/0.2/0.4**	**0.4/0.3/0.5**	**0.3/0.2/0.4**	**0.4/0.3/0.5**	**0.3/0.3/0.5**	**0.4/0.3/0.6**	**0.3/0.3/0.5**	**0.5/0.4/0.6**	**0.4/0.3/0.6**	**0.4/0.3/0.5**

Table 5.7 The number of required corners for the previous best methods to attain a comparable MAE level when compared against the proposed accurate version of the prediction strategy, evaluated in the sub-10-nm process node (adopted from [28])

	Proposed	Previous best							
Known corner	#1	#1	#2	#3	#4	#5	#6	#7	# minimum required corner
0	**0.30**	1.62	0.83	0.5	0.38	0.40	0.39	**0.29**	**7**
1	**0.41**	1.36	0.62	0.47	0.50	0.50	0.53	**0.27**	**7**
2	**0.36**	1.06	0.64	0.52	0.51	0.53	0.38	**0.28**	**7**
3	**0.43**	0.88	0.60	0.60	0.61	**0.36**			**5**
4	**0.37**	1.12	0.62	0.63	**0.34**				**4**
5	**0.49**	0.92	0.74	0.50	**0.38**				**4**
6	**0.38**	0.89	0.63	0.41	**0.38**				**4**
7	**0.50**	0.83	0.77	0.71	0.76	**0.36**			**5**
8	**0.48**	1.18	0.93	0.76	**0.36**				**4**
9	**0.43**	1.02	0.77	**0.40**					**3**
Average									**5**

runtime (approximately 10 seconds) across all designs. The fast version achieves over a 2× speedup with minimal accuracy degradation of less than 0.5 ps on average.

In addition to the aforementioned approaches, [22] proposes a graph learning-based model to predict signoff interconnect delay under a single corner. However, this prior work poses challenges in terms of reproducibility due to limited information provided in the paper. Nevertheless, we made our best effort to compare against it by assuming that the total runtime in their case is the product of the number of corners and the time required for a single-corner prediction. It should be noted that the technology nodes in our evaluation and [22] are different. In cases where processes can run in parallel, the runtime metric can also be translated into the total amount of computing resources required. Table 5.9 summarizes the comparison results, demonstrating that our proposed framework outperforms [22] in terms of both accuracy and runtime.

5.3.6.4 Cross-Corner Interconnect Slew and Load Prediction

Figure 5.9 highlights the critical importance of acquiring accurate interconnect slew and net capacitance values for effective cell delay prediction using library lookup tables. In Table 5.10, we compare the prediction accuracy of our proposed method with prior approaches, all evaluated on a 28 nm process. The prior approaches used in this comparison are the same as those employed for examining interconnect delays. Our proposed prediction method showcases a remarkable decrease in MAE, yielding substantial improvements in both slew predictions and net capacitance estimates. Specifically, our method achieves a significant reduction in MAE,

Table 5.8 A summary of the predicted MAE (ps) (mean) of the unobserved corners when using c_i as the observed corner in the 28 nm node. The runtime (s) is also listed (adopted from [28])

Benchmark	Corner	0	1	2	3	4	5	6	7	8	9	Runtime (s)[a]
B19	bp	0.122	0.133	0.162	0.169	0.105	0.110	0.151	0.162	0.111	0.116	161
	fa	0.464	0.483	0.529	0.540	0.457	0.466	0.552	0.548	0.501	0.519	**53**
	acc	**0.005**	**0.005**	**0.005**	**0.006**	**0.005**	**0.005**	**0.006**	**0.006**	**0.005**	**0.006**	172
PCI_BRIDGE	bp	0.240	0.268	0.258	0.269	0.185	0.197	0.254	0.261	0.192	0.190	115
	fa	0.801	0.860	0.857	0.860	0.752	0.811	0.869	0.865	0.810	0.827	**50**
	acc	**0.011**	**0.011**	**0.011**	**0.011**	**0.011**	**0.011**	**0.012**	**0.012**	**0.011**	**0.011**	126
DES_PERT	bp	0.088	0.084	0.107	0.094	0.078	0.078	0.108	0.088	0.063	0.067	59
	fa	0.450	0.459	0.625	0.600	0.425	0.508	0.572	0.600	0.499	0.517	**42**
	acc	**0.011**	**0.012**	**0.012**	**0.013**	**0.011**	**0.012**	**0.013**	**0.013**	**0.013**	**0.012**	69
AES-128	bp	0.240	0.268	0.258	0.269	0.185	0.197	0.254	0.261	0.192	0.190	96
	fa	0.802	0.854	0.891	0.920	0.818	0.839	0.930	0.927	0.854	0.876	**47**
	acc	**0.009**	**0.010**	**0.010**	**0.010**	**0.009**	**0.009**	**0.010**	**0.010**	**0.010**	**0.010**	107
SALSA	bp	0.341	0.391	0.407	0.442	0.287	0.289	0.388	0.419	0.273	0.281	150
	fa	1.436	1.475	1.502	1.560	1.397	1.455	1.569	1.610	1.467	1.518	**54**
	acc	**0.012**	**0.012**	**0.012**	**0.012**	**0.012**	**0.012**	**0.012**	**0.013**	**0.012**	**0.012**	161
OPENGFX	bp	0.147	0.162	0.205	0.220	0.135	0.135	0.218	0.223	0.135	0.138	123
	fa	0.620	0.646	0.682	0.706	0.609	0.647	0.692	0.723	0.650	0.671	**49**
	acc	**0.007**	**0.007**	**0.007**	**0.008**	**0.007**	**0.007**	**0.008**	**0.008**	**0.007**	**0.007**	134

[a] This runtime encompasses the total time required for predicting interconnect timing for all unobserved corners. Notably, it does not include preprocessing time, which is deemed negligible and therefore excluded for all conditions when making comparisons

Table 5.9 Comparison against a state-of-the-art work (DATE 2023 [22]) targeting single-corner interconnect delay prediction (adopted from [28])

			B19	PCI_BRIDGE	DES_PERT	AES-128	SALSA	OPENGFX
MAE (ps)	16 nm	[22]	–	–	0.43	1.14	–	2.54
	28 nm	acc	**0.005**	**0.011**	**0.011**	**0.009**	**0.012**	**0.007**
		fa	0.457	0.752	0.450	0.802	1.397	0.620
Runtime (s)	16 nm	[22]	–	–	274	567	–	976
	28 nm	acc	172	126	69	107	161	134
		fa	**53**	**50**	**42**	**47**	**54**	**49**

lowering slew predictions from 3 to 0.2 ps and enhancing net capacitance estimates from 0.1 to 0.03 fF.

5.3.6.5 ECO Runtime Improvement

Traditionally, ECO flows involve invoking the STA tools and parasitic extraction tools for each corner, which adds complexity and duration to each ECO iteration. However, the proposed framework offers a significant reduction in runtime and can be broken down into three primary components: extraction time and signoff time for a single observed corner (in the accurate version) and cross-corner prediction time. Table 5.11 presents a comparison between the proposed framework and a traditional ECO flow that solely relies on commercial tools. It demonstrates that the proposed framework achieves a runtime improvement of over ten times in an ECO iteration. This remarkable improvement holds great promise for the application of this framework in the signoff of designs. It is worth noting that while the fast version of the framework may not provide absolute accuracy, it can still be utilized in the initial design phase to assess the design's coverage. Typically, designs are implemented and optimized under only one or a few corners, making the fast version valuable for evaluating design coverage during this stage.

5.3.6.6 Known Corner Selection Accuracy

The selection accuracy was assessed by creating a dataset consisting of 600 labeled Blurred MAE hot graphs. The sample size, denoted as S, was varied from 100 to 500. The experiment was conducted specifically on the 28 nm node, and the training and testing datasets were split at a ratio of 7:3. The "mixed" label indicates a dataset that was mixed from the previous three samples. The results presented in Table 5.12 show that even with a small sample size of 500, the test accuracy surpasses 75%. Moreover, similar experiments conducted on sub-10 nm processes achieved a test accuracy of over 90%. These findings highlight the effectiveness and accuracy of the selection process, especially for sub-10 nm processes.

Table 5.10 MAE from the cross-corner slew (ps)/net load cap (ff) prediction in 28 nm (adopted from [28])

Benchmark	Corner	0	1	2	3	4	5	6	7	8	9	Avg.
B19	bp	2.64/0.06	2.85/0.07	2.59/0.07	2.73/0.06	2.88/0.07	2.88/0.06	2.32/0.06	2.75/0.07	2.42/0.06	2.67/0.06	2.67/0.07
	acc	**0.19/0.02**	**0.18/0.02**	**0.18/0.02**	**0.18/0.02**	**0.18/0.02**	**0.18/0.02**	**0.19/0.02**	**0.18/0.02**	**0.18/0.02**	**0.19/0.02**	**0.18/0.02**
PCI_BRIDGE	bp	3.03/0.08	3.34/0.10	3.10/0.10	3.09/0.08	3.91/0.10	3.35/0.08	2.70/0.08	3.28/0.10	2.90/0.09	3.22/0.09	3.19/0.09
	acc	**0.18/0.02**	**0.19/0.02**	**0.18/0.02**	**0.18/0.02**	**0.18/0.02**	**0.18/0.02**	**0.19/0.02**	**0.18/0.02**	**0.18/0.02**	**0.18/0.02**	**0.18/0.02**
DES_PERT	bp	1.34/0.05	1.21/0.05	1.28/0.05	1.21/0.05	1.53/0.05	1.21/ 0.05	1.21/0.05	1.13/0.05	1.20/0.04	1.07/0.05	1.24/0.05
	acc	**0.19/0.02**	**0.20/0.02**	**0.20/0.03**	**0.21/0.03**	**0.20/0.02**	**0.19/0.02**	**0.20/0.03**	**0.21/0.03**	**0.20/0.02**	**0.20/0.02**	**0.20/0.02**
AES-128	bp	4.12/0.06	4.91/0.07	3.92/0.07	4.32/0.07	4.65/0.07	4.70/ 0.07	3.47/0.06	4.08/0.07	3.87/0.06	4.43/0.06	4.25/0.07
	acc	**0.22/0.02**	**0.22/0.02**	**0.22/0.02**	**0.23/0.02**	**0.22/0.02**	**0.22/0.02**	**0.22/0.02**	**0.22/0.02**	**0.23/0.02**	**0.22/0.02**	**0.22/0.02**
SALSA	bp	3.72/0.21	3.60/0.23	3.75/0.23	3.53/0.22	4.36/0.22	3.79/ 0.21	3.79/0.20	3.63/0.22	3.54/0.20	3.53/0.21	3.73/0.22
	acc	**0.17/0.06**	**0.16/0.06**	**0.17/0.06**	**0.17/0.06**	**0.17/0.06**	**0.17/0.06**	**0.17/0.06**	**0.17/0.06**	**0.17/0.06**	**0.17/0.06**	**0.17/0.06**
OPENGFX	bp	3.43/0.10	3.61/0.12	3.37/0.12	3.11/0.10	3.94/0.11	3.75/ 0.10	3.13/0.10	3.14/0.11	3.25/0.10	3.37/0.10	3.41/0.11
	acc	**0.23/0.03**	**0.22/0.03**	**0.23/0.03**	**0.23/0.03**	**0.23/0.03**	**0.23/0.03**	**0.23/0.03**	**0.23/0.03**	**0.24/0.03**	**0.23/0.03**	**0.23/0.03**

Table 5.11 Runtime (s) for two ECO iterations (28 nm) (adopted from [28])

	Iteration #1			Iteration #2		
Benchmark	fa	acc	Traditional ECO	fa	acc	Traditional ECO
B19	53	169	1574	53	172	1608
PCI_BRIDGE	70	147	1354	50	126	1154
DES_PERT	43	79	697	42	69	588
AES-128	42	112	1009	47	107	962
SALSA	53	127	1582	54	161	1502
GFX	50	100	1285	49	134	1228
Average	**51.83**	**122.33**	**1250.17**	**49.17**	**128.17**	**1173.67**

Table 5.12 Corner selection accuracy w.r.t. sample size (28 nm) (adopted from [28])

Sample size	Training accuracy	Test accuracy
100	0.663	0.611
200	0.805	0.744
500	0.771	0.767
Mixed	0.814	0.8

5.3.7 Summary

Section 5.3 introduces a learning-based cross-corner timing signoff framework. This framework takes an "one-for-all" approach, meaning that it only requires one RC corner to predict interconnect timing for the remaining unobserved corners. The proposed framework undergoes comprehensive evaluations on two mainstream technology nodes and is compared to various state-of-the-art methodologies. The results demonstrate significant accuracy improvements with an acceptable runtime penalty. Additionally, a fast version of the framework is introduced for swift timing estimation. This fast version provides a quicker estimate of timing but may sacrifice some accuracy compared to the accurate version. Furthermore, a learning-based known corner selection approach is introduced, which achieves commendable accuracy in identifying known corners. By integrating this framework into the ECO flow, each iteration is substantially accelerated, achieving a runtime improvement of over ten times. This enhancement in runtime efficiency is a valuable contribution to the ECO process.

5.4 Conclusions and Outlook

In this chapter, a novel graph learning-based timing estimation model is introduced, recognizing the increasing importance of timing signoff throughout the design cycle. This model, leveraging the capabilities of graph learning, provides a detailed perspective on interconnect topology. The contributions of this work can be summarized in two main aspects. Firstly, we showcase the effectiveness of graph learning in estimating interconnect delay at a single RC corner. Secondly, we extend our contributions by introducing a framework that integrates an enhanced graph learning model. This framework efficiently predicts interconnect timing across corners using data from only one known corner. In both instances, we present ultrafast and ultra-accurate versions, accommodating diverse needs along the accuracy efficiency trade-off curve.

Looking ahead, our next steps involve the development of an automatic ECO generation step based on the predicted timing from the proposed framework. Furthermore, we are actively engaged in making this research available to the broader community by taking open-source initiatives, aiming to contribute to the collaborative efforts of researchers in the field.

Acknowledgments This work is partially supported by the National Science and Technology Major Project under Grant No. 2021ZD0114701 and by an open fund from the State Key Laboratory of Integrated Chips and Systems (SKLICS).

References

1. Yang, T., Zhao, Z., Han, A., & Liu, G. (2022). Automatic timing eco using stage-based path delay prediction. In *2022 20th IEEE Interregional NEWCAS Conference (NEWCAS)* (pp. 455–459). IEEE.
2. Golshan, K. (2020). *The art of timing closure*. Springer.
3. Kahng, A. B. (2015). New game, new goal posts: A recent history of timing closure. In *Proceedings of the 52nd Annual Design Automation Conference* (pp. 1–6).
4. Cheng, H. H., Jiang, I. H. R., & Ou, O. (2020). Fast and accurate wire timing estimation on tree and non-tree net structures. In *Proceedings - Design Automation Conference*.
5. Zhu, L., Gu, Y., & Guo, X. (2023). RC-GNN: Fast and accurate signoff wire delay estimation with customized graph neural networks. In *2023 IEEE 5th International Conference on Artificial Intelligence Circuits and Systems (AICAS)* (pp. 1–5). IEEE.
6. Elmore, W. C. (1948). The transient response of damped linear networks with particular regard to wideband amplifiers. *Journal of Applied Physics, 19*(1), 55–63.
7. Alpert, C. J., Devgan, A., & Kashyap, C. (2000). A two moment rc delay metric for performance optimization. In *Proceedings of the 2000 international symposium on Physical design* (pp. 69–74).
8. Huang, G., Hu, J., He, Y., Liu, J., Ma, M., Shen, Z., Wu, J., Xu, Y., Zhang, H., Zhong, K., et al. (2021). Machine learning for electronic design automation: A survey. *ACM Transactions on Design Automation of Electronic Systems (TODAES), 26*(5), 1–46.
9. Ren, H., & Hu, J. (2023). *Machine learning applications in electronic design automation*. Springer Nature.
10. Xie, Z., Liang, R., Xu, X., Hu, J., Chang, C.-C., Pan, J., & Chen, Y. (2022). Pre-placement net length and timing estimation by customized graph neural network. *IEEE Transactions on Computer-Aided Design of Integrated Circuits and Systems, 0070*(c), 1–1.
11. Barboza, E. C., Shukla, N., Chen, Y., & Hu, J. (2019). Machine learning-based pre-routing timing prediction with reduced pessimism. In *Proceedings of the 56th Annual Design Automation Conference 2019* (pp. 1–6).
12. Kahng, A. B., Kang, S., Lee, H., Nath, S., & Wadhwani, J. (2013). Learning-based approximation of interconnect delay and slew in signoff timing tools. In *2013 ACM/IEEE International Workshop on System Level Interconnect Prediction (SLIP)* (pp. 1–8). IEEE.

13. He, X., Fu, Z., Wang, Y., Liu, C., & Guo, Y. (2022). Accurate timing prediction at placement stage with look-ahead RC network. In *Proceedings - Design Automation Conference* (pp. 1213–1218).
14. Han, S. S., Kahng, A. B., Nath, S., & Vydyanathan, A. S. (2014). A deep learning methodology to proliferate golden signoff timing. In *Proceedings -Design, Automation and Test in Europe, DATE*.
15. Garyfallou, D., Vagenas, A., Antoniadis, C., Massoud, Y., & Stamoulis, G. (2022). Leveraging machine learning for gate-level timing estimation using current source models and effective capacitance. In *Proceedings of the Great Lakes Symposium on VLSI 2022* (pp. 77–83).
16. Chen, T., & Guestrin, C. (2016). Xgboost: A scalable tree boosting system. In *Proceedings of the 22nd ACM Sigkdd International Conference on Knowledge Discovery and Data Mining* (pp. 785–794).
17. Kahng, A. B., Luo, M., & Nath, S. (2015). SI for free: Machine learning of interconnect coupling delay and transition effects. In *International Workshop on System Level Interconnect Prediction, SLIP, 2015-July*.
18. Lopera, D. S., Servadei, L., Kiprit, G. N., Hazra, S., Wille, R., & Ecker, W. (2021). A survey of graph neural networks for electronic design automation. In *2021 ACM/IEEE 3rd Workshop on Machine Learning for CAD (MLCAD)* (pp. 1–6). IEEE.
19. Lu, Y. C., Nath, S., Pentapati, S., & Lim, S. K. (2023). Eco-gnn: Signoff power prediction using graph neural networks with subgraph approximation. *ACM Transactions on Design Automation of Electronic Systems, 28*(4), 1–22.
20. Lopera, D. S., & Ecker, W. (2022). Applying gnns to timing estimation at rtl. In *Proceedings of the 41st IEEE/ACM International Conference on Computer-Aided Design* (pp. 1–8).
21. Guo, Z., Liu, M., Gu, J., Zhang, S., Pan, D. Z., & Lin, Y. (2022). A timing engine inspired graph neural network model for pre-routing slack prediction. In *Proceedings - Design Automation Conference* (pp. 1207–1212).
22. Ye, Y., Chen, T., Gao, Y., Yan, H., Yu, B., & Shi, L. (2023). Fast and accurate wire timing estimation based on graph learning. In *Proceedings -Design, Automation and Test in Europe, DATE* (pp. 1–6).
23. Ye, Y., Chen, T., Gao, Y., Yan, H., Yu, B., & Shi, L. (2023). Graph-learning-driven path-based timing analysis results predictor from graph-based timing analysis. In *Proceedings of the 28th Asia and South Pacific Design Automation Conference* (pp. 547–552). IEEE.
24. Zhang, C., Song, D., Huang, C., Swami, A., & Chawla, N. V. (2019). Heterogeneous graph neural network. In *Proceedings of the 25th ACM SIGKDD International Conference on Knowledge Discovery & Data Mining* (pp. 793–803).
25. Wang, M., Zheng, D., Ye, Z., Gan, Q., Li, M., Song, X., Zhou, J., Ma, C., Yu, L., Gai, Y., et al. (2019) Deep graph library: A graph-centric, highly-performant package for graph neural networks. *arXiv preprint arXiv:1909.01315*.
26. Kim, K., Joo, B., Park, Y. M., Jeong, T., Kim, K. T., & Chung, E.-Y. (2021). Cross-corner delay variation model for standard cell libraries. *IEEE Access, 9*, 72299–72315.
27. Golshan, K. (2020). *Multi-mode multi-corner analysis* (pp. 65–76). Springer International Publishing.
28. Zhu, L., Cai, Y., & Guo, X. (2024). One-for-all: An unified learning-based framework for efficient cross-corner timing signoff. In *Proceedings of ACM/IEEE International Conference on Computer-Aided Design (ICCAD)*. IEEE.
29. Onaissi, S., Taraporevala, F., Liu, J., & Najm, F. (2011). A fast approach for static timing analysis covering all pvt corners. In *Proceedings of the 48th Design Automation Conference*, DAC '11 (pp. 777–782). Association for Computing Machinery.
30. e Silva, L. G., Miguel Silveira, L., & Phillips, J. R. (2007). Efficient computation of the worst-delay corner. In *Proceedings of the Conference on Design, Automation and Test in Europe*, DATE '07 (pp. 1617–1622). EDA Consortium.
31. Onaissi, S., & Najm, F. N. (2008). A linear-time approach for static timing analysis covering all process corners. *IEEE Transactions on Computer-Aided Design of Integrated Circuits and Systems, 27*(7), 1291–1304.

32. Kahng, A. B., Mallappa, U., Saul, L., & Tong, S. (2019). "unobserved corner" prediction: Reducing timing analysis effort for faster design convergence in advanced-node design. In *2019 Design, Automation & Test in Europe Conference & Exhibition (DATE)* (pp. 168–173). IEEE.
33. Zhao, Z., Zhang, S., Liu, G., Feng, C., Yang, T., Han, A., & Wang, L. (2022). Machine-learning-based multi-corner timing prediction for faster timing closure. *Electronics, 11*(10), 1571.
34. Xing, W. W., Xing, Z., Lu, R., Wang, Z., Xu, N., Cheng, Y., & Zhao, W. (2023). Total: Multi-corners timing optimization based on transfer and active learning. In *2023 60th ACM/IEEE Design Automation Conference (DAC)* (pp. 1–6). IEEE.
35. Ma, Y., Liu, X., Shah, N., & Tang, J. (2023). *Is homophily a necessity for graph neural networks?*.
36. Feng, A., You, C., Wang, S., & Tassiulas, L. (2022). KerGNNs: Interpretable graph neural networks with graph kernels. In *Proceedings of the 36th AAAI Conference on Artificial Intelligence, AAAI 2022* (Vol. 36, No. 1, pp. 6614–6622).

Chapter 6
AI-Enabled Placement for 2D and 3D ICs

Abdullah Mansoor, Malgorzata Chrzanowska-Jeske, and Marcin Jeske

6.1 Introduction

6.1.1 Overview of 2D Placement Problem in VLSI Design

The 2D placement problem is an NP-complete problem [1], and despite decades of research, the optimality of results is still an open dilemma.

NP describes a class of problems that require polynomial time for a nondeterministic Turing machine to find a solution or for a deterministic Turing machine to verify that a solution is correct. In short, they are computationally complex problems where no (known) efficient algorithm can guarantee optimal solutions. All NP problems can be transformed efficiently into a subset of NP problems that are termed NP-complete, which by definition are as hard to solve as any other problem in NP. In that light, figuring out how to approach an optimal solution with an efficient algorithm for 2D placement is a challenging and critical issue in VLSI design.

Problem Formulation The inputs of placement are a set of cells, width and height for each cell, and a netlist. Placement aims to find a location $L(x, y)$ for each cell to minimize a given cost function while meeting all constraints. Figure 6.1 shows an example of 2D placement.

One or more objectives can be optimized for in the cost function. Common placement objectives for optimization to achieve a good overall design include:

– Performance: Every design must meet its performance requirements. The placement of the components can affect the performance of the design, so it is

A. Mansoor (✉) · M. Chrzanowska-Jeske · M. Jeske
Portland State University, Portland, OR, USA
e-mail: mansoor4@ece.pdx.edu; jeske@ece.pdx.edu; jeskem@ece.pdx.edu

A. Iranmanesh, H. Sayadi (eds.), *AI-Enabled Electronic Circuit and System Design*,
https://doi.org/10.1007/978-3-031-71436-8_6

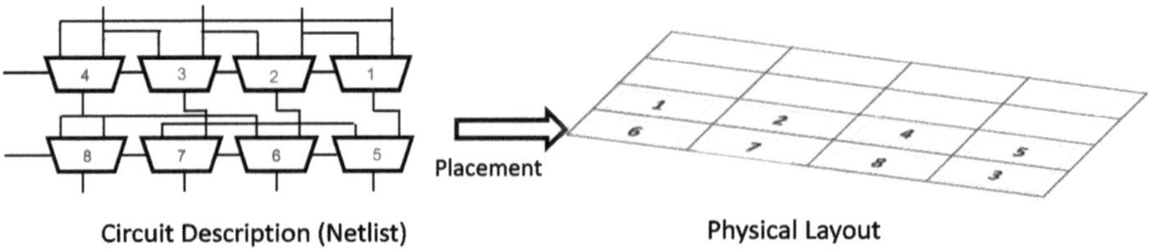

Fig. 6.1 Example of 2D placement of 4-bit Mux-Shifter (Muxs) Circuit

important to co-optimize placement and performance to achieve a good overall design.

- Manufacturability: A feasible design must be fabricated using existing manufacturing processes. It may be difficult or impossible to manufacture certain arrangements of components, or a given placement may lead to poor yields.
- Power distribution: The power distribution network (PDN) delivers power to all components. Component placement within a design can affect the PDN's performance. Therefore, co-optimizing placement and power distribution are important for a good overall design.

Placement may need to adhere to a set of constraints to ensure good quality. Here is the spectrum of placement constraints:

- Geometric constraints: Geometric constraints directly affect the physical layout of the design. Such constraints include the size and shape of the components, the minimum spacing between components, and the location of fixed components.
- Performance constraints: Performance constraints set the performance requirements of the design. These constraints include the design's maximum clock frequency, the design's maximum power consumption, and the design's density.
- Electrical constraints: Electrical constraints derive from the electrical properties of the design, including the design's timing requirements, the design's power requirements, and the design's thermal requirements.
- Manufacturability constraints: Manufacturability constraints define the constraints that must be satisfied to fabricate the design using existing manufacturing processes. These constraints include the minimum feature size, the maximum feature size, and allowed types of routing patterns.

6.1.2 Role of AI in Placement Algorithms

Since AI-enabled placement algorithms are at an early stage, we will review breakthroughs in the ImageNet Challenge to suggest a potential role of AI in VLSI placement.

The ImageNet Large Scale Visual Recognition Challenge (ILSVRC) is a competition that was held annually from 2010 to 2017 [2]. The challenge required participants to train machine learning models to classify images into 1000 different categories. The ImageNet Challenge helped to popularize deep learning and contributed to the rapid advances in computer vision that have been made in recent years.

The 2018 Turing Award was given to Yoshua Bengio, Geoffrey Hinton, and Yann LeCun "for conceptual and engineering breakthroughs that have made deep neural networks a critical component of computing" [3]. A large portion of that work is related to the ImageNet Challenge.

In 2012, Hinton's team won the ILSVRC by submitting a deep convolutional neural network (CNN) with a top-5 error rate of 15.3%, a measure of the percentage of test images where the model's top-5 ranked predictions do not include the correct label. This was a significant advance over the previous state of the art, which had a top-5 error rate of 26.4% [4].

Bengio and LeCun also made significant contributions to the ImageNet Challenge. Hinton's team used Bengio's research on deep learning algorithms to develop the winning CNN [5]. LeCun's early work on CNNs also laid the foundation for Hinton's success [6]. While the ImageNet Challenge and the development of CNN played a crucial role in advancing computer vision, their work extends beyond this specific area. CNNs with deep learning allow high accuracy and transfer of learning to unseen cases.

Overall, today's leading approaches to the ImageNet Challenge are very promising. Deep learning models have achieved near-human accuracy on the task of image classification and in recent years have become much faster. The current state-of-the-art model on ImageNet is BASIC-L (Lion, fine-tuned), which achieved a top-1 accuracy of 91.1% on the ImageNet validation set in 2023 [7]. This is only slightly lower than the accuracy of human experts, who typically achieve a top-1 accuracy of around 95% [2].

This progress is driven by several factors, including the ever-greater volume of accessible data, more powerful computers, and new algorithms and techniques. Breakthroughs emerging from the ImageNet Challenge could revolutionize many industries and improve quality of life for billions of people.

6.1.3 Motivation for AI-Enabled Placement

We will discuss the potential benefits of AI-enabled placement algorithms [8].

- **Optimality gaps:** The difficulty of finding optimal placement solutions increases with the complexity of the design. Machine learning algorithms can improve the optimality of placement algorithms by (a) learning from past data to identify good placements for different types of designs and (b) using this information to develop more efficient and effective placement algorithms.
- **Need of predictability and stability:** Predictability and stability are key gaps in traditional placement algorithms. Predictability is important because it allows

designers to estimate a placement algorithm's performance and choose the best suited for their needs. Stability is important because it ensures the placement algorithm will not produce different results for the same input. However, predictability and stability are important properties of AI algorithms.
- **Need of co-optimization and cooperation:** Co-optimizations are needed because placement and other design factors such as timing, power distribution, and manufacturability must be considered. Cooperation is needed because placement can affect other design steps, such as routing, and vice versa. Co-optimization and cooperation are challenging for traditional placement algorithms, but AI algorithms can often better accommodate those needs.

6.2 Fundamentals of AI Techniques for Placement

6.2.1 Overview of Machine Learning Algorithms

In VLSI design, machine learning algorithms can be designed to assist in optimizing the physical layout of components. AI-based placement strategies using machine learning can potentially create powerful tools to solve complex placement challenges. These algorithms can be of three types: supervised, unsupervised, and reinforcement learning. One way to differentiate these algorithms is based on the dataset (examples of placements from one or more benchmarks) with a known label (also known as a target value or ground truth). Supervised learning would use labeled data to guide the placement process and predict performance metrics. In contrast, unsupervised learning explores data without labels—clustering components or reducing dimensionality to uncover hidden structures in the placement. Reinforcement learning introduces agents that learn from interactions with the environment, making sequential decisions to optimize placement over time. VLSI designers can create more efficient and effective placement strategies by combining these machine learning approaches.

6.2.2 Supervised, Unsupervised, and Reinforcement Learning

Supervised Learning is a machine learning style that involves training an algorithm on a dataset with labels attached to it. The dataset consists of input data and their corresponding output labels. The main goal of supervised learning is to develop a predictive model that can accurately predict the output label of new input data that it has never seen before. The algorithm accomplishes this by analyzing the labeled training data and identifying patterns and relationships that exist between the input data and the output labels.

An example of supervised learning is to predict an estimate of the average delay of the placement or the routed wirelength. Delay simulations and routing of placement are time-consuming steps in the VLSI flow. Exact delay or placement-routability calculations will significantly increase the runtime of the placement flow. However, supervised learning can be used to provide an approximate estimate much faster. Since machine learning algorithms for VLSI design are in the early stages, there are no readily available datasets for supervised learning. However, creating a dataset to train supervised learning is feasible, and we will discuss various algorithms useful for supervised learning in Sect. 6.2.3.

Unsupervised Learning is a machine learning approach where the algorithm finds hidden structures, patterns, or relationships in a dataset by analyzing the data's inherent characteristics without the need for labels or ground truths. It primarily encompasses techniques for clustering data points into groups or reducing the dimensionality of the data for further analysis.

An example of unsupervised learning is the clusters of cells based on the placement or based on the connectivity of cells. Cluster data can help guide the configuration of commercial or academic placers.

Reinforcement Learning is a machine learning framework where an agent interacts with an environment, learns to take a sequence of actions, and receives feedback in the form of rewards or penalties. The agent aims to learn a policy (action selected based on the current state) that maximizes the cumulative expected reward over time. For placement, the rewards or penalties usually derive from the change in the output of the cost function that considers factors such as area, wire length, and temperature. For example, an RL placer algorithm would create various placements interactively and identify a policy that efficiently leads to good placement.

6.2.3 Machine Learning Algorithms

Machine learning algorithms can take various forms, including Linear Regression, Decision Trees, Neural Networks, and Ensembles. The description of these algorithms is part of books on ML [9, 10, 11]. Nowadays, these algorithms are available as off-the-shelf components within machine learning libraries. The developer's responsibility is to tune the hyperparameters—parameters that can significantly impact the algorithm's performance. In Sect. 6.6, we provide a detailed framework on how to tune these algorithms for placement and compare their performance. Next, we will review key hyperparameters of each algorithm type:

Linear Regression is a simple yet effective supervised learning algorithm. Here are important hyperparameters of the linear model:

- Family: the type of probability distribution for the response variable (e.g., Gaussian, Poisson, and Binomial)

- Link Function: the function that relates the linear predictor to the mean of the distribution (e.g., identity, logit, and log)
- Regularization: hyperparameters for regularization techniques (e.g., L1/L2 penalty)

Decision Trees are versatile ML models with several hyperparameters that can influence their performance and behavior. Here are a few important hyperparameters for decision trees:

- Maximum Depth: This hyperparameter determines the maximum depth or height of the decision tree.
- Minimum Samples per Leaf: The minimum number of samples to form a leaf node in the decision tree.
- Minimum Samples per Split: This hyperparameter governs the minimum number of samples for a node before it can be split.
- Criterion for Splitting: Decision trees use a criterion to determine the best feature to split on at each node. Common criteria include Gini impurity and entropy.

Neural Networks refer to the specific layouts or structures of neurons, including the arrangement of layers, the number of neurons in each layer, and the connections between them.

- Number of Layers: the neural network's architecture, including any hidden layers
- Number of Neurons: each layer's neurons (nodes)
- Activation Functions: the activation functions used in each layer (e.g., ReLU, sigmoid, and tanh)
- Learning Rate determines the step size during gradient descent optimization.

Ensembles such as Random Forest, Gradient Boosting (e.g., XGBoost and AdaBoost), and Stacking, may have a range of hyperparameters, but here are a few important ones:

- Number of Base Models: Ensemble algorithms typically combine the predictions of multiple base models (e.g., decision trees).
- Shrinkage: This parameter controls the contribution of each base model to the ensemble.
- Base Model Hyperparameters: Each base model within an ensemble may have its own set of hyperparameters (e.g., tree depth, minimum samples per leaf, and learning rate for gradient boosting).
- Subsampling: Subsampling controls the fraction of the training dataset used for building each base model in the ensemble.

6.2.4 Deep Learning and Convolutional Neural Networks

Neural networks can be further classified into Multilayer Perceptrons (MLPs), Deep Learning Networks, Convolutional Neural Networks, and Graph Convolutional Neural Networks. We will discuss the above categories next.

Multilayer Perceptrons and Deep Learning MLPs effectively extract global features from 1D input data. The output multilayer perceptron, V', for a given input vector s and weight parameters w with three layers is given by

$$V' = h\left\{\sum_{k=1}^{K} w_{lk}^{(3)} h\left[\sum_{j=1}^{J} w_{kj}^{(2)} h\left(\sum_{i=1}^{I} w_{ji}^{(1)} s_i\right)\right]\right\} \tag{6.1}$$

Figure 6.2 shows the network of MLP given by the above equation.

- The MLP in Eq. 6.1 has three layers, denoted as I, J, and K, and each layer is parameterized by its weight matrix: $w^{(1)}$ for the input to hidden layer J, $w^{(2)}$ for hidden layer I to hidden layer K, and $w^{(3)}$ for hidden layer K to the output layer.
- $\sum_{i=1}^{I} w_{ji}^{(1)} s_i$ computes a weighted sum of the input features s_i for each neuron in the first hidden layer. This is a linear combination of the inputs, and it is typically followed by an activation function.

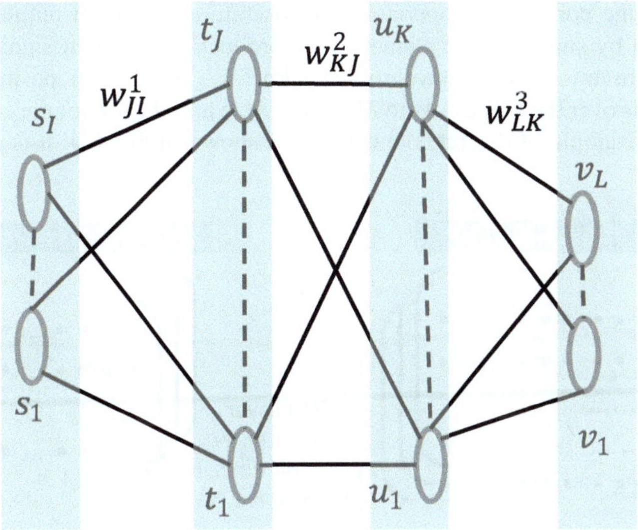

Fig. 6.2 Three-layered MLPs

- $\sum_{j=1}^{J} w_{kj}^{(2)} h(\ldots)$ computes a similar weighted sum for the hidden layer J after applying an activation function h.
- Finally, $\sum_{k=1}^{K} w_{lk}^{(3)} h(\ldots)$ computes the weighted sum for the output layer K after applying another activation function.

In summary, this equation represents the forward pass of an MLP, where the input s is transformed through multiple layers with weight parameters and nonlinear activation functions to produce the final output V'. The specific choice of activation functions and the dimensions of the weight matrices will determine the behavior and capabilities of the MLP for a given task.

When the number of layers is 4 or more, such neural networks are termed Deep Learning. Adding more layers to MLP is expected to improve the performance at the expense of more training data.

Convolutional Neural Networks CNNs effectively extract local features from 2D input data with shared weights. This allows them to capture relevant patterns in the data.

Let us consider an RGB placement image represented by a matrix X, W is the learnable convolutional kernel/filter, h is the activation function, and b is the bias term. An example of a CNN layer is shown in Fig. 6.3.

The output of the convolution layer, V', can be denoted as

$$V' = h(X * W + b) \tag{6.2}$$

where $*$ is the convolution operator. The discrete convolution output at index n is computed by summing the element-wise product of the input signal X and the flipped (or time-reversed) convolutional kernel W at different positions m. The summation is over the range $-M$ to M, where M is half the size of the convolutional kernel. An example of discrete convolution is shown in Fig. 6.4. It is given by the

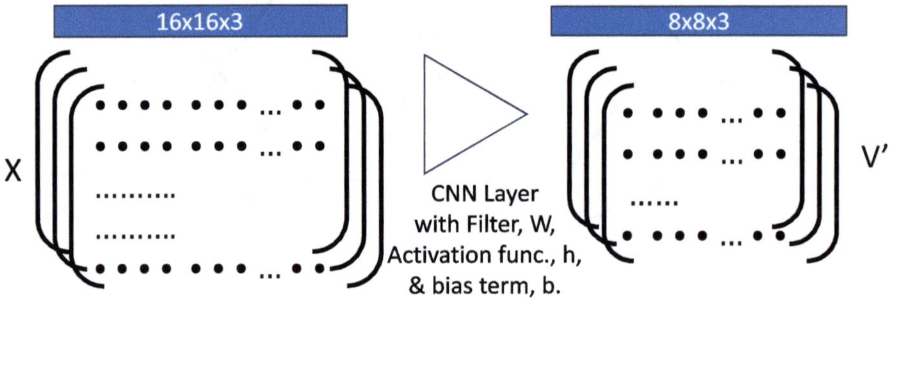

Fig. 6.3 Example of CNN layer with X (16X16X3 image data as input) and V' (8x8x3 data at the output); W the convolutional filter, h is the activation function, and b is the bias term

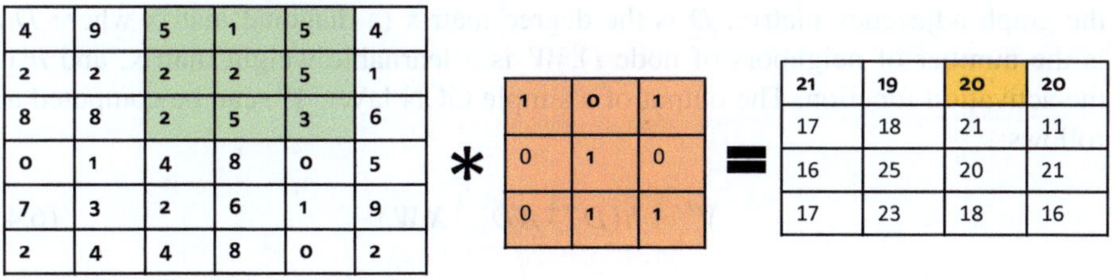

Fig. 6.4 Example of discrete convolution with X (6 × 6), W (3 × 3), and results (4 × 4)

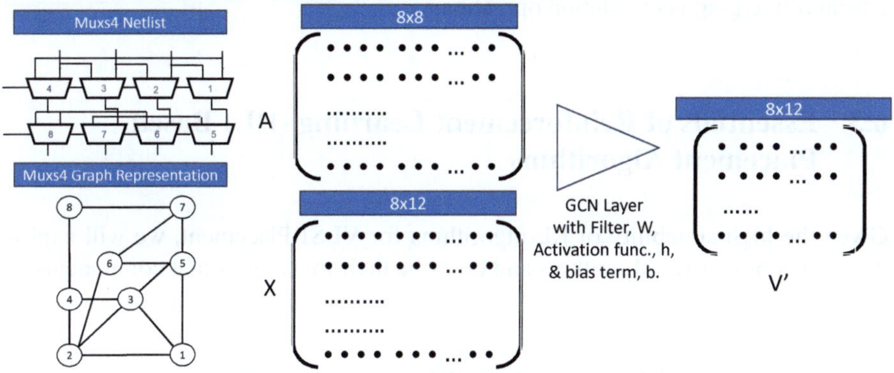

Fig. 6.5 Example of GCN layer with A (8 × 8 adjacency matrix), X (8 × 12 features matrix), and V' (8 × 12 data at the output); W is the convolutional filter, h is the activation function, and b is the bias term. The adjacency matrix identifies the neighbors of a node and aggregates data from them

following equation:

$$(X * W)[n] = \sum_{m=-M}^{M} X[n-m]W[m] \tag{6.3}$$

During training, the network adjusts these filters to automatically learn features or patterns from the input data relevant to the task at hand, such as placement feature representation.

Graph Convolutional Networks (GCNs) VLSI netlists typically lack a fixed orientation and can have a variable number of neighboring components or elements. This variability in the input data can pose a challenge for traditional convolutional networks, which are typically designed for regular grid-like structures. GCNs are a type of convolutional network that can extract local features from graphs when there are a variable number of neighbors (Fig. 6.5).

For GCN, we represent netlist as a graph, $G(V, E)$, such that V is the set of cells, E is the set of edges (nets that connect the cells), X is the features set, A is the graph adjacency matrix, D is the degree matrix (a diagonal matrix where D_{ii} is the number of neighbors of node i), W is a learnable weight matrix, and h is the activation function. The output of a simple GCN layer, V', can be computed as follows:

$$V' = h(D^{-\frac{1}{2}} A D^{-\frac{1}{2}} X W) \tag{6.4}$$

This formula incorporates the graph structure by normalizing the adjacency matrix using the degree matrix. The multiplication $D^{-\frac{1}{2}} A D^{-\frac{1}{2}}$ can be treated as a form of the graph convolution operation.

6.3 Essentials of Reinforcement Learning (RL)-Based Placement Algorithms

Given the high suitability of RL algorithms for VLSI Placement, we will explore the philosophy of RL algorithms and describe their most important components.

6.3.1 RL Algorithms for VLSI Placement and Their Classification

Formally, reinforcement learning is defined as the finite Markov Decision Process (MDP). It is a paradigm of interactive learning where an agent takes action $a \in A$ according to policy $\pi(a|s)$ and changes the state of the environment from s to s', where $s \in S$ and $s' \in S$. Reward $r \in [-R_{max}, R_{max}]$ or some finite R_{max} is obtained during the transition from s to s' by observing changes in the cost function from iteration to iteration. S is the set of all possible states, called the state space, and A is the finite set of actions, called the action set (Fig. 6.6). The MDP and agent generate a trajectory sequence like this $s_0, a_0, r_1, s_1, a_1, r_2, s_2, a_2, r_3, ...$ The goal of the RL is to sample trajectories and generalize the learning to find the optimum policy π^*, which maximizes the discounted reward G. The reward discounting factor is given by gamma, γ. The agent follows the policy π with $1 - \epsilon$ probability and deviates from the policy with ϵ probability. The discounted reward when the agent deviates from the policy and thereby follows the policy π is called q_π, and it is given by

$$q_\pi(s) = \sum_{s',r} p(s', r \mid s, a)(r + \gamma v_\pi(s')) \tag{6.5}$$

The discounted reward is estimated using value function $v_\pi(s)$, also known as Bellman equation [12], and is given by

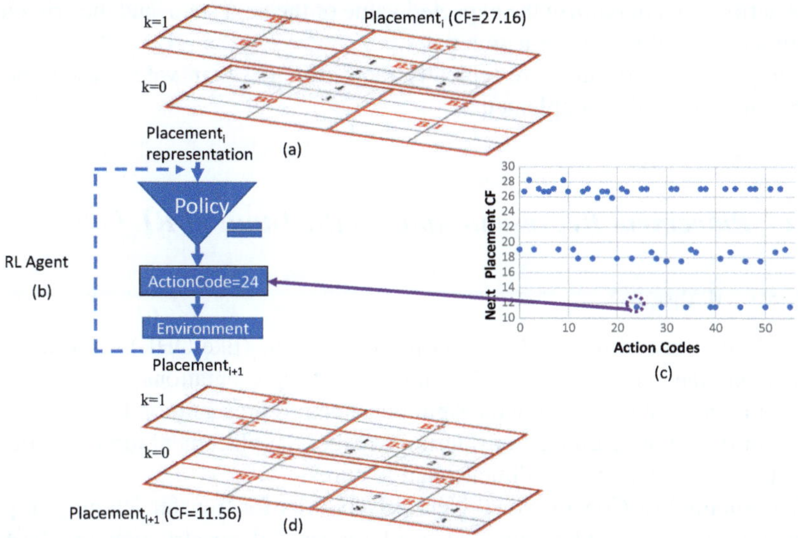

Fig. 6.6 Examples of reinforcement learning agent, placement representation, action, and reward based on cost function [13]

$$v_\pi(s) = \sum_a \pi(a \mid s) \sum_{s',r} p(s', r \mid s, a)(r + \gamma v_\pi(s')) \tag{6.6}$$

Classification of RL Algorithms The reinforcement learning algorithms in the book *Reinforcement Learning: An Introduction* by Sutton and Barto [12] can be classified into two main categories: value-based algorithms and policy-based algorithms.

Value-based algorithms learn a value function, which is a mapping from states to expected returns. The value function is then used to select actions that maximize the expected return. Value-based algorithms include:

- **Dynamic programming (DP)** algorithms learn a value function by solving a set of recursive equations. DP algorithms are guaranteed to find the optimal value function, but they can be computationally expensive.
- **Monte Carlo (MC) value estimation** algorithms learn a value function by estimating the expected return for each state.
- **Temporal difference (TD)** algorithms learn a value function by updating the value of states based on the current reward and the estimated value of the next state. TD algorithms include Q-learning and SARSA.

Policy-based algorithms learn a policy, which maps states to actions. The policy is then used to select actions in the environment. For example, **actor-critic** algorithms simultaneously learn a policy and a value function. The actor learns to select actions that maximize the expected value of the next state, and the critic learns to estimate the value of state-action pairs.

Policy-based algorithms are typically more efficient than value-based ones but can be more difficult to implement.

6.3.2 Placement Representations as the Input of RL Policy

Next, we will discuss:

- The limitations of linear placement representations (like OHE) for VLSI placement and the need for nonlinear, feature-based representations
- The importance of choosing the right placement representation for policy design in reinforcement learning, considering whether the placement representation is a vector of features or multidimensional features
- The potential of Convolutional Neural Networks for transfer learning in placement scenarios, emphasizing the need for special construction of placement representations to fit the multidimensional input requirements of CNNs.

Linear Versus Nonlinear Placement Representations Linear placement representations, such as One Hot Encoding (OHE), encode the placement as a binary vector, with each bit representing a possible placement location. This approach is simple to implement, but it has several drawbacks. First, it can become very large and unwieldy for complex VLSI designs with many possible placement locations. Second, it does not capture the spatial relationships between placement locations, which is important for many VLSI placement objectives.

Nonlinear placement representations, on the other hand, offer more flexibility and better capture the nuances of VLSI placement. Some examples of nonlinear placement representations include:

- Neural Networks: Multilayer network using CNNs and MLPs can be used to represent the VLSI placement layout image shown in Fig. 6.1 into feature vectors using techniques discussed in Sect. 6.2.4, which can be used as input to the RL policy.
- Coordinate or Sequence Encoding: Coordinate encoding is a simple but effective way to encode the spatial relationships between placement locations. Coordinate encoding represents each placement location as a vector of coordinates, which can then be used to compute distances and other spatial relationships between placement locations.

Representation Techniques for MLPs MLPs are a popular neural network type often used in RL policies. MLPs expect a vector of numbers as input, so the placement representation must be converted to a vector before it can be used as input to the MLP, illustrated in Fig. 6.7.

Some techniques for constructing placement representations that are suitable for MLPs include:

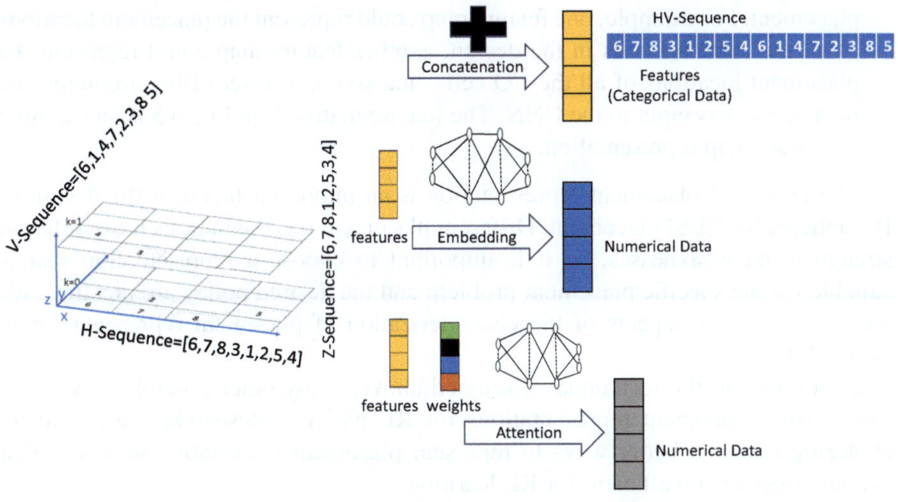

Fig. 6.7 Examples of representation techniques for MLPs

- Concatenation: Concatenation is a simple technique that combines multiple feature vectors into a single vector. For example, the feature vectors for each placement location could be concatenated to form a single vector representing the entire placement.
- Embedding: Embedding is a technique that converts categorical data into numerical data. For example, the placement locations could be embedded into a numerical space using CNNs. Multi-view embeddings can be used to combine categorical data from multiple sources, such as congestion data, technology information, design rules, etc.
- Attention: Attention is a technique that creates vector by focusing on different parts of the placement representation. For example, attention mechanism can be used to focus on the placement locations of certain critical cells in the design.

Representation Techniques for CNN CNNs are a type of neural network that is well suited for processing spatial data. CNNs expect multidimensional data as input, so the placement representation must be converted to a multidimensional tensor before it can be used as input to the CNN.

Some techniques for constructing placement representations that are compatible with CNNs include:

- Grid representation: The placement can be represented as a grid, where each cell in the grid represents a placement location. The grid can then be used as input to the CNN. The RGB placement image represented by matrix X in Fig. 6.3 is an example of grid representation.
- Feature map representation: The placement can be represented as a set of feature maps, where each feature map represents a different type of information about the

placement. For example, one feature map could represent the placement locations of all the standard cells in the design, another feature map could represent the placement locations of all the I/O cells, and so on. The set of feature maps can then be used as input to the CNN. The feature matrix X in Fig. 6.5 is an example of feature map representation.

The choice of placement representation is an important factor in the design of RL policies for VLSI placement. Different placement representations have different strengths and weaknesses, so it is important to choose a representation that is suitable for the specific placement problem and the desired policy architecture. We will share various aspects of the characterization of placement representation in Sect. 6.6.5.

In addition to the techniques discussed above, many other possible ways exist to construct placement representations for RL policies. Researchers are actively exploring new and better ways to represent placement information in a way that is both effective and efficient for RL learning.

6.3.3 Action as the Output of the RL Policy

– Sequential versus concurrent actions: The RL algorithm can learn to predict the placement of cells one by one in a sequential manner or to perturb the placement of more than one cell.
– Linear versus nonlinear actions: The output of the policy can encode actions using linear methods like OHE or multi-output encoding with nonlinear characteristics.

6.3.4 RL Policy for Placement Algorithms

The policy is the heart of the RL algorithm, and in this section, we will discuss three different types of RL policies.

– Policy Without ML: RL algorithm can be designed to learn the policy dynamically without using neural networks. However, such learning cannot be stored or reused for future runs.
– Policy with ML: Policy can be designed using machine learning that learns global features of the placement. Such a policy can save learning in the form of structure and weights of the ML model. This allows the reuse of historical learning in future runs.
– Policy with Generalized Learning: Policy with machine learning model that can learn based on local features. For example, properties could be based on the connectivity of neighboring cells in the netlist. This allows the policy to

generalize learning and transfer knowledge across different circuits and is a powerful benefit of RL algorithms over traditional algorithms.

6.4 AI-Enabled 2D Placement Algorithms

The VLSI 2D placement problem strives to find the most optimal cell positions to minimize area and/or the total wire length. We will review solutions to the 2D placement problem using various RL approaches that have been proposed and can be broadly categorized as GPU acceleration methods [14, 15], placement of macros (floorplanning) [16], initial solution [17], parameters tuning [18, 19], PPA optimization [20], and routing cooperation [21, 22].

We will categorize the above algorithms into the following categories:

- AI-enabled algorithm using GPU acceleration
- AI-enabled algorithm for placement of macros (floorplanning)
- AI-enabled algorithm for initial solution
- AI-enabled algorithm for parameter tuning
- AI-enabled algorithm with PPA co-optimization
- AI-enabled algorithm with routing cooperation.

6.4.1 AI-Based Algorithm Using GPU Acceleration

If we treat placement optimization as a neural network training problem [9], we can construct Directed Acyclic Graphs (DAGs) that represent the netlist as a neural network. The DAG is used to compute the network's predictions, loss, and gradients. We then update the network's weights based on those gradients to minimize the loss function. The gradient descent algorithms are computationally expensive, but GPU acceleration can be utilized to improve efficiency.

VLSI circuits can be represented as a hypergraph $H = (W, E)$, where W is the set of cells represented by vertices, and E is the set of nets represented by hyperedges. Let $w(x, y) \in W$ hold the location of the cell; such a graph is equivalent to a computational graph in neural networks. The use of GPUs can improve the efficiency of determining neural network weights using gradient descent algorithms since this optimization involves matrix multiplications [23].

DREAMPlace [14] is a deep learning toolkit-enabled GPU acceleration framework for modern VLSI placement [14] compatible with state-of-the-art analytical placers. This framework uses equations from the ePlace analytical placement algorithm [24] to build the graph representation and leverage available CPUs and GPUs to accelerate training. DREAM-GAN [15], an extension of the DREAMPlace algorithm, uses generative adversarial learning to compare intermediate results against placements generated by commercial tools. The idea of GPU acceleration is used in many other VLSI placement algorithms.

Placement Representation Placement is represented using a Graph Neural Network (GNN) to embed the netlist into a low-dimensional latent space. This latent space representation captures the structural and electrical properties of the netlist and can be used to efficiently compute various placement metrics, such as wirelength and congestion.

Optimization Algorithm The network inputs the netlist embedding and the current placement solution and computes loss based on feedforward results. The loss can be used to generate the updated placement solution as it happens in gradient descent algorithms. The feedforward and backpropagation passes happen repeatedly until algorithms converge to an optimum solution.

6.4.2 AI-Enabled Algorithm for Placement of Macros (Floorplanning)

Reinforcement learning can be used during the floorplanning stage to optimize the location of macros [16]. Figure 6.8 shows the high-level flow of such a two-stage algorithm. An actor-critic policy-based RL algorithm (Sect. 6.3.1) places the macros from a chip netlist onto a chip placement layout one at a time, while other algorithms are used for the cell-level placement.

Placement Representation Multi-view embeddings (Sect. 6.3.2) are generated to represent the netlist, chip layout, and related features. Graph Convolutional Networks are used to embed netlist information. MLPs (Sect. 6.2.4) are used to represent information related to the technology process.

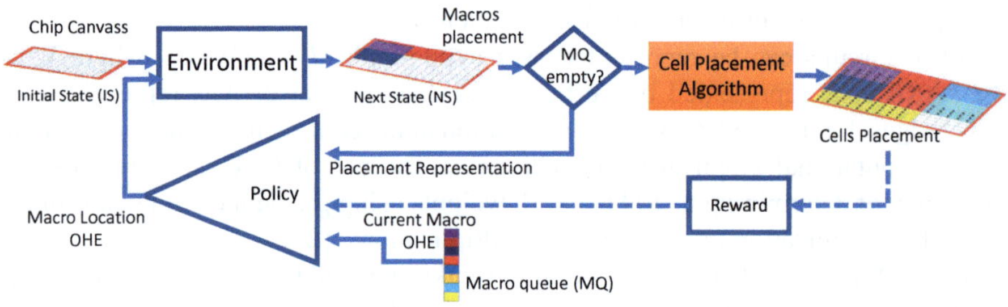

Fig. 6.8 Reinforcement learning for placement of macros. The policy input is a multi-view placement embedding and the current macro to be placed. The policy decides the location of the current macro. The RL stage ends when all macros are placed one by one. The cell placement algorithm is used to place all cells. The reward is then calculated based on the change in the placement quality compared to the previous iteration. This process repeats until RL training is completed

Action Representation A sequential action approach means macros are placed one by one. After placing all the macros, the final placement is generated with a force-directed method [25], and the reward is calculated.

Policy Network The deconvolutional layer in the policy network works by first upsampling the input representation of the netlist. The deconvolutional layer can be considered a special case of convolutional networks. In deconvolution, the filter (or kernel) typically has a larger spatial extent than the input patch it operates on. This larger filter size contributes to the expansion of the output feature map and is one of the reasons why deconvolution is used for upsampling. This upsampling increases the representation's spatial resolution, allowing the network to learn more fine-grained features of the netlist.

The output of the policy network is a probability distribution over the possible placements for each node in the netlist. The RL agent uses this probability distribution to select a placement for the node.

6.4.3 AI-Enabled Algorithm for Initial Solution

The method in [17] leverages RL to quickly generate a good initial solution but leaves the final optimization to simulated annealing (SA), as shown in Fig. 6.9.

Placement Representation Sequence pair is a concatenation technique that is very commonly used in traditional placement algorithms. Though it is a very compact and memory-efficient representation technique, it is not a well-suited technique for machine learning models since it is not inductive and contains no information about the chip layout or the density of cells.

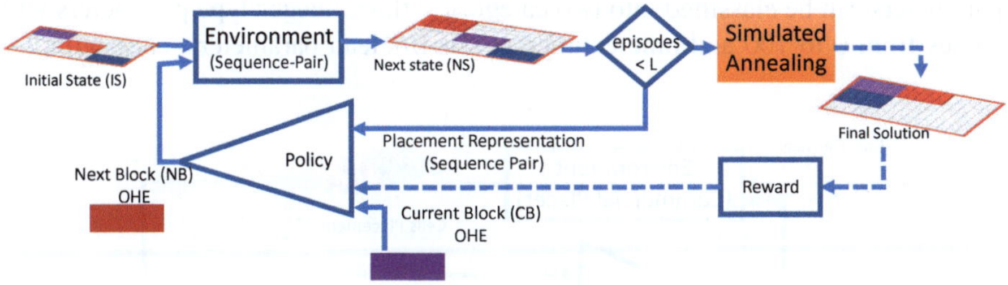

Fig. 6.9 Reinforcement Learning for Initial Placement used by Simulated Annealing to generate the final solution, the quality of which is used as feedback to update the policy. The policy input is the current block and the sequence pair for placement representation. The policy predicts the next block to be exchanged, which later becomes the current block in subsequent iterations. RL flow takes L episodes to generate the initial solution

Action Representation This algorithm allows only one type of action: switching the current block with another block. For a benchmark with 625 blocks, the policy network can select from one of the 624 other blocks.

Policy Network The authors have not disclosed what type of machine learning model is used for the policy network. However, the policy network can be treated as a multi-class classification problem. Generally, the accuracy of such networks is challenging with the increasing number of classes/actions.

6.4.4 AI-Enabled Algorithm for Parameters Tuning

AI-enabled algorithm for placer guidance or parameters tuning is adopted by Lu et al. [19] and Agnesina et al. [18]. GNN is used by Lu et al. to guide the human for placer configuration [19], whereas Agnesina et al. use GNN in the RL loop to guide the placer without human intervention, and we will discuss its main ideas next. Figure 6.10 shows their "deep reinforcement learning (RL) framework to optimize the placement parameters of a commercial EDA tool [using] a mixture of handcrafted features [and] graph embeddings" [18].

Placement Representation Embedding techniques (Sect. 6.3.2) are used to feed relevant data into the policy network. The GraphSAGE algorithm [26] is used to create embedding as Graph Neural Networks. Inputs are a graph representation of the netlist, features representing netlist properties like cell count, number of nets, etc., and the current placer parameters. Neural networks are used to reduce the dimensionality of the data such that it is ready for the policy network.

Action Representation While commercial placers have thousands of configuration parameters, the authors have selected 12 for automated tuning. The selected parameters can be classified into two categories: three integer-type parameters with values from 0 to 100 and nine enumerated or Boolean parameters, each with 2–3

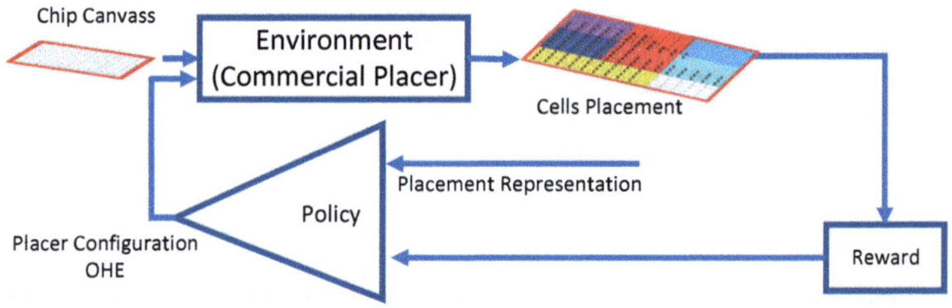

Fig. 6.10 Reinforcement learning flow for placement optimization using parameters tuning. The input of the RL policy is the graph-based embedding, and it predicts values of the selected configuration parameters. The commercial placer used the RL-generated placement parameters to generate the 2D placement. The reward is calculated based on the change in the cost function compared to the previous iteration

possible values. Despite the limited number of parameters, the search space is still in the hundreds of billions. To navigate this search space, the authors have defined 11 operators, for example, “flip Boolean parameters.”

Policy Network The policy’s goal is to create the trajectory of actions that lead the algorithm to arrive at an optimal configuration of parameters. The policy selects one of the 11 operators represented using OHE at each iteration. A policy network is created using neural network layers.

6.4.5 AI-Enabled Algorithm with PPA Co-optimization

Power-Performance-Area (PPA) co-optimization with an unsupervised algorithm [20] is used for netlist graph clustering. The clustering helps suggest commercial placer parameters. Such optimization is often a missing aspect of traditional algorithms.

6.4.6 AI-Enabled Algorithm with Routing Cooperation

RL is used for the placement of macros with routing cooperation [21, 22]. As discussed earlier (Sect. 6.3.2), multi-view embedding placement representation and sequential placement of macros are used.

6.4.7 Comparison of AI-Enabled 2D Placement Techniques

A common theme with all these algorithms is to leverage advancement in ML algorithms to better solve the optimization problem, using the methods and techniques described in greater detail in Sects. 6.2 and 6.3. But due to the challenges of optimizing a large search space using ML, current approaches mostly use these algorithms and techniques to help guide traditional methods or produce a partial placement solution.

Table 6.1 compares these 2D placement algorithms using the following metrics:

- Number of Stages (NoS): how many total stages are used before arriving at the final placement
- Learning Reuse (LRU): if learning is stored as weights or in the structures of the ML algorithm for use in future runs, to help speed up algorithms
- Machine Learning Algorithm Type (MLAT): what is the approach or what is the model used in the case of a model-based approach.

Table 6.1 Comparison of AI-enabled 2D placement techniques

	NoS	LRU	MLAT	ACD
GPU acceleration [14, 15]	2	No	Gradient descent	Netlist
Placement of macros [16]	2	Yes	Multi-view	Grid
Initial solution [17]	2	N/A	N/A	Grid
Parameters tuning [18, 19]	2	Yes	Multi-view	Tool
PPA co-optimization [20]	2	Yes	GCN	Grid
Routing co-optimization [21, 22]	2	Yes	Multi-view	Grid

– Action-space Complexity Dependency (ACD): Large action-space will affect model accuracy, so to ensure consistent results, it is important to manage that complexity in relation to netlist or benchmark size.

One clear observation from Table 6.1 is that all the algorithms use more than one stage. This reflects that the initial wave of ML algorithms for placement produce partial solutions that are further optimized in a second stage. Second, some algorithms suffer a key limitation by not reusing learning in future runs. MLAT can be model-free (e.g., a gradient descent-based approach) or model-based, like CNN or GCN. The ACD is related to either the netlist or the grid size. Often the action space complexity grows exponentially with grid size and poses challenges to the scalability of such approaches. The research challenges of AI-enabled 2D algorithms are summarized in Sect. 6.7.

6.5 AI-Enabled 3D Placement Algorithms

6.5.1 3D Integration Technologies

Three-dimensional (3D) integration enables reduced footprint, more bandwidth, faster performance, and lower power requirements for system-on-chips (SOCs) [27]. Today's 3D integration process involves stacking of dies fabricated in separate processes, known as 3D IC with through-silicon vias (TSVs) [28]. To better benefit from 3D technology, next generation research is focused on sequential fabrication of two or more active layers in a monolithic die, also known as Monolithic 3D (M3D) IC [29].

Monolithic 3D IC allows three different design styles: transistor-level, gate-level, and block-level [30]. In the transistor-level design style, the PMOS and NMOS within each cell is divided into different tiers. Interconnects in the bottom layer are mainly used for cell local connections. The gate-level design style allows gates to be fabricated either in the top or in the bottom layer with inter-cell connection between the tiers utilizing monolithic inter-tier vias (MIVs). In block-level design style, all of the cells of a design block are assigned to the same layer. The three design styles can be mixed in a single chip to produce a hybrid design style.

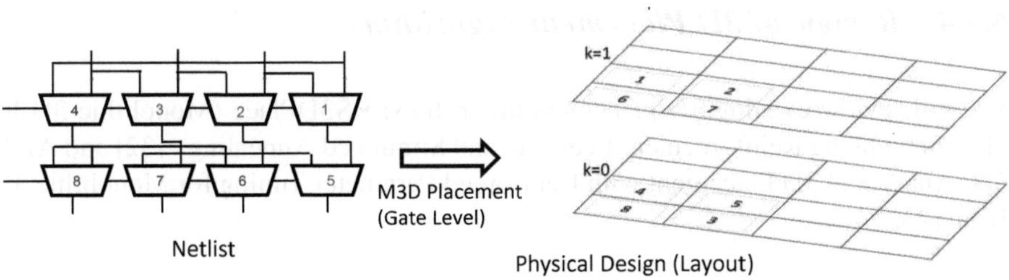

Fig. 6.11 Example of 3D placement (gate-level) using 4-bit Mux-shifter circuit

With minimal changes, standard EDA tools can design partitions for TSV-based 3D IC and M3D with transistor- and block-level styles. For gate-level 3D technology, cell placement must use an approach customized for M3D [31, 30] because gates of a design block may be spread out over multiple tiers.

6.5.2 3D Placement Problem

Given a set of cells, a netlist, the dimensions of the cells (width and height), and the count (k) of 3D tiers, M3D placement aims to find a location $L(x, y, z)$ for each cell to minimize a given cost function while meeting all constraints.

The method of estimating area and total wirelength (TWL) used in the cost function differs in 2D versus 3D Placement. A simple method for estimating area is to calculate the area of the minimum bounding box for the locations of all the cells if projected on a 2D surface. Similarly, the x-y portion of the TWL can be estimated using the Half-Perimeter method if net locations are translated to a 2D surface. The z portion of the TWL can be estimated using a weighted sum of the count of MIVs (Fig. 6.11).

6.5.3 3D Placement Strategies

3D placement strategies can be divided into 2D-to-3D transformation-based algorithms and 3D placement algorithms.

2D-to-3D transformation-based algorithms first generate a 2D placement using existing 2D algorithms. The 2D placement is then transformed to a 3D placement, either simultaneously optimizing the 3D placement (dynamic) or through an additional optimization step for each 3D tier (static, sometimes called pseudo-3D algorithms).

True 3D placement algorithms consider the multiple tiers and possible interconnects between them from the beginning of the placement process.

6.5.4 Review of 3D Placement Algorithms

We will review existing 3D placement algorithms: RS3DPlace (Monolithic 3D IC placement using Reinforcement Learning and Simulated Annealing) [32] and ART-3D (Analytical 3D Placement with Reinforced Parameter Tuning for Monolithic 3D ICs) [33, 34].

6.5.5 AI-Enabled 3D Placement for Initial Solution

As discussed in Sect. 6.4.3, an application of AI-enabled algorithms is to quickly generate a decent initial solution that can be further refined by traditional methods. The RS3DPlace algorithm [32] uses this approach as illustrated in Fig. 6.12. The placement representation is generated by the concatenation of cell names arranged in a manner to represent the relative positioning of cells. Additionally, a concurrent action approach is used to perturb placement layout with all cells. The cells are placed in a gridded layout, as illustrated in Fig. 6.13. To simplify action space, a coarse grid is superimposed on the original grid, creating large bins, as shown in Fig. 6.14. This approach has shown good results using cells with uniform size and relatively small-sized benchmarks.

Placement Representation RS3DPlace represents the placement as a sequence of cells, in the order the representation algorithm encounters each cell as it walks through the M3D grid layout. Using the 3D Cartesian coordinate system, the M3D layout is aligned to the origin at k=0 (Fig. 6.13). The algorithm walks the grid, always incrementing coordinates. The columns of the grid are numbered in the x-direction and rows are numbered in the y-direction, while the z-direction counts successively higher tiers. There are two distinct walk operations that determine the path of the walk in the gridded layout.

1. A local walk covers one layer in either the x- or y-direction.
2. A global walk recursively explores the tiers 0 to $k - 1$, in the x- or y-direction.

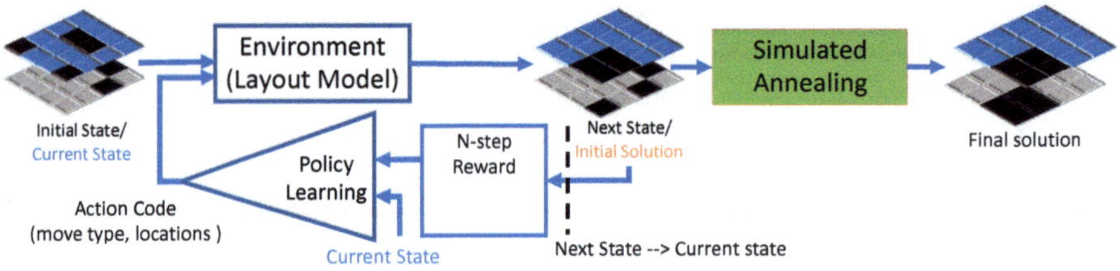

Fig. 6.12 RS3DPlace flow diagram. The basic cycle of RL is represented by the blue section of the diagram on the left. After the RL loop, RS3DPlace proceeds to the SA stage (in green) on the right, where the RL output becomes an initial solution for further optimization

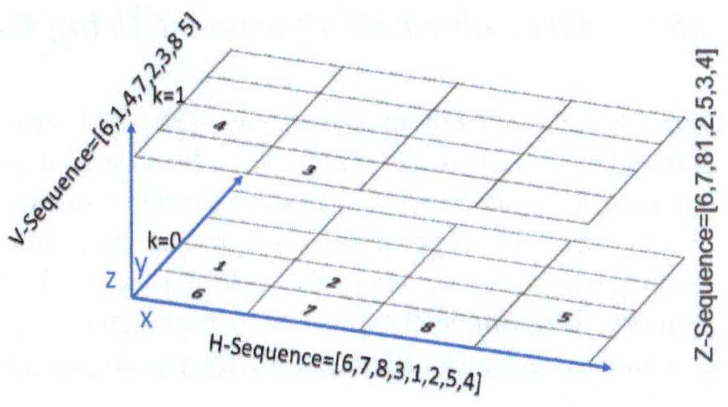

Fig. 6.13 Placement representation for 2D/3D

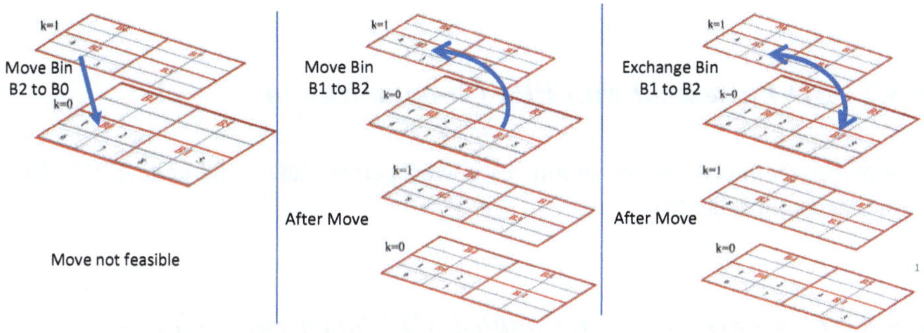

Fig. 6.14 Examples for different types of actions in action scheme BA1 (**a**) bin move that is not feasible, (**b**) bin move that is feasible, and (**c**) exchange bins

The H-Sequence refers to the path generated by a global walk performed in the x-direction layer by layer for a given row. Similarly, the V-Sequence represents the path generated by a global walk conducted in the y-direction layer by layer for a given column. Lastly, the Z-Sequence records a global walk performed layer by layer with a local walk in the x-direction carried out on each row of that layer before going to the next layer.

Action Representation RS3DPlace uses two actions: (a) exchange bins by swapping the content of the two bins, and (b) if feasible, merge the content of two randomly chosen bins. The merge move is possible if the destination bin is not occupied or if there is no overlap between the cells of the origin bin and the cells present at the destination bin. Figure 6.14 depicts several different scenarios.

Policy Network RS3DPlace uses MLPs to predict action given approximate placement representation. This approach generates a rough solution as a starting point for the Simulated Annealing algorithm. By reusing historical learning in the form of model structure and weights, this ML model shows success in making fast predictions. However, RS3DPlace struggles to predict the best action, primarily due to its relatively simple ML network.

6.5.6 AI-Enabled 3D Placement Using Parameters Tuning

The ART-3D placement algorithms [33] and similar work [34] rely on placer parameter optimization, which was discussed in Sect. 6.4.4. Instead of utilizing machine learning models that store learning in the form of a network's structure or weights, this work utilizes a dynamic programming, a type of reinforcement learning (Sect. 6.3.1) where the results of optimization experiments are stored in memory to optimize the placement parameters. OpenTuner [35] was used as a dynamic programming framework for parameter tuning of an academic 3D pacer, NL3D [36].

6.5.7 3D Placement with PPA Co-optimization

The placer method [34] co-optimizes power, performance, and area (PPA) along with placement objectives.

6.5.8 Comparison of AI-Enabled 3D Placement Techniques

A comparison of these 3D placement algorithms is shown in Table 6.2. The research challenges of AI-enabled 2D and 3D algorithms are summarized in Sect. 6.7.

6.6 How to Design and Characterize RL Components?

Characterizing the three main components: (a) model, (b) placement representation, and (c) action representation outside the cyclic loop is necessary for the best-performing RL algorithm, especially for the key metrics of placement quality and runtime.

Table 6.2 Comparison of AI-enabled 3D placement techniques: NoS—Number of Stages, LRU—Learning Reuse in future runs, MLAT—Machine Learning Algorithm Type, ACD—Action-Space Complexity Dependency

	NoS	LRU	MLT	ACD
RS3DPlace	2	Yes	MLP	Grid
ART-3DPlace [31]	2	No	N/A	Tool

6.6.1 Problem Formulation

The ideal model predicts the best perturbation action to enhance the current placement. We treat the best action as one of the placement attributes, the same way we treat wirelength and area. We call the model, placement representation, and action representation "good" if and only if they contribute to better predicting one or more attributes.

Formally, let P represent the current placement and A be the set of attributes of P, which includes wirelength, area, and the best perturbation action. Additionally, let M denote the model, M. The model, M, is considered "good" if it can predict at least one specific attribute a_1 of the current placement, where a_1 is a chosen attribute from the set of placement attributes, A.

$$M \text{ is good } \Leftrightarrow M(P) = a_1 \tag{6.7}$$

A placement representation, R_P, is considered "good" if it enables the policy model to predict the chosen attribute a_1.

$$R_P \text{ is good } \Leftrightarrow M(R_P) = a_1 \tag{6.8}$$

An action representation, R_A, is considered "good" if, given the placement representation R_P, the policy model can predict the best action for a_1.

$$R_A \text{ is good } \Leftrightarrow M(R_P, R_A) = \text{best action for } a_1 \tag{6.9}$$

Ideally, the goodness of the model, M, should be tied to the perturbation action attribute, A, for placement, P. This is a complex problem since measurements of goodness would involve multiple variables, for example, with, i.e., M, R_p, and R_a, the resulting $M(R_p, R_a)$ will be "bad" if even one of M, R_p, or R_a is "bad." A careful and independent evaluation of goodness requires a simplified approach (involving at least two variables, like M and Rp), with "goodness" specifically based on the ability to predict the chosen attribute a_1.

The design and characterization of RL algorithm components are complicated by their cyclic nature. A common approach with similar cyclic systems (such as feedback amplifiers or phase-locked loops in some control systems) is to transfer the data into another domain where analysis is easier and then bring the results back into the original domain. Similarly, it is important to analyze the RL loop in a way that makes it easier to see the cause and effect of various components of the RL algorithm. This is achieved by measuring goodness in terms of the chosen attribute, which may not always be the perturbation action.

So far, RL has been mainly used in game-based applications where the solution space is small and linear methods are used to represent states. Linear state representations such as One-Hot Encoding (OHE) have a low dimension and allow one-to-one mapping between states and weights of the neural networks. The solution space of the VLSI placement problem, in contrast, is in the factorial range, and linear methods cannot be used for placement representation due to large memory requirements. Unfortunately, using nonlinear methods with high dimensionality to represent placement adds challenges to the analysis of goodness.

Nonlinear methods of placement representation avoid the explosion of the number of variables required for representation, but that comes at the cost of complexity in the design and analysis. A simple nonlinear representation using concatenation requires the vector size on the scale of $q * n$ to represent placement, where n is the number of cells and q is a constant. Also, the placement representation is fed into neural networks in the n^p size range for MLPs where p is the number of layers. Hand calculations (estimating value functions using feedforward and backpropagation passes) are cumbersome even for the simplest two-cell problem. A single iteration with a four-layered MLP requires 2,304 multiplication operations for each forward propagation and backpropagation step. Hence, there is a need for a simple automatic framework for design and analysis.

6.6.2 RL Design and Characterization Framework (RLDCF)

RLDCF is a systematic framework for the design and the characterization of M, R_p, and R_a for RL VLSI placement algorithm is outlined below. The original idea of RL characterization is published in [13]. The RLDCF framework reuses open-source regression libraries and allows independent evaluation of each RL component. Evaluation involves the creation of relevant datasets, model selection, model hyperparameters tuning, and validation of models. We use mean absolute error or root mean squared error metrics for the characterization of M, R_p, and R_a independently.

Dataset The first step is to create datasets to enable the utilization of open-source regression analysis libraries for selecting ML models outside the feedback loop. The datasets comprise placement examples from multiple benchmarks with placement representation and the chosen attribute, a_1. The ground truth or label for a_1 is calculated based on placement P.

Model Selection Model selection involves choosing the type or architecture of the ML model to use for a specific task. With RLDCF, possible candidates are Linear Regression, Ensemble Methods, Decision Trees, and Neural Networks (Fig. 6.15).

There is no one-size-fits-all ML algorithm [37]. The prevailing approach for model selection remains one of experimentation and comparison, ultimately exploring various algorithms to find the best fit for a specific problem. To facilitate this process, open-source libraries offering diverse algorithms with standardized input, output, and procedures are invaluable tools for researchers and practitioners in the field.

We used Generalized Linear Model (GLM) [38] for linear regression, Gradient Boosting Machine (GBM) [39], Extreme Gradient Boosting (XGBoost) [40],

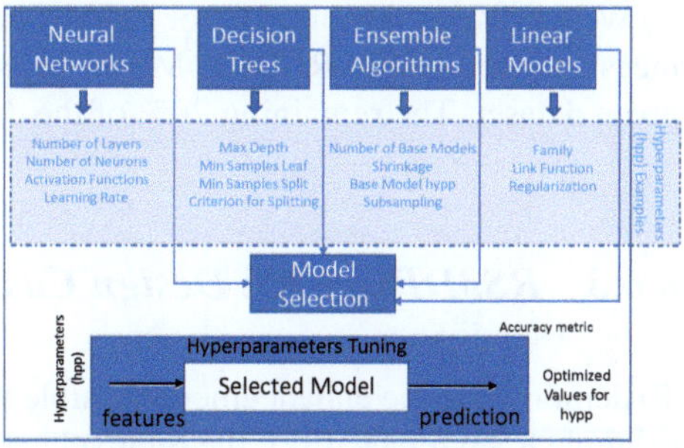

Fig. 6.15 Model selection and tuning

Stacked Ensemble for Ensemble Learning algorithms, and Distributed Random Forest (DRF) [41] and Extremely Randomized Trees (XRT) [41] for Decision Tree/Ensemble algorithms through h2o [42]. We use Neural Network models like MLP and Deep Learning models using Fully Connected layers through Keras library [43], using the TensorFlow library back end for backpropagation using automatic differentiation [44].

Hyperparameter Tuning Hyperparameter tuning involves finding a chosen ML model's optimal configuration or settings. The process of hyperparameter tuning is essential for optimizing complex machine learning models with diverse hyperparameters.

Grid search [45] is a systematic hyperparameter tuning technique that divides the hyperparameter space into discrete grids. Each grid point represents a specific combination of hyperparameter values. The grid search algorithm systematically traverses all possible combinations of hyperparameters to find the point that maximizes the average performance score. This point indicates the best possible combination of hyperparameter values for the model.

Off-the-shelf ML algorithms come with a wide range of hyperparameters to play with. Default values may work well in some cases, and a comprehensive search of the best hyperparameters for a given problem can become computationally challenging. Just like with model selection, the prevalent approach remains iterative exploration. We have reviewed some of the hyperparameter examples in Sect. 6.2.3.

Hyperparameter optimization of neural network models is implemented with KerasTuner [46] using the Hyperband algorithm [47]. The hyperparameter optimization of the rest of the models is implemented using h2o.ai [42].

Model Validation We use cross-validation [48], a statistical procedure used for assessing and comparing ML algorithm performance. In the case of k-fold cross-validation, the dataset is divided into k roughly equal-sized folds. We train the model on $k-1$ folds and test with the remaining folds in each iteration. We go through this process k times, with each fold being the test set once. The model evaluation scores obtained in each iteration are then summarized to provide an overall assessment of the model's performance. Cross-validation helps ensure the model's performance is reliable and not overly dependent on a random data split.

We use Mean Absolute Error (MAE) and Root Mean Squared Error (RMSE) to measure goodness in RLDCF. Model M is trained on randomly selected 80% of a given dataset. The remaining 20% of the data is used for testing.

6.6.3 RS3DPlace RL Design Case-Study

To demonstrate the enhancement possible with the RLDCF framework, we use as an example RS3DPlace [32], the relatively straightforward approach discussed earlier in Sect. 6.5.5 that is well suited for academic research and development purposes. The model placement representation and action representation were analyzed using the RLDCF framework and then tested using five benchmarks of different sizes, including RLcase1 with 18 cells [49, 32], and Mux shifter4 (Muxs4), Muxs8, Muxs64, and Muxs128 [50, 32] with 24, 64, 384, and 896 cells, respectively.

6.6.4 RS3DPlace Policy Model Selection

We select the model by learning the relationship between placement and its Cost Function (CF), the weighted sum of proxy total wirelength (TWL) and area. Although TWL calculations require routing and are time-consuming, proxy TWL is relatively easy to estimate using methods like Half-Perimeter. The proxy TWL is also easier to determine than the placement relationship to the best action. We use the model to predict CF in the placement-CF domain and use it to reveal RL policy behavior in the placement-action domain. We do this by comparing the Predicted CF (PCF) against the ground truth values (best solutions based on proxy TWL) of CF (GCF) obtained from our dataset. Although the accuracy of GCF is questionable, the regression-based selection method of ML models is still meaningful.

Figure 6.16 shows the comparison of ML models. Overall, the MAE of Stacked Ensemble (SE) is 2.5X better than the MLP used in RS3DPlace [32]. Table 6.3 shows the hyperparameters tuning of a sample of hyperparameters for the selected ML models.

6.6.5 Characterization of RS3DPlace Placement Representation

Sequence representation [32] for placement has a one-to-many relationship with the placement layout. However, we propose enhancing sequence representation by adding spatial features of the physical design to help learning algorithms better differentiate placements and learn more accurately.

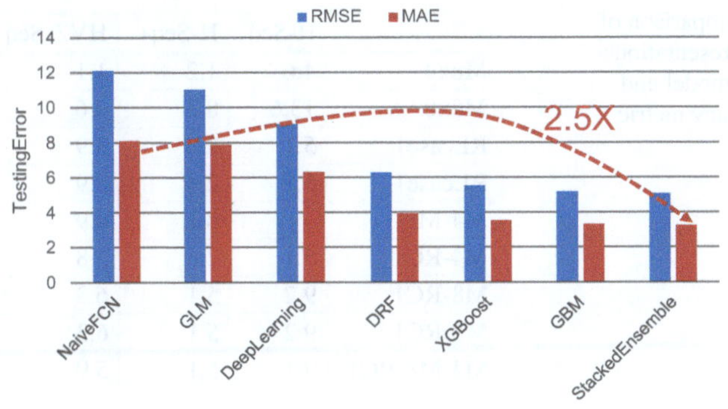

Fig. 6.16 Comparison of policy ML models/algorithm using RLcase1 benchmark and HVZ-Sequence placement representation

Table 6.3 Hyperparameters tuning of a sample of parameters for policy models

Model	Param	Options	Selected value
MLP	Layers	{2,3}	3
	Units	{73, 292,..., 1022}	1022
DL	Layers	{4,5,6}	6
	Activation	{relu, sigmoid}	relu
	Learning Rate	{1e-4, 1e-2, LOG}	1e-4
GLM	Family	{gaussian, gamma,}	gaussian
	Link Function	{Identity, Log, Inverse}	identity
XRT	Max Depth	{15,20,25,30}	20
	Min Samples Leaf	{1,2,3}	1
	Min Samples Split	{1e-03,1e-05,1e-10 }	1e-05
GBM	Number of trees	{40,140,240,340}	340
	Shrinkage	{0.1,0.2,0.3,0.4}	0.1
	Subsampling	{0.6, 0.8, 1.0}	0.8
XGBoost	Number of trees	{40,50,60,70}	70
	Shrinkage	{0.1,0.2,0.3,0.4}	0.3
	Subsampling	{0.6, 0.8, 1.0}	0.6
SE	Base Models Count	[3,4,5,6,7,8]	8

The spatial features are defined by the concatenation of area, total wirelength, and center of mass (CM) features with sequence features. Spatial features are calculated on a per-layer basis for all layers. Equations 6.10–6.12 summarize the method of calculations for the center of mass.

$$xCM_k = \frac{1}{N}\sum^{i} x_{ik} \tag{6.10}$$

Table 6.4 Comparison of placement representations using SE ML model and MAE as accuracy metric

	H-Seq	H-Seq+	HVZ-Seq	HVZ-Seq+
Mux4	4.6	1.2	1.1	1.4
Mux8	13.6	8.3	9.6	7.6
RLcase1	5.1	2.8	3.9	2.1
RLcase1	5.1	2.8	3.9	2.1
M4-M8	–	8.1	8.9	7.2
M4-RC1	5.1	2.6	3.8	1.99
M8-RC1	9.2	5.1	6.3	4.5
M8-RC1	9.2	5.1	6.3	4.5
M4-M8-RC1	9.1	5.1	5.9	4.4

$$yCM_k = \frac{1}{N}\sum^{i} y_{ik} \tag{6.11}$$

$$CM = \frac{1}{(\sqrt{(xCM_1 - xCM_0)^2 + (yCM_1 - yCM_0)^2} + 0.01)} \tag{6.12}$$

We define four different variations of placement representation methods for comparison using this framework:

- H-Seq: horizontal sequence using sequence method
- H-Seq+: concatenation of H-Seq with spatial features
- HVZ-Seq: concatenation of horizontal (H), vertical (V), and inter-tier (Z) sequences
- HVZ-Seq+: HVZ-Sequence concatenated with spatial features

Table 6.4 compares these four placement representations. The MAE is best for the HVZ-Seq+ placement representation method on an average basis. Since HVZ-Seq+ has the most information related to placement, we can conclude that the accuracy of SE ML is related to the amount of information in the R_p.

6.6.6 *Characterization of RS3DPlace Action Representation*

We use RLDCF to evaluate two action schemes: (a) RS3DPlace action scheme BA1 with two action types: bin-move and bin-exchange (as shown in Fig. 6.14) and (b) action scheme BA2 with four types: bin-column-exchange, bin-row-exchange, bin-column-transpose, and bin-row-transpose (as illustrated in Fig. 6.17). RLDCF evaluation of $M(R_p, R_a)$ = best action for a_1 requires a dataset with placement examples and ground truth or label that defines the best action. Table 6.5 shows action count and MAE accuracy using Random Decision Forest (RDF), a type of Ensemble Learning algorithm. Results show that the BA2 action scheme has better accuracy than BA1. Although BA2 has four action types, there are fewer

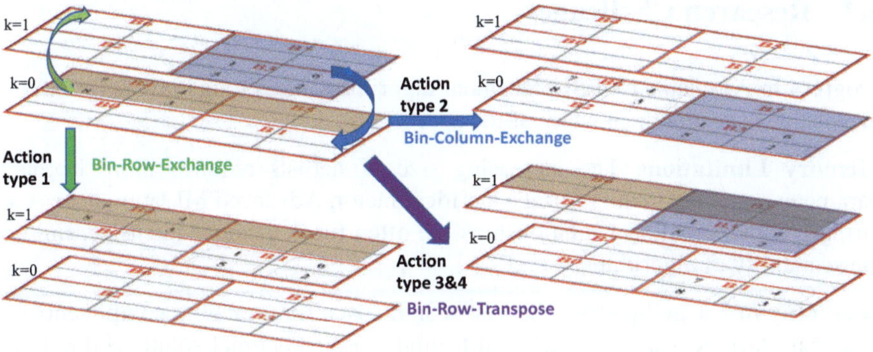

Fig. 6.17 BA2 action scheme highlighting four action types used in SERS-3DPlace [13]: bin row exchange (green), bin column exchange (blue), and bin row/column transpose (purple). The benchmark Muxs4 has 56 actions possible on each placement. Black lines show the base grid, and red lines show the binned grid [32]

Table 6.5 Comparison of action schemes for Muxs32 using RDF

Action scheme	Actions count	Model accuracy
BA1	19012	54%
BA2	728	66%

permutations than with BA1, because the actions of BA2 manipulate rows and columns rather than bins as in BA1. Treated as a multi-class classification, higher accuracy is harder to achieve as the total number of actions or categories increases. Hence, an ideal action scheme would have a smaller set of actions but of various types, making it easy to search for the optimal solution.

6.6.7 Application to 2D Placement Problem

We earlier presented a detailed case-study of 3D placement using the RS3DPlace algorithm. The 2D placement problem, though less complex than the 3D problem, can be evaluated using a similar framework. The main difference in extending RS3DPlace to the 2D placement problem is that fewer types of actions are required to explore the search space. Placement can be represented using HV-Sequence instead of HVZ-Sequence. So both the representation input to the RL Policy and action representations output by the policy are changed. But the machine learning algorithms themselves remain unchanged other than their parameters. Thus, the ideas discussed above for characterizing ML models are equally applicable to the 2D placement problem.

6.7 Research Challenges

Progress in AI-enabled VLSI placement algorithms faces tough challenges due to increasing scale and the novelty of the approach.

Memory Limitations The increasing size of netlists requires more learnable parameters, making memory usage a critical factor. Advanced ML approaches capturing complex nonlinear relationships are often more memory-intensive, limiting the scalability of algorithms.

Searching in Large Spaces The vast search space for arranging components in a practical placement makes it difficult to find close-to-optimal solutions due to the sheer number of possibilities. Algorithms must often sacrifice optimality or runtime, and sometimes some of both, to achieve acceptable placements in reasonable time.

Training Data Generation Another challenge is the lack of ML-friendly datasets tailored for VLSI 2D and M3D placement problems. Such datasets should include diverse placement examples, different circuit types, and labels/metrics created using known cost functions.

Interpretability and Explainability It is crucial that human engineers be able to understand the decisions made by AI models. Industry users must trust that these tools are delivering what they promise and that includes being able to analyze the output, detect errors, and track down the underlying cause. Numerous examples can be found of AI models learning incorrectly from inadvertently biased datasets. Further, if researchers cannot properly interpret and explain decisions made by the model, they will not have a solid basis for improvements and future research directions.

6.8 Summary

AI algorithms are often classified into supervised, unsupervised, and reinforcement learning algorithms based on the role of datasets. Well-known supervised and unsupervised algorithms cannot be used to solve placement problems because datasets for benchmarks are unavailable and there are significant challenges creating them. Conversely, Reinforcement Learning (RL) algorithms interactively create variable-sized datasets and are a natural choice for solving optimization problems like VLSI placement. We briefly described the basic components of RL algorithms, including placement representation, action representation, and policy. We reviewed published AI approaches to placement in 2D and 3D systems and highlighted their strengths, shortcomings, and challenges. We also shared a framework for the design and characterization of RL components and discussed various trade-offs. AI-enabled algorithms for VLSI placement are at an early stage of development and therefore face numerous challenges in terms of memory constraints, sampling efficiency, datasets, and interpretability of models.

References

1. Sait, S. M., & Youssef, H. (1995). *VLSI physical design automation—theory and practice*. New York: IEEE.
2. Russakovsky, O., Deng, J., Su, H., Krause, J., Satheesh, S., Ma, S., Huang, Z., Karpathy, A., Khosla, A., Bernstein, M., Berg, A. C., & Fei-Fei, L. (2015). ImageNet Large Scale Visual Recognition Challenge. https://doi.org/10.48550/arXiv.1409.0575
3. ACM (2019). *The Turing Award: Conceptual and Engineering Breakthroughs That Have Made Deep Neural Networks a Critical Component of Computing*. New York: Springer.
4. Krizhevsky, A., Sutskever, I., & Hinton, G. E. (2012). ImageNet Classification with Deep Convolutional Neural Networks. In *Proceedings of the 25th International Conference on Neural Information Processing Systems—Volume 1*, ser. NIPS'12 (pp. 1097–1105). Red Hook, NY, USA: Curran Associates Inc.
5. Bengio, Y. (2009). Learning deep architectures for AI. *Foundations and Trends in Machine Learning, 2*(1), 1–127.
6. LeCun, Y., & Bengio, Y. (1995). Convolutional networks for image recognition. In *Proceedings of the IEEE international conference on neural networks* (vol. 3, pp. 95–100).
7. Chen, X., Liang, C., Huang, D., Real, E., Wang, K., Liu, Y., Pham, H., Dong, X., Luong, T., C.-J. Hsieh, Lu, Y., & Le, Q. V., Symbolic Discovery of Optimization Algorithms. arXiv:2302.06675 [cs.LG] (2023). https://doi.org/10.48550/arXiv.2302.06675
8. Kahng, A. B. (2021). Advancing Placement. In *Proceedings of the 2021 International Symposium on Physical Design* (pp. 15–22). New York: ACM. https://dl.acm.org/doi/10.1145/3439706.3446884
9. Goodfellow, I. J., Bengio, Y., & Courville, A. (2016). *Deep Learning*. Cambridge, MA, USA: MIT Press. http://www.deeplearningbook.org
10. Geron, A. (2017). *Hands-on machine learning with Scikit-Learn and TensorFlow: Concepts, tools, and techniques to build intelligent systems*. Sebastopol, CA: O'Reilly Media.
11. Delen, D., & Safari, an O'Reilly Media Company (2020). *Predictive Analytics: Data Mining, Machine Learning and Data Science for Practitioners, 2nd Edition*, 2nd ed. New York: FT Press, oCLC: 1224592041.
12. Sutton, R. S., & Barto, A. G. (2018). *Reinforcement learning: an introduction*, 2nd edn., ser. Adaptive computation and machine learning series. New York: The MIT Press.
13. Mansoor, A., & Chrzanowska-Jeske, M. (2024). SERS-3DPlace: Ensemble Reinforcement Learning for 3D Monolithic Placement. In *2024 IEEE International Symposium on Circuits and Systems (ISCAS)*.
14. Lin, Y., Dhar, S., Li, W., Ren, H., Khailany, B., & Pan, D. Z. (2019). Dreamplace: Deep learning toolkit-enabled GPU acceleration for modern VLSI placement. In *Proceedings of the 56th Annual Design Automation Conference 2019* (pp. 1–6).
15. Lu, Y.-C., Ren, H., Hsiao, H.-H., & Lim, S. K. (2023). DREAM-GAN: Advancing DREAMPlace towards Commercial-Quality Using Generative Adversarial Learning. In *Proceedings of the 2023 International Symposium on Physical Design*, ser. ISPD '23 (pp. 141–148). New York, NY, USA: Association for Computing Machinery. https://doi.org/10.1145/3569052.3572993
16. Mirhoseini, A., Goldie, A., Yazgan, M., Jiang, J., Songhori, E., Wang, S., Lee, Y.-J., Johnson, E., Pathak, O., Bae, S., Nazi, A., Pak, J., Tong, A., Srinivasa, K., Hang, W., Tuncer, E., Babu, A., Le, Q. V., Laudon, J., Ho, R., Carpenter, R., & Dean, J. (2020). Chip Placement with Deep Reinforcement Learning. arXiv:2004.10746 [cs]. http://arxiv.org/abs/2004.10746

17. Vashisht, D., Rampal, H., Liao, H., Lu, Y., Shanbhag, D., Fallon, E., & Kara, L. B. (2020). Placement in Integrated Circuits using Cyclic Reinforcement Learning and Simulated Annealing. arXiv:2011.07577 [cs]. http://arxiv.org/abs/2011.07577
18. Agnesina, A., Chang, K., & Lim, S. K. (2020). VLSI Placement Parameter Optimization using Deep Reinforcement Learning. In *2020 IEEE/ACM International Conference On Computer Aided Design (ICCAD)* (pp. 1–9).
19. Lu, Y.-C., Pentapati, S., & Lim, S. K. (2020). VLSI placement optimization using graph neural networks. In *Proceedings of the 34th Advances in Neural Information Processing Systems (NeurIPS) Workshop on ML for Systems, Virtual* (pp. 6–12).
20. Lu, Y.-C., Yang, T., Lim, S. K., & Ren, H. (2022). Placement Optimization via PPA-Directed Graph Clustering. In *Proceedings of the 2022 ACM/IEEE Workshop on Machine Learning for CAD*, ser. MLCAD '22 (pp. 1–6). New York, NY, USA: Association for Computing Machinery. https://doi.org/10.1145/3551901.3556482
21. Cheng, R., & Yan, J. (2021). On Joint Learning for Solving Placement and Routing in Chip Design. In *35th Conference on Neural Information Processing Systems (NeurIPS)*. https://arxiv.org/abs/2111.00234
22. Lai, Y., Mu, Y., & Luo, P. (2022). MaskPlace: Fast chip placement via reinforced visual representation learning. In *36th Conference on Neural Information Processing Systems (NeurIPS)*.
23. Povey, D., Zhang, X., & Khudanpur, S. (2014). Parallel training of deep neural networks with natural gradient and parameter averaging. arXiv preprint arXiv:1410.7455. https://citeseerx.ist.psu.edu/document?repid=rep1&type=pdf&doi=a14ef7c2907ab77dc835909bb690c8860b3fe889
24. Lu, J., Chen, P., Chang, C.-C., Sha, L., Huang, D. J.-H., Teng, C.-C., & Cheng, C.-K. (2015). ePlace: Electrostatics-Based Placement Using Fast Fourier Transform and Nesterov's Method. *ACM Transactions on Design Automation of Electronic Systems, 20*(2). https://doi.org/10.1145/2699873
25. Spindler, P., Schlichtmann, U., & Johannes, F. M. (2008). Kraftwerk2—A Fast Force-Directed Quadratic Placement Approach Using an Accurate Net Model. *IEEE Transactions on Computer-Aided Design of Integrated Circuits and Systems, 27*(8), 1398–1411.
26. Hamilton, W. L., Ying, R., & Leskovec, J. (2018). Inductive Representation Learning on Large Graphs. arXiv:1706.02216 [cs, stat]. http://arxiv.org/abs/1706.02216
27. Mallik, A., Vandooren, A., Witters, L., Walke, A., Franco, J., Sherazi, Y., Weckx, P., Yakimets, D., Bardon, M., Parvais, B., Debacker, P., Ku, B. W., Lim, S. K., Mocuta, A., Mocuta, D., Ryckaert, J., Collaert, N., & Raghavan, P. (2017). The impact of sequential-3D integration on semiconductor scaling roadmap. In *2017 IEEE International Electron Devices Meeting (IEDM)* (pp. 32.1.1–31.1.4). New York: IEEE. http://ieeexplore.ieee.org/document/8268483/
28. Chang, K., Koneru, A., Chakrabarty, K., & Lim, S. K. (2017) Design automation and testing of monolithic 3D ICs: Opportunities, challenges, and solutions: (invited paper). In *2017 IEEE/ACM International Conference on Computer-Aided Design (ICCAD)* (pp. 805–810). New York: IEEE. http://ieeexplore.ieee.org/document/8203860/
29. Lee, Y.-J., Limbrick, D., & Lim, S. K. (2013). Power benefit study for ultra-high density transistor-level monolithic 3D ICs. In *Proceedings of the 50th Annual Design Automation Conference on—DAC '13* (p. 1). New York: ACM Press. http://dl.acm.org/citation.cfm?doid=2463209.2488863
30. Panth, S. A., Samadi, K., Du, Y., & Lim, S. K. (2014) Design and CAD methodologies for low power gate-level monolithic 3D ICs. In *Proceedings of the 2014 International Symposium on Low Power Electronics and Design* (pp. 171–176). New York: ACM. https://dl.acm.org/doi/10.1145/2627369.2627642
31. Guler, A. & Jha, N. K. (2018) Hybrid Monolithic 3-D IC Floorplanner. *IEEE Transactions on Very Large Scale Integration (VLSI) Systems, 26*(10), 1868–1880. https://ieeexplore.ieee.org/document/8374061/
32. Mansoor, A. & Chrzanowska-Jeske, M. (2022). RS3DPlace: Monolithic 3D IC placement using Reinforcement Learning and Simulated Annealing. In *2022 IEEE International Symposium on Circuits and Systems (ISCAS)* (pp. 394–398).

33. Murali, G., Shaji, S. M., Agnesina, A., Luo, G., & Lim, S. K. (2022). ART-3D: Analytical 3D placement with Reinforced Parameter Tuning for Monolithic 3D ICs. In *Proceedings of the 2022 International Symposium on Physical Design*, ser. ISPD '22 (pp. 97–104). New York, NY, USA: Association for Computing Machinery. https://doi.org/10.1145/3505170.3506725
34. Murali, G., Agnesina, A., & Lim, S. K. (2023). A PPA Study of Reinforced Placement Parameter Autotuning: Pseudo-3D vs. True-3D Placers. *ACM Transactions on Design Automation of Electronic Systems*. https://doi.org/10.1145/3582007
35. Ansel, J., Kamil, S., Veeramachaneni, K., Ragan-Kelley, J., Bosboom, J., O'Reilly, U.-M., & Amarasinghe, S. (2014). OpenTuner: An Extensible Framework for Program Autotuning. In *International Conference on Parallel Architectures and Compilation Techniques (PACT)*, Edmonton, Canada. http://groups.csail.mit.edu/commit/papers/2014/ansel-pact14-opentuner.pdf
36. Luo, G., Shi, Y., & Cong, J. (2013). An Analytical Placement Framework for 3-D ICs and Its Extension on Thermal Awareness. *IEEE Transactions on Computer-Aided Design of Integrated Circuits and Systems, 32*(4), 510–523.
37. Wolpert, D. H., & Macready, W. G. (1997). No free lunch theorems for optimization. *IEEE Transactions on Evolutionary Computation, 1*(1), 67–82.
38. Breslow, N. E. (2007). Generalized linear models: Checking assumptions and strengthening conclusions.
39. Friedman, J. H. (2000). Greedy Function Approximation: A Gradient Boosting Machine. *Annals of Statistics, 29*, 1189–1232.
40. Chen, T., & Guestrin, C. (2016). XGBoost: A Scalable Tree Boosting System.. In Krishnapuram, B., Shah, M., Smola, A. J., Aggarwal, C., Shen, D., & Rastogi, R., (Eds.), *KDD* (pp. 785–794). New York: ACM. http://dblp.uni-trier.de/db/conf/kdd/kdd2016.html#ChenG16
41. Geurts, P., Ernst, D., & Wehenkel, L. (2006). Extremely Randomized Trees. *Machine Learning, 63*(1), 3–42. https://doi.org/10.1007/s10994-006-6226-1
42. H2O.ai (2022). *H2O: Python Interface for H2O*, python package version 3.38.0.2. https://github.com/h2oai/h2o-3
43. Chollet, F., et al. (2015). Keras. https://github.com/fchollet/keras
44. Abadi, M., Agarwal, A., Barham, P., Brevdo, E., Chen, Z., Citro, C., Corrado, G. S., Davis, A., Dean, J., Devin, M., Ghemawat, S., Goodfellow, I., Harp, A., Irving, G., Isard, M., Jia, Y., Jozefowicz, R., Kaiser, L., Kudlur, M., Levenberg, J., Manevich, D., Monga, R., Moore, S., Murray, D., Olah, C., Schuster, M., Shlens, J., Steiner, B., Tucker, I., Vasudevan, V., Warden, P., Wei, M., Yu, Q., Zhang, X., & Zheng, X. (2016). TensorFlow: Large-scale machine learning on heterogeneous distributed systems. arXiv preprint arXiv:1603.04467.
45. Bergstra, J., & Bengio, Y. (2011). Algorithms for hyperparameter optimization. *Journal of Machine Learning Research, 13*(1), 2563–2595.
46. O'Malley, T., Bursztein, E., Long, J., Chollet, F., Jin, H., Invernizzi, L., et al. (2019). Keras Tuner. https://github.com/keras-team/keras-tuner
47. Li, L., Jamieson, K., & Deisenroth, M. P. (2018). Hyperband: A bandit-based approach to hyperparameter optimization. *Journal of Machine Learning Research, 18*(1), 6566–6606.
48. James, G., Witten, D., Hastie, T., & Tibshirani, R. (2013). *An Introduction to Statistical Learning: with Applications in R*. Berlin: Springer. https://faculty.marshall.usc.edu/gareth-james/ISL/
49. He, X., Wang, Y., Guo, Y., & Cotofana, S. (2017). A Mixed-Size Monolithic 3D Placer with 2D Layout Inheritance. In *Proceedings of the on Great Lakes Symposium on VLSI 2017*, ser. GLSVLSI '17 (pp. 29–34). New York, NY, USA: Association for Computing Machinery. https://doi.org/10.1145/3060403.3060411
50. Zhuang, H., Lu, J., Samadi, K., Du, Y., & Cheng, C.-K. (2013). Performance-driven placement for design of rotation and right arithmetic shifters in monolithic 3D ICs. In *2013 International Conference on Communications, Circuits and Systems (ICCCAS)* (pp. 509–513). New York: IEEE. http://ieeexplore.ieee.org/document/6765395/

Chapter 7
Enhancing FPGA CAD Flow with AI-Powered Solutions

Chunxiao Lin and Yang Yi

7.1 What Is FPGA?

Field-Programmable Gate Arrays (FPGAs) are a category of programmable logic devices that use electronic components to build reconfigurable digital circuits. Among these, Read-Only Memories (ROMs) and Mask Programmable Gate Arrays (MPGAs) belong to factory-programmable devices. They are programmed during the manufacturing process. This programming is permanent and aims at high-volume and application-specific design. Another type of device, called field-programmable devices, allows users to reprogram the devices after fabrication. Simple programmable logic devices (SPLDs), complex programmable logic devices (CPLDs), and FPGAs all belong to this type. SPLDs are suitable for the implementation of small-scale logic circuits, with limited flexibility and complexity. CPLDs are made up of several SPLDs with programmable switch matrices in between. This provides better flexibility, which is, however, still restricted by the architecture of sea of gates. The development of FPGAs solved the problem by adopting the RAM-based lookup table (LUT) structure. Compared to all other programmable devices, FPGAs can achieve great flexibility and support for large-scale circuit design.

Figure 7.1 shows the basic architecture of an FPGA. There are three main elements: configurable logic blocks (CLBs), switch matrices, and programmable I/O blocks. CLBs are the primary building blocks of FPGAs, including LUTs and flip-flops (FFs) for the implementation of combinational logic and sequential logic. Switch matrices are used to route signals between different CLBs and other components within the FPGA. This supports the implementation of complex circuits. The interaction between the FPGA and external components, such as

C. Lin (✉) · Y. Yi
Bradley Department of Electrical and Computer Engineering, Virginia Tech, Blacksburg, VA, USA
e-mail: chunxiaol@vt.edu; yangyi8@vt.edu

A. Iranmanesh, H. Sayadi (eds.), *AI-Enabled Electronic Circuit and System Design*,
https://doi.org/10.1007/978-3-031-71436-8_7

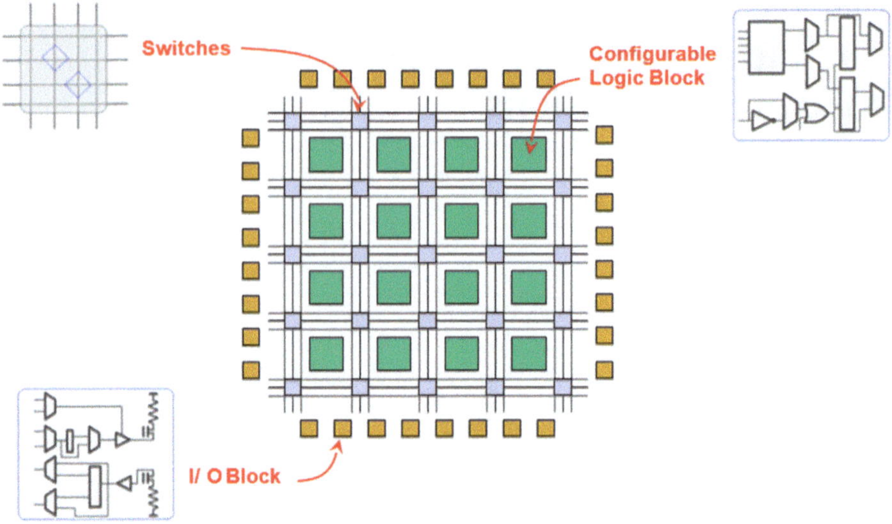

Fig. 7.1 The FPGA structure with I/O block, logic block, and switches

the processor, is enabled by programmable I/O blocks. Moreover, there are more specialized programmable function blocks like ALUs (arithmetic logic units), block RAMs, and DSP (digital signal processing) slices, which are distributed throughout the board and not displayed in Fig. 7.1. Therefore, all these components enable the ultrahigh programmability of FPGAs.

FPGA, as a hardware development platform, also shows its irreplaceability in comparison with other platforms. Unlike CPUs with powerful scheduling capability, full programmability but low computational complexity and high power consumption, GPUs with high parallelism but only within a single application, and ASICs with ultralow cost, low power consumption, but time-consuming fabrication, FPGAs achieve a balance between the programmability and the power consumption, with both pipeline parallelism and data parallelism for higher concurrent processing, and a relatively shorter development cycle.

Due to their ability to create complex digital circuits directly in hardware, FPGAs are widely used in several fields. This includes data centers [1, 2], the automobile [3, 4], telecommunications [5–8], aerospace [9–11], and defense industries [12–14]. The capacity for high-speed parallel processing has also helped FPGAs become better known in recent years in the fields of artificial intelligence and machine learning [15–19].

Currently, the top FPGA manufacturers are listed below, with more details listed in Table 7.1.

- **Xilinx** currently holds the dominant position in the global market for FPGAs, with a market share exceeding 50%. The target market of Xilinx includes Communications, Industrial, Data Center, and Aerospace and Defence markets.

Table 7.1 The latest FPGA products from the top companies

Company	FPGA products	Main markets	Design tools
Xilinx	UltraScale+, UltraScale, Virtex-7	Communications, aerospace and defence, data center	Vivado
Intel (Altera)	Stratix, Arria, MAX, Cyclone	Communications, military, automotive	Quartus
Lattice	Avant, CertusPro, CrossLink, MachXO5	Communications, computing, industrial, consumer	lattice diamond
Microchip	RTG4, SmartFusion2	Aerospace and defence, medical, industrial	Libero
Achronix	Speedster7t, Speedcore	High-performance computing, networking and telecom	ACE
QuickLogic	EOS S3, ArcticLink3, PolarPro3	Displays, audio and voice, mobile and IoT	Sensor development kit
Flex logic	EFLX	Embedded apps	Inference compiler

The company provides a wide range of FPGA options, including Versal ACAP (Adaptive Compute Acceleration Platform), UltraScale+, UltraScale, 7-series, Spartan-7, and previous generations, making it the most extensive portfolio in the industry. For the latest Versal ACAP portfolio, which utilizes the TSMC 7-nm FinFET process technology, is the first platform to combine software programmability and domain-specific hardware acceleration. The portfolio provides six device series, each uniquely designed to offer scalability and AI inference capabilities for a wide range of applications, from cloud and networking to wireless communications and edge computing.

- **Intel** acquired Altera in 2015 and now offers FPGAs under the Intel brand. As the second largest FPGA supplier with around 30% market share, Intel shows a strong presence in the communications infrastructure and military/aerospace markets.

  This company also provides a wide variety of FPGA series, including Intel Agilex, Stratix, Arria, MAX, and Cyclone. The Intel Agilex FPGA Portfolio, for example, is built on Intel 10 SuperFin and Intel 7 technology, which greatly enables customized acceleration.
- **Lattice Semiconductor** is another influential FPGA company, focusing on the low-power FPGA market, where the problems are across the network, from the Edge to the Cloud, mainly in communications, computing, industrial, automotive, and consumer markets.

  In Lattice's product lines, there are Avant, CertusPro-NX as general-purpose FPGAs, CrossLink-NX for video connection FPGAs, MachXO5-NX for control and security FPGAs, and iCE40 UltraPlus, iCE40 Ultra/UltraLite for ultralow power FPGAs. With an ultra-low power advanced process with static current as low as 75 μA and 1–10 mA active current, iCE40 UltraPlus FPGA can achieve low power connectivity and computing.

- In addition to the companies mentioned above, there are other important players in the FPGA market. **Microchip (formerly Microsemi)** is a leading supplier of FPGAs for the defense and aerospace industries. The company's FPGAs for space applications are manufactured and tested using specialized processes to ensure the highest standards of quality and reliability for satellite payloads and launch vehicles. **Achronix Semiconductor** offers the Speedster7t FPGA family, which is optimized for high-bandwidth workloads and eliminates performance bottlenecks associated with traditional FPGAs. Speedster7t FPGAs feature a revolutionary new 2D network-on-chip (2D NOC), which can deliver ASIC-level performance while retaining the full programmability of FPGAs. **QuickLogic** develops low-power FPGAs for the portable electronics market. Their products, the EOS S3, ArcticLink, and PolarPro 3 FPGAs, target energy-efficient and battery-powered devices.

While FPGAs offer significant advantages in terms of customization, low latency, and power efficiency, as shown above, they also come with certain limitations. One of the main challenges is the complexity of design and programming. Unlike CPUs and GPUs, which have mature and user-friendly software ecosystems, FPGAs require a deep understanding of hardware design and low-level programming. This potentially can be a big barrier for developers in other fields. Additionally, while FPGAs can be reprogrammed to perform specific tasks efficiently, they generally do not match the raw computational power of GPUs, which show extreme performance in data computation with regular parallelism and high floating-point compute throughput. Therefore, FPGAs are often not the best choices for general-purpose computing that requires high computational throughput. Finally, the cost of FPGA devices can be higher compared to other hardware options, such as ASIC chips. This makes them less accessible for small-scale projects with tight budgets. Despite these challenges, FPGAs continue to be used in a variety of applications where their unique strengths provide a competitive advantage.

To meet computational requirements in the AI era, the integration of AI modules in FPGA has become a significant trend in the current technological landscape. Companies like AMD Xilinx, Intel, and Achronix have introduced FPGA boards that incorporate AI modules, enhancing their performance and computational capabilities. Xilinx's Versal AI core and Alveo AI Acceleration Platforms provide high-performance AI inference and adaptable computing. Intel introduces the Agilex 7 FPGA and FPGA AI Suites with high-bandwidth memory and AI-optimized DSP blocks. Achronix Speedster7t features advanced interconnect for high-speed data transfer, AI/ML optimization, and a large number of logic elements and DSP blocks.

These AI-integrated FPGAs offer autonomous decision-making, real-time adaptation to new data, and intelligent information processing. The AI modules boost FPGA performance by allowing for faster response times and enabling more specialized, innovative operations. With the ability to customize processing functions at the basic logic port level, AI-integrated FPGAs can become better accelerator options than GPUs in specific scenarios.

7.2 Overview of the FPGA CAD Flow

FPGA CAD (Computer-Aided Design) flow is a series of steps that designers need to follow to implement or program any circuit on the FPGA chips. The mainstream FPGA synthesis tools from the companies above are all designed to assist designers based on the steps. Figure 7.2 illustrates the typical FPGA CAD flow, from the initial Design Entry to the final Bitstream Generation.

In Design Entry, designers will conceptualize their ideas about the function and structure of their circuits and transform them into a computerized representation, which is called RTL (Register-Transfer Level) design. RTL is a level of abstraction in digital circuit design where the behavior of the circuit is described in terms of the flow of data between registers. In this process, specialized computer languages called Hardware Description Languages (HDLs) are adopted, such as VHDL, Verilog, and SystemVerilog. As their names imply, HDLs are only tools for circuit description instead of tools for circuit design. When it comes to a large and complex system, HDL programming can be difficult and time-consuming. Therefore, high-level synthesis (HLS) was proposed to automate the process, which allows designers to build and verify their designs more efficiently by coding only in

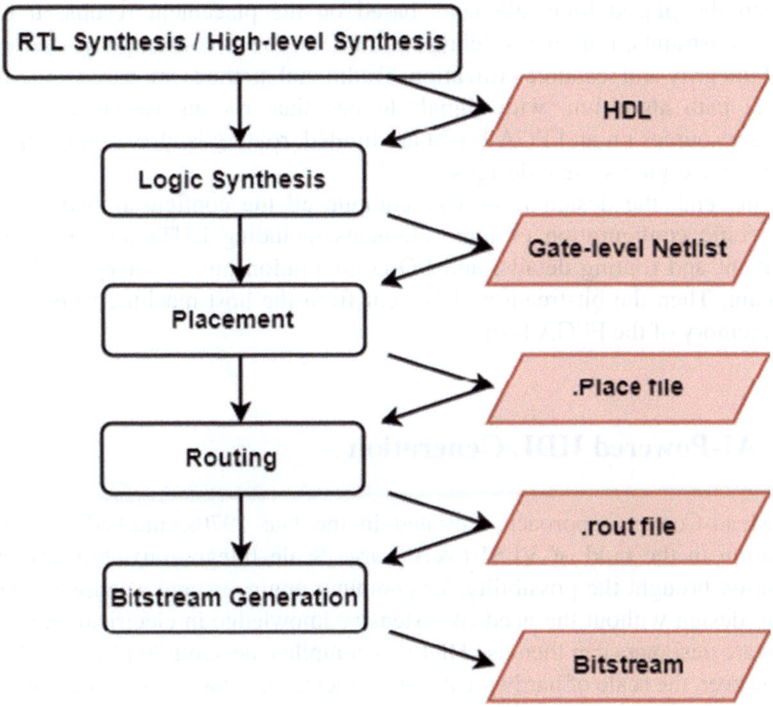

Fig. 7.2 An overview of the FPGA CAD flow

high-level programming languages such as C and C++. By enabling a higher level of abstraction, HLS in FPGA CAD can simplify algorithm expression, accelerate development cycles, and facilitate easier verification.

In Logic Synthesis, the design at the RTL level will be forwarded to the gate level. With the synthesis tool of the selected FPGA vendor and model, designers can feed the HDL files to it and generate the netlist that is built with the primitives provided by the vendor. There are usually additional built-in functions in the synthesis tools, such as logic optimization and timing analysis. These techniques can further improve the performance of the generated netlist.

After generating the synthesized netlist, the next task is to assign all the primitives to discrete slots on the FPGA board, which is called placement. Placement is one of the most time-consuming steps in the whole FPGA CAD flow. Designers need to decide on the physical location of each logic component using synthesis tools to optimize wire length and timing. For example, there are 485,760 logic cells and 37,080 kb memory on the Xilinx Virtex-7 VC 707 FPGA board, which a designer needs to select from. Two main tasks in placement are congestion estimation and routability prediction, focusing on the management of the routing resources. Smart placement will reduce the work load of the subsequent routing task and lead to a final design with better performance.

Routing is the step in the FPGA CAD flow to establish interconnections between the placed logic elements based on the placement results. It ensures timing constraints, minimizes delays, minimizes power consumption, and ensures signal integrity and resource utilization. Traditional methods are based on Dijkstra's shortest path algorithm, with signals to negotiate routing resources. Since the routing resources on an FPGA board are limited, routing is also a time-consuming task in large digital system designs.

In the end, the design tools will compile all the configuration data such as the specific configuration of logic elements including LUTs and flip-flops, the placement and routing details, and I/O configuration into a binary file known as bitstream. Then the bitstream will be sent from the host machine to program the flash memory of the FPGA board.

7.3 AI-Powered HDL Generation

The Mead-Conway approach, initiated in the late 1970s, marked a significant revolution in the field of VLSI (Very Large-Scale Integration) chip design. The revolution brought the possibility for computer engineers and scientists to engage in chip design without the need for extensive knowledge in electrical engineering. Hardware designers can then use HDLs to compile a design into physical chips.

However, the scale of hardware design is increasing at an exponential rate. Better abstraction of programming languages has been explored to reduce workload. One of the design abstraction improvements is the register-transfer level (RTL). It allows designers to describe the circuit based on the flows of digital signals between registers and the logic performance based on these signals.

High-level synthesis is another abstraction improvement after RTL. Instead of working on individual registers and operations in each clock cycle, HLS operates on a higher level abstraction. Hardware can be described using higher level programming languages, such as C, C++, and SystemC, which further simplifies the design process.

7.3.1 Conventional High-Level Synthesis

A typical HLS process is shown in Fig. 7.3. It begins with the hardware description in high-level programming language. The code is processed by the front-end compilation and then by the back-end hardware generation. In this way, the finite-state machine (FSM) and datapath components are generated, which will be used to produce the low-level HDL code.

According to Cadence's assessment, their HLS product line has the potential to cut down design and verification costs by 25–50%. This has led major companies such as Synopsys, Mentor, Intel, and Xilinx to create their own HLS design automation tools to address power and timing issues, thus increasing efficiency in custom hardware design [20, 21]. In [22], an overview of academic and commercial HLS tools is provided, in which the HLS tools are classified according to the input language. We can differentiate between two primary types: HLS tools that use domain-specific languages (DSLs) and tools that are based on general-purpose programmable languages (GPLs). DSLs can be divided into newly created languages specifically designed for a particular tool flow and dialects of GPLs,

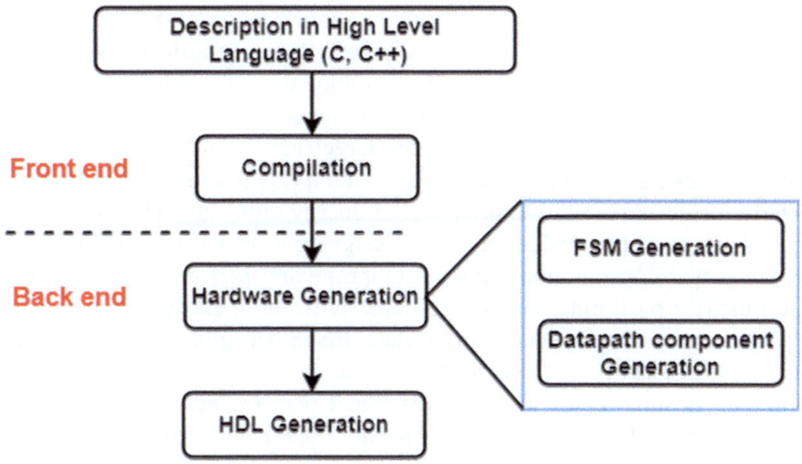

Fig. 7.3 Conventional High-Level Synthesis (HLS) process

which are languages derived from a GPL (e.g., C) and augmented with specific elements to communicate hardware information to the tool.

Although HLS shows great improvements in the development speed of the HDL generation, the generated circuits cannot provide the same quality as the manually generated ones. This is due to the following limitations that impact the hardware quality:

1. The high-level design process of HLS conceals potential optimization opportunities.
2. The design space exploration is timing-consuming, and the current HLS tools are not fully developed for the prediction of all the performance metrics.
3. The standard sets of code optimization in CPU, like LLVM, are not suitable for an FPGA back-end.

Therefore, the top two objectives of AI-based research in HLS are to (1) achieve better accuracy in the prediction of the performance and resource utilization and (2) explore the design space.

7.3.2 ML-Based Metric Prediction Algorithms in HLS

High-level synthesis tools have been powerful enough today after decades of development. For widely used HLS tools, such as the Vitis HLS from Xilinx [23] and Intel HLS [24], they can already reduce design efforts and time to market significantly. In addition to HDL generation, metric estimation is another necessary function for each HLS tool. The key metrics to evaluate the FPGA design are timing, latency, resource utilization, and power consumption. Timing analysis provides insight into the clock cycle performance, ensuring that the design meets the necessary speed requirements. Latency estimation reveals the number of clock cycles required for a particular operation or a set of operations, which is crucial for applications with strict performance constraints. Resource utilization metrics offer a detailed amount of the available resources onboard—such as logic elements, DSP blocks, and memory—that are being used by the design. This information is vital for ensuing that the design fits within the target FPGA and for identifying areas where resources can be optimized. Power consumption is another important metric, especially for designs with requirements of energy efficiency. HLS tools estimate the power usage of the design, allowing designers to make informed decisions to minimize power consumption without compromising performance. By providing detailed metrics on timing, latency, resource utilization, and power consumption, HLS tools empower designers to make data-driven decisions to meet their design requirements efficiently and effectively.

However, in most cases, the HLS-estimated performance metrics will deviate significantly from the actual quality of results (QoR) achieved by FPGA-targeted designs. The inaccurate estimation prevents designers from evaluating their design

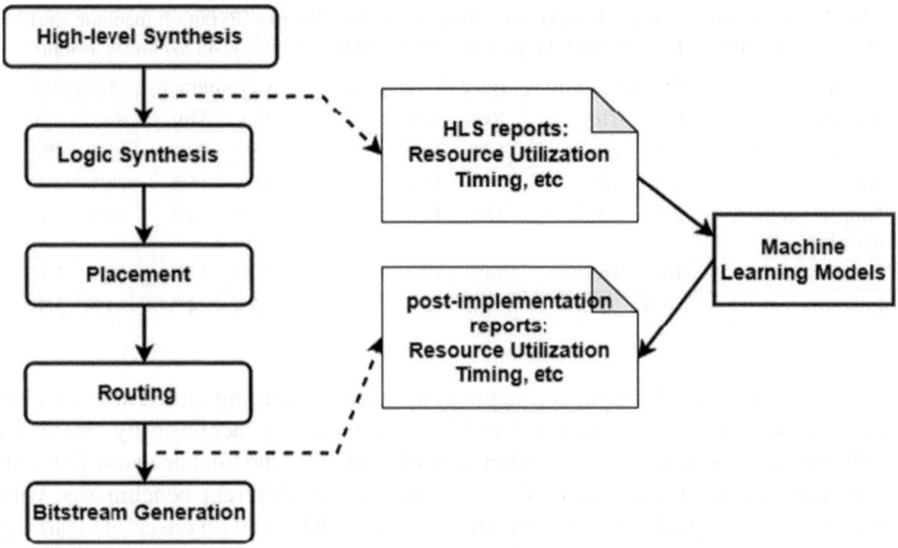

Fig. 7.4 A typical ML-based metric estimation based on FPGA HLS reports

and further exploring the design space. To address this challenge, HLS metric estimation algorithms have been explored with the strength of Artificial Intelligence.

A typical ML-based QoR estimation method is discussed in [25]. Dai et al. built a large collection of C-to-FPGA results from a diverse set of realistic HLS applications and identified relevant features from HLS reports for estimating post-implementation metrics. Then he leveraged these features and data to train and compare a number of promising machine learning models to effectively and efficiently bridge the accuracy gap. As shown in Fig. 7.4, the models are trained by estimates of resource utilization, timing, and HDL details in the HLS reports. The trained models can then be used to predict those metrics for post-implementation. In this work, they focus on regression models, including the linear model, the artificial neural network, and gradient tree boosting. It is provided in the article that their ML models can reduce the errors of the HLS estimations by up to 138%, which proves the effectiveness of AI in the estimation of the HLS metrics.

Similarly, [26] proposed a framework Pyramid based on the ensemble machine learning model for the estimation of the throughput or throughput to area. Features are also extracted from HLS reports, but an automated hardware optimization tool called Minerva is adopted for close-to-optimal results in this work, which employs a customized heuristic algorithm to find the maximum frequency. In this way, they create a comprehensive dataset (2700 samples) with all features from the HLS report, using Pearson's correlation coefficient calculations to remove redundant ones. For ML models, one model from each branch of ML is selected, including the regression model, ANN, support vector machine (SVM), and random forest (RF). In addition, this work proposes an ensemble ML model based on a stacking

Table 7.2 Error of different ML techniques for different benchmarks (Pyramid: machine learning framework to estimate the optimal timing and resource usage of a high-level synthesis design)

Benchmark	Machine learning		Image/video processing		Cryptography		Mathematical	
Targets	Res	Tim	Res	Tim	Res	Tim	Res	Tim
Linear regression	29%	25%	17%	16%	38%	22%	11%	8%
ANN	17%	14%	13%	11%	19%	14%	8%	7%
Support vector machine	22%	19%	18%	17%	23%	18%	10%	7%
Random forest	16%	16%	14%	12%	20%	15%	9%	7%
Ensemble	**6%**	**5%**	**4%**	**3%**	**5%**	**4%**	**4%**	**3%**

approach. Ensemble learning is a segment of machine learning aimed at enhancing the precision and efficacy of standard ML models. This is achieved by creating a collection of base learners and combining their output for the final decision [27, 28]. The performance of the ML models is compared for different benchmarks, with two optimization goals. For all benchmarks, this model based on ensemble learning demonstrates an accuracy of over 95%, as shown in Table 7.2.

Besides the estimation of timing, utilization, and throughput, there are also ML-based estimation methods for power, delay, and routing congestion.

Power consumption is critically important in hardware design, particularly for portable and embedded systems. Recent studies [29–31] have developed power modeling methods to expedite power analysis in hardware design. However, these methods have notable limitations: (1) They produce design-specific power models that are not transferable to other designs, (2) their modeling requires multiple power characterization iterations, following the time-consuming RTL implementation process, and (3) adapting these techniques to new platforms is challenging due to their reliance on specialized hardware modeling knowledge.

To solve this problem, Lin et al. [32] introduce HL-Pow, a power modeling framework for FPGA HLS based on regression models with supervised learning. In this work, an automated feature construction flow is introduced for rapid identification and extraction of features that are most related to power consumption from HLS reports. The HL-Pow model can then offer faster power prediction compared to the time-consuming RTL-based power estimation flow. Lin et al. further propose another framework called PowerGear in [33], which is an improved power estimation method based on graph neural networks (GNNs). The HLS designs are transformed into graph-structured data, where both intraoperation micro-architectures and inter-operation interconnects are encoded and annotated with switching activities. The novel power-aware heterogeneous edge-centric GNN model can effectively learn heterogeneous edge semantics and structural properties of the constructed graphs via edge-centric neighborhood aggregation and fit the formulation of dynamic power. This new ML-based power estimator achieves errors of 3.6 and 8. 81% in estimating the total and dynamic power, along with a speedup of $4\times$ over the Vivado power estimator.

GNNs are a class of neural networks designed to operate directly on graph-structured data, which can perform much better than traditional neural networks in graph-related tasks. Besides the GNN-based power estimation proposed in [33], a GNN-based delay estimation is carried out in [34]. In this work, Ecenur Ustun et al. focus on issues about operation mapping of hardened blocks, such as DSP blocks and carry blocks. Existing HLS tools are found to always fail to capture operation mapping patterns, leading to inaccurate estimation of resource utilization and delay. The proposed GNN model is trained on microbenchmarks directly to realistic designs through inductive learning. The approach leads to a 72% reduction in root mean square error compared to the delay estimation in Vivado HLS.

In addition to traditional metrics that indicate circuit performance, there are also studies on predicting routing congestion. Usually, the routability is difficult to evaluate in the HLS step, which lacks the post-implementation details after Place and Route. This can lead to a much longer development time in those subsequent steps. To solve this problem, Zhao et al. developed an ML-based method to resolve routing congestion in the early design stage [35]. The training dataset is collected by backtracking the congestion metrics of CLBs and linking them with the HLS IR (intermediate representation). And with their trained model, the highly congested parts can be located in the source code, which is a great improvement in developing efficiency.

7.3.3 ML-Based Design Space Exploration in HLS

One distinct advantage of raising the abstraction level to HLS is that it allows the generation of different microarchitectures of equivalent functions without changing the detailed behavioral description. Instead, this is done by setting synthesis options in the form of pragmas or constraints of the number and type of functional units (FUs). To be more specific, designers are able to control the synthesis of arrays (RAM or registers), loops (unroll, partially unroll, no unroll, or pipeline), and functions (inline or not). In this way, designers can generate their circuits with unique areas, performance, and power. The process of achieving Pareto-optimal designs by changing the pragma combinations is called Design Space Exploration (DSE).

One of the main problems of HLS DSE is the large design space to explore, which grows supralinearly with the number of pragma settings. Therefore, the task of ML-based algorithms is to achieve faster and more accurate exploration in the HLS design space.

The state-of-the-art HLS DSE techniques can be basically divided into two categories: synthesis-based approaches and AI-based methods.

As shown in Fig. 7.5, the synthesis-based methods refer to the methods that generate new pragma combinations and let HLS evaluate the effect of these new combinations. This category of DSE methods can be further divided into meta-heuristics, dedicated heuristics, and transfer learning.

Fig. 7.5 Overview of the HLS DSE techniques

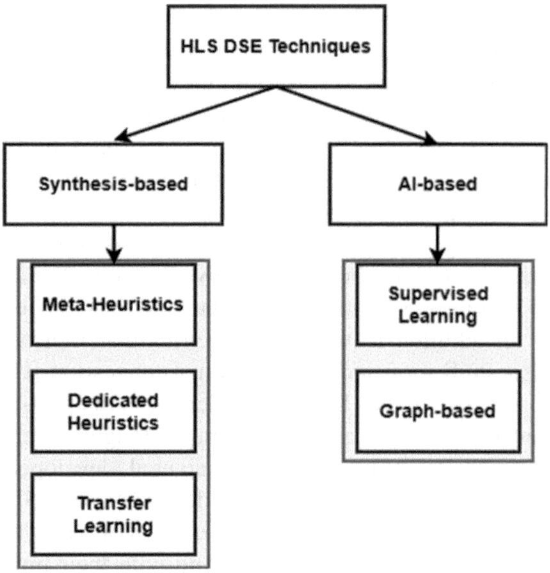

Meta-heuristics are general strategies that are usually based on natural processes, such as simulated annealing (SA), genetic algorithms (GAs), and ant-colony optimizations (ACOs). These techniques have been demonstrated to be highly effective when applied to multi-objective optimization issues like HLS DSE. And they have been proven to be effective in the HLS DSE task [36–40]. However, these methods are limited by their sensitivity to the hyper-parameters. To address this problem, dedicated heuristics are proposed for better performance [41–48]. In addition, transfer learning-based methods are adopted to provide more efficient exploration by making use of previously generated exploration results [49–51]. Instead of testing random pragma combinations, transfer learning can reduce the exploration time with an initiation based on the previous exploration results, which helps to make DSE more efficient and accurate.

AI-based DSE methods are based on predictive models of different performance metrics, which have been discussed in Sect. 7.3.2. Instead of invoking the HLS process every time for pragma combination evaluation, AI-based DSE methods use the generated predictive models for faster evaluation. As shown in Fig. 7.5, AI-based methods can be further classified into supervised learning DSE [52–56] and graph-based DSE [57–59], according to the ML algorithms adopted.

A typical supervised learning-based genetic algorithm for DSE is proposed in [52]. Details of the GA-ML (genetic algorithm machine learning) algorithm are provided in Fig. 7.6. First, the high-level C code is parsed to find all the constructs for exploration. A random set of pragmas for these constructs is generated and then evaluated by the HLS tool. The above step is then repeated to train the predictive ML model. Once the accuracy of the model reaches the given threshold, it can continue the exploration by replacing the HLS tool.

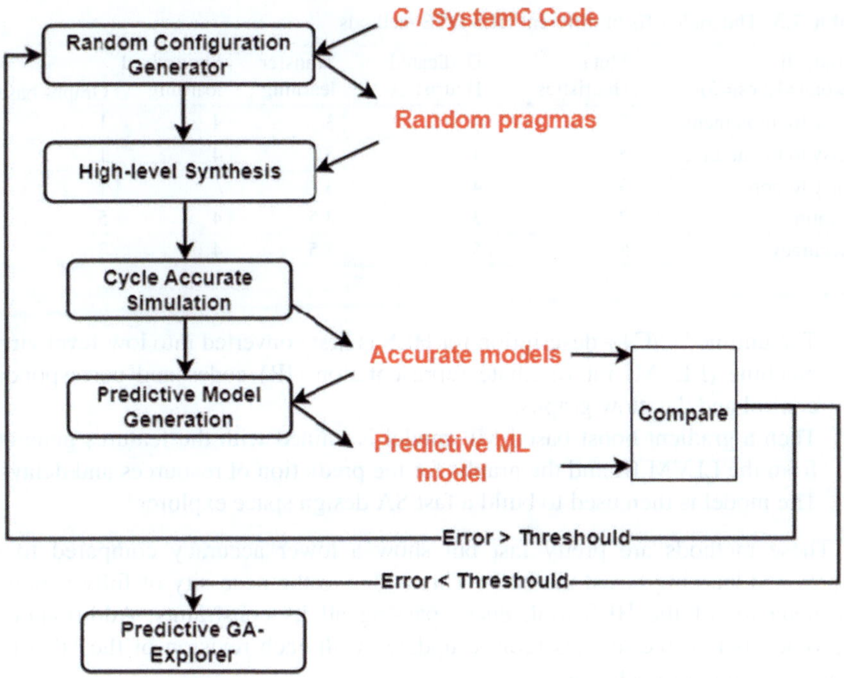

Fig. 7.6 A supervised learning-based algorithm for DSE

Considering that learning-based methods in works like [52] still rely on local search techniques, Liu et al. [53] proposed a better model, especially for HLS, along with an effective training scheme. In their exploration, the Random Forest (RF) is consistently more accurate than the other seven advanced learning models in the prediction of the area and effective latency of a Discrete Fourier Transform (DFT) design. Random Forest (RF) is an ensemble approach that involves numerous decision trees. It works by randomly dividing the training data to train each tree separately. The ultimate prediction is determined by aggregating the predictions from all individual trees through a voting mechanism. The two steps allow RF to minimize both the generalization error and prediction variance and hence achieve high accuracy. In addition to improving the learning model, a transductive experimental design and randomized selection were also included in the proposed method to help find an approximate Pareto set of RTL designs.

Unlike the supervised learning-based DSE methods, graph-based DSE methods generate predictive models and estimate area and performance with features extracted from the Control and Data Flow Graph (CDFG). An example of graph-based DSE methods is the work of Md Imtiaz Rashid in [59]. In this work, a hybrid approach was developed by integrating the conventional metaheuristic technique, Simulated Annealing (SA), with machine learning, resulting in a fast and robust exploration of HLS. Design details are as follows:

Table 7.3 The trade-offs of different HLS DSE methods

Tradeoff (worst=1,best=5)	Meta-Heuristics	Dedicated Heuristics	Transfer learning	Supervised learning	Graph- based
Easy to implement	5	4	3	4	1
Easy to maintain	5	4	3	4	1
Easy to port	5	4	3	/	1
Runtime	2	3	3.5	4	5
Accuracy	5	5	3.5	4	3

(1) The untimed C/C++ description for HLS is first converted into low-level virtual machine (LLVM) intermediate representation (IR) codes and corresponding control and dataflow graphs.
(2) Then a gradient boost-based ML model is trained with the features generated from the LLVM IR and the graphs for the prediction of resources and delays.
(3) The model is then used to build a fast SA design space explorer.

These methods are pretty fast but show a lower accuracy compared to the supervised learning-based methods. This is due to the necessity of fully capturing the behavior of the HLS tool, encompassing all its constraints. Additionally, a drawback is that the models require updates with each revision of the HLS tool, as its behavior could change.

Moreover, it is then found that the area and performance estimation reported after HLS do not always match the results after logic synthesis very well [60, 61]. To build a more accurate ML prediction model, researchers are considering using the results of post-logic synthesis as training data [59]. This has been used by Md Imtaize Rashid to compensate for the low accuracy of their graph-based HLS explorer, which helps to achieve a great improvement in the prediction results.

In conclusion, all methods show different trade-offs between their running time and the accuracy of prediction models. The summary of the trade-offs is provided in Table 7.3.

7.4 AI-Powered Logic Synthesis

Logic synthesis, as suggested by the name, is about transforming the design in HDL format into a netlist of logic gates. Specifically, the RTL code we wrote or generated by HLS in the last step will be converted into the design at the gate level by a synthesis tool. Synthesis tools are vendor-specific computer software that has been integrated into the FPGA design tools, as shown in Table 7.1.

The automation of logic design and synthesis processes has significantly simplified electronic design. By employing high-level programming languages for logic design and synthesis tools for generating logic gates, a level of standardization is achieved among different designers and electronics manufacturers. This not only

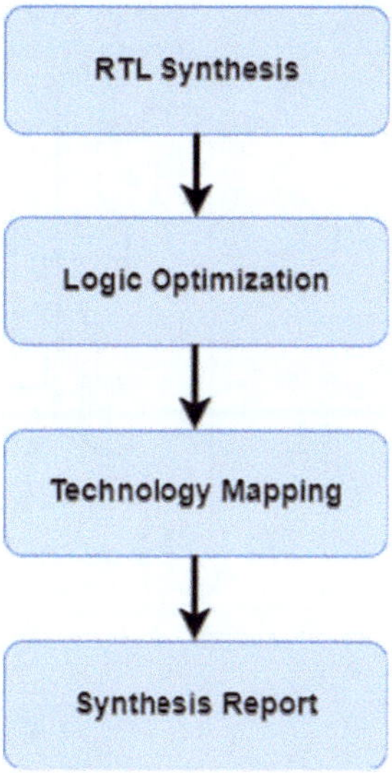

Fig. 7.7 The procedure of logic synthesis

minimizes the likelihood of errors and mistakes but also facilitates the reuse and modification of designs for future applications.

7.4.1 Procedure of FPGA Synthesis

The built-in logic synthesis of an FPGA design tool usually consists of the following four steps: RTL synthesis, logic optimization, technology mapping, and synthesis report generation, which is displayed in Fig. 7.7.

In RTL synthesis, there are three subtasks: syntax analysis, semantic analysis, and modeling. Syntax analysis checks whether the code complies with the syntax rules. Semantic analysis checks the correctness of the code's semantics. Then modeling is about inference and processing of the high-level structures, such as adders, multipliers, RAMs, and FSMs, into gate-level logic. Here, the tool identifies specific patterns within the Verilog logic design or description and processes them according to user-defined instructions.

However, typical logic synthesis involves more than just gate-level translation. It also includes two stages of optimization to improve the quality of the generated circuits. The first optimization is called technology-independent circuit optimization. The circuit is optimized without considering any technology-related metric at this stage. Instead, the optimization objective is to universally enhance the circuit, for instance, by reducing the depth and the quantity of nodes in the Boolean network of the circuit. The conventional approach is to apply a series of optimization steps to the circuit in a predetermined sequence, which has been demonstrated to be effective for typical circuits.

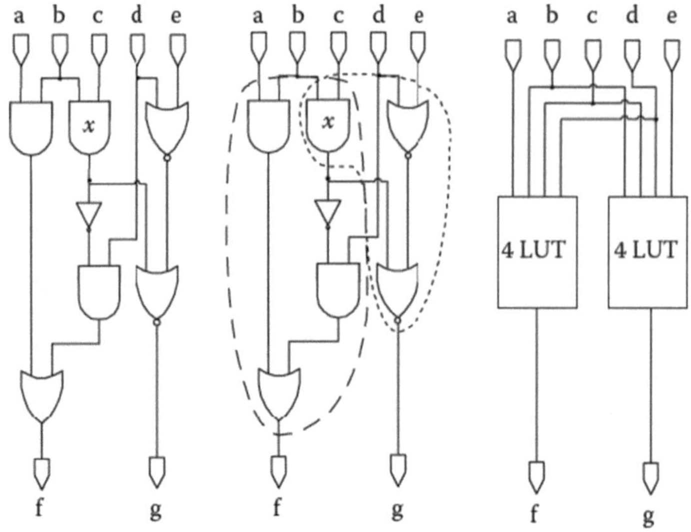

Fig. 7.8 An example of technology mapping in FPGA synthesis (from Hutton et al. "FPGA synthesis and physical design")

The second stage of optimization is technology-dependent optimization, which is also called technology mapping. The network of primitive gates is further mapped into a netlist of primitives that are compatible with the available technology, such as LUTs in FPGA synthesis. LUT-based technology mapping is best seen as a covering problem since it is common and necessary to cover some gates by multiple LUTs for an efficient solution. A more detailed illustration is provided by Ling et al. [62] in Fig. 7.8, which shows how the original netlist is covered and replaced by the four-input LUTs. The optimization goal of this stage is to minimize both the unit depth and the number of cells in the mapped network.

Before synthesis, some setups are also required, such as adding constraint files, specifying the FPGA model, and configuring the synthesis options. Constraint files are essential files in the FPGA implementation process, containing information about time constraints, pin assignments, and clocks. The technology and available resources on different FPGAs can vary greatly, which requires users to provide the FPGA model before synthesis. Synthesis options refer to the settings used by the synthesizer when synthesizing the code, such as optimization levels and timing constraints. Finally, the synthesis tools will generate a synthesis report, which includes information about the synthesis results, resource utilization, timing constraints, etc. The synthesis report contains gate-level implementation details, timing analysis, and constraint verification, allowing you to examine the circuit's implementation and timing performance.

However, conventional logic synthesis methods show restrictions in several aspects. In conventional logic synthesis, a fixed set of optimization algorithms for netlist mapping is applied to different circuits, which is deemed to work well in most cases. This can now be powered by ML methods to achieve better flexibility in the selection of unique algorithms. Another restriction of conventional logic synthesis is the precision of the estimation of expected performance, resource usage, and timing. The generated HLS reports usually do not show accurate results, which can be significantly improved with ML-based algorithms. Moreover, the problem of slow synthesis time is gaining attention as well. ML is adopted to help deliver timing closure of FPGA synthesis and hence speed up the CAD flow.

7.4.2 Reinforcement Learning-Based Logic Optimization and Technology Mapping

As discussed in the last section, conventional logic synthesis uses only one set of optimization algorithms for different circuits, which shows pretty low flexibility and cannot be auto-adjusted to fit each circuit to its best. In recent research, reinforcement learning (RL) has been explored for improving the two optimization stages in LS.

RL is a state-of-the-art ML paradigm that has shown high performance in sequential decision-making problems. RL model adopts an agent who continually interacts with the environment and is trained by reward and punishment to take correct actions. Considering that the conventional LS approach is made up of a sequence of optimization, RL can be used to help optimize the performance of LS.

In [63], an RL-based logic synthesis is proposed, along with a feature-importance analysis utilizing a random forest technique to trim down the feature space. The overall flow of their proposed framework is displayed in Fig. 7.9, which is a very typical structure for the exploration of LS with RL.

In the framework, they use a popular logic synthesis tool called ABC [64] that uses an AND-inverter graph (AIG) as the subject graph and a feature extractor called CIRC [65] to extract features from the circuit netlist. The RL agent receives the

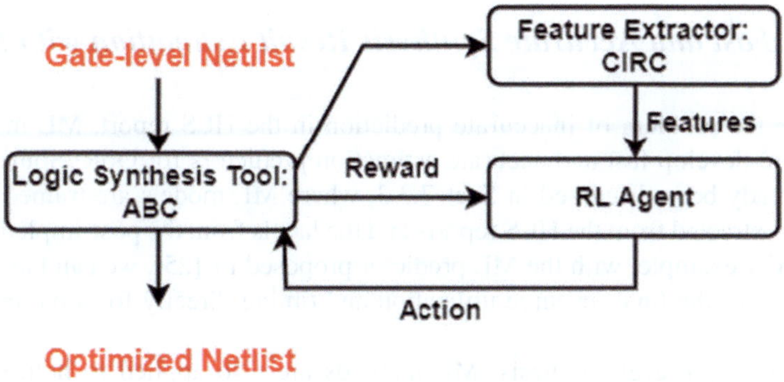

Fig. 7.9 Enhanced logic synthesis with reinforcement learning

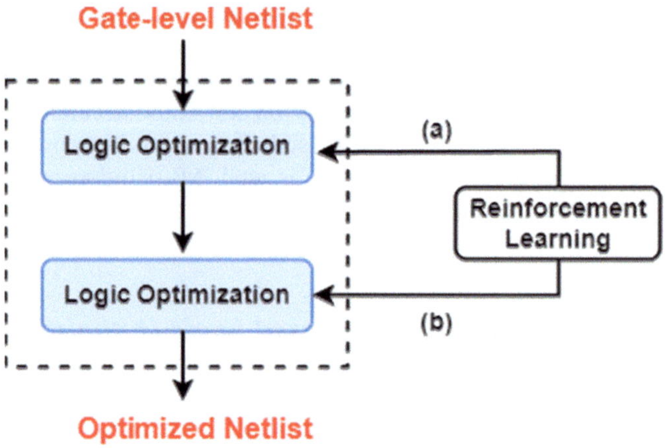

Fig. 7.10 Exploration of logic optimization and technology mapping with RL

observation from both the CIRC tool and the ABC synthesis tool. For every action, the environment provides the agent with observations and rewards. Subsequently, the circuit undergoes technology mapping to attain the area optimization objective. The proposed RL agent proves to finally achieve a 10.5% LUT count improvement and shows its generalization in the optimization of different circuits.

Another RL-based framework called ESE (Efficient Synthesis Exploration) is proposed to automatically explore the logic synthesis process in [66]. In this work, two RL environments are formulated to model the logic optimization and technology mapping, respectively, as Fig. 7.10 shows. The reward functions and termination conditions in RL environments are devised to offer balanced guidance to optimize both metrics and execution time. Specifically, for tackling the FPGA mapping task, logic optimization and technology mapping are merged to enable flexible learning.

7.4.3 *Fast and Accurate Synthesis Result Estimation with ML*

To solve the problem of inaccurate prediction in the HLS report, ML models are applied to develop fast and accurate estimation predictors for QoR. Some of them have already been discussed in Sect. 7.3.3, where ML models are trained with the features extracted from the HLS reports and the labels from the post-implementation results. For example, with the ML predictor proposed in [25], we can have a better estimation of the final resource utilization and timing directly from the input HLS reports.

Besides high-level synthesis, ML methods are also applied to help designers make early decisions on logic synthesis. At the stage of technology mapping, gate-level netlists will be mapped into circuits with FPGA hardware primitives.

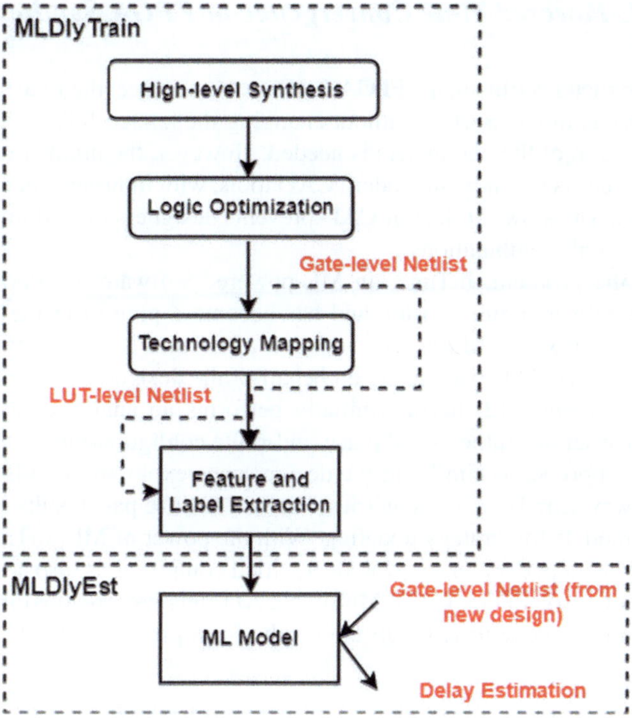

Fig. 7.11 The proposed framework for delay prediction of technology mapping

Considering that early decisions can greatly affect the final timing QoR, it will speed up the synthesis process if we can predict the delay before the time-consuming technology mapping stage. Therefore, in [67], Hu et al. proposed an ML-based delay predictor for quick timing analysis before mapping.

As shown in Fig. 7.11, the proposed framework consists of two parts, MLDlyTrain for training and MLDlyEst for inference. For the training dataset, 12 features are chosen and categorized into four groups, in order to encompass the gate-level netlist's topology and the behavior of technology mapping. These features include the count of Primary Inputs (PI), the count of gates, the count of paths, and the path length. Labels are extracted from the LUT-level netlist after mapping in the format of LUT logic levels. In this work, a supervised learning-based model called Light Gradient Boosting Machine (LGBM) is adopted. LGBM is based on gradient boosting and works well for nonlinear relationships, which makes it a good option for the delay estimation task. The final experiment results also prove its capability with an error of less than 2% for most of the selected benchmarks and an 8× faster runtime.

7.4.4 ML-Powered Time Convergence in FPGA Synthesis

Since the parameter setting in the FPGA CAD tools can affect the final timing results of the circuits as much as 70% without changing the source RTL code, a fast and accurate selection of the parameters is needed. However, the number of parameters and option values is too large in modern CAD tools, with thousands of combinations to select from. The slow runtime of CAD prevents designers from doing a thorough investigation of all combinations.

To solve the problem, InTime, an ML-powered software, is proposed to help FPGA CAD deliver timing closure and has become a plug-in in the FPGA CAD tools of both Xilinx and Altera. The function of InTime is to automatically select and tune the FPGA CAD parameters on behalf of the designers.

As shown in Fig. 7.12, InTime initially performs an analysis and generates a comprehensive set of synthesis and place-and-route configurations using predefined recipes. This approach is similar to the design space exploration tools provided by vendors. However, InTime has a metrics database to save past results and use them to train ML models for strategy updating. With the power of ML, InTime surpasses other DSE tools in performance. Moreover, cloud computing is also adopted for the parallel execution of FPGA CAD. Multiple CAD runs are executed in each round in parallel and can be adjusted with the results from previous rounds. In this way,

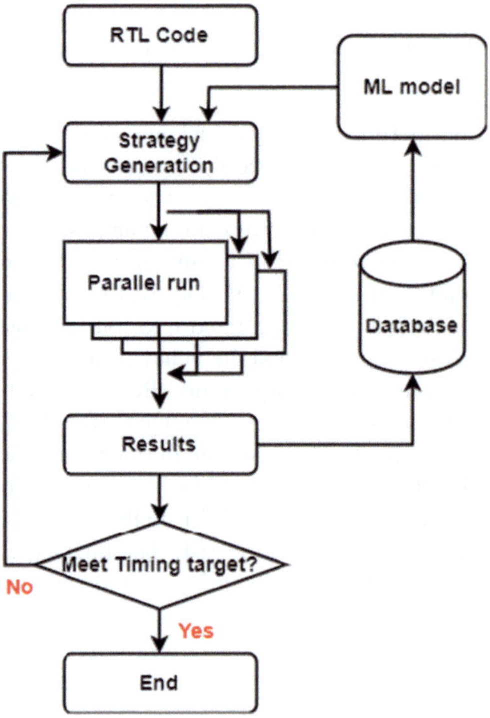

Fig. 7.12 Overview of the structure of InTime tool

InTime is capable of reducing the exploration space and achieving a great speedup in the FPGA CAD time convergence.

Although the invention of InTime greatly accelerates the FPGA CAD, including logic synthesis, one key challenge is the accuracy of ML-based prediction on potentially good combinations of CAD parameters. A more accurate prediction makes CAD much easier to achieve time convergence, which reduces not only the development time but also the cloud computing costs for parallel CAD runs.

To further speed up time convergence in FPGA CAD, Yanghua et al. [68] built an accurate estimation flow to help identify parameter combinations that are more likely to improve timing. In this work, a broad set of classification algorithms is explored in terms of their ability to reduce InTime timing scores, including logic regression, bagging, random forest, support vector machine, neural network, stack, and ensemble learning. By configuring a stack-based approach that combines multiple classifiers, they finally achieved a 2.7× reduction in time in timing closure.

7.5 AI-Powered Placement

Placement is part of the physical design flow of FPGA, which consists of packing, placement, and routing. Packing is a unique step for the placement of FPGA, which is usually included in the placement step. In this step, the logic elements are pre-clustered into more compact groups called logic clusters. These logic clusters will be mapped onto specific blocks called CLB, which is the basic logic resource on FPGA, as mentioned in Sect. 7.1. Then in placement, these CLBs will be assigned to physical locations on the FPGA, making this a slot assignment problem, where a CLB must be placed on a location with CLB fabrication.

7.5.1 Conventional Placement Algorithms

In placement, the main goal is to optimize the total wirelength and the timing. For different types of FPGA architecture, different placement algorithms have been developed to achieve the goal.

The Simulated Annealing Algorithm (SA) is a metaheuristic algorithm commonly used to solve combinational optimization problems such as the traveling salesman problem, layout problems, etc. The SA-based placement algorithm is the most widely used FPGA placement algorithm and is suitable for different architectures, especially island-style architectures. The CLBs are kept moving for an optimized location based on a cost function. This method shows great flexibility and can accommodate most objectives, such as the wirelength or timing performance. Random perturbations are applied to continuously optimize the objective function. The CLBs are then required to move until the desired optimization goal is achieved. An example of SA-based placement is shown in Fig. 7.13, which sets the wirelength

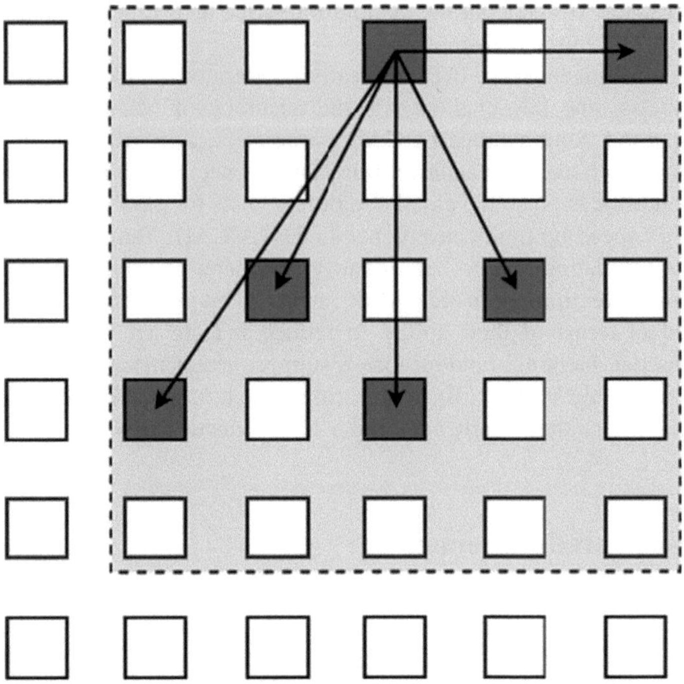

Fig. 7.13 Simulated annealing based placement

as the objective. There are still challenges for SA-based placement. It relies on random perturbations to escape local optima, but there is no guarantee of finding the global optimum. Since the algorithm relies on a sequential exploration of the solution space, the parallelization of it is also challenging. Although there have been improvements in SA, such as multi-CPU acceleration and the VPR algorithm, SA-based placement is still time-consuming compared to the following algorithms.

The partitioning-based placement algorithm is another method for hierarchical architectures. Netlist instances are grouped and used to create clustered hypergraphs based on interconnections. Top-down partitioning is then applied to divide the hypergraphs recursively until all instances are smaller than a predefined threshold. A simple illustration of this process is provided in Fig. 7.14. In this way, partitioning-based placement can transform the original placement problem into mutually orthogonal subproblems, which allows for pretty fast time convergence. However, this method may also easily get trapped in local optima.

The analytical placement algorithm is another popular method in FPGA placement, which models the problem into a continuous optimization problem in ML. The optimal solution to the problem can then be achieved by gradient descent. Based on the objective function, the algorithms can be divided into quadratic placement techniques and nonlinear placement techniques. Compared to the previous two algorithms, analytical placement can deliver great speedup and

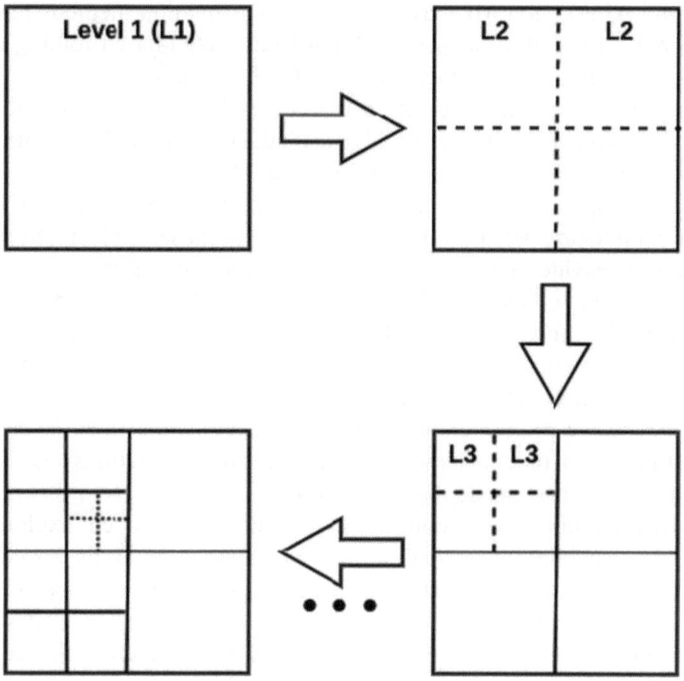

Fig. 7.14 Partitioning-based placement

performance improvement. There are still limitations to this method. The solvers used in Analytical placement require the objective function to be both continuous and differentiable, which does not fit the discrete nature of FPGA placement. Therefore, enhanced analytical placement techniques have been explored to target more complicated FPGAs [69].

However, as one of the most time-consuming steps in CAD, placement has two main tasks, congestion estimation, and routability prediction, in order to prevent potential problems in the routing step. To achieve accurate evaluations, ML techniques are also considered in these tasks.

7.5.2 ML-Based Congestion Estimation

Both placement and routing can consume a large amount of time in the overall CAD flow. A failed routing attempt will lead to iterative placement steps to create congestion-free regions for wiring. Therefore, an accurate post-placement estimation for congestion is necessary for CAD acceleration. Routing congestion is measured as the ratio of the number of routing tracks used to the number of routing tracks available.

Before the AI era, non-ML-based congestion estimation methods have already been explored. Liu et al. proposed a Global Route (GR)-based routing estimation tool for both intercell and intracell nets [70]. In some research, the congestion estimation problem is converted into a graph-based model and solved by routing demand calculation [71]. There are also statistical and probabilistic model-based methods for congestion estimation mentioned in [72].

With technology development and design rule checking (DRC) rules with increasing complexity, ML-powered methods are proposed to learn from previous experience and provide more accurate estimation for congestion.

In [73, 74], the congestion estimation is formed as a binary classification problem. The ML model is trained by features extracted from post-placement or post-routing netlist and will predict whether there will be DRC violations caused by routing congestion. These methods show better performance in modeling the relationship between routing congestion and DRC violations than the conventional global routing-based model. However, there are also limitations, as they do not predict exact congestion inside the global cells.

Supervised learning-based congestion estimation has been explored for an increasing accuracy and lower runtime. In [75], Maarouf et al. improved an SVM-based linear regression model that achieves 95% precision in comparison with the estimates of the Vivado Design Suite, but only 60% in comparison with actual congestion. In this work more features are tested and carefully selected, allowing all regression models in comparison to produce a high R^2 accuracy of more than 85%. The proposed linear regression model for congestion prediction is displayed in Fig. 7.15. This framework is commonly used and widely applicable, serving as the foundation for other machine learning methods [76, 77].

On the basis of the work in [75], Al-Hyari proposed a new congestion estimation framework called MLCong. The training dataset is generated from two academic placers, GPlace3.0 and Pripple. Four features are selected, and all the combinations of them are explored for the accuracy measurements. The researchers also compared different regression models, including linear regression, KNN, ANN, and random decision forests. The final model based on random forest algorithms achieves a 19% reduction in runtime, with a satisfactory level of estimation.

7.5.3 ML-Based Routability Prediction

Traditional non-ML prediction of routability is usually based on the estimated congestion map during placement [77–79]. Then ML-based methods have proven to deliver much better results with the cost of dataset generation and model training.

In [80], a supervised learning-based routability prediction model is proposed. An overview of the ML training procedure is shown in Fig. 7.16. A fast global router is adopted to generate the global routing features, which will be sent to train the ML model along with the features generated from placement results. DRC violations are selected as the label in the training, which is obtained from the post-routing results.

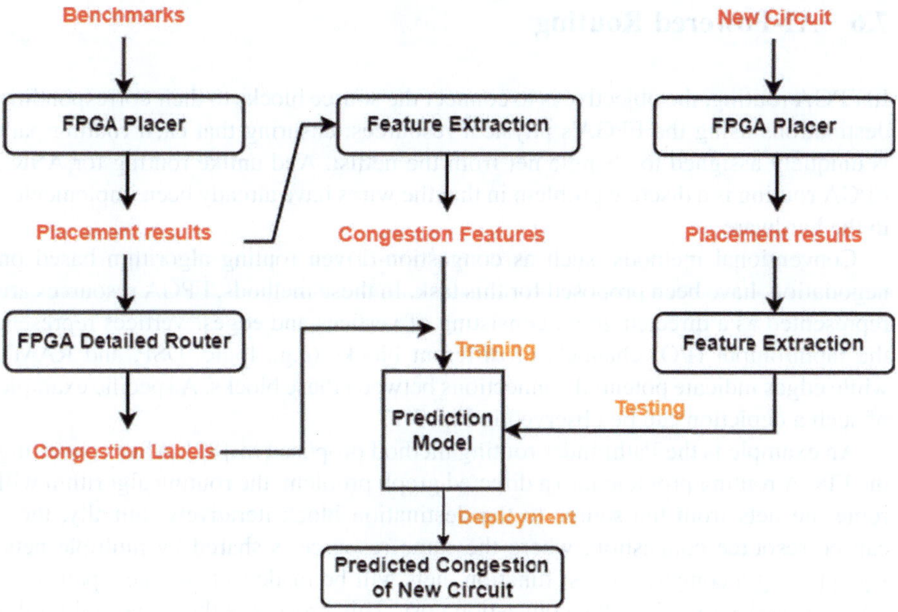

Fig. 7.15 Machine learning-based congestion estimation framework

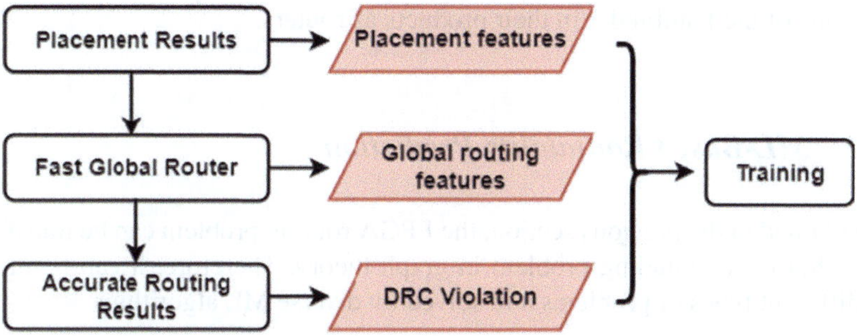

Fig. 7.16 The training of proposed supervised learning-based routability predictor

With the training dataset prepared, a multivariate adaptive regression is adopted to train the model. Then the trained model can be used to predict the DRC violation before the real detailed routing. The proposed model achieves an average accuracy of 79.8%, and it requires keeping the router and design rules unchanged for a high prediction performance.

Al-Hyari, in the work mentioned above [75], proposed another ML model called MLRoute to predict the routability based on the congestion information generated from MLCong, which shows an accuracy in the range of 93–98%. And later in [81], they upgraded the MLRoute to DLRoute, using CNN to achieve high prediction accuracy within an inference time at the millisecond level.

7.6 AI-Powered Routing

In FPGA routing, the objective is to connect the source blocks to their corresponding destinations using the FPGA's physical resources, ensuring that each routing path is uniquely assigned to a single net from the netlist. And unlike routing for ASIC, FPGA routing is a discrete problem in that the wires have already been implemented in the hardware.

Conventional methods, such as congestion-driven routing algorithm based on negotiation, have been proposed for this task. In these methods, FPGA resources are represented as a directed graph consisting of vertices and edges. Vertices represent the input/output (I/O) channels of different blocks (e.g., logic, DSP, and RAM), while edges indicate potential connections between these blocks. A specific example of such a depiction can be observed in Fig. 7.17.

An example is the PathFinder routing method proposed in [82]. After converting the FPGA routing problem into a directed graph problem, the routing algorithm will route the nets from the source to the destination block iteratively. Initially, there can be resource congestion, where the same resource is shared by multiple nets. By applying a congestion cost function, nets will be made to find a new path with different nodes to reduce the congestion cost. This process is then repeated until a congestion-free routing plan is achieved.

Currently, the two main FPGA vendors, Xilinx and Intel Altera, both use variations of the PathFinder in their products as routers.

7.6.1 *ML-Based Routability Prediction*

As explained in the previous section, the FPGA routing problem can be transformed into a shortest pathfinding problem in graph theory. Therefore, it can be modeled into different types of problems and solved by diverse ML algorithms.

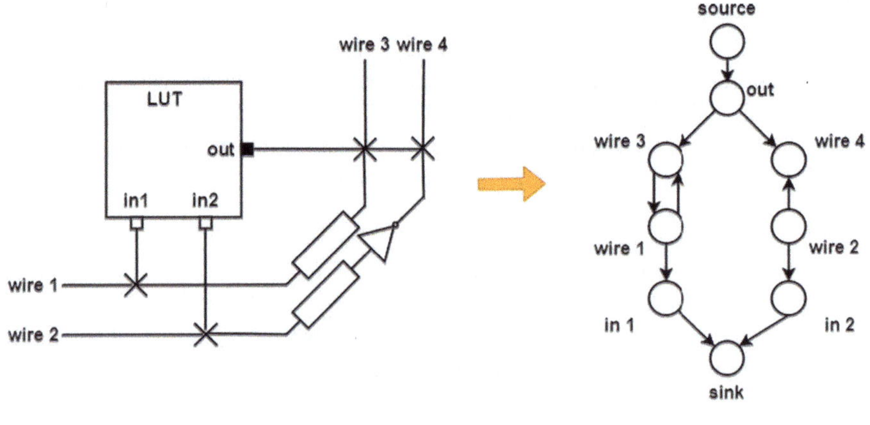

Fig. 7.17 A representation of the FPGA routing resource in the form of directed graph

In [83–86], the routing problem is formulated into continuous decision-making problems. RL-based methods and deep RL (DRL)-based methods are applied to solve it. Haiguang et al. present a DRL-driven global routing method in [83].

In this chapter, a Q-network with a proposed QDN formulation can model the routing resources into a single network and can be solved in a non-heuristics method. The Q-network learns from all previous experience and makes decisions with a reward function. As shown in Fig. 7.18, a problem is adopted to generate problems with specific requirements. Problems are stored separately in text files and then parsed. Each net will be decomposed into a set of two-pin problems using a minimum spanning tree algorithm. Two routers, the A* router and the proposed DQN router, are employed to resolve these two-pin problems. Here, the A* router will provide the memory burn-in to the DQN. The final performance for each router is generated by merging solutions for individual nets and will be evaluated and compared in the format of total congestion. Here, an overview of the DQN router is also provided in Fig. 7.19, showing how it interacts with the environment.

Similarly, Farooq et al. chose to use RL techniques to speed up the routing process [84]. In this work, they adopt a move-based RL model, which changes the reward at each move, making it a dynamic RL problem. The results demonstrate that this method can achieve a 30% reduction in runtime. In [87], however, Jain models the routing as a binary segmentation problem and proposed a Convolutional Neural

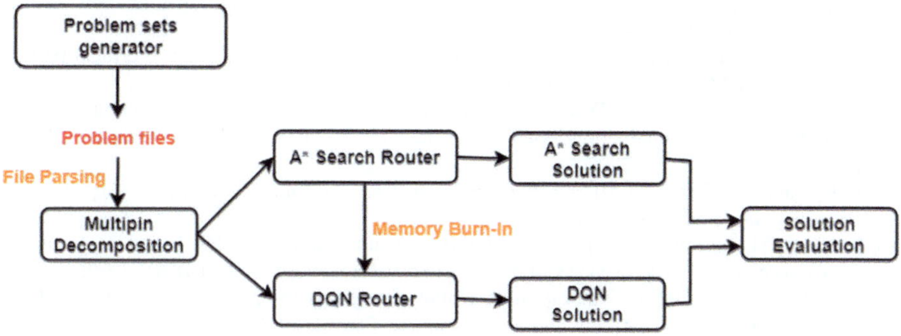

Fig. 7.18 Deep Q-network-based approach for global routing

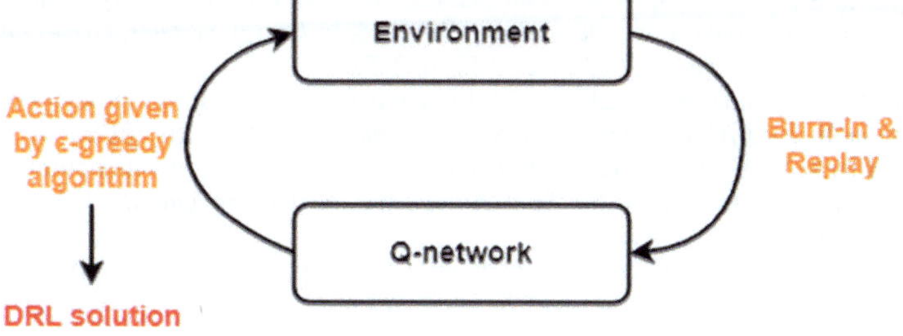

Fig. 7.19 DQN-based router

Network (CNN)-based router. The model is trained to learn the implicit design rules. In this work, they created their own datasets in the format of layout images with eight layers (for pins, metal layers, and vias). Each layer is binary encoded at the pixel level. The pixels in each layout layer are classified by the CNN router as on or off, which is used for the routing decision.

References

1. Tarafdar, N., et al. (2017). Enabling flexible network FPGA clusters in a heterogeneous cloud data center. In *Proceedings of the 2017 ACM/SIGDA International Symposium on Field-Programmable Gate Arrays*.
2. Weerasinghe, J., et al. (2016). Network-attached FPGAs for data center applications. In *2016 International Conference on Field-Programmable Technology (FPT)*. IEEE.
3. Guillot, A., et al. (2017). Multi-FPGA platform for smart automobiles embedded electronics design and prototyping. In *Proceedings of the International Conference on Embedded Systems, Cyber-physical Systems, and Applications (ESCS). The Steering Committee of The World Congress in Computer Science, Computer Engineering and Applied Computing (WorldComp)*.
4. Bhavani, M., & Devi, D. A. (2019). Design of smart Monitor for automobiles using FPGA based data logger. In *2019 International Conference on Communication and Electronics Systems (ICCES)*. IEEE.
5. Liu, L., et al. (2012). Downlink MIMO in LTE-advanced: SU-MIMO vs. MU-MIMO. *IEEE Communications Magazine, 50*(2), 140–147.
6. Liu, L., Zhang, J., & Pi, Z. (2012). Inter-cell interference avoidance for downlink transmission. *U.S. Patent No. 8,238,954*.
7. Shafin, R., et al (2020). Artificial intelligence-enabled cellular networks: A critical path to beyond-5G and 6G. *IEEE Wireless Communications, 27*(2), 212–217.
8. Chang, H.-H., et al. (2018). Distributive dynamic spectrum access through deep reinforcement learning: A reservoir computing-based approach. *IEEE Internet of Things Journal, 6*(2), 1938–1948.
9. Kastensmidt, F., Rech, P. (2016). *FPGAs and parallel architectures for aerospace applications. Soft errors and fault-tolerant design*.
10. Lattuada, M., Ferrandi, F., & Perrotin, M. (2016). Computer assisted design and integration of FPGA accelerators in aerospace systems. In *2016 IEEE Aerospace Conference*. IEEE.
11. Curzel, S., et al. (2023). Exploration of synthesis methods from simulink models to FPGA for aerospace applications. In *Proceedings of the 20th ACM International Conference on Computing Frontiers*.
12. Hosking, R. (2006) New FPGAs tackle real-time DSP tasks for defense applications. In *Boards & Solutions Magazine (November 2006)*.
13. Soumya, D. S., Aishwarya, C., & Vasavi, S. (2023). FPGA-based military vehicles detection and classification from drone videos using YOLOV5. In *Energy Systems, Drives and Automations: Proceedings of ESDA 2021* (Vol. 1057, p. 265).
14. Pillutla, N., & Kumar, S. (2023). FPGA implementation of multi-RT protocol for avionics simulation. In *2023 IEEE International Systems Conference (SysCon)*. IEEE.
15. Lin, C., Liang, Y., & Yi, Y. (2022). FPGA-based reservoir computing with optimized reservoir node architecture. In *2022 23rd International Symposium on Quality Electronic Design (ISQED)*. IEEE.
16. Gan, V. M., et al. (2021). A cost-efficient digital ESN architecture on FPGA for OFDM symbol detection. *ACM Journal on Emerging Technologies in Computing Systems (JETC), 17*(4), 1–15.

17. Shears, O., et al. (2021). A hybrid FPGA-ASIC delayed feedback reservoir system to enable spectrum sensing/sharing for low power IoT devices ICCAD special session paper. In *2021 IEEE/ACM International Conference On Computer Aided Design (ICCAD)*. IEEE.
18. Lin, C., Azmine, M. F., & Yi, Y. (2023). Accelerating next-G wireless communications with FPGA-based AI accelerators. In *2023 IEEE/ACM International Conference on Computer Aided Design (ICCAD)*. IEEE.
19. Yi, Y., et al. (2016). FPGA based spike-time dependent encoder and reservoir design in neuromorphic computing processors. *Microprocessors and Microsystems, 46*, 175–183.
20. Platforms, X. V. (2022). Available online: https://www.xilinx.com/products/design-tools/vitis/vitis-platform.html (Accessed on December 9, 2022).
21. EC2, A. (2022). Available online: https://aws.amazon.com/ec2/instance-types/f1/ (Accessed on December 9, 2022).
22. Nane, R., et al. (2015). A survey and evaluation of FPGA high-level synthesis tools. *IEEE Transactions on Computer-Aided Design of Integrated Circuits and Systems, 35*(10), 1591–1604.
23. Xilinx (2016). *Vivado High-level Synthesis UG902*. http://www.xilinx.com/
24. Intel (2016). Intel FPGA SDK for OpenCL. http://www.altera.com/.
25. Dai, S., et al. (2018). Fast and accurate estimation of quality of results in high-level synthesis with machine learning. In *2018 IEEE 26th Annual International Symposium on Field-Programmable Custom Computing Machines (FCCM)*. IEEE.
26. Makrani, H. M., et al. (2019). Pyramid: Machine learning framework to estimate the optimal timing and resource usage of a high-level synthesis design. In *2019 29th International Conference on Field Programmable Logic and Applications (FPL)*. IEEE.
27. Sayadi, H., et al. (2018). Ensemble learning for effective run-time hardware-based malware detection: A comprehensive analysis and classification. In *Proceedings of the 55th Annual Design Automation Conference*.
28. Sayadi, H., et al. (2019). 2smart: A two-stage machine learning-based approach for run-time specialized hardware-assisted malware detection. In *2019 Design, Automation & Test in Europe Conference & Exhibition (DATE)*. IEEE.
29. Lee, D., et al. (2015). Learning-based power modeling of system-level black-box IPs. In *2015 IEEE/ACM International Conference on Computer-Aided Design (ICCAD)*. IEEE.
30. Lin, Z., Sinha, S., & Zhang, W. (2018). An ensemble learning approach for in-situ monitoring of FPGA dynamic power. *IEEE Transactions on Computer-Aided Design of Integrated Circuits and Systems, 38*(9), 1661–1674.
31. Liang, Y., Wang, S., & Zhang, W. (2018). FlexCL: A model of performance and power for OpenCL workloads on FPGAs. *IEEE Transactions on Computers, 67*(12), 1750–1764.
32. Lin, Z., et al. (2020). HL-Pow: A learning-based power modeling framework for high-level synthesis. In *2020 25th Asia and South Pacific Design Automation Conference (ASP-DAC)*. IEEE.
33. Lin, Z., et al. (2022). PowerGear: Early-stage power estimation in FPGA HLS via heterogeneous edge-centric GNNs. In *2022 Design, Automation & Test in Europe Conference & Exhibition (DATE)*. IEEE.
34. Ustun, E., et al. (2020). Accurate operation delay prediction for FPGA HLS using graph neural networks. In *Proceedings of the 39th International Conference on Computer-Aided Design*.
35. Zhao, J., et al. (2019). Machine learning based routing congestion prediction in FPGA high-level synthesis. In *2019 Design, Automation & Test in Europe Conference & Exhibition (DATE)*. IEEE.
36. Schafer, B. C., Takenaka, T., & Wakabayashi, K. (2009). Adaptive simulated annealer for high level synthesis design space exploration. In *2009 International Symposium on VLSI Design, Automation and Test*. IEEE.
37. Ferrandi, F., et al. (2008). A multi-objective genetic algorithm for design space exploration in high-level synthesis. In *2008 IEEE Computer Society Annual Symposium on VLSI*. IEEE.

38. Wang, Z., & Schafer, B. C. (2020). Machine learning to set meta-heuristic specific parameters for high-level synthesis design space exploration. In *2020 57th ACM/IEEE Design Automation Conference (DAC)*. IEEE.
39. Wang, G., et al. (2006). Design space exploration using time and resource duality with the ant colony optimization. In *Proceedings of the 43rd annual Design Automation Conference*.
40. Mishra, V. K., & Sengupta, A. (2014). PSDSE: Particle swarm driven design space exploration of architecture and unrolling factors for nested loops in high level synthesis. In *2014 Fifth International Symposium on Electronic System Design*. IEEE.
41. Schafer, B. C., & Wakabayashi, K. (2009). Design space exploration acceleration through operation clustering. *IEEE Transactions on Computer-Aided Design of Integrated Circuits and Systems, 29*(1), 153–157.
42. Xydis, S., et al. (2010). Efficient high level synthesis exploration methodology combining exhaustive and gradient-based pruned searching. In *2010 IEEE Computer Society Annual Symposium on VLSI*. IEEE.
43. Pham, N. K., et al. (2015). Exploiting loop-array dependencies to accelerate the design space exploration with high level synthesis. In *2015 Design, Automation & Test in Europe Conference & Exhibition (DATE)*. IEEE.
44. Cilardo, A., & Gallo, L. (2015). Interplay of loop unrolling and multidimensional memory partitioning in HLS. In *2015 Design, Automation & Test in Europe Conference & Exhibition (DATE)*. IEEE.
45. Ferretti, L., Ansaloni, G., & Pozzi, L. (2018). Cluster-based heuristic for high level synthesis design space exploration. *IEEE Transactions on Emerging Topics in Computing, 9*(1), 35–43.
46. Ferretti, L., Ansaloni, G., & Pozzi, L. (2018). Lattice-traversing design space exploration for high level synthesis. In *2018 IEEE 36th International Conference on Computer Design (ICCD)* (pp. 509–512). IEEE.
47. Schafer, B. C. (2015). Probabilistic multiknob high-level synthesis design space exploration acceleration. *IEEE Transactions on Computer-Aided Design of Integrated Circuits and Systems, 35*(3), 394–406.
48. Xydis, S., et al. (2014). SPIRIT: Spectral-aware Pareto iterative refinement optimization for supervised high-level synthesis. *IEEE Transactions on Computer-Aided Design of Integrated Circuits and Systems, 34*(1), 155–159.
49. Wang, Z., Chen, J., & Schafer, B. C. (2020). Efficient and robust high-level synthesis design space exploration through offline micro-kernels pre-characterization. In *2020 Design, Automation & Test in Europe Conference & Exhibition (DATE)*. IEEE.
50. Ferretti, L., et al. Leveraging prior knowledge for effective design-space exploration in high-level synthesis. *IEEE Transactions on Computer-Aided Design of Integrated Circuits and Systems, 39*(11), 3736–3747.
51. Wang, Z., & Schafer, B. C. (2022). Learning from the past: Efficient high-level synthesis design space exploration for FPGAs. *ACM Transactions on Design Automation of Electronic Systems (TODAES), 27*(4), 1–23.
52. Schafer, B. C., & Wakabayashi, K. (2012). Machine learning predictive modelling high-level synthesis design space exploration. *IET Computers & Digital Techniques, 6*(3), 153–159.
53. Liu, H.-Y., & Carloni, L. P. (2013). On learning-based methods for design-space exploration with high-level synthesis. In *Proceedings of the 50th Annual Design Automation Conference*.
54. Zhong, G., et al. (2017). Design space exploration of FPGA-based accelerators with multi-level parallelism. In *Design, Automation & Test in Europe Conference & Exhibition (DATE), 2017*. IEEE.
55. Meng, P., et al. (2016). Adaptive threshold non-pareto elimination: Re-thinking machine learning for system level design space exploration on FPGAs. In *2016 Design, Automation & Test in Europe Conference & Exhibition (DATE)*. IEEE.
56. Mahapatra, A., & Schafer, B. C. (2014). Machine-learning based simulated annealer method for high level synthesis design space exploration. In *Proceedings of the 2014 Electronic System Level Synthesis Conference (ESLsyn)*. IEEE.

57. Zhong, G., et al. (2016). Lin-analyzer: A high-level performance analysis tool for FPGA-based accelerators. In *Proceedings of the 53rd Annual Design Automation Conference*.
58. Zhao, J., et al. (2017). COMBA: A comprehensive model-based analysis framework for high level synthesis of real applications. In *2017 IEEE/ACM International Conference on Computer-Aided Design (ICCAD)*. IEEE.
59. Goswami, P., Schaefer, B. C., & Bhatia, D. (2023). Machine learning based fast and accurate high level synthesis design space exploration: From graph to synthesis. *Integration, 88*, 116–124.
60. Liu, D., & Schafer, B. C. (2016). Efficient and reliable high-level synthesis design space explorer for FPGAs. In *2016 26th International Conference on Field Programmable Logic and Applications (FPL)*. IEEE.
61. Makrani, H. M., et al. (2019). Pyramid: Machine learning framework to estimate the optimal timing and resource usage of a high-level synthesis design. In *2019 29th International Conference on Field Programmable Logic and Applications (FPL)*. IEEE.
62. Ling, A., Singh, D. P., & Brown, S. D. (2005). FPGA technology mapping: A study of optimality. In *Proceedings of the 42nd Annual Design Automation Conference*.
63. Zhou, G., & Anderson, J. H. (2023). Area-driven FPGA logic synthesis using reinforcement learning. In *Proceedings of the 28th Asia and South Pacific Design Automation Conference*.
64. Mishchenko, A. (2007). *ABC: A system for sequential synthesis and verification*. http://www.eecs.berkeley.edu/alanmi/abc.
65. Hutton, M. D., et al. (1998). Characterization and parameterized generation of synthetic combinational benchmark circuits. *IEEE Transactions on Computer-Aided Design of Integrated Circuits and Systems, 17*(10), 985–996.
66. Qian, Y., et al. (2022). Efficient reinforcement learning framework for automated logic synthesis exploration. In *2022 International Conference on Field-Programmable Technology (ICFPT)*. IEEE.
67. Hu, H., et al. (2022). Machine-learning based delay prediction for FPGA technology mapping. In *Proceedings of the 24th ACM/IEEE Workshop on System Level Interconnect Pathfinding*.
68. Yanghua, Q., et al. (2016). Improving classification accuracy of a machine learning approach for fpga timing closure. In *2016 IEEE 24th Annual International Symposium on Field-Programmable Custom Computing Machines (FCCM)*. IEEE.
69. Gort, M., & Anderson, J. H. (2012). Analytical placement for heterogeneous FPGAs. In *22nd International Conference on Field Programmable Logic and Applications (FPL)*. IEEE.
70. Shojaei, H., Davoodi, A., & Linderoth, J. T. (2011). Congestion analysis for global routing via integer programming. In *2011 IEEE/ACM International Conference on Computer-Aided Design (ICCAD)*. IEEE.
71. Kannan, P., Balachandran, S., & Bhatia, D. (2001). fGREP-fast generic routing demand estimation for placed FPGA circuits. In *Field-Programmable Logic and Applications: 11th International Conference, FPL 2001 Belfast, Northern Ireland, August 27–29, 2001 Proceedings 11*. Springer.
72. Saeedi, M., Zamani, M. S., & Jahanian, A. (2007). Evaluation, prediction and reduction of routing congestion. *Microelectronics Journal, 38*(8–9), 942–958.
73. Chan, W.-T. J., et al. (2017). Routability optimization for industrial designs at sub-14nm process nodes using machine learning. In *Proceedings of the 2017 ACM on International Symposium on Physical Design*.
74. Tabrizi, Fakheri, A., et al. A machine learning framework to identify detailed routing short violations from a placed netlist. In *Proceedings of the 55th Annual Design Automation Conference*.
75. Maarouf, D., et al. (2018). Machine-learning based congestion estimation for modern FPGAs. In *2018 28th International Conference on Field Programmable Logic and Applications (FPL)*. IEEE.
76. Al-Hyari, A., et al. (2019). Novel congestion-estimation and routability-prediction methods based on machine learning for modern FPGAs. *ACM Transactions on Reconfigurable Technology and Systems (TRETS), 12*(3), 1–25.

77. Maarouff, D., et al. (2020). A deep-learning framework for predicting congestion during FPGA placement. In *2020 30th International Conference on Field-Programmable Logic and Applications (FPL)*. IEEE.
78. Yu, C., Zhang, Z. (2019). Painting on placement: Forecasting routing congestion using conditional generative adversarial nets. In *Proceedings of the 56th Annual Design Automation Conference 2019*.
79. Alawieh, M. B., et al. (2020). High-definition routing congestion prediction for large-scale FPGAs. In *2020 25th Asia and South Pacific Design Automation Conference (ASP-DAC)*. IEEE.
80. Zhou, Q., et al. (2015). An accurate detailed routing routability prediction model in placement. In *2015 6th Asia Symposium on Quality Electronic Design (ASQED)*. IEEE.
81. Al-Hyari, A., et al. (2021). A deep learning framework to predict routability for FPGA circuit placement. *ACM Transactions on Reconfigurable Technology and Systems (TRETS), 14*(3), 1–28.
82. McMurchie, L., & Ebeling, C. (1995). PathFinder: A negotiation-based performance-driven router for FPGAs. In *Proceedings of the 1995 ACM Third International Symposium on Field-Programmable Gate Arrays*.
83. Liao, H., et al. (2020). A deep reinforcement learning approach for global routing. *Journal of Mechanical Design, 142*(6), 061701.
84. Farooq, U., et al. (2021). Efficient FPGA routing using reinforcement learning. In *2021 12th International Conference on Information and Communication Systems (ICICS)*. IEEE.
85. He, Y., et al. (2022). Circuit routing using Monte Carlo tree search and deep reinforcement learning. In *2022 International Symposium on VLSI Design, Automation and Test (VLSI-DAT)*. IEEE.
86. Baig, I., & Farooq, U. (2022). Efficient detailed routing for FPGA back-end flow using reinforcement learning. *Electronics, 11*(14), 2240.
87. Jain, S. R., & Okabe, K. (2017). Training a fully convolutional neural network to route integrated Circuits. *arXiv preprint arXiv:1706.08948*.

Chapter 8
AI-Enabled 3D Integration

Vandana Kumari and Manoj Kumar Majumder

8.1 Introduction to 3D Integration

In the rapidly evolving landscape of semiconductor technology, the quest for smaller, faster, and more powerful electronic devices has been relentless. Over the years, we have witnessed the miniaturization of transistors and the integration of billions of these tiny switches onto a single silicon wafer. However, as we approach the physical limits of Moore's law, which predicted the doubling of transistors on a chip every 2 years [1], innovation has taken on a new dimension—one that extends into the third dimension. This chapter serves as a gateway into the intriguing world of 3D integration, a disruptive technology that has the potential to reshape the semiconductor industry and propel electronic devices to new heights of performance and efficiency. The 3D integration represents a paradigm shift from traditional 2D (planar) IC (integrated circuit) design and manufacturing, offering an alternative path to overcome the limitations of scaling down feature sizes. Instead of cramming more transistors onto a 2D plane, 3D integration stacks multiple layers of ICs on top of each other, creating a new dimension for packing functionality into a smaller footprint. The motivation behind 3D integration is twofold: the relentless demand for more computational power and the need for improved energy efficiency. As our society becomes increasingly reliant on data-driven applications, such as machine learning, artificial intelligence, and high-performance computing, the challenges of conventional 2D scaling become more apparent. Dennard scaling, the phenomenon in which shrinking transistors led to lower power consumption and improved performance, has been slowing down since the early 2000s [2, 3]. To continue the

V. Kumari (✉) · M. K. Majumder
Departments of Electronics and Communication Engineering, International Institute of Information Technology, Naya Raipur, Chhatisgarh, India

A. Iranmanesh, H. Sayadi (eds.), *AI-Enabled Electronic Circuit and System Design*,
https://doi.org/10.1007/978-3-031-71436-8_8

trajectory of Moore's law and meet the demands of next-generation technology, a fresh approach was necessary. This is where 3D integration comes into play.

Throughout this chapter, we will embark on a journey to explore the fundamental concepts, principles, and technologies that underpin 3D integration. We will delve into the key advantages it offers, the challenges it poses, and the various approaches to its implementation. From through-silicon vias (TSVs) to interposer technologies, and from heterogeneous integration to system-in-package (SiP) solutions, this chapter will provide a comprehensive foundation for understanding the exciting field of 3D integration. Moreover, we will discuss the historical context and the evolution of this technology, looking at pioneering developments, as well as current applications and emerging trends. As we traverse the intricate world of 3D integration, you will gain a deeper appreciation for how this innovative approach can revolutionize the electronics industry, enabling a new era of compact, powerful, and energy-efficient electronic devices.

So, fasten your seatbelts and prepare to embark on a journey through the third dimension of electronics. Welcome to the world of 3D integration.

8.1.1 Advantage of 3D Integration Compared to 2D/2.5D IC Packaging Technology

Before delving into the intricacies of 3D integration, it is essential to grasp the foundational concepts of 2D chip packaging, which has long been a cornerstone of semiconductor technology, providing a reliable and cost-effective means of assembling electronic components. In the world of semiconductor packaging, 2D chip packaging has been the go-to solution for integrating individual chips onto substrates. This conventional approach, while effective, is limited by its planar nature, which constrains the density of components and interconnects. This conventional approach involves mounting individual semiconductor chips onto a printed circuit board (PCB) or substrate, typically made of materials like silicon or ceramic. These chips, containing ICs and various electronic components, are arranged side by side on a PCB or substrate that interconnected via wire bonding, or flip-chip techniques as shown in Fig. 8.1a, b. The used substrate provides mechanical support and electrical connections for the chips, facilitating their integration into larger electronic systems. The 2D chip package configuration has long been favored for its simplicity, reliability, and cost-effectiveness. Continuously, as transition from 2D to 3D in packaging involves a shift from flat configurations to layered structures. This evolution addresses challenges related to size, power consumption, and signal transmission in semiconductor technology, marking the advent of a new era. Remarkably, 2.5D IC packaging represents a transitional phase beyond traditional 2D IC packaging. In contrast to the flat arrangement of chips in 2D packaging, 2.5D packaging involves positioning two or more active semiconductor chips alongside each other on a silicon interposer as shown in Fig. 8.1c. This interposer facilitates connectivity between the chips, enabling a

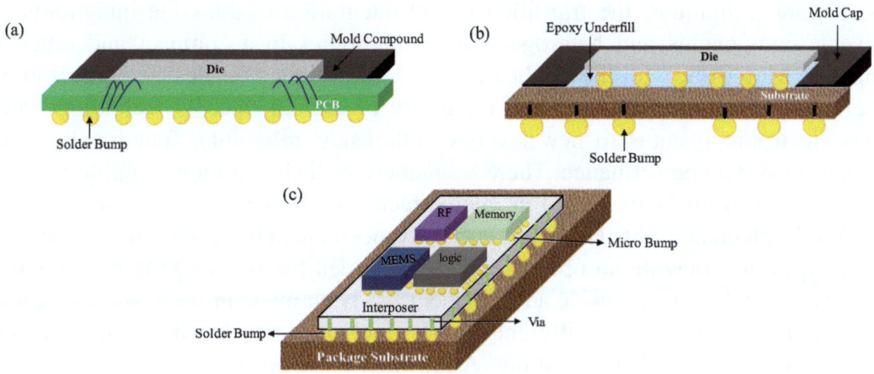

Fig. 8.1 The 2D IC packaging: (**a**) wire-bond BGA, (**b**) flip-chip BGA, (**c**) 2.5D IC

notably high die-to-die interconnect density and finer lines and spaces compared to 2D IC packaging. While 2D packaging disperses chips across a single plane, 2.5D packaging begins to layer components. This approach provides an intermediary between 2D and 3D packaging. Conceptually, 2.5D IC packaging resembles a cityscape with buildings of uniform height connected by bridges. Each "building" corresponds to a chip with a distinct function, while the "bridges" symbolize silicon interposers facilitating swift and efficient communication between the chips. However, with increasing demand of improved performance, better functionality, and miniaturization continues to grow, the limitations of traditional 2D packaging are becoming more apparent. But, as the demand for higher performance and miniaturization continues to escalate, there arises a need for innovative solutions that can overcome the limitations of traditional packaging methods. This is where 3D integration comes into play; by stacking multiple layers of chips vertically and employing advanced interconnect technologies, the 3D integration offers a promising alternative to conventional 2D packaging. In the following sections, we will delve into the advantages of 3D integration compared to 2D and 2.5D IC packaging technology, exploring how it addresses the limitations of traditional approaches and unlocks new possibilities for cutting-edge electronic systems. In traditional electronics packaging, 2D chip packaging has been a cornerstone of semiconductor technology for decades. In the context of advancing technologies and emerging applications, such as artificial intelligence (AI), there is a growing need for innovative packaging solutions that can deliver higher performance and improved efficiency. This has led to the exploration of 3D integration techniques, which offer opportunities to overcome the limitations of traditional 2D packaging and unlock new possibilities for AI-enabled systems.

In the recent three-dimensional (3D) technology, ICs are vertically stacked and interconnected, allowing for denser integration and shorter interconnection lengths. This leap in technology offers numerous advantages, including improved system performance, reduced power consumption, and enhanced form factors. By leveraging through-silicon vias (TSVs), advanced interposer technologies and innovative packaging techniques, the transition to 3D integration enables the integration of diverse components and heterogeneous technologies in a compact and efficient manner. Moreover, with the application of AI techniques for design optimization, thermal management, testing, and reliability prediction, the 3D technology propels electronic systems to new levels of efficiency, reliability, functionality, and enhanced system performance. There are numerous 3D integration technologies that enable small form factor (SFF) in microstructures, allowing for compact wireless system implementations. However, such advancements also come with associated challenges in achieving an optimal balance between the cost and the performance [4]. Additionally, an extensive advancement has been made in integrating a variety of heat TSV models with different pitches, including through holes and circular truncated cones, into 3D packaging. Altogether, these advancements have conclusively demonstrated the superior performance of TSV technology compared to the traditional wire bonding methods [5]. As we delve deeper into the world of 3D integration, it becomes increasingly evident that this innovative technology offers a host of advantages when compared to traditional 2D and 2.5D IC packaging approaches. In this section, we will explore these advantages in greater detail, shedding light on why 3D integration is garnering widespread attention and adoption in the semiconductor industry.

A. *Improved performance*

One of the most compelling primaries of 3D integration is the substantial improvement in performance it offers when contrasted with 2D and 2.5D technologies. In a typical 2D IC, interconnections between different components can be relatively lengthy, introducing signal latency and limiting data transfer rates. In stark contrast, 3D integration drastically shortens these interconnects, resulting in faster signal propagation. This, in turn, leads to vastly improved data transfer rates and significantly reduced communication delays, making 3D integrated devices ideally suited for high-performance computing applications, data centers, and cutting-edge consumer electronics.

B. *Higher energy efficiency*

In an age where energy efficiency is paramount, 3D integration addresses this concern admirably. As transistors have continued to shrink and densely populate 2D designs, issues like power density and thermal management have emerged as critical challenges. 3D integration effectively mitigates these problems by reducing the distance signals needed to travel, that minimizes power consumption and heat generation. Moreover, 3D integration enables more efficient power delivery and management, further enhancing the energy efficiency of electronic systems.

C. *Compact footprint*

Traditional 2D and 2.5D packaging technologies often necessitate larger footprints to accommodate various components and interconnects. In contrast, 3D integration leverages the power of vertical stacking, allowing for a remarkably compact form factor. This attribute is particularly valuable in applications where space is at a premium, such as mobile devices, wearables, and aerospace systems. The ability to pack more functionality into a smaller space stands as a central driving force behind the adoption of 3D integration.

D. *Enhanced heterogeneous integration*

3D integration facilitates the integration of diverse technologies and materials within the same package. This capability is pivotal in the realization of heterogeneous systems that amalgamate different functionalities, such as analog, digital, and MEMS, into a single device. The flexibility of 3D integration in accommodating a variety of technologies and materials renders it a versatile choice for applications that demand diverse functionalities within a compact package.

E. *Bandwidth and memory advantages*

The vertical orientation of 3D integration is key to its efficient use of space and the significant boost in memory bandwidth it offers. In applications where memory access speed and bandwidth are of paramount importance, such as graphics processing, artificial intelligence, and high-performance computing, 3D integration stands head and shoulders above its 2D/2.5D counterparts. Stacking memory on top of logic units significantly reduces data transfer distances and substantially enhances overall system performance.

F. *Enhanced yield and reliability*

The manufacturing process for 3D integrated devices often involves a "known good die" approach, where individual dies are rigorously tested before stacking. This approach substantially improves yield, as any defective dies can be identified and replaced before the final assembly. Additionally, the shorter interconnects in 3D integration considerably reduce the likelihood of signal degradation, further bolstering overall system reliability.

In summary, the advantages of 3D integration, including improved performance, energy efficiency, compact form factors, heterogeneous integration, memory bandwidth, and reliability, make it an extremely compelling choice for a wide range of applications. In the subsequent sections of this chapter, we will delve deeper into each of these advantages, providing a comprehensive understanding of the unique capabilities and potential that 3D integration technology holds.

8.1.2 Types and Architectures of 3D Integration

In the realm of 3D integration, there are various types and architectural approaches that cater to different application requirements and design considerations. Here, we will explore the primary types and architectures of 3D integration, providing an overview of each and discussing their advantages and used cases.

A. *Through silicon vias (TSVs)*

One of the most foundational forms of 3D integration is through-silicon vias (TSVs). TSVs are vertical interconnections that pass through the substrate, connecting active layers of integrated circuits. This approach enables the stacking of multiple ICs, creating a true three-dimensional structure. TSVs can be categorized into two main types: filled TSVs, which are typically filled with conductive material such as Copper (Cu), Tungsten (W), Cobalt (Co), Nickel (Ni), and unfilled TSVs, wherein the vias is left empty [6]. TSV-based 3D integration is well-suited for applications that require high interconnect density and improved performance, such as advanced microprocessors and memory devices. In 2.5D integration, a silicon or organic interposer serves as an intermediary layer that connects multiple dies. These interposers are equipped with TSVs and passive components that facilitate communication and power distribution between the stacked dies. The use of interposers allows for a degree of vertical integration without directly stacking active silicon layers, making it a more cost-effective and less complex solution compared to full 3D integration. This approach is commonly employed in high-performance computing, networking, and graphics applications.

B. *Wafer-on-wafer (WoW) and die-on-wafer (DoW)*

In wafer-on-wafer (WoW) integration, entire wafers are bonded together, creating a true 3D structure at the wafer level. There are several methods for achieving 3D stacking, including wire bonding, monolithic integration, and TSVs. Although wire bonding presents a simple and cost-effective approach, it is primarily suitable for ICs with lower power and frequency requirements, featuring fewer inter-die connections. Wire bonding is commonly used for staggered dies, as depicted in Fig. 8.2a. While this technique allows for extensive stacking, it comes with a higher cost. Die-on-wafer (DoW) integration involves bonding individual dies to a wafer, resulting in a mixed 2D and 3D structure. Both approaches are advantageous for applications that require a high level of integration and compactness, such as RF (radiofrequency) and millimeter-wave devices, where the reduced interconnect length is critical for performance.

C. *Monolithic 3D integration*

Monolithic 3D integration is a form of 3D stacking in which multiple active layers are built on the same substrate using various technologies like advanced CMOS (complementary metal-oxide-semiconductor) processes. The layers are interconnected through vias that are formed within the same substrate, as shown in Fig. 8.2b. Monolithic 3D integration is particularly useful in applications that demand high levels of integration, such as image sensors, wherein multiple processing layers are required to enhance the sensitivity and functionality [7].

D. *Heterogeneous integration*

Heterogeneous integration, as the name suggests, involves combining different types of devices and materials within a single 3D package. TSV-based 3D inte-

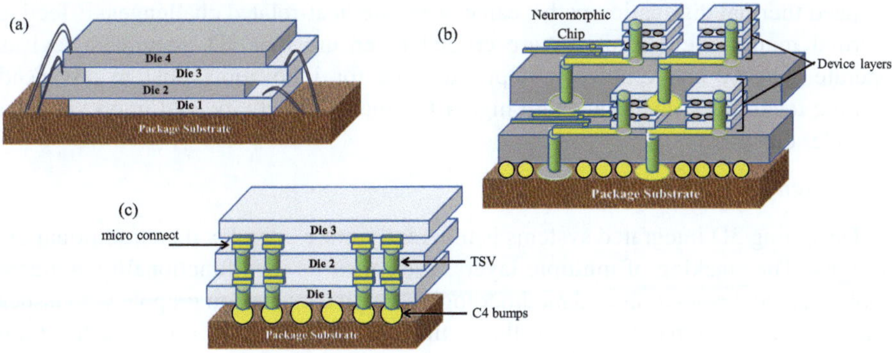

Fig. 8.2 The 3D Integration techniques: (**a**) wire bonding, (**b**) monolithic 3D IC, (**c**) TSV integration

gration shown in Fig. 8.2c, involves through-silicon vias and offers an alternative approach to achieve a 3D stacking.

This approach is essential for applications like the Internet of Things (IoT), where sensors, microcontrollers, and communication components need to be integrated into a single device. Heterogeneous integration enables the incorporation of various technologies, such as MEMS, RF, and power devices, to create versatile and multifunctional systems.

E. *Chip-on-chip and package-on-package*

Chip-on-chip (CoC) and package-on-package (PoP) are variations of 3D integration that are typically used in consumer electronics. CoC involves stacking one chip on top of the other, allowing for compact designs in applications like mobile phones. PoP, on the other hand, stacks separate packages on top of one another, providing a modular and easily upgradable solution for devices like laptops and gaming consoles. Each of these types and architectures of 3D integration offers unique advantages and trade-offs, making them suitable for different applications and design requirements.

8.1.3 Challenges and Opportunity in 3D Integration

The world of 3D integration, while promising and transformative, is not without its fair share of challenges. Here, we will examine the key challenges that engineers and researchers face when developing and implementing 3D integration technologies, as well as the exciting opportunities that lie ahead.

A. *Heat dissipation*

As electronic devices become more densely packed, managing heat has become a paramount concern. The 3D integration, with its higher transistor density and reduced thermal dissipation paths, can exacerbate heat-related challenges. Effective thermal management solutions are crucial to ensure that 3D integrated devices operate reliably and efficiently. Opportunities for innovation exist in advanced cooling techniques, materials with higher thermal conductivity, and improved heat spreader designs.

B. *Design complexity*

Designing 3D integrated systems is inherently more complex than traditional 2D designs. The stacking of multiple layers, each with its own functionality, requires sophisticated design tools and methodologies. Engineers need to grapple with issues like optimizing signal routing in three dimensions, managing power distribution efficiently, and ensuring that different components do not interfere with each other. Opportunities arise in the development of design automation tools and methods that simplify the creation of 3D integrated systems.

C. *Test and quality control*

Ensuring the reliability and quality of 3D integrated devices is a multifaceted challenge. Each layer in a 3D stack must be tested and verified before final assembly, increasing testing complexity. Furthermore, diagnosing and repairing faults in stacked devices can be challenging. Opportunities for advancements exist in test strategies, including pre-bond and post-bond testing, as well as innovations in fault isolation and repair techniques.

D. *Cost considerations*

While 3D integration offers numerous advantages, it often comes with higher manufacturing costs, stemming from the added complexity of processes like TSV fabrication and wafer bonding. Reducing the cost of 3D integration is a significant challenge, especially for consumer electronics and cost-sensitive applications. Opportunities lie in optimizing fabrication techniques, exploring cost-effective materials, and scaling up production volumes.

E. *Material and process compatibility*

Integrating various materials and technologies in a single 3D package can lead to compatibility issues. Differing coefficients of thermal expansion, material stress, and other factors must be carefully managed to prevent delamination or other reliability problems. Opportunities exist in developing materials with better compatibility and refining processes to mitigate material-related challenges.

F. *Standardization*

The lack of industry-wide standards for 3D integration can hinder its widespread adoption. Standardization efforts are essential to ensure interoperability between different devices and to promote collaboration across the semiconductor industry. Opportunities lie in establishing common interfaces, design rules, and testing standards to create a more harmonious ecosystem.

G. *Emerging applications*

The opportunities for 3D integration extend beyond overcoming challenges. Emerging applications in fields such as artificial intelligence, augmented reality, and edge computing are poised to benefit significantly from 3D integration. The ability to combine diverse functionalities and technologies within compact packages opens doors to novel products and services that were previously unattainable.

H. *Customization and specialization*

3D integration allows for greater customization and specialization of devices. This can be particularly advantageous in applications where unique functionalities are required. Opportunities exist for tailoring 3D integrated solutions to specific industry needs, resulting in high-performance, application-specific systems.

I. *Sustainability and environmental benefits*

3D integration can offer environmental benefits through reduced energy consumption and smaller form factors. Opportunities for green technology advancements can result from the energy efficiency of 3D integrated devices and their potential to reduce electronic waste through compact designs.

In conclusion, the challenges and opportunities in 3D integration coexist, with each challenge presenting a chance for innovation and improvement. As the field of 3D integration continues to evolve, engineers and researchers are poised to overcome obstacles, harness the potential of this technology, and usher in a new era of highly integrated, energy-efficient, and compact electronic devices.

8.1.4 Possibility of AI Application in 3D Integration

Artificial intelligence (AI) is rapidly transforming numerous industries, and the field of 3D integration is no exception. In this section, we will explore the exciting possibilities and emerging applications of AI in the context of 3D integration. Table 8.1 shows different opportunities in the AI market, with leaders those emerging various applications of AI in 3D integration as follows:

A. *Design optimization*

The AI-driven algorithms have the potential to revolutionize the design and optimization of 3D integrated systems. Machine learning models can analyze vast design spaces, considering factors such as power consumption, heat dissipation, and signal integrity. By iteratively generating and evaluating designs, AI can identify optimal 3D integration configurations that human designers might overlook. This not only accelerates the design process but also leads to more efficient and innovative solutions. Figure 8.3 depicts the utilization of an AI approach to optimize the microsystem prospect. The optimization process starts with identifying the optimized design parameters and performance parameters. The algorithm is then

Table 8.1 Opportunities of AI in 3D integration-based application

Application	Opportunities	
	Existing	New potential
Memory	On-chip memory, high bandwidth memory	Evolving nonvolatile memory (NVM)
Data storage	Potential growth in demand for current systems as more data is retained	AI-optimized storage systems, emerging NVM as storage solution
Compute	Parallel processing accelerators, such as GPU[a] and FPGA[b]	Workload-specific AI accelerators
Network	Architecture for data centers	High speed interconnect, programmable switches
Solutions	Computer vision technology, automation tools	Enhance machine learning models, robust security
Sustainability	Energy-efficient appliance	Recyclable packaging
Optimization	Optimizing inventory level, supply chain optimization	Improving logistics and distribution processes
Yielding	Defects prevention	3D IC yield enhancement

Source: McKinsey analysis
[a]Graphics processing units
[b]Field programmable gate array

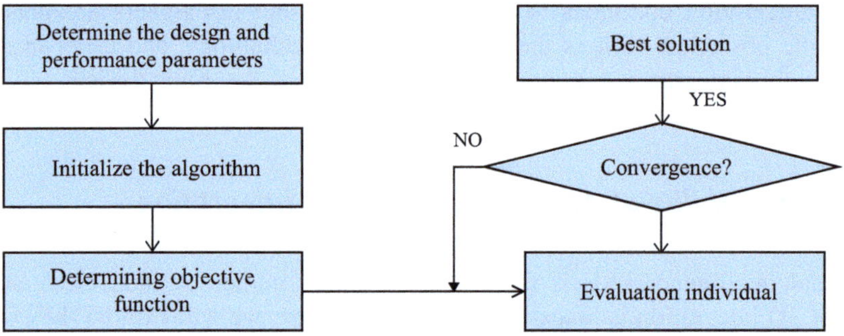

Fig. 8.3 Utilizing an AI approach to optimize the microsystem prospect

initialized, and an optimized target function is formulated based on the performance parameters slated for optimization [8]. This optimized target function serves as the fitness function for the optimization algorithm. Subsequently, the algorithm employs this fitness function to assess an individual performance quality. The optimization continues iteratively, with the algorithm evaluating and refining the design parameters using the fitness function the process ends either when the error between the fitness function for optimized performance and the global optimal fitness function falls below a predetermined accuracy threshold, or when the number of iterations reaches a set value. This ensures that the algorithm converges to a solution that meets the specified optimization criteria.

B. *Predictive maintenance*

AI can play a crucial role in ensuring the reliability and longevity of 3D integrated devices. Through the analysis of sensor data and real-time monitoring, AI algorithms can predict when components may fail or when performance may degrade. This enables proactive maintenance, reducing downtime and enhancing the lifespan of 3D-integrated systems in critical applications like data centers and aerospace.

C. *Thermal management*

Heat management is a significant challenge in 3D integration, and AI can provide dynamic solutions. AI offers several avenues to improve thermal management and heat dissipation in 3D integration. By analyzing thermal data collected from sensors embedded within the IC stack, AI algorithms can predict potential hotspots and thermal performance under various operating conditions [9]. This proactive approach enables designers to optimize the layout and configuration of the 3D IC to improve heat dissipation and prevent thermal issues before they arise. Additionally, dynamic thermal management systems adjust component parameters in real-time, redistributing tasks to mitigate localized heating and overall thermal stress, thereby enhancing the overall performance and reliability of the system. Altogether, topology optimization is a key aspect of 3D integration that can benefit from AI-driven design techniques. By leveraging AI algorithms to explore a vast design space, designers can identify the most thermally efficient configurations for the IC stack. This optimization process involves intelligently adjusting the placement and connectivity of components to minimize thermal resistance and enhance heat dissipation pathways, ultimately improving the overall thermal performance of the device [10]. Furthermore, AI-driven material selection accelerates the discovery of materials with superior thermal properties, improving overall thermal performance [11, 12]. Finally, AI-based fault detection and diagnosis continuously monitors thermal behavior, promptly identifying and mitigating thermal anomalies to ensure the system reliability. AI-based anomaly detection algorithms can continuously monitor the thermal behavior of the IC stack and detect deviations indicative of thermal issues or component failures. By promptly identifying and diagnosing thermal anomalies, AI systems can facilitate proactive maintenance and prevent catastrophic thermal events, ensuring the long-term reliability and performance of 3D integrated systems. Some of the examples are given below:

Example 8.1
When it comes to 3D IC design integration, thermal management hits a big challenge. The primary issue lies in the increased power consumption due to thermal concerns like current crowding and elevated temperatures resulting from heat generation within the stacked layers. To address these challenges effectively, an efficient machine learning (ML) model has been developed to optimize wire length and temperature [13]. Specifically, a linear regression model is favored for its ability to significantly enhance TSV layer assignment performance. This linear regression approach employs a gradient-based methodology, predicting errors at each step by

tracing the gradient cost function. Through this adaptable ELRM (error-based linear regression model), optimized TSV layer assignments are achieved. Performance analysis indicates that this ELRM-based TSV assignment method outperforms alternatives in terms of wire length reduction and temperature management. The main contribution of this work is to implement an enhanced linear regression algorithm to optimize the 3D IC design, minimizing thermal and wire length for TSV-based integration, and analyzing its performance. In this study, an enhanced gradient-based linear regression algorithm is proposed to improve the optimization of the 3D IC method, specifically focusing on reducing wire length and temperature. To assess the suitability of the machine learning linear regression algorithm for layer assignment in achieving effective TSV optimization, we aim to prevent thermal mismatches. The linear regression model, using gradient descent, minimizes errors by monitoring the gradient of the cost function at each training instance. Additionally, a fitness function is calculated to ensure effective TSV layer assignment. As a result, effective linear regression reduces temperature rise and optimizes wire length within the TSV layer as anticipated.

Used Methodology

The workflow of the proposed TSV approach methodology, discussed in Fig. 8.4, employs the modified stochastic-based genetic algorithm, and describes the techniques used for effective linear regression. In this method, gradient descent calculations and a fitness function are employed to detect optimal nodes.

The fitness calculation considers both thermal considerations and wire length for random TSV placements. During the selection process, TSV positions are updated by adjusting chromosome positions. Following this, the cost function evaluates the

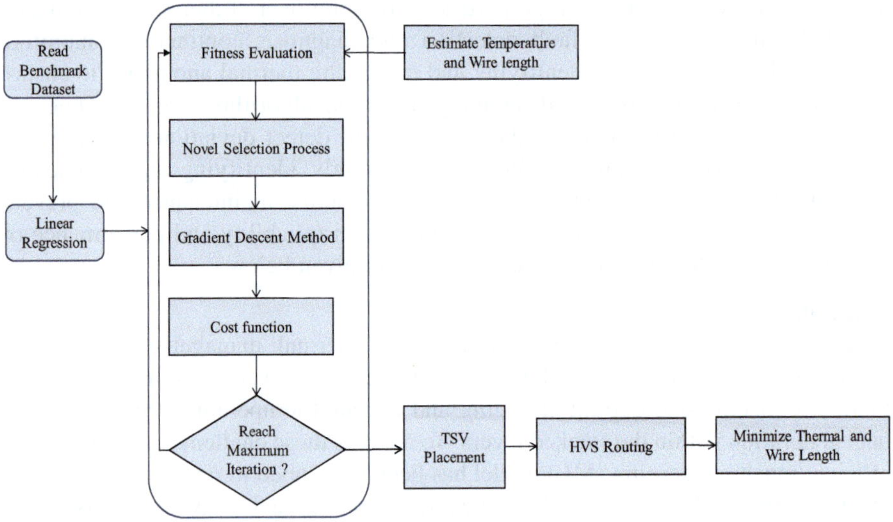

Fig. 8.4 Workflow of described methodology

movement of TSVs under thermal conditions. Utilizing a novel linear regression model significantly reduces both wire length and temperature.

Example 8.2
In this example, an intelligent design method of thermal through-silicon via (TTSV) for thermal management of a chiplet-based system is proposed based on the finite element method (FEM), back propagation neural network (BP-NN) model, and particle swarm optimization (PSO) algorithm. The study utilizes FEM simulations to analyze the impact of TTSV design parameters (TTSV pitches, radius of TTSV unit, and thickness of oxide layer) on thermal distribution. BP-NN models are then employed to map the relationships between these parameters and thermal distribution indices based on simulation data. Subsequently, a multi-objective optimization criterion is formulated to optimize design parameters using a modified PSO algorithm, aligning with desired chiplet-based system performance. This intelligent design method effectively optimizes TTSV parameters, ensuring efficient thermal management in complex chiplet-based systems [14]. Here, the developed intelligent design method effectively manages the thermal distribution indexes of the chiplet-based system, with errors between optimized and simulated thermal distribution indexes consistently below 1 K.

Used methodology
To fabricate the chiplet-based system with the desired thermal distribution, an intelligent optimization method is proposed to optimize the design parameters of the TTSV array, as presented in Fig. 8.5.

According to the desired performance criteria for a chiplet-based system with a TTSV array, multi-objective optimization criteria are formulated, and design

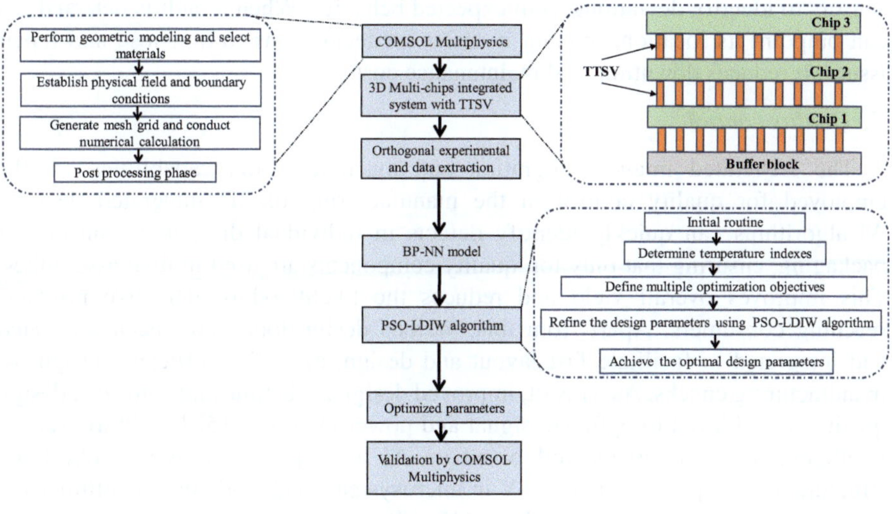

Fig. 8.5 Flowchart of the intelligent design method

parameters are optimized using a modified PSO algorithm. The parameters of the TTSV array in the chiplet-based system are then obtained and this research adopts PSO to find the optimal design parameters for the TTSV array. An intelligent design approach is developed using a modified PSO algorithm with linearly decreasing inertia weight (PSO-LDIW). Initially, parameters for the PSO-LDIW algorithm and multi-objective optimization criteria are initialized. Next, BP-NN models trained on thermal distribution indices predict the system's thermal behavior based on TTSV array design parameters. Subsequently, multi-objective optimization criteria are determined using predicted and desired thermal distribution indices. The PSO-LDIW algorithm is then used to finalize the TTSV array's design parameters. Ultimately, the optimal design parameters for the TTSV array are derived. This research establishes a detailed finite element method (FEM) model for a chiplet-based system featuring a TTSV array, revealing a complex and irregular relationship between design parameters and thermal distribution indices. The study employs backpropagation neural network (BP-NN) models to describe mapping relationships between these parameters, achieving remarkable accuracy with average absolute relative errors (AAREs) below 0.000045% across three performance metrics (PTs). An intelligent design method is proposed, utilizing the particle swarm optimization (PSO) algorithm to effectively manage thermal conditions in the chiplet-based system. The achieved thermal distribution indices for three chips (340.81 K, 334.76 K, and 314.85 K) closely align with target values (340 K, 335 K, and 315 K), demonstrating the efficacy of the approach in practical thermal management applications.

D. *Fault detection and isolation*

AI-driven fault detection and isolation techniques can enhance the reliability of 3D-integrated devices. Machine learning models can analyze data from various sensors to identify deviations from expected behavior. When a fault is detected, AI can pinpoint its location, making it easier for technicians to diagnose and repair issues. It reduces downtime and maintenance costs.

E. *Quality control*

The AI-related image recognition and machine vision capabilities can be employed for quality control in the manufacturing of 3D integrated devices. AI algorithms can quickly identify defects in individual dies, interconnects, or packaging, ensuring that only top-quality components are used in final assemblies. This improves overall yield and reduces the likelihood of defective products reaching consumers. Apart from this, various design tools have been innovated and integrated to facilitate fast layout and design, as well as efficient design for manufacturing checks. As a result, improved design cycle time and enhanced design quality are achieved to optimize signal and power integrity [15, 16]. Moreover, an intelligent optimization method based on AI is proposed to swiftly adjust the structural design parameters of TSV in microsystems and modeling to optimize the thermal-mechanical coupling problem [17, 18].

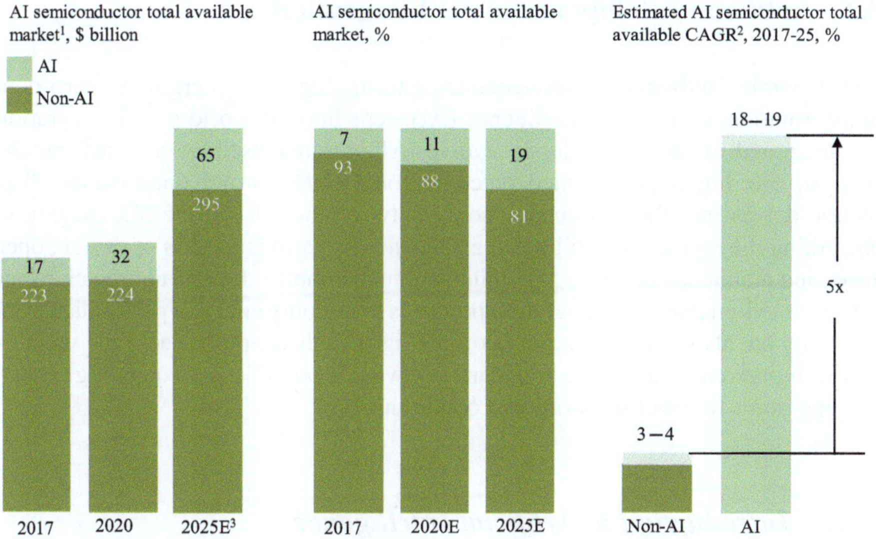

[1]Total available market includes processors, memory, and storage; excludes discretes, optical, and micro-electrical-mechanical systems.
[2]compound annual growth rate.
[3]E=estimated.
Source: Bernstein; Cisco System; Gartner; IC Insights; Machina Research; McKinsey analysis

Fig. 8.6 Growth for semiconductors related to artificial intelligence

F. *Data security*

AI can be used to enhance the security of 3D-integrated devices by analyzing data patterns and identifying potential security threats. Machine learning algorithms can monitor data traffic within the device and detect anomalies that may indicate a security breach. Figure 8.6 represents the growth for semiconductors related to AI and is anticipated to be five times higher than growth in the rest of the market [19]. This is particularly important in applications where data integrity and confidentiality are critical, such as IoT devices and cloud servers.

G. *Autonomous assembly*

In the manufacturing process of 3D-integrated systems, AI-driven robotic systems can take on a more prominent role. These robots equipped with AI vision and control algorithms can autonomously handle the intricate assembly process, aligning and bonding the various components with precision. This not only reduces human labor but also increases the efficiency and accuracy of the assembly process.

In summary, AI presents a wide array of possibilities in the field of 3D integration, from design optimization and predictive maintenance to fault detection and supply chain management. The synergy between AI and 3D integration has the potential to yield more efficient, reliable, and intelligent electronic systems, further accelerating the adoption and advancement of this transformative technology.

8.2 Artificial Intelligence in 3D Integration

The dynamic landscape of semiconductor technology is undergoing a profound transformation as artificial intelligence (AI) steps into the world of 3D integration. As integrated circuits relentlessly evolve, AI presents itself as a transformative catalyst, offering unprecedented potential for innovation and optimization. This section delves into the profound synergy between the AI and the 3D integration, unearthing the myriad possibilities it affords across various aspects of design, operation, and maintenance. From the optimization of complex designs to proactive fault detection, AI-enabled 3D integration promises to not only enhance performance and reliability but also usher in a new era of intelligent, self-adaptive electronic systems. In this digital renaissance, the boundaries of what is possible are expanding rapidly, opening doors to novel solutions and opportunities.

8.2.1 Introduction to Artificial Intelligence

Artificial intelligence (AI) stands at the forefront of technological advancement, promising a future where machines emulate human intelligence and perform tasks that once belonged exclusively to the realm of science fiction. This section serves as a gateway into the realm of AI, providing an overview of its history, key concepts, and potential applications. In an age defined by data-driven decision-making and automation, understanding AI's core principles is vital. As we embark on this journey, we will explore the foundations of AI, the historical context that has shaped its evolution, and the myriad ways it is transforming industries and reshaping our daily lives. Consequently, the emerging novel AI technologies primarily incorporate logic, computing, networking, and memory functions. On the other hand, an increasing demand for high computing power and bandwidth amplified the breakthrough for the semiconductor industry. Table 8.2 outlines the technology stack comprising nine distinct layers that facilitate the two essential activities in AI applications includes both training and inference. When developers seek enhancements in these processes, they are frequently facing challenges linked to the hardware layer. This layer encompasses storage, memory, logic, and networking. Semiconductor companies have the potential to overcome these hurdles by introducing advanced accelerator architectures, enhancing computational efficiency, and streamlining the transfer of extensive datasets across memory and storage.

The fusion of artificial intelligence (AI) and 3D integration presents a compelling synergy at the intersection of semiconductor technology and intelligent systems. In this section, we embark on a journey to unravel the multifaceted landscape of AI-enabled 3D integration. Before delving into the depths of this transformative union, it is essential to grasp the fundamental principles of AI, the evolution of 3D integration, and the profound implications that arise when these two technological giants merge their capabilities. In the twenty-first century, the essence of artificial

Table 8.2 Technology stack for artificial intelligence

Technology	Stack	Definition
Services	Solution and application	All-inclusive solutions including hardware, models, training data, and other parts like voice recognition systems
Training	Data types	Data fed into AI systems for analysis
Platform	Methods	Techniques for optimizing weights given to model inputs
	Architecture	Structured approach to extract features from data
	Algorithm	A set of rules that iteratively adjusts the weights assigned to specific model inputs within the neural network during training to optimize inference
	Framework	Software tools for defining architectures and executing algorithms on hardware through the interface
Interface	Interface systems	Systems within framework that establish and facilitate communication pathways between software and underlying hardware
Hardware	Head node	A hardware unit responsible for orchestrating and coordinating computations among accelerators
	Accelerator	A silicon chip engineered to execute highly parallel operations demanded by AI, enabling simultaneous computations

intelligence (AI) lies in its dynamic evolution. AI in this modern era becomes more than just a concept but a transformative force.

Machine learning plays a central role in this AI renaissance, propelling it to new heights by enabling systems to learn and adapt autonomously. Furthermore, the horizons of AI applications continue to expand, touching every facet of our lives, from health care to transportation, showing the boundless potential of this technology. This exploration sets the stage for understanding the profound impact and possibilities of AI in our rapidly evolving world. The evolution of 3D integration from the historical transition of computing technology from 2D to 3D, it becomes evident that the promises and challenges of stacking components are central to advancing the field. The quest for higher performance and miniaturization has fueled this evolution, and now, we stand at the cusp of a new era with the convergence of AI and 3D integration. AI's role in optimizing 3D designs is paramount, ensuring efficiency and functionality in complex structures. Moreover, the synergy of AI-enabled dynamic thermal management is vital for sustained performance, and predictive maintenance and reliability enhancement complete the picture, guaranteeing the long-term success of 3D integration technologies. These interconnected aspects form the foundation of a promising future for 3D integration in the digital age. In the dynamic landscape of 3D integration, a multitude of critical aspects come into play. From fault detection, isolation, and the reduction of downtime to AI-driven supply chain and quality control, the potential for optimizing supply chains with AI and ensuring quality assurance in manufacturing is immense. To achieve product excellence, AI extends its reach to autonomous assembly and data security. AI-enhanced robotic assembly techniques further enhance efficiency.

This section sets the stage for a comprehensive journey through the possibilities, challenges, and frontiers of AI in the world of 3D integration.

8.2.2 AI Techniques and Algorithms for 3D Integration

In the realm of 3D integration, the role of AI is becoming increasingly pivotal. AI techniques and algorithms are the building blocks that empower 3D integration to realize its full potential. In this section, we will explore the various AI methodologies and algorithms that underpin the advancements in 3D integration, exploring how they optimize designs, enhance reliability, and revolutionize the semiconductor landscape.

A. *Machine Learning in 3D Integration*

In the realm of 3D integration, a spectrum of machine learning techniques plays pivotal roles. Supervised learning stands as the foundation for predictive modeling, harnessing labeled data to make informed predictions. On the other hand, unsupervised learning takes center stage in anomaly detection, revealing concealed patterns and irregularities within complex structures. Reinforcement learning, with its ability to adapt and learn from interactions, finds its place in adaptive control systems for optimized 3D integration processes.

Moreover, the emerging fields of semi-supervised and self-supervised learning contribute to the efficient utilization of limited labeled data, offering innovative solutions to the challenges of integrating AI into the 3D landscape. These diverse approaches form the bedrock of AI in 3D integration, enabling enhanced performance, precision, and adaptability. During the TSV fabrication process, various defects like voids, cracks, misaligned chips and impurity failure formations in TSVs or bump defects are identified [20, 21]. Each defect or potential defect contributes uniquely to signal degradation [22]. Figure 8.7 illustrates different defect types, such as TSV voids, RDL shorts, deformed or missing bumps, and Cu-pillars, on a typical TSV structure with vertically stacked heterogeneous ICs. Testing TSV interconnects for manufacturing defects is challenging, requiring innovative design-for-test techniques. Here, we specifically analyze the electrical characteristics of TSV channels with defects by introducing varying amounts of void defects. Subsequently, a machine learning (ML) technique is applied to create models for accurately classifying voids, shorts, and open defects within a TSV.

B. *Defect detection and classification by machine learning*

Here, we simplify the TSV (through-silicon via) interconnection by modeling it as a cylindrical structure with a circular cross-section. Passivation layers shield the upper and lower sides of the silicon substrate, while an annular oxide liner fills the gap between the conductor and silicon, creating an insulating capacitance that inhibits direct current flow. To evaluate the impact of defects on the electrical properties of the TSV channel, we employ 3D electromagnetic field solvers,

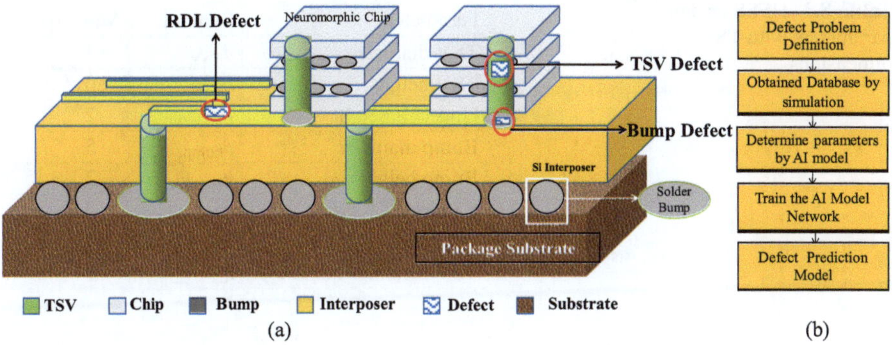

Fig. 8.7 Defects and its prediction model: (**a**) various kind of defect in TSV (**b**) flow chart of defects prediction model in 3D integration micro system

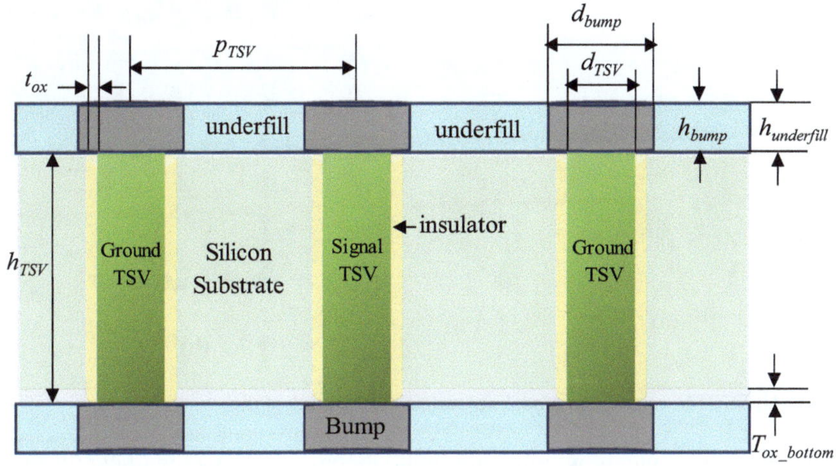

Fig. 8.8 Cross-sectional view of a TSV geometry

specifically HFSS simulators, to analyze electrical characteristics and transmission performance. Figure 8.8 depicts a physical structure with void defects in HFSS using a graphical interface, suitable for both single-die and stacked-die configurations. Table 8.3 details cross-sections of the TSV geometry utilized in this study, applicable to both single-die and stacked-die configurations.

The void content within the TSV is quantified as a percentage, termed as the void ratio (e). It can be computed using the formula:

$$e = \frac{V_{\text{void}}}{V_{\text{tsv}}} \tag{8.1}$$

Where, V_{void} is the volume of the voids, and V_{tsv} is the volume of the TSV.

Table 8.3 HFSS design parameters with TSV dimensions

Parameter	Symbol	Value (μm)
TSV diameter	d_{TSV}	3
TSV height	h_{TSV}	30
Underfill height	$h_{underfill}$	2
Bump diameter	d_{bump}	5
Bump height	h_{bump}	2
TSV to TSV pitch	p_{TSV}	7
Bottom SiO_2 thickness	T_{ox_bottom}	4
TSV insulator thickness	t_{ox}	1

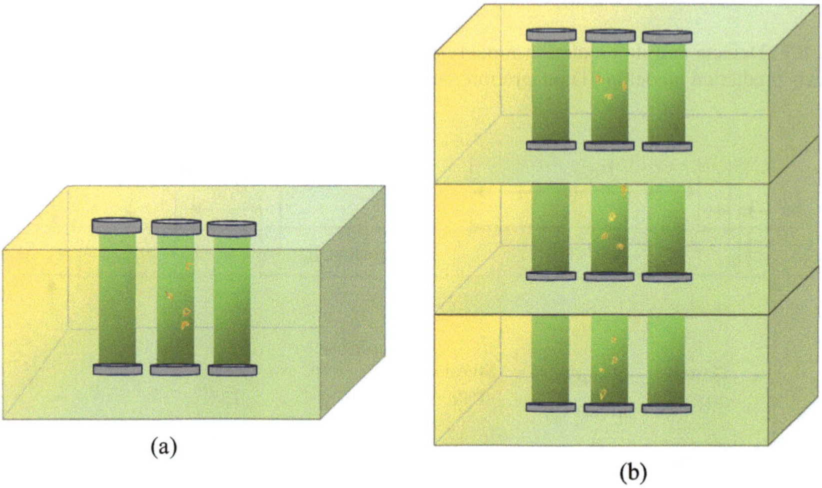

Fig. 8.9 HFSS TSV void defect model: (**a**) single die (**b**) stacked die

Various void contents within a through-silicon via (TSV) significantly impact signal transmission quality, as illustrated in Fig. 8.9 with a 3D model of a typical physical structure containing voids in both single-die and stacked-die configurations. The modeling methodology is adaptable for high-frequency structure simulator (HFSS) simulations, enabling the analysis of different void contents in these structures. Simulated variations of return loss (S_{11}) and insertion loss (S_{21}) for various void ratios inside TSV and nonvoided samples are depicted in Fig. 8.10a.

The degradation in Cu TSV signal integrity is observed to increase with higher void ratios, attributed to increased void area acting as scattering centers. The insertion losses (S_{21}) also escalate with rising defect ratios due to increased local impedance caused by voids, as shown in Fig. 8.10a. The evaluation extends to a stacked-die structure, as seen in Fig. 8.10b, where larger insertion losses are noted compared to a single-die structure. These distinctions arise from diverse structures surrounding the TSV channel and extended path lengths. Additionally, the existence of void defects across the TSV channel within stacked-die structures leads to a greater series capacitance. Both single- and stacked-die configurations

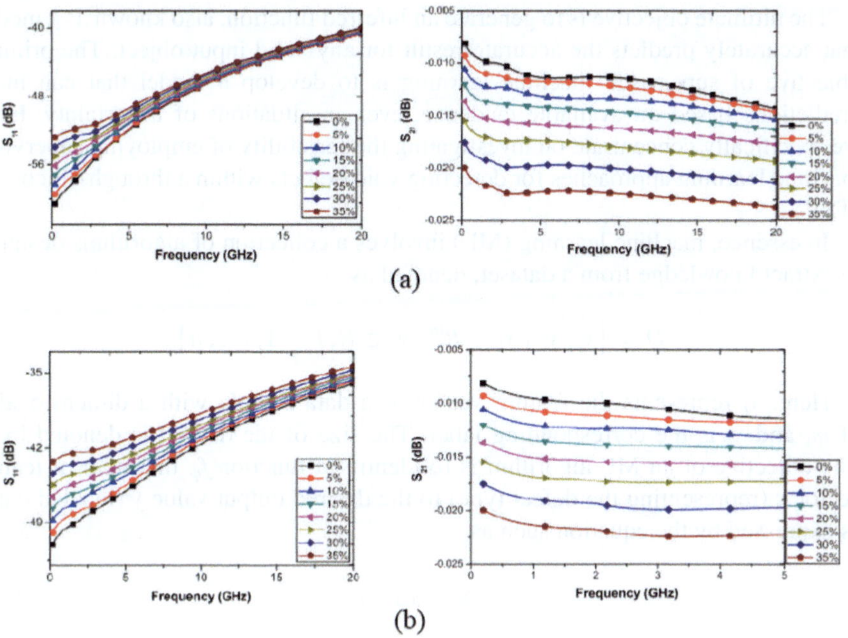

Fig. 8.10 Simulation results of various void ratios within TSV die as S11 and S21: (**a**) single die (**b**) stacked die

exhibit decreasing magnitudes of S_{11} and S_{21} with increasing frequency, linked to the decreasing impedance of the series capacitance due to disconnection at higher frequencies. In the broader context, return loss reflects signal loss from reflection coefficients, while insertion loss indicates the quality of signal transfer through the path. Higher void content in voided samples correlates with increased return losses compared to nonvoided samples. Defects, especially voids, serve as scattering centers and are recognized to deteriorate electrical performance in conductors. The 3D TSV structure incorporates redistribution layers (RDLs), and their impact on S-parameters is also explored. HFSS models for GSG-type TSVs and RDLs demonstrate simulation results for S_{11} and S_{21}, showcasing the effects of RDL short and open scenarios.

Supervised learning is a machine learning (ML) task that involves deducing a function from provided training data. The training dataset comprises a collection of instances, each composed of an input object, usually characterized by a set of features, along with the corresponding desired output value, commonly known as "labels." In the context of supervised learning, algorithms utilize training data with associated responses to learn the mapping function, which establishes the relationship between input samples and their respective outputs. The output can be discrete (in which case the inferred function is termed a "classifier") or continuous (resulting in a "regression function").

The ultimate objective is to generate an inferred function, also known as a model, that accurately predicts the accurate result for any valid input object. The primary objective of supervised machine learning is to develop a model that can make predictions based on available evidence, even in situations of uncertainty. Here, we specifically concentrate on investigating the feasibility of employing supervised machine learning approaches for detecting void defects within a through-silicon via (TSV).

In essence, machine learning (ML) involves a collection of algorithms designed to extract knowledge from a dataset, denoted as

$$D = \left\{x_i, y_i \mid x_i \in R^m, y_i \in R, i = 1, \ldots, n\right\}.$$

Here, x_i represents the feature vector of a data sample with a dimensionality of m, and y_i is the corresponding label. The size of the dataset is denoted by n. The objective of an ML algorithm is to identify a function f_ϕ that maps a feature vector x (representing the defect type) to the desired output value $\hat{y}$ (labeled data), as expressed by the equation such as

$$\hat{y} = f_\phi(x) \tag{8.2}$$

where ϕ is the model's parameter. Supervised learning specifically focuses on predicting missing information based on observed information. In this context, the function $f_\phi(x)$ is learned from a set of feature vectors and labeled samples $\{(x_i, y_i),\ldots, (x_n, y_n)\}$, known as training data, which serves as input for the learning algorithm. The supervised learning process involves training an ML model in the first step using the learning algorithm and training data, followed by testing the model's accuracy on unseen test data in the second step [23]. Accuracy is determined by calculating the number of correctly classified instances divided by the total number of instances.

This employs high-frequency structure simulator (HFSS) simulation data for various void ratios inside TSV and nonvoided samples. The dataset is partitioned into testing and training sets. A defect detection model is constructed from HFSS-simulated scattering parameters (S_{11} and S_{21}) using ML techniques. The supervised learning model for TSV void defect prediction is illustrated in Fig. 8.11. The performance of $f_\phi(x)$ (average classification error) is assessed using a distinct set of labeled data known as test data. The difference between prediction and observation is quantified by a loss functional L, and Fig. 8.12 demonstrates the initial chunking of the dataset D into K disjoint subsets (or blocks) of the same size $m = n/K$. Let us denote test block T_i for the i th such block, and D_i the training set obtained by excluding the elements in T_i from D. For a training set $D_i = \{z_1,\ldots, z_n\}$ with $z_i = (x_i, yi)$, the cross-validation estimator is defined as the average of the errors on test block T_i obtained when the training set is derived from T_i [24].

$$\mathrm{CV}(D) = \frac{1}{K}\sum_{k=1}^{K}\frac{1}{m}\sum_{x_i \in T_i} L\left(f\left(D_i\right), z_i\right) \tag{8.3}$$

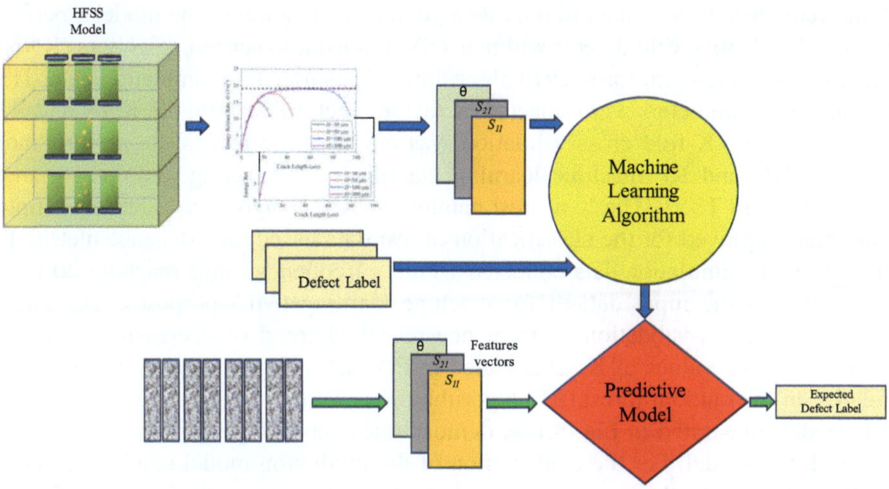

Fig. 8.11 Supervised learning model for TSV defect detection

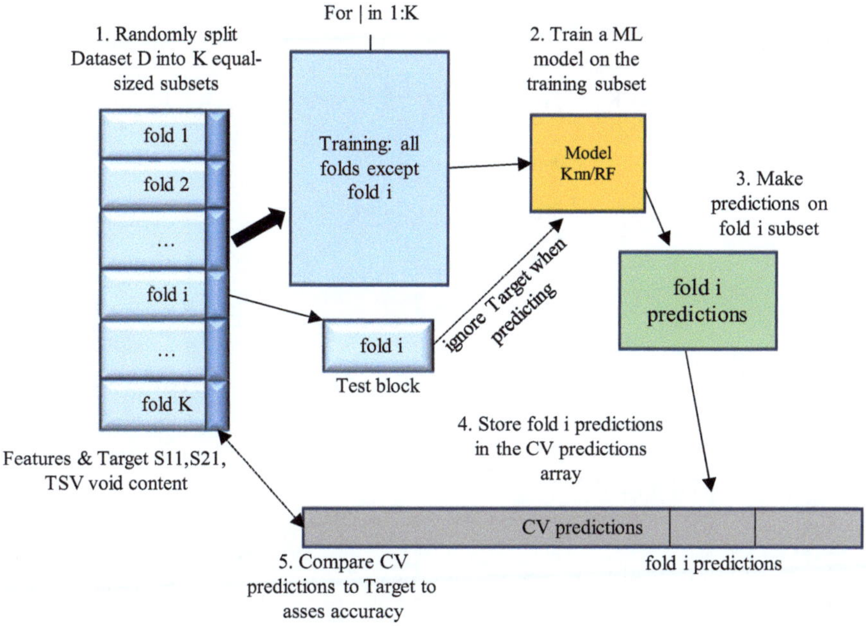

Fig. 8.12 Holdout method of cross-validation

The precision of the defect detection and classification model is subsequently assessed through k-fold cross-validation. This method is frequently employed to gauge the effectiveness of a predictive model on a specific training set while also preventing overfitting. By conducting n iterations of k-fold validation, the average of n accuracies is calculated to provide a genuine evaluation of the model's performance. To identify void defects within a TSV, trained a K-nearest neighbors (KNN) classifier and a random forest (RF) algorithm [22] using HFSS simulation data. The S_{11} and S_{21} parameters were selected as feature vectors for both the KNN and RF classifiers. The K-fold cross-validation was employed to assess the performance of the KNN and RF machine learning classifiers in detecting and categorizing void defects in TSVs. The K-nearest neighbors (KNN) algorithm is an impromptu classifier employed for the classification of test data based on a distance metric. In this research, simulation data collected within a frequency range reaching 20 GHz is utilized as the input dataset for machine learning (ML) purposes. The KNN algorithm's cross-validation performance reveals a trend of decreasing accuracy with increasing values of K, as depicted in Fig. 8.13. To evaluate its performance relative to the random forest (RF) algorithm, K is set to 5 for KNN algorithm. The RF model, illustrated in Fig. 8.14a, demonstrates superior performance compared to the KNN model. For the construction of the prediction model employing KNN and RF algorithms, the input dataset within a lower frequency range of up to 5 GHz is chosen. The cross-validation results, as shown in Fig. 8.14b, indicate that the

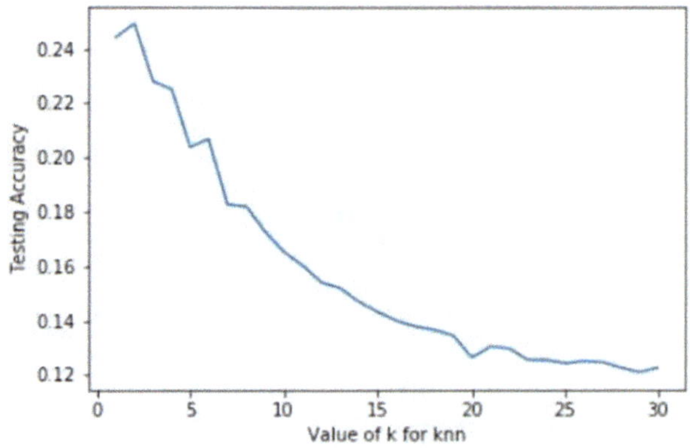

Fig. 8.13 Comparative analysis of KNN model performance across various K values

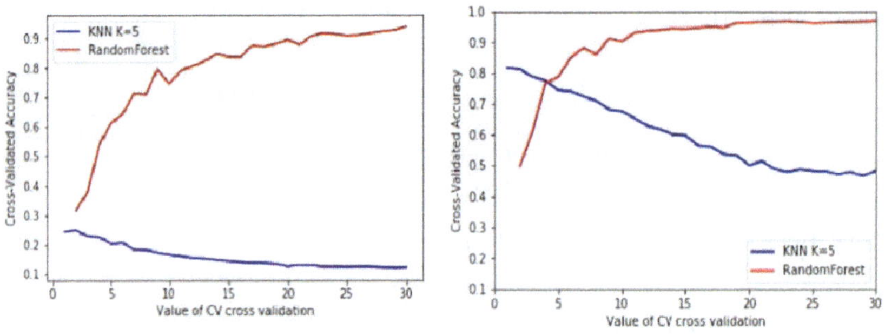

(a) (b)

Fig. 8.14 Performance comparison: (**a**) KNN and RF using input data set up to 20 GHz (**b**) KNN and RF

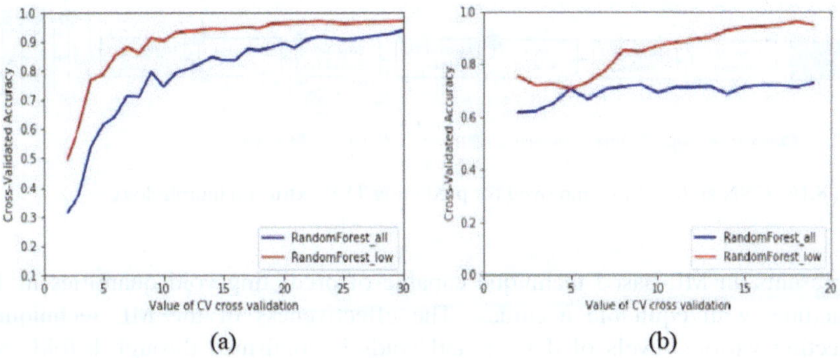

Fig. 8.15 Performance comparison of RF classifiers (**a**) across diverse quantities of void defects data set (**b**) for 3D TSV structure encompassing RDLs defect

KNN model does not exhibit significant improvements in performance. However, the accuracy of the RF model is noteworthy. The accuracy of the random forest (RF) model demonstrates superior classification results with reasonable accuracy for void detection. Furthermore, the accuracy tends to enhance with an increasing number of cross-validation (CV) folds for the RF algorithm. However, it becomes apparent that the accuracy levels off when approximately 20 CV folds are reached. The effectiveness of classifying voids in through-silicon vias (TSV) is influenced by the frequency range, primarily due to the MOS device characteristics exhibited by TSV [25, 26]. The capacitance of the oxide layer encasing a TSV plays a dominant role, particularly in low frequencies, determining the electrical behavior. Defects within the TSV, such as voids, change the electrical characteristics, and the impact of series impedance caused by disconnections in void defects manifests differently across frequency ranges. As illustrated in Fig. 8.15a, comparing the RF algorithm's performance with training data spanning the frequency range of 20–5 GHz reveals that lower frequency data leads to superior classification outcomes. This frequency-dependent classification efficiency also applies to the 3D TSV structure, which includes redistribution layers (RDLs) with open/short defects. Figure 8.15b demonstrates a similar comparison of RF algorithm performance, highlighting that classification results for detecting RDL defects improve notably with lower frequency range data.

The above discussion has highlighted the necessity for an in-line metrology method to characterize voids in TSV structures. By utilizing S-parameters, different levels of voids within TSVs exhibit distinct frequency response characteristics. However, accurately measuring the void quantity requires additional experimental data. While various methods like SAM and LIT rely on image processing algorithms for high-resolution imaging to detect TSV voids, ML-based approach involves training algorithms to analyze data and make predictions. The distinctive advantage of our method is the elimination of the need for image processing steps, leading to potential reductions in cost and time associated with void defect detection.

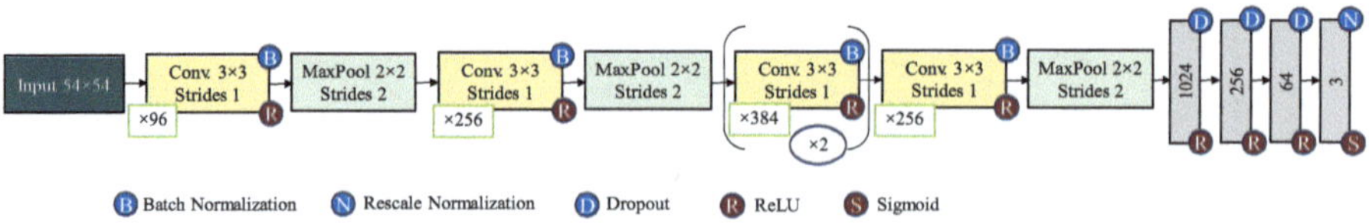

Fig. 8.16 CNN architecture employed for predicting TSV extrusion morphology

It presents an ML-based technique capable of predicting void quantities in TSV structures with equitable accuracy. The effectiveness of the ML technique in detecting various levels of defects and voids is confirmed through k-fold cross-validation. Additionally, the RF technique demonstrates its ability to forecast the quantities of shorts, voids, and opens in TSV-based 3D-stacked-die structure.

C. *Neural networks and deep learning*

Artificial neural networks (ANNs) serve as the backbone of modern machine learning, simulating the human brain's interconnected neurons to process complex data. Convolutional neural networks (CNNs), specialized for computer vision, excel at recognizing patterns and objects in images, revolutionizing fields like autonomous vehicles and facial recognition [27]. The CNN architecture model recognition [28] presented in Fig. 8.16 proposes a method wherein each 54 × 54 input image undergoes a transformation, resulting in 256 feature maps of size 3 × 3. These feature maps are then flattened and supplied to the fully connected (FC) layers within the network. Subsequently, the model generates class probability predictions for the input TSV image. Recurrent neural networks (RNNs) are tailored for temporal data, making them essential for applications such as speech recognition and time series forecasting. Meanwhile, the emergence of transformers has revolutionized natural language processing (NLP), leading to significant advances in machine translation, chatbots, and text generation. Each of these neural network architectures plays a unique and indispensable role in the diverse landscape of artificial intelligence applications.

D. *Evolutionary algorithms for optimization*

Designing in the realm of 3D integration involves harnessing a diverse array of optimization techniques. Genetic Algorithms serve as powerful tools for exploring design spaces, allowing for the evolution of innovative solutions. Particle swarm optimization (PSO) comes into play for efficient component placement, optimizing spatial arrangements. Simulated Annealing and Tabu Search are indispensable for layout optimization, fine-tuning the physical design for maximum efficiency. Ant colony optimization and swarm intelligence, inspired by nature's collaborative systems, finds application in 3D integration, offering novel approaches for complex problem-solving. These optimization methods collectively form a toolkit that empowers engineers and designers in the pursuit of cutting-edge solutions within the world of 3D integration.

E. *Natural language processing (NLP) in 3D design*

In the ever-evolving landscape of design and collaboration, language models are becoming invaluable tools for documenting and communicating ideas. Language models for design documentation streamline the process, offering efficient ways to convey complex concepts and specifications. Sentiment analysis in collaborative environments enhances team dynamics by providing insights into the emotional aspects of interactions, fostering better communication and decision-making. Moreover, NLP for automatic code generation revolutionizes software development, automating portions of the coding process and increasing efficiency. These applications of natural language processing are transforming the way we design, collaborate, and innovate in the digital age.

F. *Computer vision in quality control*

In the domain of quality control and precision engineering, image recognition plays a crucial role in defect detection, allowing for swift and accurate identification of imperfections in products. Object detection, on the other hand, is vital for precise component placement, ensuring that each element is correctly positioned within the complex systems. In microscopy and inspection processes, image segmentation proves indispensable by dividing images into meaningful segments, enabling detailed analysis and examination. These image processing techniques collectively contribute to the high precision and reliability required in various industries, from manufacturing to health care.

G. *Reinforcement learning in robotic assembly*

Reinforcement learning agents are at the forefront of modern robotics, seamlessly integrating AI into robotic systems. They enable autonomous assembly and precision alignment in complex 3D integration processes, ensuring components are accurately positioned and secured. Learning-based robotic control further enhances the efficiency and adaptability of these systems, enabling them to evolve and optimize their performance over time. This convergence of AI and robotics represents a significant leap forward in achieving precision and reliability in the intricate world of 3D integration.

As we journey through this section, we uncovered the intricate interplay of AI techniques and algorithms in the realm of 3D integration. From optimizing designs and enhancing reliability to enabling autonomous assembly and predictive maintenance, these AI methodologies are paving the way for a future where 3D integration achieves new heights of efficiency and innovation.

8.2.3 Application of AI in 3D Integration

Artificial intelligence (AI) has numerous applications in 3D integration, which is the process of stacking and interconnecting multiple integrated circuits (ICs) or dies to create complex systems. Here are some key applications of AI in 3D integration:

A. *Design optimization*

AI algorithms can explore vast design spaces and optimize the layout of components in a 3D structure. They can consider factors like power consumption, heat dissipation, and signal integrity to create more efficient and innovative designs.

B. *Predictive maintenance*

AI can monitor the health of 3D-integrated devices in real time by analyzing sensor data. This enables predictions of potential failures or performance degradation, supporting proactive maintenance and extending the device's life span.

C. *Dynamic thermal management*

Heat management is a significant challenge in 3D integration. AI algorithms can continuously monitor temperature sensors and predict thermal hotspots. By adjusting power delivery and cooling systems in real time, AI optimizes heat dissipation.

D. *Fault detection and isolation*

AI-driven fault detection techniques can analyze data from various sensors to identify deviations from expected behavior. When a fault is detected, AI can pinpoint its location, reducing downtime and maintenance costs.

E. *Supply chain optimization*

AI can optimize the entire supply chain for 3D integration, from materials sourcing to production and distribution. Machine learning algorithms can analyze historical data and market trends to predict demand, streamline inventory management, and optimize production schedules.

F. *Quality control*

AI's image recognition and machine vision capabilities can be employed for quality control in the manufacturing of 3D-integrated devices. AI algorithms can quickly identify defects in individual dies, interconnects, or packaging, ensuring high-quality components.

G. *Autonomous assembly*

AI-driven robotic systems equipped with vision and control algorithms can autonomously handle the assembly process, aligning and bonding components with precision. This reduces human labor and increases assembly efficiency.

H. *Data security*

AI can enhance the security of 3D-integrated devices by analyzing data patterns and identifying potential security threats. This is crucial for applications where data integrity and confidentiality are critical, such as IoT devices and cloud servers.

I. *Emerging applications*

AI-enabled 3D integration has applications in various industries, including AI hardware, edge computing, health care, aerospace, and automotive. AI can improve the performance, efficiency, and reliability of 3D-integrated systems in these domains.

J. *Ethical considerations*

With the use of AI in 3D integration, ethical and societal considerations come into play. Addressing issues related to privacy, bias, and fairness in AI systems is essential to ensure responsible and secure implementation. These applications highlight the transformative potential of AI in 3D integration, offering opportunities to create more efficient, reliable, and intelligent electronic systems across a wide range of industries.

The application of AI in 3D integration extends seamlessly into the realm of testing and quality assurance. By leveraging AI-driven algorithms, real-time fault detection, and predictive maintenance become integral components of ensuring the reliability and performance of 3D-integrated systems. The adaptive nature of AI facilitates not only innovative design but also enhances the precision and efficiency of quality assurance processes. As 3D integration evolves, the symbiotic relationship between AI applications and testing methodologies becomes increasingly crucial in achieving optimal functionality and reliability in complex, densely packed structures.

8.3 AI-Enabled Testing and Quality Assurance in 3D Integration

The quality and reliability of 3D-integrated devices are paramount, and achieving these standards demands innovative approaches to testing and quality assurance. Artificial intelligence (AI) has emerged as a powerful ally in this endeavor, offering the capability to enhance the efficiency and effectiveness of testing processes and ensuring the highest level of quality in 3D integration. In this section, we explore the application and significance of AI in the domain of testing and quality assurance, shedding light on the transformative potential it brings to this essential aspect of semiconductor technology.

Testing 3D-integrated devices presents a unique set of challenges that demand innovative solutions. The complex structures and high interconnect density of 3D integration require testing procedures that can navigate intricate configurations and assess multiple layers effectively. Thermal challenges further complicate the testing process, as the concentration of components in a compact space can lead to elevated temperatures, potentially affecting device performance. Comprehensive fault detection and diagnosis are imperative, as the dense interconnects and intricate architectures necessitate a thorough understanding of potential issues. Additionally, the need for speed and precision in testing is paramount, as these devices often operate at high speeds and demand meticulous attention to ensure accurate results. Balancing these challenges with effective testing strategies is essential to guarantee the reliability and performance of 3D-integrated devices in the ever-advancing field of electronics and technology.

AI-driven testing strategies are at the forefront of innovation in the testing of complex 3D-integrated devices. AI-powered test pattern generation is a game changer, as it enables the automatic creation of test patterns that can efficiently assess intricate configurations and high-density interconnects. The integration of automated test equipment (ATE) with AI further enhances the testing process, allowing for intelligent, data-driven decision-making during the assessment. Real-time monitoring and sensor data analysis play a pivotal role, enabling on-the-fly adjustments and providing insights into the device performance. The collective result is not only improved test coverage but also enhanced efficiency, ensuring that these advanced devices are rigorously and accurately tested, thus guaranteeing their reliability and performance in the ever-evolving landscape of technology.

AI contributes significantly to data security in testing by offering robust encryption and anomaly detection mechanisms, safeguarding sensitive information, and preventing unauthorized access. However, with great power comes great responsibility, and privacy and ethical concerns must be addressed. The collection and use of data in AI-enabled testing raises important questions about the privacy of individuals and the ethical implications of data handling. Furthermore, the potential for bias in AI testing systems demands scrutiny and measures to ensure fairness and equity. As AI continues to revolutionize testing processes, striking a balance between security, privacy, and ethical considerations is essential to uphold the highest standards of integrity and trust in these advanced technologies. The landscape of AI-enabled testing is rapidly expanding, particularly in high-reliability sectors such as aerospace, automotive, and medical devices. These critical industries benefit from AI's precision and efficiency in ensuring the reliability and safety of their products. The applications span from advanced quality control in aerospace to rigorous testing of autonomous systems in the automotive sector, as well as meticulous quality assurance in medical device manufacturing. Looking to the future, the prospects for advancements in AI-optimized testing and quality assurance are promising. The synergy between artificial intelligence and high-reliability sectors is poised to drive innovation, enhance safety, and streamline manufacturing processes. As AI continues to advance and adapt to the unique needs of these industries, it promises to play a pivotal role in shaping the future of quality assurance and testing. As we venture deeper into the realm of AI-enabled testing and quality assurance in 3D integration, it becomes clear that AI has the potential to revolutionize how we ensure the reliability, performance, and safety of 3D-integrated devices. With its capacity for fault detection, real-time monitoring, and efficiency improvements, AI brings a transformative edge to quality assurance in the semiconductor industry.

8.3.1 *AI-Based Testing Approaches and Strategies*

In the dynamic landscape of 3D integration, where the quality and reliability of stacked integrated circuits are paramount, Artificial intelligence (AI) is making significant inroads into testing processes. AI-driven testing approaches and strategies are redefining how we ensure the highest level of quality in 3D-integrated devices. This section explores the application and significance of AI in testing, shedding light on the innovative approaches and strategies that AI brings to the forefront of semiconductor technology.

3D integration testing presents a set of intricate challenges that necessitate innovative solutions. The complex structures and high interconnect density found in these systems demand testing procedures that can navigate intricate configurations and assess multiple layers effectively. Thermal challenges during testing are also crucial, as densely packed components can lead to elevated temperatures, potentially affecting device performance. Comprehensive fault detection and diagnosis are imperative in these complex architectures, requiring a thorough understanding of potential issues. Furthermore, finding the right balance between speed and precision in testing is vital, as these devices often operate at high speeds and demand meticulous attention to ensure accurate results. In response to these challenges, AI-powered test pattern generation emerges as a groundbreaking solution. Leveraging machine learning algorithms, it enables the automatic creation of test patterns that efficiently assess intricate configurations and high-density interconnects. Moreover, the optimization of test coverage and efficiency through AI ensures that testing is both comprehensive and streamlined. Additionally, AI plays a pivotal role in adapting test patterns for varying device configurations, promoting flexibility in testing procedures. This integrated approach marks a significant leap forward in the world of electronic testing, where AI-driven innovations are revolutionizing performance and cost-effectiveness, ensuring the reliability and efficiency of 3D-integrated devices in an ever-evolving landscape of technology.

The integration of artificial intelligence (AI) into the realm of testing and quality assurance is poised to revolutionize a broad array of industries. In the context of automated test equipment (ATE), AI plays a transformative role by enhancing ATE capabilities, offering real-time monitoring and adaptive testing, and enabling autonomous test execution. This not only optimizes testing efficiency but also ensures real-time adjustments and precise fault detection. Machine learning for anomaly detection, predictive maintenance, and AI-enabled isolation of fault locations further contribute to minimizing downtime and maintenance costs, ensuring the reliability of intricate systems. Furthermore, the advancement of technology has introduced various challenges, including different faults that arise in TSV-RDL structures. To address these faults, AI-based algorithms such as K-nearest neighbors (KNN) and feed-forward neural networks (FNN) have been employed, leveraging their capabilities to enhance accuracy and precision. In addition to this paradigm of machine learning, an appropriate artificial neural network is crafted, trained, and used for the identification of type and position defects in the daisy chain 3D structure of TSV. Moreover, this contribution explores both the benefits and constraints of this method, quantifying the disparities between the actual and estimated defect positions [29, 30].

Quality control and inspection, powered by AI-enhanced image recognition and machine vision, improve manufacturing yield, and uphold product excellence by automating defect detection in stacked dies and interconnects. Meanwhile, in performance testing, real-time monitoring, AI-driven performance optimization, and the identification of thermal hotspots guarantee efficient and reliable device operation. Yet, as AI becomes integral to these processes, security and ethical considerations are paramount, focusing on data protection, privacy, and fairness to prevent bias in AI testing systems. These advances have far-reaching applications across high-reliability sectors such as aerospace, automotive, and medical devices, wherein AI-optimized testing, and quality assurance enhance safety and streamline manufacturing processes, promising a future marked by innovation, safety, and efficiency.

As we delve into the realm of AI-based testing approaches and strategies in 3D integration, it becomes evident that AI has the potential to revolutionize the way that ensures the reliability, performance, and safety of 3D-integrated devices. AI brings a transformative edge to quality assurance and testing in the semiconductor industry by addressing complex challenges and offering innovative solutions.

8.3.2 Quality Assurance Technique in 3D Integration

In the realm of 3D integration, where the stacking and interconnection of integrated circuits yield advanced electronic systems, ensuring quality and reliability is of paramount importance. This chapter explores the various quality assurance techniques and strategies applied in 3D integration. With a particular focus on the role of artificial intelligence (AI), we delve into how AI-driven quality assurance is reshaping the landscape of semiconductor technology.

Quality assurance in the context of 3D integration presents unique challenges that must be met with innovative solutions. The complexity of structures and high interconnect density inherent to these systems demands new approaches. Additionally, considering thermal and electrical performance in multilayered designs is crucial to ensure the reliability of these intricate systems. Conventional quality assurance methods have included visual inspection and manual testing, non-destructive testing (NDT) techniques, reliability testing (e.g., HTOL, HAST), and functional testing and simulation. However, as 3D integration technologies evolve, the need for more advanced and efficient quality assurance processes becomes apparent. AI-enhanced quality assurance takes center stage, with AI playing a pivotal role in defect detection and classification. Predictive maintenance powered by AI further enhances the reliability of these systems, ensuring their longevity. Real-time monitoring of performance allows for immediate intervention, while AI-optimized quality control and inspection streamlines the process of identifying defects and maintaining the high standards of product excellence essential in a world driven by advanced technologies. Machine learning has emerged as a powerful tool for anomaly detection, with AI models trained to recognize deviations from the norm. This technology allows for real-time anomaly detection during operation, enabling proactive fault identification and diagnosis. In the realm of thermal management and reliability, AI-driven dynamic thermal management and predictive analysis work together to prevent thermal hotspots and extend the lifespan of devices, ensuring their longevity and performance. However, as AI continues to play a central role in these processes, security and ethical considerations are paramount. AI's involvement in data security and privacy is essential to safeguard sensitive information and ensure responsible data handling. Ethical and fairness considerations are equally critical in AI-powered quality assurance, as data integrity and unbiased decision-making are essential. Addressing potential bias and maintaining data integrity are ongoing concerns in the ever-evolving landscape of AI-enhanced quality assurance, where technology and ethics intersect.

As we journey through this chapter, we recognize the multifaceted nature of quality assurance in 3D integration. AI, with its capabilities in defect detection, predictive maintenance, and performance optimization, is reshaping the approach to quality assurance. The combination of traditional methods and AI-driven techniques is paving the way for a future where quality, reliability, and innovation coexist seamlessly in the world of 3D-integrated devices.

8.3.3 Case Studies

The choice of filler material in 3D integration plays a pivotal role in achieving structural integrity, thermal performance, and device reliability. In this chapter, we delve into real-world case studies that highlight the significance of filler materials, with a specific focus on how artificial intelligence (AI) is employed to optimize their selection and application. Table 8.4 provides a clear and concise overview of the challenges, solutions, and outcomes in various case studies related to AI-driven 3D integration in different industries.

These case studies demonstrate the multifaceted role of filler materials in 3D integration and the transformative impact of AI in optimizing their selection and application. They highlight the critical importance of selecting the right filler material and the potential for AI to enhance efficiency, reliability, and innovation in a variety of industries. Building upon quality assurance techniques in 3D integration, the chapter seamlessly transitions into the realm of AI-enhanced reliability and failure analysis. AI, having been integral to quality assurance, now takes center stage in proactively identifying potential failures through real-time monitoring and advanced defect detection. This shift signifies a comprehensive approach, utilizing AI algorithms to not only ensure quality but also to predict and address potential points of failure, ultimately enhancing the overall reliability and robustness of 3D-integrated systems.

Table 8.4 Case studies related to AI-driven 3D integration

Sl. No.	Case study	Challenge	Solution	Outcome
1.	High-performance data center cooling	Data centers demand efficient cooling solutions to maintain optimal operating conditions for stacked servers and storage systems	AI-driven optimization of filler materials, including graphene-based composites, for improved heat dissipation and reduced cooling costs	Enhanced data center performance, reduced energy consumption, and lower operating expenses
2.	AI-enhanced aerospace electronics	Aerospace applications require lightweight, reliable 3D-integrated systems capable of withstanding extreme conditions	Integration of AI-powered design optimization with aluminum filler materials to ensure structural integrity and efficient thermal management.	Robust and high-performance avionics systems with increased reliability and reduced weight
3.	Medical device miniaturization	Miniaturizing complex medical devices while maintaining functionality and reliability	Utilizing benzocyclobutene (BCB) filler material combined with AI-driven design optimization to achieve compact yet reliable medical devices	Smaller, lightweight medical devices with enhanced performance for improved patient care
4.	AI-optimized automotive electronics	Ensuring the reliability of automotive electronics in the face of temperature fluctuations and vibrations	AI-enhanced copper filler materials in automotive control units to improve thermal stability and fault tolerance	Enhanced automotive electronics with extended lifespans and greater resistance to environmental stressors
5.	Telecommunications infrastructure	Maintaining signal integrity and thermal efficiency in high-frequency telecommunications equipment	AI-optimized BCB filler materials for signal routing and heat management	Improved signal quality, reduced heat-induced issues, and greater network reliability
6.	Ethical considerations and bias mitigation	Addressing ethical considerations and bias in AI-driven filler material selection and application	Implementing AI governance frameworks to ensure fairness, transparency, and responsible AI use	Ethical and unbiased AI-driven decisions in filler material selection and quality assurance

8.4 AI-Enhanced Reliability and Failure Analysis in 3D Integration

The reliability and performance of 3D-integrated devices are paramount in the ever-evolving landscape of semiconductor technology. In this section, we explore how artificial intelligence (AI) is transforming reliability assessment and failure analysis in 3D integration. With AI as our guide, we navigate through innovative approaches to ensuring device longevity and reliability. Reliability and failure analysis in the context of complex 3D structures pose a significant challenge, exacerbated by high interconnect density and variability in device manufacturing. Rapid and accurate failure analysis is essential to maintain the integrity and performance of these intricate systems. Traditional reliability assessment methods have included accelerated life testing (ALT) and stress testing, failure modes and effects analysis (FMEA), reliability prediction models, and postmortem failure analysis. However, as technology advances, the demand for more efficient and proactive reliability assessment becomes apparent.

AI-enhanced reliability assessment takes center stage, with machine learning algorithms contributing to predictive maintenance and real-time monitoring through sensor data analysis. These algorithms enable the prediction of failure patterns and anomalies, offering a proactive approach to reliability enhancement. AI-driven failure analysis automates defect detection and classification, predicting failures through AI models, isolating fault locations, and conducting root cause analysis, ultimately reducing downtime and maintenance costs. This integrated approach marks a significant advancement in the world of reliability and failure analysis, where AI-driven innovations are revolutionizing performance and cost-effectiveness, ensuring the longevity and dependability of 3D-integrated devices in an ever-evolving landscape of technology.

The future is marked by AI's profound influence on reliability assessment, particularly in high-reliability sectors such as aerospace, automotive, and medical devices. AI-optimized reliability analysis is increasingly vital in these industries, promising to enhance safety, streamline manufacturing processes, and advance innovation. As the synergy between AI and high-reliability sectors continues to evolve, the prospects for advancements in AI-optimized reliability analysis are exciting, with the promise of improved safety and efficiency in critical industries. In this section, we uncover the transformative impact of AI in the domain of reliability and failure analysis for 3D integration. AI-driven predictive maintenance, real-time monitoring, and automated failure analysis are reshaping the semiconductor landscape, ensuring enhanced device longevity, performance, and reliability.

8.4.1 Reliability Challenges in 3D Integration

The pursuit of advanced 3D integration is marked by numerous innovations, but it also presents unique challenges in the realm of reliability. Here, we probe into the multifaceted reliability challenges encountered in 3D integration and how artificial intelligence (AI) is emerging as a key player in addressing these issues. Table 8.5 provides a clear overview of the multifaceted reliability challenges in 3D integration and the corresponding AI-driven solutions, emphasizing the role of artificial intelligence in addressing these issues.

As we navigate through the challenges of reliability in 3D integration, it becomes evident that AI is a transformative force, offering solutions to complex problems and ensuring that the technology continues to evolve with a focus on reliability, performance, and innovation.

Table 8.5 Reliability challenges with AI solution in 3D integration

Reliability challenge	Description	AI solution
Complex structures and interconnect density	Stacking multiple dies in three-dimensional structures increases the complexity and interconnect density, leading to thermal management and signal integrity challenges	AI-driven thermal and signal routing optimization to balance performance and reliability
Variability in device manufacturing	Variations in manufacturing processes can result in device-to-device discrepancies, impacting reliability	Machine learning algorithms for predictive maintenance and real-time monitoring to adapt to manufacturing variability
Thermal challenges	Managing heat within tightly stacked 3D structures is critical for reliability and performance	AI-optimized dynamic thermal management to prevent hot spots and extend device lifespan
Reliability prediction and analysis	Accurate reliability prediction models and timely failure analysis are essential for device maintenance and optimization	AI-enhanced failure analysis, predictive maintenance, and anomaly detection for proactive reliability improvement
Security and ethical considerations	Ensuring data security and addressing ethical concerns in AI-powered reliability assessments	Implementation of AI governance frameworks to maintain data integrity and ethical AI use
Emerging applications and future prospects	Adapting to reliability demands in high-stakes applications such as aerospace, automotive, and medical devices	AI-empowered reliability assessments for critical sectors with an eye on future advancements

8.4.2 AI-Based Reliability Prediction and Analysis

In the intricate landscape of 3D integration, the reliability of devices is not just a goal but a necessity. The ability to predict and analyze reliability is pivotal in ensuring the sustained performance of 3D-integrated systems. In this section, we explore how artificial intelligence (AI) is redefining the field of reliability prediction and analysis, offering innovative approaches to enhancing the durability and dependability of 3D-integrated devices. Reliability prediction and analysis in the context of complex 3D structures present a multifaceted challenge, particularly due to the intricacies of interconnect density and variability in manufacturing processes. The timely and accurate prediction of reliability is essential to ensure the longevity and performance of these intricate systems.

Traditional methods of reliability assessment have included accelerated life testing (ALT) and stress testing, failure modes and effects analysis (FMEA), reliability prediction models, and postmortem failure analysis. However, the ever-advancing technology landscape calls for more efficient and proactive reliability assessment.AI-powered reliability prediction takes center stage, with machine learning algorithms facilitating predictive maintenance, real-time monitoring through sensor data analysis, and the capability to predict failure patterns and anomalies, offering a proactive approach to reliability enhancement. Furthermore, AI-driven failure analysis automates defect detection and classification, utilizes AI models for predictive failure analysis, isolates fault locations, and conducts root cause analysis, ultimately reducing downtime and maintenance costs. This integrated approach marks a significant advancement in the world of reliability prediction and analysis, where AI-driven innovations are revolutionizing performance and cost-effectiveness, ensuring the longevity and dependability of 3D-integrated devices in an ever-evolving landscape of technology.

As AI takes on a pivotal role in reliability prediction and analysis, security and ethical considerations are of paramount importance. AI's contribution to data security and privacy in reliability analysis safeguards sensitive information, reducing the risk of unauthorized access and ensuring responsible data handling. Ethical and fairness considerations in AI-powered predictive maintenance and reliability assessment are crucial, as they ensure that decision-making is unbiased and data handling aligns with ethical standards. Additionally, addressing potential bias and data integrity concerns is an ongoing challenge in the ever-evolving landscape of AI-driven predictive maintenance and reliability assessment.

The future holds great promise for AI-enabled reliability assessment, particularly in high-reliability sectors like aerospace, automotive, and medical devices. AI-optimized reliability prediction and analysis is increasingly vital in these industries, promising to enhance safety, streamline manufacturing processes, and drive innovation. As the synergy between AI and high-reliability sectors continues to evolve, the prospects for advancements in AI-optimized reliability prediction and analysis are exciting, with the potential for improved safety and efficiency in critical industries. As we embark on this exploration, it becomes clear that AI-based reliability prediction and analysis is not just a technological advancement but a fundamental shift in ensuring that 3D-integrated devices remain dependable, efficient, and capable of withstanding the rigors of modern applications. In this section, we witness the transformative power of AI in the domains of failure analysis and prognostics. AI-driven approaches enable a proactive stance toward device reliability, allowing for early detection and mitigation of potential issues, ultimately enhancing the performance, safety, and durability of 3D-integrated devices.

8.5 AI-Driven Manufacturing and Assembly in 3D Integration

The manufacturing and assembly of 3D-integrated devices pose intricate challenges that demand precision, efficiency, and adaptability. In this section, we explore how artificial intelligence (AI) is revolutionizing the manufacturing and assembly processes, reshaping the way 3D integration is achieved.

Manufacturing and assembly in the realm of 3D integration bring forth a set of multifaceted challenges, particularly due to the complexity of structures, high interconnect density, and the necessity for microscale assembly precision. Managing variability in manufacturing processes while maintaining scalability and adaptability in production is also imperative to meet the demands of rapidly evolving technology. Traditional manufacturing and assembly methods have encompassed manual assembly and inspection, semi-automated processes, quality control, and defect detection within batch production models. However, the ever-evolving landscape of 3D integration requires more advanced and efficient methodologies.

The future presents promising applications for AI-driven manufacturing and assembly, particularly in high-reliability sectors such as aerospace, automotive, and medical devices, where safety and reliability are imperative. AI-optimized manufacturing and assembly are increasingly vital in these industries, promising advancements that will enhance safety, streamline manufacturing processes, and drive innovation. As AI continues to evolve and adapt to the unique needs of high-reliability sectors, the prospects for advancements in AI-optimized manufacturing and assembly are exciting, marking a significant leap forward in safety and efficiency in critical industries.

As we navigate through this section, it becomes clear that AI is not just a technological advancement but a fundamental shift in the way 3D integration devices is manufactured and assembled. AI-driven processes improve efficiency, reduce errors, and enhance adaptability, ultimately leading to a more innovative and reliable 3D-integrated future.

8.5.1 AI-Based Manufacturing Process in 3D Integration

The manufacturing process in 3D integration is a complex endeavor, demanding precision, speed, and adaptability. Artificial intelligence (AI) has emerged as a transformative force in optimizing and streamlining the manufacturing process, revolutionizing how 3D-integrated devices are created. Manufacturing processes in the realm of 3D integration are confronted with a unique set of challenges, primarily driven by the intricacies of complex stacked structures, high interconnect density, and the demand for microscale precision in assembly. Managing variability within manufacturing processes while simultaneously ensuring scalability and adaptability for efficient production becomes pivotal in the ever-evolving landscape of technology. Conventional manufacturing methods have historically involved manual assembly and inspection, semi-automated processes, and adherence to batch production models. These methods, while effective, have their limitations in addressing the intricacies of 3D integration manufacturing. As technology advances, the call for more advanced and efficient manufacturing methodologies becomes apparent to meet the demands of rapidly evolving technology. Innovative approaches, often empowered by AI, automation, and real-time quality control, are essential for achieving precision and efficiency in the production of 3D-integrated devices, ultimately upholding quality, and product excellence in this dynamic technological landscape. AI is taking manufacturing processes in the realm of 3D integration to the next level. With AI-powered robotics and automation, precision and efficiency are significantly enhanced, as machines are equipped with learning capabilities that optimize their performance. Real-time quality control and inspection ensure that manufacturing processes are continuously monitored for accuracy and adherence to quality standards. Adaptive manufacturing, empowered by AI feedback, is a game changer. This approach allows for dynamic adjustments during the manufacturing process, ensuring that the end product meets the highest standards of quality and performance. Moreover, machine learning for adaptive assembly, driven by AI algorithms for error detection and correction, predictive maintenance for manufacturing equipment, and real-time adjustments based on sensor data, guarantees that the production of 3D-integrated devices is streamlined, efficient, and responsive to the ever-evolving technological landscape. The result is an improvement in product quality, precision, and cost-effectiveness, which is crucial in meeting the demands of complex 3D integration manufacturing. AI-optimized quality control stands as a pivotal element in the manufacturing of 3D integration devices, where precision and efficiency are paramount. AI-enhanced image recognition and machine vision enable automated defect detection and classification, ensuring the continuous enhancement of manufacturing yield and product excellence.

As AI continues to evolve and adapt to the unique needs of high-reliability sectors, the prospects for advancements in AI-optimized manufacturing are exciting, marking a significant leap forward in safety and efficiency in critical industries. As we explore the impact of AI in the manufacturing process of 3D integration, it becomes clear that AI is reshaping how these intricate devices are made. AI-driven manufacturing processes enhance precision, reduce errors, and improve efficiency, ultimately leading to more innovative and reliable 3D-integrated systems.

8.5.2 Assembly Techniques and Optimization with AI

The assembly of 3D-integrated devices is a critical phase in the journey toward high-performance systems. In the realm of 3D integration assembly, numerous challenges emerge, each demanding innovative solutions for optimal performance. The first challenge involves the creation of complex stacked structures with high interconnect density. This intricate process of stacking multiple dies in three-dimensional configurations introduces difficulties in thermal management and signal integrity. Tables 8.6 and 8.7 show the conventional and AI-enhanced assembly techniques, respectively. Here, AI-driven solutions play a pivotal role by optimizing thermal and signal routing, thereby enhancing overall performance and reliability. The second challenge revolves around achieving microscale precision and alignment as devices become increasingly compact. The intricacies of this task are met with real-time adjustments facilitated by AI, ensuring the precise alignment of components during the assembly process. Handling variability in component dimensions constitutes the third challenge. Inherent variations in manufacturing processes can lead to device-to-device discrepancies, impacting overall reliability. To address this, machine learning algorithms have come into play, offering predictive maintenance and real-time monitoring that adapt to the inherent variability in manufacturing.

Finally, scalability for mass production stands as a critical challenge in meeting the demands of large-scale manufacturing. AI-enabled robotics and automation,

Table 8.6 Conventional assembly techniques

Challenge	Technique
Manual assembly and alignment	Human-led assembly with manual precision
Pick-and-place automation	Automated processes for component placement
Challenges in component matching and orientation	Addressing alignment issues in automation
Quality control and defect detection	Inspection processes to ensure product quality

Table 8.7 AI-enhanced assembly techniques

AI solution	Description
AI-powered robotics and automation	Integration of AI with robotic assembly systems
Real-time precision alignment	Immediate adjustments for precise alignment
Adaptive assembly with AI feedback	Dynamic assembly process based on AI insights
Scalable and efficient integration	Capability to adapt and scale for mass production

complemented by adaptive assembly techniques, serve as effective solutions, ensuring the efficiency and scalability of integration processes.

These challenges collectively underscore the significance of advanced technologies, particularly AI and machine learning, in navigating the complexities of 3D integration assembly to achieve reliable and scalable outcomes. In this section, we explore how artificial intelligence (AI) is playing a pivotal role in revolutionizing assembly techniques and optimizing the entire process, ensuring precise and efficient integration.

In the realm of adaptive assembly, the integration of machine learning and artificial intelligence (AI) is instrumental in enhancing precision, efficiency, and overall product excellence. AI algorithms dedicated to component matching and alignment play a pivotal role in ensuring precise alignment during the assembly process. This is complemented by predictive maintenance strategies, where anticipatory maintenance based on AI predictions ensures the continuous reliability of assembly equipment. Real-time adjustments based on sensor data further contribute to the adaptability of the assembly process, allowing immediate and informed modifications guided by sensor feedback. On the quality control front, AI-optimized techniques come into play with enhanced image recognition and machine vision. These technologies improve the accuracy of recognition and inspection processes. Automated defect detection and classification, facilitated by AI, contribute to the identification of defects during assembly. The overarching goal is to achieve an enhanced manufacturing yield and product excellence, where AI-driven methodologies lead to improved product quality and increased yield, particularly beneficial in industries demanding precision and reliability. Artificial intelligence (AI) assumes a pivotal role in ensuring robust data security and privacy throughout the assembly process. This involves the conscientious handling of data generated during assembly, safeguarding sensitive information from any form of unauthorized access. Beyond security, ethical considerations take center stage in the context of AI-enabled assembly. Measures are implemented to address potential biases inherent in AI algorithms, ensuring fair treatment and preventing unintended discriminatory outcomes. Upholding ethical standards in the utilization of AI for assembly processes is emphasized, promoting responsible and equitable use. Additionally, addressing concerns related to bias and data integrity is integral. Proactive measures are taken to identify and rectify biases in AI algorithms, coupled with a steadfast commitment to maintaining the integrity and accuracy of data throughout the assembly process. These conscientious efforts underscore the commitment to not only technological excellence but also ethical responsibility in the integration of AI within the assembly domain. The integration of Artificial intelligence (AI) in assembly processes has proven particularly impactful in sectors demanding utmost reliability, notably aerospace, automotive, and medical devices. AI-driven assembly techniques are instrumental in elevating the robustness and reliability of systems within these critical domains. In applications specific to aerospace, automotive, and medical devices, AI brings forth advancements that transcend traditional assembly methods. The implementation of AI enhances efficiency, precision, and overall reliability in the assembly of devices crucial to these sectors, ensuring optimal performance and safety standards. Looking ahead, the prospects for further advancements in AI-optimized assembly techniques are promising. The continuous evolution of AI techniques is anticipated to lead to innovative approaches, refining assembly processes and contributing to the ongoing enhancement of reliability and efficiency across diverse industries. The synergy of AI and assembly in high-reliability sectors not only marks a current technological milestone but also paves the way for a future characterized by continuous advancements and improvements in assembly methodologies.

As we navigate through this chapter, it becomes evident that AI is not just a technological advancement but a fundamental shift in how 3D-integrated devices is assembled. AI-driven assembly techniques enhance precision, reduce errors, and improve efficiency, ultimately contributing to more innovative, reliable, and high-performance 3D-integrated systems.

8.5.3 Robotics and Automation in 3D Integration

The field of 3D integration is constantly evolving, demanding precision, speed, and adaptability in the assembly and manufacturing processes. In this section, we explore the pivotal role of robotics and automation, empowered by artificial intelligence (AI), in reshaping the landscape of 3D integration.

The assembly and manufacturing of 3D integration devices present a series of complex challenges, driven by the intricacies of complex stacked structures, high interconnect density, and the necessity for microscale precision and alignment. Handling variability in component dimensions and ensuring scalability for mass production are crucial considerations in this dynamic field.

Conventional assembly and manufacturing techniques have involved manual assembly and alignment, as well as pick-and-place automation. However, these methods face challenges in component matching and orientation, particularly in the context of intricate 3D structures.

AI-powered robotics and automation are transforming the landscape of assembly and manufacturing. AI-enhanced robotic arms and automated systems enable real-time precision alignment and component placement, ensuring microscale accuracy. Adaptive assembly with AI feedback allows for dynamic adjustments during the assembly process, streamlining integration and upholding the highest standards of quality and precision. The scalability and efficiency offered by AI-powered assembly and manufacturing are essential in meeting the demands of complex 3D integration, ultimately ensuring cost-effectiveness and product excellence in this rapidly evolving technological landscape.

Machine learning is at the forefront of adaptive robotics, transforming the field of component matching and alignment. AI algorithms facilitate precision in component matching and alignment, ensuring that components are seamlessly integrated into complex 3D structures. Predictive maintenance for robotic systems ensures that the automation process remains efficient, minimizing downtime and costly repairs, while real-time adjustments based on sensor data enable robots to adapt dynamically to the changing requirements of the assembly process.

AI-optimized quality control is a critical element in ensuring the manufacturing excellence of 3D integration devices. AI-enhanced image recognition and machine vision technologies allow for automated defect detection and classification, enhancing the overall manufacturing yield by identifying and addressing defects in real time. This approach upholds the highest standards of product excellence and quality in the dynamic landscape of 3D integration assembly and manufacturing. As AI takes center stage in robotics and automation, security and ethical considerations are of paramount importance. AI's role in data security and privacy safeguards sensitive information, reducing the risk of unauthorized access and ensuring responsible data handling in the field of robotics and automation. Ethical and fairness considerations in AI-enabled assembly ensure that decision-making aligns with ethical standards, and addressing potential bias and data integrity concerns remains an ongoing endeavor in the dynamic landscape of AI-driven assembly and manufacturing.

The future holds exciting potential for AI-driven robotics and automation, particularly in high-reliability sectors such as aerospace, automotive, and medical devices, where precision and reliability are imperative. AI-optimized robotics and automation are increasingly vital in these industries, promising advancements that will enhance safety, streamline manufacturing processes, and drive innovation. As AI continues to evolve and adapt to the unique needs of high-reliability sectors, the prospects for advancements in AI-optimized robotics and automation are compelling, marking a significant leap forward in safety and efficiency in critical industries. Here, we witness the transformative power of AI-driven robotics and automation in the domain of 3D integration. These technologies enhance precision, reduce errors, and improve efficiency, ultimately contributing to the development of more innovative, reliable, and high-performance 3D-integrated systems.

8.6 Future Directions and Emerging Trends

The landscape of 3D integration continues to evolve, driven by technological advancements and the transformative potential of Artificial intelligence (AI). In this section, we explore the future directions and emerging trends that are set to influence the course of 3D integration in the years ahead. In the realm of quality assurance and reliability, AI-driven quality control and defect detection have eliminated the need for manual inspection, ensuring product excellence. Predictive maintenance and real-time monitoring play a pivotal role in maintaining optimal performance and device longevity. AI-driven testing and prognostics further enhance the reliability of 3D integration devices. Real-time fault detection and correction through AI algorithms guarantee the seamless operation of devices, while predictive maintenance and device health monitoring minimize downtime. The integration of AI across these critical aspects of 3D integration manufacturing, quality assurance, and testing marks a significant advancement in technology, promising unprecedented efficiency, reliability, and product excellence.

In the realm of 3D integration, the incorporation of AI comes with a twofold commitment to security and ethics. AI is instrumental in ensuring data security and privacy, fortified with robust safeguards to protect against cyber threats, safeguarding sensitive information and averting unauthorized access. Moreover, ethical and fairness considerations are integral to AI systems, actively working to mitigate bias and promote responsible AI use, ensuring that the technology operates with the highest ethical standards. The future of 3D integration is a diverse and expanding landscape. Technology is expanding into high-reliability sectors like aerospace, automotive, and medical devices, where precision and reliability are paramount. Simultaneously, 3D-integrated systems are proliferating across various industries, from telecommunications to consumer electronics, fundamentally transforming the way technology is applied in these sectors. This expansion heralds a new era of innovation and reliability, driven by AI and 3D integration, promising to reshape the technological landscape across diverse fields.

The prospects for research and innovation in the field of 3D integration are incredibly promising. Researchers are actively exploring AI's role in achieving sustainable and environmentally friendly 3D integration, aiming to reduce the ecological footprint of technology. This quest for eco-friendly solutions aligns with the growing emphasis on sustainability in the tech industry. Collaboration between academia, industry, and research institutions is a key driver of these innovations, as it brings together diverse expertise and resources to pioneer new AI-driven solutions. This collaborative approach fosters an environment where cutting-edge research can be translated into practical applications, ultimately leading to the development of more efficient, environmentally friendly, and innovative 3D integration technologies.

In this concluding section, we envision a future where AI-driven 3D integration plays a central role in shaping the technological landscape. The integration of AI in materials, design, manufacturing, quality assurance, and more opens the door to unprecedented innovation and reliability, heralding an era of cutting-edge 3D-integrated systems.

8.6.1 Promising Trends and Technologies

The world of 3D integration is marked by continuous innovation and progress. AI-enhanced manufacturing and assembly techniques are enabling the creation of flexible and wearable 3D devices that adapt to our dynamic needs. Real-time AI plays a pivotal role in ensuring adaptive functionality and sensing, enhancing the user experience and the range of potential applications.

Quantum computing integration within 3D structures marks a quantum leap in computing power. AI takes the forefront in optimizing quantum computing devices and algorithms, bringing unparalleled computational capabilities to the forefront.

These integrated innovations are propelling us into a future where computing is not only more intelligent but also more adaptable and closely integrated into our daily lives, offering transformative possibilities in various industries and applications. These promising trends and technologies represent the frontiers of 3D integration. AI is crucial for enhancing their capabilities and ensuring practicality, enabling the development of innovative and efficient 3D-integrated systems that can address the evolving needs of various industries and applications. In the ever-evolving landscape of technology, heterogeneous integration is a key innovation, marking the convergence of various technologies, materials, and functions into a single 3D package. This approach not only enhances the compactness of systems but also enables multifunctionality, addressing the demand for more versatile and efficient devices. Alongside this, quantum dot technology is transforming the realms of display and lighting technology, offering improved visual quality and energy-efficient lighting solutions. To further optimize the applications of quantum dots in 3D displays, AI-driven algorithms play a central role. These algorithms are instrumental in enhancing the performance and efficiency of quantum dot-based displays, ensuring that the integration of quantum dot technology in 3D displays aligns with the highest standards of quality and environmental sustainability, ultimately reshaping the technological landscape.

In the dynamic world of technology, innovation knows no bounds, and nanotechnology stands as a testament to this truth. It involves the utilization of nanoscale materials and structures to enhance 3D integration, pushing the boundaries of what's possible in miniaturization and precision. Guided by AI, the precision assembly of nanoscale components becomes a reality, ensuring that even the tiniest elements come together seamlessly and with unparalleled accuracy.

8.6.2 Research Challenges and Opportunities

The realm of AI-enabled 3D integration offers fertile ground for both challenges and exciting opportunities. In this section, we delve into the key research challenges that need to be addressed and the promising opportunities that lie ahead in the pursuit of advancing 3D integration with AI. Table 8.8 provides a structured view of the research challenges and opportunities across various domains of 3D integration.

Here, we recognize that research challenges are the building blocks of innovation, and the opportunities are the rewards of visionary efforts. AI-enabled 3D integration holds the promise of transforming various industries and addressing global challenges, provided researchers rise to meet these challenges with creativity, ingenuity, and perseverance.

Table 8.8 Challenges and opportunities across different areas related to AI-driven 3D integration

Integration domain	Research challenge	Opportunity
Heterogeneous integration challenges	Developing AI algorithms for optimizing heterogeneous integration processes, ensuring compatibility and efficient performance	Pioneering innovative applications by integrating diverse technologies and functions within compact 3D packages
AI-driven design challenges	Developing AI models for real-time design adaptation and optimization	Enabling AI to drive continuous design enhancements, leading to more efficient, reliable, and adaptable 3D systems
Material and nanotechnology challenges	Exploring the integration of nanomaterials and the role of AI in precise nanoscale assembly	Unveiling new frontiers in 3D integration by harnessing the potential of nanotechnology and AI
Biological integration challenges	Ensuring the seamless integration of biological materials and AI for health care and biotechnology applications	Pioneering medical breakthroughs and biotechnological advancements through AI-driven 3D integration
Energy-efficient computing challenges	Developing AI algorithms for energy-efficient computing and power management in 3D systems	Reducing energy consumption and environmental impact while enhancing performance
Thermal management challenges	Addressing thermal challenges in densely packed 3D structures through AI-optimized designs	Enhancing device reliability and performance while mitigating thermal issues
Edge computing and IoT challenges	Enabling AI-driven real-time processing and decision-making in 3D-integrated edge computing and IoT devices	Revolutionizing data processing at the device level for greater efficiency and responsiveness
Neuromorphic computing challenges	Advancing the capabilities of AI algorithms for neuromorphic computing in 3D systems	Harnessing the potential of brain-inspired computing for a wide range of applications
Flexible and Wearable Electronics Challenges	Optimizing manufacturing and AI-driven adaptability in flexible and wearable 3D devices	Empowering wearable technology for health, fitness, and beyond
Quantum computing integration challenges	Exploring the integration of quantum computing elements within 3D structures and AI for quantum applications	Unleashing the transformative power of quantum computing within 3D systems

8.6.3 Ethical Challenges and Implications

The fusion of artificial intelligence (AI) and 3D integration brings forth not only technological advancements but also ethical considerations that need to be carefully addressed. Here, we explore the ethical challenges and implications associated with AI-enabled 3D integration [31]. Organizations must create ethical guidelines and frameworks to direct the development and implementation of AI in 3D integration. Table 8.9 organizes the information into a clear format, making it easier to understand the challenges and implications across different areas related to AI-driven 3D integration. A multidisciplinary approach involving technologists, policymakers, ethicists, and stakeholders is required to tackle these ethical challenges. It involves incorporating ethical considerations into the design and deployment of AI systems in 3D integration, as well as ensuring ongoing monitoring, evaluation, and accountability for their ethical implications.

As we explore the ethical challenges and implications, it becomes evident that the ethical and responsible use of AI in 3D integration is vital. A proactive approach to these challenges can ensure that AI-driven 3D systems are not just technologically advanced but also aligned with societal values and expectations.

8.6.4 Privacy and Security Concerns

The incorporation of artificial intelligence (AI) in 3D systems ushers in numerous benefits, but it also raises important concerns about the privacy and security of data and devices. Here, we delve into the privacy and security concerns that arise with AI-enabled 3D integration. Table 8.10 provides a clear overview of the challenges, implications, and aspects related to data privacy in AI-enhanced 3D systems.

The privacy and security concerns related to AI-enabled 3D integration are multifaceted and require vigilant attention. A comprehensive approach, encompassing robust encryption, secure communication, ethical AI practices, and regulatory compliance, is essential to safeguard the privacy and security of AI-driven 3D systems and the data they handle.

8.6.5 AI-Driven 3D Integration in Industry

The fusion of AI with 3D integration has brought forth a new age of creativity and productivity across multiple sectors. Here, we delve into the significant impact of AI-driven 3D integration in diverse industrial sectors. Table 8.11 helps to quickly grasp the advancements and benefits across various industries related to AI-driven 3D integration.

Table 8.9 Challenges and implication across different areas related to AI-driven 3D integration

Area	Challenge	Implication
Data privacy and security	Safeguarding sensitive data in AI-driven 3D-integrated systems, particularly in applications where privacy is paramount	Potential for data breaches and security vulnerabilities if not properly managed
Bias and fairness	Detecting and mitigating biases in AI algorithms used in 3D integration, ensuring fair treatment and equitable outcomes	Unintentional discrimination and unequal opportunities if bias is not addressed
Transparency and explain ability	Ensuring that AI algorithms used in 3D integration can be transparently understood and explained	Trust issues and difficulties in decision-making if AI processes are seen as "black boxes"
Accountability	Defining responsibility in the event of errors or malfunctions in AI-enabled 3D systems	Legal and ethical accountability for AI-driven actions and decisions
Impact on employment	Addressing the potential displacement of human workers by AI-driven automation in 3D manufacturing	Need for retraining and reskilling of the workforce to adapt to changing job requirements
Ethical AI governance	Developing and implementing ethical frameworks for AI-enabled 3D integration	Establishing a foundation for responsible and ethical AI use in 3D systems
Environmental impact	Assessing and mitigating the environmental impact of AI-enhanced 3D integration processes	Establishing a foundation for responsible and ethical AI use in 3D systems
Intellectual property and ownership	Defining ownership and intellectual property rights in AI-generated designs and innovations	Reducing carbon footprints and waste associated with advanced manufacturing
Consumer awareness and education	Ensuring that consumers and end-users are aware of AI's role in 3D-integrated devices	Ethical concerns and expectations regarding AI use in consumer products

In these diverse industries, AI-driven 3D integration is revolutionizing the way businesses operate. From improved product performance to cost savings and enhanced user experiences, the integration of AI in 3D systems is a driving force behind innovation and competitiveness across the industrial landscape. AI-enabled 3D integration is not just a theoretical concept; it is a tangible reality with numerous successful implementations across various industries. The above real-world examples demonstrate the transformative power of AI-driven 3D integration in enhancing efficiency, reliability, and user experiences across a wide range of industries. The successful implementations serve as a testament to the potential of AI in reshaping the technological landscape.

Table 8.10 Challenges, implications, and aspects related to data privacy in AI-enhanced 3D systems

Aspect of data privacy in AI-enhanced systems	Challenge	Implication
Protecting sensitive data	Protecting sensitive data collected and processed by AI in 3D-integrated systems, particularly in applications where privacy is paramount	The risk of unauthorized access and data breaches, which can compromise personal and confidential information
Security of AI algorithms	Ensuring the security of AI algorithms used in 3D integration, preventing tampering or unauthorized modifications	Possible weakness that could be leveraged by malicious individuals to disrupt the functioning of devices
Secure communication in AI-connected devices	Securing communication channels between AI-enabled devices to prevent eavesdropping and data interception	Risks of data leakage and unauthorized access to communication streams
Device authentication and authorization ethical AI use and data handling	Implementing robust device authentication and authorization mechanisms to prevent unauthorized access to AI-driven 3D devices. Ensuring ethical AI use, which includes fair treatment, unbiased decision-making, and responsible data handling	Ethical concerns and societal backlash if AI are perceived as infringing on privacy or making unfair decisions Unauthorized control or manipulation of devices, potentially leading to adverse consequences
Protection against malware and cyberattacks	Developing safeguards to protect AI-enhanced 3D systems from malware and cyberattacks	Device malfunctions, data loss, and potential security breaches if not adequately protected
Data ownership and control	Defining ownership and control over the data generated and processed by AI in 3D integration	Disputes over data ownership and control, especially in collaborative or shared environments
User consent and transparency	Ensuring that users are well-informed and provide informed consent for AI use in 3D-integrated devices	Ethical concerns and potential legal issues if users are not adequately informed or involved in decision-making
Regulatory compliance	Adhering to evolving privacy and security regulations and standards	Legal and financial penalties for noncompliance, as well as reputational damage

8.7 Summary

As we reach the conclusion of this journey into the realms of AI-enabled 3D integration, it is evident that we stand on the precipice of a new technological era. From health care to aerospace, from manufacturing to space exploration, the

Table 8.11 Advancements and benefits across various industries related to AI-driven 3D integration

Industry	Advancements	Benefits
Aerospace and defense	AI-optimized thermal management and fault detection in critical aerospace components	Enhanced reliability, reduced maintenance costs, and improved performance in aircraft and defense systems
Automotive	AI-enhanced in-vehicle systems with 3D integration for advanced driver assistance and infotainment	Safer and more autonomous driving experiences, improved fuel efficiency, and enhanced connectivity
Health care and medical devices	AI-embedded medical devices with 3D integration for real-time patient monitoring and diagnostics	Early disease detection, remote patient care, and improved treatment outcomes
Telecommunications	AI-driven optimization of 3D-integrated network infrastructure for low latency and high bandwidth	Enhanced network performance, reduced downtime, and improved data transfer speeds
Consumer Electronics	AI-enhanced 3D integration in smartphones, wearables, and smart home devices	Smarter and more responsive devices, improved battery life, and enhanced user experiences
Energy and utilities	AI-optimized energy management systems with 3D integration for efficient power distribution	Reduced energy wastage, improved grid stability, and cost savings
Manufacturing and Industry 4.0	AI-driven smart factories with 3D integration for automated and adaptive manufacturing	Increased production efficiency, reduced errors, and lower operational costs
Environmental monitoring	AI-enabled sensors with 3D integration for real-time environmental data collection	Enhanced insights into climate change pollution control, and disaster prediction
Space exploration	AI-optimized 3D-integrated spacecraft systems for autonomous navigation and exploration	Improved space missions reduced human intervention, and cost-efficient exploration
Finance and banking	AI-driven 3D-integrated financial systems for risk assessment, fraud detection, and customer service	Enhanced security, improved decision-making, and personalized services

successful implementations, and real-world examples we have encountered are testimony to the transformative potential of AI-driven 3D systems. The future of AI-enabled 3D integration is a canvas, awaiting the brushstrokes of innovative minds and the collaborative efforts of researchers, engineers, and visionaries. It is a future where we can realize even more efficient, reliable, and adaptable systems, where AI optimizes not just technology but also human experience. So, as we conclude this exploration, let us recognize that AI-enabled 3D integration is not just a concept; it is an evolving reality, where the interplay of AI and 3D integration holds the power to redefine our world. It is a journey without an end, an ever-evolving narrative where each discovery, each innovation, is a steppingstone toward a more advanced and interconnected future.

References

1. Mack, C. A. (2011). Fifty years of Moore's law. *IEEE Transactions on Semiconductor Manufacturing, 24*(2), 202–207.
2. Dennard, H. R., et al. (1974). Design of ion - implanted MOSFET's with very small physical dimensions. *IEEE Journal of Solid-State Circuits, 9*, 256–268.
3. Johnsson, L., & Netzer, G. (2016). The impact of Moore's law and loss of Dennard scaling: Are DSP SoCs an energy efficient alternative to x86 SoCs? *Journal of Physics: Conference Series, 762*(1), IOP Publishing.
4. Choudhury, D. (May 2010). 3D integration technologies for emerging microsystems. In *Proc. IEEE MTT-S international microwave symposium* (pp. 1–4).
5. Tian, W., Wang, W., & Wang, H. (2012, Aug). TSV modeling and thermal analysis based on 3D package. In *Proc. 13th international conference on electronic packaging technology & high-density packaging* (pp. 546–548).
6. Selvanayagam, C. S., Lau, J. H., Zhang, X., Seah, S. K. W., Vaidyanathan, K., & Chai, T. C. (2009). Nonlinear thermal stress/strain analyses of copper filled TSV (through silicon via) and their flip-chip microbumps. *IEEE Transactions on Advanced Packaging, 32*(4), 720–728.
7. Khan, N., & Hassoun, S. (2012). *Designing TSVs for 3D integrated circuits*. Springer Science & Business Media.
8. Shan, G., Guangbao, L., Guoliang, Y., Wang, C., Xing, Y. Z., & Yang, Y. (2023). Application and prospect of artificial intelligence methods in signal integrity prediction and optimization of microsystems. *Micromachines, 14*(2), 344.
9. Liu, H., Aljbri, A., Song, J., Jiang, J., & Hua, C. (2022). Research advances on AI-powered thermal management for data centers. *Tsinghua Science and Technology, 27*, 303–314.
10. Chen, X., Tu, R., & Yang, X. (2023). Parameter prediction optimization of data center's heat dissipation system using machine learning algorithms. *Applied Thermal Engineering, 232*, 121047.
11. Guo, K., Yang, Z., Yu, C. H., & Buehler, M. J. (2021). Artificial intelligence and machine learning in design of mechanical materials. *Materials Horizons, 8*(4), 1153–1172.
12. Cecen, A., Dai, H., Yabansu, Y. C., Kalidindi, S. R., & Song, L. (2018). Material structure-property linkages using three-dimensional convolutional neural networks. *Acta Materialia, 146*, 76–84.
13. Pandiaraj, K., Sivakumar, P., & Prakas, K. J. (2021). Machine learning based effective linear regression model for TSV layer assignment in 3DIC. *Microprocessors and Microsystems, 83*, 103953.
14. Wang, X., Yang, Y., Chen, D., & Li, D. (2023). Intelligent design method of thermal through silicon via for thermal management of Chiplet-based system. *IEEE Transactions on Electron Devices, 70*, 5273–5280.
15. Chang, K. T., et al. (2019, May). Ultra high-density IO fan-out design optimization with signal integrity and power integrity. In *IEEE 69th electronic components and technology conference (ECTC)* (pp. 41–46).
16. Li, Y., Yu, H., & Li, E. P. (2023). Signal integrity analysis of neuronal spike signal in 3-D packaging. In *IEEE transactions on signal and power integrity* (pp. 1–10).

17. Shan, G., et al. (2023). Thermodynamic multi-field coupling optimization of microsystem based on artificial intelligence. *Micromachines, 14*, 411.
18. Takawadekar, V., & Vaidya, Y. M. (2021). Neuromorphic computing: Modelling of 3D integrated circuit components using TSV. In *Proc. IEEE 12th international conference on computing communication and networking technologies (ICCCNT)* (pp. 1–7).
19. Batra, G., Jacobson, Z., Madhav, S., Queirolo, A., & Santhanam, N. (2019). *Artificial-intelligence hardware: New opportunities for semiconductor companies*. McKinsey and Company.
20. Kim, D. W., & Hwang, T. (2019). The future of advanced package solutions. In *Proc. IEEE 2019 symposium on VLSI technology* (pp. T48–T49).
21. Hsieh, M.-C., Wu, S.-T., Wu, C.-J., & Lau, J. H. (2014). Energy release rate estimation for through silicon vias in 3-D IC integration. *IEEE Transactions on Components, Packaging and Manufacturing Technology, 4*(1), 57–65.
22. Dunne, R., et al. (2012, May). Development of a stacked WCSP package platform using TSV (through silicon via) technology. In *Proc. IEEE 62nd Electron. Compon. Technol. Conf. (ECTC)* (pp. 1062–1067).
23. Liu, D., & Park, S. (2014). Three-dimensional and 2.5-dimensional interconnection technology: State of the art. *Journal of Electronic Packaging, 136*(1), 014001-1–014001-9.
24. Kumari, V., Chandrakar, S., Verma, S., & Majumder, M. K. (2023). Reliability concerns of TSV based 3D integration: Impact of interfacial crack. In *IEEE transactions on components, packaging and manufacturing technology* (Vol. 13, p. 1734–1742).
25. Kumari, V., Chandrakar, M., & Majumder, M. K. (Apr 2023). Performance analysis of cylindrical through silicon via with interfacial crack. In *International symposium on quality electronic design, USA*.
26. Marro, J., Okoro, C., Obeng, Y., & Richardson, K. (2014). Defect and microstructural evolution in thermally cycled Cu through-silicon vias. *Microelectronics Reliability, 54*(11), 2586–2593.
27. Huang, Y. J., Pan, C. L., Lin, S. C., & Guo, M. H. (2018). Machine-learning approach in detecttion and classification for defects in TSV-based 3-D IC. *IEEE Transactions on Components, Packaging and Manufacturing Technology, 8*(4), 699–706.
28. Ndip, I., et al. (2011). High-frequency modeling of TSVs for 3-D chip integration and silicon interposers considering skin-effect, dielectric quasi-TEM and slow-wave modes. *IEEE Transactions on Components, Packaging and Manufacturing Technology, 1*(10), 1627–1641.
29. Bengio, Y., & Grandvalet, Y. (2004). No unbiased estimator of the variance of K-fold cross-validation. *Journal of Machine Learning Research, 5*, 1089–1105.
30. Biau, G. (2012). Analysis of a random forests model. *Journal of Machine Learning Research, 13*, 1063–1095.
31. Huang, Y.-J., Pan, C.-L., & Guo, M.-H. (2017). Defect detection for 3D through silicon via based on machine learning approach. *Nanoscience and Nanotechnology Letters, 9*(8), 1175–1179.

Chapter 9
AI-Enabled Hardware Security

Emanuele Strieder, Johanna Baehr, and Matthias Hiller

9.1 Introduction

Electronic devices are fundamental building blocks for today's world. Our technologically driven society is based on tiny physical components that process our daily data. As a result, it is clear that the information security of these hardware devices is a critical, non-optional requirement.

Core principles of information security are confidentiality, integrity, availability, and accountability:

- *Confidentiality* ensures that only authorized entities have access to sensitive information that is mainly achieved by cryptographic algorithms and protocols.
- Information *integrity* refers to the protection of data against unauthorized alteration or manipulation.
- *Availability* incorporates methods that guarantee reliable and uninterrupted information access.
- *Accountability* is a major concern for information security that can be achieved by logging and monitoring of accesses or manipulations of information.

All aforementioned principles have to be ensured during the whole life cycle of a hardware component, such that the software can rely on a trust anchor. This involves strategies to secure the design process, the production and supply chain, the usage, and finally retirement of a device.

Hardware security covers a wide range of research and engineering domains that create methods or primitives to achieve these goals. As examples that will be further discussed in this work, Side-Channel Analysis (SCA) can be used to compromise

E. Strieder (✉) · J. Baehr · M. Hiller
Fraunhofer Institute AISEC Garching, Garching, Germany
e-mail: Emanuele.Strieder@aisec.fraunhofer.de; Johanna.Baehr@aisec.fraunhofer.de; Matthias.Hiller@aisec.fraunhofer.de

A. Iranmanesh, H. Sayadi (eds.), *AI-Enabled Electronic Circuit and System Design*,
https://doi.org/10.1007/978-3-031-71436-8_9

the confidentiality of secret keys, security primitives such as Physical Unclonable Functions (PUFs) can validate the integrity of the system, and reverse engineering can detect if the security of the system is authentic.

Initially, the community was hesitant to adopt Artificial Intelligence (AI) in their tool chains. However, the development and availability of powerful AI methods makes it a crucial complement to traditional statistical algorithms. For example, classic Machine Learning (ML) methods like Support Vector Machines (SVMs) or linear discriminant analysis (LDA) have been extensively researched, successfully applied, and integrated into Common Criteria (CC) certification. However, in recent years, Convolutional Neural Networks (CNNs) are increasingly being employed in SCA and image recognition during netlist reverse engineering. The growing popularity of such Neural Network (NN)-based approaches is accompanied by a computationally costly and thus economically demanding hyperparameter search. This can be a concern in certification contexts when budgets are limited and an exhaustive search is not a viable solution.

On a larger scale, AI reinforces the problem of how to achieve the highest feasible security level that eliminates as many assumption and estimation errors about the security of a system, device, or algorithm without being able to test each possible attack path a priori. However, it is always difficult to ensure that there cannot be a more efficient method to compromise a system, which is in particular relevant to also address the explainability of AI.

As the hardware landscape is constantly changing, e.g., with the rise of the ARM instruction set and recent work on RISC-V, HW/SW co-design and CPUs with integrated FPGA fabric, new architectures like ZnyQ or Versal, this change requires to constantly adapt the technologies, designs, and tools. AI pose the chance that the security technologies can keep the pace with advance of the underlying system design. Going beyond the specific system and architecture, this chapter aims to provide generic approaches and technologies that can be applied to various use cases.

The remainder of this chapter is organized as follows: Sect. 9.2 addresses AI methods in the context of side-channel analysis. Section 9.3 applies AI methods to reverse engineer circuits. Section 9.4 introduces AI methods to analyze PUFs and countermeasures to limit the exposure of information. Finally, Sect. 9.5 gives conclusion and an outlook on the topic.

9.2 AI-Enabled Side-Channel Analysis

Side-Channel Analysis (SCA) summarizes a collection of measuring methods and data analytical algorithms that enable the extraction and exploitation of unintentionally observable information of otherwise internally processed data. The algorithms create a relationship between the observable information and a hardware device's internal state. Because of this, it is a constant threat to devices that store and process secret information like keys, certificates, or intellectual property due to their implementation, even if the underlying cryptographic theory and device specification are secure. An internally processed key, for instance, might be extractable solely by observing the power consumption of a microcontroller if there are no countermeasures that disguise the algorithms' operations. Therefore, it is critical during the iterative design process of a security-relevant chip, to plan ahead, include algorithmic and analog countermeasures and evaluate these by a formal verification, and last and most importantly perform a thorough penetration test.

Over the last years, AI algorithms have gained a lot of attention in the SCA community. The successes in other fields like image recognition or AI-based language translators sparked the fear that commonly used data analytical methods are inferior to the new approaches. This could have severe implications if adversaries have a better tool set at hand to exploit leaking side-channel information.

First, a short introduction into the basic concept of SCA is presented, displaying challenges during leakage assessment and exploitation. We next demonstrate the adaption of ML approaches in different steps during a SCA evaluation by describing and referring to existing literature.

9.2.1 Workflow

The SCA workflow consists out of a number of typical steps. This includes the acquisition of side-channel traces of a target algorithm and a feature engineering step where noise is reduced, sample points are aligned, and traces are transformed. This includes the detection and assessment of leakages. If leakage is found in the feature engineering phase, it is evaluated if it is exploitable. We use this standard workflow to sketch classical methods that are used to evaluate cryptographic implementations. After this we will discuss and introduce multiple contributions that used ML as an alternative approach.

9.2.2 Sketching the Classic Approach

To perform a successful SCA, there has to be a physical side channel that leaks information about the internally processed data. Commonly exploited physical side channels are the execution time, the power consumption, or the electromagnetic (EM) dissipation of parts of an algorithm that run on an accessible device. While power or EM measurements typically need to be taken from close physical proximity, timing side channels can also be observable through remote connections. The following paragraphs sketch the origin of the most basic source of a physical side channel, the power consumption. The power consumption is directly related to the technology in the digital circuits.

9.2.2.1 CMOS Technology

Microcontrollers, Field Programmable Gate Arrays (FPGAs), or application-specific integrated circuits (ASICs) play a major role in today's electronics. They are built from large digital circuits, manufactured in Complementary Metal-Oxide-Semiconductor technology (CMOS) technology, and enriched with some analog components. CMOS relies on semiconducting materials that can be used to build transistors. The semiconductors are doped with elements such that two types of complementary transistors can be created: One is conducting when there is a high control voltage and one if there is a low control voltage. With the help of boolean algebra, these complementary transistors can be assembled to digital circuits and on a global scale any Turing machine.

Figure 9.1 shows a simplified inverter circuit that represents a fundamental building block in CMOS-based devices. The complementary transistors are arranged in a pull-up and a pull-down network. Depending on the control voltage applied in parallel to both transistors, either the NMOS or PMOS transistor will be conducting. In case of an inverter, this translates to a logical 1 if the input receives a logical 0 and a 0 if the input receives a logical 1.

The power consumption of such CMOS transistor circuits can be approximated by three sources: the power consumption due to short circuits, static power consumption, and dynamic power consumption as shown in Eq. (9.1).

$$P_{tot} = P_{static} + P_{short} + P_{dyn}. \tag{9.1}$$

The static power consumption is a result of small leakage currents, heat dissipation, and other capacitive loads that are maintaining their logical values. The small leakage currents are a result of imperfections in the CMOS design itself. The short dynamic power consumption is depicted in Fig. 9.1a and b. Based on the applied control voltage at the input of the circuit, the load capacitance has to be either

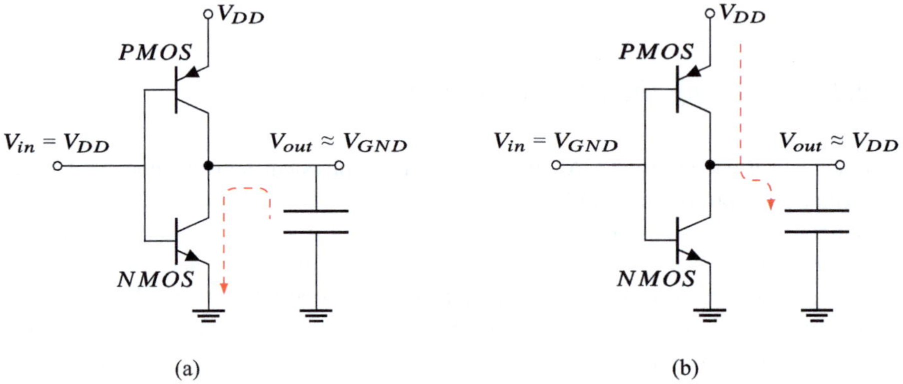

Fig. 9.1 Simplified inverter circuit during charge and discharge of the load capacitance. The current flow is indicated by dashed arrows. Charge and discharge of the load capacitance is a source for the dynamic power consumption. (**a**) Inverter during discharge. (**b**) Inverter during charge

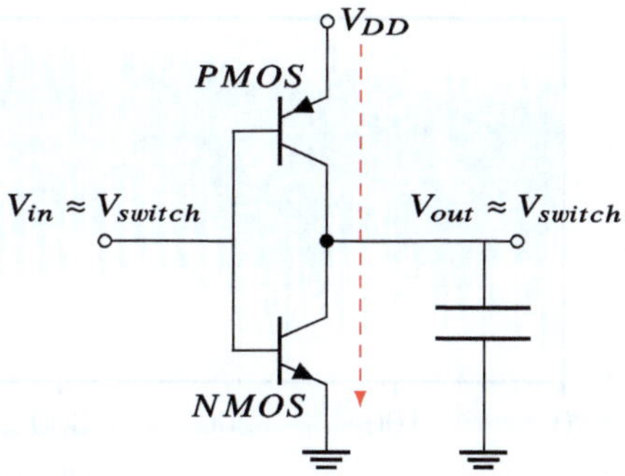

Fig. 9.2 Simplified inverter circuit during value change. The short circuit current flow is indicated by a dashed arrow

charged or discharged. This results in a current flow and power consumption. Note that this power consumption is data dependent as the control voltage encodes the data. The short circuit power consumption is a result of the switching process of the complementary transistors. Figure 9.2 shows the inverter during the switching process. During the switching of the control voltage, both transistors are conducting in parallel. This produces a short circuit and therefore a power consumption.

This simplified model shows that the processed data in a CMOS-based devices is mirrored in its power consumption. Especially the dynamic power consumption is prone to be exploited as a source of side-channel leakage. However, also the static power consumption has been shown to leak information [59]. Although power consumption is the most prevalent leakage source, there are others, for instance, EM emanation. Maxwell's equations reveal the connection between the EM field with the current flow in a circuit. Therefore the previous discussed emanation of the power consumption is also observable in the EM field. For a detailed discussion about EM-based leakage, we refer to [52].

9.2.2.2 Example Implementation

Listing 9.1 displays a software loop that sequentially loads an 8-bit value from the array `origin`, copies the value, and stores it into the array `destination` to illustrate the relation between processed data and power consumption of a CMOS-based device.

```c
uint8_t destination[16] = {0};

void copy_data(uint8_t *origin)
{
    for (uint8_t i = 0; i < 16; i++)
    {
        destination[i] = origin[i];
    }
    return;
}
```

Listing 9.1 A software loop written in the programming language C. The function copies 16 bytes from the array `origin` to the array `destination`.

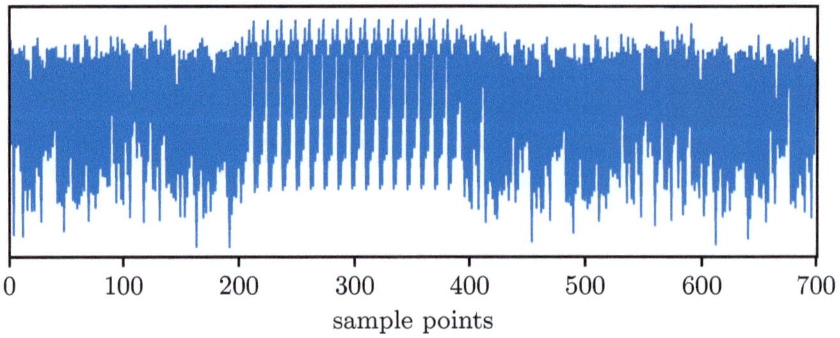

Fig. 9.3 Averaged power consumption of a SAM4S microcontroller during execution of the software loop described in Listing 9.1

Figure 9.3 depicts a mean trace that shows the operation-dependent power consumption of such a loop. The side-channel traces were captured with a NewAE ChipWhisperer Husky oscilloscope targeting a SAM4S 32-bit microcontroller. The mean trace is the result of averaging 90000 power traces, while `origin` contains random data. There are sixteen distinct peaks in the sample point range of 200 to 400, revealing the distinguishable power consumption of the loop operations. The process of analyzing the raw traces is called visual inspection and is an important step during a SCA evaluation.

9.2.2.3 Leakage Model

Visual inspection of a power trace is commonly the first step during a side-channel investigation. Figure 9.3 shows 16 distinct power consumption peaks while executing the software loop described in Listing 9.1. In certain cases, visual inspection can be enough to compromise cryptographic implementations: This is particularly true if secret-dependent information can be visually distinguished by power consumption patterns or execution times. This however is rare in security products that handle sensitive information. However, a failed visual inspection should not be confused with the device's security.

A few of the averaged traces are shown in Fig. 9.4, which appears to show just random noise. The dynamic power consumption of a device can be split into three parts: the operation-dependent part, the data-dependent part, and a noise term. The operation-dependent part of the analyzed loop was visualized in Fig. 9.3. The power consumption of the data-dependent part and the noise term at each sample point s_t can be formalized as

$$s_t = \theta(x)_t + Z, \tag{9.2}$$

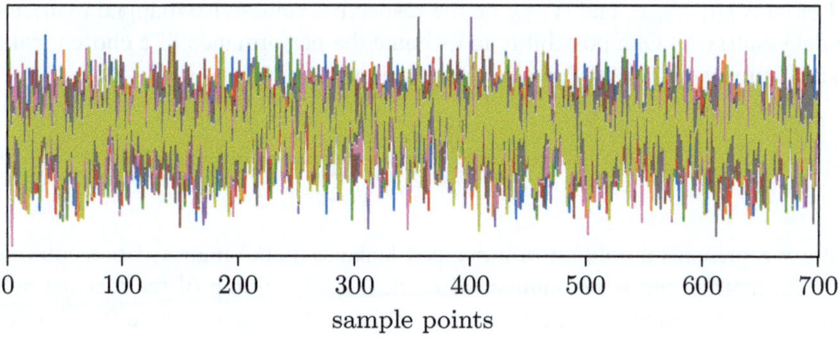

Fig. 9.4 Random traces of the power consumption of a SAM4S microcontroller during execution of the software loop described in Listing 9.1

where x is the currently process intermediate value. The noise term Z is commonly sampled from a centered normal distribution $Z \sim \mathcal{N}(0, \sigma^2)$ and θ is called leakage model. Its purpose is to optimally map a processed intermediate value to the observed power consumption. This could, for instance, be a simple one-to-one mapping, which is called the identity model. Another possibility is an approximation of the leakage model $\theta(x)$ by a weighted sum of the bits of the binary representation of x:

$$\hat{\theta}(x)_t = \sum_{i \in [1,N]} a_i \cdot g_i(x), a_i \in \mathbb{R}, \tag{9.3}$$

where $g_i(x)$ returns the value of the i-th bit. In the presented example loop, the number of bits N equals 8 as 8-bit values are processed. Leakage models are a major research field in SCA because they influence the success quality of all following postprocessing algorithms. Let T be a matrix that holds all captured traces—each row represents one trace. Let X be the intermediate value matrix. Each row of X contains the sixteen 8-bit values that are processed by the discussed loop. We introduce three basic leakage models to exemplify their influence: the identity model, the Hamming weight (HW) model, and the least significant bit model. They are given by

$$\theta_{ID}(x) = x, \tag{9.4}$$

$$\theta_{HW}(x) = \sum_{i \in [1,8]} g_i(x), \text{ and} \tag{9.5}$$

$$\theta_{LSB}(x) = g_8(x). \tag{9.6}$$

Please note that all a priori selected leakage models are based on simplifying assumptions about the underlying hardware.

Let be X_{ID}, X_{HW}, and X_{LSB} be the respective value-wise mapped matrices of the data matrix X. One possibility to estimate the performance of a chosen leakage model is the evaluation of Pearson's correlation given by

$$\rho(t,x) = \frac{\sum_{i=1}^{n}(t_i - \bar{t})(x_i - \bar{x})}{\sqrt{\sum_{i=1}^{n}(t_i - \bar{t})^2}\sqrt{\sum_{i=1}^{n}(x_i - \bar{x})^2}}, \tag{9.7}$$

where x represents a column of a mapped leakage model matrix, t one column of the trace matrix, and n the number of entries. Each column of the leakage model matrices is correlated with each column of the trace matrix. Figure 9.5 shows the correlation of each of the 16 loops with the respective leakage models. For all correlations, 100000 traces were used. Among the three leakage models, the HW model has the highest correlation given the provided number of samples. This is to be expected given that microcontrollers modify register bits all at once, which is well approximated by the HW assumption. This example shows the importance of

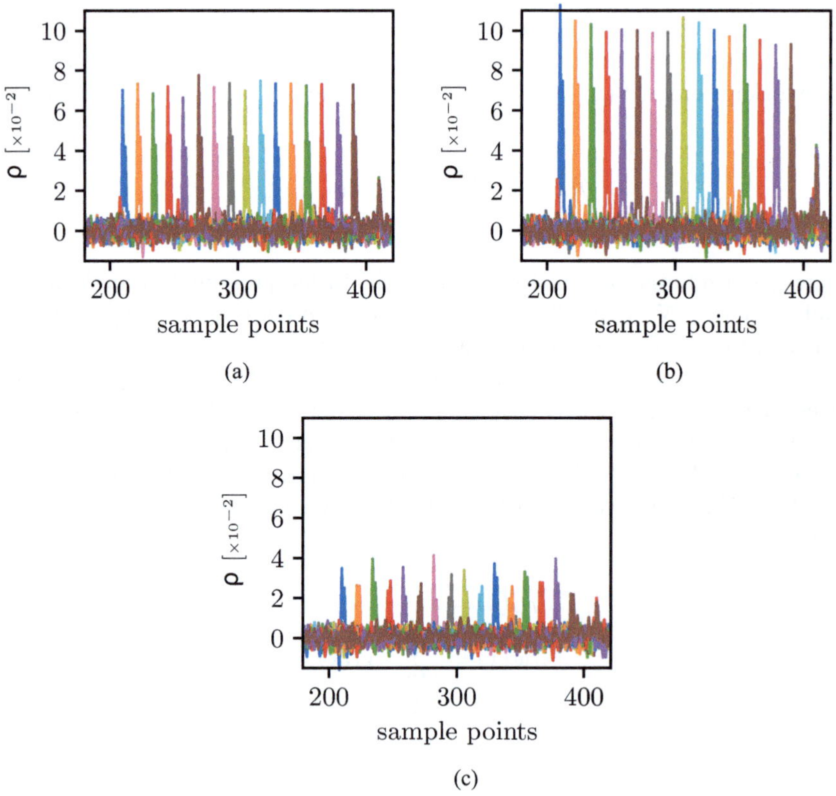

Fig. 9.5 Correlation of traces with three leakage models. (**a**) Correlation with ID model. (**b**) Correlation with HW model. (**c**) Correlation with LSB model

leakage models during side-channel evaluations. The presented leakage models are commonly used examples. However, leakage models could also be rather complex, nonlinear, and consist of multiple sample points.

9.2.2.4 Feature Engineering and Countermeasures

Feature engineering is a crucial part of SCA as in all data-driven research and development fields. Section 9.2.2.2 shows that a correlation of the power consumption of a microcontroller with a fitting leakage model can reveal information about the location in time of the processed data. This is only feasible if the information-carrying sample points of all captured traces are aligned vertically and the noise level still allows to distinguish the signal of the process intermediate value from noise. Feature engineering aims to increase data quality with approaches such as alignment and filtering. On the other side, there are numerous ways to mitigate information leaks by deliberately misaligned sample points or introduction of noise. These deliberately implemented methods are called countermeasures and are applied at several levels of a cryptographic chip. We differentiate four levels: the transistor-level-, the program-level-, the algorithmic-level-, and the protocol-level-based countermeasures.

The *transistor-level*-based countermeasures are the most expensive and, in terms of chip design, the most intriguing. They involve the engineering of physical countermeasures that are integrated in the digital circuits with the goal of mitigating the information leakage a priori. This could include the engineering of noise engines, special jittery clock circuits, dual-rail logic, or securely designed memory. Clock jitter refers to the timing variance of the clocking signal. This is a side effect of imperfections during production of CMOS devices; however, it can also be introduced intentionally by design. This impedes the alignment of sample points.

Program-level-based countermeasures try to alter the execution flow of an algorithm to mitigate information leakage. This can be achieved by randomly introduced dummy instructions or the randomization of the execution time of an instruction. The latter is called shuffling that is a hiding method. In this case, the hiding method does not reduce the leakage, but the leakage occurs at an unexpected point in time. Shuffling is only possible if the randomized instructions are independent of each other. In the presented software loop, this might be implemented by randomizing the order in which the indices are processed. This can be achieved by random selection of the start index or a computational costly randomization of all 16 indices. The functionality is the same in both cases, but the alignment of the sample points is random.

Algorithmic-level-based countermeasures are tightly coupled to the operations in an algorithm. The most prominent algorithmic countermeasure is masking. In a masked implementation, secret-dependent values are split into multiple parts, called shares. Let x be a secret-dependent intermediate value that should be masked and m a randomly chosen mask with the same bit width than it holds that:

$$x = (x \oplus m) \oplus m = x_m \oplus m,$$

where $\oplus$ defines the binary XOR operation. A masked implementation performs all operations that involve the processing of secret-dependent data with masked intermediate values. The masked value x_m should be at least uncorrelated but in best case statistical independent of the original value x. Thus, an emanation from a captured side-channel signal like the power consumption is also uncorrelated. Note that this is only one high-level concept of masking that depends on the binary XOR operation. There other implementation possibilities are dependent on the concrete algorithm and implementation. We will later discuss that if an adversary has access to side-channel emanations of all shares, the secret can still be inferred.

The last group of countermeasure are *protocol-based* countermeasures. This group of countermeasures has the goal of reducing the number of side-channel observations an attacker can exploit. This could be, for example, an enforced secret switch after a number of encryptions.

All stated countermeasures have to be investigated for leaks, after they have been implemented. Often, the research community considers a worst-case adversary that is in line with Kerckhoffs's principle. According to Kerckhoff's principle, cryptographic systems should be secure even if an adversary is aware of every detail of the systems except the secret. This includes the chance that an adversary has gained knowledge about the implemented countermeasures. Feature engineering helps to estimate the effort required to reverse these countermeasures. We will later see that ML can ease this process.

9.2.2.5 Leakage Exploitation

Section 9.2.2.2 shows that the execution of a simple software loop is observable in side-channel traces. Section 9.2.2.4 discusses a number of countermeasures that mitigate the risk of information leakage. A major measure during SCA evaluation is the number of traces that are needed to extract the correct value of the processed value. The strongest attack path are profiled SCA methods that allow to extract an intermediate value in only a few traces. In extreme cases, only a single trace is necessary. The basic idea of profiled SCA methods is to create a typical signal distribution for each representation of an intermediate value while having full control over a profiling device. The knowledge about the distribution of each representation of the intermediate value can be used to match a target trace of a black-box device. An example for this are Gaussian template attacks [11, 67]. To create a Gaussian template during profiling, many side-channel traces are recorded, while the target algorithm processes random data. This is necessary to have enough samples that model the whole variation in the side-channel signal. The traces are grouped by using a leakage model of the target intermediate value, e.g., the HW. The Gaussian profile of each group g is based on a mean vector $\boldsymbol{\mu}_g$ and a covariance matrix $\boldsymbol{\Sigma}_g$. In the attack phase, n attack traces are matched with the profile of each

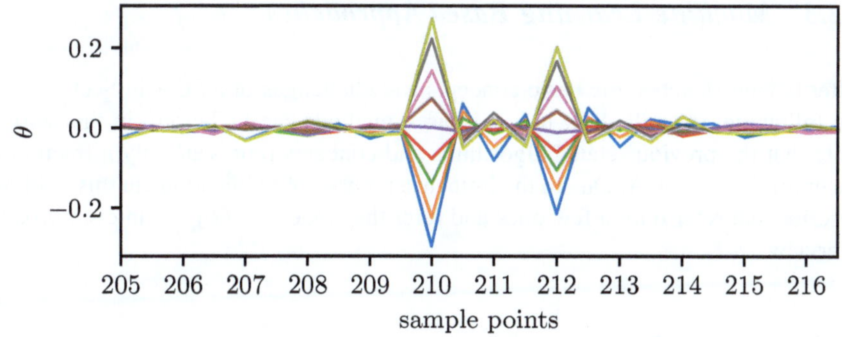

Fig. 9.6 Mean vector of all profiles of each of the 9 possible HW groups. Profiles were derived by using 800k traces

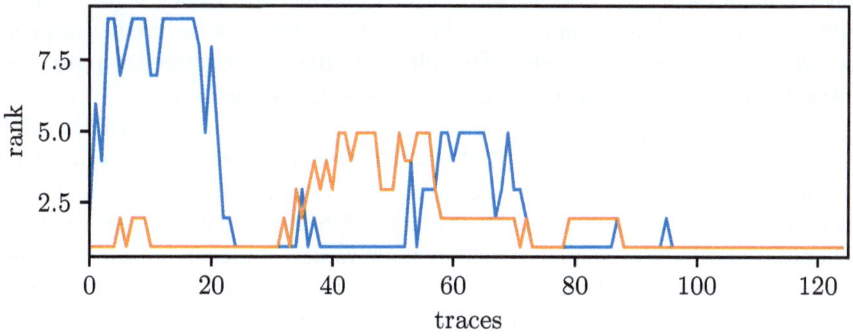

Fig. 9.7 Rank evolution of two attack sets, per number of used attack traces

of the pre-calculated group profiles:

$$s_g = \prod_{i=1}^{n} \mathcal{N}(\boldsymbol{t}_i, \boldsymbol{\mu}_g, \boldsymbol{\Sigma}_g), \tag{9.8}$$

where s_g is called score value. Figure 9.6 shows the mean vectors of each of the nine possible HW groups for the first loop iteration given by Listing 9.1. The first statistical moment of the profiled groups is showing a distinguishable pattern at several sample points. The second step is to match attack traces with the given profiles. Figure 9.7 shows two matching processes of sets of attack traces. Although the profiles seemed to be clear in the first statistical moment, there are up to 100 traces necessary that the matching converges to the best rank. Rank is a metric that measures the distance between the correct value and the current guess. The convergence process is of course caused by noisy attack measurements.

9.2.3 Machine-Learning-Based Approaches

After having sketched the basic concepts and challenges of a SCA in Sect. 9.2.2 in the following, we will show how AI-based methods can be helpful in this setting. Note that the previous stated algorithms and concepts represent only a fraction of the entire field of SCA. Due to the immense number of publications in this field, we describe and refer only a few ones and refer the reader to [66] for an even broader overview.

9.2.3.1 Leakage Detection and Leakage Assessment

Section 9.2.2.3 discusses the methodology to detect leakage by selecting a fitting leakage model and correlate the transformed intermediate values with the traces. Although this is a feasible option, it relies on the accuracy of the predictions about the processed intermediate value. Hypothesis testing is a common way to detect unspecific leakage to mitigate the risk of an incorrect assumption.

Hypothesis testing does not require knowledge about the intermediate values, making it the de facto standard for leakage detection. The main concept is to capture two disjunctive sets of traces. One set contains only traces that are captured during the execution of a fixed input value to the target algorithm and another set containing either traces with the execution of random input values or a second fixed input value. Hypothesis testing checks if the distributions of these two disjunctive sets have similar statistical properties. If the distributions are distinguishable by statistical measures, the implementation is leaking unspecifically. Unspecifically in this setting means that it is not clear if this leakage allows an exploitation like discussed in Sect. 9.2.2.5. One common example is Welch's t-test that tests for the equality of the mean of the sets or the χ^2-test. For a more detailed overview about leakage detection in the field of SCA, we refer to [29, 74, 61]. The present correlation method as well as hypothesis testing relies on aligned traces and leakage in the first statistical moments.

Moos, Wegener and Moradi [60] introduced a Deep Learning (DL)-based leakage detection method called Deep Learning Leakage Assessment (DLLA). An NN-based binary classifier is trained using the same disjunctive trace sets as described above. If the binary classifier reaches a significantly higher prediction rate than guessing, it can be inferred that the trace groups contain features that allow the separation of the two trace groups. In contrast to other leakage detection methods, there is no need for an assumption about the underlying distribution of the sample points. Further, depending on the used NN, DLLA mitigates the need for preprocessing steps like the alignment of traces. For instance, this can be achieved by training a NN-based binary classifier that contains convolutional layers. The convolution over multiple sample points allows for a certain degree of shift invariance of features. Leakage tests that use univariate leakages as discussed in Sect. 9.2.2.3 have the inherent problem of false negatives: Leakage that is only detectable by a combination of sample points will not be detected. DLLA is a valuable addition in the SCA toolbox to mitigate this risk, because NNs are inherently multivariate. However, a wrong assumption would be to be sure that no leakage is present if no DL was found. There is currently no way to rule out the possibility of leakage with 100% certainty.

If a DL model is found that has a better-than-guessing accuracy, it is relevant to extract the sample points that were decisive for the information gain, which is solved by the authors using a sensitivity analysis. Moos, Wegener and Moradi demonstrate by several case studies that DL is a handy addition to classical leakage detection tests even in the presence of countermeasure as discussed in Sect. 9.2.2.4.

After a successful unspecific leakage detection and the identification of the leaking sample points, a leakage model has to be found. An NN-based approach to this problem was introduced by Cristiani, Lecomte and Maruine [16]. They use a Generative Adversarial Network (GAN) structure that contains three NNs to estimate linear and nonlinear leakage models. GANs are unsupervised NN structures where one network, the discriminator, is trained to distinguish between correct and incorrect samples that are created by a generator. The authors use real SCA measurements of a cryptographic algorithm as well as simulations to support their methodology. These findings are a useful first step in reducing the impact of the error-prone choice of a leakage model, even though they still need to be further assessed in more complicated scenarios with multivariate leakages, alternative implementations, and a range of noise levels.

The examples provided demonstrate that AI-based methods may be utilized for leakage detection and leakage assessment, which is a critical stage in SCA. In the next sections, we explore various scenarios in which leakage exists and is exploited utilizing DL.

9.2.3.2 Leakage Exploitation Using ML

Since the seminal work of Hospodar et al. [45] in 2011, a significant portion of the publications in the field of Machine Learning based Side-Channel Analysis (MLSCA) focused on the search of NNs that allow the exploitation of leakage if countermeasures are present [66]. We introduced a few of the possible countermeasures to enhance the resistance of cryptographic implementations in Sect. 9.2.2.4. In the following, we will show that non-profiled as well as profiled SCA attacks against implementations with countermeasures are possible with AI methods.

Section 9.2.2.3 discusses the usage of Pearson's correlation coefficient to evaluate the performance of different leakage models. The usage of correlation as a distinguisher can be extended to the so-called Differential Power Analysis (DPA) that is a non-profiled SCA attack [57]. Instead of correlating a known vector of intermediate values with the captured traces, multiple hypothetical intermediate value vectors are correlated, and the best fit reveals the secret value, eventually. The hypothesis is selected based on the interaction of a known input value and an unknown secret value within the algorithm. Let $s^\star$ be a fixed secret 8-bit value that interacts with an 8-bit known input value p in a known way $f(p, s^\star)$. During a SCA, X side-channel traces are captured, while X input values p_i are processed. Let $c_{i,j}$ be the output of $f(p_i, \hat{s}_k)$, where p_i is the i-th input value and $\hat{s}_k$ the k-th hypothetical value of the secret s^*. Based on all possible values of $\hat{s}$, we build $K = 256$ hypothetical vectors representing the output of the interaction function. This can be written as

$$\begin{bmatrix} c_{1,1}, c_{1,2}, \ldots, c_{1,K} \\ c_{2,1}, c_{2,2}, \ldots, c_{2,K} \\ \vdots \\ c_{X,1}, c_{X,2}, \ldots, c_{X,K} \end{bmatrix}, \tag{9.9}$$

where each column represents a hypothetical SCA emanation vector. An attacker now correlates each column of the hypothesis matrix with each column of the trace matrix as described in Sect. 9.2.2.3. The hypothesis vector that correlates the best reveals the value of the secret s^*.

Timon has introduced a DL-based approach to non-profiled SCA that is called Differential Deep Learning Analysis (DDLA) [80]. Instead of applying Pearson's correlation or another classical distinguisher like Mutual Information (MI), Timon uses the training and validation loss as an indicator of the goodness of the used hypothesis vector. He demonstrates that a Multi-Layer Perceptron (MLP) as well as a CNN can in certain cases outperform the classical methods. Even in settings where traces are desynchronized as described in Sect. 9.2.2.4, this method is robust. A sensitivity analysis of the first layer is used to identify which sample points were decisive. This again shows that an ML-based approach is added as valuable tool to evaluate the potential risks of side-channel leakages of a cryptographic implementation.

A more intuitive approach to use DL in SCA is the profiled scenario that is seen as the worst-case scenario. The classical approach of a Gaussian template attack was discussed in Sect. 9.2.2.5. The idea is to create a profile with as much traces as are available while having control over the target implementation. In the attack phase, only a few traces are exploited to predict a single or multiple intermediate values. The main portion of publications are attacking symmetric ciphers [42, 58, 60]; however, also asymmetric ciphers [83, 10] have been investigated. Different architectures are tested like MLPs, CNNs [96], or Residual Neural Networks (ResNets) [30]. GANs [28] or Autoencoders (AEs) have been tested for noise reduction [91]. On the other side some authors have used artificial noise to make NN predictions more robust [50]. There is a lot of research with respect to hyperparameter search [90, 68, 31] and other details like loss functions [95]. The number of publications in this field exploded in the last five to ten years. For an even broader overview, we refer to Picek et al. [66].

9.2.4 *Summary*

We have sketched the classical SCA approach and showed how AI methods are increasingly researched and introduced into SCA. Especially in a black-box setting where the points of interest are unknown or countermeasures are involved, traces are unaligned or algorithmic dependencies can be learned, and ML is a valuable extension of the SCA toolbox. The new findings in this field have to be considered during chip design, and countermeasures have to be introduced if secret dependent data is processed.

9.3 Hardware Reverse Engineering

Both integrated circuits (ICs) and FPGAs have become the target of more and more hardware-based attacks, both by the end user and during design, fabrication, and commercialization. While attacks such as side-channel and probing attacks attack a concrete implementation of a chip post-fabrication, with the aim to extract secret information contained within specific chips, attacks against the Intellectual Property (IP) of the chip often aim for a better understanding of the chip, to copy or attack the design during fabrication, or prepare for an attack on a specific chip post-fabrication. Such attacks can be classed into the creation of counterfeits, including IP theft and the resale of recycled hardware, the insertion of malicious hardware (e.g., Hardware Trojans (HTs)), or the discovery of attack vectors. Although such hardware attacks are more complex and costly to perform than software attacks, countermeasures against such attacks are also far more difficult to implement, and in the worst case the attacker is able to attack many millions of devices at once.

Hardware reverse engineering is the process of analyzing, decomposing, and modeling a black or gray box in order to reconstruct its functionalities as closely as feasible. The reasons for reverse engineering are manifold and can be malicious or constructive. Most commonly, hardware reverse engineering is used to identify counterfeits and possible hardware alterations. However, it is also required to insert hardware Trojans, and for many subsequent attacks. Reverse engineering frequently requires the analysis of large amounts of data that would be computationally intractable without AI. We will discuss how AI methods are used during reverse engineering of digital circuits to aid, e.g., Trojan detection and how they can help mitigate safety and security issues of recycled ICs and FPGAs.

9.3.1 *Digital Circuits Reverse Engineering*

The reverse engineering workflow of an IC, starting from an unknown chip, is a complex process. The most common approach relies on Scanning Electron

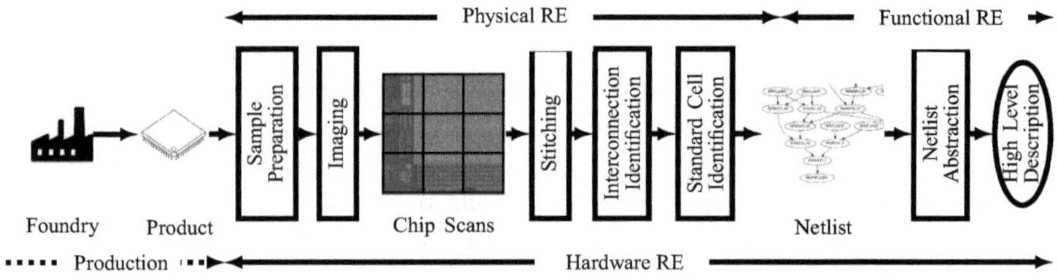

Fig. 9.8 Process Flow for Hardware Reverse Engineering of Integrated Circuits

Microscope (SEM) imagining of the chip to reconstruct the gate-level netlist. Very generally, as illustrated in Fig. 9.8, the chip must first be depackaged, delayered, and imaged. Image processing methods are used to stitch the resulting images, and to identify interconnects and standard cells, to reach a gate-level netlist [85, 54, 14]. In the final step, known as netlist analysis, a gate-level netlist is abstracted into a high-level description, i.e., into a human-readable form.

Both physical and functional reverse engineering heavily rely on ML methods. However, all these AI approaches suffer from similar issues. The datasets used for training and testing are often small, with only limited open-source datasets available, and the data itself is frequently unlabeled, requiring manual labeling efforts. Furthermore, the choice and design of AI methods are often based on convenience rather than a comprehensive analysis of the problem. Finally, the improper choice of features to represent the problem contributes to suboptimal performance [9].

Complex image processing is a critical component of physical reverse engineering, characterized by the intricate task of stitching together images with various types of offsets and overlaps. This complexity is further exacerbated by defects resulting from physical imperfections on the chip surface, as well as the presence of deformed and warped images. Additionally, managing the number and size of images to be stitched adds to the complexity of the process. Usually, thousands of images are taken per layer, with seven to ten layers being the norm. Images must be aligned horizontally as well as vertically, to ensure a correct identification of design elements [54]. This is not a task unique to chip reverse engineering, for example, the horizontal stitching is comparable to stitching images for Google Earth. However, there are some differences: Often many pictures are taken of the same area, physical defects are not an issue, and large amounts of training data are available. Furthermore, the effect of imperfect images for Google Earth are minor, while small errors in the images of a chip can cause large problems for subsequent reverse engineering steps.

Once the chip has been imaged, stitched, and aligned, the next type of problem deals with identifying the wires and via in the metal layers, and the logic gates in the poly silicon layer [54, 14, 51, 55]. Complexities include bad contrast, many repeating structures, and poor resolution of the images, which can be compensated by using AI methods. To counteract possible defects, it is possible to use machine learning to classify encountered defects, for example, using neural networks [4, 8]. Logic gates are usually more easily identified than wires and vias, as their structure is unique for each type of logic gate, and the placement is aligned along rows and columns, but these may be flipped or mirrored. Additionally, the technology libraries used to generate the logic gates are often not available for comparison, and so such libraries must first be manually built by extracting all possible logic cells. Finally, especially for small technology nodes, it becomes more difficult to distinguish the borders between logic gates. This process usually results in a gate-level netlist with less than 1% error.

To recover a human-readable version of the gate-level netlist, the final step of the process is carried out: the interpretation of the gate-level netlist. This step is often broken into a two-stage process. First, it is necessary to partition a large gate-level netlist into smaller submodules, and then, in a second step, the functionality of each submodule is identified by comparison to a known golden model. Early works in netlist reverse engineering accomplished such a comparison between an unknown netlist and a known golden model by functional comparison, i.e., the functionality was compared using formal verification, with a match between two functionally identical netlist, and a mismatch if there were any differences.

In 2018, it was first proposed to use structural features as a basis for such design interpretation, with the analysis of the structure of cryptographic circuits [85]. However, it was not until 2019 that the idea of more large-scale structural comparison using machine learning methods was considered [5]. Here structural features of netlists were aggregated and compared against a golden model library using k-nearest neighbors (kNNs). This method is based on the idea that in hardware design, form follows function. This means that specific functionalities will very commonly be synthesized by a design tool in a very similar way, resulting in similar structural features, even when the high-level implementations show some differences. When considering a gate-level netlist as a graph, where each node repents a logic gate, and each wire represents a connection between them, it becomes possible to visualize these similarities. As an example, Fig. 9.9 shows four different implementations of a multiplier, where the structural similarities are clearly visible.

More recent methods have been published that utilize Graph Neural Networks (GNNs) for netlist comparisons. All works use structural graph-based features to compare the functionality of a candidate submodule to a golden model. However, both the features used to train the GNN, the structure of the GNN, and the training and testing data sets differ. The first of these works [44] focused on adder designs and could differentiate between different architectures. Later work combined netlist partitioning and functionality identification for comparators, multipliers and adders [2], or focuses on arithmetic blocks [82, 37].

Unfortunately, many of the machine-learning-based methods used to identify the functionality of a netlist are trained and tested on limited data sets. Some methods only consider a few designs of small size. To better understand the most useful NN and GNN architecture, larger and more realistic datasets are needed. Additionally,

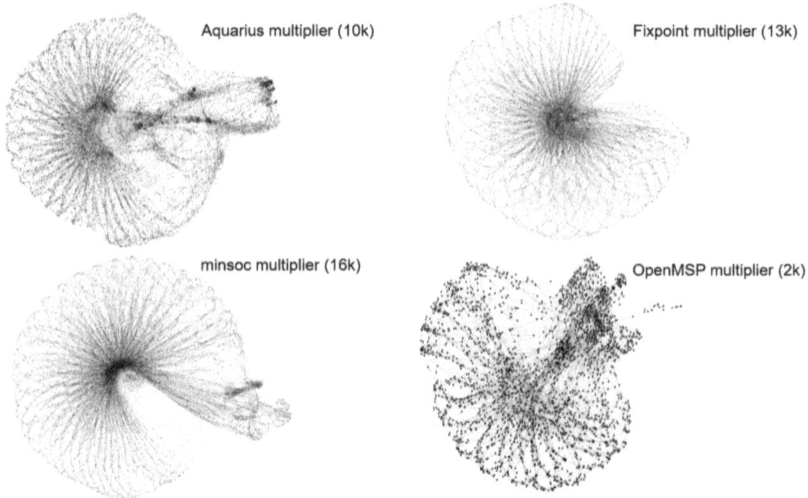

Fig. 9.9 Graph visualization of different multiplier implementations

feature selection is currently optimized for these small designs, and it is unclear how these features will perform for larger designs or if new features will be necessary.

Furthermore, many of these methods fail to consider the impact of the synthesis tool and the chosen technology library. As mentioned earlier, the design structure remains largely unchanged for the same functionality, regardless of the implementation method. This also applies to the same design, synthesized with different synthesis tools, optimizations, and technology libraries. However, certain differences do arise, particularly with the choice of technology library, resulting in gate count variations of over 50%. Figure 9.10 illustrates this issue by presenting a graph visualization of the same design synthesized using two different synthesis tools and three different technology libraries. Again, it remains unclear how and if this issue will affect machine-learning-based methods. For example, the usability of a GNN trained on a specific synthesis tool and technology library for designs implemented using a different tool and technology library remains unclear.

While there are still some uncertainties surrounding the application of machine learning in functionality recovery and netlist analysis, these methods have shown remarkable results in their current use cases. Additionally, while not proven for netlists, these methods have demonstrated scalability in other areas, by effectively handling large graphs consisting of millions of nodes. It is worth noting that alternative methods based on formal verification face significant challenges due to errors in the netlist resulting from previous reverse engineering attempts [5]. Therefore, machine learning methods hold great promise for the future, but further research is necessary to fully explore their potential.

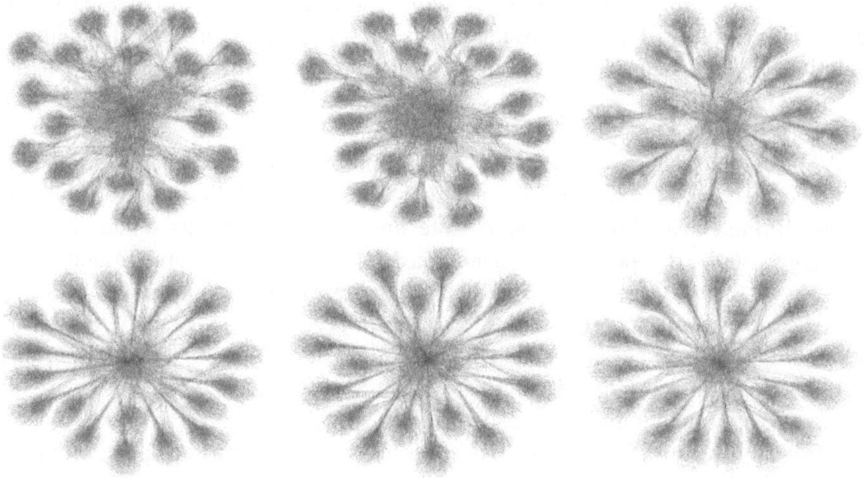

Fig. 9.10 Graph visualization of two different synthesis tools (vertical) and three different technology libraries (horizontal) for same design

9.3.2 Hardware Trojan Detection

As with intrusion detection and malicious software detection, AI methods can facilitate HT detection, especially because they are often unable to be detected using traditional verification methods [64]. We can differentiate between detection methods that work directly on the image data, those that work on the gate-level netlist, and those that consider side-channel information to classify benign and malicious logic.

HT detection on imaging data is often evaluated using relatively simple AI approaches, such as SVMs [6, 81], although new work in the last years has used more innovative methods, for example, using deep-learning-based methods to automatically create GDSII data [53]. Differences in methods tend to arise in the choice of image type, and whether a golden sample is available for comparison, or if the focus is on detecting suspicious changes in the netlist. In cases where a golden sample exists, some works prefer to use deterministic methods instead of machine learning approaches to avoid introducing uncertainties or compromising interpretability [56]. Research into the creation of explainable and predictable AI architectures for this purpose would greatly improve these efforts. The previously mentioned issue of creating NNs capable of handling multiple technology nodes, and the difficulty of exchanging pretrained neural networks due to the constraints imposed by non-disclosure agreements (NDAs), also creates additional challenges.

When it comes to netlist-based methods to identify HTs, the focus lies on identifying HT-specific functional or structural features, which can be aggregated over the entire netlist, and then classified using ML methods, to identify malicious logic gates. Initially, SVM methods with a limited number of Trojan-specific features were utilized [33]. Subsequent research explored various ML techniques including kNNs, decision trees, random forests, and different architectures of NNs [63, 35, 34, 22]. These works also included new potential Trojan features, with some using as many as 50 features.

While many of these approaches yield promising results on the tested datasets, their accurate evaluation and thus their real-world applicability are again hindered by the limited availability of large-scale and realistic datasets containing HTs. Although there are open-source benchmarks [71, 73], these are limited in size and consist of only a few design types with repeated implementations of HTs. Subsequent research has focused on creating larger and more realistic datasets by developing tools that automatically insert HTs into designs, using both deterministic methods [38] and AI-based methods [17]. Such datasets will enable a more systematic approach to selecting and verifying specific neural network and machine learning architectures. Additionally, researchers have addressed the challenge of defining the entire possible HT space by using more general features (i.e., less Trojan-specific features), thus enabling the detection of HTs outside of the training set [38, 24, 36].

HTs can also be detected by comparing the side-channel signals of infected designs to uninfected designs or by analyzing the signals of an infected design based on the activation status of the HT. Over the years, numerous machine learning algorithms have been evaluated on different types of side-channel signals for this purpose. These include power and EM side channels, as well as thermal maps. Commonly employed techniques to classify designs as Trojan-free or determine whether a Trojan is currently active include SVMs and various types of neural networks [49, 47, 84, 77, 92, 23, 18].

Although these approaches usually yield good results for the validation data, they are often trained and validated on very limited datasets, sometimes involving only a few designs or even a single design. Additionally, these methods frequently rely on simulated data rather than real world, measured side-channel signals, which limits their ability to account for noise [77]. Therefore, to properly evaluate these methods, it is crucial to have a more extensive training and validation dataset, along with real-world measurements obtained from infected ICs and FPGAs [39, 77].

9.3.3 Recycled FPGA Detection

Another example of the advantages of AI for hardware security is the ability to distinguish between recycled chips and newly manufactured chips. Recycled FPGAs and ICs are a big concern for the safety and reliability of products and are occasionally sold as new. To differentiate between recycled and new products, the aging of the FPGA or IC can be considered. Ageing occurs in electronics due to various reasons and causes the device to slow down [48]. In the case of ICs, this commonly requires adding circuitry capable of measuring aging during the chip's design phase, resulting in an area overhead [48]. For FPGAs, such circuits can be programmed during the verification of the device, and thus no overhead is incurred. The simplest version of such a circuit is a ring oscillators (ROs)-based design, where multiple XOR and XNOR gates are chained together. The frequencies of these ROs can be compared against a known new device; if the device is not new, the frequencies should have significantly degraded [21, 15]. However, if a known new device is unavailable, the frequencies of all possible variations of ROs through the paths of the LookUp Tables (LUTs) of the FPGAs can be compared against each other. In a new device, the frequency variation between different paths should be minimal, while in aged devices, there should be a significant difference in frequencies between frequently used paths and rarely used paths, indicating that the device is not new [1].

Depending on the availability of a known new device, supervised (e.g., SVMs, NNs) or unsupervised (e.g., k-means clustering) ML approaches can be used to evaluate the aging measurements [1, 3]. Both approaches give good results, with supervised ML being able to distinguish devices that are more than a few hours old. To classify recycled FPGAs, ML offers distinct advantages over more common traditional classifiers. In the case of supervised machine learning methods, they excel at handling large amounts of data without the need to identify specific points of interest [32]. Even in situations where a known new device is unavailable, unsupervised ML can effectively identify hidden patterns in the data, indicating possible aged devices. Additionally, both supervised and unsupervised approaches demonstrate robustness in handling noisy data, making them especially advantageous for inherently noisy datasets.

9.3.4 Outlook and Discussion

AI has emerged as a valuable tool for enhancing IC and FPGA security. It offers significant advantages in distinguishing between recycled and newly manufactured chips, detecting HTs, and aiding in reverse engineering processes.

Machine learning methods are heavily utilized in the area of netlist generation and interpretation, both for image processing of the delayered chip, and for analysis of the netlist graph to identify the functionality of the chip. To improve their efficacy, the development of comprehensive NN and ML architectures that can handle multiple technology nodes is imperative. Furthermore, larger and more diverse datasets, as well as optimized feature selection, are needed.

Detection of HTs is another area where AI demonstrates its usefulness. It enables the identification of suspicious logic by analyzing imaging data or gate-level netlists, or comparing side-channel signals of infected designs with design without an active HT. ML algorithms, including SVMs and various neural network architectures, have shown promise in HT detection. However, the evaluation of these methods would benefit from more extensive training and validation datasets, including real world, and thus noisy, measurements obtained from infected ICs and FPGAs. Furthermore, the development of explainable and predictable AI architectures, as well as a more sophisticated methodology in selecting these AI architectures, could enhance the reliability of HT detection approaches. Researchers should also focus on addressing challenges related to limited datasets, simulated data, and the impact of noise in side-channel signal analysis.

In the case of recycled chips, AI-based methods enable the identification of aging patterns that differentiate aged devices from new ones. ML approaches, both supervised and unsupervised, can effectively evaluate aging measurements and classify devices, even in the absence of a known new device. AI methods excel at handling large amounts of data and are robust in the presence of noise, making them well suited for inherently noisy datasets encountered in hardware security.

In summary, AI has shown great potential in enhancing hardware security by enabling the distinction between recycled and new chips, facilitating HT detection, and aiding in reverse engineering processes. Further advancements in dataset availability, neural network architectures, and the integration of real-world designs and measurements will drive the progress of AI in this field and contribute to the development of more robust and reliable hardware security solutions.

9.4 AI-Enabled Analysis of Physical Unclonable Functions

PUFs evaluate physical variation of objects and can be used especially within integrated circuits for identification, authentication, or cryptographic key generation [40]. Typically, secure key storage for integrated circuits is expensive as it requires Non-Volatile Memories (NVMs) that can withstand the physical attacks, such as optical analysis, also while the device is powered down. However, embedded flash technology is only available in larger technology nodes. Going to new technologies, like for example MRAM, FRAM, or RRAM, limits the available manufacturing technologies and requires access to additional circuit IP [12].

In contrast, PUFs either evaluate external factors that can be evaluated from outside a device, such as optical properties [65], or internal properties that are only available in the powered device that has a higher resistance against physical attacks. However, as PUFs evaluate physical measures, they have an inherent noisiness that needs to be addressed by dedicated error correction schemes for PUFs [43] or algorithms based on a challenge-response interface [93].

AI and ML have been used for over a decade for analyzing PUFs, especially since the attack paper published by Rührmair et al. in 2010 [69] starting a line of research.

9.4.1 Physical Unclonable Functions

Silicon PUFs evaluate manufacturing variations in integrated circuits that can be observed, e.g., through timing behavior or mismatches between transistors. Popular PUFs include the arbiter PUF, [27] ring oscillator PUF [76], or bistable ring PUF [13, 41] among others. An overview over more recent PUF types can be found here [79].

PUF designers are faced with the challenge of contradicting requirements as one design goal tries to limit the impact noise spreading through the circuit and a high resilience against ML through a high complexity and especially nonlinear elements in the circuit.

For electronics design, and especially AI-based design, it is important to design PUF structures as regularly as possible to avoid bias and correlation that introduce structure into the random PUF responses.

9.4.2 Attacks

PUFs-based authentication relies on the security of the protocol and the underlying hardware security primitives. The different approaches come with specific advantages and pitfalls on protocol level when building secure lightweight protocols [19]. If protocols withstand the first analytical analysis, they can be attacked by ML methods, introduced in [69] or physical attacks, discussed, e.g., in [78]. In the following, we will focus on the first.

As introduced by Rührmair et al. [69], the first step involves to query the device to collect a large number of hundreds to millions of Challenge Response Pairs (CRPs) and analyze them with SVMs, evolution strategies (ess), or logistic regression (LR), whatever captures the structure of the PUF best. Going to NNs further increases the efficiency, as demonstrated, e.g., in [89]. For example, the model for the runtimes of the PUF in LR closely relates to the weights in an NN. As the protocol is inherently noise-resistant, the model needs to be close enough to the PUF but not exactly accurate. In addition to the models used for arbiter PUFs [46], also dedicated attack can be performed, e.g., on the bistable ring PUF [72].

A typical countermeasure is the XORing of PUF response bits to increase the resilience against ML at the expense of an increase complexity and error probability. It was shown in [87] that the reliability against stochastic errors can be significantly increased if multiple readouts are performed and an internal majority vote is performed for each PUF before the XOR is applied. However, this does not allow deterministic errors, for example, due to aging or temperature drift.

The combination of multiple PUF primitives, e.g., in composite PUFs can increase the ML resistance of the designs [70]. However, the paper [20] shows that multiple protocols, namely, PolyPUFs, OB-PUFs, RPUFs, LHS-PUFs, and PUF–FSMs, can be attacked mathematically by ML methods to allow impersonation attacks. In addition, even though the protocols include a True Random Number Generator (TRNG), they still do not offer an effective resistance against side-channel attacks. Also using only input transformation does not increase the learning significantly in comparison to plain XOR arbiter PUFs [86].

One recent example to counter ML attacks is the interpose PUF [62]. Here, the outputs of a PUF are fed into the next PUF to increase the learning complexity, which was however attacked successfully 1 year later [88].

If no ML model is possible, generalized attack can also be performed with no mathematical model [25, 26]. If the PUF responses are not directly accessible, helper data of PUF-based error correction can be attacked to model the behavior of the PUF [75].

In sum, the number of successful attacks shows the vulnerability of PUF-based authentication and clearly stresses that it is necessary to have a suitable model for the problem and model the PUF with the correct AI-based methods, whereas only the lack of the model often leads to the erroneous assumption that no attacks are possible.

9.4.3 Attack-Resistant Lockdown Protocols

To counter ML- and AI-based attacks of PUF-based authentication using CRPs, Yu et al. [94] introduced two lockdown protocols that limit the exposure of exploitable CRPs.

9.4.3.1 Definition and Notation

Basic strong PUF authentication of a device i (prover) toward the server (verifier) builds upon a strong PUF (SPUF) with a challenge-response interface that is used for d unilateral authentication events of the device. The challenge $\mathbf{c}$ is fed into a PRNG to derive a list of sub-challenges $\langle \mathbf{c} \rangle$. The sub-challenges are fed into the SPUF to generate single response bits that are concatenated to a response vector $\widetilde{\mathbf{r}}$.

During an initial one-time enrollment before deployment of the device, the device i is queried d times with random challenges, and the d CRPs $(\mathbf{c}_{ij}, \mathbf{r}_{ij})$ are stored in the server. Note that the device identifier $\mathbf{id}_i$, e.g., a serial number, is stored on chip, e.g., in a one-time programmable storage (OTP-S), as public information that does not need to be kept secret.

The authentication is initiated by the server by querying the identifier $\mathbf{id}$ of the device. Then, challenge $\mathbf{c}_{ij}$ is transmitted from the server to the device, and the response $\widetilde{\mathbf{r}_{ij}}$ is returned, both in clear, so that CRPs $(\mathbf{c}_{ij}, \widetilde{\mathbf{r}_{ij}})$ is observable by the attacker. The authentication succeeds if the fractional Hamming distance (FHD) between the device's response $\widetilde{\mathbf{r}}$ and the response $\mathbf{r}$ stored in the database exceeds an authentication threshold τ. Subsequently, challenge $\mathbf{c}_{ij}$ is removed from the server's list to prevent replay attacks.

For an ideal PUF, the attacker cannot derive unknown responses for other challenges or other PUFs from the exposed CRPs. In addition, the security of the protocol relies on the unpredictability of the challenges in the database of the server so that an attacker cannot query the PUF in advance and replay the response later.

The protocol can be attacked by analyzing the exposed CRPs with AI. As the server is not authenticated by the devices, an active attacker can initiate the communication and apply new challenges to the PUF to generate an arbitrary number of CRPs. Therefore, it is not possible to upper bound the number of CRPs, and thus the training set for the modeling, to limit the capabilities of the attacker.

9.4.3.2 Lockdown Protocol I

The lockdown protocol shown in Fig. 9.11 addresses this issue by authenticating the server and thus limiting the availability of training material.

Device i		**Server**
	Enrollment (1×)	
OTP-S: $\mathbf{id} \leftarrow \mathbf{id}_i$	$\xleftarrow{\mathbf{id}_i}$	$\mathbf{id}_i \leftarrow \mathsf{SerialNumber}()$
	$\longleftrightarrow$	$(\mathbf{r}_1, \mathbf{r}_2)_{ij}$ with $j \in [1, d]$
		$j_i \leftarrow \mathbf{0}$
	Authentication (d×)	
	$\xleftarrow{\mathsf{init}}$	
	$\xrightarrow{\mathbf{id}}$	Abort if $\forall i : \mathbf{id} \neq \mathbf{id}_i$
		$\mathbf{c} \leftarrow j_i$
		$j_i \leftarrow j_i + 1$
		Abort if $j_i > d$
		$(\mathbf{r}_1, \mathbf{r}_2) \leftarrow (\mathbf{r}_1, \mathbf{r}_2)_{ij}$ with $j = j_i$
	$\xleftarrow{\mathbf{c} \Vert \mathbf{r}_1}$	
$\widetilde{\mathbf{r}}_1 \Vert \widetilde{\mathbf{r}}_2 \leftarrow \mathsf{WPUF}(\mathbf{c})$		
Abort if $\mathsf{FHD}(\widetilde{\mathbf{r}}_1, \mathbf{r}_1) > \tau$	$\xrightarrow{\widetilde{\mathbf{r}}_2}$	
		Abort if $\mathsf{FHD}(\widetilde{\mathbf{r}}_2, \mathbf{r}_2) > \tau$

Fig. 9.11 Lockdown protocol I, providing unilateral authentication with a weak PUF. A strong PUF could also be used by replacing $\widetilde{\mathbf{r}}_1 \Vert \widetilde{\mathbf{r}}_2 \leftarrow \mathsf{WPUF}(\mathbf{c})$ with $\widetilde{\mathbf{r}}_1 \Vert \widetilde{\mathbf{r}}_2 \leftarrow \mathsf{SPUF}(\mathsf{PRNG}(\mathbf{c}))$ [94]

Enrollment

During enrollment, the server sends d deterministic challenges $\mathbf{c}$ to the devices and obtains responses that are split into $\mathbf{r}_1$ and $\mathbf{r}_2$ for each challenge. The tuples $(\mathbf{r}_1, \mathbf{r}_2)$ are stored in the database of the server.

Authentication

Again, authentication is initiated by the server by querying the identifier **id** of the device to send a packet containing $\mathbf{c}$ and $\mathbf{r}_1$ to the device. The device generates response $\widetilde{\mathbf{r}}_1$ and compares it to $\mathbf{r}_1$. If FHD exceeds the preset threshold τ, the authentication of the server fails and the device aborts the process. Otherwise, the second part of the response $\widetilde{\mathbf{r}}_2$ is generated and sent to the server so that the server can compare it to the enrolled version $\mathbf{r}_2$.

Remarks

- Introducing an interface inside the device with the PUF facilitates to limit the training material of the attacker and thus create a defined attacker model, within the PUF can be secure against ML attacks, or not.
- As the device will not respond to challenges with incorrect data $\mathbf{r}_1$, the challenges can be created in a deterministic manner, e.g., through a counter that seeds an expansion function.
- Note that the protocol limits the number of exposed CRPs but still allows the attacker to query the PUF multiple times with the same challenge. This allows to generate reliability information of the PUF bits that permit, e.g., the attack in [7].

9.4.3.3 Lockdown Protocol II

Protocol II, also introduced in [94] and shown in Fig. 9.12, extends Protocol I by introducing freshness to prevent noise side channels that occur when the same challenge is applied multiple times. It requires that the PUF has two states: one state where it is easy to learn and one with a high complexity. This can be implemented for example through an XOR arbiter PUF where the arbiter PUFs can be characterized independently. Later, the XOR is activated permanently, e.g., through a fuse, to create a high PUF with a high resistance against machine learning.

Enrollment

Instead of storing dedicated responses or CRPs, a single authentication verification model $\widehat{\mathsf{SPUF}}_i$ is learned from an unlocked PUF and stored on the server. Then, the PUF is locked irreversibly.

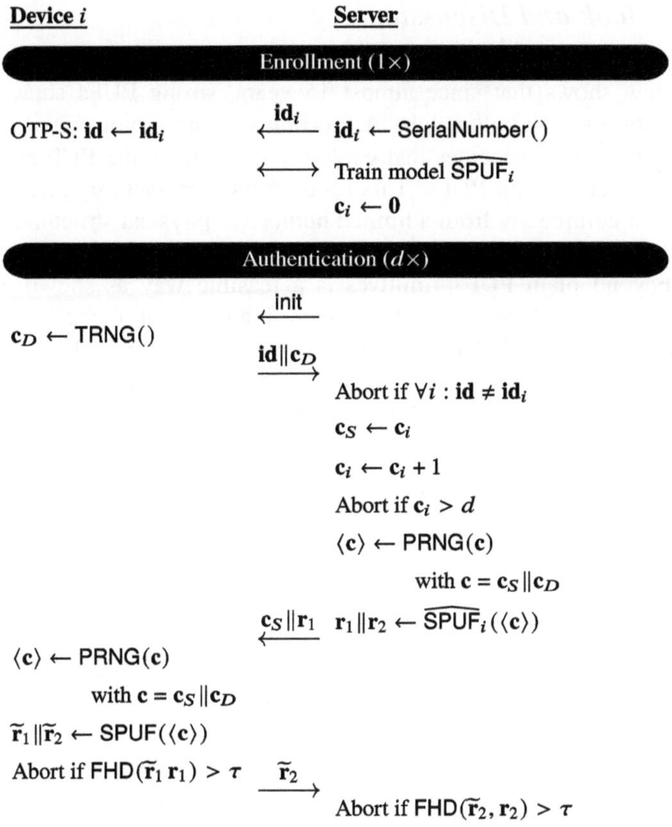

Fig. 9.12 Lockdown protocol II, providing mutual authentication. A device-side nonce $\mathbf{c}_D$ enables a noise side-channel countermeasure [94]

Authentication

After triggering the authentication, the device generates a device-side challenge $\mathbf{c}_D$ and sends it together with its device identifier **id** to the server. The server generates $\mathbf{c}_S$ and computes $\mathbf{r}_1$ from the emulated software model of the PUF $\widehat{\mathsf{SPUF}}_i$ with $\langle \mathbf{c} \rangle$ that is a function of both $\mathbf{c}_S$ and $\mathbf{c}_D$. Then, $\mathbf{c}_S$ and $\mathbf{r}_1$ are sent to the device. Similarly to Protocol I $\widetilde{\mathbf{r}}_1$ is generated and compared to $\mathbf{r}_1$. If it succeeds, $\widetilde{\mathbf{r}}_2$ is generated and released for the comparison against the $\mathbf{r}_2$ that was emulated using $\widehat{\mathsf{SPUF}}_i$.

Remarks

- The device-side challenge allows a *challenge exchange*, so neither the device nor the server can unilaterally determine all the bits of $\langle \mathbf{c} \rangle$.
- For a detailed security and reliability analysis, we refer to [94].

9.4.4 Outlook and Discussion

This overview shows that since almost 15 years, strong PUFs show a constant arms race between new lightweight PUF primitives that promise resilience against ML attacks and attack schemes that exploit a structure in the PUF and sometime also generalize attacks on PUFs. This roots in the constant struggle to achieve a high degree of complexity from a limited number of physical structures, while also maintaining reliability.

Going beyond plain PUF primitives is a feasible way as shown exemplarily through the discussed protocols Lockdown I and II, at the expense of extra implementation overhead. However, this facilitates to use PUFs in lightweight authentication and still offers questions and room for research how more advanced AI methods can be used to improve modeling of PUFs and thus also increase the security and attack resilience.

9.5 Conclusion

Hardware security is a constant arms race between efficient designs and analysis methods to protect a system and assess its security. AI methods can ease the analysis of embedded systems, while the explainability of the methods remain challenging. They cover, for example, side-channel analysis of cryptographic implementations, reverse engineering of circuits, and systems, and PUF as physical hardware security primitives. However, the impact is not limited to the presented use cases such that we expect further analyses methods, but also more generative work using AI.

With new methods such as LLMs revolutionizing other areas of IT and technology, we also expect that the analysis and introduction of new countermeasures will also benefit. While this chapter focuses on the analysis side, also new design techniques and the integration of countermeasures through AI that will need to be further investigated in the future.

References

1. Alam, M. M., Tehranipoor, M., & Forte, D. (2019). Recycled FPGA Detection Using Exhaustive LUT Path Delay Characterization and Voltage Scaling. *IEEE Transactions on Very Large Scale Integration (VLSI) Systems, 27*(12), 2897–2910.
2. Alrahis, L., Sengupta, A., Knechtel, J., Patnaik, S., Saleh, H., Mohammad, B., Al-Qutayri, M., & Sinanoglu, O. (2021). GNN-RE: Graph Neural Networks for Reverse Engineering of Gate-Level Netlists. In *IEEE Transactions on Computer-Aided Design of Integrated Circuits and Systems* (pp. 1–1).

3. Arvin, K. & Jha, R. (2021). Aging prediction of integrated circuits using ring oscillators and machine learning. In *2021 IEEE Physical Assurance and Inspection of Electronics (PAINE)* (pp. 1–8).
4. Asadizanjani, N., Tehranipoor, M., & Forte, D. (2017). Counterfeit Electronics Detection Using Image Processing and Machine Learning. *Journal of Physics: Conference Series, 787*(1), 012023.
5. Baehr, J., Bernardini, A., Sigl, G., & Schlichtmann, U. (2019). Machine learning and structural characteristics for reverse engineering. In *Proceedings of the 24th Asia and South Pacific Design Automation Conference* (pp. 96–103), Tokyo Japan: ACM.
6. Bao, C., Forte, D., & Srivastava, A. (2014). On application of one-class SVM to reverse engineering-based hardware Trojan detection. In *Fifteenth International Symposium on Quality Electronic Design* (pp. 47–54).
7. Becker, G. T. (2015). The gap between promise and reality: On the insecurity of XOR arbiter PUFs. In Güneysu, T., & Handschuh, H. (Eds.), *Cryptographic Hardware and Embedded Systems* (pp. 535–555). Berlin, Heidelberg: Springer Berlin Heidelberg.
8. Bette, A.-C., Brus, P., Balazs, G., Ludwig, M., & Knoll, A. (2021). Automated Defect Inspection in Reverse Engineering of Integrated Circuits.
9. Botero, U. J., Wilson, R., Lu, H., Rahman, M. T., Mallaiyan, M. A., Ganji, F., Asadizanjani, N., Tehranipoor, M. M., Woodard, D. L., & Forte, D. (2021). Hardware Trust and Assurance through Reverse Engineering: A Tutorial and Outlook from Image Analysis and Machine Learning Perspectives. *ACM Journal on Emerging Technologies in Computing Systems, 17*(4), 62:1–62:53.
10. Carbone, M., Conin, V., Cornelie, M., Dassance, F., Dufresne, G., Dumas, C., Prouff, E., & Venelli, A. (2019). Deep learning to evaluate secure RSA implementations. *IACR Transactions on Cryptographic Hardware and Embedded Systems, 2019*(2), 132–161.
11. Chari, S., Rao, J. R., & Rohatgi, P. (2002). Template attacks. In Kaliski Jr., B. S., Koç, Ç. K., & Paar, C. (Eds.), *Cryptographic Hardware and Embedded Systems—CHES 2002, 4th International Workshop, Redwood Shores, CA, USA, August 13–15, 2002, Revised Papers. Lecture Notes in Computer Science* (Vol. 2523, pp. 13–28). Berlin: Springer.
12. Chen, A. (2016). A review of emerging non-volatile memory (NVM) technologies and applications. *Solid-State Electronics, 125*, 25–38.
13. Chen, Q., Csaba, G., Lugli, P., Schlichtmann, U., & Rührmair, U. (2011). The bistable ring PUF: A new architecture for strong physical unclonable functions. In *IEEE International Symposium on Hardware-Oriented Security and Trust (HOST)* (pp. 134–141).
14. Cheng, D., Shi, Y., Lin, T., & Gwee, B.-H. (2023). Domain-Integrated Machine Learning for IC Image Analysis. In Hatzilygeroudis, I. K., Tsihrintzis, G. A., & Jain, L. C. (Eds.), *Fusion of Machine Learning Paradigms: Theory and Applications*, Intelligent Systems Reference Library, pp. 129–151. Cham: Springer International Publishing.
15. Couch, J., & Arkoian, J. (2016). An investigation into a circuit based supply chain analyzer for FPGAs. In *2016 26th International Conference on Field Programmable Logic and Applications (FPL)* (pp. 1–9).
16. Cristiani, V., Lecomte, M., & Maurine, P. (2023). The EVIL machine: Encode, visualize and interpret the leakage. In Hong, J., Lanperne, M., Park, J. W., Cerný, T., & Shahriar, H. (Eds.) *Proceedings of the 38th ACM/SIGAPP Symposium on Applied Computing, SAC 2023, Tallinn, Estonia, March 27–31, 2023* (pp. 1566–1575). New York: ACM.
17. Cruz, J., Gaikwad, P., Nair, A., Chakraborty, P., & Bhunia, S. (2022). Automatic Hardware Trojan Insertion using Machine Learning. arXiv:2204.08580 [cs].
18. Dakhale, B., Vipinkumar, K., Narotham, K., Kadam, S., Bhurane, A. A., & Kothari, A. G. (2023). Automated Detection of Hardware Trojans using Power Side-Channel Analysis and VGG-Net. In *2023 2nd International Conference on Paradigm Shifts in Communications Embedded Systems, Machine Learning and Signal Processing (PCEMS)* (pp. 1–5).
19. Delvaux, J. (2017). Security analysis of PUF-based key generation and entity authentication. Ph. D. dissertation.

20. Delvaux, J. (2019). Machine-learning attacks on polypufs, ob-pufs, rpufs, lhs-pufs, and puf-fsms. *IEEE Transactions on Information Forensics and Security, 14*(8), 2043–2058.
21. Dogan, H., Forte, D., & Tehranipoor, M. M. (2014). Aging analysis for recycled FPGA detection. In *2014 IEEE International Symposium on Defect and Fault Tolerance in VLSI and Nanotechnology Systems (DFT)* (pp. 171–176).
22. Dong, C., Chen, J., Guo, W., & Zou, J. (2019). A machine-learning-based hardware-Trojan detection approach for chips in the Internet of Things. *International Journal of Distributed Sensor Networks, 15*(12), 155014771988809.
23. Faezi, S., Yasaei, R., Barua, A., & Faruque, M. A. A. (2021). Brain-Inspired Golden Chip Free Hardware Trojan Detection. *IEEE Transactions on Information Forensics and Security, 16*, 2697–2708.
24. Gaikwad, P., Cruz, J., Chakraborty, P., Bhunia, S., & Hoque, T. (2023). Hardware IP Assurance against Trojan Attacks with Machine Learning and Post-processing. In *ACM Journal on Emerging Technologies in Computing Systems* (p. 3592795).
25. Ganji, F., Tajik, S., Fäßler, F., & Seifert, J.-P. (2016). Strong machine learning attack against PUFs with no mathematical model. In Gierlichs, B. & Poschmann, Y. A. (Eds.), *International Conference on Cryptographic Hardware and Embedded Systems (CHES)* (pp. 391–411). Berlin: Springer Berlin Heidelberg.
26. Ganji, F., Tajik, S., Fäßler, F., & Seifert, J.-P. (2017). Having no mathematical model may not secure PUFs. *Journal of Cryptographic Engineering, 7*(2), 113–128.
27. Gassend, B., Clarke, D., Dijk, M. v., & Devadas, S. (2002). Silicon physical random functions. In *ACM Conference on Computer and Communications Security (CCS)* (pp. 148–160).
28. Genevey-Metat, C., Heuser, A., & Gérard, B. (2021). Trace-to-trace translation for SCA. In Grosso, V., & Pöppelmann, T. (Eds.), *Smart Card Research and Advanced Applications—20th International Conference, CARDIS 2021, Lübeck, Germany, November 11–12, 2021, Revised Selected Papers. Lecture Notes in Computer Science* (Vol. 13173, pp. 24–43). New York: Springer.
29. Gilbert, B., Goodwill, J., Jaffe, J., Rohatgi, P., et al. (2011). A testing methodology for side-channel resistance validation. In *NIST Non-invasive Attack Testing Workshop* (Vol. 7, pp. 115–136).
30. Gohr, A., Jacob, S., & Schindler, W. (2019). CHES 2018 side channel contest CTF—solution of the AES challenges. *IACR Cryptology ePrint Archive* (pp. 94).
31. Gupta, P., Drees, J. P., & Hüllermeier, E. (2023). Automated side-channel attacks using black-box neural architecture search. In *Proceedings of the 18th International Conference on Availability, Reliability and Security, ARES 2023, Benevento, Italy, 29 August 2023–1 September 2023* (pp. 5:1–5:11). New York: ACM.
32. Gustin, L.-H., Durvaux, F., Kerckhof, S., Standaert, F.-X., & Verleysen, M. (2014). Support Vector Machines for Improved IP Detection with Soft Physical Hash Functions. In Prouff, E. (Ed.), *Constructive Side-Channel Analysis and Secure Design* (pp. 112–128). Cham: Springer International Publishing.
33. Hasegawa, K., Oya, M., Yanagisawa, M., & Togawa, N. (2016). Hardware Trojans classification for gate-level netlists based on machine learning. In *2016 IEEE 22nd International Symposium on On-Line Testing and Robust System Design (IOLTS)* (pp. 203–206).
34. Hasegawa, K., Shi, Y., & Togawa, N. (2018). Hardware Trojan Detection Utilizing Machine Learning Approaches. In *2018 17th IEEE International Conference on Trust, Security and Privacy in Computing and Communications/ 12th IEEE International Conference on Big Data Science and Engineering (TrustCom/BigDataSE)* (pp. 1891–1896).
35. Hasegawa, K., Yanagisawa, M., & Togawa, N. (2017). Hardware Trojans classification for gate-level netlists using multi-layer neural networks. In *2017 IEEE 23rd International Symposium on On-Line Testing and Robust System Design (IOLTS)* (pp. 227–232).
36. Hassan, R., Meng, X., Basu, K., & Dinakarrao, S. M. P. (2023). Circuit Topology-aware Vaccination-based Hardware Trojan Detection. In *IEEE Transactions on Computer-Aided Design of Integrated Circuits and Systems* (pp. 1–1).

37. He, Z., Wang, Z., Bail, C., Yang, H., & Yu, B. (2021). Graph Learning-Based Arithmetic Block Identification. In *2021 IEEE/ACM International Conference On Computer Aided Design (ICCAD)* (pp. 1–8).
38. Hepp, A., Baehr, J., & Sigl, G. (2022). Golden Model-Free Hardware Trojan Detection by Classification of Netlist Module Graphs. In *2022 Design, Automation Test in Europe Conference Exhibition (DATE)* (pp. 1317–1322).
39. Hepp, A., & Sigl, G. (2021). Tapeout of a RISC-V crypto chip with hardware trojans: a case-study on trojan design and pre-silicon detectability. In *Proceedings of the 18th ACM International Conference on Computing Frontiers* (pp. 213–220). Virtual Event Italy: ACM.
40. Herder, C., Yu, M., Koushanfar, F., & Devadas, S. (2014). Physical unclonable functions and applications: A tutorial. *Proceedings of the IEEE, 102*(8), 1126–1141.
41. Hesselbarth, R., & Sigl, G. (2016). Fast and reliable PuF response evaluation from unsettled bistable rings. In *Euromicro Conference on Digital System Design (DSD)* (pp. 82–90).
42. Heuser, A., Picek, S., Guilley, S., & Mentens, N. (2020). Lightweight ciphers and their side-channel resilience. *IEEE Transactions on Computers, 69*(10), 1434–1448.
43. Hiller, M., Kürzinger, L., & Sigl, G. (2020). Review of error correction for PuFs and evaluation on state-of-the-art FPGAs. *Journal of Cryptographic Engineering, 10*(3), 229–247.
44. Hong, X., Lin, T., Shi, Y., & Gwee, B. H. (2021). ASIC Circuit Netlist Recognition Using Graph Neural Network. In *2021 IEEE International Symposium on the Physical and Failure Analysis of Integrated Circuits (IPFA)* (pp. 1–5).
45. Hospodar, G., Gierlichs, B., Mulder, E. D., Verbauwhede, I., & Vandewalle, J. (2011). Machine learning in side-channel analysis: A first study. *Journal of Cryptographic Engineering, 1*(4), 293–302.
46. Hospodar, G., Maes, R., & Verbauwhede, I. (2012). Machine learning attacks on 65nm arbiter PUFs: Accurate modeling poses strict bounds on usability. In *IEEE International Workshop on Information Forensics and Security (WIFS)*. New York: IEEE.
47. Hu, T., Wu, L., Zhang, X., Yin, Y., & Yang, Y. (2019). Hardware Trojan Detection Combine with Machine Learning: An SVM-based Detection Approach. In *2019 IEEE 13th International Conference on Anti-counterfeiting, Security, and Identification (ASID)* (pp. 202–206).
48. Huang, K., Carulli, J. M., & Makris, Y. (2012). Parametric counterfeit IC detection via Support Vector Machines. In *2012 IEEE International Symposium on Defect and Fault Tolerance in VLSI and Nanotechnology Systems (DFT)* (pp. 7–12).
49. Jap, D., He, W., & Bhasin, S. (2016). Supervised and unsupervised machine learning for side-channel based Trojan detection. In *2016 IEEE 27th International Conference on Application-specific Systems, Architectures and Processors (ASAP)* (pp. 17–24).
50. Kim, J., Picek, S., Heuser, A., Bhasin, S., & Hanjalic, A. (2019). Make some noise. Unleashing the power of convolutional neural networks for profiled side-channel analysis. *IACR Transactions on Cryptographic Hardware and Embedded Systems, 2019*(3), 148–179.
51. Kimura, A., Scholl, J., Schaffranek, J., Sutter, M., Elliott, A., Strizich, M., & Via, G. D. (2020). A decomposition workflow for integrated circuit verification and validation. *Journal of Hardware and Systems Security, 4*, 34–43.
52. Koç, Ç. K. (Eds.), *Cryptographic Engineering*. New York: Springer (2009).
53. Lin, T., Shi, Y., & Gwee, B. H. (2023). SEM2GDS: A deep-learning based framework to detect malicious modifications in IC layout. In *2023 IEEE International Symposium on Circuits and Systems (ISCAS)* (pp. 1–5).
54. Lippmann, B., Werner, M., Unverricht, N., Singla, A., Egger, P., Dübotzky, A., Gieser, H., Rasche, M., Kellermann, O., & Graeb, H. (2019). Integrated flow for reverse engineering of nanoscale technologies. In *Proceedings of the 24th Asia and South Pacific Design Automation Conference*, ASPDAC '19 (pp. 82–89), New York, NY, USA: Association for Computing Machinery.
55. Liszewski, K., McDonley, T., Delozier, J., Elliott, A., Jones, D., Sutter, M., & Kimura, A. (2021). Netlist Decompilation Workflow for Recovered Design Verification, Validation, and Assurance. Technical Report 780. ZSCC: 0000001.

56. Ludwig, M., Bette, A.-C., & Lippmann, B. (2021). ViTaL: Verifying Trojan-free physical layouts through hardware reverse engineering. In *2021 IEEE Physical Assurance and Inspection of Electronics (PAINE)* (pp. 1–8). Washington, DC: IEEE.
57. Mangard, S., Oswald, E., & Popp, T. (2007). *Power Analysis Attacks—Revealing the Secrets of Smart Cards*. Berlin: Springer.
58. Masure, L., Dumas, C., & Prouff, E. (2020). A comprehensive study of deep learning for side-channel analysis. *IACR Transactions on Cryptographic Hardware and Embedded Systems, 2020*(1), 348–375.
59. Moos, T., Moradi, A., & Richter, B. (2020). Static power side-channel analysis—an investigation of measurement factors. *IEEE Transactions on Very Large Scale Integration Systems, 28*(2), 376–389.
60. Moos, T., Wegener, F., & Moradi, A. (2021). DL-LA: deep learning leakage assessment A modern roadmap for SCA evaluations. *IACR Transactions on Cryptographic Hardware and Embedded Systems, 2021*(3), 552–598.
61. Moradi, A., Richter, B., Schneider, T., & Standaert, F. (2018). Leakage detection with the x2-test. *IACR Transactions on Cryptographic Hardware and Embedded Systems, 2018*(1), 209–237.
62. Nguyen, P. H., Sahoo, D. P., Jin, C., Mahmood, K., Rührmair, U., & van Dijk, M. (2019). The interpose PUF: Secure PUF design against state-of-the-art machine learning attacks. *IACR Transactions on Cryptographic Hardware and Embedded Systems* (pp. 243–290).
63. Noor, N. Q. M., Sjarif, N. N. A., Azmi, N. H. F. M., Daud, S. M., & Kamardin, K. (2017). Hardware Trojan identification using machine learning-based classification. *Journal of Telecommunication, Electronic and Computer Engineering (JTEC), 9*(3-4), 23–27. ZSCC: 0000005.
64. Pan, Z. & Mishra, P. (2022). A survey on hardware vulnerability analysis using machine learning. *IEEE Access, 10*, 49508–49527.
65. Pappu, R., Recht, B., Taylor, J., & Gershenfeld, N. (2002). Physical one-way functions. *Science, 297*(5589), 2026–2030.
66. Picek, S., Perin, G., Mariot, L., Wu, L., & Batina, L. (2023). SoK: Deep learning-based physical side-channel analysis. *ACM Computing Surveys, 55*(11), 227:1–227:35.
67. Rechberger, C. & Oswald, E. (2004). Practical template attacks. In Lim, C. H. & Yung, M., (Eds.), *Information Security Applications, 5th International Workshop, WISA 2004, Jeju Island, Korea, August 23–25, 2004, Revised Selected Papers. Lecture Notes in Computer Science* (Vol. 3325, pp. 440–456). Berlin: Springer.
68. Rijsdijk, J., Wu, L., Perin, G., & Picek, S. (2021). Reinforcement learning for hyperparameter tuning in deep learning-based side-channel analysis. *IACR Transactions on Cryptographic Hardware and Embedded Systems, 2021*(3), 677–707.
69. Rührmair, U., Sehnke, F., Sölter, J., Dror, G., Devadas, S., & Schmidhuber, J. (2010). Modeling attacks on physical unclonable functions. In *ACM Conference on Computer and Communications Security (CCS)* (pp. 237–249).
70. Sahoo, D. P., Nguyen, P. H., Mukhopadhyay, D., & Chakraborty, R. S. (2015). A case of lightweight PUF constructions: Cryptanalysis and machine learning attacks. *IEEE Transactions on Computer-Aided Design of Integrated Circuits and Systems, PP(99)*, 1–1.
71. Salmani, H., Tehranipoor, M., & Karri, R. (2013). On design vulnerability analysis and trust benchmarks development. In *2013 IEEE 31st International Conference on Computer Design (ICCD)* (pp. 471–474).
72. Schuster, D. & Hesselbarth, R. (2014). Evaluation of bistable ring PUFs using single layer neural networks. In Holz, T. & Ioannidis, S., (Eds.), *Trust and Trustworthy Computing (TRUST). LNCS* (Vol. 8564, pp. 101–109). Berlin: Springer International Publishing.
73. Shakya, B., He, T., Salmani, H., Forte, D., Bhunia, S., & Tehranipoor, M. (2017). Benchmarking of Hardware Trojans and Maliciously Affected Circuits. *Journal of Hardware and Systems Security, 1*(1), 85–102.
74. Standaert, F. (2018). How (not) to use Welch's t-test in side-channel security evaluations. In Bilgin, B. & Fischer, J., (Eds.), *Smart Card Research and Advanced Applications, 17th*

International Conference, CARDIS 2018, Montpellier, France, November 12–14, 2018, Revised Selected Papers. Lecture Notes in Computer Science (Vol. 11389, pp. 65–79). Berlin: Springer.
75. Strieder, E., Frisch, C., & Pehl, M. (2021). Machine learning of physical unclonablefunctions using helper data. *IACR Transactions on Cryptographic Hardware and Embedded Systems, 2021*(2), 1–36.
76. Suh, G. E. & Devadas, S. (2007). Physical unclonable functions for device authentication and secret key generation. In *ACM/IEEE Design Automation Conference (DAC)* (pp. 9–14).
77. Sun, C., Cheng, L. Y., Wang, L. W., Huang, Q., Huang, Y., & Feng, G. L. (2021). A machine learning method for hardware Trojan detection on real chips. *AIP Advances, 11*(5), 055006.
78. Tajik, S. (2019). *On the Physical Security of Physically Unclonable Functions*. T-Labs Series in Telecommunication Services. Berlin: Springer International Publishing.
79. Tehranipoor, M. H., Pundir, N., Vashistha, N., & Farahmandi, F. (2023). *Hardware Security Primitives*. Berlin: Springer.
80. Timon, B. (2019). Non-profiled deep learning-based side-channel attacks with sensitivity analysis. *IACR Transactions on Cryptographic Hardware and Embedded Systems, 2019*(2), 107–131.
81. Vashistha, N., Lu, H., Shi, Q., Rahman, M. T., Shen, H., Woodard, D. L., Asadizanjani, N., & Tehranipoor, M. (2018). Trojan Scanner: Detecting Hardware Trojans with Rapid SEM Imaging Combined with Image Processing and Machine Learning, Phoenix, Arizona, USA (pp. 256–265).
82. Wang, Z., He, Z., Bai, C., Yang, H., & Yu, B. (2022). Efficient arithmetic block identification with graph learning and network-flow. In *IEEE Transactions on Computer-Aided Design of Integrated Circuits and Systems* (pp. 1–1).
83. Weissbart, L., Picek, S., & Batina, L. (2019). One trace is all it takes: Machine learning-based side-channel attack on EDDSA. In Bhasin, S., Mendelson, A., & Nandi, M., (Eds.), *Security, Privacy, and Applied Cryptography Engineering—9th International Conference, SPACE 2019, Gandhinagar, India, December 3–7, 2019, Proceedings. Lecture Notes in Computer Science* (Vol. 11947, pp. 86–105). Berlin: Springer.
84. Wen, Y. & Yu, W. (2020). Combining thermal maps with inception neural networks for hardware Trojan detection. *IEEE Embedded Systems Letters,13*(2), 45–48.
85. Werner, M., Lippmann, B., Baehr, J., & Gräb, H. (2018). Reverse engineering of cryptographic cores by structural interpretation through graph analysis. In *2018 IEEE 3rd International Verification and Security Workshop (IVSW)* (pp. 13–18).
86. Wisiol, N., Becker, G. T., Margraf, M., Soroceanu, T. A. A., Tobisch, J., & Zengin, B. (2020). Breaking the lightweight secure PUF: Understanding the relation of input transformations and machine learning resistance. In Belaid, S., & Güneysu, T. (Eds.) *Smart Card Research and Advanced Applications* (pp. 40–54). New York: Springer International Publishing.
87. Wisiol, N. & Margraf, M. (2019). Why attackers lose: design and security analysis of arbitrarily large XOR arbiter PUFs. *Journal of Cryptographic Engineering, 9*(3), 221–230.
88. Wisiol, N., Mühl, C., Pirnay, N., Nguyen, P. H., Margraf, M., Seifert, J.-P., van Dijk, M., & Rührmair, U. (2020). Splitting the interpose PUF: A novel modeling attack strategy. *IACR Transactions on Cryptographic Hardware and Embedded Systems, 2020*(3), 97–120.
89. Wisiol, N., Thapaliya, B., Mursi, K. T., Seifert, J.-P., & Zhuang, Y. (2022). Neural network modeling attacks on arbiter-puf-based designs. *IEEE Transactions on Information Forensics and Security, 17*, 2719–2731.
90. Wu, L., Perin, G., & Picek, S. (2020). I choose you: Automated hyperparameter tuning for deep learning-based side-channel analysis. *IACR Cryptology ePrint Archive* (p. 1293).
91. Wu, L. & Picek, S. (2020). Remove some noise: On pre-processing of side-channel measurements with autoencoders. *IACR Transactions on Cryptographic Hardware and Embedded Systems, 2020*(4), 389–415.
92. Yang, S., Chakraborty, P., & Bhunia, S. (2021). Side-channel analysis for hardware Trojan detection using machine learning. In *2021 IEEE International Test Conference India (ITC India)* (pp. 1–6).
93. Yu, M. (2018). *PUF Constructions with Limited Information Leakage*. Dissertation.

94. Yu, M., Hiller, M., Delvaux, J., Sowell, R., Devadas, S., & Verbauwhede, I. (2016). A lockdown technique to prevent machine learning on PUFs for lightweight authentication. *IEEE Transactions on Multi-Scale Computing Systems, 2*(3), 146–159.
95. Zaid, G., Bossuet, L., Dassance, F., Habrard, A., & Venelli, A. (2021). Ranking loss: Maximizing the success rate in deep learning side-channel analysis. *IACR Transactions on Cryptographic Hardware and Embedded Systems, 2021*(1), 25–55.
96. Zaid, G., Bossuet, L., Habrard, A., & Venelli, A. (2019). Methodology for efficient CNN architectures in profiling attacks. *IACR Cryptology ePrint Archive* (p. 803).

Chapter 10
On AI-Enabled Cybersecurity: Zero-Day Malware Detection

Hossein Sayadi and Zhangying He

The rapid technological advancements in cloud and edge computing, digital transformation, ubiquitous connectivity, and generative artificial intelligence in recent years have led to new security vulnerabilities, making advanced computer systems prime targets for increasingly sophisticated cyber threats. The rise in frequency and complexity of high-profile cyberattacks and intrusions has garnered intense media and consumer attention, prompting the widespread implementation of common cybersecurity practices in both business and personal computing and networking systems. In response, hackers have adapted their tactics to exploit any exposed security vulnerabilities, complicating cyber defense efforts.

According to SonicWall's 2023 Cyber Threat Report [44], there were 6.3 trillion cyber threat attempts in 2022, a 19% increase from 2021. Malware attacks continue to dominate, reaching 5.5 billion incidents, while IoT attacks surged to 12.3 million, an 87% increase from the previous year. Among these attacks, cybercriminals increasingly exploit new vulnerabilities, especially through zero-day malware attacks. The same SonicWall report highlights a significant rise in never-before-seen zero-day attacks in 2022, with nearly one million new malware variants and approximately 1,300 new malware variants detected per day. The dramatic increase in malware attacks underscores the growing vulnerabilities and potential drawbacks of technological advancements. This chapter explores the emerging field of AI-enabled cybersecurity, with a particular focus on the critical aspect of zero-day malware detection. This topic is of immense significance and has a profound impact on the overall security landscape.

H. Sayadi (✉)
Computer Engineering and Computer Science Department, California State University, Long Beach (CSULB), Long Beach, CA, USA
e-mail: Hossein.Sayadi@csulb.edu

Z. He
California State University Long Beach, Long Beach, CA, USA

A. Iranmanesh, H. Sayadi (eds.), *AI-Enabled Electronic Circuit and System Design*,
https://doi.org/10.1007/978-3-031-71436-8_10

10.1 Security Vulnerabilities

A security vulnerability refers to a susceptible aspect, such as a bug, flaw, or error, within a computer system. Cyberattackers strategically identify these vulnerabilities as potential entryways, utilizing them to infiltrate a system and implant malicious code, ultimately compromising its integrity. Examples of common security vulnerabilities encompass unauthorized system access, SQL injection, cross-site request forgery, security misconfiguration, remote code execution, security feature bypass, path traversal, arbitrary file reading, and social engineering tactics like phishing attacks.

To launch a cyber threat, hackers first must perform vulnerability scanning to identify weaknesses in a computer system and then plot an exploit to mount an attack. An exploit is a piece of code or a sequence of commands crafted by attackers. It can be embedded in the applications or software running in the target computer system through the security weak point to enter the system and launch attacks. Security vulnerabilities can exist in various layers in a computer system, including the underlying hardware layer, networking, infrastructure layers (e.g., SAAS), operating system, software, and applications running in the system.

Aiming at reporting and sharing the vulnerabilities to encourage patching, several government agencies and corporations report the Common Vulnerabilities and Exposure (CVE) as public knowledge. This helps warn of potential threats and urges software updates and patches. For instance, MITRE Corporation maintains the platform to update newly found CVEs from publicly released software to warn related companies and developers about common weaknesses. Figure 10.1 shows the top 15 vulnerabilities in 2021 reported by the US's Cybersecurity & Infrastructure Security Agency (CISA) [1]. As depicted in Fig. 10.1a, hackers exploited diverse vulnerabilities, enabling them to assume control and execute malicious codes on the targeted victim system. Figure 10.1b highlights that Microsoft Exchange Servers stand out as a particularly susceptible point, providing attackers with a significant opportunity to elevate privileges and execute malicious code on the system.

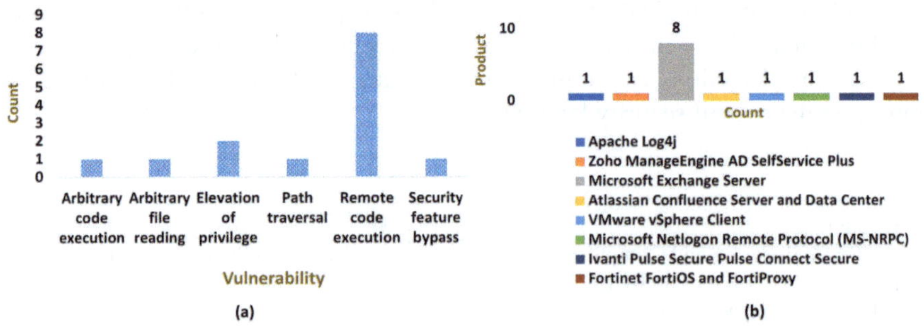

Fig. 10.1 Top 15 vulnerabilities reported in 2021 by CISA [1]. (**a**) Count by vulnerability; (**b**) count by product exposed to vulnerability

10.2 Malware Attacks

Malware, short for malicious software, is a general term for any malicious software developed by cyber criminals to perform harmful activities like stealing data or disrupting or damaging computers and computer systems [3, 36]. To hide their true harmful intentions, attackers typically blend malicious code into seemingly benign and legitimate applications to lure unwary users into downloading malicious programs on the target victim's system [16, 17]. A malware attack is the first step for hackers to take control of the victim's system. Common examples of malware include viruses, worms, Trojans, spyware, adware, ransomware, and more.

10.2.1 Get Familiar with Malware Types

10.2.1.1 Virus

A virus is a self-propagating malware through code injecting itself into other software, data, or applications within the infected victim system. When the victim file is opened, it executes the virus code and spreads the virus code to other files, software, and attached hardware of the system. The victim system can suffer data loss, interruption, and damage.

10.2.1.2 Worm

A worm is also a self-propagating malware, but its designed purpose is to inject code into other systems beyond the current victim system. In some cases, users do not take any actions before their systems are infected by a worm malware.

10.2.1.3 Trojans

Trojans conceal themselves as legitimate application such as antivirus software, legitimate links in emails, or pop-ups to users. It relies on users to click on it to download to their system. Trojans can cause data loss, leakage, and stealing.

10.2.1.4 Spyware

Spyware is designed to spy on the victim's computer system and gather information like login credentials, banking details, transaction details, and more. Spyware is installed in the victim's system via users' unintentional clicking. This stolen information can cause data and financial loss. Attackers gathering businesses' information via spyware can cause consumer data leakage and reputation damage.

10.2.1.5 Adware

Adware is a type of malware that is often embedded in digital advertisements delivered to users. Users thought it was a legitimate ad, and once it is clicked, the adware is installed quietly into their computer system. Adware often redirects users to malicious links that expose users to malicious programs and advertisements.

10.2.1.6 Ransomware

Ransomware often employs social engineering techniques to trick users into clicking on a link. The type of trick can be an email from a person you know, or a seemingly legitimate action you should click. Through users' clicking, ransomware is downloaded to the target victim's computer and can take effect in minutes. Users are locked out of their computers. The ransomware encrypts users' file systems with an encryption key (known to attackers), and users lose their access to their data. Attackers offer to sell the encryption key to the victim users to regain access to their data. In recent years, governments and businesses have been the most targeted by ransomware attacks.

10.2.1.7 Rootkit

Rootkit is designed to give hackers control over a target device. It can infect software, operating systems, computer hardware, and firmware. It is one of the stealthy malware that can conceal itself from being detected.

10.2.1.8 Botnet

Botnet is a type of malware attack in which hackers first infect a network of computers and devices, and then they control these computers/devices without the knowledge of the owners to conduct further attacks such as flooding other servers (with spam, launching DoS attacks), and sending phishing emails. Hackers often trick users by clicking on malicious links to flood the botnet malware to the victim's computer. Without user's awareness, their computers/devices are joined as part of the infected botnet network to carry out tasks by their design or launch nefarious attacks.

10.2.1.9 Backdoor

Backdoor malware creates a hidden way to access a computer system, bypassing customary security mechanisms. Hackers use the backdoor to gain remote access, often in conjunction with a worm or virus, to further control and damage the target system.

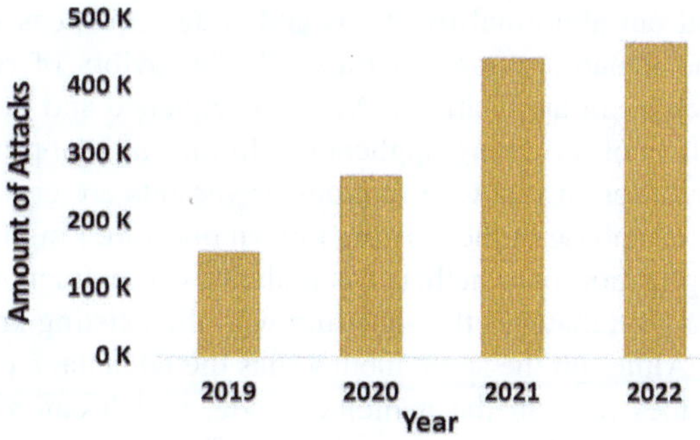

Fig. 10.2 Zero-day malware attacks keep dominating the landscape of cyber threats in 2022 [44]

10.2.2 Unknown (Zero-Day) Malware Attacks

Unknown, zero-day malware attacks represent a category of cyber threats characterized by their novelty and absence of prior detection within the cybersecurity community. These attacks have gained prominence due to their association with high-profile breaches in recent years. Zero-day malware typically capitalizes on vulnerabilities that are newly discovered, exploiting novel malware variants and initiating the attack before effective defense mechanisms can be implemented. Consequently, these malicious programs swiftly exploit targeted vulnerabilities, introducing the malicious code into the target computer system.

According to the SonicWall 2023 cyber threat report [44], the year 2022 witnessed a record-breaking surge in zero-day malware attacks. The report highlights the identification and mitigation of over one million new malware variants, underscoring the evolving and dynamic nature of the threat landscape. Figure 10.2 demonstrates the yearly new malware variants from 2019 to 2022.

Detecting and mitigating zero-day malware attacks pose significant challenges [14, 40, 18], primarily due to the fact that the vulnerabilities used for penetrating the attacks are discovered in a short time, which leaves minimal time for the development and deployment of patches or software updates. Traditional defense and mitigation solutions are generally built on existing knowledge of known malware variants, rendering them ineffective against never-before-seen malware attacks. It is crucial to note that the defense against malicious cyber threats varies depending on the underlying system, such as Linux or Windows. Customized strategies tailored to the specific system characteristics are essential for achieving seamless and effective protection. In the upcoming sections, we will explore a spectrum of malware detection techniques, encompassing both traditional and emergent methods.

10.3 Malware Detection Techniques

There are two systematic approaches to detecting malware: static and dynamic [2, 45]. The static detection method analyzes the malware code's internal structure to find out abnormal functions and code sequences without really running the code. The dynamic approach traces the behaviors of running the malware executable, such as an application's API call sequence and the hardware performance counter values of a running application. In the static approach, the abnormal code patterns for different malware families or variants are converted into signatures and stored in a database of the existing known malware's signatures. When analyzing a sample application on whether it is malicious, it extracts the code pattern to the signature and then matches the signature with the existing known signature in the database.

Although the static method has the advantage of fast detection due to the nature it does not run the malicious code, which can save time when scanning a large number of files, this method is not effective in detecting unknown zero-day malware, which is a prominent threat in recent years. Furthermore, depending on the operating system's specific structure, some code execution behaviors cannot be invoked until apps or software are executed at run time. Thus, it limits the effectiveness of the static approach and is the main motivation for the emergence of a more dynamic approach in recent years by the community of cybersecurity.

Dynamic malware detection approach has improved performance in detecting unknown zero-day malware, largely because the intent and goal of malicious attacks are exposed during running the applications and are captured by dynamic feature extractions. Machine learning has been widely studied and integrated with dynamic feature analysis and extractions in malware detection. The machine-learning-based dynamic approach requires neither a deep dive into each malware variant's internal code structure nor a concrete mathematics model to find out the patterns of abnormality, and it has shown promising results in defending against malware attacks. In the next section, we will review several essential, state-of-the-art malware detection approaches. Each technique can embed static, dynamic, or a hybrid of static and dynamic approaches.

10.3.1 Signature-Based Malware Detection

The signature-based approach employs the malware's signature in the database to check and examine whether a sample is malware or benign. There are different ways to create a signature. In a nutshell, signatures are created by encapsulating the binary executable's instructions and summarizing them into a binary sequence. Since an executable can contain a large number of instructions, the common practice is to scan the essential part potentially related to the malware detection section in the instructions. This process requires expertise and knowledge of how malware works internally.

There exist various ways to create such signatures, including string scanning, top-and-tail scanning, entry point scanning, and integrity checking [2]. For instance, entry point scanning only scans the entry point of the malware's code since malware usually changes the entry point at the beginning of the code. Most antivirus tools use signature-based detection methods to scan the potential harmful intrusions. It is worthwhile to mention that generating signatures needs experts who understand deeply how malware works [45]. It also demands much man power to analyze the code instructions and extract useful signatures. While security mechanisms work around the clock to catch up with generating more signatures for defending evolving malware variants, malicious hackers develop more advanced malware techniques to conceal their intentions and evade being detected by security mechanisms. More complex malware techniques disrupt the sequence of byte instructions by intentionally breaking the established patterns, resulting in a mismatch with the stored signatures in the database, thus evading security systems.

10.3.2 Behavior-Based Malware Detection

The behavior-based approach analyzes the running applications' behavior to determine whether the application is malicious or not. It first uses monitoring tools to observe and collect behavioral data, like system call sequences, registry snapshots, file changes, process monitoring, performance counters' data, and network activities. It leverages the fact that running applications leave footprints in all layers of a computer system. A machine learning approach can be employed to analyze and detect the discrepancy of the behavioral differences between malware and benign software. Such an approach first explores what behaviors are effective features that can effectively distinguish such discrepancy. After that, a monitoring tool is involved and behaviors are traced and monitored. Selected top features are extracted from the traced behaviors using various feature extraction methods. At last, a machine learning classifier that will make decisions is developed and trained.

10.4 Machine Learning Algorithms

Machine learning (ML), a branch of Artificial Intelligence (AI), aims to teach computers to program themselves using data to accomplish various tasks. Machine learning algorithms are typically categorized into four major types: supervised learning, unsupervised learning, semi-supervised learning, and reinforcement learning. Supervised learning deals with labeled data, unsupervised learning handles unlabeled data, and semi-supervised learning uses a combination of both labeled and unlabeled data. Reinforcement learning, on the other hand, does not directly use data but learns from interactive experiences with the environment, often tackling complex classification or control tasks.

Tables 10.1, 10.2, and 10.3 list various popular algorithms for supervised, unsupervised, and reinforcement learning techniques. These AI algorithms span various categories and can be applied to address diverse challenges including cybersecurity problems.

Table 10.1 A list of supervised machine learning algorithms

Algorithm	Description
Naive Bayes (NB)	Probabilistic classifier based on Bayes' theorem
Linear discriminant analysis (LDA)	Dimensionality reduction and classification
Logistic regression (LR)	Classification
K-nearest neighbors (KNN)	Traditional ML for classification or regression
Support vector machine (SVM)	Using a hyperplane in high-dimensional space to separate data points into different classes, for classification or regression
Decision tree (DT)	Hierarchical tree-like structures used for both classification and regression tasks
Random forest (RF)	Ensemble learning method combining multiple decision trees for improved performance, for classification and regression
Adaptive boosting (AdaBoost)	Meta algorithm for classification or regression
Extreme gradient boosting (XGBoost)	Builds an ensemble of weak learners, for classification or regression
Stochastic gradient descent (SGD)	Traditional ML for classification
Multilayer perceptron (MLP)	Fully connected multilayer NN
Convolutional neural network (CNN)	DNN for classification, identification, etc.
Recurrent neural networks (RNNs)	DNN for time series and sequential data
Long short-term memory networks (LSTMs)	DNN for time series and sequential data
Autoencoder (AE)	DNN for dimensionality reduction, noise reduction, anomaly detection, can be semi-supervised, supervised, unsupervised learning
Restricted Boltzmann machines (RBM)	DNN for dimensionality reduction, classification, feature learning, etc.
Deep belief network (DBN)	DNN for classification of image or nonimage data
Generative adversarial network (GAN)	DNN for generating synthesized data
Simple and multiple linear regression	Regression
Polynomial regression	Regression
LASSO and ridge regression	Regression

10.4.1 Classical Machine Learning

This subsection briefly describes some traditional/classical machine learning classifiers that are commonly employed for cybersecurity applications. We selected these machine learning models because they are from different branches of ML covering a diverse range of learning algorithms.

Decision tree (DT) is a sequential supervised learning model (a.k.a. divide and conquer algorithm), which logically combines a sequence of simple tests where a numerical attribute is compared against a threshold value or a set of possible values.

Table 10.2 A list of unsupervised machine learning algorithms

Algorithm	Description
K-means clustering	Dividing data points into clusters based on their similarity
Mean-shift clustering	Clustering
Density-based spatial clustering of applications with noise (DBSCAN)	Density-based clustering
Gaussian mixture models (GMMs)	Data clustering
Agglomerative hierarchical clustering	Clustering
Self-organizing Maps (SOMs)	DNN for clustering
Autoencoder (AE)	DNN for dimensionality reduction, noise reduction, and anomaly detection, can be semi-supervised, supervised, unsupervised learning
Latent Dirichlet Allocation (LDA)	Topic modeling, document clustering/summarization, information retrieval, feature extraction
Principal component analysis (PCA)	Dimensionality reduction

Table 10.3 A list of reinforcement learning algorithms

Algorithm	Description
Monte Carlo methods	Model free
Q-learning	Off-policy, model free
Deep Q-learning	Off-policy, model free
Actor critic	Off-policy, model free
State-action-reward-state-action (SARSA)	On-policy, model free
Proximal policy optimization (PPO)	On-policy, model free
Deep deterministic policy gradient (DDPG)	Off-policy, model free
AlphaGo	Model based

It is essentially a flow chart-like structure where each internal node denotes a test on an attribute with each branch representing an outcome of the test and each leaf holding a class label, as shown in Fig. 10.3. A decision tree is a strong algorithm but tends to overfit data when the dataset has many features, so feature selection and dimensional reduction of a dataset are essential to avoid overfitting for DT. Meantime, DT is sensitive to the data, and different inputs result in distinctive tree structure.

Random Forest (RF) is an ensemble machine learning algorithm that grows multiple randomized decision trees, which are then merged to produce a more accurate prediction. It builds the tree by splitting on a random subset of features as shown in Fig. 10.4. At each split, the model selects only a small subset of features randomly with replacement to create a sub-tree, a process known as bagging or bootstrap aggregation. The randomly created sub-trees have no correlations among each other, ensuring model diversity among sub-trees. All sub-trees together produce a final prediction, so its model tends to have a low bias and a moderate variance. RF is a robust machine learning algorithm that can properly handle imbalanced datasets.

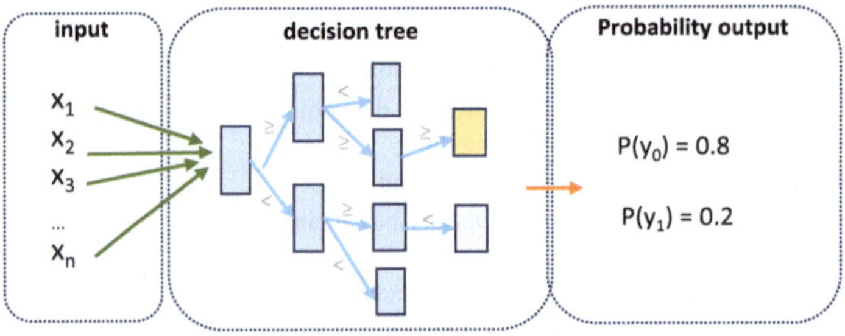

Fig. 10.3 Architecture of decision tree

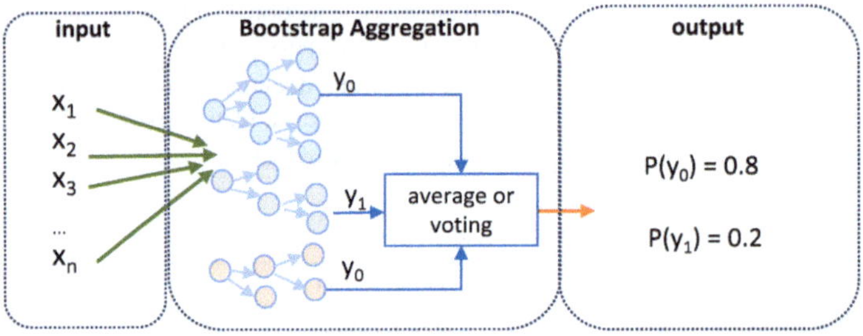

Fig. 10.4 Architecture of random forest

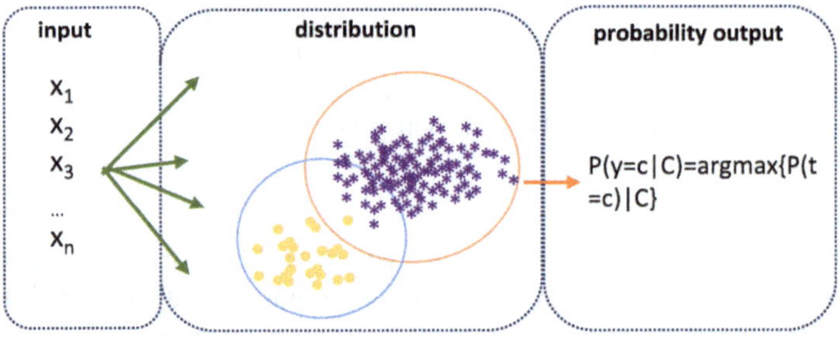

Fig. 10.5 Architecture of GNB

Gaussian Naïve Bayes (GNB) is a simple classification technique based on the Bayes theorem. It can handle a high-dimensional dataset with a smaller model size. Complex classification problems can also be implemented by using the Gaussian Naive Bayes Classifier. GNB supports continuous data, assuming that the feature sets associated with each class are distributed according to a Gaussian distribution. As shown in Fig. 10.5, a class prediction probability is calculated by the class prior and class conditional density considering the class-specific mean and covariance.

Stochastic Gradient Descent Classifier (SGD) is a linear classifier optimized using stochastic gradient descent, an efficient method for fitting linear classifiers and regressors under convex loss functions, including (linear) Support Vector Machines

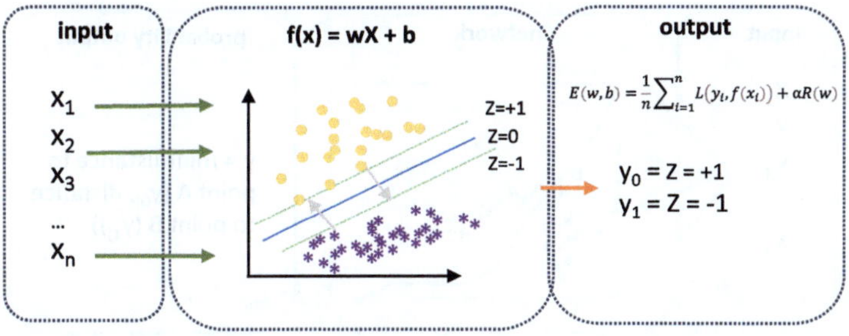

Fig. 10.6 Architecture of SGD

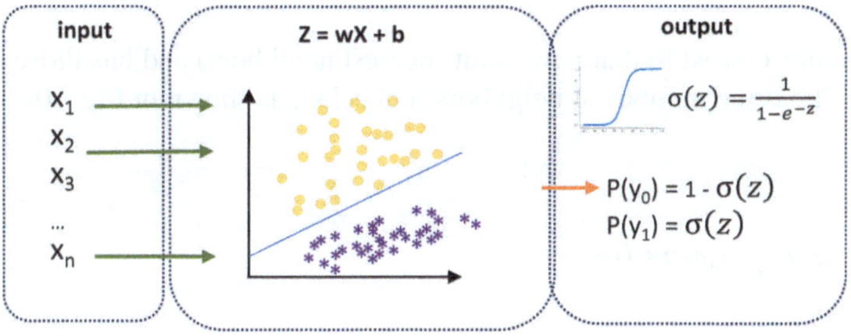

Fig. 10.7 Architecture of logistic regression

(SVM) and Logistic Regression (LR). In stochastic gradient descent, the algorithm randomly selects a batch of data points from the entire dataset to compute the gradient descents. By default, Scikit Learn implements a stochastic gradient descent learning routine over a linear SVM, which supports various loss functions and penalties for classification. This method allows mini-batch learning, where each batch of samples is used to estimate the gradient of the loss function. The model is continuously updated with a scheduled decreasing learning rate. SGD is an efficient machine learning algorithm, though it is sensitive to feature scaling. It relies on several parameters, such as regularization parameters, iterations, mini-batch sizes, and learning rates. Its architecture is briefly represented in Fig. 10.6.

Logistic Regression (LR) is a linear statistical regression-based algorithm used for analyzing a dataset in which there are one or more independent variables that determine an outcome. The goal of this algorithm is to find the best-fitting model to describe the relationship between a dependent variable (response or outcome variable) and a set of independent (predictor or explanatory) variables as shown in Fig. 10.7. In particular, the logistic regression algorithm models and refines the relationship between features and labels iteratively using error measures and output probabilities of classes.

K-Nearest Neighbor (KNN) is a supervised classification algorithm that takes labeled points and uses them to learn how to classify other points by looking at the

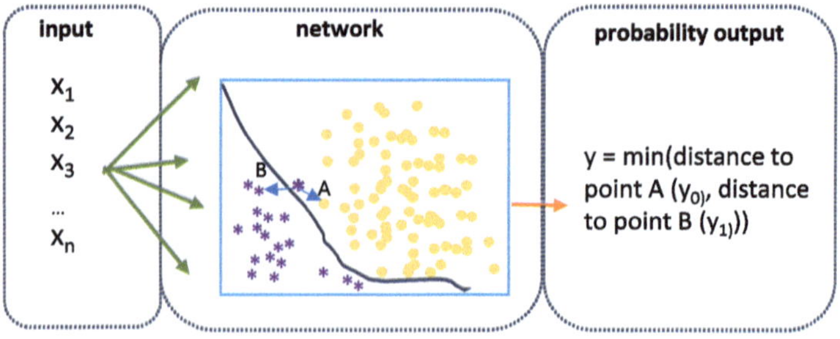

Fig. 10.8 Architecture of KNN

labeled points closest to that new point (nearest neighbors) and has those neighbors vote (the "k" is the number of neighbors it checks), as shown in Fig. 10.8.

10.4.2 Deep Learning

Traditional ML classifiers are primarily used to train on structured data such as tabular data stored in CSV files or relational databases. On the other hand, Deep Neural Networks (DNNs) are most commonly used on unstructured data such as images and natural language processing. Computer vision-based DNN models can recognize hierarchical relationships in analyzing simple to complex features and characterize the visual system as a hierarchical and feed-forward system. While the neurons in the early layers of a DNN have small receptive fields and are sensitive to local features, they can capture more generalized patterns in deeper layers.

Deep learning achieves great success in image-based and language processing applications with a growing number of new model architectures and learning techniques introduced frequently. In general, DL can be further categorized into several types. The following sections will describe the most popularly used deep learning algorithms.

10.4.2.1 Multilayer Perceptron

Multilayer Perceptron (MLP) [29] is one type of Neural Network, often used as a baseline for more complex deep learning architectures. MLP refers to such an architecture that has one or more hidden layers between input and output layers, and nodes are fully connected from the prior layer to its next layer to allow the information flow from lower to higher layers, noticing neurons in the same layer are not connected. Figure 10.9 shows an MLP architecture with three hidden layers, and one input and one output layer. Learning of an MLP is through a four-step process explained as follows to adapt the weights of the connections to obtain a minimal

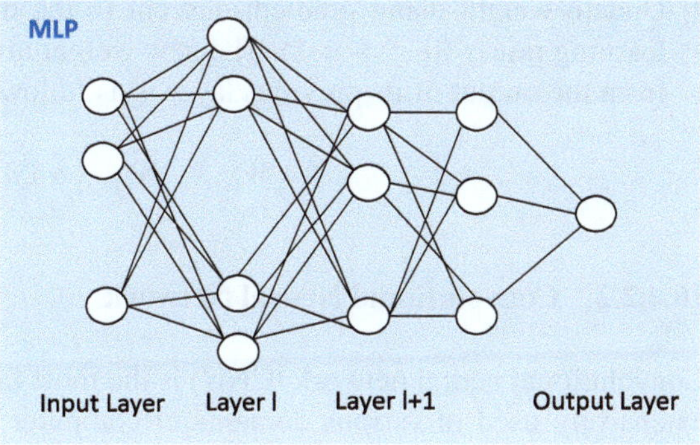

Fig. 10.9 An MLP architecture with three hidden, one input, and one output layer

difference (loss) between the network output (prediction) and the desired output (label):

1. Forward pass: It applies input data to each hidden layer with its weight matrix (W) and bias (b) in a forward direction until the output layer generates the output result. Each hidden layer computes its output (z^l) and then applies an activation function (f(z)), which is an input to its succeeding hidden layers. A simple notation can be written as follows:

$$z^l = W^l a^{l-1} + b^l a^l = f(z^l), \tag{10.1}$$

where f is the activation function (e.g., Sigmoid, Tanh, ReLU, leakyReLU).
2. Backpropagation: While forward pass is to compute from the input layer to the output layer, backpropagation is to compute the error ($E(w) = \sum_{n=1}^{\infty}(J(y_i, o_i))$ between the predictive output (o_i, from forward pass) and the label (y_i) back to the input layer. The error is defined as the loss function, which differs in regression (e.g., Mean Square Error) and classification problem (e.g., Entropy). The loss function can be selected from the existing algorithms or a tailored one to solve a specific problem. For the binary classification problem, the loss function (if using binary cross-entropy) is defined as follows:

$$J(y, o) = -y \log o - (1 - y) \log(1 - o). \tag{10.2}$$

3. Finding gradient descent: This step calculates the partial derivatives of the error for each node ($\frac{\partial(E)}{\partial W_{i,j}^l} = \frac{J(y,o)}{\partial W_{i,j}^l} = \delta_i^l a_j^{l-1}$), starting from the output layer to the input layer, layer by layer; δ for node i is calculated as follows, and f' is the activation function used for layer l.

$$\delta_i^l = f'(z_i^l) \sum_j (W_{ji}^{l+1} \delta_j^{l+1}). \tag{10.3}$$

4. Update weight using gradient descent in the next round of forward pass by a learning rate α ($0 < \alpha < 1$). The new weight after applying the gradient descent from the output of its previous layer is as follows:

$$W_{ij}^{l} \leftarrow W_{ij}^{l} - \alpha \delta_i^l a_j^{l-1}. \tag{10.4}$$

10.4.2.2 Convolutional Neural Network

Convolutional neural network (CNN) is the most famous deep learning architecture extensively used in various domains of computer vision and language processing applications. Compared with traditional machine learning and MLP, its main advantage resides in its ability to automatically extract relevant features without human intentions. Meantime, CNNs have shown to be easy to build a new architecture by stacking customized layers of its building blocks that encourages various experiments, and thus state-of-the-art results in the ImageNet Large-Scale Visual Recognition Challenge (ILSVRC) [35]. Several most famous CNN architectures include LeNet-5 [25], AlexNet [24], VGG-16, Inception-v3, ResNet [13], Xception, and SqueezeNet. The following section will briefly review its building blocks for CNN architectures.

A typical CNN architecture contains the following layers:

- Input layer: It is formed of a multidimensional matrix. For image-based input, it contains a batch (e.g., 32 images per batch) of images. Each image is an RGB image or a grayscale image. The RGB image is a 3-dimensional matrix, and the grayscale image only has a 1-dimensional matrix. Considering the batch size of 32, the input of 32 RGB images with 256 resolution is a 4-dimensional matrix of [32, 256, 256, 3].
- Convolutional layer: The convolutional layer is the core building block of a convolutional neural network (CNN) and is where most computations take place. It involves a few key components: input data, a filter, and a feature map. To detect features in the input images, a feature detector (also known as a kernel or filter) moves across the receptive fields of the image. By performing convolution operations (essentially dot products), the feature map is generated as the output from these dot products. The dot product of the kernel function is applied to an input (e.g., an RGB image represented as a 3D matrix), and the results are aggregated to produce the feature map in a 2D format.

  In many well-known CNN architectures, it is common to see multiple convolutional layers stacked together. This stacking creates a hierarchical structure, where later layers can observe the pixels within the receptive fields of the preceding layers. The earlier layers detect lower-level patterns in the input image, while the later layers perform higher-level feature extraction, combining the lower-level features from previous layers. Ultimately, the convolutional layer transforms the image into numerical values, enabling the neural network to interpret and extract relevant patterns.

- Pooling layer: Pooling layers, also known as downsampling layers, perform dimensionality reduction, decreasing the number of parameters in the input. Similar to convolutional layers, pooling layers use a filter that sweeps across the entire input. However, unlike convolutional filters, pooling filters do not have weights. Instead, the kernel applies an aggregation function to the values within the receptive field to populate the output array. The two main types of pooling are Max Pooling and Average Pooling.
- Fully connected (FC) layer: The fully connected layer typically appears in the final layers before the last activation function in CNNs. Adding an FC layer is a cost-effective way to learn nonlinear combinations of high-level features represented by the output of the convolutional layers. In the fully connected layer, each node in the output layer is directly connected to every node in the preceding layer.
- Classification layer: Following the FC layers, the network leverages an activation function (e.g., a Softmax activation function) to output a probability distribution across the number of output classes. The Softmax function turns a vector into a vector of n class probability distributions that sum to 1. For k output classes where z_i is the input vector (output of the last FC layer), the Softmax function can be expressed as follows:

$$\text{Softmax}(z_i) = \frac{e^{z_i}}{\sum_{j=1}^{k} e^{z_i}}. \tag{10.5}$$

10.4.2.3 Recurrent Neural Network

Similar to a feed-forward network, a Recurrent Neural Network (RNN) [41] contains input, hidden, and output layers. Its difference resides in its sequence-embedded structure that contains repeated same structured layers for processing sequential data, and these layers can hold memory from the previous layer(s) and then pass to the succeeding layer(s). This embedded structure is designed for processing the sequential data that can be suitable for a range of applications, such as voice processing and natural language processing. However, a regular RNN architecture has a main issue. It is sensitive to the exploding gradient and vanishing problems, resulting in the network not learning as designed from long sequence data due to the derivatives either exponentially exploding or decaying during the backpropagation process that caused its inability to learn from the sequence data.

To address this issue, several improved versions of RNN architectures are proposed like Long Short-Term Memory (LSTM). It modifies the standard RNN architecture to have four neural network layers in each cell instead of one to better handling of long-term dependencies. The implementation in LSTM adds input, forget, and output gates to manage the output at each layer neither exploded nor decayed. A forget gate is added at the first Sigmoid function that takes in the previous hidden layer's output h_{t-1} and the input sequence data at step t x_t to output forget (0) or keeps (1) the information. The information flows through input, forget,

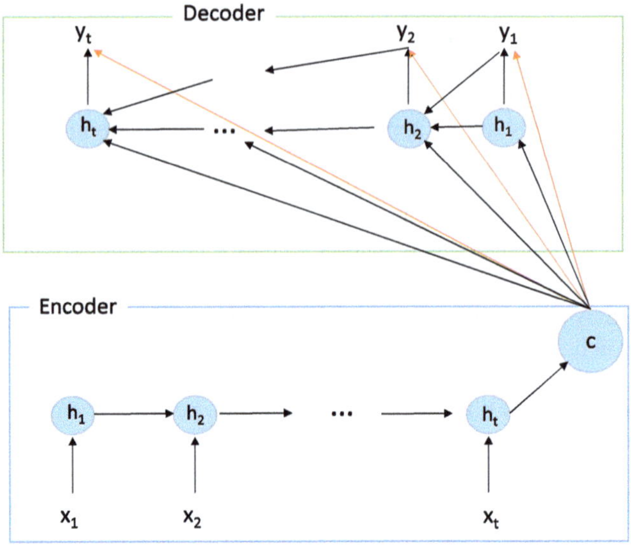

Fig. 10.10 An encoder-decoder architecture [5]

and output gates to decide what information is read, forgotten, and stored in the cell state. At the last layer, the Tanh function pushes the output value to be with -1 and 1 to maintain the gradient descent within a valuable range (not exploded or vanished).

Encoder-Decoder There are several variants of encoder-decoder architectures like large deep LSTM and bidirectional-RNN-based encoder-decoder to improve the translation performance. Encoder-decoder network builds foundations for many later sequential data processing architectures. As depicted in Fig. 10.10, the network employs an encoder and a decoder each is a separate network based on RNN or LSTM to process the sequence data. The encoder reads the input $x = x_1, x_2, ..., x_t$ sequentially. It contains hidden states (h) and outputs, notated a c, a summary of the sequential output of the encoder. Specifically, the output of h is updated by $h(t) = f(h(t-1), x_t)$ at each time step that f is the activation function such as a sigmoid function or an LSTM cell. The encoder outputs a vector that represents the input as a c.

The decoder employs a similar approach but is trained to predict the next word based on the current word (output of the current hidden state). It takes the input sequence summary c, the last step's hidden state, and output from the last step's hidden state to output its current step's word, updated by $h(t) = f(h(t-1), f_{t-1}, c$.

10.4.2.4 Large Language Models and Transformers

Large Language Models (LLMs), such as ChatGPT, are powerful applications of deep learning designed to handle and generate natural language. The architecture of these models is primarily based on Transformers, which employ a self-attention mechanism that significantly enhances the practicality and efficiency of natural language processing tasks. This section reviews the Transformer architecture and its applications in LLMs.

The transformer model was first introduced by Google Brain in 2017 [49]. Unlike previous models that relied heavily on Convolutional Neural Networks (CNNs), Recurrent Neural Networks (RNNs), or Long Short-Term Memory (LSTM) networks, the Transformer architecture uses an encoder-decoder structure that is solely based on attention mechanisms to handle the encoding and decoding processes. Recall that an attention function is to map a query and key-value pairs to an output, which is a pure matrix computation process. Among the several attention structures, their work uses the multi-head attention mechanism that linearly projects queries, keys, and values h times (a parameter to tune) to dimensions of d_k, d_k, and d_v, respectively, followed by applying a Scaled Dot-Product Attention that follows the formula defined in Eq. 10.6. After computing the attention for each head, the resulting matrices are concatenated into a single matrix and linearly transformed to produce the final multi-head attention output. This output matrix contains the weighted sum for each value, with each weight reflecting the relatedness of the query to the corresponding key.

$$\text{Attention}(Q, K, V) = \text{Softmax}\left(\frac{QK^T}{\sqrt{d_k}}V\right). \tag{10.6}$$

The transformer architecture contains six layers of a multi-head attention mechanism for the encoder and six layers of a modified version of the multi-head attention mechanism (by adding a feed-forward layer) for the decoder. Notice that the attention mechanism itself does not recognize the positions of each word in the sentence. To know each word's position, the Transformer architecture implements position encoding in the input embedding of the encoder and the output embedding of the decoder. A Transformer network using a self-attention mechanism reduces the computation complexity per layer from $O(n \cdot d^2$ to $O(n^2 \cdot d$, where n is the sequence length, and d is the representation (input) dimension. Other benefits include enabling training parallelization (due to the h-head nature) and shortening the path length between long-range dependencies. The Transformer model's attention mechanism allows it to efficiently handle relationships in data, making it particularly effective for natural language processing tasks. In recent years, various Transformer architectures have been developed including ChatGPT. Their applications range from language translation, sentiment understanding, question and answering, content recommendation, code generation, and many more.

10.5 Machine Learning-Enabled Malware Detection

Table 10.4 provides a comparison between traditional malware detection approaches and ML-based methodologies. Traditional approaches, such as signature analysis,

Table 10.4 Comparison of traditional malware detection approach with ML-based malware detection approach

Metric	Traditional approach	ML-based approach
Cost	Man power to find signature patterns, collect and extract new signatures for storing in the database	Time to train and continuously update the ML classifier with new data, complex classifier takes longer training time
Knowledge	Domain knowledge of how malware is created, knowledge is harder to transfer among malware variants	Domain knowledge of how ML works for finding patterns of malware, knowledge is easy to transfer among malware variants
Defense time	Can detect known malware attacks; usually defense after zero-day malware attacks	Advanced ML detectors can guard the system before known and zero-day attacks
Detector update pattern	Linear, continuously expand the signature database with new malware variants	Iterative, continuous improvement of ML detector with new malware variants and amount of data
Resistance to stealthy malware	Analyze stealthy techniques, create new ways to extract meaningful signatures, usually ineffective	Analyze root causes of stealthy techniques, implement advanced ML classifiers in the detection pipeline, can be effective

API call sequence graph analysis, and Opcode analysis, are effective for detecting known malware but struggle with unknown, zero-day threats. Signature-based analysis, in particular, requires significant man power to analyze and collect signatures for new malware variants. Given the rapid mutation of malware, with approximately 1,300 new variants emerging daily [44], this method is increasingly inadequate. The signature-based approach is inherently reactive, always trying to catch up with hackers, which makes it insufficient for protecting critical computer systems in real time.

Machine-learning-based methodologies can overcome the limitations of traditional conventional malware detection methods due to their capabilities of detecting more complex, unknown malware variants. If implemented effectively, ML-based methods can protect systems even before new malware variants are created and deployed by attackers [7, 47]. Although updating ML classifiers with new data can be costly, especially for complex models like deep learning models, the knowledge gained from creating and enhancing ML detectors is typically easier to transfer between different malware families and variants. This makes the ML-based approach ideal for defending against today's large volume of rapidly evolving malware threats. Engineers with expertise in ML can continuously develop high-performance anti-malware detectors with some understanding of malware's general behavior. Additionally, the ML-based malware detection approach can be combined with both static and dynamic feature extraction methods.

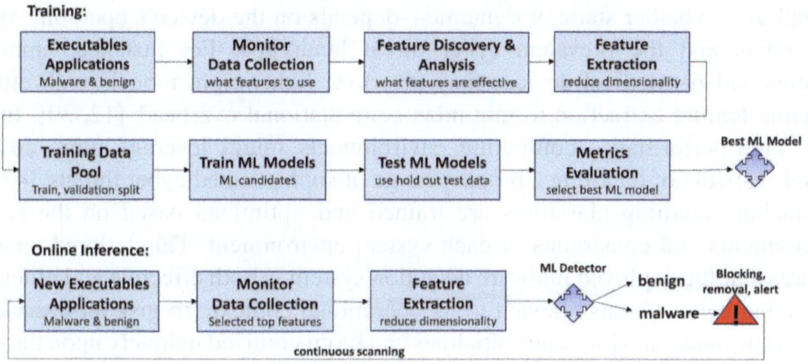

Fig. 10.11 Top: process for training ML classifiers; bottom: process for ML online inference

10.5.1 Process of ML-Enabled Malware Detection Approach

ML-based malware detection approach typically involves two phases as shown in Fig. 10.11. The top of the figure depicts the ML training process, while the bottom demonstrates the online defense process after a well-trained ML detector is deployed to the system.

In the training process, multiple machine learning classifiers are trained based on existing known malware and benign applications. Depending on the feature extraction approach to be used, either static or dynamic, features can be monitored and collected from the executables or the running applications. These features are analyzed with the most relevant features to the label of malware or benign. High-dimensional data are further reduced to lower dimensions to reduce the resources needed to train the model, particularly reducing the training time for the initial model and the subsequent model updating phase. A hold-out test set is established together with the training set. A list of ML classifiers is trained and evaluated over the hold-out test data. The best-performed ML model is selected as the online defender.

In the phase of online inference (defense), the best-performed ML classifier is deployed to the operational environment. Following a similar process as training (without feature analysis as it has been done during training), incoming monitored data and dimension-reduced data (through feature extraction) are produced in real time. The data is constantly fed to the ML defender to scan whether it is a potential intrusion. If malware is detected, the system can be alerted, and a removal or blocking process can be initiated to defend against potential attacks.

While the aforementioned ML-assisted malware detection process provides a general framework of a machine learning pipeline, it can be implemented differently to fit the unique circumstances of the system environment. This adaptation involves considering specific characteristics of the targeted applications, such as malware families and their stealthiness, as well as the operational platforms such as Android, Windows, Linux, and IoT devices [26, 36, 37, 38, 20, 32]. The choice of features to analyze—whether static or dynamic—depends on the device's operating system constraints and the prevalent cyber threat landscape. For instance, embedded systems and biomedical devices may prioritize lightweight models with efficient dynamic feature extraction to minimize computational overhead [12, 39]. In contrast, high-performance computing environments might leverage more complex models capable of handling a broader range of sophisticated cyber threats [48].

Machine learning classifiers are trained and optimized based on the specific requirements and constraints of each system environment. This tailored approach ensures that the deployed malware detection system is both effective and efficient in mitigating cyber threats relevant to its operational context. To give an overview of how ML training and inference pipelines can be customized uniquely upon the target system, the following section will briefly review various integrated approaches running in three different operation systems, combined with static or dynamic feature analysis methods and various machine learning algorithms.

10.5.2 ML-Based Malware Detection in Windows Systems Using Strings Extracted from PE Files

Figure 10.12 illustrates the ML-assisted malware detection process focused on identifying malicious Windows Portable Executable (PE) files [50]. This method employs static feature collection and analysis techniques. Initially, it scans PE files to extract API function calls from the Import Table, forming interpretable strings that include API calls and strings with semantic meaning. These interpretable strings serve as static signatures stored in a database for training the ML classifier.

During the training phase, a subset of hold-out validation data is reserved from the training pool to evaluate model performance. The training data are sampled using a bootstrapping technique, where each iteration randomly selects 70% of the total training data to train diverse classical ML models. This bootstrapping, or bagging, approach ensures that each subset of sampled data is independent, enhancing the robustness of the ML models. In the inference phase, the ensemble of trained models employs a majority voting mechanism to classify new sample data as either malware or benign, leveraging the diversified detection capabilities of the individual ML models.

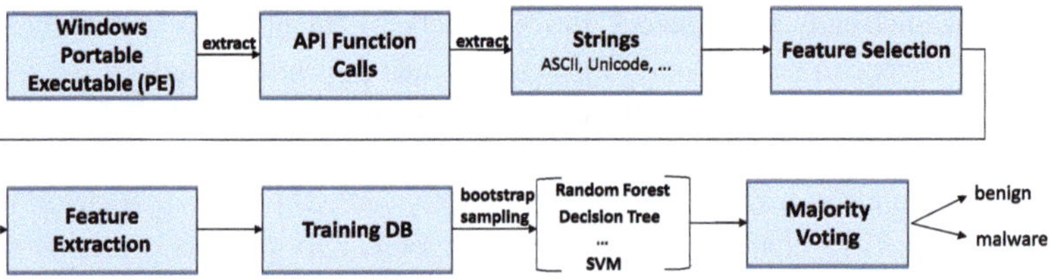

Fig. 10.12 Illustration of Window PE files malware detection combined with static, signature-based feature analysis

10.5.3 Android Malware Detection Using Machine Learning and a Hybrid Feature Analysis Approach

Android applications have become a key target for plotting malicious attacks due to their open-source nature and the large number of devices used worldwide. Hence, the ability to detect an app to be installed initially and during its life cycle is essential for safeguarding this massive device consumer community. Early prior works usually use static features in the malware code to detect any abnormalities, and such static feature analysis methods include known suspicious information flow, calling structure, code patterns, requests for excessive permissions, and API invocations usually used by malicious code [4, 27]. However, this static method could not obtain a full picture of the Android application due to the characteristics of Android OS. Android OS handles the application's behavior based on the events needed when running the application. Some functions such as runtime control, data flow, permissions, app life cycle, and GUI handling can only occur at run time and are heavily dependent on runtime context, which is not able to be estimable solely on static analysis of code itself. As such, dynamic analysis and malware detection using machine learning are more suitable for Android malware detection.

During run time, it is important to collect impactful features on a device while initially installing an application and continuously scan the running apps (after installation) for detecting potential harmful malware (hackers can plot the malicious behavior at a later time). Malware classifiers can be trained offline and are updated to the on-device malware detector/protector on Android phones. The top section in Fig. 10.13 demonstrates one example of an Android on-device malware detection process. Generally speaking, an ML detector suits well for an on-device detector. This can be set as an automated process by the manufacturer, and the user can have the option to turn it off.

Meanwhile as shown in the bottom section of Fig. 10.13, the process can be a combination of static and dynamic, on-device, and cloud-based solutions. At first, on-device apps monitor can scan and collect the predefined impactful features (e.g., API calls) when apps are running. It then feeds the features to an on-device ML defender to scan for potentially harmful code alerts. Meantime, more diversified features can be collected in the background and sent to cloud-based scanning [4, 27]. Finally, a scan report can be combined from both on-device and cloud-based solutions to the device manager, which can block an application or warn the user about potential harm.

10.5.4 Arm-Based IoT Malware Detection Using Byte Sequences from ELF Files

ARM-based processors are widely used in IoT and embedded devices, including ARM Cortex-A processors, Cortex-M processors, and Cortex-R processors. To train

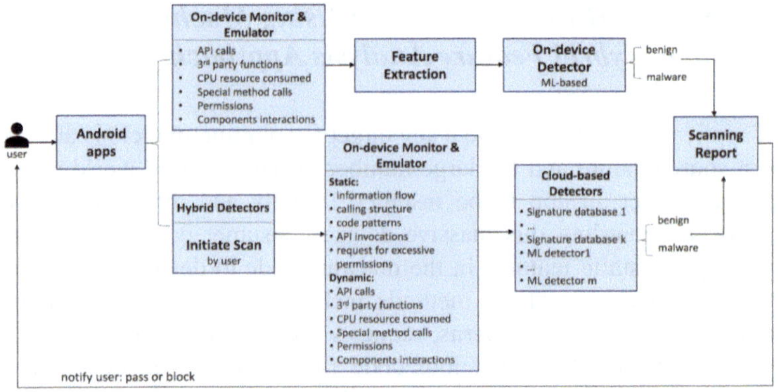

Fig. 10.13 A hybrid approach of combining on-device and cloud scanning for Android malware detection using dynamic features

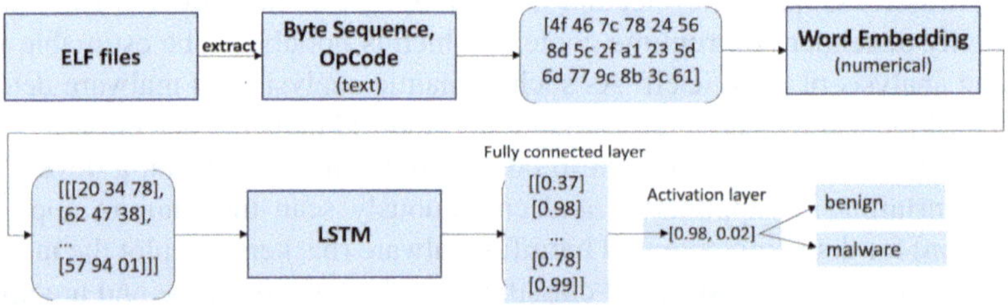

Fig. 10.14 IoT malware detection using ELF and LSTM algorithm

a corresponding malware detector for IoT, using malware and benign applications run on ARM-based processors is a suitable data collection method. For offline ML model training purposes, malware data can be collected through the Virus Total Threat Intelligence platform.

Figure 10.14 shows the process of extracting features in the code structure (e.g., Opcode, byte sequence, etc.) when executing malware and benign applications on ARM-based platforms. The features are extracted using static analysis. The byte sequences or Opcode are considered as string type data, while machine learning often needs numerical data. Considering the unique situation of sequential data of byte sequences (or Opcode), word embedding techniques are used to convert the string format of byte sequences into high-dimensional, numerical data that is suitable for the input of a Long Short-Term Memory (LSTM) model [10]. The LSTM model outputs a sequence output. To make it a classification problem, this method adds one or more fully connected layers at the end of the LSTM layers. Finally, an activation function is applied to perform binary or multiclass classification [34].

10.5.5 ML Performance Evaluation Metrics

Assessing the performance of machine learning classifiers is a crucial step in developing effective ML-based malware detection techniques. In machine learning and statistics, there are various measures available to evaluate the performance of a detection method, providing insights into its detection accuracy. Table 10.5 summarizes the standard evaluation metrics used for performance analysis in malware detection and classification. In this section, we briefly describe each evaluation metric.

In the field of statistical machine learning, a confusion matrix is a specific table that represents the prediction performance of a machine learning classifier. It consists of two dimensions, "actual" and "predicted," each containing identical sets of "classes." Each row of the confusion matrix represents the instances in a predicted class, while each column represents the instances in an actual class, or vice versa.

A confusion matrix is constructed from the four outcomes generated by a binary classification. A binary classifier predicts each data instance in a test dataset as either positive or negative. This classification (or prediction) results in the following four outcomes:

- True Positive (TP): Correct positive prediction; malicious applications that are correctly classified as malware
- False Positive (FP): Incorrect positive prediction; benign applications that are incorrectly classified as malware
- True Negative (TN): Correct negative prediction; benign applications that are correctly classified as benign
- False Negative (FN): Incorrect negative prediction; malicious applications that are incorrectly classified as benign

In malware detection performance evaluation, malicious application samples are typically considered positive instances. Consequently, the True Positive Rate (TPR), or hit rate, represents sensitivity, which is the proportion of correctly identified positives. Essentially, it measures how effectively the classification model identifies malware samples. The True Negative Rate (TNR), also known as specificity, measures the proportion of correctly identified negatives. Additionally, the False Positive Rate (FPR) indicates the rate at which benign files (i.e., negative instances) are incorrectly classified as malware.

The F-measure (F-score) in machine learning is a weighted average of precision (p) and recall (r). Precision is the ratio of true positives to the total predicted positives, while recall is the ratio of true positives to all actual positives. Precision and recall alone are insufficient for demonstrating detection performance and can sometimes be contradictory, as they do not account for all results and samples. To address this limitation, the F-score (or F-measure) is calculated based on both precision and recall, providing a more balanced evaluation. This makes it particularly resilient to class imbalances in the dataset, which is relevant to our

Table 10.5 Performance metrics used for evaluating ML-based malware detection techniques

Evaluation metric	Equation/Description
True positive (TP)	Count of correct positive predictions (e.g., malware is predicted as malware).
False positive (FP)	Count of incorrect positive predictions (e.g., benign is predicted as malware).
True negative (TN)	Count of correct negative predictions (e.g., benign is predicted as benign).
False negative (FN)	Count of incorrect negative predictions (e.g., malware is predicted as benign).
Specificity: True negative rate (TNR)	$TNR = TN/(TN + FP), TNR = 1 - FPR$, defined as the proportion of genuinely negative samples predicted as negative results among all negative samples.
False negative rate (FNR)	$FNR = FN/(FN + TP), FNR = 1 - TPR$, defined as the proportion of genuinely positive samples predicted as negative results among all positive samples.
False positive rate (FPR)	False alarm rate, defined as the proportion of genuinely negative samples predicted as positive results among all negative samples, $FPR = FP/(FP + TN), FPR = 1 - TNR$.
Recall/sensitivity: True positive rate (TPR)	$TPR = TP/(TP + FN), TPR = 1 - FNR$, defined as the proportion of genuinely positive samples predicted as positive results among total positive samples. It refers to the proportion of correctly identified positives (the rate of malware samples) by the model.
Precision (P)	$P = TP/(FP + TP)$, Defined as the ratio of true positive predictions to total predicted positives, indicating the confidence level of attack detection.
F-Measure (F1-Score)	$Fmeasure = 2 \times (P \times R)/(P + R)$, Represents a weighted average of precision (P) and recall (R). It offers a more comprehensive evaluation than accuracy, considering both precision and recall, and is robust for evaluating both balanced and imbalanced datasets.
Detection accuracy (ACC)	$ACC = (TP + TN)/(TP + FP + TN + FN)$, The ratio of correctly classified samples to total samples. Accuracy is a suitable metric when the dataset is balanced.
Error rate (ERR)	$ERR = (FP + FN)/(P + N)$, where P, N represent total positive and negative samples, respectively. It is the number of all incorrect predictions divided by the total number of the dataset.
Receiver operating characteristic (ROC) curve	Visually depicts the trade-offs between true positive rate and false positive rate, providing insights into detection performance with varying discrimination thresholds. Each prediction result corresponds to a point in the ROC space, and the upper-left corner (coordinate (0, 1)) signifies optimal detection, denoting 100% sensitivity and 100% specificity.
Area under the curve (AUC)	$AUC = \int_0^1 TPR(x)dx = \int_0^1 P(A > \tau(x))dx$, where τ is the thresholds on the decision function used to compute FPR and TPR. AUC quantifies the entire two-dimensional area beneath the ROC curve, spanning from (0,0) to (1,1).

experiments. Detection Accuracy (ACC) measures the rate of correctly classified positive and negative samples.

The Receiver Operating Characteristic (ROC) curve is a statistical plot that illustrates binary detection performance across varying discrimination thresholds.

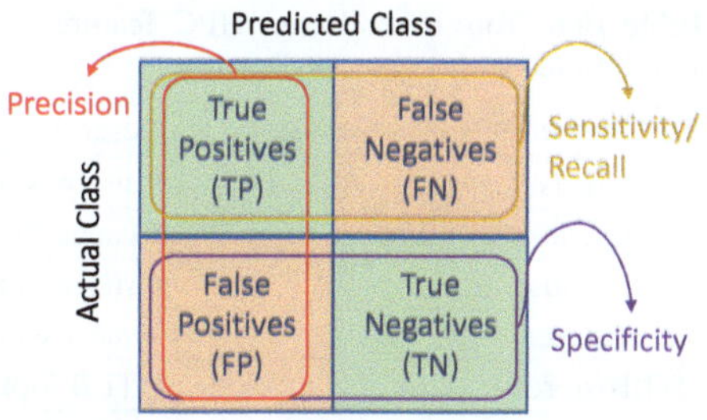

Fig. 10.15 Confusion matrix layout for a machine learning classifier

The ROC space is defined by the False Positive Rate (FPR) on the x-axis and the True Positive Rate (TPR) on the y-axis, helping to identify trade-offs between true positives and false positives. Since TPR and FPR correspond to sensitivity and (1-specificity), respectively, each prediction result is represented as a point in the ROC space. The point at the upper-left corner of the ROC curve represents the best detection result, signifying 100% sensitivity and 100% specificity, also known as perfect detection. The Area Under the Curve (AUC) typically ranges from 0.5 to 1.0, with higher values indicating better detection performance. Different metrics can conflict with each other, making it challenging to achieve high precision and recall simultaneously. Therefore, achieving a balance between precision and recall is crucial, and the F-score serves as a valuable indicator of overall detection performance.

10.6 Image-Based Zero-Day Malware Detection Using Deep Transfer Learning

In this section, we dive into a case study that utilizes multiple techniques of machine learning to assist zero-day malware detection. These techniques at a glance include traditional machine learning and image-based convolutional neural networks, transfer learning, and various performance metrics to evaluate an ML model's performance. We begin by introducing the datasets and detailing the feature engineering process.

10.6.1 Dataset Characteristics

This case study focuses on hardware-assisted malware detection using data from an Intel Xeon X5550 machine. To address non-determinism and overcounting issues in HPC registers, as discussed in recent works [6, 52], we utilized low-level CPU events extracted with the *Perf* tool, employing a static performance monitoring approach. HPC events were sampled every 10ms within Linux Containers (LXC) for isolation. Numerous benign and malware applications were executed, collecting data from real-world applications (MiBench [9], SPEC2006 [19]), Linux

Table 10.6 Top 16 collected HPC features from the Perf tool in this case study and their descriptions

HPC event	Description
LLC-loads	# successful memory load operations in L3 cache
L1-dcache-load-misses	# cache lines brought into L1 data cache
Node-loads	# successful load operations to DRAM
Mem-stores	# successful store in main memory
dTLB-stores	# TLB lookup for a data memory store
cpu/branch-instructions/	# cpu's branch instructions
dTLB-loads	# TLB lookup for a data memory load
cpu/cache-references/	# of references cached
cpu/branch-misses/	# branches mispredicted
cpu/instructions/	# instructions retired
Cache-references	# last level cache references
Instructions	# branch instructions retired
Branch-instructions	# of processor's branch instructions
Branch-loads	# successful branches
Branch-misses	# all retired mispredicted branches
msr/tsc/	# time-stamp counter

programs, browsers, text editors, and nine malware types (VirusShare, VirusTotal) including worms, viruses, botnets, ransomware, spyware, adware, Trojan, rootkit, and backdoor. Each application was run 11 times to capture 44 HPC events, divided into batches of four due to the processor's four available counter registers. Containers were destroyed after each run to prevent contamination, resulting in 16K malware and 30K benign samples. The HPC data monitored and collected are numerical values, which are traces left by the malware or benign applications on the Hardware Performance Counters. Table 10.6 reports a subset of sixteen deployed low-level features captured by HPC registers from the Perf tool under Linux in our experiments and their descriptions.

Once traced HPC data are collected, the next step is to preprocess them to extract meaningful features from them to train effective machine learning models. This is an important step for the ML-assisted malware detection approach, and we introduce this process in the section on Feature Engineering. A brief mention of its benefits is to reduce the redundancy of features, remove useless features, and reduce training time, and this is particularly influential to complex models that retraining the model takes larger computing resources and a longer time. Training can occur in an online or offline manner. In this case study, we apply training in an offline manner and inference in an online manner. Once all candidate models are well trained and tested on a hold-out validation dataset, the best-performed ML model is deployed online to defend against potential attacks. In the online inference scenario, real-time HPC data is monitored, processed in an online, automated preprocessing pipeline to extract useful features, and then fed to the deployed ML model to identify whether it is malicious or not.

10.6.2 Threat Model: Zero-Day Malware Detection

Recent ML-assisted malware detection methods using HPC events have mainly considered two major validation methods including cross-validation and percentage splits to examine the effectiveness of their models. The cross-validation method splits the dataset into $K(1, ..., n)$ folds and selects one of them as a target testing dataset, while the rest of the folds are used for the training dataset. And in the percentage split method, the dataset is divided into two sections based on the percentage setting allocated to training and the other to the testing set. However, the major issue with these validation techniques is that the testing data is split from the large dataset and is part of the same data type used in the training dataset. Hence, such validation methods could not imitate the zero-day or unknown testing scenarios occurring in real-world applications in which the trained machine learning classifiers should have never seen the testing dataset.

To model the zero-day malware threat type in our experiments, among all nine malware types, all four types of malware from rootkit, backdoor, virus, and ransomware are held out as the target zero-day test data. These four types of malware are not presented in the training and known test datasets, and thus, the zero-day malware set is unknown from the training dataset. For benign, 30% of all benign data are set aside as a zero-day test benign dataset. This process of generating the zero-day test dataset imitates the zero-day testing in real-world scenarios where the malware is undocumented in the training database of the detection mechanism. The rest of the five types of malware including Trojan, spyware, botnet, worm, and adware as well as the rest of the benign samples are considered for training and known test purposes, and they are randomly split into 70% for training and 30% for known testing.

The difference between the known test and zero-day test in this study is that the known test data contains the same malware types as the training dataset but with different unseen data and the zero-day test data contains different malware types from the training dataset that are considered as new unknown attacks. After data is split, all types of malware are relabeled as malware and left benign as benign. Notably, the image-based method uses the same datasets during training, known test, and zero-day test same as all classifiers. Also, classical ML models use tabular format data, and the image-based framework employs the image data converted from corresponding tabular data.

10.6.3 Feature Engineering

10.6.3.1 Feature Engineering Workflow

Deciding which features to use for malware detection is the first step in a successful malware detection process. As discussed in previous sections, various features, whether static or dynamic, can be employed individually or collectively in a malware detection mechanism. Feature exploration and analysis are typically conducted offline to identify the most suitable features and feature sets. Once a specific set of features is proven effective for the corresponding detection approach, the process can then be implemented online.

Figure 10.16 illustrates the feature engineering process for both offline and online stages. The primary goal of the offline process is to select and determine the targeted feature set that is effective in enhancing the malware detection approach. Once these feature sets are validated, core features can be extracted using feature dimensionality reduction and extraction techniques. Training machine learning (ML) algorithms are resource-intensive, but by applying feature dimensionality reduction, the training time and cost can be significantly lowered, allowing the ML classifier to be updated more frequently.

During the online inference process, certain components of feature engineering are integrated into the detection mechanism's pipeline before the core features are fed into the ML detector. In the following section, we discuss both the offline and online stages of feature engineering using the traces collected from the underlying hardware performance counters.

As previously mentioned, feature analysis and selection (e.g., evaluating the importance of hardware events) is a crucial step in developing accurate ML models for hardware-based malware detection. Various popular methods for feature importance and selection are available. This study first employs the Mutual Information (MI) method in information theory [23] to analyze and rank the importance of HPC events relative to the target label Y (benign/malware). Next, it examines the correlations among the top 16 features to filter out redundant ones, thereby reducing training costs and improving model performance. Finally, it selects the four most significant features from the top-ranked set to meet the constraint of having only four HPCs available on most processors during run time.

10.6.3.2 Feature Normalization

Feature normalization is a crucial step in scaling tabular data along its column or row values to ensure that features with large values do not dominate the learning process.

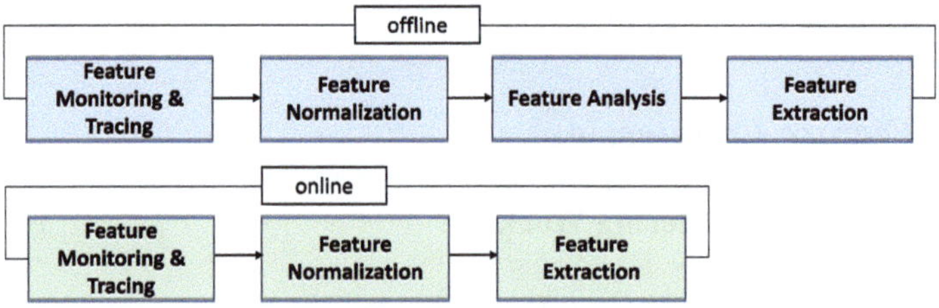

Fig. 10.16 Feature engineering process from feature monitoring, and feature preprocessing to feature extraction. Top: feature engineering process used in the training process, conducted offline; bottom: feature engineering process used in online inference, conducted online

This technique is effective across all machine learning (ML) algorithms and is particularly important for algorithms sensitive to feature distance values. Generally, column-based normalization is more common than row-based normalization, as it helps models converge faster and often results in more effective performance. Scikit-Learn ($sklearn.preprocessing$) is a widely used library for performing feature normalization, offering L1, L2, Max, and MinMax normalization functions [30].
L1 Normalization: is used to scale each row of the data so that the sum of absolute values of each row is 1. The formula is as follows:

$$|x| = \sum_{i=1}^{n} |x_i|. \tag{10.7}$$

L2 Normalization: is to scale the data so that the sum of squares of each row is 1.

$$|x| = \sqrt{\sum_{i=1}^{n} |x_i|^2}. \tag{10.8}$$

Max Normalization: is to scale the data based on the maximum value of each row or column.

$$x = \frac{x_i}{\max(x)}. \tag{10.9}$$

MinMax Normalization: is to scale the data based on the minimum and maximum values of each row or column.

$$x = \frac{x_i - \min(x)}{\max(x) - \min(x)}. \tag{10.10}$$

In this chapter, we process normalization over features (columns) using MinMax normalization ($sklearn.preprocessing.MinMaxScaler$).

10.6.3.3 Feature Importance Analysis

Mutual Information (MI) measures the dependency between two variables [23]. Regarding features X and label Y, the MI measure $I(X, Y)$ is obtained by estimating the marginal entropies $H(X)$, $H(Y)$, and the joint entropy $H(X, Y)$ as follows:

$$I(X, Y) = H(X) + H(Y) - H(X, Y). \tag{10.11}$$

For each data point i, the MI method computes I^i based on its neighboring data points. It first finds the k-closest neighbors falling inside of the distance to point i. Using $\psi(\cdot)$ as the digamma function, N is the total samples, Nx_i is the data sample falling within the distance d with k neighbors, and m_i is the total number of neighbors in the dataset. The estimated MI is defined as below:

$$I^i = \psi(N) - \psi(Nx_i) + \psi(k) - \psi(m_i). \tag{10.12}$$

In this chapter, mutual information (MI) is employed to assess the dependency between each HPC feature and the target label Y. MI quantifies how much each HPC feature contributes to predicting label Y. To achieve this, the Scikit Learn library's *mutual_info_classif* algorithm [30] is used. This algorithm estimates MI based on k-nearest neighbor statistics. The resulting feature importance values for each HPC with respect to Y are then ranked. Specifically, the top 16 features that contribute the most to distinguishing between malware and benign samples are identified. These important features are visualized in a heatmap, as depicted in Fig. 10.17. As demonstrated, mutual information (MI) provides feature importance weight for each feature based on its contribution to the target label Y. A higher weight indicates greater importance of the feature in predicting Y.

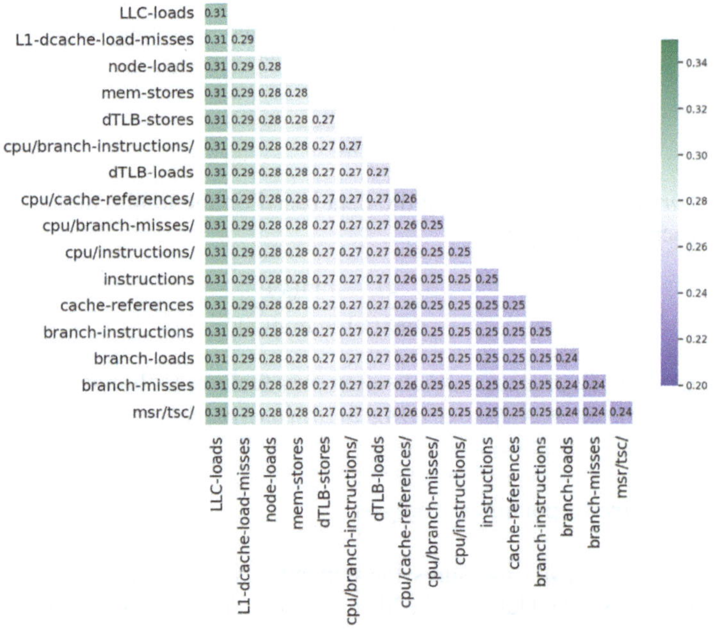

Fig. 10.17 Top HPC features' accumulated mutual information gain to Y. The more green highlights, the greater the contribution of the feature to the label Y

10.6.3.4 Feature Correlations Analysis

The next step in the Feature Engineering process involves a thorough analysis of feature correlations, aiming to elucidate the relationships between pairs of hardware performance counter features. In instances where redundancy becomes apparent, action is taken to eliminate such features from consideration within the feature list. This process begins with the application of Mutual Information, as detailed in Sect. 10.6.3.3, which allows us to identify and prioritize the top 16 features. These top features are those that exhibit the most substantial contributions to the target variable (Y label), rendering an exhaustive examination of all features unnecessary.

Among the top 16 features, we employ Panda's Pearson's correlation coefficient method [28] to conduct a comprehensive correlation analysis, which measures the linear relationship between each pair of HPC features for all data. Less significant HPC features that displayed a correlation coefficient exceeding a threshold of 0.9 with another HPC feature are dropped. Figure 10.18 depicts the heatmap of the HPC features and their correlations. Notably, a vivid red color (excluding the diagonal) on the heatmap signifies the presence of redundant HPC features. For example, "dTLB-loads" was identified as redundant to "dTLB-stores," and given its lesser contribution to the target label Y in comparison to "dTLB-stores," "dTLB-loads" is removed from the feature list. Similarly, "cache-references" was deemed redundant to "node-loads" and was considered less significant, leading to its exclusion from consideration.

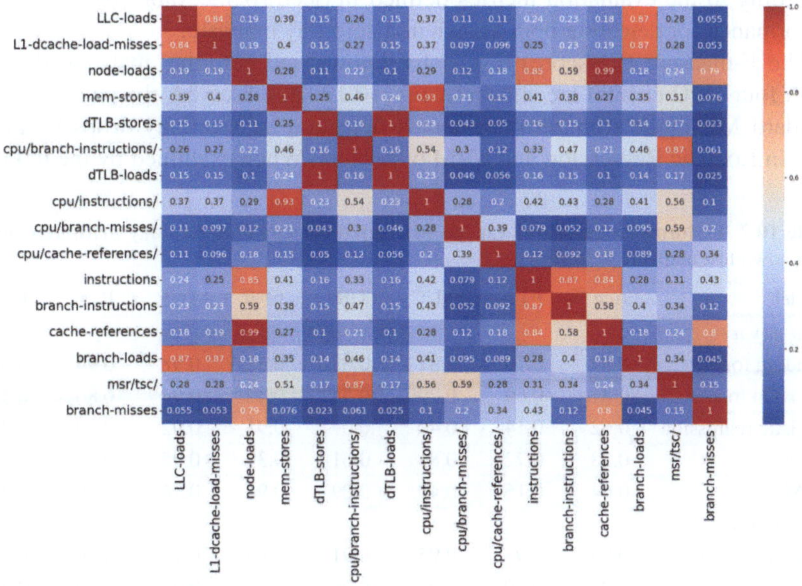

Fig. 10.18 Top 16 HPC features' correlations heatmap

10.6.3.5 Feature Selection

Modern microprocessors can only monitor a limited number of events simultaneously during application execution time [36]. We prioritize the top four hardware events based on accumulated information gains for training a model. As the last step of feature analysis, among the top prioritized features as shown in Fig. 10.17, the top four hardware events that show significant accumulated information gains (regarding label Y) to train a model are selected. If these top features are highly correlated, we drop the least important one and consider adding the fifth top feature. In this dataset, the selected four hardware events include $node-loads$, $LLC-loads$, $L1-dcache-load-misses$, and $mem-stores$. Next, we extract the top four numerical features from the dataset, which can then be processed using Pandas DataFrames [28] and Numpy multidimensional arrays and matrices [11].

10.6.4 Standard ML-Based Malware Detectors

Various ML models are implemented and tested using both known and unknown zero-day datasets to explore the feasibility of the standard learning classifiers in detecting malware based on hardware events. We experimented with and evaluated several widely used ML classifiers across different categories. We used Scikit Learn's library to implement these ML algorithms and evaluated their performance according to the evaluation metrics defined in Sect. 10.5.5. Table 10.7 reports the performance for detecting both known and zero-day malware.

The F-measure (F1-score) results for known and zero-day malware detection (with four HPC events) are shown in Fig. 10.19. As observed, the performance of standard ML models on zero-day attack detection substantially drops by around 50% in Logistic Regression and SVM classifiers. When examined by the unknown

Table 10.7 Performance results of various ML-based detectors for zero-day malware detection and known test malware detection

Model	Acc	F1	AUC	TPR	FNR	TNR	P	R
Zero-day test								
Random forest	0.87	0.79	0.87	0.68	0.32	0.98	0.94	0.68
Decision tree	0.81	0.69	0.79	0.58	0.42	0.94	0.86	0.58
Logistic regression	0.62	0.14	0.40	0.09	0.91	0.92	0.39	0.09
KNN	0.83	0.72	0.81	0.61	0.39	0.95	0.87	0.61
SVM	0.64	0.15	0.49	0.09	0.91	0.96	0.56	0.09
Known test								
Random forest	0.95	0.93	0.95	0.91	0.09	0.98	0.96	0.91
Decision tree	0.93	0.90	0.91	0.90	0.10	0.94	0.89	0.90
Logistic regression	0.78	0.63	0.74	0.53	0.47	0.92	0.78	0.53
KNN	0.93	0.90	0.92	0.89	0.11	0.95	0.91	0.89
SVM	0.81	0.65	0.78	0.52	0.48	0.96	0.88	0.52

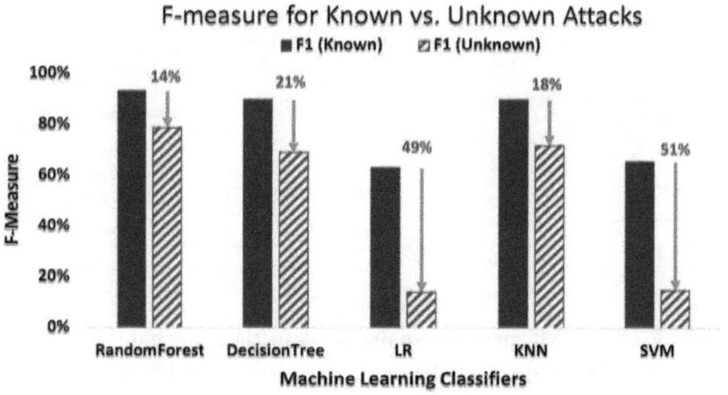

Fig. 10.19 Evaluation of standard ML classifiers for known and unknown (zero-day) malware detection

(zero-day) test data, the trained machine learning classifiers have never seen the testing malware types. Even for the most robust ML model, Random Forest, the F-measure for zero-day malware drops by 14% as compared with the scenario of detecting known malware. The results confirm the limitation of standard ML algorithms in recognizing the signatures of unknown malware using HPC events and further highlight the importance of proposing an effective mechanism to enhance the detection rate of the hardware-based zero-day malware detection process.

10.6.5 Advanced Framework: Image-Based Deep Transfer Learning-Enabled Method: Deep-HMD

To further enhance the detection performance for zero-day malware detection, we introduce a framework called *Deep-HMD* [15], a multilayer intelligent and scalable framework that achieves an accurate and robust zero-day malware detection performance.

First, we examine the general overview of *Deep-HMD*, illustrated in Fig. 10.20. In the initial level, the framework converts tabular data of malware and benign hardware events in numerical format (monitored from the underlying processor) into image-formatted data using an effective 2D embedding image features conversion method. The detailed algorithm, code, and visual effects of this conversion method are provided in the following section. The detailed algorithm, code, and visual effects of this conversion method are provided in the following section. While a deep neural network (DNN) can be trained on both tabular and image data, DNN-based architectures have shown exceptional performance in visual recognition tasks, achieving high top-1 and top-5 accuracy with low top-5 error rates [13]. In contrast, studies suggest that DNN models trained on tabular data may not perform as well [43].

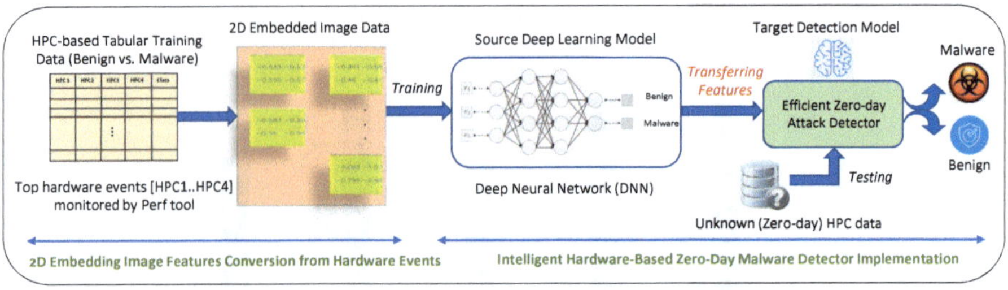

Fig. 10.20 Overview of the ML-assisted malware detection mechanism, *Deep-HMD*: the proposed image-based zero-day malware detection framework utilizing deep transfer learning [15]

After generating the image data, *Deep-HMD* leverages a high-performance ResNet architecture [13] and utilizes its pretrained model parameters from a large image dataset, ImageNet. Through transfer learning techniques, *Deep-HMD* recognizes zero-day malicious images in real time.

10.6.5.1 Transfer Learning Approach

Recent studies have reported that a well-trained DNN could transfer its knowledge of generalized features and feature extraction ability from one domain to another. Kornblith et al. [22] highlight that transfer learning can also share the architecture-related parameters in a new field while maintaining a high performance rate. Although transfer learning has been widely successful in computer vision, its application to tabular data, particularly in cybersecurity, remains relatively unexplored. This chapter aims to fill this gap by investigating transfer learning methods on tabular data using just four HPC numerical values.

Deep Neural Networks, while state-of-the-art, have been reported to be less effective than classical ML models for tabular data [43]. However, studies have shown that architectures like Convolutional Neural Networks (CNNs) can transfer their knowledge from domains like ImageNet to new domains with high accuracy and stability. ImageNet, known for its diverse image categories, contrasts sharply with cybersecurity datasets that include numerical features extracted from processor HPC events. Motivated by these findings, in this chapter we explore a transfer learning strategy combined with a DNN model for zero-day hardware-based malware detection.

In this case study, the dataset comprises tabular-based HPC data containing both malware and benign samples. We explore whether leveraging transfer learning and DNN can improve knowledge transfer to boost a higher detection performance than traditional MLs. He et al. [15] were the first in the field to explore the functionality and effectiveness of leveraging the DNN network transferred from ImageNet to the tabular-like zero-day malware detection domain based on only a few hardware tracing events monitored at run time. Building upon this foundation, we investigate state-of-the-art deep learning model architectures trained over the ImageNet Large-Scale Visual Recognition Challenge (ILSVRC). ILSVRC uses the smaller portion of the ImageNet that consists of 1000 categories with a total of 1.3 million training images, 50,000 validation images, and 100,000 testing images.

Using transfer learning offers several advantages. Firstly, pretrained models have already learned to recognize patterns from millions of images. A convolutional neural network acts as a crucial feature detector and extractor within its convolutional layers, followed by classification based on extracted features in subsequent layers. Secondly, training a model from scratch requires a larger dataset, significant training time, and iterative parameter tuning. In contrast, a pretrained model combined with transfer learning can maintain high accuracy when fine-tuned for a new task. According to a survey on transfer learning [46], ResNet is a particularly suitable choice for network-based deep transfer learning among the many popular DNN architectures.

ResNet employs residual blocks with "skip connections" to address the vanishing gradient problem, providing an alternative path for gradients to flow through. This architecture also facilitates identity learning, ensuring that higher layers do not degrade in performance compared to lower layers. ResNet's simplicity ensures that residual blocks do not introduce significant complexity, making it compatible with standard optimization techniques for training. Therefore, this study implements a transfer learning approach based on ResNet18, which consists of 18 layers, to enhance the detection performance of zero-day malware using tabular-based HPC data.

10.6.5.2 Embedding Tabular Data to Images

As we discussed in the previous sections, nonimage data like text, byte, and numerical data saved in CSV files can be preprocessed and converted into image format using the technique of embedding. Embedded data are more suitable as input to Deep Learning (DL)-based machine learning algorithms to find hidden spacial patterns, or apply transfer learning techniques to improve detection performance. DL-based algorithms can be more capable of tackling complex, obfuscated, stealthy malware than traditional ML algorithms.

Several prior works use the technique of embedding to convert nonimage-based data (text, byte) into image-like embedding as a preprocessing process to leverage Deep Neural Networks (DNNs) to assist in detecting cybersecurity threats, thanks to DNN's known performance in image processing and Nature Language Processing (NLP).

The work in [33] employs a static analysis directly on the byte program from Microsoft Windows Portable Executable (PE) without running it, to convert the raw byte sequences into high-level representations before feeding them to an LSTM model. They treat each raw byte sequence as a long sequence classification problem on the order of two million time steps. Similar to this approach, [42] converts malware binaries into grayscale images and then transforms them into sequence embeddings to feed to a Recurrent Neural Network (RNN) for stealthy malware detection. They first convert malware binaries into gray images for pattern matching, which employs Cosine similarity to evaluate all images that fall into several distance scopes and label the malware binaries with various similarity distances with unique IDs. Then, these grayscale images are transformed into sequence embedding to meet the input restriction for RNN that requires input sequences of equal length for training the RNN for malware detection. [31] transforms Android applications' API call sequence derived from Android applications' API call graph structure into a low-dimension numeric vector feature set (embedding) to feed a DNN, which limits the malware detection to those having to call malicious API(s) to execute malicious code. Meantime, notice this work extracts the API call graph representation from a malicious code in a static approach without running it, which does not represent a runtime scenario. The work in [21] is based on prior opcode sequence embedding work to solve the challenge of long opcode sequence problem, and they convert sequential opcodes using low-dimensional opcode embeddings to discover the malicious patterns.

This study leverages a dynamic analysis approach to collect HPC tracing data independently from the running applications when malicious codes are running and then presents a lightweight algorithm to embed HPC data directly to image embedding using only four numerical HPC values (features). Compared with PE byte sequences, opcode sequences, or applications' API call graphs, using only four numerical HPC values is a greatly very small amount of information, which is a challenging task to train an effective DNN model. Directly training such a DNN model poses a challenge, and thus this work employs a transfer learning approach to a transferable DNN architecture to counter such a challenge.

Meanwhile, compared to prior works utilizing RNN and LSTM architectures, which, despite being state-of-the-art, involve larger model architectures that introduce significant overhead alongside the already large data embeddings, this work investigates the effectiveness of embedding only four numerical HPC events using the proposed two-stage methodology.

The embedding method converts each row of four HPC tabular data into one image data represented by a two-dimensional Numpy array with the size of 256×256 and three color channels (red, green, and blue), as shown in Algorithm 2. It first normalized all rows of tabular HPC data using the standard scalar in Scikit Learn, which removes the mean and scales the data to unit variance. Next, it uses functions from the OpenCV library to evenly project the four numeric data to a $256 \times 256 \times 3$ resolution image with equal spacing and no overlapping as shown in Fig. 10.21. The example on the right side shows that nine rows of tabular data are converted into 9 two-dimensional image data. Each image on the right side is associated with a target label of either malware or benign. Overall, in this stage, the four features are drawn with equal sizes on the image with even spacing and ensure that they are not overlapped. Such images contain spatial relationships and hidden features among malware and benign applications resulting from a limited number of performance counter data.

Algorithm 2: Converting tabular data into 2D images

Input: HPC features in tabular format X = $\{x1, x2, x3, x4\}$
Output: Image data equivalent to X
while *tabular data* **do**
    for *each row of X={x1, x2, x3, x4}* **do**
        - Normalize HPCs
        - Set font size as 50, resolution as 256 × 256 × 3
        - Set 2 columns 2 rows per image
        - Apply OpenCV's cv2.putText() to draw the HPC number on the image
        - Save the converted image to a folder
    end
end

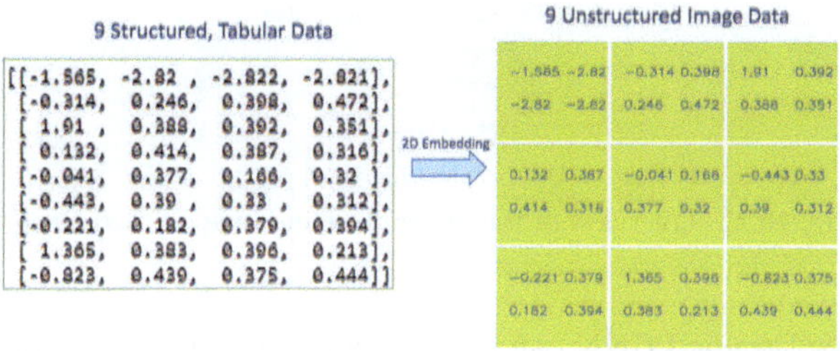

Fig. 10.21 Nine sample structured tabular data projected to 9 two-dimensional images using 2D embedding image features conversion from hardware events

10.6.5.3 Training and Testing of ML Detector

Figure 10.22 demonstrates the overall training and testing process of *Deep-HMD* in four steps. Steps 1 and 2 are dedicated to the training process and steps 3 and 4 are for the testing phase.

During **Step 1**, it first examines how the baseline model of the *Deep-HMD* performs on the known and zero-day test datasets. To this aim, it loads the ImageNet parameter of *Deep-HMD* in fast.ai, runs two tests, and calculates various metrics. Fast.ai is a deep learning wrapper library with PyTorch as its underlined backbone that can quickly train high-performance deep learning models and supports application domains in computer vision, natural language processing, and tabular models. The baseline of *Deep-HMD* tests how much knowledge it learns from ImageNet on generic features such as lines and strokes. It is found the baseline can detect 92% of benign applications but only 6% of actual malware. This experiment indicates that the front layers of the pretrained transferrable model trained on a sizeable ImageNet can be used as a feature extractor, and the extracted features are versatile according to Tan et al.'s study on transfer learning [46]. Notice the mentioned "transferrable" means that there is a relationship between model architecture and transferability [51]. The front layers of such pretrained models learn the morphology of images in general, and the morphology knowledge can be transferred to other visual recognition tasks with or without labeled data [8]. As shown in this baseline experiment where there is no training occurred, the model can detect 92% of benign as benign that further supports the understanding that a

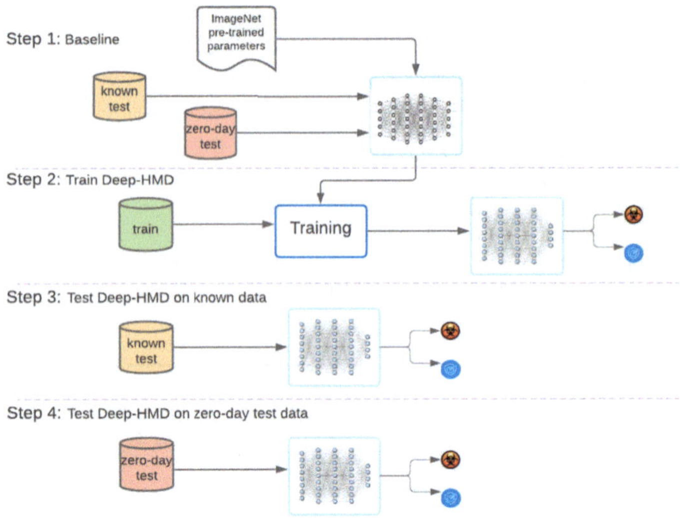

Fig. 10.22 Overview of training and testing image-based *Deep-HMD*

well-trained CNN can be an efficient feature extractor for unknown contexts. With fine-tuning model parameters on malware datasets, it continues to learn domain knowledge, particularly the pattern of malware from benign. In particular, this work examines the possibility of transferring the knowledge of pattern recognition to a new domain with a limited number of training samples and fast training time.

During **Step 2**, it loads the transferable pretrained baseline model with the ImageNet parameter, removes the last layer of the network, and replaces it with a Softmax function that outputs binary classification and trains the model.

In this training stage, the generated two-dimensional images are resized to $224 \times 224 \times 3$ and fed into the *Deep-HMD* network initialized with the ImageNet pretrained weights. Firstly, five steps with a batch size of 64 are trained to find an optimized learning rate for the next training cycle. In this step, generic features learned from ImageNet are transferred to the *Deep-HMD* network. In the customized training, cyclical learning rates, oversampling, and weight decay techniques are applied. No augmentation was applied to the images. Batch normalization is used right after each convolution and before activation. The cross-entropy loss function with a mini-batch size of 64 is used. The learning rate starts from 0.0001, and then fast.ai's learning rate finder for every 7–20 epochs to find the optimized learning rate to apply to the next training steps. A weight decay of 0.01 and a momentum of [0.95, 0.85, 0.95] are applied across the whole training process and train the models up to 30–80 iterations. Training loss and validation loss decrease until the validation loss is close to the training loss. Checkpoints are saved periodically, and the best model is selected for the testing phase.

10.6.6 Experimental Results and Evaluation

Table 10.8 presents the performance results of *Deep-HMD* for detecting zero-day and known malware using four HPC events. It is important to note that *Deep-HMD* is trained on image-based HPC data, whereas classical machine learning models are trained on the same data in tabular format. Our analysis includes various ML-based detectors commonly used in existing HMD techniques for comparison.

As the results indicate, the proposed *Deep-HMD* achieves 97% in F-measure, 97% in Area Under the Curve (AUC), and 98% accuracy for zero-day (unknown) malware detection. In addition, it obtains 96% in precision and 99% in recall. Recall that the best performing standard ML classifier, Random Forest, can only achieve 79% in F-measure, 87% in AUC and accuracy, and 68% in recall when used for detecting unknown malware. Random Forest (also other MLs) experiences a huge detection performance drop from detecting known malware to unknown malware, which is analyzed in the previous section. This is the main weakness of classical MLs when it comes to detecting diversified unknown zero-day malware. In contrast, *Deep-HMD* demonstrates strong resilience in unknown zero-day malware detection.

Figure 10.23 demonstrates the enhancement in overall F1-scores achieved by *Deep-HMD* (green bar) compared to classical ML models (blue bars). Specifically, *Deep-HMD* improves the F1-score by 17% over Random Forest, 28% over Decision Tree, 25% over KNN, and 82% over both Logistic Regression and SVM.

Overall, the results demonstrate that the suggested intelligent hardware-based technique for detecting zero-day malware, known as *Deep-HMD*, is the most precise model out of all the classifiers tested. *Deep-HMD* not only achieves an F-measure of 97% on the unknown zero-day test but also provides a true positive rate of 99%

Table 10.8 Performance results of *Deep-HMD* for zero-day malware detection and known test malware detection

Model	Acc	F1	AUC	TPR	FNR	TNR	P	R
Zero-day test								
Deep-HMD	0.98	0.97	0.97	0.99	0.01	0.97	0.96	0.99
Known test								
Deep-HMD	0.99	0.99	0.99	0.99	0.01	0.99	0.99	0.99

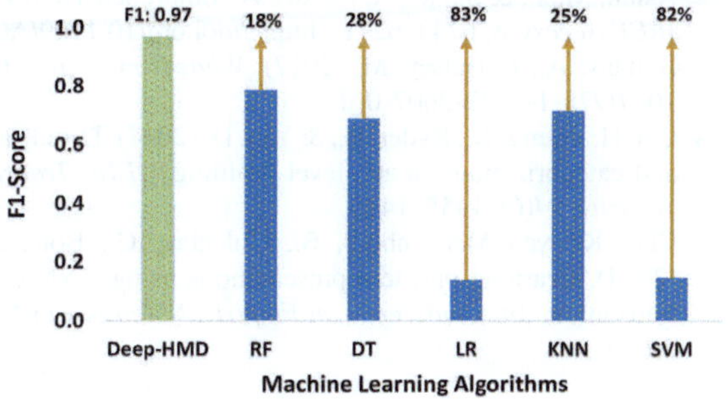

Fig. 10.23 The F1-score is boosted up (marked in orange line) by *Deep-HMD* (green bar) compared with standard ML models (blue bars) for zero-day malware detection. RF: Random Forest, DT: Decision Tree, LR: Logistic Regression

and a false positive rate of only 1%. This is a significant improvement compared to the results obtained by the best standard ML (Random Forest classifier), which only delivers a true positive rate of 68% and a false positive rate of 32%.

10.7 Conclusion

This chapter has conducted a thorough examination of AI-enabled cybersecurity methods developed to protect modern computing systems against malicious software, with a specific emphasis on the detection of zero-day (unknown) attacks. Emphasizing the significance of feature analysis in machine-learning-based detection approaches, we closely examined two categories: static and dynamic approaches. While machine-learning-based malware detection, combining static and dynamic feature processing, has proven highly effective in aiding the identification of known malware, it encounters limitations in detecting never-before-seen zero-day malware. Our work initially delved into classical and deep learning algorithms, outlining a typical ML-based malware detection process. Subsequently, we conducted an in-depth review of AI-enabled malware detection methodologies tailored for Windows, Android, and Arm-based IoT systems. The chapter further presents an AI-enabled methodology in the context of hardware security, which leverages deep learning and transfer learning in both training and inference processes for known and unknown image-based malware detection. It showcased an effective strategy for detecting zero-day malware using a minimal set of four numerical hardware-based data points converted into 2D images. The results demonstrated that such an AI-enabled approach could achieve significant detection performance, highlighting the potential of AI solutions in enhancing cybersecurity measures against sophisticated zero-day threats.

References

1. 2021 Top Routinely Exploited Vulnerabilities. https://www.cisa.gov/news-events/cybersecurity-advisories/aa22-117a (2022).
2. Aslan, M.A., & Samet, R. (2020). A comprehensive review on malware detection approaches. *IEEE Access, 8*, 6249–6271. https://doi.org/10.1109/ACCESS.2019.2963724
3. Bettany, A., & Halsey, M. (2017). *What is malware?* (pp. 1–8). Apress. https://doi.org/10.1007/978-1-4842-2607-0_1
4. Cai, H., Meng, N., Ryder, B., & Yao, D. (2018). DroidCat: Effective android malware detection and categorization via app-level profiling. *IEEE Transactions on Information Forensics and Security, 14*(6), 1455–1470.
5. Cho, K., van Merrienboer, B., Gulcehre, C., Bougares, F., Schwenk, H., & Bengio, Y. (2014). Learning phrase representations using RNN encoder-decoder for statistical machine translation. In: *Conference on Empirical Methods in Natural Language Processing (EMNLP 2014)*.

6. Das, S., Werner, J., Antonakakis, M., Polychronakis, M., & Monrose, F. (2019) Sok: The challenges, pitfalls, and perils of using hardware performance counters for security. In: *2019 IEEE Symposium on Security and Privacy (SP)* (pp. 20–38). IEEE.
7. Demme, J., Maycock, M., Schmitz, J., Tang, A., Waksman, A., Sethumadhavan, S., Stolfo, S.: On the feasibility of online malware detection with performance counters. *ACM SIGARCH Computer Architecture News, 41*(3), 559–570.
8. George, D., Shen, H., Huerta, E. A.: Classification and unsupervised clustering of LIGO data with deep transfer learning. *Physical Review D, 97*, 101501. https://doi.org/10.1103/PhysRevD.97.101501
9. Guthaus, M., Ringenberg, J., Ernst, D., Austin, T., Mudge, T., & Brown, R. (2001). MiBench: A free, commercially representative embedded benchmark suite. In *Proceedings of the Fourth Annual IEEE International Workshop on Workload Characterization. WWC-4 (Cat. No.01EX538)* (pp. 3–14). https://doi.org/10.1109/WWC.2001.990739
10. HaddadPajouh, H., Dehghantanha, A., Khayami, R., & Choo, K. K. R. (2018). A deep recurrent neural network based approach for internet of things malware threat hunting. *Future Generation Computer Systems, 85*, 88–96.
11. Harris, C. R., Millman, K. J., van der Walt, S. J., Gommers, R., Virtanen, P., Cournapeau, D., Wieser, E., Taylor, J., Berg, S., Smith, N. J., Kern, R., Picus, M., Hoyer, S., van Kerkwijk, M. H., Brett, M., Haldane, A., del Río, J. F., Wiebe, M., Peterson, P., Gérard-Marchant, P., Sheppard, K., Reddy, T., Weckesser, W., Abbasi, H., Gohlke, C., & Oliphant, T. E. (2020). Array programming with NumPy. *Nature, 585*(7825), 357–362. https://doi.org/10.1038/s41586-020-2649-2
12. He, Z., & Sayadi, H. (2023). Image-based zero-day malware detection in IoMT devices: A hybrid AI-enabled method. In: *2023 24th International Symposium on Quality Electronic Design (ISQED)* (pp. 1–8). https://doi.org/10.1109/ISQED57927.2023.10129348
13. He, K., Zhang, X., Ren, S., & Sun, J. (2016). Deep residual learning for image recognition. In: *Proceedings of the IEEE Conference on Computer Vision and Pattern Recognition* (pp. 770–778).
14. He, Z., Miari, T., Makrani, H. M., Aliasgari, M., Homayoun, H., & Sayadi, H. (2021). When machine learning meets hardware cybersecurity: Delving into accurate zero-day malware detection. In: *2021 22nd International Symposium on Quality Electronic Design (ISQED)* (pp. 85–90). https://doi.org/10.1109/ISQED51717.2021.9424330
15. He, Z., Rezaei, A., Homayoun, H., & Sayadi, H. (2022). Deep neural network and transfer learning for accurate hardware-based zero-day malware detection. In: *Proceedings of the Great Lakes Symposium on VLSI 2022, GLSVLSI '22* (pp. 27–32). Association for Computing Machinery.
16. He, Z., Fernandes, C. W., & Sayadi, H. (2024). ObfusGate: Representation learning-based gatekeeper for hardware-level obfuscated malware detection. In: *2024 Design, Automation & Test in Europe Conference & Exhibition (DATE)* (pp. 1–2).
17. He, Z., Homayoun, H., & Sayadi, H. (2024). Beyond conventional defenses: Proactive and adversarial-resilient hardware malware detection using deep reinforcement learning. In *ACM/IEEE Design Automation Conference (DAC)*.
18. He, Z., Homayoun, H., & Sayadi, H. (2024). Guarding against the unknown: Deep transfer learning for hardware image-based malware detection. *Journal of Hardware and Systems Security, 8*, 1–18.
19. Henning, J. L. (2006). Spec cpu2006 benchmark descriptions. *ACM SIGARCH Computer Architecture News, 34*(4), 1–17.
20. Hussain, A., Asif, M., Ahmad, M. B., Mahmood, T., & Raza, M. A. (2022). Malware detection using machine learning algorithms for windows platform. In A. Ullah, S. Anwar, Á. Rocha, & S. Gill (Eds.), *Proceedings of International Conference on Information Technology and Applications* (pp. 619–632). Springer Nature Singapore.
21. Kakisim, A. G., Gulmez, S., & Sogukpinar, I. (2022). Sequential opcode embedding-based malware detection method. *Computers & Electrical Engineering, 98*, 107703.

22. Kornblith, S., Shlens, J., & Le, Q. V. (2019). Do better imagenet models transfer better? In: *Proceedings of the IEEE/CVF Conference on Computer Vision and Pattern Recognition (CVPR)*.
23. Kraskov, A., Stögbauer, H., & Grassberger, P. (2004). Estimating mutual information. *Physical Review E—Statistical, Nonlinear, and Soft Matter Physics, 69*, 066138. https://doi.org/10.1103/PhysRevE.69.066138
24. Krizhevsky, A., Sutskever, I., & Hinton, G. E. (2017). ImageNet classification with deep convolutional neural networks. *Communications of the ACM, 60*(6), 84–90. https://doi.org/10.1145/3065386
25. Lecun, Y., Bottou, L., Bengio, Y., & Haffner, P. (1998). Gradient-based learning applied to document recognition. *Proceedings of the IEEE, 86*(11), 2278–2324. https://doi.org/10.1109/5.726791
26. Liu, X., Lin, Y., Li, H., & Zhang, J. (2020). A novel method for malware detection on ML-based visualization technique. *Computers & Security, 89*, 101682. https://doi.org/10.1016/j.cose.2019.101682
27. Massarelli, L., Aniello, L., Ciccotelli, C., Querzoni, L., Ucci, D., & Baldoni, R. (2017). Android malware family classification based on resource consumption over time. In: *2017 12th International Conference on Malicious and Unwanted Software (MALWARE)* (pp. 31–38). IEEE.
28. McKinney, W. (2010). Data structures for statistical computing in python. In: S. van der Walt & J. Millman (Eds.), *Proceedings of the 9th Python in Science Conference* (pp. 56–61). https://doi.org/10.25080/Majora-92bf1922-00a
29. Nielsen, M. (2015). *Neural networks and deep learning*. Determination Press. https://books.google.com/books?id=STDBswEACAAJ
30. Pedregosa, F., Varoquaux, G., Gramfort, A., Michel, V., Thirion, B., Grisel, O., Blondel, M., Prettenhofer, P., Weiss, R., Dubourg, V., Vanderplas, J., Passos, A., Cournapeau, D., Brucher, M., Perrot, M., & Duchesnay, E. (2011). Scikit-learn: Machine learning in python. *Journal of machine Learning research, 12*, 2825–2830.
31. Pektaş, A., & Acarman, T. (2020). Deep learning for effective android malware detection using API call graph embeddings. *Soft Computing, 24*(2), 1027–1043 (2020). https://doi.org/10.1007/s00500-019-03940-5
32. Putrevu, M. A., Putrevu, V. S. C., & Shukla, S. K. (2023). Early detection of ransomware activity based on hardware performance counters. In: *Proceedings of the 2023 Australasian Computer Science Week, ACSW '23* (pp. 10–17). Association for Computing Machinery. https://doi.org/10.1145/3579375.3579377
33. Raff, E., Barker, J., Sylvester, J., Brandon, R., Catanzaro, B., & Nicholas, C. (2017). *Malware detection by eating a whole exe*. https://arxiv.org/abs/1710.09435
34. Ravi, V., Pham, T. D., & Alazab, M. (2022). Attention-based multidimensional deep learning approach for cross-architecture IoMT malware detection and classification in healthcare cyber-physical systems. *IEEE Transactions on Computational Social Systems, 10*(4), 1597–1606.
35. Russakovsky, O., Deng, J., Su, H., Krause, J., Satheesh, S., Ma, S., Huang, Z., Karpathy, A., Khosla, A., Bernstein, M., Berg, A. C., & Fei-Fei, L. (2015). ImageNet large scale visual recognition challenge. *International Journal of Computer Vision, 115*(3), 211–252. https://doi.org/10.1007/s11263-015-0816-y
36. Sayadi, H., Patel, N., Sasan, A., Rafatirad, S., & Homayoun, H. (2018). Ensemble learning for effective run-time hardware-based malware detection: A comprehensive analysis and classification. In *Proceedings of the 55th Annual Design Automation Conference, DAC '18*. Association for Computing Machinery. https://doi.org/10.1145/3195970.3196047
37. Sayadi, H., Makrani, H. M., Pudukotai Dinakarrao, S. M., Mohsenin, T., Sasan, A., Rafatirad, S., & Homayoun, H. (2019). 2smart: A two-stage machine learning-based approach for run-time specialized hardware-assisted malware detection. In *2019 Design, Automation & Test in Europe Conference & Exhibition (DATE)* (pp. 728–733). https://doi.org/10.23919/DATE.2019.8715080

38. Sayadi, H., Gao, Y., Mohammadi Makrani, H., Mohsenin, T., Sasan, A., Rafatirad, S., Lin, J., & Homayoun, H. (2020). StealthMiner: Specialized time series machine learning for run-time stealthy malware detection based on microarchitectural features. In *Proceedings of the 2020 on Great Lakes Symposium on VLSI, GLSVLSI '20* (pp. 175–180). *Association for Computing Machinery*. https://doi.org/10.1145/3386263.3407585
39. Sayadi, H., He, Z., Fernandes, C. W., & Miari, T. (2023). Cyber-immunity at the core: Securing biomedical devices through hardware-level machine learning defense. In *2023 IEEE Biomedical Circuits and Systems Conference (BioCAS)* (pp. 1–5). https://doi.org/10.1109/BioCAS58349.2023.10388581
40. Sayadi, H., He, Z., Makrani, H. M., & Homayoun, H. (2024). Intelligent malware detection based on hardware performance counters: A comprehensive survey. In *2024 25th International Symposium on Quality Electronic Design (ISQED)* (pp. 1–10). https://doi.org/10.1109/ISQED60706.2024.10528369
41. Sherstinsky, A. (2020). Fundamentals of recurrent neural network (RNN) and long short-term memory (LSTM) network. *Physica D: Nonlinear Phenomena, 404*, 132306. https://doi.org/10.1016/j.physd.2019.132306
42. Shukla, S., Kolhe, G., & Rafatirad, S. (2019). Microarchitectural events and image processing-based hybrid approach for robust malware detection: work-in-progress. In: *Proceedings of the International Conference on Compliers, Architectures and Synthesis for Embedded Systems Companion, CASES '19*. Association for Computing Machinery. https://doi.org/10.1145/3349569.3351538
43. Shwartz-Ziv, R., & Armon, A. (2022). Tabular data: Deep learning is not all you need. *Information Fusion, 81*, 84–90.
44. Sonicwall Cyber Threat Report: Charting Cybercrime's Shifting Frontlines. https://www.sonicwall.com/news/2023-sonicwall-cyber-threat-report-casts-new-light-on-shifting-front-lines-threat-actor-behavior/ (2023).
45. Supriya, Y., Kumar, G., Sowjanya, D., Yadav, D., & Kameshwari, D. L.: Malware detection techniques: A survey. In *2020 Sixth International Conference on Parallel, Distributed and Grid Computing (PDGC)* (pp. 25–30). https://doi.org/10.1109/PDGC50313.2020.9315764
46. Tan, C., Sun, F., Kong, T., Zhang, W., Yang, C., & Liu, C. (2018). A survey on deep transfer learning. In V. Kůrková, Y. Manolopoulos, B. Hammer, L. Iliadis, & I. Maglogiannis (Eds.), *Artificial Neural Networks and Machine Learning – ICANN 2018* (pp. 270–279). Springer International Publishing.
47. Tang, A., Sethumadhavan, S., & Stolfo, S. J. (2014). Unsupervised anomaly-based malware detection using hardware features. In A. Stavrou, H. Bos, & G. Portokalidis (Eds.), *Research in attacks, intrusions and defenses* (pp. 109–129). Springer International Publishing.
48. Tian, D., Ying, Q., Jia, X., Ma, R., Hu, C., & Liu, W. (2021). MDCHD: A novel malware detection method in cloud using hardware trace and deep learning. *Computer Networks, 198*, 108394. https://doi.org/10.1016/j.comnet.2021.108394
49. Vaswani, A., Shazeer, N., Parmar, N., Uszkoreit, J., Jones, L., Gomez, A. N., Kaiser, L., & Polosukhin, I. (2023). *Attention is all you need.*
50. Ye, Y., Chen, L., Wang, D., Li, T., Jiang, Q., & Zhao, M. (2009). SBMDS: An interpretable string based malware detection system using SVM ensemble with bagging. *Journal in Computer Virology, 5*, 283–293.
51. Yosinski, J., Clune, J., Bengio, Y., & Lipson, H. (2014). How transferable are features in deep neural networks? In Z. Ghahramani, M. Welling, C. Cortes, N. Lawrence, & K. Weinberger (Eds.), *Advances in Neural Information Processing Systems* (Vol. 27). Curran Associates.
52. Zhou, B., Gupta, A., Jahanshahi, R., Egele, M., & Joshi, A. (2018). Hardware performance counters can detect malware: Myth or fact? In *Proceedings of the 2018 on Asia Conference on Computer and Communications Security, ASIACCS '18* (pp. 457–468). Association for Computing Machinery. https://doi.org/10.1145/3196494.3196515

Chapter 11
Machine Learning Applications and Attacks Using Side Channel Analysis

Chaitanya Bhure, Geraldine Shirley Nicholas, and Fareena Saqib

11.1 Introduction

Side channel analysis is a physical implementation vulnerability of cryptographic algorithms and does not rely on the algorithm or key strength. Mathematically, the security strength of an encryption engine is directly proportional to the key size, whereas the side channel attack targets the physical parameters such as power dissipation and EM radiation to study and break the transitions of data during execution to extract secret information. The side channel provides additional information that makes the process of breaking secret in linear time by applying fewer combinations as otherwise requires exponential effort. Side channel analysis is a class of noninvasive attack, where the attacker needs to be in the close proximity of the victim device to eavesdrop and capture the side channels to break in secret. Such attacks can go undetected, and the victim machine might never know the device has been compromised, thus posing a serious threat to the various implementations of the symmetric, asymmetric algorithms, and hash functions.

The physical sources of side channel leakage include signal propagation, register latching, power analysis, electromagnetic variations, timing variations, etc. The taxonomy of side channel attacks includes:

C. Bhure (✉) · F. Saqib (✉)
Department of Electrical and Computer Engineering, University of North Carolina at Charlotte, Charlotte, NC, USA
e-mail: cbhure@charlotte.edu; fsaqib@charlotte.edu

G. S. Nicholas
Department of Electrical and Computer Engineering, High Point University, High Point, NC, USA
e-mail: gnichola@highpoint.edu

A. Iranmanesh, H. Sayadi (eds.), *AI-Enabled Electronic Circuit and System Design*,
https://doi.org/10.1007/978-3-031-71436-8_11

1. Power and Electromagnetic Analysis: Power analysis captures the transient effects of the voltage drop because of the transitions of signals on the power rail. These signals can be captured using probes and traces are collected. Several side-channel-based statistical attacks can be applied, such as correlation power analysis (CPA) and differential power analysis (DPA) that help in extracting the cryptographic secret. Similar to the power analysis, electromagnetic emanations can be captured to reveal/reverse engineer the secrets. The attacks include Simple Electromagnetic Analysis (SEMA), CEMA, and DEMA.
2. Direct Memory Access (DMA): Direct Memory Access (DMA) attack is a type of side channel attack that targets the memory to exploit during runtime execution. Direct Memory Access (DMA) is a hardware feature that allows peripherals or devices to access a computer's memory without involving the CPU (Central Processing Unit) and is commonly used for high-speed data transfers between peripherals. In side channel, an attacker may use DMA to scan contents of the memory potentially locating cryptographic keys and other sensitive data.
3. Timing Attacks: Timing attacks exploit variations in the time it takes for a system to perform cryptographic operations. By measuring these variations, an attacker can infer valuable information about the cryptographic keys or data being processed. Broadly, timing attacks can also be classified into two distinct categories including Simple Timing Attacks and Differential Timing Attacks.
4. Fault Injection: Fault injection attacks involve deliberately inducing faults in a cryptographic device's operation. These faults can lead to the extraction of sensitive information or the compromise of cryptographic keys.
5. Cache Attacks: Cache attacks exploit the behavior of CPU caches during cryptographic operations. By measuring cache access patterns and timing, the secret information can be inferred [1].

This chapter focuses on the AI attacks on Power, EM, and DMA attacks. The details on architecture-based side channel attacks including fault attacks, timing attacks, and cache attacks are not covered in this chapter.

11.2 Machine Learning Overview

Machine learning is a branch of artificial intelligence (AI) that enables computers to learn from data and make predictions without explicit programming. Deep learning is a category of machine learning that utilizes multiple layers to model data, excelling in tasks like image and speech recognition. Despite achieving breakthroughs, deep learning demands significant computational resources and extensive fine-tuning. In contrast, machine learning is versatile but may require feature engineering and struggle with complex relationships. The choice depends on the problem, resources, and the balance between interpretability and performance. The hierarchical architecture allows deep learning models to excel in tasks like image classification, natural language processing, and speech recognition. Despite

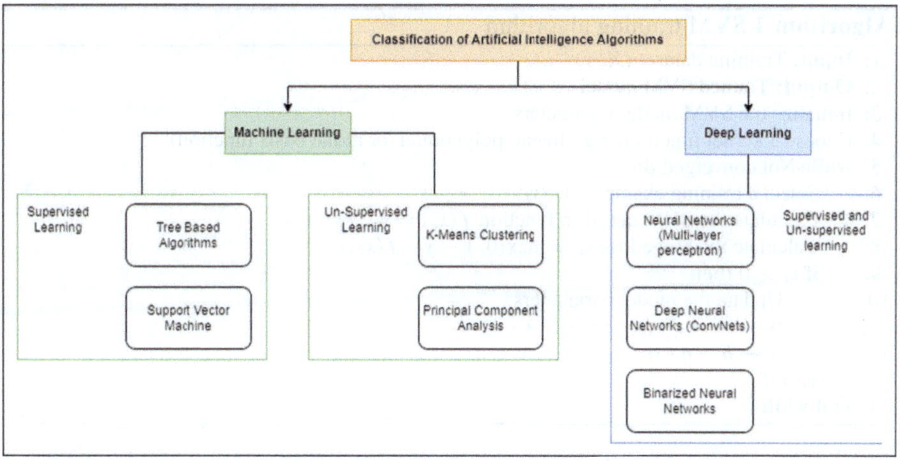

Fig. 11.1 Taxonomy of Artificial Intelligence models used in this chapter

its remarkable achievements, deep learning often requires substantial computational resources, large datasets, and meticulous fine-tuning. The choice between deep learning and other approaches depends on the specific problem, available resources, and the desired balance between model interpretability and predictive performance.

The machine learning process includes data preprocessing, model fitting, and optimization approaches that may be considered in one of the categories of supervised, semi-supervised, unsupervised, and reinforcement learning.

In this section, we will explore the fundamental categories of machine learning, supervised and unsupervised learning, and introduce some well-known algorithms and delve into key concepts in deep learning (Fig. 11.1).

11.2.1 Supervised Machine Learning

Supervised machine learning, a foundational paradigm in artificial intelligence, has gained prominence for training computer systems using labeled data. In this approach, a training dataset with input-output pairs guides the algorithm to build predictive models mapping inputs to outputs. Common algorithms like linear regression, decision trees, and neural networks are chosen based on application requirements. The iterative training phase optimizes the model by adjusting parameters to minimize prediction differences from actual labels. Supervised learning has applications in diverse domains, from image recognition to medical diagnoses. Supervised learning has a set of challenges that includes data label quality and model interpretability that are driving ongoing research and development. As machine learning advances, supervised learning remains vital, empowering AI practitioners to develop intelligent systems with transformative potential.

Algorithm 1 SVM training algorithm

Input: Training dataset (X, y)
Output: Trained SVM model
Initialize the SVM model parameters
Choose a kernel function (e.g., linear, polynomial, or radial basis function)
while Not converged **do**
    Select a training example (x_i, y_i)
    Calculate the SVM decision function $f(x_i) = \mathbf{w}^T \cdot \mathbf{x}_i + b$
    Calculate the hinge loss $\ell_i = \max(0, 1 - y_i \cdot f(x_i))$
    if $\ell_i > 0$ **then**
        Update the model parameters:
        $\mathbf{w} \leftarrow \mathbf{w} + \eta \cdot (y_i \cdot \mathbf{x}_i - 2 \cdot \lambda \cdot \mathbf{w})$
        $b \leftarrow b + \eta \cdot y_i$
    end if
end while

Support Vector Machines (SVMs) SVM is particularly popular for its effectiveness in handling both linear and nonlinear data by finding the optimal hyperplane that best separates data points into different classes. The SVM algorithm aims to find a hyperplane that maximally separates data points into different classes. The objective is to maximize the margin while minimizing the classification error. If a linear separation is not possible, SVM can use a kernel trick to transform the data into a higher-dimensional space, where a hyperplane can linearly separate the classes. The SVM training algorithm, as described below, iteratively selects training examples, computes the decision function for each example, calculates the hinge loss (a measure of classification error), and updates the model parameters. The training process continues until convergence, where the model has learned the optimal hyperplane or decision boundary. Regularization terms, such as L1 or L2 regularization, can be incorporated to prevent overfitting. SVMs are versatile and have applications in various fields, including image classification, text categorization, and bioinformatics, making them a valuable tool in machine learning [2].

Naive Bayes The Naive Bayes algorithm is a probabilistic classification technique based on Bayes' theorem. It is well suited for text classification and spam detection but can be applied to a wide range of classification tasks. The "naive" assumption in Naive Bayes is that the features used for classification are conditionally independent given the class labels. This simplifying assumption makes the algorithm computationally efficient and easy to implement. Given a new data point, the algorithm calculates the probability of it belonging to each class and assigns it to the class with the highest probability [3].

Algorithm 2 Naive Bayes classification algorithm

Input: Training dataset (X, y), New data point x_{new}
Output: Class label for x_{new}
Compute class prior probabilities $P(y_i)$ for each class y_i
for each feature f_j **do**
  Compute feature likelihoods $P(x_{\text{new},f_j}|y_i)$ for each class y_i
end for
for each class y_i **do**
  Compute the posterior probability $P(y_i|x_{\text{new}})$ using Bayes' theorem
end for
Assign x_{new} to the class with the highest posterior probability

11.2.2 Unsupervised Machine Learning

Unsupervised machine learning, a subset of artificial intelligence, focuses on discovering patterns in data without labeled outcomes. This approach explores data to identify inherent structures, making it valuable for uncovering hidden information. One application is clustering, where algorithms group data points based on similarity, like k-means and hierarchical clustering. Another aspect is dimensionality reduction, aiming to simplify high-dimensional data. PCA and t-SNE are notable techniques for this purpose. Unsupervised learning finds applications in diverse domains, including document clustering, image segmentation, and cybersecurity for anomaly detection. It plays a crucial role in revealing hidden patterns and relationships, providing valuable insights for further analysis.

K-Means Clustering K-Means clustering is an unsupervised machine learning algorithm that is designed to partition a dataset into K distinct, nonoverlapping clusters. This method aims to group data points into clusters based on their similarity, with the number of clusters, K, specified by the user. It starts by initializing K cluster centers randomly. It then iteratively assigns each data point to the nearest cluster center and updates the center's position based on the mean of all points assigned to it. The process continues until the algorithm converges, which typically occurs when the cluster assignments and centers no longer change significantly. The result is a set of K clusters, each characterized by its centroid, which serves as a representative point for that cluster [4].

Principal Component Analysis (PCA) Principal Component Analysis (PCA) is a dimensionality reduction technique while retaining as much variance as possible. This is achieved by emphasizing the principal components (orthogonal axes) that capture the most significant information in the data. These principal components are ordered by their variance, with the first principal component explaining the most variance in the data. PCA is particularly useful for data compression, visualization, and feature selection [5].

Algorithm 3 K-means clustering

Input: Data points X, Number of clusters K
Output: Cluster assignments and cluster centers
Initialize cluster centers randomly
while Not converged **do**
  for each data point x_i **do**
    Assign x_i to the nearest cluster center
  end for
  for each cluster center **do**
    Update the center as the mean of its assigned data points
  end for
end while
Return: Cluster assignments and cluster centers

Algorithm 4 Principal component analysis (PCA)

Input: Data matrix X, Number of principal components k
Output: Transformed data matrix X_{PCA}, Principal components
Center the data by subtracting the mean from each feature
Compute the covariance matrix $C = \frac{1}{m}X^T X$
Perform eigenvalue decomposition of C
Sort the eigenvectors by decreasing eigenvalues
Select the top k eigenvectors as the principal components
Project the data onto the selected principal components
Return: Transformed data matrix X_{PCA} and principal components

11.2.3 Deep Learning Concepts

Deep learning has revolutionized artificial intelligence in recent years. It focuses on the use of artificial neural networks that are inspired by the structure and functioning of the human brain, to solve complex tasks and automatically learn representations of data. There are a few key concepts in deep learning that are explained below:

Neural Networks Neural networks consist of layers of interconnected artificial neurons, each with its own set of weights and biases. Input data is passed through the network, and the weights are adjusted during training to make accurate predictions.

Deep Architectures Deep learning networks are characterized by their depth, comprising multiple hidden layers. This depth allows them to learn hierarchical and abstract representations of data.

Activation Functions Activation functions, such as ReLU (Rectified Linear Unit) and Sigmoid, introduce nonlinearity into neural networks. They help the network capture complex patterns and relationships in data.

Loss Functions Loss functions, like Mean Squared Error (MSE) for regression tasks and Cross-Entropy for classification tasks, quantify the error between predicted and actual values. Training aims to minimize this error, adjusting network parameters accordingly.

Optimization Algorithms Optimization techniques like Stochastic Gradient Descent (SGD) and its variants are used to update weights and biases to minimize the loss function. These algorithms play a crucial role in training deep neural networks.

Backpropagation Backpropagation is the process of computing gradients of the loss function with respect to the network's parameters, allowing for efficient weight updates. It propagates errors backward through the network.

Accuracy Metrics For classification tasks, accuracy is a common metric, representing the proportion of correctly classified instances. Precision, recall, and F1-score are also used to evaluate the trade-offs between true positives, false positives, and false negatives.

Deep learning has applications in image and speech recognition, natural language processing, recommendation systems, and autonomous driving. It continues to drive advancements in technology and is a fundamental component of many cutting-edge AI systems. There are a couple of algorithms that are rendered important in the field of hardware security, and they are discussed below.

Binarized Neural Networks (BNNs) Binarized Neural Networks (BNNs) are a specialized class of neural networks designed to perform computations using binary values, which are either -1 or 1. BNNs aim to reduce memory and computational requirements while maintaining competitive accuracy, making them particularly suitable for resource-constrained environments and applications such as edge devices and embedded systems. The key innovation in BNNs lies in quantizing both the model's weights and activations into binary values, thus significantly reducing the model's memory footprint and computational complexity. This simplification allows BNNs to be more energy efficient and faster for inference tasks, without compromising accuracy. The core of a Binarized Neural Network is its binary activation and weight quantization [6].

Deep Neural Networks (DNNs) Deep Neural Networks (DNNs) have revolutionized a wide range of applications, including image and speech recognition, natural language processing, and reinforcement learning. Training deep neural networks

---

Algorithm 5 Binarization of neural network parameters

Input: Neural network weights W, Activations A
Output: Binarized weights $\hat{W}$, Binarized activations $\hat{A}$
for each weight matrix W_i in W **do**
  Compute binary weights $\hat{W}_i$ by thresholding W_i at 0
end for
for each activation matrix A_i in A **do**
  Compute binary activations $\hat{A}_i$ by thresholding A_i at 0
end for
Return: Binarized weights $\hat{W}$ and Binarized activations $\hat{A}$

Algorithm 6 Deep neural network training

Input: Training data (X, y), Network architecture, Hyperparameters
Output: Trained neural network model
Initialize network weights and biases randomly or using pre-trained weights
for each epoch in training **do**
    Initialize gradient accumulators for weights and biases
    for each mini-batch in training data **do**
        Perform forward pass to compute predictions
        Compute loss using a specified loss function (e.g., Cross-Entropy or Mean Squared Error)
        Perform backward pass to compute gradients
        Accumulate gradients for weights and biases
    end for
    Update weights and biases using an optimization algorithm (e.g., Stochastic Gradient Descent)
end for
Return: Trained neural network model

involves forward and backward passes to compute predictions and gradients, respectively, followed by weight updates using optimization algorithms.

Machine learning models have several applications in the field of hardware security, offering innovative solutions to address various challenges. One such application involves the use of machine learning for enhancing the security of Physical Unclonable Functions (PUFs). PUFs are hardware security primitives that exploit the inherent variations in physical properties of integrated circuits to generate unique identifiers. Machine learning can be employed to enhance the reliability of PUF-based authentication systems [7], mitigating noise and environmental variations, ultimately bolstering their effectiveness.

In the realm of counterfeit Integrated Circuit (IC) detection, deep learning has emerged as a powerful tool. Image processing techniques, in conjunction with deep learning models, can detect counterfeit ICs by analyzing their physical features [8, 9]. This technology can differentiate genuine ICs from counterfeit ones, reducing the risks associated with fraudulent hardware components infiltrating critical systems.

Furthermore, machine learning plays a vital role in defending against Direct Memory Access (DMA) attacks on embedded systems. DMA attacks exploit vulnerabilities to gain unauthorized access to a system's memory. By leveraging machine learning, security measures can be implemented to detect and mitigate such attacks in real time [10]. Advanced anomaly detection algorithms can recognize unauthorized memory access patterns, thus enhancing the security of embedded systems against DMA threats.

Overall, machine learning models demonstrate their versatility in addressing hardware security concerns, offering robust solutions to safeguard the integrity and reliability of critical hardware components and systems. The following section covers side channel attacks, its applications in hardware security, and machine learning modeling used to exploit hardware system vulnerabilities.

11.3 Side Channel Attacks

11.3.1 Power and EM Analysis

Power and electromagnetic side channel attacks are techniques used to extract sensitive information from electronic devices by exploiting unintended emissions occurring during device operation. Power side channel attacks concentrate on fluctuations in power consumption, utilizing equipment like oscilloscopes or power analyzers to measure power traces, which are then analyzed to uncover patterns or correlations between device operations and power consumption. In contrast, electromagnetic side channel attacks focus on capturing and analyzing the electromagnetic radiation emitted by devices using specialized equipment like antennas or field sensors. These attacks differ not only in the nature of the emissions they exploit but also in the equipment required and their potential range, making countermeasures vital to protect against them in electronic systems.

11.3.1.1 Simple Power Analysis

Simple Power Analysis (SPA) is a foundational technique in side channel attacks on cryptographic systems. It involves monitoring a device's power consumption during cryptographic operations to detect information leakage. SPA analyzes power consumption patterns, exploiting variations in the power trace to interpret operations and data transitions. The fluctuating power consumption is influenced by the processed data, algorithm operations, and secret keys. SPA leverages this information leakage, observing power consumption changes corresponding to specific bits of secret keys during cryptographic operations. This power leakage can be detected and analyzed (Fig. 11.2) [11].

Attackers exploit variations in power consumption during cryptographic operations, particularly in the processor and memory components. SPA attacks target symmetric cryptographic algorithms like block ciphers and stream ciphers, aiming to disclose crucial information about secret keys. Visual inspection in SPA involves capturing power traces over time using specialized equipment such as

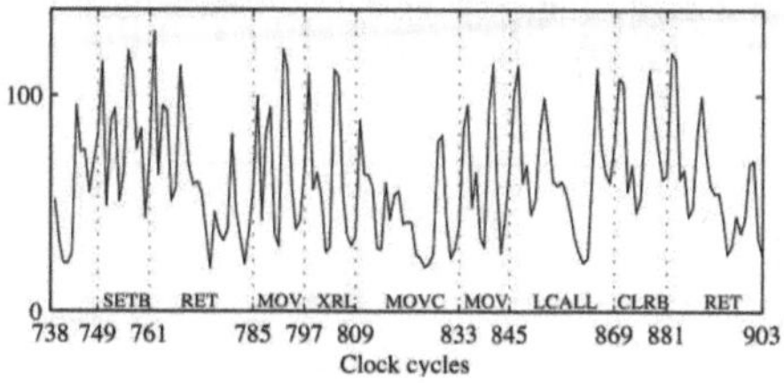


```
LCALL SET_ROUND_TRIGGER
MOV A,ASM_input + 0       ; load a0
XRL A,ASM_key + 0         ; add k0
MOVC A,@A + DPTR          ; S-box look-up
MOV ASM_input, A          ; store a0
LCALL CLEAR_ROUND_TRIGGER
```

Fig. 11.2 Annotated Sequence of AddRoundKey, SubBytes, and ShiftRow Operations and the corresponding sequence of assembly-level instructions [11]

oscilloscopes, measuring voltage or current fluctuations during multiple executions of a cryptographic algorithm with varying inputs or keys. While the process can be meticulous, successful visual inspection of power traces poses a significant threat to cryptographic security.

11.3.1.2 Differential Power Analysis

Differential Power Analysis (DPA) is an advanced power analysis attack that analyzes power consumption differences during cryptographic computations, correlating them with specific operations or intermediate values in the algorithm. By statistically analyzing these differences, DPA attackers can deduce sensitive information like secret keys. DPA identifies data-dependent correlations within power consumption measurements, involving partitioning traces into subsets. If this partitioning is independent of the data, subset averages tend toward zero; otherwise, deviations indicate correlations with the processed data values.

The DPA technique applies averaging process at various offsets or points within the traces, effectively examining how power consumption varies over times [12].

DPA attack results are visualized on a graph with the X-axis representing trace offset (time) and the Y-axis showing discrepancies between subset averages at that offset. Significant deviations from zero, often appearing as spikes, indicate regions where power consumption correlates with the internal operations of the target device, revealing valuable information about the secret key. Conversely, regions unrelated to data-dependent operations show statistically similar distributions, resulting in flat areas as more traces are analyzed. DPA attacks are powerful and have been successful in compromising cryptographic implementations in real-world scenarios (Fig. 11.3) [13].

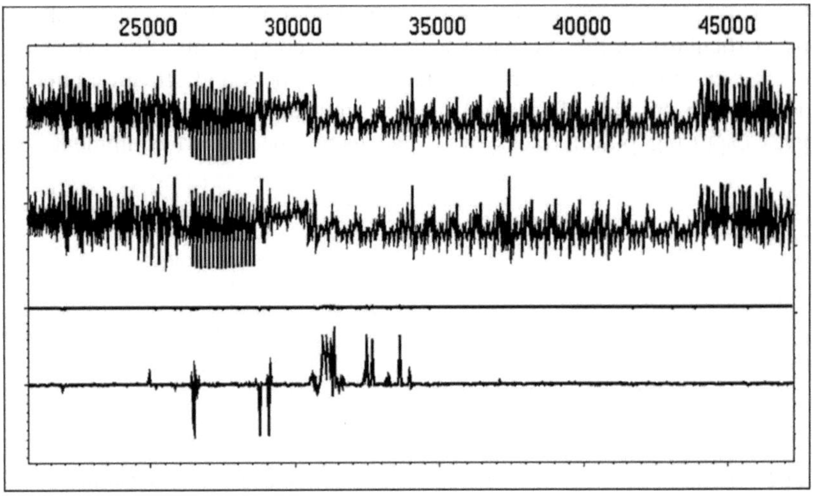

Fig. 11.3 Result of DPA analysis [13]

Table 11.1 S-box hypotheses and minimum sample requirements results from [14]

S-box and hypothesis	Min samples (mean diff.)	Min samples (max. likl.)
S1, B3	640	350
S2, B3	630	210
S7, B3	110	40
S8, B3	130	90

DPA is a powerful tool for attackers, especially when power variations are subtle. The typical attack process for DPA involves the following steps [13]:

Algorithm 7 Typical DPA attack stages

Stage 1: Device Instrumentation
Input: Access to target device, measurement apparatus
Output: Instrumented device ready for data collection
Description: Prepare to interface with the target device for cryptographic operations. Connect measurement equipment like a PC-controlled digital oscilloscope.
Stage 2: Measurement
Input: Instrumented device, cryptographic operations
Output: Captured power traces with associated cryptographic data
Stage 3: Signal Processing
Input: Captured traces
Output: Processed traces (if required)
Stage 4: Prediction and Selection Function Generation
Input: Processed traces, hypotheses about a portion of the key
Output: Selection functions for analysis
Stage 5: Averaging
Input: Processed traces, selection function outputs
Output: Averages of input trace subsets
Stage 6: Evaluation
Input: Averaged data
Output: Most likely candidate key guesses

The improved techniques for side channel analysis [14] detail an improved version of the DPA attack using a better metric that shows a significant reduction in the minimum number of signals required for a DPA attack.

For the DPA attack, obtaining the maximum likelihood estimate of the test parameters involves solving a set of nonlinear coupled equations. Instead of using these maximum likelihood estimates, sample estimates are employed. The application of this method to attacking the S-box lookup for a DES implementation is presented in Table 11.1 from [14]. The table indicates the predicted bit, the number of samples required for the correct key hypothesis to emerge as the winner using the traditional DPA metric, and the number of samples needed with the new metric. The improved metric significantly reduces the number of minimum signals required.

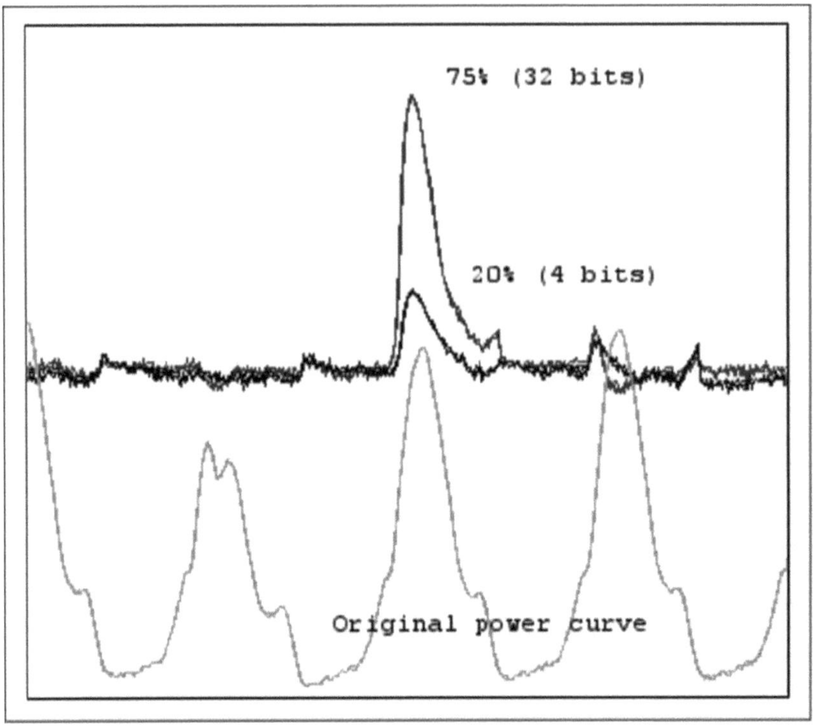

Fig. 11.4 Two correlation peaks for full word (32 bits) and partial (4 bits) predictions [15]

11.3.1.3 Correlation Power Analysis

Correlation Power Analysis (CPA) identifies correlations between the power consumption and hypotheses about the secret key or intermediate values during cryptographic operations. CPA relies on advanced statistical methods to measure the strength of these correlations. This technique helps attackers recover cryptographic keys or sensitive data by analyzing the power traces. CPA is particularly effective when the power leakage is less obvious or even in case of countermeasures that can be circumvented with the strong correlation analysis (Fig. 11.4) [15].

A linear model implies specific relationships among the variances of various factors treated as random variables [15]:

$$\sigma_W^2 = a^2\sigma_H^2 + \sigma_b^2$$

In classical statistics, a correlation factor (ρ_{WH}) is introduced to gauge the fitting quality of the linear model between the Hamming distance (W) and the measured power (H). Assuming uncorrelated noise, this definition can be expressed as

$$\rho_{WH} = \frac{\text{cov}(W, H)}{\sigma_W \sigma_H} = \frac{a\sigma_H}{\sigma_W} = \frac{a\sigma_H}{a^2\sigma_H^2 + \sigma_b^2}$$

This equation adheres to the widely recognized property: $-1 \leq \rho_{WH} \leq 1$. When the model is perfect, the correlation factor approaches ± 1 as the noise variance approaches 0, with the sign determined by the linear gain (a). If the model is applicable to only l independent bits out of m, a partial correlation still exists:

$$\rho_{WH}^{l/m} = \frac{a\sqrt{l}}{\sqrt{m}\sqrt{ma^2 + 4\sigma_b^2}}$$

This can be simplified to:

$$\rho_{WH}^{l/m} = \frac{\rho_{WH}^{l}}{\sqrt{m}}$$

The CPA attack process emphasizes the measurement and analysis of correlations:

Algorithm 8 Correlation Power Analysis (CPA) attack process

1: **Measurement:** An attacker measures the power consumption of the cryptographic device while it performs cryptographic operations, capturing multiple power traces.
2: **Multiple Traces:** CPA requires multiple power traces for the same cryptographic operation, each with variations in the data or secret key bits.
3: **Correlation Analysis:** The attacker focuses on the correlation between the power traces and potential values of secret key bits or intermediate data. Specifically, they look for patterns where certain power consumption variations consistently align with specific hypotheses.
4: **Hypothesis Testing:** Attackers formulate hypotheses about the values of secret key bits or data and test them against the observed correlations. They seek to identify the most likely values based on the strength of these correlations.

Some assumptions with DPA mentioned in [15] make it difficult for application in real-world scenarios. These assumptions are listed as follows:

1. Even when the attacker guesses the correct sub-key, DPA peaks can appear in the power consumption traces, even if the targeted bit is not directly manipulated. This contradicts one of the assumptions that power consumption should not depend on the targeted bit unless explicitly handled.
2. DPA can produce peaks for incorrect sub-key guesses, known as "ghost peaks." These can make it difficult for the attacker to make accurate decisions and contradict the assumption that the predicted value for the targeted bit should be independent of the sub-key guess.
3. The genuine DPA peak associated with the correct guess may sometimes be smaller than certain ghost peaks, or it may even be nonexistent or negative.

This can be confusing for the attacker. These anomalies are partly due to the oversimplification of the initial assumption that non-targeted bits have a consistent influence.

In essence, these practical issues challenge the assumptions made when applying DPA, making the attack less straightforward than initially anticipated. On the contrary, CPA is highly effective in situations where power leakage is less pronounced or even in the case of countermeasures is applied to protect the design from side channel attacks. It is adaptable to various cryptographic algorithms and implementations. CPA can reveal secrets even when encryption algorithms are secure against mathematical attacks.

11.3.2 ML Attacks on Power and EM

Machine learning (ML) techniques have gained prominence in the field of side channel attacks, offering novel approaches for extracting sensitive information from electronic devices. In the first study of machine learning in side channel analysis [16], the authors present a powerful technique called Least Squares Support Vector Machine (LS-SVM) and its applications in side channel analysis. The study demonstrates power side channel analysis (SCA), utilizing LS-SVM classifiers. The LS-SVM parameter choice significantly influenced results, highlighting the need for automated parameter tuning. In situations with more complex and nonlinear data, the RBF kernel proves more suitable. The number of power traces used for training notably impacted success rates, especially with a small trace count. Increasing components in classifier inputs led to worse results. Preprocessing techniques and feature selection shows limited improvements. LS-SVM classifiers with RBF kernels outperformed TAs in the interpolated approach, but TAs performed similarly to LS-SVM classifiers with linear kernels, yielding comparable (poor) results in that specific scenario (Table 11.2) [16].

Table 11.2 Success rates (%) for bit(4) approach [16]

Method	Number of components	Success rate (%)
LS-SVM: RBF	3	74.7
LS-SVM: RBF	4	67.1
LS-SVM: RBF	6	52.7
LS-SVM: Linear	3	75.5
LS-SVM: Linear	4	75.5
LS-SVM: Linear	6	75.1
Template Attack	3	73.0
Template Attack	4	75.8
Template Attack	6	75.0

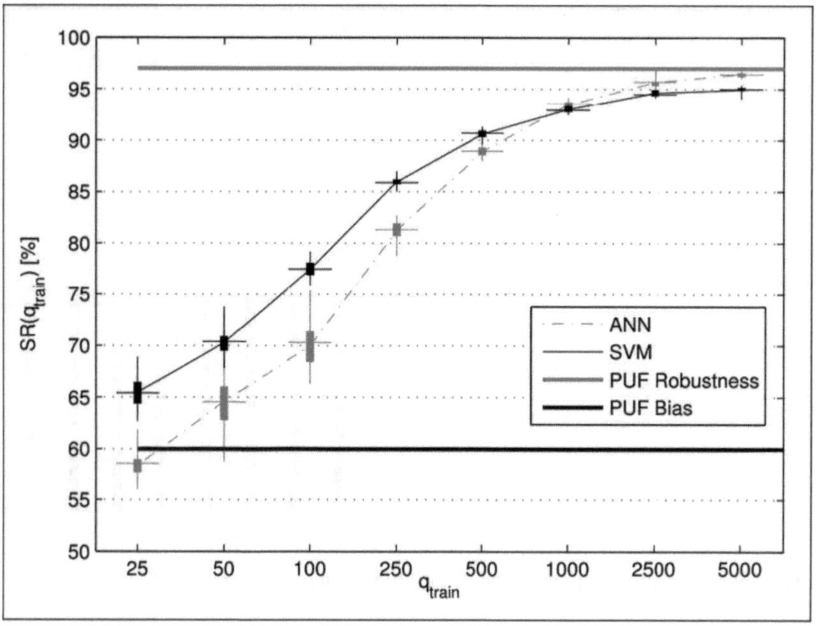

Fig. 11.5 Box plots of obtained success rates (%) for different ML attacks for Simple Arbiter PUF [17]

Machine learning models were first explored as a way to implement modeling attacks on a 65nm CMOS Arbiter PUF; [17]. With minimal training data comprising of a few dozen Challenge-Response Pairs (CRPs), the model was able to accurately predict responses from simple Arbiter PUFs with a success rate significantly higher than random guessing. The results of the modeling attacks on the simple Arbiter PUF using Artificial Neural Networks (ANN) and Support Vector Machines (SVM) are depicted in Fig. 11.5 [17]. On average, SVM tends to yield more accurate models for the Arbiter PUF when *qtrain* is less than or equal to 500, but for larger training sets, ANN outperforms SVM. Specifically, SVM achieves an SR of approximately 70% for *qtrain* equal to 50, and when *qtrain* reaches 500, both SVM and ANN exhibit the capability to predict responses with an accuracy close to 90%. Notably, when *qtrain* exceeds 5,000, ANN becomes proficient at modeling the Arbiter PUF with success rates approaching the PUF's inherent robustness. The decreasing height of the box plots suggests that as the value of *qtrain* increases, the estimation of SR(*qtrain*) becomes more precise.

Hyperparameter optimization is a critical process in machine learning that holds paramount importance in achieving the best performance from a model [18]. There is absence of comprehensive benchmarks for side channel analysis, especially in terms of electromagnetic traces. Focusing on crucial time-domain properties such as Mean and Kurtosis and frequency-domain properties such as Median Amplitude of Spectrum and Mean, Median and Frequency ratio distribution as

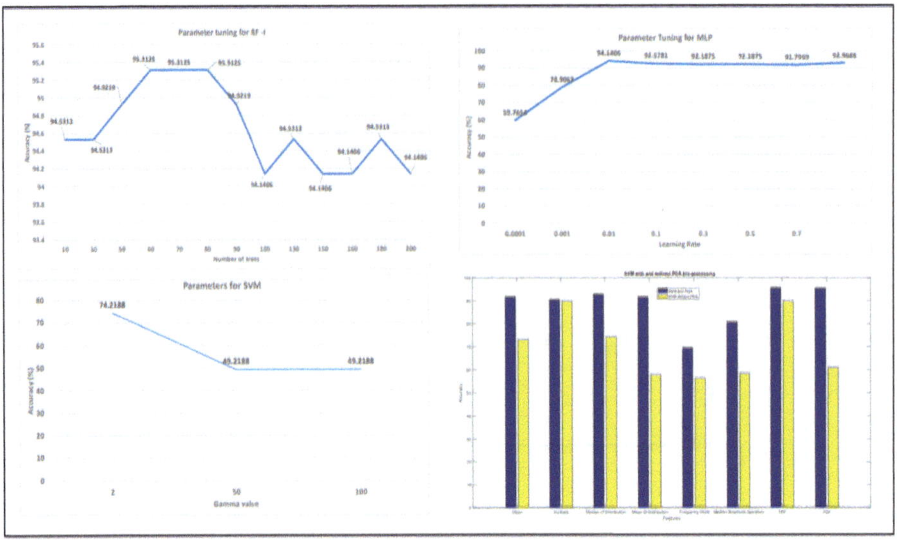

Fig. 11.6 Hyperparameter optimization and its effects on different models [18]

features, various combinations of models to features are tested to assess changes in accuracy. Employing Random Forest, Naive Bayes, Support Vector Machine, and a Multilayer Perceptron with k-fold cross-validation to mitigate biases and PCA for dimensionality reduction. Significant improvement is observed when PCA is used with the SVM algorithm. Other parameters such as the number of trees, gamma values, and learning rates in Random Forest, SVM, and MLP models were also varied, demonstrating the resulting accuracy fluctuations shown in Fig. 11.6 [18].

The supervised learning approach can also be used to infer from applications running on an Android platforms using EM side channel emissions [19]. Machine learning methods exhibit superiority over profiling-based approaches for complex System-on-Chip (SoC) architectures. Experimental results on a Quad-core Snapdragon 820 SoC demonstrate early detection of these apps within 700 milliseconds, with an accuracy rate exceeding 85% for unknown applications. The Snapdragon 820 SoC features two clusters of Kyro cores, providing a comprehensive view of DVFS state information, enhancing the accuracy of the experimental analysis Fig. 11.7 [19]. The hardware setup for this experiment is described in Fig. 11.8 [19].

While the prior work demonstrates attacks on sequentialized hardware, a novel deep-learning-based SCA on lattice-based key-exchange protocols targeting parallelized hardware was proposed in [20]. The technique targets parallelized implementations of post-quantum key-exchange (PQKE) protocols, specifically Frodo and NewHope. Results demonstrate the model's superiority over conventional techniques, achieving up to a 1.5× improvement in recovering a 100% success rate with less data compared to deep learning with 1-D input. Machine learning

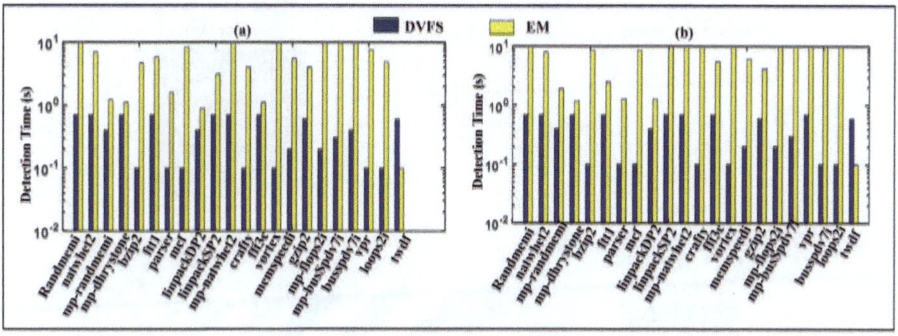

Fig. 11.7 Detection time for applications with (**a**) SVM inference model and (**b**) RF inference model using features from DVFS and EM side channel [19]

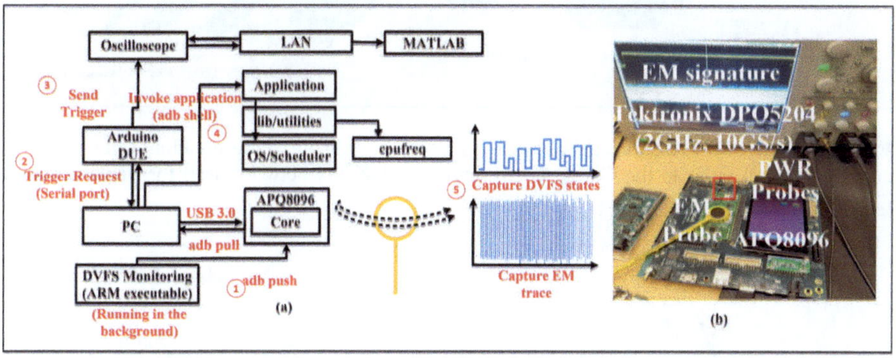

Fig. 11.8 Measurement setup to capture DVFS and EM signatures (**a**) and Open-Q system on module development platform (**b**) [19]

techniques outperform template attacks by up to 1.25×. The evaluation setup involves a Sakura-G board with a Xilinx Spartan-6 XC6SLX75 FPGA, measuring voltage drop across a 1-ohm shunt resistor, and utilizing a PicoScope-3206D oscilloscope for power measurements. Preprocessed raw power traces and model training took approximately 10 minutes, yielding an accuracy rate exceeding 99%. The input processing required to convert 1D input into 2D reconstructions utilizing Gramian angular sum field (GASF) and using that as input to the 2D-CNN network is shown in Fig. 11.9 [20].

The results suggest that 2D-CNN is a promising and robust choice, particularly when addressing the challenges posed by algorithmic noise and cross-device attacks as shown in Fig. 11.10 [20].

The potential challenges and opportunities pertaining to the use of machine learning techniques for hardware security are detailed in [21]. The two defense mechanisms—dummy state insertion and clock jitter-based hiding—are used to counteract classical attacks. Dummy state insertion strategically introduces false or dummy operations during the execution of cryptographic algorithms, which

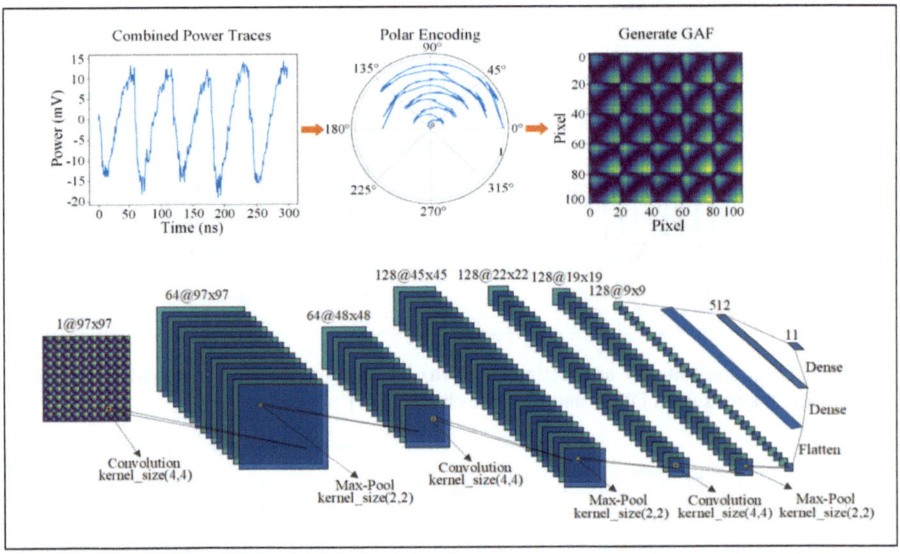

Fig. 11.9 Conversion of five subtraces to a polar coordinate representation followed by a conversion from polar to the GASF. GASF input to 2D-CNN network used for implementation [20]

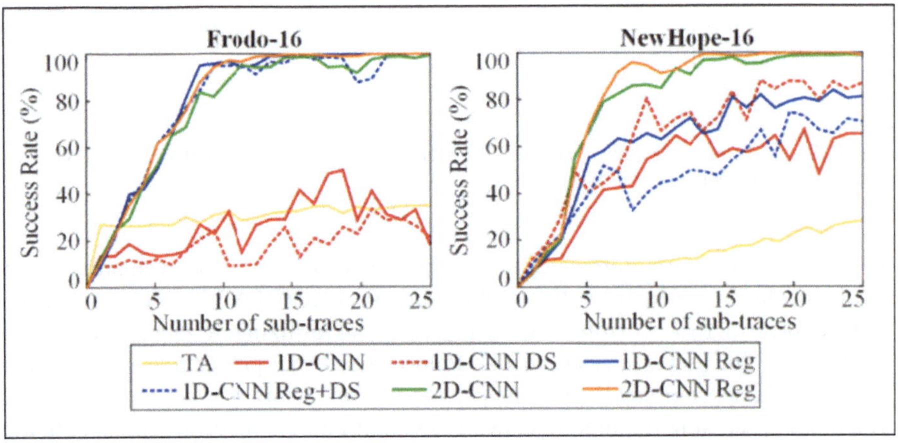

Fig. 11.10 Comparison of cross-device attacks on Frodo-16 and NewHope-16 [20]

makes it challenging for adversaries to extract meaningful information from the side channels. Similarly, clock-based jitter serves as a countermeasure by introducing intentional variations in the timing of cryptographic operations. Deep-learning-based attacks, specifically using a convolutional neural network, prove adaptable and robust against these countermeasures, maintaining a 100% success rate in contrast to horizontal DPA and template attacks Fig. 11.11 [21].

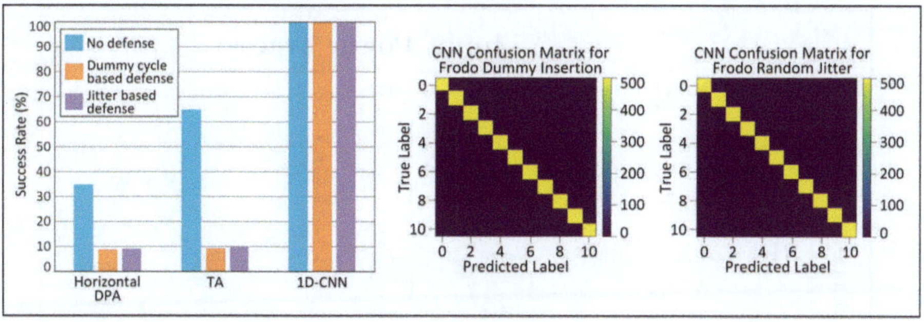

Fig. 11.11 Comparison of SCA defenses using all available subtraces (i.e., 84 subtraces) of Frodo and the corresponding confusion matrix for CNN [21]

Emphasizing model training via a multiphase approach, the work in Deep K-TSVM [22] presents a pioneering deep learning model that integrates deep neural networks and Twin Support Vector Machine (TSVM). This model employed pretraining with the Restricted Boltzmann Machine (RBM) method and subsequent fine-tuning using gradient descent. Efficient model building attacks on cryptographic implementations, even with a single trace, include CNNs, NeuroSVM, and MLP-based attacks. The impact of hyperparameter selection on the model's generalization is highlighted on the positive influence of pretraining. The identification of the correct key candidate by examining sorted probabilities from the Identity (ID) leakage model offers a robust metric for evaluating deep-model-based side channel attacks Fig. 11.12 [22]. Model results demonstrate the capability to recover secret keys with 99% success rate Fig. 11.13 [22].

To ensure confidentiality, integrity, and authenticity of startup configuration files, the secure boot mechanism in ZU+ devices is evaluated [23]. Using Electromagnetic (EM) measurements, the first side channel analysis of the ZU+ encryption unit is conducted. The AES encryption rounds and Deep Learning (DL) methods to attack the 256-bit secret key are employed, providing valuable insights for configuring the key rolling parameter and enhancing device security. The study uses a dataset of 200,000 traces with random keys and IVs, allocating 75% for training, 24% for validation, and 1% for attack traces. Linear Discriminant Analysis (LDA) and Correlation Power Analysis (CPA) are applied for baseline key recovery attempts. Two DNN models, Multilayer Perceptron (MLP) and Convolutional Neural Network (CNN), are employed in the analysis, using trace standardization for the MLP and Adam optimizer for training (Figs. 11.14 [23] and 11.15) [23].

Machine-learning-based side channel attacks represent a powerful and evolving approach in the field of hardware security. They offer automated and adaptable methods for extracting sensitive information from electronic devices, making them a significant concern for device manufacturers and a subject of ongoing research for robust countermeasures.

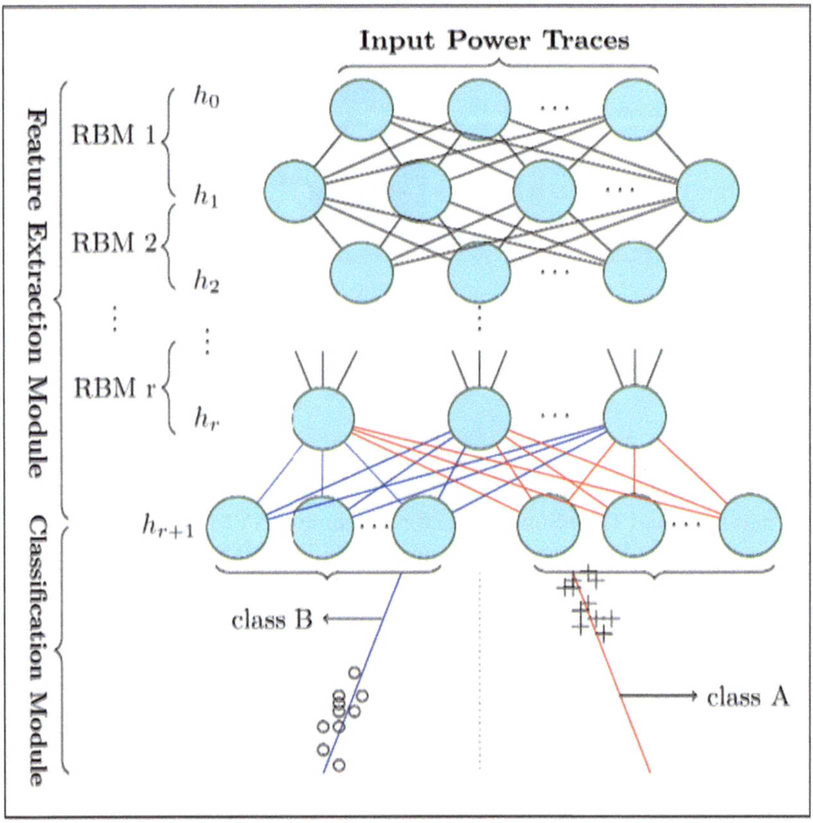

Fig. 11.12 Deep K-TSVM architecture [22]

11.3.3 Direct Memory Access (DMA) as a Side Channel Attack

Direct Memory Access (DMA) is a hardware feature that allows peripheral devices to access a computer's memory without involving the central processing unit (CPU). While DMA is essential for efficient data transfer, it can also be exploited as a side channel attack vector.

The seminal work in [24] describes a sophisticated data processing system incorporating Direct Memory Access (DMA) controllers connected to a bus architecture. These controllers are responsible for managing memory access, each generating dual addresses—one for its own memory space and another for a memory space controlled by a different DMA controller. This enables simultaneous transfer of data between these memory spaces within a single clock cycle. When a DMA controller assumes the role of a bus master, it signals its intended access to the system, user, or external memory space. Crucially, each controller includes comparison circuits to verify address validity against predefined limits, triggering an error if

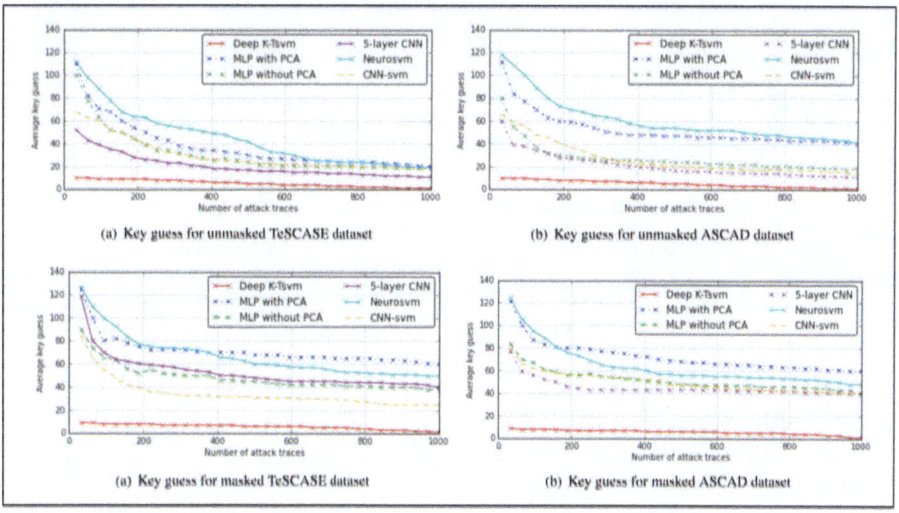

Fig. 11.13 Average key rank for third key byte on masked and unmasked implementation of AES [22]

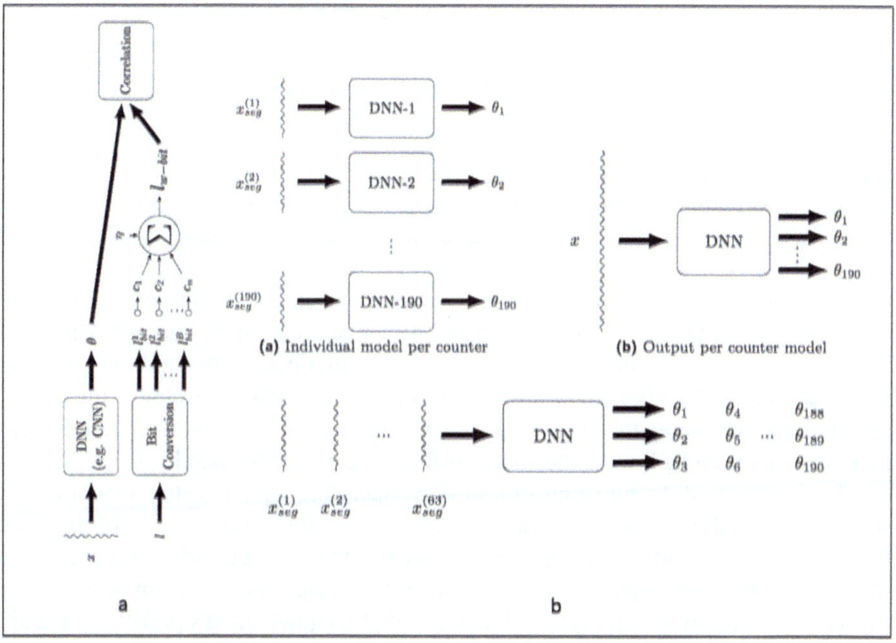

Fig. 11.14 Correlation optimization using small MLP model to approximate bitwise weighting (**a**) and different approaches to attack the ZU+ encryption engine (**b**) [23]

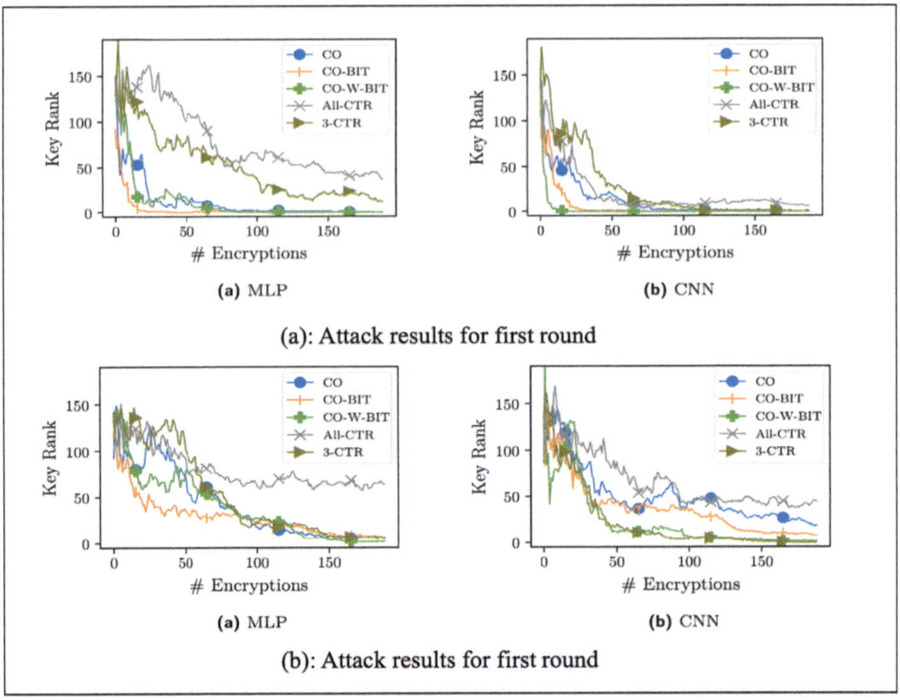

Fig. 11.15 Attack results for different models on first and second rounds of AES [23]

unauthorized access is detected. Furthermore, the controllers feature registers for interrupt addresses, constant values, and status, facilitating automatic transfer of this information over the bus for storage at the interrupt address upon exception detection.

The resilience against physical security threats facilitated by Firewire technology was evaluated on Windows 7 platform [25]. An authentication-bypass attack method leverages Firewire's read/write memory access to manipulate the code responsible for password verification. Through an in-memory patching strategy, the attack modifies the authentication process, allowing the use of any password for system login or unlocking following the successful application of the patch.

The securing of feature-rich commodity operating systems due to numerous security vulnerabilities is an ongoing concern. To address this issue, [26] explores the concept of excluding the operating systems from the trusted computing base of applications while maintaining their benefits through the use of hypervisors or trusted hardware. A novel threat called controlled-channel attacks pose a significant risk to shielding systems like Overshadow, Ink Tag, or Haven. These attacks enabled an untrusted operating system to extract sensitive information from protected applications, such as complete text documents and outlines of JPEG images, compromising the envisioned protection of unmodified legacy applications in the presence of legacy operating systems on off-the-shelf hardware.

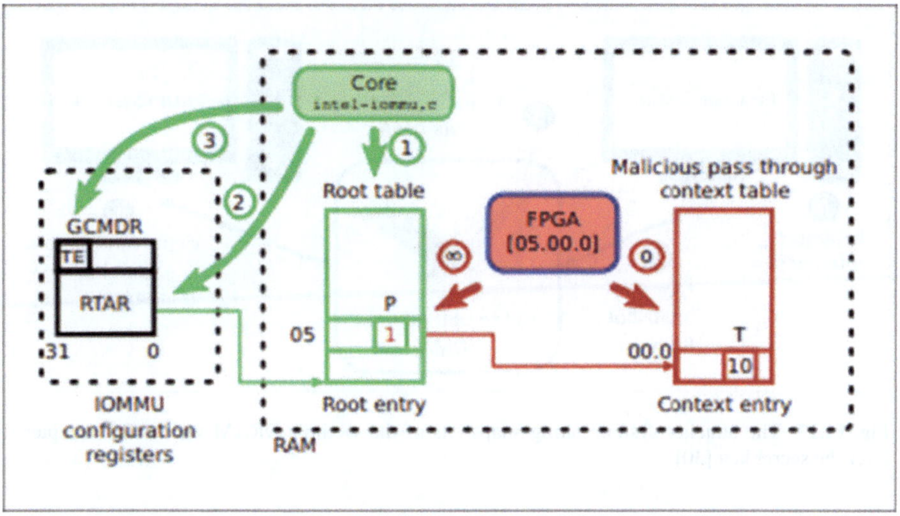

Fig. 11.16 Linux IOMMU driver attack implementation [27]

A design weakness in the configuration of an IOMMU by the Intel IOMMU Linux driver was discovered in [27] leading to a potential exploitation scenario allowing a malicious peripheral to bypass the underlying protection mechanism Fig. 11.16 [27]. The successful exploitation of a vulnerability related to the configuration of the Linux IOMMU kernel driver was demonstrated in the experiment, enabling unauthorized read and write accesses in main memory by malicious peripherals and bypassing the embedded protection mechanisms in the IOMMU. Other operating systems, including Windows and BSD systems, were investigated to assess the exclusivity of this vulnerability to the Linux kernel. Additionally, scrutiny extended to the boot process when Intel TXT was activated, ensuring the technology did not exhibit similar weaknesses.

Similarly, Thunderbolt, a proprietary I/O protocol promoted by Intel and integrated into various laptops, desktops, and systems, served as an external interconnect exposing the internal PCI Express (PCIe) domain to external devices [28]. Thunderbolt, an Intel-promoted I/O protocol integrated into various systems, served as an external interconnect exposing the internal PCIe domain. "Evil Maid" attacks exploiting Thunderbolt's DMA-enabled I/O, providing complete access to a PC's state and system memory with minimal physical access. Revised Thunderbolt controllers introduced software-based access control measures to authorize trusted devices, addressing the Thunderspy vulnerabilities. In computer security, systems with memory over 4GB face vulnerabilities as Kernel Modules (KMD) with unrestricted access can exploit PCIe DMA, searching for code signatures and patching identified code [29]. Robust defenses against Direct Memory Attacks on the Kernel are crucial to safeguard systems from potential exploits.

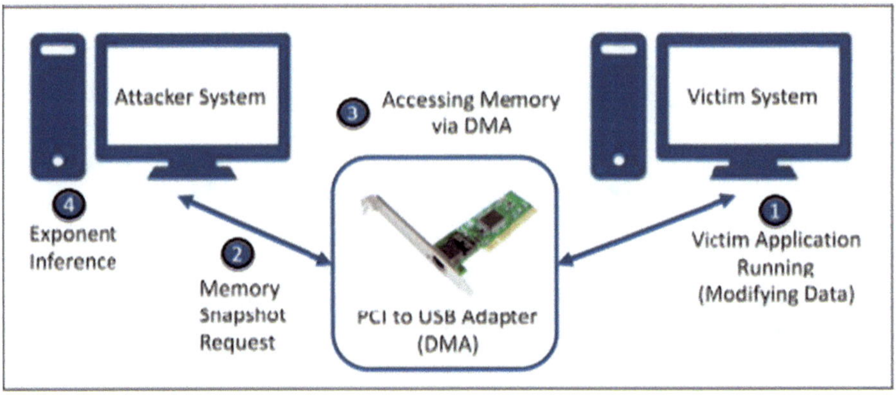

Fig. 11.17 The attacker system taking snapshots of the victim's DRAM via the PCI adapter to infer the secret key [30]

The exposure of sensitive user data through data-dependent access patterns to untrusted storage systems is a well-known concern. A study presented an actual attack on the power-side-channel-resistant Montgomery's ladder-based modular exponentiation algorithm, widely utilized in public-key cryptography [30]. To understand the victim algorithm's write access patterns in realistic settings, a compromised DMA device was leveraged to capture frequent snapshots of the application's address space as shown in Fig. 11.17 [30]. The study also explores the potential of a similar attack on the McEliece public-key cryptosystem, exploiting write access patterns to unveil the secret key.

The widespread adoption of the Direct Memory Access (DMA) technique has significantly enhanced data transfer efficiency between peripherals and host machines. Nevertheless, these efficiency improvements come at the cost of heightened security vulnerabilities, particularly susceptibility to DMA attacks exploiting direct access features to pilfer data from the live memory of the target system. The proposed approach in [31] involves constructing a unique identifier for each DMA-supported PCIe device based on profiling time and establishing a trusted database for authentication purposes. Experimental results demonstrate the scheme's efficacy in reducing noise during the identifier construction process, achieving a 100% success rate in device authentication across all tested devices. Following up on this paper, a machine learning model was also utilized to do a comparative analysis as well as preparing a countermeasure for DMA attack mitigation that will be discussed in the later sections (Fig. 11.18) [31].

11.3.4 ML Attacks on DMA

The study in [10] outlines a lightweight process that creates resilience against DMA attacks with minimal modifications to the DMA protocol configuration. Results indicate the generation of unique identifiers for trusted devices and successful

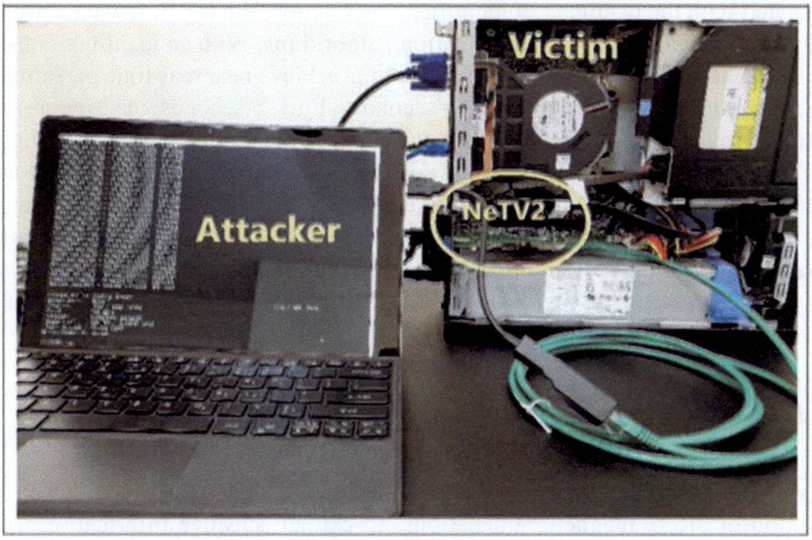

Fig. 11.18 Experimental attack setup for DMA [31]

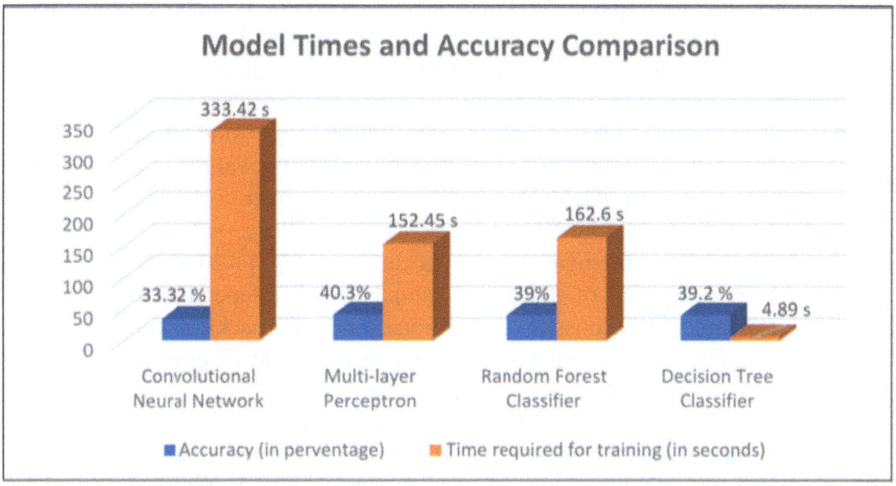

Fig. 11.19 Time required for training versus accuracy for each of the four models [10]

authentication. The paper explores various machine learning models such as Decision Trees, Random Forest, multilayer perceptrons and Convolutional Neural Networks to enhance authentication, emphasizing the success of the proposed security measures in authenticating PCIe devices utilizing DMA without physical modifications to system hardware or protocols. The machine learning models demonstrate device authentication (40% accuracy) using raw data, despite some overlap in profile times for the devices Fig. 11.19. While there is a limited accuracy associated with the overhead of identifier construction, the use of raw data provides a significant advantage to machine learning algorithms. With an identifier construction overhead of 4.89 seconds, these algorithms achieve near real-time performance, authenticating devices within a few seconds (1 to 2 seconds on average). The continual improvement of algorithms with the expansion of the dataset suggests promising results for machine learning applications in device authentication. Further exploration involves machine learning algorithms to improve time efficiency and accuracy, considering environmental changes and real-time resource utilization's impact on profile time variations. Ongoing studies aim to refine the authentication process over time, addressing factors like aging and environmental fluctuations.

11.4 Side Channel Attacks on ML Models

In cybersecurity, side channel attacks exploit unintended signals, like power consumption or electromagnetic emanations, to extract sensitive information from a target system. Machine learning model-based attacks can unveil key parameters, compromising model confidentiality. Adversaries use side channel attacks to steal the structure and hyperparameters of machine learning models, posing threats to intellectual property. As machine learning advances and models are deployed widely, safeguarding against side channel attacks becomes crucial for protecting the integrity and security of these assets.

The reverse engineering of neural networks through side channel analysis techniques is explored in [32]. The attack methodology is validated on the Atmel ATmega328P and on the ARM Cortex-M3 microcontroller. Side channel activity is captured using the Lecroy WaveRunner 610zi oscilloscope, with each known input generating a measurement or trace as shown in Fig. 11.20 [32]. The number of traces, each composed of numerous samples, varies depending on the target operation, ranging from thousands to millions. The chip package is mechanically opened to enhance measurement quality, and a near-field electromagnetic (EM) probe is employed to collect EM measurements, functioning as an antenna for spying on the EM side channel leakage from the underlying processor running machine learning (ML). This setup allows for both side channel and timing analysis. The methodology's relevance extends to GPU and FPGA platforms, with known leakage models for these environments. While the low signal-to-noise ratio (SNR) of ARM Cortex-M3 measurements may require increased sample sizes and signal preprocessing, the core principles of the analysis remain applicable. The attacks, more generic and potent than previous works, are able to successfully recover sensitive parameters of the neural network.

The GUINNESS (GUI-based binarized neural network synthesizer) serves as an open-source tool flow facilitating the implementation of binarized deep neural networks on FPGAs [33]. The tool includes both GPU-based training and FPGA-based inference through a GUI, eliminating the need for software designers to script neural network structures or training behaviors. The tool flow addresses challenges

(a) Target 8-bit microcontroller mechanically decapsulated (b) Langer RF-U 5-2 Near Field Electromagnetic passive Probe

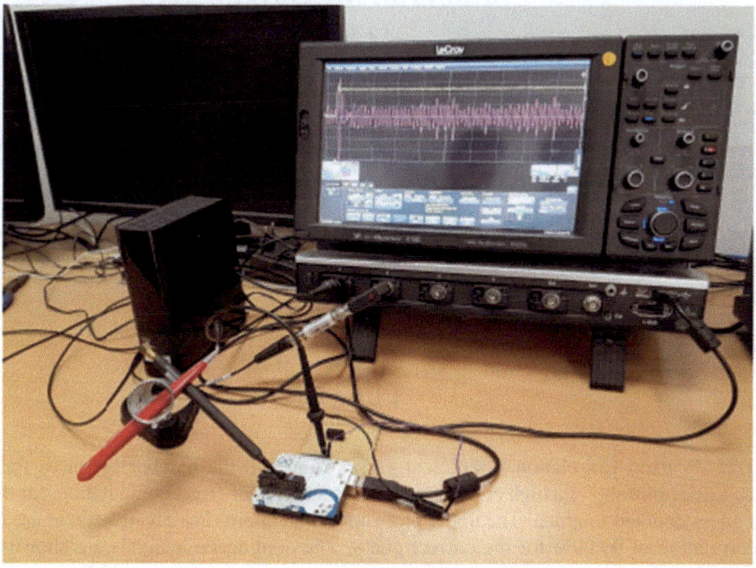

Fig. 11.20 The complete measurement setup [32]

in binarized CNN hardware, modifying training algorithms to handle minimal bias, and eliminates the need for additional hardware in inference. Benchmarking against conventional FPGA-based binarized implementations, the tool showcases comparable classification accuracy but excels in performance per power efficiency (5.1 times), performance per area efficiency (8.0 times), and performance per memory (8.2 times). Comparative analyses against CPU and GPU designs reveal the FPGA-based design's significant speed improvement (1776.3 times faster than ARM Cortex-A57, 11.5 times faster than Maxwell GPU), lower power dissipation (3.0 times lower than ARM Cortex-A57, 7.3 times lower than Maxwell GPU), and superior performance per power efficiency (5706.3 times better than ARM Cortex-

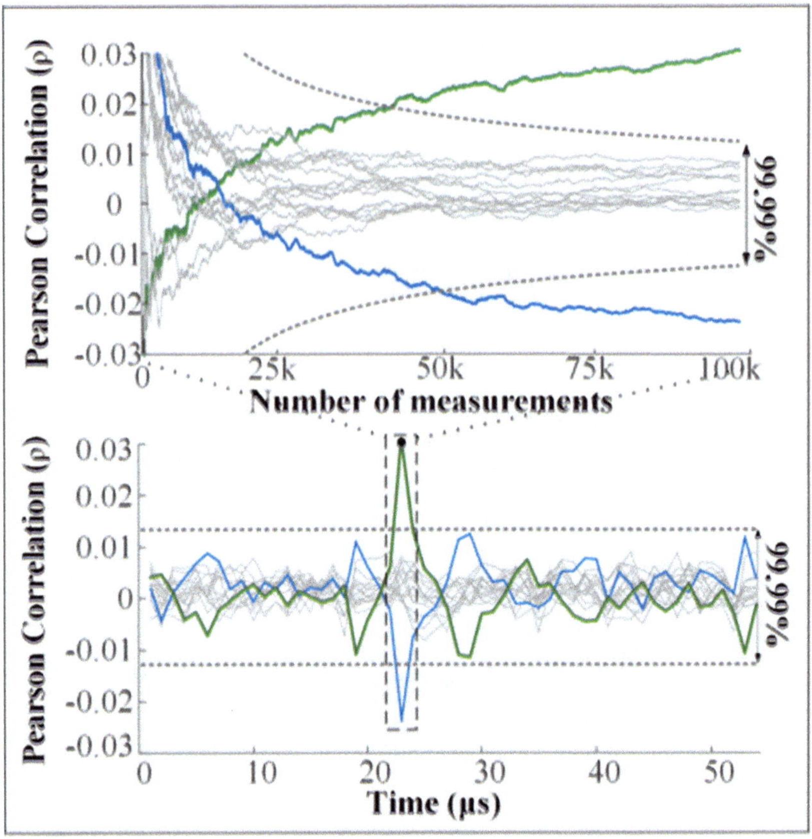

Fig. 11.21 Pearson Correlation Coefficient versus time and the number of traces for DPA on weights. Lower plot shows a high correlation peak at the time of target computation, for the correct weight guess denoted in green. The upper plot shows that approximately 40k traces are needed to get a correlation of 99.99% for the correct guess. The confidence intervals are shown in dotted lines. The blue plot denotes the 2's complement of the correct weight guess [34]

A57, 83.0 times better than Maxwell GPU). However, a notable drawback is the additional time required for FPGA executable code synthesis, with a total design duration of 75 hours, including three hours for synthesis. Despite this, considering the dominant role of CNN training, the overhead remains reasonable.

Dubey et al. [34] demonstrates DPA attacks during inference, extracting secret model parameters such as weights and biases of a neural network. This work highlights the need for new avenues in side channel analysis research, specifically tailored to deep learning inference engines, and emphasizes the importance of investing in securing deep learning frameworks given the diverse nature of neural networks and the ongoing challenges in masking (Fig. 11.21) [34].

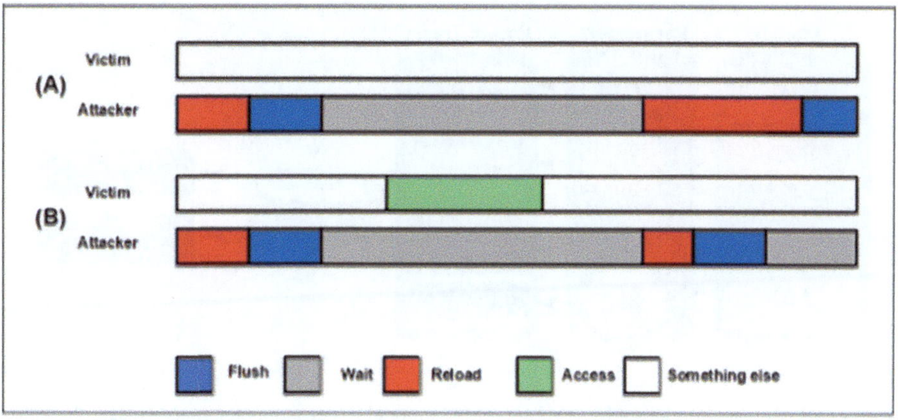

Fig. 11.22 Flush + Reload principle [1]

The efficacy of deep learning algorithms on security evaluations is explored; however, the hyperparameters required to reproduce the results often remain undisclosed. A comprehensive study of deep learning algorithms in the context of side channel analysis, establishing connections with classical template attacks and the exploration of hyperparameter choices for convolutional neural networks, is performed in [35]. The study reveals that the design principles of the VGG-16 algorithm, used for image recognition, also prove effective in defining architectures for side channel analysis. ASCAD stands as the first open database in its category, positioned to serve as a foundational resource for future research in this domain.

An experiment to extract a support vector machine (SVM) kernel function using the Flush + Reload technique via information through a cache side channel is performed in [1], posing a serious threat to the security of machine learning models. It specifically targets the scikit-learn machine learning framework, successfully requiring only two attacks to accurately infer the kernel function utilized by the victim's SVM training program. Introducing a novel approach grounded in the Flush + Reload cache side channel attack Fig. 11.22 [35], the paper demonstrates the effectiveness of this method in stealing the SVM kernel function, thereby compromising the confidentiality integral to the security requirements of machine learning models.

While cryptographic algorithms, often preferred targets, exhibit vulnerability to such attacks, machine learning algorithms, due to their ubiquity and rigidity, become susceptible to the inference of model parameters, raising significant security concerns. Utilizing the simplified process of deriving sensitive information from cryptographic keys, such as AES keys, through a $35 USD microcontroller is detailed in [36]. This investigation employs traditional side channel techniques, particularly power analysis using a ChipWhisperer Lite, to extract information from

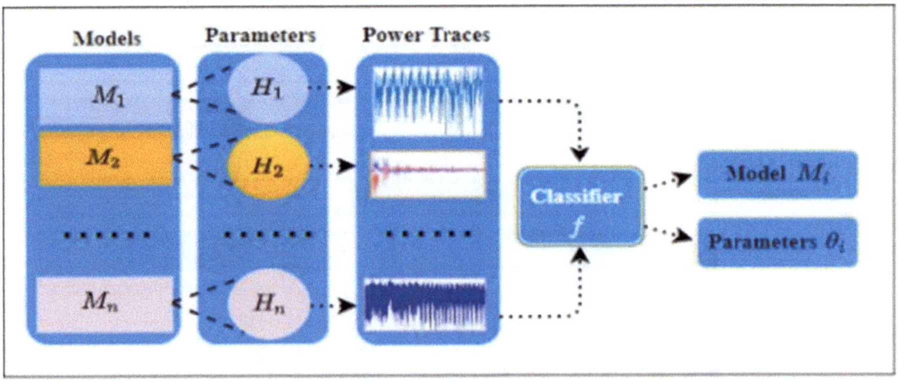

Fig. 11.23 Given a model set M = M1, · · ·, Mn, a model hyperparameter space H, sample matrix XL and the corresponding label vector YL, train a classifier to classify unlabeled power traces Xu, and further infer the running model with some specific hyperparameter values [36]

Neural Networks and Linear Regression models running on a target microcontroller. Time series classification tasks are then performed on the power traces, showcasing the ability to differentiate between various models and their hyperparameters Fig. 11.23 [36]. The accuracy of these classification tasks ranges from 100% when distinguishing between neural networks and linear regression models to 67.92% in the more challenging task of differentiating between virtually identical linear regression models. The results affirm the feasibility of inferring machine learning model parameters through power-based side channel attacks, highlighting the potential security implications. The findings underscore the distinguishability of power traces from different machine learning training cycles, paving the way for further exploration into parameter inference.

Existing approaches at the time utilize mathematical attacks or exploit hardware vulnerabilities such as side channel leakage for stealing trained machine learning (ML) models. The emphasis on escalating development costs associated with these models is studied as an emerging concern in [37]. This work introduces a novel attack vector targeting ML models on embedded devices through the abuse of scan chain infrastructure Fig. 11.24 [37]. The results highlight the effectiveness of this new threat vector for stealing ML models from hardware, specifically through the exploitation of debug ports and observation of intermediate states during inference.

Reverse engineering of model internals highlights the escalating threat of intellectual property (IP) thefts targeting trained ML models through side channel attack on inference engines [38]. Existing works indicate that a straightforward adaptation of such defenses falls short in providing comprehensive security or results in significant area overheads. To overcome these limitations, this study proposes the construction of neural networks to be inherently more compatible with masking. The implementation of fully masked neural networks on an FPGA

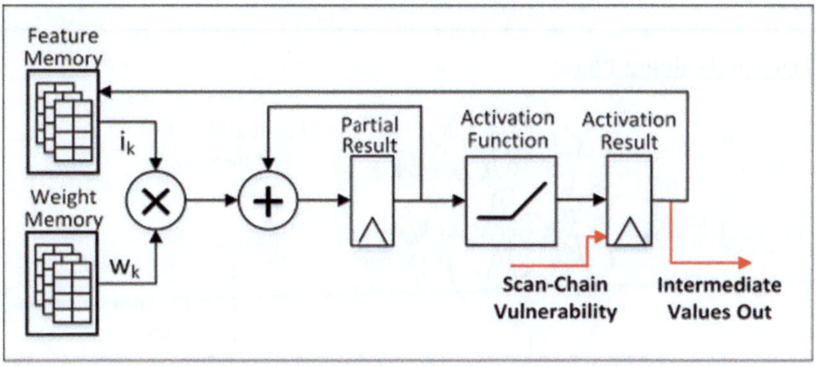

Fig. 11.24 Scan chains enable accessing internal states and leaking intermediate results, which can be used to efficiently extract parameters [37]

showcases comparable latency with a notable reduction in Flip-Flop (FF) and Look-Up Table (LUT) costs, surpassing state-of-the-art protected implementations by 34.2% and 42.6%, respectively. Furthermore, the study demonstrates the first-order side channel security of these implementations with up to one million traces.

With the rising need of safeguarding machine learning models against IP Piracy attacks, this work introduces and optimizes various side channel defenses for ML models within hardware blocks, focusing on Boolean masking as the primary defense mechanism [39]. Implementation of the masked hardware blocks is coupled with an optimized adder design to minimize area and latency overheads, complemented by a shuffle-based defense (Fig. 11.25) [39].

Machine learning and deep learning approaches exhibit advantages over earlier methods such as the template attack, particularly in scenarios with limited training traces; however, there are challenges pertaining to imbalanced datasets and evaluation metrics [40].

Methodologies such as Bayesian optimization and reinforcement learning have emerged as effective approaches in hyperparameter tuning. These techniques involve sophisticated computational processes that aim to fine-tune hyperparameters systematically, contributing to improved performance. It highlights the significance of leveraging advanced optimization strategies to enhance the efficiency and effectiveness of machine learning models. However, with the advent of targets featuring more intricate countermeasures, minimal architectures may prove inadequate, necessitating innovative deep learning approaches. This study delves into the performance of residual neural networks (ResNets) in side channel analysis and explores the construction of deeper ResNets capable of handling larger input sizes with minimal tuning requirements [41]. Notably, the findings highlight the effectiveness of ResNets, particularly in scenarios with a large number of profiling traces and features within a trace (Fig. 11.26) [41].

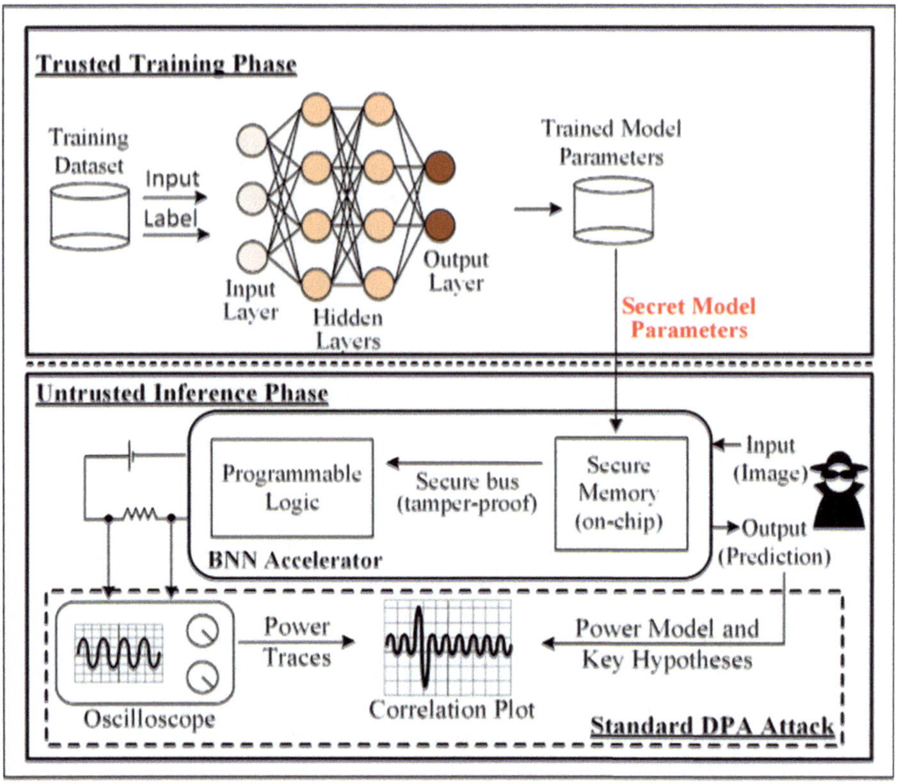

Fig. 11.25 Standard DPA threat model applied to ML model stealing where the trained neural network is deployed on an edge device running in an untrusted environment [39]

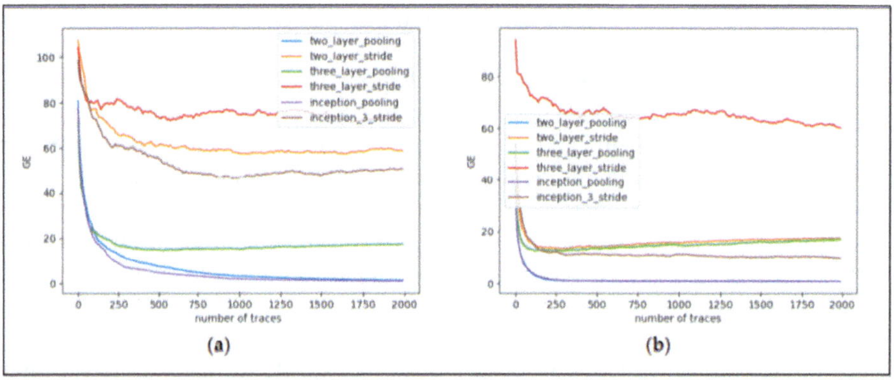

Fig. 11.26 Attacking performance of both the final model and the model that had the best validation loss. (**a**) Attack performance of the models after 100 epochs. (**b**) Attack performance of models achieving best validation loss [41]

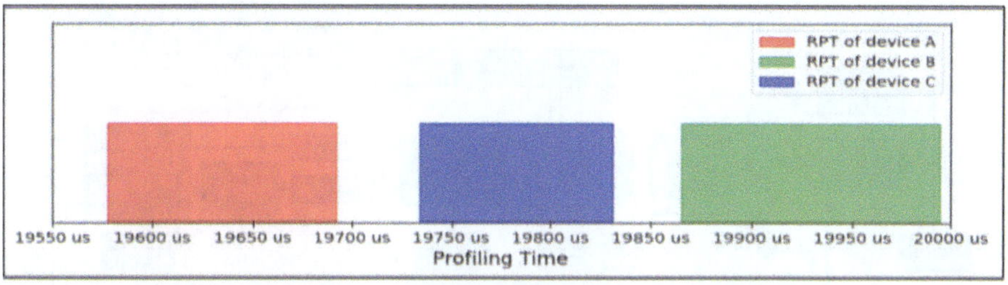

Fig. 11.27 Range of profile times for each device [10]

11.5 Novel AI-Based Countermeasures

Given the significance of protecting the architecture and weights of these models against extraction attacks, especially when adversaries have physical access to the devices, dedicated efforts have been directed toward devising specialized watermarking and fingerprinting techniques for ML hardware.

A digital watermarking technology is used as a means to ensure intellectual property protection and the detection of infringement in the utilization of trained models [42]. The approach embeds the watermarks into deep neural networks. The method does not compromise the performance of the host networks, as the watermark is seamlessly embedded during the training process. Its efficiency in embedding watermarks during the training lifecycle of a DNN encompasses both initial training and subsequent fine-tuning and distillation stages, without detrimentally affecting performance. The embedded watermark persists even post fine-tuning or parameter pruning, showcasing resilience even after a substantial reduction of up to 65% of parameters.

Furthermore, this watermark can be applied to secure authentication of DMA devices as discussed in the research paper in [10] that illustrates a secure authentication process for DMA devices based on ROI (Region of Interest) and machine learning techniques. The proposed security measures establish an authentication framework for PCIe devices utilizing DMA, employing unique profiling times for each device Fig. 11.29 [38]. Importantly, this lightweight scheme necessitates no physical alterations to the system hardware or protocols and has negligible impact on the overall system performance.

Other AI-based countermeasures include securing the neural networks using a masking technique. As discussed in [34] the countermeasure involved by securing neural networks through the masking technique. The countermeasure development process addresses the intricacies of masking neural networks and outlines the complete implementation of neural network inference. The fundamental concept behind masking involves ensuring that all intermediate computations remain independent of the secret key, hindering adversaries from correlating leaked information with the secret value. Notably, masking is implementation-agnostic, allowing its

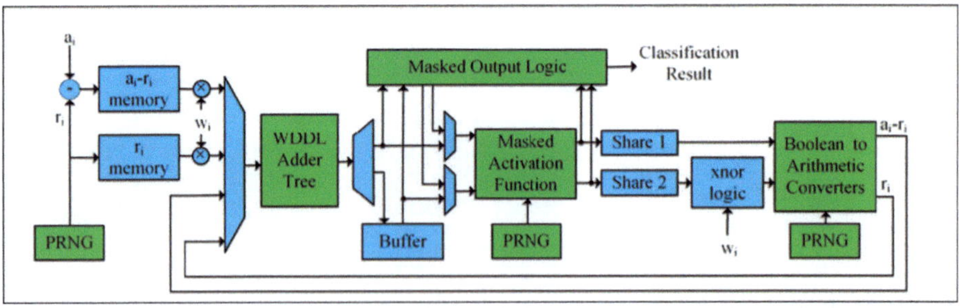

Fig. 11.28 Components of a masked BNN. The blocks in green represent the proposed masking blocks [34]

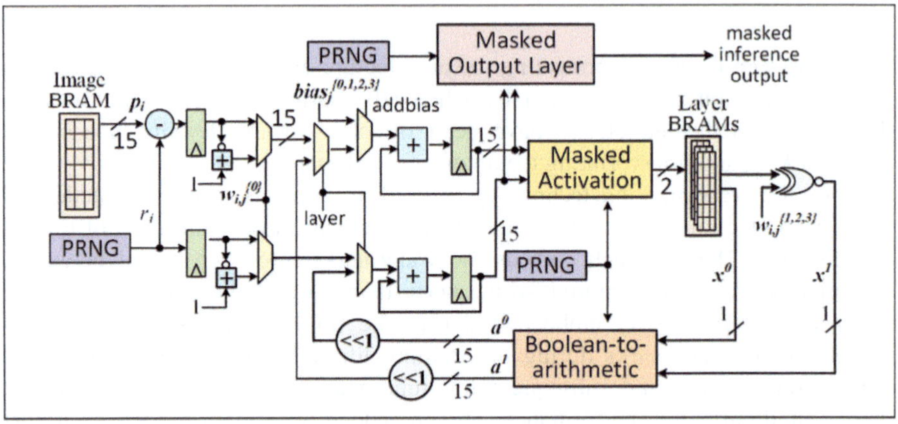

Fig. 11.29 Hardware design of the fully masked ModuloNET inference showing various top-level masked components [38]

application on various circuit styles such as FPGAs or ASICs without necessitating manipulation of back-end tools.

Similarly the masking technique can be deployed in ML applications by alternating the target algorithm itself that is traditionally not done in cryptography [38]. The technique constructs efficient masking schemes, empirically validating their security through practical side channel tests and theoretically validating them with proofs. The neural networks are constructed to make use of masking optimizations for the given network style. The complexity of such optimizations is demonstrated through the masking of ConvNets, extending to other network types.

The first fully masked neural network was introduced in [39], demonstrating its security against up to two million traces and providing a quantification of the potential overheads associated with countermeasures. A significant challenge addressed masking integer addition through Boolean masking, extending this approach to tackle the unique linear and nonlinear computations inherent in fully connected neural networks, which diverge from traditional cryptographic applications. Furthermore, the coupling of a first-order masked design with a lightweight shuffle countermeasure enhances second-order side channel security. The combined effect of these defenses improves the side channel resilience.

The masking technique can be applied on the BNN networks where the conventional continuous-valued weights of a neural network are replaced with binary values, typically $+1$ or -1. This binary representation significantly reduces the memory footprint and computational requirements of the network, making it more efficient for deployment on resource-constrained devices. The presented ideas in the study serve as a benchmark for analyzing vulnerabilities in neural network computations and constructing more robust and efficient countermeasures.

The increasing integration of machine learning (ML) algorithms in security-sensitive applications has raised concerns regarding the vulnerability of these systems to side channel attacks. ML-based side channel attacks exploit subtle information leakage from a system's physical implementation, such as power consumption or electromagnetic emanations, to infer sensitive information.

11.6 Side Channel Attacks for Post-quantum Cryptography

We know now that side channel attacks exploit physical leakages from cryptographic implementations, such as timing information, power consumption, electromagnetic emissions, or acoustic signals, to extract secret keys or other sensitive data. The power of Artificial Intelligence and Machine Learning makes these side channel attacks more formidable than ever.

As cryptographic paradigms evolve to resist the threats posed by quantum computing, post-quantum cryptography (PQC) algorithms are being developed to ensure security against quantum adversaries. However, the resilience of these PQC algorithms to side channel attacks remains a critical concern. While PQC algorithms are designed to withstand theoretical attacks from quantum computers, their practical implementations must also be scrutinized and fortified against side channel vulnerabilities.

One study delves into the vulnerabilities of PQC implementations to physical leakages, proposing countermeasures to enhance the resilience of these cryptosystems in practical scenarios [43]. Another research effort focuses on error detection schemes for lattice-based key encapsulation mechanisms (KEMs), specifically applied to FrodoKEM, Saber, and NTRU, which are finalists in the NIST PQC standardization process. These schemes involve recomputing with shifted, negated, and scaled operands, and their implementation on FPGA devices demonstrates acceptable overhead and high error coverage [44]. Further, another paper presents fault injection countermeasures for the McEliece cryptosystem, concentrating on different composite field arithmetic units. The proposed architectures utilize tailored error detection signatures, and FPGA implementations are used to benchmark the overhead and performance impact, highlighting their suitability for constrained embedded systems [45]. Additionally, novel error detection schemes for number-theoretic transform (NTT) architectures are introduced, crucial for polynomial multiplication in post-quantum cryptosystems. These schemes detect both permanent and transient faults by recomputing with negated, scaled, and swapped operands. Implementation on application-specific integrated circuits (ASICs) shows acceptable hardware overhead, making these schemes suitable for deeply embedded architectures [46]. These advancements in securing cryptographic implementations against side channel attacks, particularly in the context of post-quantum cryptographic systems, are also active areas of research for researchers and practitioners.

Similarly, lightweight cryptography (LWC) is designed to provide robust security for resource-constrained devices, such as those found in the Internet of Things (IoT), smart cards, and embedded systems. These cryptographic schemes must balance stringent resource limitations with the need for adequate security, making them potential targets for side channel attacks (SCAs). Active side channel attacks, which involve inducing errors during cryptographic operations to reveal secret information, pose significant threats to LWC implementations. Error detection schemes are critical countermeasures in this context, as they can identify and mitigate erroneous operations that result from intentional fault injections. By integrating robust error detection mechanisms, lightweight cryptographic systems can enhance their resilience against active SCAs, thereby maintaining the integrity and confidentiality of the protected data even in the face of sophisticated physical attack strategies.

This chapter addresses the challenge of providing robust security at low cost for these constrained devices by introducing error detection schemes for the nonlinear subblocks of the WAGE cipher. These signature-based schemes, applicable to both logic- and LUT-based implementations, demonstrate acceptable overhead and practical error coverage when benchmarked on FPGA platforms [47]. Another study highlights fault diagnosis schemes for the Midori lightweight block cipher, which is optimized for low energy consumption and minimal latency. This work is pioneering in proposing fault diagnosis architectures for both the 64-bit and 128-bit variants of Midori, enhancing the cipher's reliability through comprehensive fault injection simulations and FPGA benchmarks [48]. Additionally, the ASCON cipher, a finalist in the NIST lightweight cryptography standardization competition, is examined in terms of error detection mechanisms for its secure hardware implementations. The proposed signature, interleaved signature, and cyclic redundancy check approaches achieve high error coverage and are shown to have acceptable overheads on Spartan-7 and Kintex-7 FPGA families [49]. Furthermore, another study presents fault detection architectures for the Pomaranch stream cipher, used in sensitive smart infrastructures. These architectures, designed for low-power operation and assessed through error simulations, provide resiliency against false alarms and are validated via application-specific integrated circuit realizations, thus proving their feasibility for intelligent, emerging applications [50]. Lightweight cryptography and error detection schemes for LWC are also another developing area of research along with Machine Learning and Post-quantum Cryptography. As these areas evolve, there might be a convergence from these domains to strengthen the security of the modern world.

11.7 Conclusion

The chapter covers a comprehensive overview of machine learning applications in side channel analysis. It covers the vulnerabilities introduced by machine learning models on different side channels including power analysis, electromagnetic emanations, and system-level side channel attacks on interfaces such as direct memory access interface. The comprehensive literature review covers the machine learning attacks on side channels, model inversion and extraction techniques, and the potential risks associated with deploying such models in security-sensitive applications. Furthermore, machine learning use in countermeasures to side channel attacks is explored in the later section that also outlines potential strategies to fortify defenses against adversarial side channel and machine-learning-based attacks. By integrating these insights into the development and deployment of machine learning techniques, their security can be bolstered to create a more resilient and trustworthy system for the future.

References

1. Yan, H. (2021). Stealing kernel function of support vector machine through cache side channel. In *2021 International Conference on Computer, Blockchain and Financial Development (CBFD)* (pp. 162–167). https://doi.org/10.1109/CBFD52659.2021.00038.
2. Kecman, V. (2005). Support vector machines–an introduction. In: *Support vector machines: Theory and applications* (pp. 1–47). Springer.
3. Rish, I., et al. (2001). An empirical study of the naive bayes classifier. In: *IJCAI 2001 Workshop on Empirical Methods in Artificial Intelligence* (Vol. 3, pp. 41–46).
4. Sinaga, K. P., & Yang, M.-S. (2020). Unsupervised k-means clustering algorithm. *IEEE Access, 8*, 80716–80727.
5. Bro, R., & Smilde, A. K. (2014). Principal component analysis. *Analytical Methods, 6*(9), 2812–2831.
6. Simons, T., & Lee, D.-J. (2019). A review of binarized neural networks. *Electronics, 8*(6), 661.
7. Wen, Y., & Lao, Y. (2017). Enhancing PUF reliability by machine learning. In: *2017 IEEE International Symposium on Circuits and Systems (ISCAS) pp. 1–4*. IEEE.
8. Bhure, C. M., Nicholas, G. S., Ghosh, S., Zhong, Y., & Saqib, F. (2022). Automated transfer learning model for counterfeit IC detection. In: *2022 IEEE Physical Assurance and Inspection of Electronics (PAINE)* (pp. 1–7). IEEE.
9. Bhure, C., Nicholas, G. S., Ghosh, S., Asadi, N., & Saqib, F. (2024). AutoDetect: Novel autoencoding architecture for counterfeit IC detection. *Journal of Hardware and Systems Security, 8*, 1–20.
10. Gui, Y., Bhure, C., Hughes, M., & Saqib, F. (2021). A delay-based machine learning model for DMA attack mitigation. *Cryptography, 5*(3), 18. https://doi.org/10.3390/cryptography5030018.
11. Mangard, S., Oswald, E., & Popp, T. (2008). *Power analysis attacks: Revealing the secrets of smart cards* (Vol. 31). Springer.
12. Kocher, P., Jaffe, J., & Jun, B. (1999). Differential power analysis. In: *Advances in Cryptology–CRYPTO'99* (pp. 789–789).

13. Kocher, P., Jaffe, J., Jun, B., & Rohatgi, P. (2011). Introduction to differential power analysis. *Journal of Cryptographic Engineering, 1*, 5–27
14. Rohatgi, P. (2009). Improved techniques for side-channel analysis. In: *Cryptographic engineering* (pp. 381–406, Chap. 14).
15. Brier, E., Clavier, C., & Olivier, F. (2004). Correlation power analysis with a leakage model. In: *CHES* (Vol. 3156, pp. 16–29)
16. Hospodar, G., Gierlichs, B., De Mulder, E., Verbauwhede, I., & Vandewalle, J. (2011). Machine learning in side-channel analysis: A first study. *Journal of Cryptographic Engineering, 1*(4), 293.
17. Hospodar, G., Maes, R., & Verbauwhede, I. (2012). Machine learning attacks on 65nm arbiter PUFs: Accurate modeling poses strict bounds on usability. In: *2012 IEEE International Workshop on Information Forensics and Security (WIFS)* (pp. 37–42). IEEE.
18. Mukhtar, N., & Kong, Y. (2018). Hyper-parameter optimization for machine-learning based electromagnetic side-channel analysis. In: *2018 26th International Conference on Systems Engineering (ICSEng)* (pp. 1–7). IEEE.
19. Chawla, N., Singh, A., Kar, M., & Mukhopadhyay, S. (2019). Application inference using machine learning based side channel analysis. In: *2019 International Joint Conference on Neural Networks (IJCNN)* (pp. 1–8). https://doi.org/10.1109/IJCNN.2019.8852124.
20. Kashyap, P., Aydin, F., Potluri, S., Franzon, P. D., & Aysu, A. (2020) . 2deep: Enhancing side-channel attacks on lattice-based key-exchange via 2-d deep learning. *IEEE Transactions on Computer-Aided Design of Integrated Circuits and Systems, 40*(6), 1217–1229. https://doi.org/10.1109/TCAD.2020.3035590.
21. Regazzoni, F., et al. (2020). Machine learning and hardware security: Challenges and opportunities -invited talk-. In: *2020 IEEE/ACM International Conference on Computer Aided Design (ICCAD)* (pp. 1–6).
22. Ghandali, S., Ghandali, S., & Tehranipoor, S. (2021). Deep K-TSVM: A novel profiled power side-channel attack on AES-128. *IEEE Access, 9*, 136448–136458. https://doi.org/10.1109/ACCESS.2021.3117761.
23. Hettwer, B., Leger, S., Fennes, D., Gehrer, S., & Güneysu, T. (2021). Side-channel analysis of the Xilinx Zynq ultrascale+ encryption engine. *IACR Transactions on Cryptographic Hardware and Embedded Systems, 2021*, 279–304.
24. Savage, S. V., & Harris, J. M. Direct Memory Access Controller for Improved System Security, Memory to Memory Transfers, and Interrupt Processing. US 4,797,853.
25. Böck, B., & Austria, S. B. (2009). Firewire-based physical security attacks on Windows 7, EFS and BitLocker. Technical report, Security Business Austria Research Lab.
26. Xu, Y., Cui, W., & Peinado, M. (2015). Controlled-channel attacks: Deterministic side channels for untrusted operating systems. In: *Proceedings of the 2015 IEEE Symposium on Security and Privacy*, San Jose, CA, USA.
27. Morgan, B., Alata, E., Nicomette, V., & Kaâniche, M. (2016). Bypassing IOMMU protection against I/O attacks. In: *Proceedings of the 2016 Seventh Latin-American Symposium on Dependable Computing*, California, CO.
28. When Lightning Strikes Thrice: Breaking Thunderbolt 3 Security. https://thunderspy.io/. Accessed 20 Sep 2020.
29. Frisk, U.: Direct Memory Attack the Kernel. https://media.defcon.org/DEF%20CON%2024/DEF%20CON%2024%20presentations/DEF%20CON%2024%20-%20Ulf-Frisk-Direct-Memory-Attack-the-Kernel.pdf. Accessed 15 Sep 2020.
30. John, T. M., Haider, S. K., Omar, H., & Dijk, M. (2020). Connecting the dots: Privacy leakage via write-access patterns to the main memory. *IEEE Transactions on Dependable and Secure Computing, 17*, 436–442.
31. Gui, Y., Siddiqui, A. S., Nicholas, G. S., Hughes, M., & Saqib, F. (2021). A lightweight delay-based authentication scheme for DMA attack mitigation. In: *2021 22nd International Symposium on Quality Electronic Design (ISQED)* (pp. 263–268). IEEE. https://doi.org/10.1109/ISQED53481.2021.9428624.

32. Batina, L., Bhasin, S., Jap, D., & Picek, S. (2019). CSINN: Reverse engineering of neural network architectures through electromagnetic side channel. In: *28th USENIX Security Symposium (USENIX Security 19)* (pp. 515–532).
33. Nakahara, H., Yonekawa, H., Fujii, T., Shimoda, M., & Sato, S. (2019). Guinness: A GUI based binarized deep neural network framework for software programmers. *IEICE TRANSACTIONS on Information and Systems, 102*(5), 1003–1011.
34. Dubey, A., Cammarota, R., & Aysu, A. (2020). MaskedNet: The first hardware inference engine aiming power side-channel protection. In: *2020 IEEE International Symposium on Hardware Oriented Security and Trust (HOST)* (pp. 197–208). https://doi.org/10.1109/HOST45689.2020.9300276.
35. Bubberman, W., Karayalcin, S., Meester, M., Braakman, O., & Picek, S. (2020). Deep learning for side-channel analysis and introduction to ASCAD database. *Journal of Cryptographic Engineering, 10*, 163–188. https://doi.org/10.1007/s13389-020-00250-7.
36. Wolf, S., Hu, H., Cooley, R., & Borowczak, M. (2021). Stealing machine learning parameters via side channel power attacks. In: *2021 IEEE Computer Society Annual Symposium on VLSI (ISVLSI)* (pp. 242–247). https://doi.org/10.1109/ISVLSI51109.2021.00052.
37. Potluri, S., & Aysu, A. (2021). Stealing neural network models through the scan chain: A new threat for ML hardware. In: *2021 IEEE/ACM International Conference On Computer Aided Design (ICCAD)* (pp. 1–8). IEEE. https://doi.org/10.1109/ICCAD54762.2021.9624702.
38. Dubey, A., Ahmad, A., Pasha, M. A., Cammarota, R., & Aysu, A. (2022). ModuloNET: Neural networks meet modular arithmetic for efficient hardware masking. *IACR Transactions on Cryptographic Hardware and Embedded Systems*, 506–556. https://doi.org/10.13154/tches.v2022.i3.506-56.
39. Dubey, A., Cammarota, R., Suresh, V., & Aysu, A. (2022). Guarding machine learning hardware against physical side-channel attacks. *ACM Journal on Emerging Technologies in Computing Systems (JETC), 18*(3), 1–31. https://doi.org/10.1145/3649921.
40. Picek, S., Heuser, A., Jovic, A., Bhasin, S., & Regazzoni, F. (2023). Tipping the balance: Imbalanced classes in deep learning side-channel analysis. *IEEE Design and Test*. https://doi.org/10.1109/MDAT.2023.3288808.
41. Karayalcin, S., & Picek, S. (2023). Resolving the doubts: On the construction and use of ResNets for side-channel analysis. *Mathematics, 11*(15), 3265.
42. Uchida, Y., Nagai, Y., Sakazawa, S., & Satoh, S. (2017). Embedding watermarks into deep neural networks. In: *Proceedings of the 2017 ACM on International Conference on Multimedia Retrieval* (pp. 269–277).
43. Sarker, A., Canto, A. C., Kermani, M. M., & Azarderakhsh, R. (2022). Error detection architectures for hardware/software co-design approaches of number-theoretic transform. *IEEE Transactions on Computer-Aided Design of Integrated Circuits and Systems, 42*, 2418–2422.
44. Canto, A. C., Sarker, A., Kaur, J., Kermani, M. M., & Azarderakhsh, R. (2022). Error detection schemes assessed on FPGA for multipliers in lattice-based key encapsulation mechanisms in post-quantum cryptography. *IEEE Transactions on Emerging Topics in Computing, 11*, 791–797.
45. Canto, A. C., Kermani, M. M., & Azarderakhsh, R. (2020). Reliable architectures for composite-field-oriented constructions of McEliece post-quantum cryptography on FPGA. *IEEE Transactions on Computer-Aided Design of Integrated Circuits and Systems, 40*(5), 999–1003.
46. Sarker, A., Mozaffari-Kermani, M., & Azarderakhsh, R. (2018). Hardware constructions for error detection of number-theoretic transform utilized in secure cryptographic architectures. *IEEE Transactions on Very Large Scale Integration (VLSI) Systems, 27*(3), 738–741.
47. Kaur, J., Sarker, A., Kermani, M. M., & Azarderakhsh, R. (2021). Hardware constructions for error detection in lightweight Welch-Gong (WG)-oriented streamcipher WAGE benchmarked on FPGA. *IEEE Transactions on Emerging Topics in Computing, 10*(2), 1208–1215.
48. Aghaie, A., Kermani, M. M., & Azarderakhsh, R. (2016). Fault diagnosis schemes for low-energy block cipher midori benchmarked on FPGA. *IEEE Transactions on Very Large Scale Integration (VLSI) Systems, 25*(4), 1528–1536.

49. Kaur, J., Kermani, M. M., & Azarderakhsh, R. (2021). Hardware constructions for error detection in lightweight authenticated cipher Ascon benchmarked on FPGA. *IEEE Transactions on Circuits and Systems II: Express Briefs, 69*(4), 2276–2280.
50. Mozaffari-Kermani, M., Azarderakhsh, R., & Aghaie, A. (2015). Reliable and error detection architectures of Pomaranch for false-alarm-sensitive cryptographic applications. *IEEE Transactions on Very Large Scale Integration (VLSI) Systems, 23*(12), 2804–2812.

Chapter 12
Adapt and Defend: Reinforcement Learning for Hardware-Assisted Security

Hossein Sayadi and Zhangying He

In the dynamic landscape of cybersecurity, the constant evolution of threats necessitates innovative approaches to fortify digital defenses. This chapter delves into a cutting-edge paradigm, exploring the integration of reinforcement learning techniques with hardware-assisted (cyber)security measures. As the relentless sophistication of cyber threats challenges conventional security mechanisms, this chapter elucidates how reinforcement learning, a form of adaptive artificial intelligence, can be harnessed to adapt and fortify defenses in real time. The convergence of hardware-assisted security and adaptive learning promises a robust and adaptive shield against emerging threats particularly malicious software, bringing in a new era of proactive and resilient cybersecurity strategies in the digital age.

12.1 Malware Detection Techniques: Background and Taxonomy

In this section, we offer a concise exploration of the taxonomy underlying the existing malware detection methods. The overarching classification is visualized in Fig. 12.1. The diversity in malware detection methods necessitates a systematic categorization from various perspectives. In our study, we consider two primary categories, as depicted in Fig. 12.1: (1) misuse-based malware detection and (2) anomaly-based malware detection methods.

H. Sayadi (✉)
Computer Engineering and Computer Science Department, California State University, Long Beach (CSULB), Long Beach, CA, USA
e-mail: Hossein.Sayadi@csulb.edu

Z. He
California State University Long Beach, Long Beach, CA, USA

A. Iranmanesh, H. Sayadi (eds.), *AI-Enabled Electronic Circuit and System Design*,
https://doi.org/10.1007/978-3-031-71436-8_12

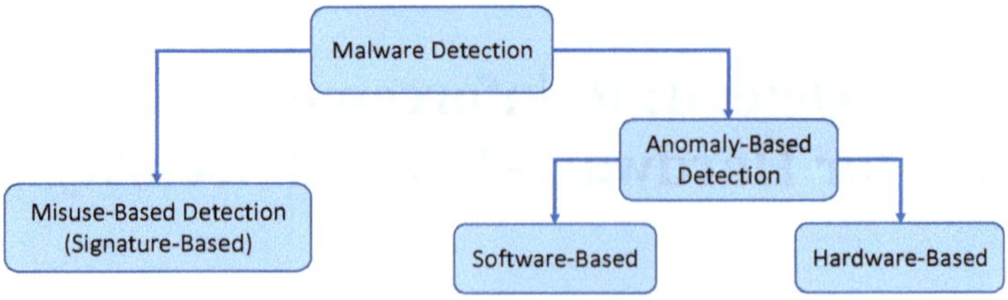

Fig. 12.1 General taxonomy of malware detection techniques

12.1.1 Misuse-Based Malware Detection

Misuse detection is a method for identifying computer security attacks by defining abnormal behaviors as an initial reference for determining malicious software. In this approach, anything that matches the predefined signatures or patterns is considered malicious [10, 32, 23]. Signature-based malware detection, a subtype of misuse detections, identifies unique signatures and malicious patterns in malware programs [25, 56]. By profiling applications on computer systems, misuse techniques recognize malware based on predetermined execution signatures and patterns extracted from known malware [59]. This category of malware detection includes signature-based, heuristic-based, and cloud-based detection methods. Off-the-shelf antivirus software (e.g., Norton, MacAfee, etc.) are considered traditional software malware detection methods that utilize the signatures of malicious patterns to detect known malware attacks [68].

Misuse malware detection methods have a notable advantage in their simplicity of incorporating known attacks into the detection model. This approach assumes that malicious behavior can be defined by specific, recognizable patterns, making it straightforward to employ. However, a significant drawback of misuse detection is its inability to identify unknown security attacks [10, 59]. These methods are vulnerable to alterations that deviate from pre-identified malware signatures. While successful in detecting known attacks, signature-based techniques are incapable of detecting new (unseen) attacks lacking predefined malicious patterns.

12.1.2 Anomaly-Based Malware Detection

Anomaly-based or behavior-based malware detection contrasts with misuse detection by defining normal system behavior first and categorizing all other behaviors as abnormal or malicious. This approach characterizes attacks based on deviations from predefined models [59] and has the potential to identify new, unknown malware [8, 59, 68, 69]. Anomaly-based detection relies on heuristics or rules, rather than patterns or signatures, aiming to detect any form of malicious activity outside normal system operation [6]. While antivirus solutions traditionally relied on signature-based methods, anomaly-based techniques are now deployed to proactively detect malicious applications, overcoming the limitations of the signature-based analysis. Unlike signature-based methods that look for specific patterns, anomaly-based detection searches for certain instructions or rules within application behaviors that are absent in normal programs [42, 41]. Thus, behavior-based malware detection methods employ weight-based or rule-based systems to identify potentially malicious patterns not previously examined by the system [68, 53].

While anomaly-based detection is effective in recognizing unknown threats for real-time protection in antivirus tools, it comes with drawbacks. The scanning and analysis process is time-consuming, degrading computer system performance [8]. Additionally, manually constructing rules/patterns for recognizing unknown malware is error-prone, limiting the effectiveness of heuristic analysis [68]. Behavior-based detection can increase false positives, blocking benign programs and restricting normal computer system operations [53, 8].

Behavior-based malware detection relies on two key phases: a training phase, involving profiling applications to capture their behavior and implementing Machine Learning (ML) solutions based on the captured information, and a testing phase, where real malware and benign applications are compared with the profiled model from the training phase [28, 6]. These detection techniques predominantly leverage Artificial Intelligence (AI) and data mining methods to identify anomalies [28, 59]. In the feature analysis step, software- and hardware-related features are extracted offline to capture file sample characteristics during training. Machine learning techniques then automatically classify file samples into "benign" or "malware" categories based on feature representations.

Anomaly-based malware detection employs various features, categorized into software-based and hardware-based approaches. Software features include permissions [70], network traffic [5, 62], system calls [25], application programming interface (API) [7], information flow [3], call stacks [26], and covert channels [11]. In recent years, there has been a shift toward hardware-based detection, utilizing information like memory access patterns, I/O interface, instruction execution, and low-level microarchitectural features. Hardware-based detectors offer real-time detection, efficient resource utilization, and resistance to infection, proving effective against emerging threats. Recent studies use ML algorithms for accurate and efficient detection of malicious attacks, leveraging low-level features captured from processors' Hardware Performance Counters (HPCs) [13, 46]. Table 12.1 provides a comparative analysis of the two primary categories of malware detection methods: misuse-based and anomaly-based including key attributes, advantages, and challenges.

Table 12.1 Comparison of misuse-based and anomaly-based detection techniques

Detection approach	Misuse-based (signature-based)	Anomaly-based (behavior-based)
Specification	– Deploys the pattern of known attacks to identify malicious applications	– Deploys the deviation from normal patterns to identify malicious applications
Advantage	– Simple mechanism	– Ability to detect unknown attacks
Challenge	– Known attacks have to be hand-coded	– Scanning takes a considerable time
	– Inability to detect new attacks	– False positives
	– Signature updates required	
	– False negatives	

12.2 Hardware-Assisted Cybersecurity: ML-Enabled Malware Detection Using Hardware Features

As the use of malicious software (commonly known as malware) continues to increase in various application domains, the development of efficient malware detection techniques has become even more critical. These techniques serve as an early protection mechanism, safeguarding the integrity and confidentiality of authenticated users' data [9, 48]. Recent advances reveal a surge in leveraging emerging vulnerabilities for malicious activities, emphasizing the critical need for robust defenses against malware threats [50]. Traditional software-based detection methods relying on static signature analysis face performance drawbacks such as static analysis limitations, inability to detect more sophisticated attacks, and sizeable computational overhead, particularly on resource-limited systems. These challenges stem from constraints in computing power and communication bandwidth within embedded system environments [34, 14, 4, 51].

In addressing these challenges, the imperative is to develop effective and cost-efficient cybersecurity countermeasures, focusing on safeguarding user information and mitigating the impact of emerging cyber threats [36]. This involves a paradigm shift toward integrating security measures into the underlying hardware, establishing as a bottom-up approach to fortify computing devices rather than treating security as an afterthought [50, 51]. Simultaneously, recent breakthroughs in artificial intelligence and machine learning, fueled by the increase in data volume across various computing systems, have yielded successful applications across various domains, especially in enhancing systems security [27, 18].

Recent research underscores the importance of identifying malicious activities at the processor hardware and architecture level due to its speed, efficiency, and lower visibility to potential attacker exploits. To this aim, Hardware-Assisted Malware

Detection (HMD) methods, specifically leveraging machine learning trained on Hardware Performance Counter (HPC) features, have emerged as a solution to the shortcomings of traditional software-based malware detection [13, 59, 37, 46, 47]. Hardware Performance Counters (HPCs) are dedicated registers found within a microprocessor's Performance Monitoring Unit (PMU). These registers track various hardware events related to application execution, such as the number of cycles executed, instructions processed, and cache misses incurred [55, 49, 12, 19]. Furthermore, machine learning techniques have demonstrated efficacy in detecting and classifying these anomalies within low-level feature spaces. By leveraging machine learning, systems can effectively discern potential threats and proactively respond to evolving behaviors in real time [52, 17].

12.3 ML-Assisted Malware Detection: Overview and Procedure

Figure 12.2 provides an overview of the process for machine-learning-driven approaches designed for enhancing cybersecurity, specifically in the context of hardware-assisted malware detection. This process encompasses stages from application monitoring for HPC data profiling, feature engineering, and training ML-based detectors and online inference. The continuous learning of ML models through the analysis of low-level microarchitectural features aims to recognize and counteract malicious patterns [52, 22]. This proactive and intelligent approach safeguards the processor architecture from potential threats, encompassing not only malicious software but also extending to microarchitectural side-channel attacks [65, 64].

As shown, hardware-assisted ML detection methods can employ dynamic feature analysis techniques in the ML training and inference process. The combination of dynamic feature collection and analysis can properly integrate with ML techniques

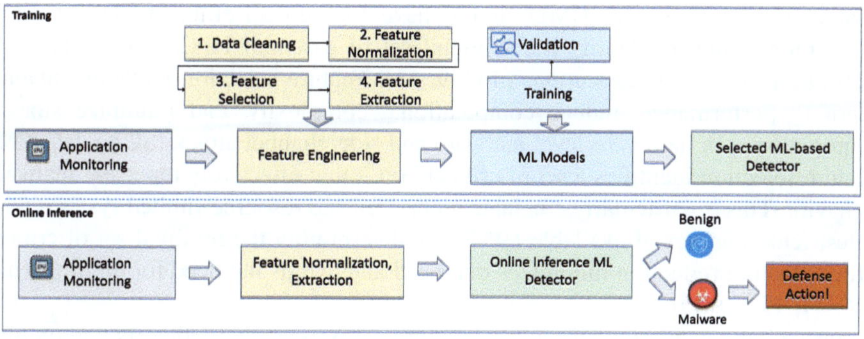

Fig. 12.2 General overview of hardware-assisted security countermeasures against malware using machine learning

during online inference [19]. During the training phase, collected hardware data can be processed with a feature engineering process offline to understand and select the most impactful features for a specific platform. Diversified ML models are trained and validated, and the most effective ML models are selected as the online defender. In the inference phase, data monitoring, collection, and feature preprocessing can be automated and processed before feeding to an ML defender. When there are malicious or abnormal activities, the system raises alarms or takes real-time defense actions.

12.4 Feature Engineering: Analysis of Key Features

Developing effective ML-based hardware-assisted malware detectors begins with crucial steps like data collection and feature selection [13, 46, 55, 52]. Modern microprocessors can collect numerous microarchitectural events, but selecting relevant low-level features is essential to avoid computational complexity and delays associated with high-dimensional datasets.

Specifically, identifying essential low-level microarchitectural features is crucial for hardware-assisted malware detection for several reasons: (a) The multitude of microarchitectural events (e.g., 100+ in Intel Xeon) results in high-dimensional data [46], (b) Handling raw datasets introduces computational complexity and causes delays [29], and (c) The selection of the most relevant microarchitectural events varies across application classes, posing challenges in specifying nontrivial events for different malware classes [47]. Modern processors contain 2–8 HPC registers, presenting a challenge for runtime malware detection to pinpoint a minimal set of HPCs that precisely capture malicious attack characteristics, minimizing unnecessary computational overhead [46, 47, 19]. This ensures the development of efficient ML-based security countermeasures with minimal impact on system performance.

In resource-constrained computing platforms like embedded systems and IoT devices, with restricted HPC registers, efficient runtime detection relies on critical feature selection. Recent HMD studies have focused on efficient runtime hardware monitoring to identify the minimal set of essential HPCs needed for data collection in a single execution [46, 49, 51]. Feature selection methods enhance learning performance, reduce computational complexity, and minimize storage requirements. In hardware-level malware and side-channel attack (SCA) detection, feature selection identifies a set of critical HPCs that effectively represent malware behavior. This is particularly crucial in embedded and resource-limited systems with a restricted number of available HPCs. By determining the minimal set of critical HPCs, feature reduction minimizes overhead and avoids the need for multiple runs [46, 44, 45, 40, 55].

Instead of considering all captured features, it is essential to identify and remove irrelevant ones using a feature reduction algorithm. A subset of HPCs is then selected to represent the most critical features for classification. These selected features train each ML-based malware detector, which aims to correlate hidden patterns with application behavior to predict whether it is benign or malware. Feature reduction also helps reduce offline learning time and hardware design complexity, especially in embedded systems and IoT devices [46].

By reviewing prior studies on ML-based malware detection, we identified several dominant feature selection techniques. These techniques include correlation attribute evaluation, principal component analysis, gain ratio evaluation, and Fisher score. Below, we briefly discuss each of these methods and compare their strengths and weaknesses at the end of this section.

12.4.1 Correlation Attribute Evaluation (CAE)

Correlation Attribute Evaluation (CAE) is a method to assess the relationship between two or more variables in statistics, measuring both strength and direction with values ranging from -1 to 1. As a feature selection method, CAE ranks features based on their importance (strength) and relevance (positive relationship) to the target classified variable. It evaluates attribute importance by measuring the correlation (e.g., Pearson correlation) between the attribute and the corresponding class, using the Pearson correlation coefficient ρ. The Pearson correlation coefficient ρ is calculated as follows:

$$\rho(i) = \frac{cov(X_i, C)}{\sqrt{(var(X_i)var(C))}}. \tag{12.1}$$

In this equation, ρ is the output Pearson correlation coefficient. Xi is the feature of the event i in the feature matrix X. C represents the classes, "Malware" or "Benign" in our case. The value of i represents the index of any feature out of all captured features X during the monitoring and extracting of the features. $cov(Xi; C)$ measures covariance between the feature Xi and the classes. The $var(Xi)$ and $var(C)$ measure the variance of both input features and output classes, respectively. Equation 12.1 can be elaborated further as shown below:

$$\rho(i) = \sum_{k=1}^{n} \frac{(x_{k,i} - \bar{x_i})(c_k - \bar{c})}{\sqrt{\sum_{k=1}^{n}(x_{k,i} - \bar{x_i})^2 \sum_{k=1}^{n}(c_k - \bar{c})^2}}. \tag{12.2}$$

In this algorithm k is the total number of input values; $x_{(k,i)}$ is kth value in the input dataset for feature i; n represents the total number of samples; c_k is kth value in the output dataset (e.g., the class). The mean of input data for feature i is denoted by $\bar{x_i}$, and that for the output classes by $\bar{c}$. CAE finds the correlation coefficient for all captured features using the above equation. Based on the ranking of ρ, the top low-level HPC features are selected for analysis and implementation of the ML classifiers used for malware detection and classification. The threshold for selecting top features according to ρvalues depends on the uniqueness of different datasets. Correlation attribute evaluation is a key technique for feature selection in hardware-assisted malware detection, as seen in recent works [46, 40, 44, 47, 45, 15].

12.4.2 Principal Component Analysis (PCA)

Principal Component Analysis (PCA) is a statistical method for dimensionality reduction, transforming observations of correlated variables into a new feature space called Principal Components (PCs) [31, 61, 54]. PCA identifies vital parameters by transforming data into new coordinates such that the principal components capture the largest variations [2, 57]. This transformation helps determine the most important features along different PC dimensions. It is essential to normalize data to the unit normal distribution to address sensitivity to the relative scaling of the original variables [47]. PCA involves applying an orthogonal transformation to convert a set of observations with possibly correlated variables into a new feature space composed of linearly uncorrelated components (PCs). These PCs are uncorrelated and represent linear combinations of the original features, capturing most of the data variations. The number of PCs can be less than or equal to the original data variables. The proposed hardware-assisted security works in [55, 45, 71, 47] have employed PCA as their feature engineering method to determine the top prominent features for malware detection.

12.4.3 Gain Ratio Evaluation (GRE)

Gain Ratio Evaluation (GRE) is an extension of the Information Gain Ratio (IGR) feature selection method introduced in [43]. IGR aims to address the bias in Information Gain (IG) by normalizing it, considering how an attribute divides input samples. The Gain Ratio method reduces bias by dividing the information gain by the entropy of the attribute relative to the class. The entropy after partitioning samples according to a given feature A is denoted as SI(S, A). The IGR is computed using equations involving SI and IG as shown in Eq. 12.4. IGR attempts to lower the bias in IG, which tends to favor features with higher entropy and greater diversity [67]. Information Gain primarily measures the uncertainty in the input data, and IG often supports features that are more diverse. IGR normalizes this by considering how an attribute divides the input samples. The normalization in Gain Ratio is performed by dividing the information gain by the entropy of the attribute with respect to the class, which reduces the bias of the information gain algorithm. Let $SI(S, A)$ be the entropy of S after partitioning the samples according to the given feature A.

$$SI(S, A) = -\sum_{v=1}^{|V|} \frac{|S_v|}{|S|} \times \log_2 \frac{|S_v|}{|S|} \tag{12.3}$$

$$IGR(S, A) = \frac{IG(S, A)}{SI(S, A)}. \tag{12.4}$$

First, let us briefly describe information gain. Suppose we partition the data samples in the training set based on their values for a given feature (e.g., attribute AA). Information gain attempts to quantify the expected reduction in entropy resulting from this partitioning, aiming to assess the feature's effectiveness in classifying file samples [67]. If S is a set of items that can be divided into C nonoverlapping subsets, where Sc denotes the c-th subset, the entropy of S is calculated as follows:

$$H(s) = -\sum_{c \in C} \frac{|S_c|}{|S|} \times \log_2 \frac{|S_c|}{|S|}. \tag{12.5}$$

Given Eq. 12.5, the IG for attribute A is calculated by the following equation:

$$IG(S, A) = H(s) = H(S) - \sum_{v \in V(A)} \frac{|S_v|}{|S|} \times H(S_v), \tag{12.6}$$

where V is the set of possible feature values of A, and Sv is the subset of samples with feature value v for the feature A.

Gain Ratio evaluation addresses a limitation of the information gain approach, particularly its bias against attributes with a high number of distinct values. In the context of building a decision tree for hardware-assisted malware detection, this bias correction is crucial. The Gain Ratio, commonly employed to determine the relevance of collected features for malware detection, assists in making informed decisions about feature selection near the root of the tree. GRE and IGR are widely adopted feature selection methods in recent HMD techniques [55, 19, 20].

12.4.4 Fisher Score (FS)

Fisher score (FS) is a widely used feature selection method aiming to measure discriminative power by minimizing redundancy and maximizing relevance to the target, such as class labels in classification [35]. In hardware-assisted malware detection, FS identifies top low-level features for malware detection using unsupervised ML techniques, as employed in [59]. FS provides a measure of average interclass distance compared to intraclass distance for a given feature, helping select features that assign similar values to data samples in the same class and different values to samples from different classes [16, 67]. As a result, the FS for the i_{th} feature denoted as Si will be calculated as follows:

$$S_j = \frac{\sum_{k=1}^{n_j} (\mu_{ij} - \mu_i)^2}{\sum_{k=1}^{n_j} (\rho_{ij})^2}, \tag{12.7}$$

where $\mu_i j$ and $\rho_i j$ are the mean and the variance of the i_{th} feature in the j_{th} class respectively, n_j is the number of instances in the j_{th} class, and μ_i is the mean of the i-th feature. Higher values of S_k imply that while members in the same class are closer together, the members belonging to different classes are further separated using the k_{th} feature. For the malware detection problem that is often a two-class problem including the positive class (malicious) or negative class (benign), the FS method simplifies [67]. FS helps select a subset of significant features for distinguishing malicious patterns, but it does not consider the correlation and mutual dependency between features, which could be a limitation [59].

The researches in [67, 59] are the recently proposed HMDs that use FS as their feature selection method to determine the prominent features.

12.4.4.1 Comparative Analysis of Feature Selection Techniques

Table 12.2 presents an overview of the discussed feature selection techniques, along with recommendations for their suitable applications. CAE is straightforward and fast, suitable for moderate to large datasets without requiring data normalization. However, it performs best with features exhibiting linear relationships, and may struggle with nonlinear correlations, imbalanced labels, or outliers.

PCA effectively handles highly correlated data by transforming feature spaces to identify patterns. It requires data normalization but may not perform well with outliers or uncorrelated features. PCA is beneficial for dimensionality reduction and noise reduction, but it does not preserve the original data structure, and may lose some information during this process. GRE offers the capability to handle both categorical and numerical data but comes with high computational costs. FS assumes data to be normally distributed. It can work effectively with high-dimension data. Besides outputting a subset of selected features, it can also be used to evaluate feature importance of a subset of features related to target classes.

Feature engineering is a crucial step in machine-learning-based security detection methodologies. It aims to improve the effectiveness of machine learning models by selecting, creating, and transforming features. In practice, multiple feature selection approaches can be employed and their results compared to choose the most effective one for the current datasets. Feature selection is often an iterative process. As datasets evolve (e.g., with new types of malware or abnormal activities), revisiting feature engineering ensures optimal problem-solving approaches. Section 12.9 explores a case study that illustrates the utilization of feature selection within the broader context of ML-based hardware malware detection.

Table 12.2 Comparison of different feature selection (FS) methods

FS	Output	Benefits	Drawbacks	Other usage
CAE	Weights	– Simple, fast	– Variance error occurs for imbalanced dataset	Feature redundancy analysis
		– Moderate or large dataset	– Not robust if outliers exist	
		– Normal and not-normal distribution	– Effective in linear features	
		– Suitable for numerical data		
PCA	Subset	– Good for highly correlated data to reduce to independent set	– The original data structure is not reserved	– Feature dimension reduction
		– Sensitive to data scaling so data need to be standardized before applying the algorithm	– Information loss during dimension reduction	– Visualize data patterns
			– Not working for uncorrelated features	– Denoise data
			– Sensitive to outliers;	
			– Not working well for nonlinear features	
GRE	Weights	– Perform better for a lower number of distinct values	– Struggle to find attributes for the high number of distinct values	– Normalize data
		– Suitable for both categorical and numerical data		
		– Can handle missing values		
Fisher score	Subset	– Computationally cheap when handling high-dimensional data	– Struggle with nonlinear data	– Evaluating a subset feature importance
		– Assume normal distribution	– Not applicable to highly redundant data	

12.5 Machine-Learning-Based Self-adaptive System

A self-adaptive system is designed to adjust its output and behavior in response to changes or optimization requirements preconfigured within the system. The definition and scope of self-adaptive systems evolve with advancements in technology and their applications [30, 21]. The demand for self-adaptability in systems has increased, leading to expanded and evolved methodologies. Machine-learning-based self-adaptive systems have garnered significant interest in various application fields [24], including cloud computing, robotics, traffic, and transportation systems, security defense software, and more. Unlike traditional machine learning

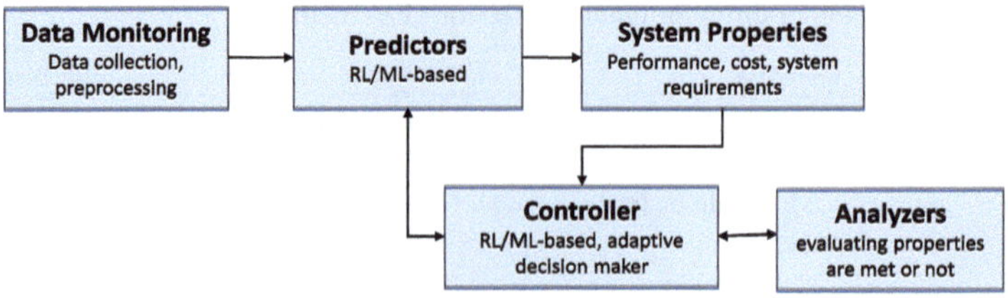

Fig. 12.3 Core components of a self-adaptive system

approaches, which are typically trained on a fixed set of data and then periodically validated and deployed, self-adaptive systems aim to reduce human involvement by enabling the system to handle adaptive functions and behaviors at runtime.

In the depicted system architecture (Fig. 12.3), a self-adaptive system actively monitors real-time data and preprocesses it. Leveraging a pretrained machine learning defense model, the system fulfills initial functional requirements, such as recommendations, detections, and identifications. The core components of self-adaptive functionality include an adaptive controller, which makes data-driven decisions related to task execution, system properties, and switching policies. Lastly, an analyzer assesses system performance and offers feedback for long-term optimization.

Self-adaptive systems can be implemented using various techniques, such as external control mechanisms, component-based software engineering, self-monitoring architectures, multiagent systems, and feedback systems. Reinforcement learning, in particular, is emerging as an effective approach for adaptive logic, and it remains a dynamic and active area of research [38]. In this chapter, we explore the process of defining, designing, and experimenting with self-adaptive systems through a reinforcement learning approach.

12.6 Reinforcement Learning: An Overview

Reinforcement learning (RL) is a distinct type of machine learning, differing significantly from supervised and unsupervised learning. In supervised learning, labeled samples in the training set are used to train models to correctly interpret new data. Unsupervised learning, on the other hand, focuses on discovering hidden structures within unlabeled data. Unlike these approaches, reinforcement learning involves a goal-directed agent interacting with an uncertain environment, considering the entire system to achieve its objective. Reinforcement learning consists of four main elements: the agent, the environment (states and actions), the reward signal, and the policy that defines the goal. Figure 12.4 illustrates these elements and their relationships within an RL system.

In addition to the agent, an RL environment includes the observation space $S = [s_1, s_2, \ldots, s_m]$ and the action space $A = [a_1, a_2, \ldots, a_k]$. The observation space

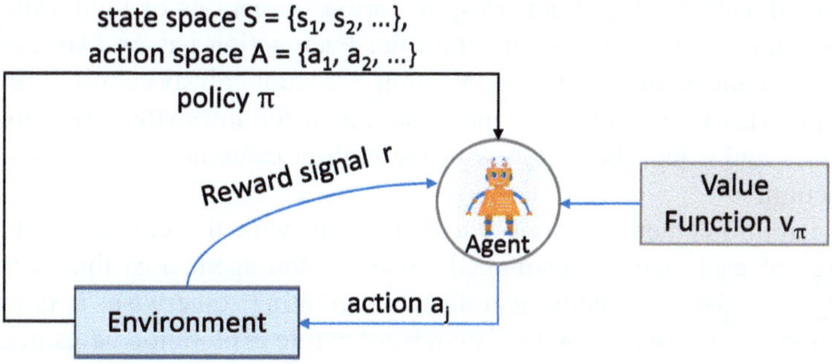

Fig. 12.4 Main components in reinforcement learning

provides contextual data points for the agent to make decisions, leading to actions chosen from the action space. The consequences of these actions are driven by the reward policy, which offers a framework of rewards and penalties, defining the goal to guide the agent at each step. This policy can range from simple lookup tables to complex computations. The value of rewards is determined by the value function, based on the reward policy and the action taken by the agent at each step. The reward signal evaluates whether an action is beneficial or detrimental to achieving the agent's goal.

Reinforcement learning can be categorized into model-free and model-based approaches. Model-based RL integrates all four elements into a comprehensive model, as seen in systems like AlphaZero and AlphaGo. This approach offers more accurate predictions and can adapt to changes in the environment, though constructing such models is complex. On the other hand, model-free RL, which is more commonly used, is simpler as it separates the agent from the RL environment. The RL environment is defined by parameters and remains stable, while the RL agent learns from its interactions with this environment. In the following sections, we will review several important reinforcement learning algorithms for agents.

12.7 Reinforcement Learning Agents

This section will introduce four unique, commonly used, model-free reinforcement learning agent algorithms.

12.7.1 Multiarmed Bandit: Upper Confidence Bound

The multiarmed bandit problem is an analogy to a slot machine, where each action is like pulling an arm of the machine, and the reward from that action is akin to the payoff of winning a jackpot. Despite various variations of multiarmed bandit problems, they all share a fundamental rule: Each action has an expected reward or value. If we knew the value of each action, we would always choose the highest-value action. However, in practice, the values are uncertain, so the goal is to estimate these values and select the action with the highest estimated value, known as the greedy action.

As the reinforcement learning agent interacts with the environment, it gains knowledge of each action's estimated values. If the agent uses this accumulated knowledge to take the greedy action, it is exploiting; otherwise, it is exploring. Exploitation maximizes the immediate reward, while exploration helps in gathering information to improve long-term rewards. Balancing these two states—exploitation and exploration—is crucial in an RL agent's learning process.

One popular method to address this balance in multiarmed bandit problems is the Upper Confidence Bound (UCB) algorithm. UCB is a deterministic algorithm that leverages a confidence boundary to decide whether to explore or exploit based on the action value formula stated in Eq. 12.8. Initially, UCB assigns a high confidence boundary for each action, allowing the RL agent to explore each action. As the agent explores and gathers rewards, it updates the confidence intervals for each action. At each step, the RL agent selects the action with the highest upper confidence bound to explore. The confidence interval of the chosen action is then updated based on the received reward, which either increases or decreases the interval. Over time, this method balances exploration and exploitation by gradually favoring actions with higher estimated rewards while still allowing for exploration to improve these estimates.

Calculate action value

$$A_t = \underset{a}{argmax}[Q_t(a) + \delta_j(t)]$$

where

$$Q_t(a) = \frac{\sum_{i=1}^{t-1} R_i | A_i = a}{N_j(a)},$$

$$N_j(a) = \sum_{i=1}^{t-1} A_i = a, \delta_j(t) = c\sqrt{\frac{\ln t}{N_j(a)}}. \tag{12.8}$$

At the end of a round, the UCB-based RL agent calculates the average reward $Q_t(a)$ and updates the upper bound boundary interval $[Q_t(a) - \delta_j(t), Q_t(a) + \delta_j(t)]$, where $Q_t(a)$ is the average reward of all rounds of N, and $\delta_j(n)$ is the knowledge gained from the current round. Delta $\delta_j(t)$ encourages more exploration, and the average reward $Q_t(a)$ facilitates exploitation. When the RL agent has little knowledge of the best branch, it favors more explorations to search through all branches. As more rounds of exploration progress, more knowledge of the possible best branch is gained, and the RL agent gradually shifts to exploitation to select the branch with the highest upper bound value. UCB greedily trials each detector on each round of exploration and evaluates the possible rewards according to the reward policy.

12.7.2 Temporal-Difference Learning: Q-Learning

Another novel idea of reinforcement learning is temporal-difference (TD) learning, which combines Monte Carlo ideas and dynamic programming (DP) [58] to update the estimated Q-value at time step t without finishing the final round based on the raw experience of the agent. TD updates the estimated value using the following formula in Eq. 12.9, where $R_{t+1} + \gamma \max Q(S_{t+1}, a)$ is the part of TD update at step $t + 1$.

$$\underbrace{Q(S_t, A_t)}_{\text{New Q-value}} \leftarrow \underbrace{Q(S_t, A_t)}_{\text{Old Q-value}} + \alpha \Big[\underbrace{R_{t+1} + \gamma \cdot \max_a Q(S_{t+1}, a) - Q(S_t, A_t)}_{\text{TD error}} \Big]. \tag{12.9}$$

Q-Learning, a popular TD learning algorithm, maintains a Q-table that gets updated after each episode. The rows correspond to states, and the columns correspond to actions. Initially, the Q-values for each state-action pair are set to zero. The RL agent randomly selects actions to explore the environment. The Q-function uses Eq. 12.9 to update the Q-value at each step. An episode ends after a set of actions is completed. In the end, the Q-table suggests the optimal policy. The Q-value considers both the current reward value $V(S_t)$ and the discounted future rewards (TD error). The current rewards are weighted more heavily than future rewards when the RL agent makes decisions. This iterative process continues until the agent has sufficiently explored the environment and the Q-values converge to optimal approximations. In summary, TD learning and Q-learning enable RL agents to learn optimal policies by continuously updating their estimates based on direct interactions with the environment, balancing the importance of immediate and future rewards.

12.7.3 Value Approximation Method: Dueling DQN

Dueling DQN (DDQN) [63] is a reinforcement learning algorithm that uses a single Q-network with two separate streams to estimate the state value function $v(s)$ and the advantage function $A(s, a)$ for each action. These two estimates are then combined to produce the state-action value $Q(s, a)$. DDQN decouples the representation and the algorithm (e.g., using two MLPs), which allows reusing any underlining architecture in the two streams. Figure 12.5 illustrates a DDQN network formed by a simple two-layer multilayer perceptron as the underlying value approximation function. This network is followed by one dense layer representing

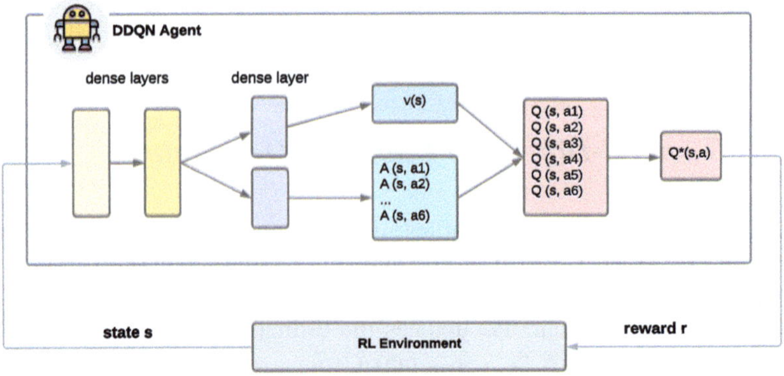

Fig. 12.5 Dueling deep Q-network

the state value function $v(s)$ and another dense layer representing the advantage function $A(s, a)$. The optimal state-action value $Q^*(s, a)$ is the maximum Q-function from all options of advantages $\{A(s, a_1), \ldots, A(s, a_6)\}$. The optimal Q-value is derived from the action that yields the highest Q-value, as shown in Eq. 12.10.

$$Q^*(s, a) = \max_{\pi} Q^{\pi}(s, a). \tag{12.10}$$

12.7.4 Policy Gradient Method: Advantage Actor Critic

A typical Advantage Actor Critic (A2C) [33] is a combination of policy gradient-based and value-based reinforcement learning algorithms that contain two deep RL networks: Actor and Critic networks. The Actor network learns to map each state S_t with a probability distribution of actions (probability of $a1_t$, $a2_t$, ... $a6_t$), and it uses a neural network with an activation function (e.g., Softmax) to output for the policy $\pi(a_t|S_t; \theta)$, which estimates the action to take. Unlike value-based method, the policy gradient method learns a parameterized reward policy $\pi(a|S, \theta)$ based on a policy gradient function $J(\theta)$, as shown in Eq. 12.11. The action and the state are input to the Critic to produce an estimated action value $V(S_t; Q_v)$, which is similar to a Q-value. The environment takes in the action and returns the next state and the reward. The actor updates the policy parameter using the policy gradient function in Eq. 12.11.

$$J(\theta) = \nabla_\theta \log \pi(a_t|s_t, \theta)(R_t - V^{\pi}(s_t)). \tag{12.11}$$

The A2C network updated the optimized action by an advantage function as shown in Eq. 12.12. In this sense, A2C integrates Q-learning with policy gradient methods.

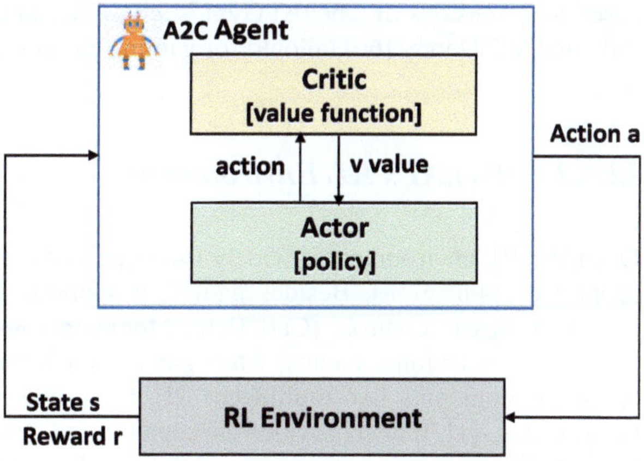

Fig. 12.6 Advantage Actor-Critic network

$$A(a_t, s_t) = Q(a_t, s_t) - V(s_t). \tag{12.12}$$

Figure 12.6 depicts one architecture of the A2C model. As the agent takes actions and moves through the environment, it learns to map the observed state of the environment to two possible outputs: (1) Recommended action output from the Actor, in which the Actor network outputs a probability distribution over actions, suggesting the best action to maximize the reward. In the context of cybersecurity, this means determining the probability for each machine learning detector in the action space. The reward is determined by the mapping between the state and the actions according to the defined reward policy in the RL environment at the current time step. (2) Estimated reward value produced from the Critic, which represents the sum of expected future rewards from the current state. This value helps to assess the quality of the action taken by the Actor. The Actor and Critic work together, with the Actor selecting actions to maximize the total expected rewards and the Critic evaluating these actions to provide feedback. The Actor updates its policy parameters to improve action selection, while the Critic updates its value estimates to better predict future rewards.

12.8 Reinforcement Learning Environment

In contrast to classical machine learning and deep learning, reinforcement learning necessitates an environment in which to train an RL agent. Designing an RL environment that accurately simulates real-world conditions encountered by an RL agent involves significant tasks such as programming, data setup, and configuration. However, there are platforms available that facilitate the creation of custom RL environments. These include widely used platforms like OpenAI Gym, which offers a diverse array of environments [1], and DeepMind OpenSpiel, particularly suitable for gaming scenarios. Additionally, customized environments can be developed based on extensions of OpenAI Gym, such as TensorTrade for trading applications [60], and VIZDoom [66] tailored for gaming environments.

12.8.1 OpenAI's RL Environment

OpenAI's RL environment is widely used due to its easy-to-customize and broadly supported applications. Besides games, it supports many existing environments including classic controls (Cart Pole, Mountain car), multi-joint dynamics used in robotics and biomechanics, Atari gaming environment, and third-party ready-to-use environments like multiagent RL, multi-objective RL, and 3D navigation environment [1]. It also provides the capability of customizing an RL environment for a specific use case, extended from the gym.Env class. Four elements need to be defined in the new RL environment class as follows:

step() defines the reward for the action taken by an agent, returning the agent with the next observations. Meantime, it also checks whether the episode and the whole process have met the termination conditions.

reset() resets the environment to the initial state, and it is required before the calling step. It returns the first agent observation for an episode and information.

render() is to visualize what the agent sees from the environment, which is an optional step in the RL environment definition. It provides visual assistance in the case of gaming and robotics, which provides a visual clue of the environment and agent-environment interactions.

close() is to terminate the environment when all episodes are finished.

Previously, we explored topics ranging from adaptive systems and feature engineering to machine-learning-based hardware malware detection and reinforcement learning. Looking ahead, we will delve into a case study that demonstrates the design process of an adaptive system for hardware-assisted security using reinforcement learning and evaluate its effectiveness.

12.9 Adaptive RL-Guided Framework for Hardware-Assisted Security

In this chapter, we explore how reinforcement learning excels at identifying suitable actions to achieve goals through dynamic environment interaction, making it ideal for self-adaptive systems. Next, we delve into a methodology titled *Reinforced-HMD* [19], where reinforcement learning techniques are properly employed to create a cost-aware, adaptive framework for hardware-assisted malware detection.

12.9.1 Dataset and Hardware Monitoring

Benign and malware programs are profiled on an Intel Xeon X5550 machine. We have extracted low-level CPU events available under the *Perf* tool using a static performance monitoring approach where we can profile applications several times measuring different events each time. HPC events are monitored using the *Perf* tool with a sampling time of 10ms by running applications in a Linux Containers (LXC) as an isolated profiling environment. LXC is an operating-system-level virtualization technique that allows developers to package and isolate applications with their entire runtime environment unlike common virtual platforms such as VMWare and provides access to actual performance counters events. We executed more than 5,000 benign and malware applications. Benign applications include real-world applications comprising MiBench and SPEC2006, Linux system programs, browsers, and text editors. Malware applications, collected from and categorized by VirusShare and VirusTotal online repositories, comprise nine types of malware including Worms, Viruses, Botnets, Ransomware, Spyware, Adware, Trojans, Rootkits, and Backdoor. After data acquisition, the HPC events are thoroughly analyzed using effective feature analysis methods.

We employ the top four HPCs to generate sub-datasets for training, validating, and testing baseline ML detectors, ensemble methods, and the RL agents. To model the zero-day malware threat type in this case study, among all nine malware types, we randomly considered four types of malware from rootkit, backdoor, virus, and ransomware as the target zero-day test data. Additionally, we reserved 30% of all benign data as a zero-day test benign dataset. Both malware and benign samples in this subset were set aside to simulate zero-day testing data, ensuring that the trained machine learning classifiers have no prior knowledge of the new malware types and have never encountered the testing dataset. The rest of the five types of malware and benign samples are considered for training and known test purposes, and we randomly split them into 70% for training and 30% for known testing. Notably, the RL-based framework adopts the same settings for training and testing.

12.9.2 Methodology

Figure 12.7 illustrates an overview of *Reinforced-HMD*, an RL-guided adaptive system comprised of four components: data acquisition, feature fusion, training of RL system, and inference of RL system for defending against zero-day malware attacks during runtime.

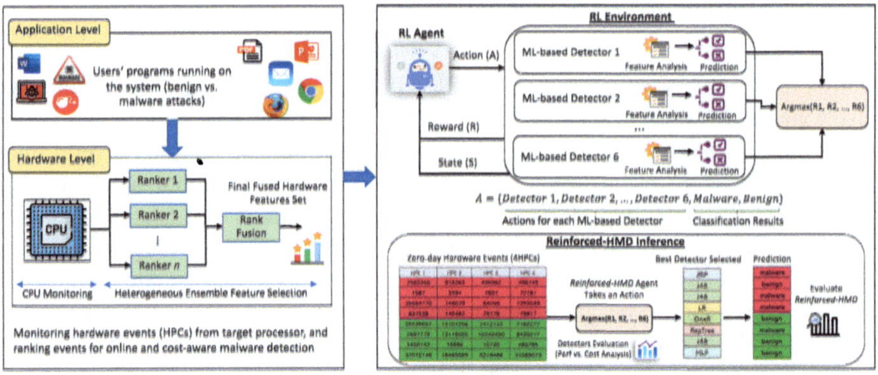

Fig. 12.7 Overview of *Reinforced-HMD*, a reinforcement-learning-guided framework for adaptive and cost-aware hardware-assisted malware detection [19]

12.9.2.1 Feature Engineering

In prior sections of this chapter, we emphasized the importance of feature engineering for creating robust machine-learning-based malware detectors that rely on hardware events. Before proceeding with feature selection, data inspection is crucial to address any missing data and outliers. Subsequently, we delve into data analysis to inform feature selection and choose appropriate ML algorithms. Additionally, conducting feature importance and correlation analyses helps us understand potential feature redundancy.

During feature selection, we can employ single or multiple feature selection methods to compare results and gain insights. In this section, our focus is on introducing a unified two-step feature fusion approach. This method utilizes a heterogeneous ensemble feature selection technique, incorporating various feature selection methods. Subsequently, an effective feature fusion step is applied to identify the most prominent hardware events.

Feature Selection (FS):
To investigate the impact of various feature selection methods on the prominence of top hardware events, we initially applied three feature selection techniques using Scikit Learn. For each method, we identify and record the top four features. The feature selection methods examined in this study are as follows:

- *Recursive Feature Elimination (RFE):* RFE is a feature analysis method that fits an ML model and removes the weakest feature(s) till it reaches a specified feature number. RFE begins by building a Decision Tree classifier on the entire dataset and computing the weights of each and all features. It then gradually eliminates less important features till the desired set is found.
- *Mutual Information (MI):* MI is a feature selection technique that calculates the mutual information between the HPC features X and labels Y. It primarily evaluates the amount of information for the target application's class (either

Table 12.3 Disparity of the four most significant HPC events across different feature selection methods (MI, RFE, and SFS) [19]

FS	Bus-cycles	Cache-references	Node-loads	Branch-loads	L1-dcache-loads	LLC-loads	L1-dcache-stores	L1-dcache load misses	iTLB-load-misses	dTLB-store-misses
MI		✓	✓			✓			✓	
RFE	✓		✓		✓			✓		
SFS				✓		✓	✓			✓

malware or benign) gained from each HPC feature in X. Mutual information between features X and label Y is calculated based on the entropy estimation from each point's k-nearest neighbors.

- *Sequential Feature Selection (SFS):* Lastly, we employ a sequential feature selection with a backward reduction in Scikit Learn. Backward-SFS is a greedy procedure that starts with all the dataset and greedily removes the less important features till meeting the desired number of features.

  Table 12.3 reports the top 4 HPC features chosen from the three tested feature selection methods. The result clearly shows the disparity of different selection methods' decisions. As observed, the top four most significant HPCs appear across ten features making the selection of the top four features uncertain. This highlights the importance of presenting a unified feature selection solution that accounts for the functionality of different methods to determine the top HPC events for online and cost-efficient HMD.

Feature Fusion (FF):

For the future fusion step, we adopt a novel fusion technique called Multi-criteria Decision-Making (MCDM) [39] that evaluates the conflicting criteria from several feature selection methods to obtain the final feature subset. MCDM is a powerful tool for making optimal decisions based on multiple criteria, including conflicting ones. In our experiments, given that three feature selection methods chose inconsistent hardware features, the MCDM method assists in structuring the goal and selecting a small set of essential features. MCDM has broad applications in many engineering problems with a rich set of algorithms. In our work, we employ a suitable one called the VIKOR algorithm as our feature fusion method.

Our implementation of the VIKOR feature fusion algorithm contains five distinct steps to determine a compromise solution that the Q measure provides a minimum of individual regrets for not selecting other features. We sort the Q measure for all HPC features in ascending order and select the top four smallest Q measure corresponding HPC features. The final top four features are L1-dcache-loads, node-stores, node-loads, and L1-dcache-stores.

In summary, our feature fusion algorithm includes the following five steps:

Step 1. Determine the best (f_j^*) and the worst (f_j^-) values of all criteria functions. The best value represents the benefit, while the worst value represents the cost of selecting the criterion.

Step 2. Compute the values S_i and R_i. S_i is the summation of the difference between the benefits and the costs among eight features' data points. R_i is the maximum benefit over costs along eight features' data points.

Step 3. Compute the values Q_i by the relation of S_i and R_i by calculating the majority of the criteria, where $Q_i = v(S_i - S^*)/(S^- - S^*) + (1-v)(R_i - R^*)/(R^- - R^*)$, $i = 1, 2, ...,$ length of p_matrix. S^* is the minimum of the sum of the benefits, S^- is the maximum of the sum of the costs, R^* is the minimum benefits for each criterion, and R^- is the maximum cost for each criterion. v is a weight for the strategy of the "majority of criteria," and 1-v is the weight of the individual regret.

Step 4. Rank the alternatives, sorting by the values S (summation of benefits), R (maximum benefits), and Q (the regrets) in ascending order. The results are three ranking lists.

Step 5. Propose a compromise solution $A^{(1)}$, which is the best ranked by the measure Q (minimum regrets) if the following two conditions are satisfied:

(1) $Q(A^{(2)}) - Q(A^{(1)}) \geq 1(m - 1)$, where $A^{(2)}$ is the second-ranked alternative by the measure Q (minimum regrets). The difference between the alternative solution $A^{(2)}$ and proposed solution $A^{(1)}$ is larger than a threshold meaning the alternative solution has a solid higher regrets than the proposed solution.

(2) The alternative $A^{(1)}$ must also be the best ranked by S and/or R so that the suggested solution $A^{(1)}$ shows stability regarding the summation of the benefits (S) and the maximum benefits (R) in step 4.

12.9.2.2 Threat Model

We utilize the top four performance counters data to create sub-datasets for training, validation, and testing. Our focus was on modeling zero-day malware threats. Among the nine malware types, we randomly selected four—rootkit, backdoor, virus, and ransomware—as the target zero-day test data. Additionally, we set aside 30% of the benign data to create a zero-day test benign dataset. This approach simulates real-world scenarios where trained machine learning classifiers encounter new malware types during testing. The remaining five malware types and benign samples were used for training and known testing, with a 70% training split and 30% for known testing.

12.9.2.3 Training Phase

As depicted in the top-right part of Fig. 12.7, the training process of the framework includes the autonomous RL agents, an RL environment, and training the classical machine learning detectors (base detectors). The RL system aims to find the best-parameterized reward function. This allows the RL agent to select accurate and cost-effective malware detectors based on user preferences and available resources, leading to high rewards (e.g., high detection rate, low hardware overhead). For malware detectors, we examine the suitability of various standard machine learning classifiers including JRIP, J48, OneR, Multilayer Perceptron, Logistic Regression (LR), and RepTree classifiers. These models are selected from different branches of machine learning, and their detection models are binary classifications aligned with the zero-day malware detection task. Algorithm 1 describes the RL-guided framework training procedure.

Algorithm 1: Training process of RL-guided *Reinforced-HMD*

Setup:
- Pre-process experience replay data;
- Initialize state S, RL agent θ, reward policy π;
- Let $e \leftarrow$ design criteria (F1, AUC, latency, area);
- Let $d \leftarrow$ the selected malware detector;
- Let $ML \leftarrow$ ML model;

while *training of RL system* **do**
- Agent gets state s_t;
- Agent predicts $y_{tk}^{-} = ML_t(s_t)$, $k = 1, 2, ..., 6$ ML detectors;
- Agent performs action $a_t \leftarrow \max\{\pi(a_{t1}|s_t, y_{t1}^{-}, e_{t1}), ..., \pi(a_{t6}|s_t, y_{t6}^{-}, e_{t6})\}$, action is selected based on detection and cost analysis of all ML detectors;
- Agent receives reward r_t and new state s_{t+1};
- Update RL agent parameters θ.

end

12.9.2.4 RL Environment

We customize the RL-guided framework's RL environment based on OpenAI's Gym [1]. OpenAI Gym is an open-source interface that provides RL environments for researchers to develop and benchmark new algorithms. It also offers a structured interface for customizing our RL environment. The RL environment consists of the following four key components and corresponding settings:

- *State S* consists of one set of experience replay data preprocessed inside the RL environment and available to the RL agent during runtime. It contains each row as a set of four HPC data, the predictions (correct/incorrect predictions) from the six malware detectors, their detection metrics (F1, AUC), and hardware overheads (latency, area). The state spaces (S) correspond to six ML-based detectors regarding their experience replay data.
- *Action Spaces A* is a set of all valid actions given to the RL agent to choose from. Our RL environment has six discrete action spaces available to the agent. Each action corresponds to one of the six tested malware detectors. Various actions will result in different consequences for getting a reward or not, and how many rewards the agent can receive. The RL agent is inclined to select the detector d_t that receives the highest rewards among the action space.

- *Reward Policy* π is a rule provided to the RL agent to decide its best action at each step. The RL-guided framework's environment provides a deterministic policy that maps $S \xrightarrow{R} A$, where S is the set of possible states from experience replay data. A is the action space containing all six detectors. The reward policy depends on the design criteria in our case. For the purpose of a comprehensive analysis of metrics trade-offs, we define five performance and design criteria ($F1, AUC, AUC/Latency, F1/Area$, and $F1 * AUC/Latency * Area$) as various target measurements to motivate the RL agent to take action. The reward policy based on F1 leads the agent to select the detector with the highest F-measure among all correct-predicted ML models at time step t. The reward policy based on $(F1 * AUC)/(Latency * Area)$ will have a more balanced view to consider selecting the detector with a high detection rate and a low hardware overhead at time step t. The goal of the reward policy is to enable the RL agent to select proper actions to maximize total future rewards for an accurate and cost-aware HMD process.
- *Experience Replay Data:* To simulate the interaction between the RL agent and the environment, the RL-guided framework uses an algorithm to collect all the possible outcomes for different malware detectors as the experience replay data. The presented pipeline reads each state, goes through each classifier for each state, and records the model detection performance (F1, AUC) and hardware overhead (Latency, Area). Then, we collect all the experience replay datasets for both known-test data (Train RL) and zero-day data (Test RL). As mentioned earlier, we consider three dataset partitions including train, known, and zero-day tests. The training dataset is used to train and validate the ML detectors. These ML detectors are placed in the branches of the RL environment so that the RL agent can adaptively decide which branch to activate and use for cost-aware malware detection. Furthermore, the unknown-test dataset is a dataset that is set aside during the initial dataset split and is used to train the RL agent. In this setting, we train the RL agent such that each branch of the ML detector has not seen the data before so that the RL agent can learn from the new dataset. Once the RL agent is well-trained, the zero-day test data is used to examine the RL agent in an unseen RL environment. This is because the zero-day test dataset contains new malware types, and both malware and benign data have never been seen by either the ML detectors or the RL agent.

12.9.2.5 RL Agents

Four model-free RL agents including UCB, Q-learning, A2C, and DDQN are implemented to interact with the environment. At each step, we trace the selected detector d_t chosen by the RL agent. We let the selected ML model defend at each time step to evaluate the HMD performance. The goal is to determine if an adaptive selection of the optimal detector would result in increasing the detection rate while reducing the resource and computational overheads.

Algorithm 2: Upper Confidence Bound (UCB)

Input: Experience replay data
Output: Agent's actions $A = [a_1, a_2, ..., a_6]$ is a set of detectors
Setup:
- Agent's actions $A = [a_1, a_2, ..., a_6]$ is a set of detectors;
- Set up state space S, action space A, reward function R;
- Initialize the number of times of detector selected N_j for each detector all as 0;
- Initialize episode length as 20;
- Initialize upper bound value as $1e^5$, let episodes $\tau = \lfloor \frac{length\ of\ training\ data}{episode\ length} \rfloor$;

while *episodes τ is not terminal* **do**
  for *steps in each episode* **do**
    for $j = 1, 2, ..., 6$ *detectors* **do**
      - calculate average reward $\bar{r}_j(n) = R_j(n)/N_j(n)$ if N_j is not zero, otherwise use the default value of $1e^5$;
      - calculate delta $\delta_j(n) = \sqrt{\frac{3}{2}\frac{\log(n)}{N_j(n)}}$;
      - update upper bound = $\bar{r}_j(n) + \delta_j(n)$;
      - $a \leftarrow$ detector j having maximum upper bound;
      - return S', r, a to the agent.
    end
    - Agent takes action a, record accumulated rewards;
    - Update current state $S \leftarrow S'$.
  end
end

- *Upper Confidence Bound (UCB):* UCB is a deterministic algorithm that leverages a maximum confidence boundary to decide whether to explore or exploit the RL environment. As shown in Algorithm 2, it firstly assigns a high maximum confidence boundary of $1e^5$ for each branch that the RL agent shall learn from the data and gradually update it for each branch later when explorations progress. Initially, the RL agent randomly selects one of the branches to explore. Depending on how much reward this branch leads to, this branch's corresponding confidence interval is either increased or reduced. In the next round, the RL agent chooses the branch with the highest upper bound to explore and update rewards in the confidence interval.

At the end of round N, the UCB-based RL agent calculates the average reward $r_j(n)$ and $\delta_j(n)$ and updates the upper bound boundary interval $[r_j(n) - \delta_j(n), r_j(n) + \delta_j(n)]$, where $r_j(n)$ is the average reward of all rounds of N, and $\delta_j(n)$ is the knowledge gained from the current round. Delta $\delta_j(n)$ encourages more explorations, and the average rewards $r_j(n)$ facilitate exploitation. When the RL agent has little knowledge of the best branch, it favors more explorations to search through all branches. As more rounds of exploration progress, more knowledge of the possible best branch is gained, and the RL agent gradually shifts to exploitation to select the branch with the highest upper bound value. UCB greedily trials each detector on each round of exploration and evaluates the possible rewards according to the reward policy. Given different base classifiers, the UCB algorithm can almost explore the maximum potential rewards among the six malware detectors, with a relatively longer computation time. Since our environment is lightweight, the time used for executing UCB is slightly longer than Q-learning but very trivial. Detailed implementation is shown in Algorithm 2.

- *Q-Learning:* As described earlier, Q-Learning is an off-policy algorithm to determine the best action given the current state by keeping track of a Q-table that gets updated after each episode with its row corresponding to the state and its column to the action. An episode ends after a set of actions is completed. In the end, the Q-table suggests the optimal policy. We implement Q-learning as shown in Algorithm 3.

Algorithm 3: Q-learning

Input: Experience replay data
Output: Agent's actions $A = [a_1, a_2, ..., a_6]$ is a set of detectors
Setup:
- Set up state space S, action space A, reward function R;
- Initialize Q table $Q[s_i, a_j] \leftarrow 0$ for 20 steps/episode for 6 detectors, with the initial value N_j, the number of detectors selected for each detector as 0;
- Initialize learning rate α, discount factor γ, epsilon ϵ; let episodes $\tau = \lfloor \frac{length\ of\ training\ data}{episode\ length} \rfloor$;

while *episodes τ is not terminal* **do**
    for *steps in each episode* **do**
        - Generate a random number $X \sim [0, 1]$;
        if $X > \epsilon$ **then**
            $a \leftarrow max(Q(s_i, a_j), j = 1, 2, ...6)$;
        end
        else
            $a \leftarrow X$;
        end
        - Agent takes action a, record accumulated rewards;
        - Update current state $S \leftarrow S'$;
        - Update Q table: $Q(s, a) \leftarrow Q(s, a) + \alpha[R + \gamma \cdot \max Q(s_{t+1}, a_{t+1}) - Q(s, a)]$;
    end
end

As highlighted in Algorithm 3, the RL agent deploys a Q-table to track the maximum reward for the action it takes at each state. Initially, the Q-value for each of the six branches is zero. The RL agent randomly selects a branch to explore. The parameter of epsilon ϵ helps the agent to decide on whether to explore or to exploit. The initial value for epsilon ϵ is a small value between [0,1]. In experiments, we found that 0.5 is the best value. The Q-function uses the Bellman equation as shown in line 6 of Algorithm 3 to update the Q-value at each step.

The Q-value considers both the current reward R and the discounted future rewards. Current rewards are more important than future rewards when the RL agent makes a decision. This is an iterative process until all training data are explored. As the RL agent starts to explore the environment, the Q-function gives optimal approximations by continuously updating the Q-values in the table. We initially set the learning rate alpha α, the discount factor gamma γ, and the epsilon each as a

small value of [0,1], and we train the RL agent to learn the most optimal values of them. Our final alpha α is 0.35, gamma γ is 0.3, and epsilon ϵ is 0.5.
- *Advantage Actor-Critic (A2C):* Advantage Actor-Critic (A2C) algorithm is a combination of policy gradient-based and value-based reinforcement learning algorithms [33]. A2C is a deterministic algorithm that contains two deep RL networks: Actor and Critic networks. The Actor network learns to map each state S_t with a probability distribution of actions (probability of $a1_t$, $a2_t$, ... $a6_t$), and it uses a neural network with an activation function (e.g., softmax) to output for the policy $\pi(a_t|S_t; \theta)$. The Critic network maps the input state S_t to a prediction value V for future reward, and it uses another neural network with a linear output (without activation) for the value function $V(S_t;\ Q_v)$. The A2C network learns from a loss function calculated by an Advantage function, and the loss value represents the difference between the Actor and Critic to update both networks to gradually align them together. Algorithm 4 shows the details of our A2C algorithm in the context of HPC-based malware detection.

In this chapter, we use two MLPs in the A2C architecture that the Critic contains five layers with three hidden layers (32*16*16), and the Actor is a four-layer MLP containing two hidden layers with 32 and 16 hidden nodes. The input layer contains the dimension of state space, and the output layer contains one action result (one optimal detector). We use the Sparse Categorical Cross Entropy loss function with Adaptive Moment Estimation (Adam) optimization algorithms for the Actor and the Mean Squared Error loss function for the Critic.

Algorithm 4: Advantage Actor Critic (A2C) [33]

Input: Experience replay data
Output: Agent's actions $A = [a_1, a_2, ..., a_6]$ is a set of detectors
Setup:
- Pre-process experience replay data; - Initialize RL agent parameters θ, θ_v discount factor γ, learning rate for Actor and Critic, and update interval;
- Initialize state S, reward policy π=F1*AUC/Latency*Area;
- Let $d \leftarrow$ the selected malware detector;
- Let $ML \leftarrow$ ML model;

while *training of RL system* **do**
- Agent reads state s_t;
- Actor predicts on state s_t, receives probabilities of all actions: $s_t \rightarrow \{P(a_{1t}), P(a_{2t}), ..., P(a_{6t})\}$;
- Agent selects the action with the highest probability a_t;
- Agent receives reward r_t and new state s_{t+1};
- Critic predicts future rewards based on current action $v(a_t)$;
- Agent computes advantage value to evaluate how good is the action, $A(s_t, a_t) = r_{t+1} + \gamma v(s_{t+1}) - v(s_t)$;
- Compute Actor loss value;
- Computer Critic loss value;
- Compute and apply gradients to update Actor and Critic networks weights w = w + $\nabla_{\theta'} log\pi(a_t|s_t; \theta') A(s_t, a_t; \theta, \theta_v)$;

end

- *Dueling DQN (DDQN):* As mentioned previously, DDQN is a single Q network with two streams: one estimating the state value function $v(s)$ and the other estimating the advantage $A(s, a)$. These outputs combine to produce the state-action value $Q(s, a)$. DDQN decouples the representation and the algorithm, allowing any underlying architecture to be reused in the two streams. We employ a simple two-layer MLP for the underlying algorithms, with one dense layer representing the state value function $v(s)$ and another dense layer representing the advantage function $A(s, a)$. The optimal $Q^*(s, a)$ is the maximum Q-function from all options of advantages $\{A(s, a_1), \ldots, A(s, a_6)\}$. The optimal Q-value comes from the action which can produce the highest Q-value $Q^*(s, a) = \max_\pi Q^\pi(s, a)$, where $a = argmax_{a' \in A} Q^*(a, a')$ and $V^*(s) = \max_a Q^*(s, a)$. At last, Q^* is aggregated followed by $Q = V + A$—mean of all A. Algorithm 5 provides further details of the Dueling DQN network developed in this work.

Algorithm 5: Dueling DQN (DDQN)

Input: Experience replay data
Output: Agent's actions $A = [a_1, a_2, ..., a_6]$ is a set of detectors
Setup:
- Pre-process experience replay data for stream 4; ▷ See Sect. 12.9.2.4
- Initialize DRL agent parameters: discount factor γ, learning rate, and epsilon;
- Initialize state S, reward policy π=F1*AUC/Latency*Area;
- Let $a \leftarrow$ the selected malware detector;
while *training of RL system* **do**
    - Agent reads state s_t;
    - Q network predicts q value on state s_t for all actions, select the action with the highest value;
    - Advantage evaluates the action based on the state;
    - Agent update Q*=v+(a-mean of all actions);
    - Agent receives reward r_t and new state s_{t+1};
    - Compute loss value;
    - Compute and apply gradients to update DDQN network weights;
end

12.9.2.6 Online Inference

Figure 12.7 (right bottom) shows a case study for the inference and evaluation process. As shown, first each row of the four zero-day tests HPCs (state s_t) is fed into the RL environment. The RL agent loads all six ML-based detectors in different branches to run predictions. Among all six models, any ML that predicts the sample HPC correctly is eligible to receive a reward from the environment. If all ML models predict wrongly, the system assigns the least-costly ML (in terms of latency and hardware overhead) as the default detector with a reward granted.

When multiple ML models perform correct predictions, the RL system will evaluate detection accuracy and cost analysis for all six ML models to select the detector with the highest reward. The selected detector is then recorded one by one for all test data until the adaptive branching selection through the RL system is complete. Lastly, to evaluate the effectiveness of the RL system, we use the recorded selection of malware detector at each time step to reproduce the defending process to obtain the result. We first load the ML model selected by the RL system at each time step and feed with the same row of four HPCs data to allow each model to run prediction. We record each prediction for all test data and calculate the RL system's evaluation metrics based on (1) the sum of rewards, (2) the RL system's malware detection accuracy, and (3) the RL system's detection and cost effectiveness. Algorithm 6 outlines the online inference procedure and evaluation process.

Algorithm 6: Online inference and system performance evaluation

```
Setup:
- Pre-process experience replay data;
- Initialize state S, RL agent θ, reward policy π;
- Let d ← the selected malware detector;
- Let ML ← ML model;
while Inference do
    for zero-day test data: 0 → n do
        - Agent gets state s_t;
        - Agent acts to select a detector d_t;
        - Agent assigns the best d_t to defend, gets predicted label y_t^- = π(a_t|s_t, θ);
        if predicted label y_t^- == y_t true label then
            - Increment detection rate;
            - Record hardware overhead, selected detector d_t;
        end
    end
    - Calculate RL system detection metrics (F1, Accuracy, AUC) and overhead (latency,
    area) for all test data.
end
```

12.9.3 Experimental Results and Evaluation

This section presents an evaluation and detailed analysis of experimental results on the effectiveness of the proposed RL-guided framework for adaptive and cost-aware zero-day malware detection.

12.9.3.1 Effectiveness Analysis of Feature Fusion

To analyze the benefits of feature fusion processing presented in Sect. 12.9.2.1, we benchmark models trained on the top four features from the feature fusion method against models trained using three other feature selection methods and then analyze their model detection performance. We compare the ML models using the top four features selected from the feature fusion method with the same models trained from

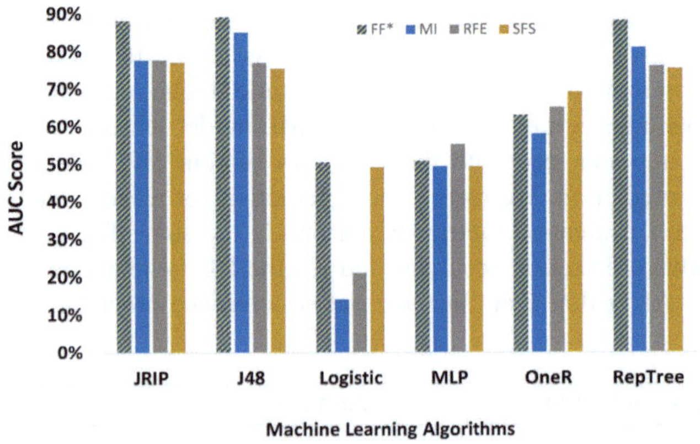

Fig. 12.8 AUC scores with vs. without Feature Fusion. FF: feature fusion, MI: mutual information, RFE: recursive feature elimination, SFS: sequential feature selection

the top four features generated from each of the three feature selection methods (MI, RFE, SFS) regarding their AUC score. Figure 12.8 displays their AUC scores.

FF represents the AUC scores of the zero-day test from the model trained on feature fusion outputted top four features. It is observed that in almost all models, their AUC scores trained from the feature fusion method outperform those same ML algorithms that are trained from various feature selection methods, indicating a disparity in the effectiveness of top feature selection by traditional methods. For instance, in the J48 algorithm, the model trained with FF-selected features outperforms the one trained with MI-selected features by 4% in AUC. The advantage of our proposed feature fusion technique is that it selects the most dominant features containing the most information gains. Consequently, the trained model can have a higher detection rate for zero-day malware overall. We also observe different ML algorithms react to important features differently. Besides FF, MI proves to be a stronger feature selection method than RFE and SFS.

12.9.3.2 Base ML-Based Malware Detectors

Table 12.4 reports the base ML-based malware detectors' detection metrics and hardware overheads. Given the importance of analyzing the hardware overhead of ML classifiers used for efficient on-device malware detection, we develop the ML-based malware detectors at the hardware level and analyze their associated area and latency overheads. We used the Vivado HLS compiler to develop the HDL implementation of the classifiers on Xilinx Virtex 7 FPGA. The latency is considered as the number of clock cycles (cycles @10 ns), and the area overhead is calculated using the total number of utilized LUTs, FFs, and DSP units in the FPGA. We evaluate ML models in two ways. Firstly, we benchmark the RL-based method with ML-based and ensemble methods. Secondly, the six pretrained MLs

Table 12.4 Base classifiers' detection rate and overhead for zero-day HMD

ML Models	Accuracy	F1-score	AUC	Latency (ms)	Area
JRIP	0.91	0.86	0.89	2	156
J48	0.91	0.87	0.89	3	584
Logistic Reg.	0.62	0.14	0.50	59	11,815
MLP	0.62	0.14	0.51	102	25,667
OneR	0.70	0.46	0.63	1	292
RepTree	0.90	0.86	0.88	3	377

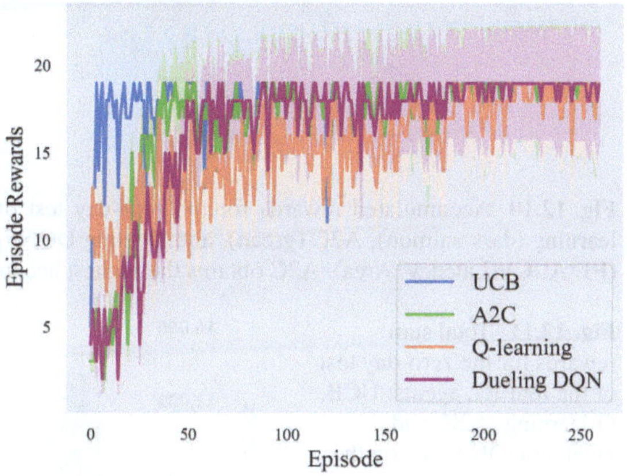

Fig. 12.9 Learning speed during training of the four RL agents: UCB (blue), Q-learning (dark salmon), A2C (green), and Dueling DQN (purple), based on the design criteria of (F1*AUC)/(Latency*Area). The maximum episode reward is 20

are available to the RL agents to select one ML at each step that returns the highest reward. This metric shows that the J48 decision tree can achieve the highest detection performance, 87% in F1-score and 89% in AUC, respectively. The results highlight the cost inefficiency of MLP and LR algorithms with the highest latency and area overhead among others. However, regarding latency, OneR is the fastest with 1 ms outperforming J48 (3 ms). Also, regarding the area, the JRIP classifier outperforms both the OneR and J48 classifiers.

12.9.3.3 RL-Guided Self-adaptive Framework Evaluation: Learning Performance

Figure 12.9 depicts the training speed in episode rewards (left) for the four RL agents used in the RL-guided framework for the design criteria of $F1 * AUC/Latency * Area$. Among four RL agents, UCB (blue) picks up the decision pattern quickly at the beginning of episodes (during episodes 0–50). The two deep RL agents A2C (green) and Dueling DQN (purple) start slower, as they spend the initial episodes exploring the RL environment to learn the optimal decision patterns based on the states (data) and the reward policy. After episode 50, both A2C and DDQN catch up with UCB regarding the episode rewards and maintain stable episode rewards afterward.

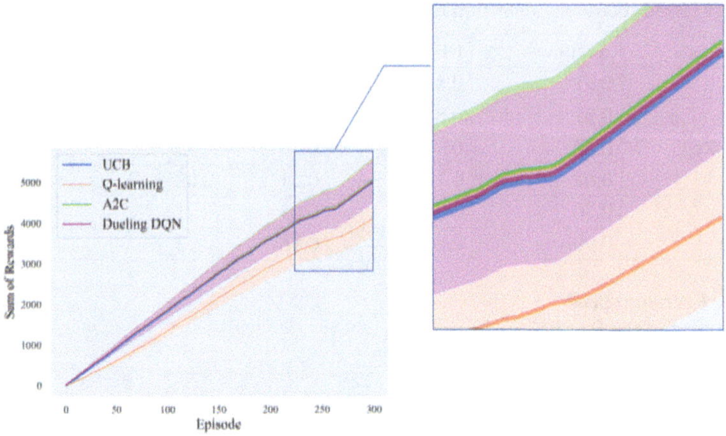

Fig. 12.10 Accumulated rewards for the zero-day test of the four RL agents: UCB (blue), Q-learning (dark salmon), A2C (green), and Dueling DQN (purple), based on the design criteria of (F1*AUC)/(Latency*Area). A2C obtains the highest accumulated rewards in the zero-day test

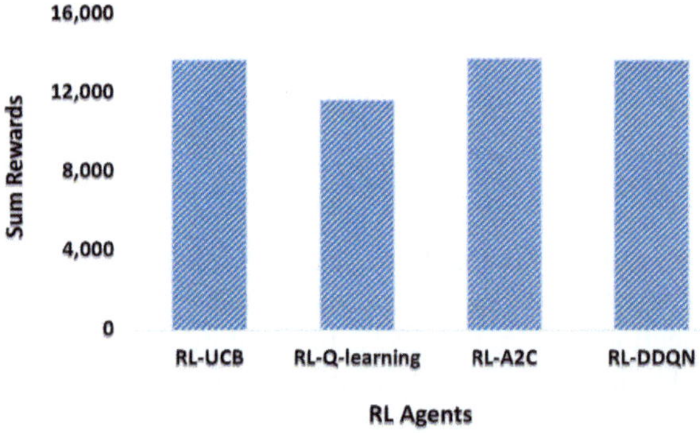

Fig. 12.11 Total sum rewards for the zero-day test of the four RL agents: UCB, Q-learning, A2C, and Dueling DQN based on the design criteria of (F1*AUC)/(Latency*Area)

12.9.3.4 RL-Guided Self-adaptive Framework Evaluation: Accumulated Rewards for Zero-Day Test

Figure 12.10 demonstrates the accumulated rewards for the four RL agents used in the experiment, over the zero-test test episodes for the design criteria of $F1 * AUC/Latency * Area$. We observe from Fig. 12.9 that in the last several episodes, the episode rewards for UCB, A2C, and DDQN are very stable (with low variance), which indicates the models are converged from exploring the environment. During the zero-day test, the accumulated rewards as shown in Fig. 12.11 indicate the A2C reaches the highest sum of rewards, followed closely by Dueling DQN and UCB algorithms.

12.9.3.5 RL-Guided Self-adaptive Framework Evaluation: Performance vs. Cost-Efficiency Analysis Against State-of-the-Art ML Models

Table 12.5 reports the detection performance and hardware overhead improvement of using our presented RL-guided framework as an adaptive and cost-aware model as compared with using the best classical ML model (J48) and widely used ensemble models. As seen, RL-UCB, RL-A2C, and RL-Dueling DQN have similar performance in selecting the optimal policy, to reach high F1 and AUC scores. RL-UCB is a deterministic, greedy policy algorithm that visits all state-action pairs at every step. It is able to select the optimal policy according to our reward function F1*AUC/Latency*Area for the highest F1 and AUC, while minimizing hardware latency and area.

RL-UCB has a nearly perfect capability (99.65%) to select the most optimal malware detector that helps the RL system reach a 96% detection rate in both F1-score and AUC, which is 1% higher than majority voting and 9% up than J48. Since the RL system considers a balanced strategy across detection metrics and low hardware overhead, RL-UCB has much less hardware latency, 12 times lower than ensemble-based majority voting, and 44% lower than the lightweight J48 algorithm. Meantime, UCB consumes 36 times less hardware area than majority voting, 2.34 times less than J48.

RL-A2C achieves the highest total accumulated rewards over zero-day test data, as seen in Fig. 12.11. A2C, a hybrid architecture combining value-based and policy-based algorithms, reduces variances during inference. This architecture can be easily saved as two separate models: one for the Actor and another for the Critic. These models are lightweight and can be imported to resource-constrained devices during inference. In this experiment, RL-A2C reaches a 99.30% probability to select the most optimal malware detector that helps the RL system obtain a 96% detection rate in both F1-score and AUC, which is 1% higher than majority voting and 9% up than J48. Its latency is slightly higher than RL-UCB by 0.13 ms, and the hardware area is 22 units higher than RL-UCB. Compared with ensemble-based majority voting, its F1-score is 1% higher, and 36% lower than the lightweight J48 algorithm. Meantime, A2C consumes 32 times less hardware area than majority voting and 1.82 times less than J48.

To accordingly account for malware detection rate and area overhead impact, in Fig. 12.12, we show the trade-off analysis between the five tested ML-based detectors regarding their detection rate (F1-score) vs. models' hardware overhead (area). We employ F1-score over area to determine the models that require a small area and yet can detect the maliciousness of the program with high performance. A classifier with a higher value in F1 and a low value in area is considered more effective. As observed, the three RL algorithms RL-UCB, RL-A2C, and RL-Dueling DQN all achieve a similar level of high detection rate and low hardware area overhead. An RL-based A2C model proves to be the best balanced solution with a high F1-score (96%) while occupying a minimal area outperforming other models. The majority voting achieves a high F1-score (95%) but with an increased area and hardware overhead that makes the solution inefficient for cost-aware on-device

Table 12.5 Analysis of the RL-guided techniques with the best detector (J48) and ensemble-based methods for adaptive and cost-aware zero-day malware detection

Approach	Selection probability	RL system F1-score	RL System AUC	HW latency (ms)	HW area
J48	N/A	0.87	0.89	3	584
Majority voting	51%	0.94579	0.9475	28	6482
Stacking	25%	0.90966	0.9135	28	6482
RL-UCB	99.65%	0.95538	0.9560	2.08	175
RL-Q-learning	89%	0.92600	0.9285	6.1	1190
RL-A2C	99.30%	0.95540	0.9560	2.21	207
RL-DDQN	99%	0.95170	0.9525	2.44	264

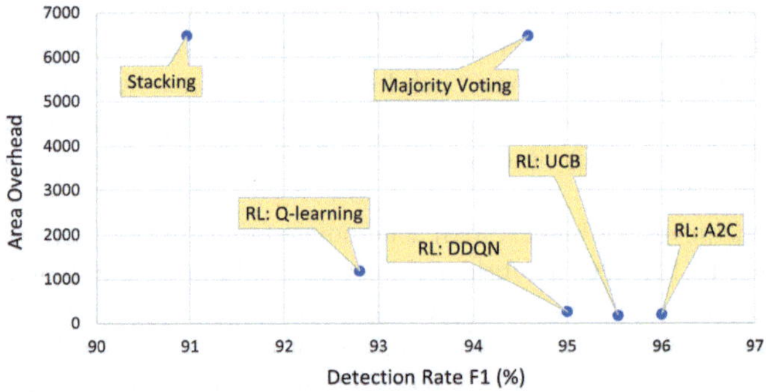

Fig. 12.12 Detection rate vs. area overhead comparison of RL (UCB, Q-learning, A2C, Dueling DQN), ensemble (majority voting, stacking), and best ML model (J48)

Fig. 12.13 The relationship between all RL and ensemble methods' selection probability of the optimal detector vs. system efficiency with the design criteria of (F1*AUC)/(Latency*Area)

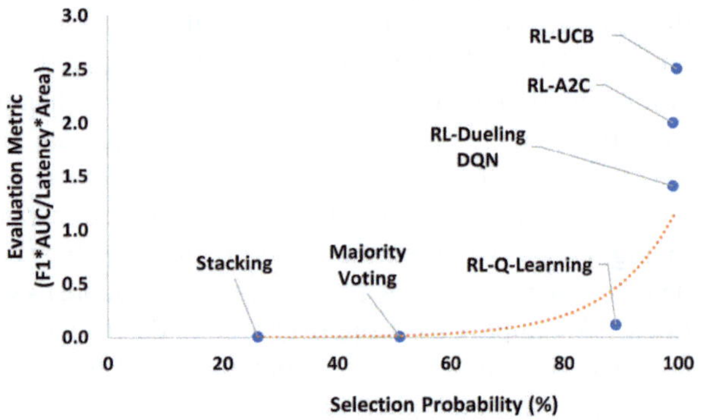

malware detection. Additionally, the stacking method is the least optimal selector with a 91% F1-score while consuming a high area on the host hardware.

Furthermore, Fig. 12.13 depicts the relationship between all methods' selection probability of the optimal detector versus the cost efficiency measured regarding the balanced design criteria of $(F1 * AUC)/(Latency * Area)$. The results show that the agents based on UCB and A2C agents lead to a higher selection probability in choosing the most optimal malware detectors during runtime which leads to a higher cost efficiency. Overall, UCB outperforms Q-learning by 11%, A2C and Dueling DQN by 0.65% in selection probability, which results in an increment of 20% efficiency measured in $(F1 * AUC)/(Latency * Area)$ than Q-learning. Moreover, RL-based UCB outperforms majority voting by nearly 49% to select the most optimal malware detector at runtime while delivering a five times higher cost efficiency. RL-based A2C outperforms majority voting by nearly 48% to select the most optimal malware detector at runtime while delivering a four times higher cost efficiency.

Considering Fig. 12.11, we observe that A2C accumulates the most rewards for the zero-day test, surpassing even UCB. This highlights the complexity of the reinforcement learning environment, where an RL agent might select suboptimal actions that still achieve high detection accuracy but incur slightly higher costs in terms of hardware latency and area overhead.

12.9.3.6 RL-Guided Self-adaptive Framework Evaluation: Inference Overhead

Table 12.6 presents the per-sample inference time and the model size for J48, ensemble learning models, and the four examined RL models. As observed, a single J48 has the lowest latency, with a moderate 77 KB model size. Ensemble learning also performs inference fast but has a relatively bigger model size than J48. RL-based UCB and Q-learning models are parameter-based, making them lightweight with negligible model size and minimal inference time. The deep RL-based A2C also has a lightweight model but requires 59 ms to complete inference per sample. Dueling DQN, while having a moderate inference time, adds 220 KB to the host device's model size.

Regarding the overall performance and overhead, different RL agents can be recommended depending on host device situations compared to ML and ensemble models. A2C offers a high detection rate (F1, AUC), a moderate hardware overhead (hardware latency and area), and a lightweight model. Although its online inference time is the longest at 59 ms, this can generally be acceptable. If inference time is the highest priority, we recommend either RL UCB or RL Dueling DQN. Dueling DQN is preferable for hardware malware detection in environments requiring better generalization to respond to changing malware intrusions. UCB may require more frequent fine-tuning to adapt to a changing environment effectively.

Table 12.6 Inference overhead

Approach	Inference time (ms)	Model size (KB)
J48	0.02	77
Majority voting	0.02	335
Stacking	0.10	172
RL-UCB	0.04	1
RL-Q-learning	0.02	1
RL-A2C	59	23
RL-DDQN	1.69	220

12.10 Conclusion

This chapter has centered around the application of AI for hardware-assisted security, specifically exploring the emerging topic of reinforcement learning with a focus on enabling adaptive hardware-assisted malware detection. It embarked on a comprehensive journey, starting with a review of the background and taxonomy of malware detection. We then delved into an exploration of hardware-assisted ML-based malware detection methodologies, providing insights into the intricate processes involved in using machine learning on hardware performance counters data for detecting malicious software. The chapter further offered a detailed overview of the machine learning process, followed by an in-depth examination of various feature selection methodologies. Notably, we discussed the innovative concepts of an ML-based self-adaptive system, elucidating the application of a reinforcement learning approach in its implementation. Detailed discussions on widely employed RL agent algorithms and the establishment of RL environments using the OpenAI Gym platform were provided. A compelling case study showcased the efficacy of an RL-guided adaptive defense system for online malware detection in hardware-assisted security, revealing that such a framework, when combined with machine learning defenders, can achieve commendable detection performance with minimal hardware overhead. Overall, this chapter has highlighted the significant potential of reinforcement learning in enhancing hardware-assisted malware detection systems, demonstrating how adaptive AI techniques can effectively respond to evolving security threats.

References

1. https://gymnasium.farama.org/api/env/. Accessed March 5, 2024.
2. Abdi, H., & Williams, L. J. (2010). Principal component analysis. *WIREs Computational Statistics, 2*(4), 433–459. https://doi.org/10.1002/wics.101
3. Abadi, M., Budiu, M., Erlingsson, U., & Ligatti, J. (2009). Control-flow integrity principles, implementations, and applications. *ACM Transactions on Information and System Security, 13*(1), 1–40. https://doi.org/10.1145/1609956.1609960

4. Adat, V., & Gupta, B. B. (2018) Security in internet of things: issues, challenges, taxonomy, and architecture. *Telecommunication Systems, 67*, 423–441.
5. Ahmed, M., Naser Mahmood, A., & Hu, J. (2016). A survey of network anomaly detection techniques. *Journal of Network and Computer Applications, 60*, 19–31. https://doi.org/10.1016/j.jnca.2015.11.016
6. Anomaly Based Intrusion Detection (2024). https://en.wikipedia.org/wiki/Anomaly based intrusion detection system. Accessed March 5, 2024.
7. Baye, G., Hussain, F., Oracevic, A., Hussain, R., & Ahsan Kazmi, S. (2021). Api security in large enterprises: Leveraging machine learning for anomaly detection. In *2021 International Symposium on Networks, Computers and Communications (ISNCC)* (pp. 1–6). https://doi.org/10.1109/ISNCC52172.2021.9615638
8. Bazrafshan, Z., et al. (2013). A survey on heuristic malware detection techniques. In *The 5th Conference on Information and Knowledge Technology* (pp. 113–120). https://doi.org/10.1109/IKT.2013.6620049
9. Bettany, A., & Halsey, M. (2017). *What is Malware?* (pp. 1–8). Apress. https://doi.org/10.1007/978-1-4842-2607-0_1
10. Cathey, R., Ma, L., Goharian, N., & Grossman, D. (2003). Misuse detection for information retrieval systems. In *Proceedings of the Twelfth International Conference on Information and Knowledge Management, CIKM '03* (pp. 183–190). Association for Computing Machinery. https://doi.org/10.1145/956863.956901
11. Chourib, M. (2019). Detecting selected network covert channels using machine learning. In *2019 International Conference on High Performance Computing & Simulation (HPCS)* (pp. 582–588). https://doi.org/10.1109/HPCS48598.2019.9188115
12. Das, S., Werner, J., Antonakakis, M., Polychronakis, M., & Monrose, F. (2019). Sok: The challenges, pitfalls, and perils of using hardware performance counters for security. In *2019 IEEE Symposium on Security and Privacy (SP)* (pp. 20–38). IEEE.
13. Demme, J., Maycock, M., Schmitz, J., Tang, A., Waksman, A., Sethumadhavan, S., & Stolfo, S. (2013). On the feasibility of online malware detection with performance counters. *ACM SIGARCH Computer Architecture News, 41*(3), 559–570.
14. Dinakarrao, S. M. P., Sayadi, H., Makrani, H. M., Nowzari, C., Rafatirad, S., & Homayoun, H. (2019). Lightweight node-level malware detection and network-level malware confinement in iot networks. In *2019 Design, Automation & Test in Europe Conference & Exhibition (DATE)* (pp. 776–781). IEEE.
15. Dinakarrao, S. M. P., et al. (2020). Cognitive and scalable technique for securing IoT networks against malware epidemics. *IEEE Access, 8*, 138508–138528. https://doi.org/10.1109/ACCESS.2020.3011919
16. Duda, R., et al. (2001). *Pattern classification*. Wiley.
17. Gao, Y., Makrani, H. M., Aliasgari, M., Rezaei, A., Lin, J., Homayoun, H., & Sayadi, H. (2021). Adaptive-HMD: Accurate and cost-efficient machine learning-driven malware detection using microarchitectural events. In *2021 IEEE 27th International Symposium on On-Line Testing and Robust System Design (IOLTS)* (pp. 1–7). IEEE.
18. Gibert, D., Mateu, C., & Planes, J. (2020). The rise of machine learning for detection and classification of malware: Research developments, trends and challenges. *Journal of Network and Computer Applications, 153*, 102526. https://doi.org/10.1016/j.jnca.2019.102526
19. He, Z., Makrani, H. M., Rafatirad, S., Homayoun, H., & Sayadi, H. (2022). Breakthrough to adaptive and cost-aware hardware-assisted zero-day malware detection: A reinforcement learning-based approach. In *2022 IEEE 40th International Conference on Computer Design (ICCD)* (pp. 231–238).
20. He, Z., Rezaei, A., Homayoun, H., & Sayadi, H. (2022). Deep neural network and transfer learning for accurate hardware-based zero-day malware detection. In *Proceedings of the Great Lakes Symposium on VLSI 2022, GLSVLSI '22* (pp. 27–32). Association for Computing Machinery.
21. He, Z., Homayoun, H., & Sayadi, H. (2024). Beyond conventional defenses: Proactive and adversarial-resilient hardware malware detection using deep reinforcement learning. In *ACM/IEEE Design Automation Conference (DAC)*.

22. He, Z., Homayoun, H., & Sayadi, H. (2024). Guarding against the unknown: Deep transfer learning for hardware image-based malware detection. *Journal of Hardware and Systems Security, 8*, 1–18.
23. Helman, P., et al. (1992). Foundations of intrusion detection (computer security). In *Computer Security Foundations Workshop V, 1992. Proceedings* (pp. 114–120). IEEE.
24. Hie, B. L., & Yang, K. K. (2022). Adaptive machine learning for protein engineering. *Current Opinion in Structural Biology, 72*, 145–152. https://doi.org/10.1016/j.sbi.2021.11.002
25. Hofmeyr, S. A., et al. (1998). Intrusion detection using sequences of system calls. *Journal of Computer Security, 6*(3), 151–180.
26. Jidiga, G. R., & Sammulal, P. (2014). Anomaly detection using smart tracing tricks on call stack. In *International Conference for Convergence for Technology-2014* (pp. 1–6). https://doi.org/10.1109/I2CT.2014.7092136
27. Kaloudi, N., & Li, J. (2020). The AI-based cyber threat landscape: A survey. *ACM Computing Surveys, 53*(1), 1–34. https://doi.org/10.1145/3372823
28. Khalkhali, I., Azmi, R., Azimpour-Kivi, M., & Khansari, M. (2011). Host-based web anomaly intrusion detection system, an artificial immune system approach. *International Journal of Computer Science Issues (IJCSI), 8*(5), 14.
29. Liu, H., & Motoda, H. (2012). *Feature selection for knowledge discovery and data mining* (Vol. 454). Springer Science & Business Media.
30. Macías-Escrivá, F. D., Haber, R., del Toro, R., & Hernandez, V. (2013). Self-adaptive systems: A survey of current approaches, research challenges and applications. *Expert Systems with Applications, 40*(18), 7267–7279. https://doi.org/10.1016/j.eswa.2013.07.033
31. Malik, M., Tullsen, D. M., & Homayoun, H. (2017). Co-locating and concurrent fine-tuning mapreduce applications on microservers for energy efficiency. In *2017 IEEE International Symposium on Workload Characterization (IISWC)* (pp. 22–31). IEEE.
32. Misuse (2024). https://en.wikipedia.org/wiki/Misuse detection. Accessed March 5, 2024.
33. Mnih, V., Badia, A. P., Mirza, M., Graves, A., Lillicrap, T., Harley, T., Silver, D., & Kavukcuoglu, K. (2016). Asynchronous methods for deep reinforcement learning. In *International Conference on Machine Learning* (pp. 1928–1937). PMLR.
34. Mosenia, A., & Jha, N. K.: A comprehensive study of security of internet-of-things. *IEEE Transactions on Emerging Topics in Computing, 5*(4), 586–602.
35. Moskovitch, R., Feher, C., Tzachar, N., Berger, E., Gitelman, M., Dolev, S., & Elovici, Y. (2008). Unknown malcode detection using opcode representation. In *Intelligence and Security Informatics: First European Conference, EuroISI 2008, Esbjerg, December 3–5, 2008. Proceedings* (pp. 204–215). Springer.
36. Nespoli, P., Papamartzivanos, D., Mármol, F. G., & Kambourakis, G. (2017). Optimal countermeasures selection against cyber attacks: A comprehensive survey on reaction frameworks. *IEEE Communications Surveys & Tutorials, 20*(2), 1361–1396.
37. Ozsoy, M., Donovick, C., Gorelik, I., Abu-Ghazaleh, N., & Ponomarev, D. (2015). Malware-aware processors: A framework for efficient online malware detection. In *2015 IEEE 21st International Symposium on High Performance Computer Architecture (HPCA)* (pp. 651–661). IEEE.
38. Palm, A., Metzger, A., & Pohl, K. (2020). Online reinforcement learning for self-adaptive information systems. In S. Dustdar, E. Yu, C. Salinesi, D. Rieu, & V. Pant (Eds.), *Advanced Information Systems Engineering* (pp. 169–184). Springer International Publishing.
39. Papathanasiou, J. P. N. (2019). *Multiple criteria decision aid: Methods, examples and python implementations*. Springer.
40. Patel, N., et al. (2017). Analyzing hardware based malware detectors. In *Proceedings of the 54th Annual Design Automation Conference 2017, DAC '17* (pp. 25:1–25:6). ACM. https://doi.org/10.1145/3061639.3062202
41. Polychronakis, M., Anagnostakis, K. G., & Markatos, E. P. (2007). Emulation-based detection of non-self-contained polymorphic shellcode. In *Recent Advances in Intrusion Detection: 10th International Symposium, RAID 2007, Gold Goast, September 5–7, 2007. Proceedings 10* (pp. 87–106). Springer.

42. Polychronakis, M., Anagnostakis, K. G., & Markatos, E. P. (2010). Comprehensive shellcode detection using runtime heuristics. In *Proceedings of the 26th Annual Computer Security Applications Conference, ACSAC '10* (pp. 287–296). Association for Computing Machinery. https://doi.org/10.1145/1920261.1920305
43. Quinlan, J. R. (1986). Induction of decision trees. *Machine Learning, 1*(1), 81–106. https://doi.org/10.1007/BF00116251
44. Sayadi, H., Houmansadr, A., Rafatirad, S., & Homayoun, H. (2018). Comprehensive assessment of run-time hardware-supported malware detection using general and ensemble learning. In *Proceedings of the 15th ACM International Conference on Computing Frontiers, CF '18* (pp. 212–215). Association for Computing Machinery. https://doi.org/10.1145/3203217.3203264
45. Sayadi, H., Makrani, H. M., Randive, O., PD, S. M., Rafatirad, S., & Homayoun, H.: Customized machine learning-based hardware-assisted malware detection in embedded devices. In *2018 17th IEEE International Conference On Trust, Security And Privacy In Computing And Communications/12th IEEE International Conference On Big Data Science And Engineering (TrustCom/BigDataSE)* (pp. 1685–1688). https://doi.org/10.1109/TrustCom/BigDataSE.2018.00251
46. Sayadi, H., Patel, N., Sasan, A., Rafatirad, S., & Homayoun, H. (2018). Ensemble learning for effective run-time hardware-based malware detection: A comprehensive analysis and classification. In *Proceedings of the 55th Annual Design Automation Conference, DAC '18*. Association for Computing Machinery. https://doi.org/10.1145/3195970.3196047
47. Sayadi, H., Makrani, H. M., Pudukotai Dinakarrao, S. M., Mohsenin, T., Sasan, A., Rafatirad, S., & Homayoun, H. (2019). 2smart: A two-stage machine learning-based approach for run-time specialized hardware-assisted malware detection. In *2019 Design, Automation & Test in Europe Conference & Exhibition (DATE)* (pp. 728–733). https://doi.org/10.23919/DATE.2019.8715080
48. Sayadi, H., Wang, H., Miari, T., Makrani, H. M., Aliasgari, M., Rafatirad, S., & Homayoun, H. (2020). Recent advancements in microarchitectural security: Review of machine learning countermeasures. In *2020 IEEE 63rd International Midwest Symposium on Circuits and Systems (MWSCAS)* (pp. 949–952). https://doi.org/10.1109/MWSCAS48704.2020.9184539
49. Sayadi, H., Gao, Y., Hosein, M. M., Lin, J., Costa, P. C., Rafatirad, S., & Homayoun, H. (2021). Towards accurate run-time hardware-assisted stealthy malware detection: A lightweight, yet effective time series cnn-based approach. *Cryptography, 5*(4), 28.
50. Sayadi, H., Aliasgari, M., Aydin, F., Potluri, S., Aysu, A., Edmonds, J., & Tehranipoor, S. (2022). Towards ai-enabled hardware security: Challenges and opportunities. In *2022 IEEE 28th International Symposium on On-Line Testing and Robust System Design (IOLTS)* (pp. 1–10). https://doi.org/10.1109/IOLTS56730.2022.9897507
51. Sayadi, H., He, Z., Fernandes, C. W., & Miari, T. (2023). Cyber-immunity at the core: Securing biomedical devices through hardware-level machine learning defense. In *2023 IEEE Biomedical Circuits and Systems Conference (BioCAS)*. IEEE.
52. Sayadi, H., He, Z., Makrani, H. M., & Homayoun, H. (2024). Intelligent malware detection based on hardware performance counters: A comprehensive survey. In *2024 25th International Symposium on Quality Electronic Design (ISQED)* (pp. 1–10). https://doi.org/10.1109/ISQED60706.2024.10528369
53. Schmall, M. (2002). *Heuristic techniques in AV solutions: An overview*.
54. Serpen, G., & Aghaei, E. (2018). Host-based misuse intrusion detection using PCA feature extraction and kNN classification algorithms. In *Intelligent data analysis*.
55. Singh, B., Evtyushkin, D., Elwell, J., Riley, R., & Cervesato, I. (2017). On the detection of kernel-level rootkits using hardware performance counters. In *Proceedings of the 2017 ACM on Asia Conference on Computer and Communications Security* (pp. 483–493).
56. Somayaji, A., & Forrest, S. (2000). Automated response using system-call delays. In *Proceedings of the 9th Conference on USENIX Security Symposium - Volume 9, SSYM'00* (pp. 14–14). USENIX Association.
57. Sorzano, C. O. S., Vargas, J., & Montano, A. P. (2014). *A survey of dimensionality reduction techniques*.

58. Sutton, R. S., & Barto, A. G. (2018). *Reinforcement learning: An introduction.* A Bradford Book.
59. Tang, A., Sethumadhavan, S., & Stolfo, S. J. (2014). Unsupervised anomaly-based malware detection using hardware features. In A. Stavrou, H. Bos, & G. Portokalidis (Eds.), *Research in attacks, intrusions and defenses* (pp. 109–129). Springer International Publishing.
60. Tensortrade: Tensortrade (2024). *Trade efficiently with reinforcement learning.* https://github.com/tensortrade-org/tensortrade
61. Trojan Horse (Computing) (2024). https://en.wikipedia.org/wiki/Principal component analysis. Accessed March 5, 2024.
62. Wang, K., Parekh, J. J., & Stolfo, S. J. (2006). Anagram: A content anomaly detector resistant to mimicry attack. In D. Zamboni & C. Kruegel (Eds.), *Recent advances in intrusion detection* (pp. 226–248). Springer.
63. Wang, Z., Schaul, T., Hessel, M., van Hasselt, H., Lanctot, M., & de Freitas, N. (2016). Dueling network architectures for deep reinforcement learning. In *International Conference on Machine Learning*.
64. Wang, H., Sayadi, H., Kolhe, G., Sasan, A., Rafatirad, S., & Homayoun, H. (2020). Phased-guard: Multi-phase machine learning framework for detection and identification of zero-day microarchitectural side-channel attacks. In *2020 IEEE 38th International Conference on Computer Design (ICCD)* (pp. 648–655. https://doi.org/10.1109/ICCD50377.2020.00111
65. Wang, H., Sayadi, H., Sasan, A., Rafatirad, S., & Homayoun, H. (2020). Hybrid-shield: Accurate and efficient cross-layer countermeasure for run-time detection and mitigation of cache-based side-channel attacks. In *Proceedings of the 39th International Conference on Computer-Aided Design, ICCAD '20*. Association for Computing Machinery. https://doi.org/10.1145/3400302.3418783
66. Wydmuch, M., Kempka, M., & Jaśkowski, W. (2019). ViZDoom competitions: Playing doom from pixels. *IEEE Transactions on Games, 11*(3), 248–259. https://doi.org/10.1109/TG.2018.2877047. The 2022 IEEE Transactions on Games Outstanding Paper Award.
67. Yan, P., & Yan, Z. (2017). A survey on dynamic mobile malware detection. *Software Quality Journal, 26*, 891–919. https://doi.org/10.1007/s11219-017-9368-4
68. Ye, Y., Li, T., Adjeroh, D., & Iyengar, S. S. (2017). A survey on malware detection using data mining techniques. *ACM Computing Surveys, 50*(3), 1–4. https://doi.org/10.1145/3073559
69. Yin, H., Song, D., Egele, M., Kruegel, C., & Kirda, E. (2007). Panorama: Capturing system-wide information flow for malware detection and analysis. In *Proceedings of the 14th ACM Conference on Computer and Communications Security, CCS '07* (pp. 116–127). Association for Computing Machinery. https://doi.org/10.1145/1315245.1315261
70. Zarni Aung, W. Z. (2013). Permission-based android malware detection. *International Journal of Scientific & Technology Research, 2*(3), 228–234.
71. Zhou, B., Gupta, A., Jahanshahi, R., Egele, M., & Joshi, A. (2018). Hardware performance counters can detect malware: Myth or fact? In *Proceedings of the 2018 on Asia Conference on Computer and Communications Security, ASIACCS '18* (pp. 457–468). Association for Computing Machinery. https://doi.org/10.1145/3196494.3196515

Chapter 13
Machine Learning-Enhanced Analysis and Design for Trustworthy Integrated Circuits

Kaixin Yang, Yinghua Hu, Dake Chen, Chunxiao Lin, Yang Yi, Peter A. Beerel, and Pierluigi Nuzzo

13.1 Introduction

Nowadays, integrated circuit (IC) manufacturing has become a collaborative process, with many companies participating in different stages, including design, fabrication, assembly, test, and distribution, due to the cost efficiency and diverse expertise offered by different companies. However, such an involvement of third-party players brings concerns regarding potential security threats, such as sensitive data leakage, intellectual property (IP) theft, design tampering, and counterfeit designs [1–3]. Establishing trust across the supply chain usually requires appropriate measures to ensure a design's security, dependability, and resilience.

As shown in Fig. 13.1, threats that affect the hardware implementation of a system can be mainly classified into five categories: (a) malicious modification, (b) IP piracy, (c) side-channel analysis, (d) reverse engineering, and (e) counterfeit

K. Yang (✉) · D. Chen · P. A. Beerel
Department of Electrical and Computer Engineering, University of Southern California, Los Angeles, CA, USA

Y. Hu
Synopsys, Inc., Mountain View, CA, USA
e-mail: yinghuah@synopsys.com

C. Lin · Y. Yi
Department of Electrical and Computer Engineering, Virginia Tech, Blacksburg, VA, USA
e-mail: chunxiaol@vt.edu; yangyi8@vt.edu

P. Nuzzo
Department of Electrical and Computer Engineering, University of Southern California, Los Angeles, CA, USA

Department of Electrical Engineering and Computer Sciences, University of California, Berkeley, Berkeley, CA, USA
e-mail: nuzzo@eecs.berkeley.edu

A. Iranmanesh, H. Sayadi (eds.), *AI-Enabled Electronic Circuit and System Design*,
https://doi.org/10.1007/978-3-031-71436-8_13

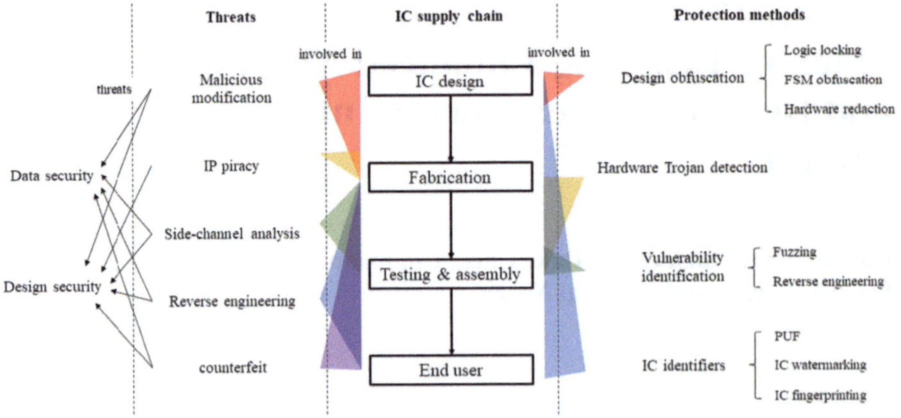

Fig. 13.1 Typical threats and possible protection methods across the IC supply chain

[4], which jeopardize the security of the hardware design itself as well as the data fed into the design. Over the years, various techniques have been proposed to mitigate risks and protect confidential information for hardware designs [3–11]. Design obfuscation techniques aim at preventing unauthorized access to the correct functionality of a digital design by either introducing additional circuitry, controlled by a key, or by replacing the original circuitry with a reconfigurable block. Hardware Trojan detection methods monitor the behaviors of the design under test (DUT) for designated test patterns to reveal malicious modifications to the original design. Vulnerability identification tools aim to broadly identify the type and location of potential vulnerabilities by simulating and analyzing the functionality of a circuit. Finally, IC identifiers can be inserted into the circuit to identify the ownership and usership. However, despite the many existing algorithms and strategies, trustworthy hardware design may remain a challenge. At design time, existing obfuscation methods often utilize heuristic evaluation metrics and rarely rely on formal foundations to define their security [12, 13]. During testing, vulnerability identification tools usually leverage heuristic methods to generate test patterns [14]. These difficulties are then compounded by scalability, both at the design time and test time, due to the size of modern integrated systems [15]. To summarize, conventional hardware design methods (1) usually have limited capability in capturing patterns in complex systems where nonlinear and high-dimensional relationships are involved, (2) often struggle to adapt to different scenarios, (3) may require a substantial amount of manual effort for specific applications.

Similar challenges have already appeared in many other domains, and some existing approaches attempt to find solutions by integrating machine learning (ML) with conventional methodologies due to ML's advantage in pattern recognition, task automation, and versatility. On the other hand, ML models may exhibit unpredictable behaviors and are difficult to interpret [16, 17]. More conventional methods can then be used to rectify their outcomes. For example, simulation can provide precise results if a physical model is known. In contrast, ML requires less understanding of the model but can give accurate predictions if provided with sufficient and unbiased data. Therefore, ML can be used to calibrate the parameters of a simulation model to improve the estimation accuracy, while simulation can assist in generating high-quality data for ML techniques [18]. In the context of computer vision (CV), for example, deep learning (DL) has been widely used in processing digital images, and its integration with more conventional CV techniques can achieve a better trade-off between performance and cost in many situations [19].

Given the success of hybrid, data-driven, and model-based design methods in other domains, this chapter explores the possibility of using ML to help build trustworthy hardware systems. A few works have already leveraged ML to improve the performance of conventional tools and address previously unsolved problems. We review the application of ML techniques in the context of hardware protection methods through the whole IC supply chain. We discuss their advantages and limitations as well as the benefits of integrating them with conventional design flows. Finally, we introduce promising directions for future research.

13.2 Design-Time Methods and Tools: Circuit Obfuscation

In the last few years, many researchers have been actively delving into research about building trustworthy hardware design, as well as formal metrics and methods to effectively and quantifiably evaluate the security level of the VLSI circuit [12, 20–25]. Modern design and manufacturing of ICs heavily rely on a global supply chain, raising concerns about the security and protection of hardware intellectual properties (IPs). To address these concerns, strategies that can obfuscate the original designs and can unlock the correct functionality only when correctly configured are proposed. Among them, a promising method called logic locking (LL) can be employed to safeguard hardware IPs from untrusted foundries. Typically, LL approaches involve incorporating additional logic that enables the introduction of secret keys. When an incorrect key is applied to the locked circuit, the circuit outputs produce incorrect results, effectively restricting access to the correct circuit functionality. Numerous LL approaches are proposed to provide protection for both combinational and sequential circuits, along with dedicated techniques for scan chains. Ideally, certain properties of logic locking, such as the correct key, should be kept as secrets from attackers. However, the implementation of logic locking may depend on these secrets. The unwanted correlation between functional/structural properties of the locked circuits and the secrets opens up a new attack surface where novel machine-learning approaches can learn from these correlations and make informative predictions on the secrets.

In this section, we focus on introducing existing applications of ML techniques to assist the attack on logic locking as a security evaluation method. We will also briefly summarize a few techniques that attempt to protect the locked design from being attacked by removing the signatures leveraged by ML. In the end, we provide a discussion on the benefits brought by ML and the potential future directions.

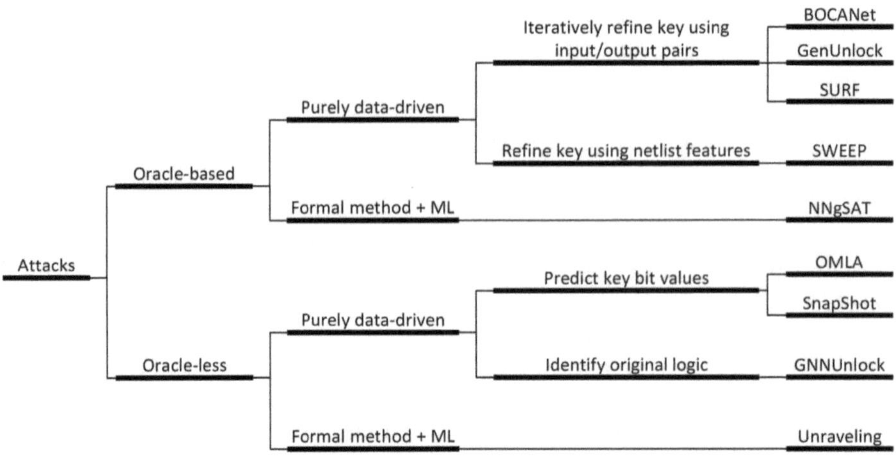

Fig. 13.2 Different types of ML-assisted attacks on logic locking techniques

13.2.1 *ML-Assisted Attacks on Logic Locking (Fig. 13.2)*

Attacks on logic locking techniques have been widely developed and used as a way to evaluate the performance of those defense techniques. The runtime or the key value prediction accuracy of successful attacks on different techniques can represent their security level under corresponding attack surfaces. Generally, there are two types of attacks for LL. The first one is oracle-based; it assumes that the attacker can access a functioning unlocked circuit's primary inputs and outputs. This category includes attacks based on satisfiability checkers (SAT) [26], sensitization [27], removal and bypass [28, 29], functional analysis [30], and GF(2) linear algebra [31]. In contrast, oracle-less attacks exploit structural signatures within the locked netlist only, associated with specific defenses. Notable examples include OMLA [32], GNNUnlock [33], Snapshot [16], among others. In the rest of this subsection, we are going to introduce several typical attacks leveraging ML techniques.

13.2.1.1 Oracle-Based Attacks

BOCANet [34] is one of the oracle-based attacks exploring the use of deep learning techniques, specifically Deep RNN, to break logic locking techniques. A Deep RNN model is trained using input/output pairs, and then the trained model is used to predict and update the key value iteratively according to the mean square error of the newly generated outputs. In this technique, the circuit is treated as a black box, and only the input/output pairs are used for refining the key value. GenUnlock [35] proposes an approach based on genetic algorithms (GAs) and particle swarm optimization (PSO) to identify the correct keys for locked circuits. It generates input vectors and queries the active IC to collect corresponding outputs as a training dataset. In the key evolution phase, a set of keys are instantiated as individuals in the population. GenUnlock performs circuit fitness evaluation, population selection, crossover, and mutation operations to evolve the population and search for the correct keys. SURF [36] is another oracle-based attack that leverages ML and refines the key value iteratively. It takes one step further as it utilizes the structural information of the netlist as well. SURF starts a correct key exploration process with a seed key generated by a machine learning-based attack called SAIL (structural analysis-based IC Locking) [37]. SAIL predicts structural changes introduced by resynthesis and reverts them to provide insight into the design before resynthesis. This information can be used for subsequent attacks, such as key retrieval or reverse engineering. The SURF algorithm starts with a seed key from SAIL and iteratively refines it by flipping key bits and measuring the output corruption. If the bit flip results in less corruption, it is kept. This process continues until the correct key or an approximate key is determined. BOCANet, GenUnlock, and SURF find candidate key values as the starting point and iteratively refine the key value. They all take the functional information as the refinement metrics, which ensure that the final predicted key can retrieve the correct functionality or approximate functionality of the original design. However, since they are all purely data-driven, they cannot guarantee the correctness of the final key they find for any new design when terminating their algorithms.

Additionally, instead of only using output information for determining the correct key, SWEEP [38] propagates constant values of 0 and 1 from key ports to collect circuit features from synthesis tools, which can disclose key bit values by analyzing discrepancies in characteristics. The features include metrics such as area reduction, number of edges, power consumption, and gate count. The correlation between these design features and the key values differs for each locking scheme, and SWEEP exploits vulnerabilities in different locking techniques. For example, the MUX-based TOC locking technique maximizes the Hamming distance between the correct and incorrect routes. Applying an incorrect key value will result in a large area reduction and changes in other design features. Therefore, the impact of key value on the correlated netlist features can be learned, and then the correct key value for any locked design can be estimated according to the obtained feature values. Similar to the techniques mentioned above, SWEEP is also data-driven and thus faces the same challenges.

Moreover, NNgSAT [39] is a neural network (NN)-guided SAT attack on circuits that are locked with or contain complex structures, making them hard to solve using traditional SAT solvers. The goal of NNgSAT is to speed up the SAT attack by using a trained neural network as a classifier to predict the satisfying assignment. NNgSAT utilizes a message-passing neural network (MPNN) with an iterative structure. It consists of two stages: in the first stage, each clause receives messages from its neighboring literals and updates its embedding based on the current embeddings. In the second stage, each literal receives messages from its neighboring clause and its complement and updates its embedding based on the current embeddings. This iterative process continues for a certain number of iterations and finally votes for the literal values that can solve the SAT problem with high confidence. Different from previous techniques, NNgSAT embeds ML techniques into the conventional SAT attack flow, which accelerates the assignment prediction step. It combines formal methods with ML techniques, which ensures the correctness of the final result as well as reduce the runtime.

13.2.1.2 Oracle-Less Attacks

OMLA [32] utilizes graph neural networks (GNNs) to extract key bit values from locked netlists without the need for an oracle or prior knowledge of the synthesis procedure. OMLA addresses the limitations of traditional machine-learning (ML) models by leveraging the graph structure of netlists. It extracts small subgraphs for each key gate from the locked netlist and uses GNNs to learn representations of these subgraphs. The labels of the subgraphs correspond to the key bit values, allowing OMLA to classify and predict these values. SnapShot [16] is also a notable attack on logic locking that utilizes neural networks to directly predict a key bit value from a locked synthesized gate-level netlist without using a golden reference. The attack aims to immediately predict the key bit value based on the extracted locality (subgraph of gates) around each key gate. Compared to other techniques like SAIL, SnapShot offers several advantages. It can be applied to various locking schemes and netlist structures, not just XOR-based locking. It allows for direct prediction of key bit values based on post-resynthesized netlists, eliminating the need for access to the original synthesis tool used in the design flow. SnapShot also requires training only one model, simplifying the attack process. This method synthesizes designs to have fixed fan-in and fan-out sizes for each node, generates the locality feature vectors for each key gate in the extracted netlist graph, and trains MLP and CNN models to predict each key bit value. Both OMLA and SnapShot predict the key bit value based on the structural information of the key port surroundings. According to their experimental results, both techniques can achieve high prediction accuracy for each single key bit. However, challenges still exist for them. For example, when the key length is large, the overall prediction accuracy for the whole key string may not be high enough. Also, these two techniques do not take the functional information into consideration, which means key values with high accuracy cannot guarantee the retrieved netlist functionality is similar to the original one.

Another oracle-less attack using the ML technique is GNNUnlock [33]. The goal of GNNUnlock is to automatically detect and remove the protection logic added by LL schemes, thereby recovering the original functionality of the locked circuit. GNNUnlock first generates a dataset of locked benchmarks by locking each benchmark multiple times with randomly generated key values and different key sizes, resulting in an extensive training set. The training process is not aware of the correct key values. Next, the locked benchmarks are transformed into graphs using a netlist-to-graph transformation. Each node is associated with a feature vector that describes its important characteristics, such as its in-degree, out-degree, and connections to primary inputs and outputs. The GNN is trained to classify the nodes in the graph as either part of the protection logic or the original design logic. GNNUnlock uses similar ML models and feature engineering with a different training target. The previous techniques train GNN models to classify the key input value as 0 or 1, while GNNUnlock trains GNN models to classify all nodes in the netlist to be part of the original design or not.

Moreover, an attack that leverages ML to pinpoint the structure signature in the latch-based logic locking (LBLL) circuit is introduced recently [40]. The attack is a two-phase attack on latch-locked circuits using a combination of deep learning, Boolean analysis, and integer linear programming (ILP). The goal of the attack is to identify the secret key used for latch locking and fully disclose the correct functionality of the circuit. In the first phase of the attack, machine learning is used to detect and remove logic decoy latches from the locked circuit. A random forest classifier is trained to identify these decoy latches based on extracted sequential graphs and feature vectors. The circuit is then simplified through constant propagation. In the second phase, the remaining latches in the circuit are classified using a combination of machine learning and ILP. Ground-truth labels are used to train a second classifier that distinguishes primary/secondary latches from delay decoys. The softmax outputs of this classifier are used as coefficients in the ILP objective function, which aims to maximize the probabilities of latch classifications. The ILP solver is configured to find the closest legal key to that identified by the machine-learning classifier. This technique is another example of combining ML techniques and formal methods. Similar to NNgSAT, this method uses ML models to obtain good prediction results to facilitate the ILP solving process.

13.2.2 Defenses Against ML-Assisted Attacks on Logic Locking

Defenses against ML attacks are also starting to appear. Since ML-based attacks mainly learn the correlation between the target, such as key value, and the sensitive information, such as the structural features, existing defenses against ML-assisted attacks focus on reducing this kind of correlation. There is not as much work for defense compared to attack, which indicates a great potential to further explore efficient defenses in the future.

UNSAIL [41] is a defense scheme designed to confuse oracle-less machine-learning attacks on logic locking. UNSAIL aims to inject adversarial samples during the learning process of ML models used in attacks like SAIL. It does this by introducing identical subgraphs in the locked netlist that can be classified as either changed or unchanged, confusing the ML models. By injecting this conflicting data, UNSAIL undermines the training efforts of the ML models and induces flawed inferences. The effectiveness of UNSAIL is evaluated against the SAIL attack, as well as other attacks like SWEEP and redundancy attacks. The results show that UNSAIL effectively reduces the accuracy of ML-based attacks on logic locking. It confuses the ML models used in the attacks and makes it difficult for them to accurately predict changes in the locked netlist. UNSAIL also demonstrates resilience against other attacks and maintains a high level of security. UNSAIL provides a promising defense against oracle-less ML attacks on logic locking, enhancing the security of integrated circuits and protecting them from unauthorized access and reverse engineering.

DECOR [42] is a recent technique that enhances existing logic locking techniques against ML-assisted attacks. This method focuses on decorrelating the design structure from the correct key value, since the oracle-less attacks heavily leverage the structural information to predict the key value, such as OMLA and SnapShot. By modifying the locked design functionality and introducing more correct key strings, a single key bit can have different correct values in different key strings. Meanwhile, different key bits that have opposite correct values can also share similar structural features. The validation results exhibit that the proposed approach confuses the OMLA attack and reduces the average key prediction accuracy by more than 30% and 10% after enhancing two logic locking techniques, XOR-based insertion and SARLock, respectively.

13.2.3 Discussion and Future Directions

As introduced in the previous subsection, existing ML techniques leverage either functional or structural information to assist in simulation result prediction, key value prediction, assignment exploration in an SAT problem, node classification, etc. The ML techniques can be integrated into the conventional attack flow as a preprocessing or post-processing step, or can be used to train a model with the information provided by the conventional methods and analysis. The combination of ML techniques and conventional methods can rectify the prediction results from the ML models to achieve higher accuracy, as well as reduce the effort and search space of conventional methods.

We notice that neural networks, especially GNN, have exhibited good capability to capture structural information. Since a netlist can be naturally converted into a graph by treating gates as nodes and connections as edges, the aggregation function of GNN can gather the information around a node, generating a locality signature. The information from the node surroundings helps improve the property prediction and node classification accuracy, as shown in OMLA and GNNUnlock. We expect to see more GNN-based techniques appear in related areas, aiding the conventional flow and improving performance.

Recently, large language models (LLMs) have become attractive, and researchers are exploring the possibility of applying LLMs in RTL designs [43, 44]. The LLMs are usually trained with a massive amount of data and thus have the capability to provide solutions for commonly used designs promptly. However, for some specific or constrained designs, LLMs may have limitations since they are heavily dependent on statistical patterns and have limited capability of rule-based reasoning. To build a trustworthy hardware design, potential vulnerabilities should

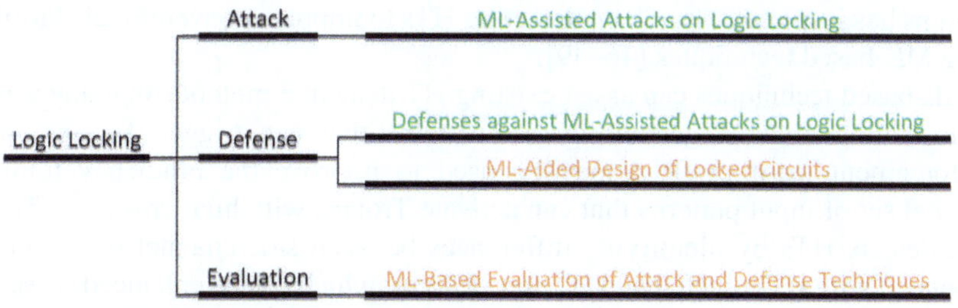

Fig. 13.3 Interesting topics related to ML-assisted logic locking

be considered, which means a design with specific security-aware constraints, such as low correlation between correct key values and the structure. In this case, we believe rule-based conventional design methods could help LLMs implement risk-free hardware designs in the future.

Furthermore, we mainly introduce the existing ML-assisted attacks on logic locking, as well as the defenses against these attacks. Except for them, we believe there are also other interesting topics that can be further explored, as listed in the Fig. 13.3. For defense methods, integrating ML techniques into the logic locking automation design flow can be another interesting direction. ML techniques can help make good decisions about the locking location, configuration, and choices of locking primitives. However, challenges exist, such as how to quantitatively define a good locked netlist for ML models to learn and design. For ML-based evaluation of attack and defense techniques, a recent paper [45] starts to discuss possible ways to estimate the SAT attack runtime given a locked netlist without running the actual attacks. Using ML-based evaluation methods can either save the evaluation runtime or provide a solution when a concrete evaluation metric is hard to explicitly express.

13.3 Test-Based Methods and Tools

Due to the distributed ICs production process, untrusted third-party companies can be involved in the supply chain. Potential adversaries from untrusted third-party companies can maliciously tamper a design by inserting hardware Trojans (HTs), which will be triggered under certain conditions, and cause multiple potential threats, including confidential information leakage, functional corruption, and even permanent fatal damage to the chip [46]. Therefore, the hardware Trojan detection mechanism is essential in realizing trustworthy IC design. Conventional HT detection methods can be classified into three main types based on the detection mechanism: (1) logic testing, (2) side-channel analysis (SCA), and (3) reverse engineering [6]. All these three types of HT detection methods share a few common challenges, including the requirement of a golden model, the scalability issue, and high cost. Recently, the success of machine learning (ML)-based techniques in many domains has inspired research on detecting HTs to improve conventional algorithms using ML-based techniques [46–49].

ML-based techniques can assist existing HT detection methods in many aspects. Logic testing aims at generating test patterns that can trigger Trojans, where reinforcement learning (RL) may be used to improve the efficiency to find a minimal set of input patterns that can activate Trojans with high coverage [50–52]. SCA detects HTs by identifying differences between side-channel measurements associated with HT-infected and HT-free designs, which can be enhanced by support vector machines (SVM) to reveal unseen signature [53]. In addition to detection techniques during the test process, there are also ML-based techniques for detecting HTs during runtime [54, 55]. Meanwhile, golden model-free HT detection is another attractive research topic. Except for some methods mentioned above, other techniques [56–59] also emerge due to the possible unavailability of a golden model and the high cost of golden model-based techniques. These techniques leverage the differences in certain features, such as side-channel properties, controllability, observability, and signal correlation, between HT-infected and HT-free designs or signals to train classifiers, and finally classify unseen designs.

In this section, we focus on applying ML techniques in logic testing and SCA-based HT detection methods. Reverse engineering can be treated as a more general technique of design vulnerability identification and will be discussed later.

13.3.1 ML-Enhanced Hardware Trojan Detection (Table 13.1)

A few works [60, 61] introduce early attempts to integrate ML methods into hardware Trojan detection techniques. These works model hardware designs as graphs and propose a few graph properties to compose the feature vectors associated with each net in the circuit. SVM and random forest (RF)-based methods are adopted to classify whether or not a net is among the ones constructing hardware Trojans. The proposed features include the number of each circuit element within a certain depth inside the input and output cones of the target net. Compared with a previous method leveraging signal correlation between Trojan nets and normal circuits, ML-based techniques may outperform the previous work regarding true positive rate (TPR) and the final precision.

AdaTest [50] is proposed to address the effectiveness, efficiency, and scalability challenges of conventional logic testing-based HT detection methods. In general, the proposed method defines HT-related reward functions and then utilizes reinforcement learning to generate high-quality HT detection test patterns iteratively. The transition probability of each signal is obtained through random logic simulation to locate rare nets, which are more probable to be leveraged by an adversary to trigger HTs. The testability values, including controllability and observability, are also calculated during the circuit profiling step of AdaTest. The transition probability and testability work as the rewards for the reinforcement learning model to generate candidate test patterns. According to the evaluation results, AdaTest can achieve

Table 13.1 Brief introduction to the ML-based hardware Trojan detection methods

HT detection method	ML technique	Information by leveraged ML	Contributions of ML	How combined with conventional methods
Trojan feature generation [60, 61]	SVM and RF	Structural information in the fan-in and fan-out cones	Reveal the correlation between the HT-affected logic and the structural features to improve detection precision	Conventional methods extract and generate structural features for training the ML model
AdaTest [50]	RL	Signal behaviors, including transition probability and testability	RL models are trained and used to select high-quality test patterns for detecting HTs	Transition probability and testability values are calculated during the conventional circuit profiling step for calculating the rewards during RL-based test pattern generation
GNN4TJ [62]	GNN	DFG of the hardware design	Identify the existence of HTs based on the different data flow behaviors	The DFG and corresponding features are generated using conventional methods for training the GNN model
Power signature analysis [53]	SVM	Power traces and PCA results	SVM captures the power signature difference between HT-infected and HT-free designs	PCA on the collected power traces extracts the signatures to feed the ML model

high Trojan coverage with fewer input patterns across various benchmarks. Similar to AdaTest, another work [51] uses RL to update test patterns by flipping bits instead of adaptively sampling from candidate test patterns according to their reward values. Furthermore, DETERRENT [52] focuses on maximizing the rare nets that can be covered by one test pattern so that each pattern is more effective in detecting HTs, and thus, the overall test pattern set could be minimized.

GNN4TJ [62, 63] proposes to utilize graph neural network (GNN) model to classify hardware Trojan-free and hardware Trojan-infected design by learning design behavior features from their data flow graphs. GNN4TJ addresses the challenge that (1) the golden reference, which is the design without any malicious modification, might not be available during the testing phase for detecting hardware Trojans, and (2) the type of the potentially inserted Trojan is unknown. The proposed method generates a data flow graph (DFG) for the given hardware design and then extracts graph-level features from the DFG. After that, the GNN model will be trained with a collection of HT-free and HT-infested designs generated using various types of hardware Trojans and employed to detect the existence of Trojans in any given design later. GNN4TJ leverages the fact that introducing hardware Trojans can bring uncommon data flow behavior into a design. GNN models can learn the behavior patterns from the training data and thus distinguish HT-infected designs based on their data flow behavior without any knowledge about their golden reference.

Another work [53] explains how ML-based techniques help improve SCA-based HT detection. This work applies principal component analysis (PCA) on the collected power traces and then provides generated power signatures for training an SVM model to distinguish between IPs with and without HTs. Compared with conventional methods, the ML-based approach helps eliminate the effects of parameter variations. Similarly, SVM models are also helpful in runtime HT detection [54, 55]. The trained models can identify the Trojan's behavior based on the collected runtime information. These approaches are more time-efficient than conventional techniques.

13.3.2 Discussion and Future Directions

Nowadays, hardware design is becoming even larger and more complex. Detecting HTs, which can be an extremely small portion of the design, will be more challenging regarding time efficiency and cost. ML-based approaches provide many opportunities to improve conventional HT detection techniques. They can help identify HTs from different perspectives, such as the circuit logic, structure, and physical properties, while they are usually efficient during the inference phase if the models are trained well. In most situations, ML models only rely on the properties of the target design itself without the necessity of a golden model, which also brings more flexibility.

However, ML techniques can also be applied oppositely. ML models can be used to learn from the features of hard-to-detect HTs and then automatically insert high-quality Trojans into the design [64]. This will bring more challenges to the existing HT detection techniques. Similar to logic locking techniques, new HT insertion techniques can focus on the design signatures leveraged by current detection methods, and hide or reduce the signature difference of inserted Trojans compared to the normal design. Recent research exhibits the possibility of utilizing the large language model (LLM) and natural language processing (NLP) to detect HTs as well [65]. They generate the natural language description of the hardware design and use GPT-4 to understand and analyze the contextual and structural intricacies. The experimental results show that GPT-4 can achieve detection performance comparable to traditional ML algorithms. This research opens another direction for ML-based HT detection and needs further exploration. For example, a fine-tuned LLM may perform better than the general GPT-4 model, and different information about the design can be described using natural language and fed into the LLM model to improve HT detection.

Therefore, we believe ML-enabled techniques can help overcome some existing challenges in HT detection and raise the design testing and verification efficiency to a higher level.

13.4 ML-Enhanced Vulnerability Identification

As the functionality of hardware design is becoming increasingly complex, detecting and locating vulnerabilities inside the design is even more complicated [66]. it is unrealistic to conduct exhaustive testing before deployment. Therefore, intelligent vulnerability identification becomes critical to avoid potential vulnerabilities introduced by careless designs. Meanwhile, the globalization of the IC supply chain also gives the adversaries chances to alter the design maliciously. Identifying the vulnerabilities brought by the malicious modification is also important [67]. In this section, we want first to introduce hardware fuzzing, which has been an attractive hardware verification technique in the last five years and contains many potential opportunities to have ML techniques involved. In the second part of this section, we want to emphasize the important role of hardware reverse engineering (RE) within the vulnerability identification flow and how ML facilitates the conventional RE methods.

13.4.1 Hardware Fuzzing

In a software development cycle, detecting potential vulnerabilities of the software is a crucial step to ensure the safety and security of the application that runs the software. In general, software vulnerabilities can be detected by two analysis methods, namely, static methods [68–70] and dynamic methods [71–73]. Static analysis methods involve analyzing the source code of the software application without directly running the code. For example, one of the commonly seen software vulnerabilities is buffer overflow [74], where the code attempts to write to a predefined buffer with data whose size is larger than what the buffer can hold. A static analysis tool can go through the source code and identify suspicious buffer write operations that fall into this category. Static analysis methods are often known for their efficiency in scanning the source code. However, not all software vulnerabilities are straightforward to be seen in the source code; some can only manifest in a specific runtime environment. For these vulnerabilities, dynamic analysis methods come into play where certain input stimuli to the software can be generated to "trigger" the vulnerability and cause abnormal functional behavior, such as unexpected function output and runtime crash, during code execution. Fuzzing, as one of the most prominent dynamic methods to detect software vulnerabilities, has received much attention in both academia and industry [75, 76].

At the high level, a fuzzing tool, also called a fuzzer, operates through a flow comprising several key components: seed generation, input mutation, execution monitoring, and vulnerability analysis. Initially, a fuzzer generates inputs or "seeds" that are fed into the target software. These seeds undergo systematic mutations to produce a wide spectrum of test cases. The mutated inputs from the seeds are then executed, and the software's behavior is closely monitored for anomalies, such as crashes or memory leaks, revealing potential vulnerabilities. In this sense, the mechanism of fuzzing is similar to that of automatic test pattern generation (ATPG) [77] in chip testing, where test patterns are incrementally generated that can trigger circuit faults and propagate them to the primary output. Unsurprisingly, there is an ongoing effort to extend fuzzing to detect vulnerabilities in hardware, by converting the RTL code of the hardware into software [7]. Despite the efficacy of traditional fuzzing, its performance can be heavily limited by the quality of the starting seeds, which requires fine-tuning by experts with domain knowledge. Recently, various machine-learning (ML) techniques have been applied to several components of fuzzing, aiming to further improve the quality and speed of fuzzing.

In this section, we aim to provide a comprehensive view of previous efforts on applying ML techniques to fuzzing. In the end, we provide our insights on the future directions of integrating ML techniques for the specific application of hardware fuzzing.

13.4.1.1 Traditional Fuzzing

A fundamental fuzzing tool is structured around four core components: test case generation, code execution, program monitoring, and vulnerability analysis.

Test case generation This initial phase often requires a seed file that adheres to the target software's input format. Incompatibility with this format can result in immediate rejection during the program's entry point, causing inconsequential failures. From this seed file, test cases are derived using three primary methods: mutation-based [76, 78, 79], generation-based [80, 81], and evolution-based strategies [82]. Mutation-based approaches are typically employed when there is limited knowledge about the software, with new test cases crafted by applying predefined mutation rules, such as bit flipping, byte substitution, and data injection. A well-known example is AFL (American Fuzzy Lop), which is renowned for its efficient algorithm that smartly mutates inputs based on code coverage feedback, quickly uncovering a wide range of bugs [76]. Conversely, generation-based methods are utilized when there is an understanding of the software's input requirements, enabling the creation of varied test cases that inherently satisfy the input criteria without the need for a seed file. Grammar-based fuzzing is a classic example of this method, where the fuzzer generates inputs based on the grammatical rules of the expected input format [80]. The mutation and generation strategies, while effective, operate somewhat blindly, without insights into the utility or redundancy of the test cases they produce. To overcome this, evolution-based methods were developed. These methods utilize feedback from the execution of earlier test cases to generate new ones with more targeted objectives, often employing genetic algorithms or evolutionary strategies. This approach allows the fuzzer to adaptively refine its test cases based on the results of previous iterations, significantly enhancing the effectiveness of the fuzzing process [82]. In the context of software fuzzing, a common aim is to enhance code coverage, encouraging the generation of test cases that collectively traverse more lines of code or unique execution paths [83].

Code execution In this stage, the generated test cases are executed in controlled environments, such as virtual machines or containers. This isolation is crucial for safety and for accurately monitoring the software's response to each input [84]. The speed of execution is a critical factor for efficient fuzzing, with faster execution allowing for more test cases to be processed in a given time, thus increasing the chances of uncovering vulnerabilities [85].

Program monitoring This component involves tracking the program's response to each test case using various metrics, such as crash reports and sanitizer outputs. The integration of sophisticated monitoring tools is crucial for effective fuzzing. These tools are designed to detect and log anomalies in program behavior, providing valuable data for identifying potential vulnerabilities [86, 87]. Studies have shown that specific monitoring techniques can significantly impact the rate of vulnerability discovery [14].

Vulnerability analysis The potential vulnerabilities identified during fuzzing are analyzed in this phase. Techniques to filter out false positives and prioritize true vulnerabilities are crucial. This analysis often involves distinguishing between exploitable bugs and noncritical anomalies, a process that can be as complex as the fuzzing itself. Post-fuzzing, static analysis methods are often employed to further scrutinize the findings, offering a deeper inspection of potential security issues revealed by the dynamic analysis [88, 89].

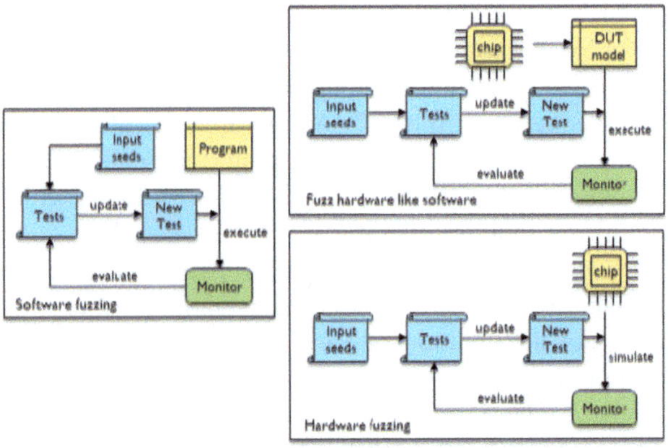

Fig. 13.4 Fuzzing for software and two basic implementations of fuzzing for hardware

Fuzzing for hardware has similar components and flow as fuzzing for software. As shown in Fig. 13.4, the first category of hardware fuzzing is to fuzz hardware like software. In this case, a software-based model will be created to capture the behavior of the hardware. This model will be executed with generated tests, and the monitor will check the behaviors to reveal potential vulnerabilities. On the other hand, another hardware fuzzing technique directly takes the hardware design and provides tests to run simulations. Based on the simulation results, the monitor will identify vulnerabilities inside the design. Due to the similarity between software fuzzing and hardware fuzzing, in the following subsections, we want to introduce the application of ML techniques in software fuzzing, and then discuss potential opportunities for hardware fuzzing.

13.4.1.2 Integrating ML into Fuzzing

As the complexity of software applications escalates, traditional fuzzing methods may face scalability issues, particularly when navigating deep state conditions and achieving extensive code coverage. For example, the selection of seed files is critical, directly influencing code coverage outcomes [90–93], and mutation operation choices further affect fuzzing efficacy [72, 94–98]. Machine learning (ML) has emerged as a powerful ally, enhancing the fuzzing process by refining components, especially during test case generation.

SmartSeed [91] uses an ML-based approach to learn from input files that have previously triggered unique crashes or new paths, and then uses this information to generate new seed files. Similarly, *Faster Fuzzing* [90] adopted two ML models, that is, Generative Adversarial Networks (GAN) and Long Short-Term Memory (LSTM), to enhance the American Fuzzy Lop (AFL) fuzzer's capabilities, yielding a higher rate of unique code path discovery. These advancements are echoed in other research efforts such as those by Cheng et al. [93] and Wang et al. [99]. The domain of sequential decision-making models, including Reinforcement Learning (RL) [72,

94–96], Q-learning [97], and Markov Chain [98], has been instrumental in refining mutation operation strategies. These models guide the fuzzing process through complex software landscapes by learning effective mutation patterns and adapting strategies based on feedback from previous fuzzing cycles. Collectively, these ML-enhanced techniques not only address the inherent limitations of traditional fuzzing methods but also pave the way for more intelligent and adaptive security testing frameworks capable of keeping pace with the rapid evolution of software systems.

Currently, not much work is being done on using ML techniques for hardware fuzzing. Although hardware fuzzing is similar to software fuzzing in terms of the flow, there are still many differences between the two domains, such as the definition of corruption. Most of the recent work focuses on how to implement the fuzzing in the context of hardware. On the one hand, since hardware and software fuzzing share a similar framework, all existing ML techniques utilized for software fuzzing can be migrated to the hardware domain to facilitate the seed input generation and the mutation operation for generating new input patterns. On the other hand, due to the complexity of hardware property verification, it is interesting to see how ML techniques can be leveraged to help generate meaningful hardware coverage points, as well as monitor and analyze hardware vulnerabilities.

13.4.1.3 Discussion and Future Directions

As we peer into the horizon of fuzzing methodologies, it becomes evident that the generation of test cases is only the prelude to the formidable challenge of their execution. The sheer volume of test cases necessitates a paradigm shift toward hardware-accelerated fuzzing solutions. By leveraging specialized hardware, such as GPUs or FPGAs, we can significantly reduce the execution time of test cases, enabling a more thorough exploration of the software's state space in less time.

Moreoever, machine learning has the potential to revolutionize the cost-function landscape of fuzzing. Traditionally, code coverage has been the de facto metric, under the presumption that higher coverage correlates with a higher likelihood of discovering unique bugs and vulnerabilities. However, this is not always the case. An ML-powered approach could entail learning a more nuanced cost function that prioritizes paths more likely to contain bugs, rather than just maximizing code coverage. For example, ML could analyze historical bug data to identify patterns or code features that are more bug-prone and adjust the cost function accordingly. ML could also be employed to predict the effectiveness of a test case before its execution, thus optimizing the allocation of computational resources. By learning from past executions, an ML model might estimate the probability of a test case revealing new vulnerabilities, allowing fuzzers to prioritize test cases with higher potential yields.

In the foreseeable future, we might witness the integration of ML models that can conduct a preliminary analysis of the code and suggest more focused areas for fuzzing, effectively creating a smarter, targeted approach to vulnerability discovery. This strategy could evolve into a semi-supervised learning model where fuzzing tools learn and adapt from both their successes and failures, continually refining their test case generation and execution strategies.

As the intersection of ML and fuzzing grows ever more intertwined, the next generation of fuzzing tools will likely be characterized by their intelligence and adaptability, heralding a new era of software security testing that is both efficient and deeply insightful.

13.4.2 IC Reverse Engineering

Reverse engineering (RE) is an essential technique in the context of hardware security and assurance, which aims to retrieve confidential and critical design information about a product. Generally, RE consists of the following steps: (1) decapsulation, (2) delayering, (3) imaging, (4) annotation, and (5) schematic creation, organization, and analysis [9]. RE is a significant threat to the security of the integrated circuit (IC) design if it is illegally applied to a chip by an adversarial user. It also plays an important role in design verification and penetration testing. For example, RE is used for gate-level netlists to reconstruct the state transition graph for detecting potential modifications to the designs [17, 100] (Fig. 13.5).

ML techniques have been leveraged to accelerate and enhance the RE process. K-nearest neighbors (KNNs) and support vector machines (SVMs) are utilized to improve the performance of RE-based hardware Trojan detection approaches [9]. Graph neural networks (GNNs) increase the classification accuracy between state and data registers in a gate-level design [17], compared with previous heuristic methods [101–103], showing the advantage of ML-based techniques in providing effective classification models. The strength of ML is also leveraged in image-based RE [104].

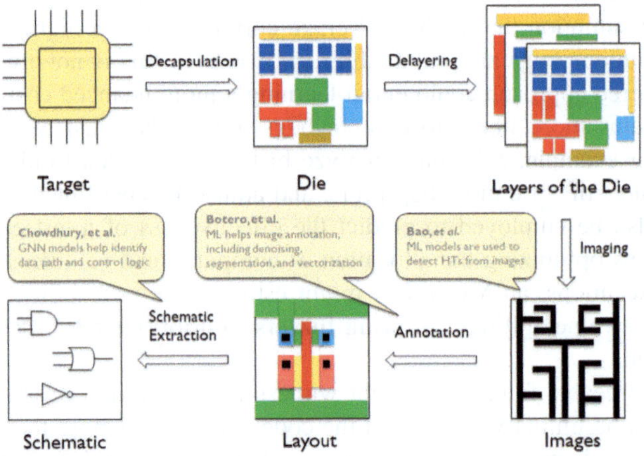

Fig. 13.5 IC reverse engineering steps

In this subsection, we want to introduce a few reverse engineering techniques with ML involved. We want to highlight the contributions of ML techniques at different steps of the reverse engineering flow. Three typical techniques are discussed below to explain how ML techniques help identify hardware vulnerabilities or how they help process the data to facilitate vulnerability identification.

13.4.2.1 ML-Enhanced IC Reverse Engineering

As introduced in Sect. 13.3, reverse engineering is one main category of approaches to detect hardware Trojans. One existing work [9] introduces their approach of utilizing ML-based classification techniques to detect HTs at the layout level. They break layout images into several nonoverlapping blocks and obtain area and centroid differences as the associated features for each block. Those features are leveraged to train SVM and K-means clustering models. With trained models, they can identify HTs in the target design using its features without knowledge about the golden design.

Another work [104] shows the application of ML-based techniques in image processing during RE. Accurate image processing can help understand functionality from the images, and thus, the potential vulnerabilities can be correctly identified in the following RE steps. In the typical RE flow, after obtaining the images of the design, the next step is to annotate the design to understand the functionality of the design from the images, which involves a lot of work of image processing. ML can be effectively utilized in this step, including denoising, segmentation, and vectorization. Thus, ML-enabled image analysis can promise high-quality RE for the target design.

ReIGNN [17] is an approach that helps accelerate the RE process, especially the schematic creation, organization, and analysis step. To identify vulnerabilities at the schematic level, it is important to reconstruct the netlist and then understand the functionality of the design, so that it can be compared to a golden model to check whether any vulnerability exists. To reconstruct the netlist, it is critical to separate the control logic and data path. In a sequential design, registers decide the current states and play a significant role in composing the control logic. Therefore, correctly identifying the control logic registers is important. Conventional approaches apply heuristic clustering methods, whose accuracy is very sensitive to the configuring parameter values. Unlike previous methods, ReIGNN converts the netlist into a graph and applies a GNN-based model to classify registers. ReIGNN provides better identification performance and prevents users from exploring the best configuration in a huge space.

13.4.2.2 Discussion and Future Directions

RE is an essential technique in the current design testing and verification processes. Within the IC RE flow, there is a significant reliance on image processing and graph analysis, where ML excels. Existing ML-based approaches may facilitate RE techniques in annotation and netlist regeneration due to their efficiency in related tasks. Therefore, developing more advanced ML-enabled RE techniques is promising. Moreover, ML-based approaches can potentially enhance the performance of techniques that rely on RE, such as hardware Trojan detection. Exploring the integration of ML in specific scenarios alongside traditional RE techniques presents intriguing opportunities for advancement.

13.5 IC Identifiers

IC counterfeiting and overbuilding are typical threats that undermine the trustworthiness of hardware design. Among the defense techniques, IC watermarking and fingerprinting can track the source of piracy and help designers claim ownership. In this section, we focus on the physical unclonable function (PUF), widely used as a unique identifier due to its physical properties.

13.5.1 Physical Unclonable Function

A physical unclonable function (PUF) is often a crucial component of a secure hardware design. PUF technology plays a key role in enhancing the security of a growing number of embedded devices. A PUF takes advantage of the inherent manufacturing variations present in embedded devices [105]. It leverages these variations to generate a device-specific random response, known as a PUF response. This response exhibits unpredictability, making it difficult to predict or replicate, and therefore serves as a distinctive fingerprint for each device. The primary applications of this fingerprint are authentication and the derivation of a device-specific secret key. By utilizing the unique PUF response, the authenticity of a device can be verified, ensuring that it is genuine and not a counterfeit or tampered device. Additionally, the PUF response can be used as a basis for generating a secret key that is tied to the specific device. This key can then be used for secure operations, such as encryption and decryption, providing an additional layer of device-specific security.

A PUF can be conceptually understood as a circuit element that produces a response, denoted as "r," when a challenge, represented as "ch," is applied to it. Currently, a diverse range of PUF primitives exists, including SRAM PUFs [106, 107], arbiter PUFs (APUF) [108], lightweight secure PUFs (LSPUF) [109], bistable ring PUF [110], multiplexer PUF (MPUF) [111], and many others. Ongoing research in this domain holds great potential for further advancements and innovations.

Table 13.2 ML-based modeling attacks on PUF designs

Attack on PUF	PUF Type	Contribution
[112]	Lightweight PUFs	This work exhibits that CRPs of lightweight PUFs can be learned and predicted by ML-based modeling attacks
[113]	XOR arbiter PUF	This paper proves that nonlinear XOR arbiter PUF is not secure
[114]	BR-PUF	The behaviors of PUFs unknown mathematical models can also be captured by ML attacks
[115]	Arbiter PUFs	The proposed learning algorithm can model CRPS of arbiter PUFs with prescribed accuracy and confidence with polynomial time complexity
[116]	Arbiter PUFs	This work improves learning processes and provides different speed options

13.5.2 ML-Based Attacks on PUF (Table 13.2)

The PUF circuit itself can be viewed as a function, and it has been shown through the universal approximation theorem that multilayer perceptrons (MLPs) have the ability to learn any input-output function. Consequently, PUFs inherently possess vulnerabilities to attacks leveraging deep learning and machine-learning techniques. In particular, machine learning-based modeling attacks are widely recognized as the most significant threat to PUF implementations [112, 117]. Numerous studies have demonstrated the effectiveness of machine learning-based attacks against PUFs, prompting extensive research efforts to develop newer attack methods. These attacks typically involve providing the adversary with access to a small portion of the challenge-response pairs (CRPs) dataset associated with a specific PUF instance. With this limited data, the adversary constructs a computational model capable of accurately predicting the PUF's response to any given challenge, achieving a high probability of success. The nonlinearity from the ReLU activation function also plays an important role in the attacks. Several ML or deep learning techniques have been commonly employed to launch attacks on PUFs, including support vector machines (SVMs), logistic regressions (LRs), and MLPs. These techniques have demonstrated their superior performance in attacking various PUF variants including various APUFs, bistable ring PUFs, and LSPUFs [108, 112, 118–121].

The paper [112] focuses on the security analysis from the perspective of ML on composite PUFs and multi-bit output lightweight secure PUFs (LSPUFs). First, the paper introduces a cryptanalysis attack that shows that composite PUFs with multiple levels of PUFs are not secure. Additionally, a machine-learning attack is developed for composite PUFs, which reduces the attack complexity. The paper highlights that the security evaluation of composite PUFs was not provided in the original proposal. Next, the paper discusses an attack on multi-bit output LSPUFs. LSPUFs rely on the design properties of input and output networks, as well as the primitive PUF instances used. The paper presents a new cryptanalysis attack that is effective for any multi-bit output LSPUF variants with practical time and data complexity. The attack demonstrates that multi-bit output LSPUFs are not secure. The paper validates the proposed attacks through experimental results on field programmable gate array (FPGA) implemented composite PUFs and multi-bit output LSPUFs. The experiments demonstrate the effectiveness of the attacks in the best-case scenario. This paper provides a comprehensive analysis of the security of composite PUFs and multi-bit output LSPUFs, highlighting their vulnerabilities and the need for careful consideration of security trade-offs in PUF design.

A machine learning-based attack on arbiter physical unclonable functions (PUFs) is presented in [113]. PUFs are used for authentication and fingerprinting purposes, but it has been shown that arbiter PUFs are susceptible to various types of attacks. ML techniques have been successful in breaking the security of arbiter PUFs by mathematically cloning them. The attack proposes a new framework for learning XOR PUFs in the presence of noisy responses. It provides a theoretical limit for ML techniques to learn XOR PUFs based on the number of arbiter PUF stages and the number of chains. It also addresses the issue of noisy responses in XOR PUFs. It introduces a noise model that accurately represents the noise observed in practice. Unlike previous approaches that rely on majority voting to deal with noisy responses, this framework does not require majority voting and can still learn XOR PUFs efficiently. The authors emphasize that the security of XOR PUFs is limited by technological constraints, such as the number of chains and the silicon area required for construction. They argue that protocols relying solely on XOR PUFs cannot be considered perfectly secure. Overall, this attack highlights the limitations and vulnerabilities of arbiter PUFs. It provides insights into the design and manufacturing of delay-based PUFs and contributes to the understanding of ML attacks on PUFs.

Most machine-learning attacks on PUFs assume that a mathematical model of the PUF functionality is known. The paper [114] addresses the scenario where the mathematical model is unknown and proposes a provable framework for machine-learning attacks against PUFs. The authors focus on the bistable ring PUF (BR-PUF) family, which is a prime example of a PUF with an unknown mathematical model. The paper presents a novel probably approximately correct learning framework that can learn the challenge-response behavior of BR-PUFs without relying on a mathematical model. The authors conduct extensive experiments on BR-PUFs implemented on FPGA to evaluate the effectiveness of their attack. The experimental results confirm the applicability and effectiveness of their machine-learning algorithm. It also discusses the existence of influential bits in PUFs and provides a constant upper bound on the number of influential bits. This result is based on empirical evidence and theoretical insights.

PAC learning attacks [115] is a learning algorithm that can model the challenge-response behavior of an arbiter PUF with prescribed levels of accuracy and confidence. The algorithm is based on a new mathematical representation of arbiter PUFs, which allows for the discretization of real-valued delay values. By mapping the delay values to a finite set of integer values, the algorithm can effectively learn the behavior of the arbiter PUF. The paper proves that the maximum number of CRPs required for the attack is polynomial in the length of the arbiter PUF. The experiments also evaluate the time complexity of the learning algorithm and show that it is polynomial in the length of the arbiter PUF, the maximum variation of delays, and the desired levels of accuracy and confidence.

Active learning attack is introduced in [116] and it aims to improve the learning process of PUFs, specifically the arbiter PUF. The researchers present two variations of the active learning technique: fast learning and slow learning. In fast learning, the goal is to quickly learn the PUF model with a small number of challenges. The technique selects challenges that are close to the current model of the hyperplane separating the PUF responses. This approach outperforms random challenge selection and achieves high accuracy with only 350 challenges for a 64-stage arbiter PUF. In slow learning, the objective is to generate a set of challenges that yield poor generalization errors, making it difficult for an attacker to accurately clone the PUF. The technique uses an initial set of challenges and a parameter k to select challenges that lie at different distances from the hyperplane. By tuning k, different learning speeds can be achieved. Slow learning with a high value of k results in challenges that are far from the hyperplane and statistically uncorrelated, making it harder to learn the PUF. The paper provides numerical results comparing the performance of the active learning technique with other learning algorithms such as LR, SVM, AdaBoost, and neural networks. The results demonstrate the effectiveness of the active learning technique in both noiseless and noisy scenarios, outperforming other methods in terms of prediction error rate and accuracy. The active learning technique offers a powerful approach to improve the learning process of arbiter PUFs, enabling faster and more accurate learning while also providing options for slower and more adversarial learning.

13.5.3 Defenses Countering ML-Based Attacks on PUF

In response, various defense mechanisms have been proposed to counter ML attacks. These defenses primarily focus on enhancing the nonlinearity of the PUF structure and challenge-response mapping [122–124]. These defense strategies aim to increase the complexity and nonlinearity of the PUF's behavior, making it more challenging for ML-based attacks to accurately model the PUF and predict its responses.

The work [118] introduces a defense mechanism called the interpose physical unclonable function (iPUF) to protect against machine-learning attacks on XOR arbiter PUFs (APUF). The iPUF is designed to be resilient against both classical machine-learning attacks and reliability-based machine-learning attacks. The defense is based on the concept of interposing a bit in the middle of the XOR APUF structure. This interposed bit disrupts the correlation between the challenge and response bits, making it difficult for machine-learning attacks to accurately model the PUF. The paper provides theoretical analysis and empirical results to demonstrate the effectiveness of the iPUF defense. To evaluate the security of the iPUF, the iPUF is compared with the XOR APUF using different machine-learning attacks. The results show that the iPUF is more resilient to attacks compared to the XOR APUF. Classical machine-learning attacks, such as LR and deep neural networks, are less efficient on the iPUF. The paper further discusses the reliability-based machine-learning attacks and proposes an enhanced reliability-based CMA-ES attack. The results demonstrate that the enhanced attack using the proposed reliability model outperforms the original attack. The iPUF is shown to have higher modeling accuracy and a greater frequency of correct models compared to the XOR APUF. The iPUF defense provides a practical solution to enhance the security of XOR APUFs against machine-learning attacks. The interposed bit disrupts the correlation between challenge and response bits, making it difficult for attackers to accurately model the PUF. The empirical results and theoretical analysis support the effectiveness of the iPUF defense, making it a promising approach for securing PUFs against machine-learning attacks.

A new lightweight PUF design called the multi-PUF (MPUF) is proposed in [124]. The MPUF design aims to enhance the security of PUFs against modeling attacks by utilizing Weak PUFs to obfuscate the challenges to a Strong PUF. The architecture of the MPUF design consists of an array of elementary 1-bit cells, each composed of a group of Weak PUFs and a Strong PUF. The proposed MPUF design demonstrates higher complexity compared to the conventional arbiter PUF, making it more difficult to attack. A mathematical model of the MPUF design is presented, showing the delay difference between the two delay paths in the APUF can be represented as the inner product of the parity vector and a constant vector. The mathematical model also shows that the actual input challenge to the APUF is obfuscated and masked, further increasing the complexity of the design. To evaluate the resistance of the MPUF design against modeling attacks, two machine learning-based attacks, LR and covariance matrix adaptation evolution strategies (CMA-ES), are used. The experimental results show that the proposed MPUF design achieves a lower prediction rate compared to the conventional Arbiter PUF, indicating its improved resistance to modeling attacks. The proposed MPUF design demonstrates higher complexity, improved resistance to modeling attacks, and good performance in terms of uniqueness and uniformity.

Another representative ML defensive PUF is proposed in [125]. A configurable tristate PUF against machine-learning attacks is introduced for IoT security. The CT PUF is designed to provide secure device authentication by generating unique responses based on challenge inputs. It incorporates obfuscation mechanisms and employs a bitwise XOR operation to hide the mapping relationship between challenges and responses. Experimental results show the resistance of CT PUF against empirical ML attack methods. The results show that CT PUF is resilient to these ML attacks. It also discusses the resistance of CT PUF to provable ML attack methods, specifically the PAC learning method. It highlights the challenges in breaking CT PUF using this method due to the obfuscation mechanisms and the difficulty in approximating CT PUF with a known Boolean function. Furthermore, the paper also presents the CT PUF-based authentication protocol, which includes an enrollment phase and an authentication phase. The CT PUF defense demonstrates its effectiveness in providing secure device authentication and resistance against various machine-learning attack methods. The experimental and mathematical analysis results support the claim that CT PUF is a robust defense mechanism for IoT security.

13.5.4 Discussion and Future Directions

The aforementioned studies demonstrate the efficacy of ML techniques in PUF modeling attacks, where the primary challenge is to accurately learn the mathematical model governing the challenge and response behaviors of PUFs. ML offers distinct advantages in handling complex, high-dimensional models, which explains its success in PUF modeling attacks. Given the critical role of PUFs in ensuring hardware trustworthiness, it is foreseeable that further research will delve into assessing the security of diverse types of PUFs under various scenarios and with attack assumptions. This will also drive the development of more resilient and secure PUF designs, thereby supporting the overall reliability of hardware systems.

Beyond modeling attacks, ML can be used in other aspects of PUF design. For example, a recent work [126] showcases the potential of ML in enhancing the authentication phase of PUFs. They build a CNN model to help improve noise tolerance of the authentication step without any more profound knowledge of the PUF structure and characteristics. The prospect of further research exploring the application of ML from the defender's perspective in PUF design or authentication processes would be intriguing. Such investigations could yield innovative approaches further to enhance the security and reliability of PUF designs.

13.6 Conclusion

In this chapter, we reviewed existing design methods that incorporate ML techniques to enhance IC security throughout the supply chain, from design, testing, and vulnerability detection, to IC identification. We discussed how IC security can benefit from ML techniques and their integration in conventional design flows. Specifically, in an integrated approach, conventional IC design methods can be used to (a) generate training datasets for building accurate ML models, and (b) as a domain-specific formal foundation to rectify the outcomes of ML models and improve on their accuracy and robustness. On the other hand, ML can help with (a) capturing important features of complex systems, as in modeling attacks on PUFs; (b) efficiently guiding decision-making toward promising directions, as in test pattern generation for hardware fuzzing; (c) efficiently detecting recurrent structural patterns in circuits, for example, in the context of IC reverse engineering; and (d) efficiently exploring large design spaces, as discussed in the context of hardware IP protection and confidentiality.

Emerging ML techniques bring further opportunities and challenges. LLMs have already demonstrated remarkable capabilities in providing innovative design solutions [127, 128]. They can rapidly generate designs in hardware description languages when the design specifications are provided in natural languages. The models can provide functionally correct codes for common designs, and the designs can be further refined with feedback either from human engineers or from the testing results. However, as for most ML techniques, the resulting models have limited interpretability and rely heavily on data. Their output still requires careful verification and validation. Overall, as hardware designs become increasingly more complex, the principled combination of ML techniques with conventional design methodologies holds the promise of ushering in a new era for secure and trustworthy hardware design.

References

1. Hassija, V., et al. (2020). A survey on supply chain security: Application areas, security threats, and solution architectures. *IEEE IoT Journal, 8*, 6222–6246.
2. Guin, U., et al. (2014). Counterfeit integrated circuits: A rising threat in the global semiconductor supply chain. *Proceedings of the IEEE, 102*, 1207–1228.
3. Liu, B., & Qu, G. (2016). VLSI supply chain security risks and mitigation techniques: A survey. *Integration, 55*, 438–448.
4. Rostami, M., Koushanfar, F., & Karri, R. (2014). A primer on hardware security: Models, methods, and metrics. *Proceedings of the IEEE, 102*, 1283–1295.
5. Kamali, H. M., et al. (2022). Advances in logic locking: Past, present, and prospects. *Cryptology ePrint Archive*.
6. Bhasin, S., & Regazzoni, F. (2015). *A survey on hardware trojan detection techniques* (pp. 2021–2024). IEEE.
7. Trippel, T., et al. (2022). *Fuzzing hardware like software* (pp. 3237–3254).
8. Cruz, J., et al. (2018). *Hardware Trojan detection using ATPG and model checking* (pp. 91–96). IEEE.
9. Bao, C., et al. (2016). On reverse engineering-based hardware Trojan detection. *TCAD, 35*, 49–57.
10. Dey, K., et al. (2021). *PUF based hardware security: A review* (pp. 1–6). IEEE.
11. Chen, X., Qu, G., & Cui, A. (2017). *Practical IP watermarking and fingerprinting methods for ASIC designs* (pp. 1–4). IEEE.
12. Hu, Y., et al. (2021). *Risk-aware cost-effective design methodology for integrated circuit locking* (pp. 1182–1185). IEEE.
13. Beerel, P., Georgiou, M., Hamlin, B., Malozemoff, A. J., & Nuzzo, P. (2022). Towards a formal treatment of logic locking. *Cryptology ePrint Archive, 2022*, 92.
14. Li, J., Zhao, B., & Zhang, C. (2018). Fuzzing: a survey. *Cybersecurity, 1*, 1–13.
15. Hu, W., et al. (2020). An overview of hardware security and trust: Threats, countermea- sures, and design tools. *TCAD, 40*, 1010–1038.
16. Sisejkovic, D., et al. (2021). Challenging the security of logic locking schemes in the era of deep learning: A neuroevolutionary approach. *JETC, 17*, 30.
17. Chowdhury, S. D., Yang, K., & Nuzzo, P. (2021). *ReIGNN: State register identification using graph neural networks for circuit reverse engineering* (pp. 1–9).
18. Pan, I., et al. (2022). Data-centric engineering: Integrating simulation, machine learning and statistics. Challenges and opportunities. *CES, 249*, 117271.

19. O'Mahony, N., et al. (2020). *Deep learning vs. traditional computer vision* (pp. 128–144). Springer.
20. Hu, Y., et al. (2023). On the security of sequential logic locking against oracle-guided attacks. *TCAD, 42*, 3628.
21. Zhang, Y., et al. (2022). *Trilock: IC protection with tunable corruptibility and resilience to SAT and removal attacks* (pp. 1329–1334). IEEE.
22. Hu, Y., et al. (2021). *Fun-SAT: Functional corruptibility-guided SAT-based attack on sequential logic encryption* (pp. 281–291).
23. Hu, Y., et al. (2020). *SANSCrypt: A sporadic-authentication-based sequential logic encryption scheme* (pp. 129–134).
24. Chowdhury, S. D., et al. (2021). *Enhancing SAT-attack resiliency and cost-effectiveness of reconfigurable-logic-based circuit obfuscation* (pp. 1–5). IEEE.
25. Hu, Y., et al. (2019). *Security-driven metrics and models for efficient evaluation of logic encryption schemes* (pp. 1–5).
26. Subramanyan, P., Ray, S., & Malik, S. (2015). *Evaluating the security of logic encryption algorithms* (pp. 137–143). IEEE.
27. Yasin, M., et al. (2015). On improving the security of logic locking. *TCAD, 35*, 1411–1424.
28. Xu, X., et al. (2017). *Novel bypass attack and BDD-based tradeoff analysis against all known logic locking attacks* (pp. 189–210). Springer.
29. Yasin, M., et al. (2017). Removal attacks on logic locking and camouflaging techniques. *Transactions on Emerging Topics in Computing, 8*, 517–532.
30. Sirone, D., & Subramanyan, P. (2020). Functional analysis attacks on logic locking. *Transactions on Information Forensics and Security, 15*, 2514–2527.
31. Chen, D., Lin, C., & Beerel, P. A. (2021). *GF-Flush: A GF(2) algebraic attack on secure scan chains*. IEEE.
32. Alrahis, L., et al. (2022). OMLA: An oracle-less machine learning-based attack on logic locking. *Transactions on Circuits and Systems II: Express Briefs, 69*, 1602–1606.
33. Alrahis, L., et al. (2021). *GNNUnlock: Graph neural networks-based oracle-less unlocking scheme for provably secure logic locking* (pp. 780–785).
34. Tehranipoor, F., et al. (2019). *Deep RNN-oriented paradigm shift through BOCANet: Broken obfuscated circuit attack* (pp. 335–338).
35. Chen, H., et al. (2019). *GenUnlock: An automated genetic algorithm framework for unlocking logic encryption* (pp. 1–8).
36. Chakraborty, P., Cruz, J., & Bhunia, S. (2019). *SURF: Joint structural functional attack on logic locking* (pp. 181–190).
37. Chakraborty, P., Cruz, J., & Bhunia, S. S. A. I. L. (2018). *Machine learning guided structural analysis attack on hardware obfuscation* (pp. 56–61).
38. Alaql, A., Forte, D., & Bhunia, S. (2019). *Sweep to the secret: A constant propagation attack on logic locking* (pp. 1–6).
39. Azar, K. Z., et al. (2020). *NNgSAT: Neural network guided SAT attack on logic locked complex structures* (pp. 1–9).
40. Chen, D., et al. (2023). Unraveling latch locking using machine learning, boolean analysis, and ILP. *arXiv preprint arXiv:2305.00107*.
41. Alrahis, L., et al. (2021). UNSAIL: Thwarting oracle-less machine learning attacks on logic locking. *Transactions on Information Forensics and Security, 16*, 2508–2523.
42. Hu, Y., et al. (2024). DECOR: Enhancing logic locking against machine learning-based attacks. *arXiv preprint arXiv:2403.01789*.
43. Lu, Y., et al. (2023). RTLLM: An open-source benchmark for design rtl generation with large language model. *arXiv preprint arXiv:2308.05345*.
44. Thakur, S., et al. (2023). *Benchmarking large language models for automated verilog rtl code generation* (pp. 1–6). IEEE.
45. Ahmed, B., et al. (2024). *Seemless: Security evaluation of logic locking using machine learning oriented estimation* (pp. 489–494).

46. Huang, Z., et al. (2020). A survey on machine learning against hardware Trojan attacks: Recent advances and challenges. *IEEE Access, 8*, 10796–10826.
47. Liakos, K. G., et al. (2019). *Machine learning for hardware Trojan detection: A review* (pp. 1–6).
48. Patnaik, S., et al. (2022). *Reinforcement learning for hardware security: Opportunities, developments, and challenges* (pp. 217–218).
49. Liakos, K. G., et al. (2020). Conventional and machine learning approaches as countermeasures against hardware Trojan attacks. *Microprocessors and Microsystems, 79*, 103295.
50. Chen, H., et al. (2023). AdaTest: Reinforcement learning and adaptive sampling for on-chip hardware trojan detection. *TECS, 22*, 1–23.
51. Pan, Z., & Mishra, P. (2021). *Automated test generation for hardware Trojan detection using reinforcement learning* (pp. 408–413).
52. Gohil, V., et al. (2022). *DETERRENT: Detecting trojans using reinforcement learning* (pp. 697–702).
53. Yang, S., Chakraborty, P., & Bhunia, S. (2021). *Side-channel analysis for hardware Trojan detection using machine learning* (pp. 1–6).
54. Kulkarni, A., et al. (2016). Real-time anomaly detection framework for many-core router through machine-learning techniques. *JETC, 13*, 1–22.
55. Kulkarni, A., Pino, Y., & Mohsenin, T. (2016). *SVM-based real-time hardware Trojan detection for many-core platform* (pp. 362–367).
56. Xue, M., Wang, J., & Hu, A. (2016). *An enhanced classification-based golden chips-free hardware Trojan detection technique* (pp. 1–6).
57. Liu, Y., Huang, K., & Makris, Y. (2014). *Hardware Trojan detection through golden chip-free statistical side-channel fingerprinting* (pp. 1–6).
58. Salmani, H. (2017). COTD: Reference-free hardware Trojan detection and recovery based on controllability and observability in gate-level netlist. *Transactions on Information Forensics and Security, 12*, 338–350.
59. Cakır, B., & Malik, S. (2015). *Hardware Trojan detection for gate-level ICs using signal correlation-based clustering* (pp. 471–476).
60. Hasegawa, K., et al. (2016). *Hardware trojans classification for gate-level netlists based on machine learning* (pp. 203–206). IEEE.
61. Hasegawa, K., et al. (2017). *Trojan-feature extraction at gate-level netlists and its application to hardware-trojan detection using random forest classifier* (pp. 1–4). IEEE.
62. Yasaei, R., Yu, S.-Y., & Al Faruque, M. A. (2021). *Gnn4tj: Graph neural networks for hardware trojan detection at register transfer level* (pp. 1504–1509).
63. Yasaei, R., Chen, L., Yu, S.-Y., & Faruque, M. A. A. (2022). *Hardware trojan detection using graph neural networks* (pp. 1–1). TCAD.
64. Cruz, J. (2022). et al, Automatic hardware trojan insertion using machine learning. *arXiv preprint arXiv:2204.08580*.
65. Saha, D., et al. (2023). Llm for soc security: A paradigm shift. *arXiv preprint arXiv:2310.06046*.
66. Chen, C., et al. (2023). *HyPFuzz: Formal-assisted processor fuzzing* (pp. 1361–1378).
67. Yu, S.-Y., Yasaei, R., Zhou, Q., Nguyen, T., & Al Faruque, M. A. (2021). *Hw2vec: A graph learning tool for automating hardware security* (pp. 13–23). IEEE.
68. Louridas, P. (2006). Static code analysis. *IEEE Software, 23*, 58–61.
69. Gomes, I., et al. (2009). An overview on the static code analysis approach in software development. *Faculdade de Engenharia da Universidade do Porto*.
70. Dahse, J., & Holz, T. (2014). *Simulation of built-in php features for precise static code analysis* (Vol. 14, pp. 23–26).
71. Fell, J. (2017). A review of fuzzing tools and methods. *PenTest Magazine*.
72. B̈ottinger, K., et al. (2018). *Deep reinforcement fuzzing* (pp. 116–122). IEEE.
73. Wang, Y., Jia, P., Liu, L., Huang, C., & Liu, Z. (2020). A systematic review of fuzzing based on machine learning techniques. *PLoS One, 15*, e0237749.
74. Foster, J. C. (2005). *Buffer overflow attacks*.

75. Nishimura, R., et al. (2016). *Implementation of the can-fd protocol in the fuzzing tool bestorm* (pp. 1–6). IEEE.
76. Wong, C. W. K. (2022). *American fuzzy lop (afl) fuzzer.*
77. Prasad, M. R., Chong, P., & Keutzer, K. (1999). *Why is atpg easy?* (pp. 22–28).
78. Cha, S. K., Woo, M., & Brumley, D. (2015). *Program-adaptive mutational fuzzing* (pp. 725–741). IEEE.
79. Smetsers, R., Moerman, J., Janssen, M., & Verwer, S. (2016). Complementing model learning with mutation-based fuzzing. *arXiv preprint arXiv:1611.02429.*
80. Godefroid, P., et al. (2008). *Grammar-based whitebox fuzzing* (pp. 206–215).
81. Liu, J., et al. (2023). Generation-based differential fuzzing for deep learning libraries. *ACM Transactions on Software Engineering and Methodology, 33*, 1–28.
82. Li, Y., et al. (2019). V-fuzz: Vulnerability-oriented evolutionary fuzzing. *arXiv preprint arXiv:1901.01142.*
83. Lou, B., & Song, J. (2020). A study on using code coverage information extracted from binary to guide fuzzing. *IJCSS, 14*, 200–210.
84. How to cyber security: Containerizing fuzzing targets. (2021). https://www.synopsys.com/blogs/software-security/effective-fuzz-testing-containerization/
85. Zhang, B., Ye, J., Bi, X., Feng, C., & Tang, C. (2018). Ffuzz: Towards full system high coverage fuzz testing on binary executables. *PLoS One, 13*, e0196733.
86. Serebryany, K., & Iskhodzhanov, T. (2009). *Threadsanitizer: Data race detection in practice* (pp. 62–71).
87. Serebryany, K., Bruening, D., Potapenko, A., & Vyukov, D. (2012). *AddressSanitizer: A fast address sanity checker* (pp. 309–318).
88. Yang, G., Feng, C., & Tang, C. (2017). *Static analysis assisted vulnerability-oriented evolutionary fuzzing* (Vol. 1820). AIP Publishing.
89. Shields, P. (2023). *Hybrid testing: Combining static analysis and directed fuzzing*. Ph.D. thesis,. Massachusetts Institute of Technology.
90. Nichols, N., et al. (2017). Faster fuzzing: Reinitialization with deep neural models. *arXiv preprint arXiv:1711.02807.*
91. Lyu, C., et al. (2018). SmartSeed: Smart seed generation for efficient fuzzing. *arXiv preprint arXiv:1807.02606.*
92. Godefroid, P., Peleg, H., & Singh, R. (2017). *Learn&Fuzz: Machine learning for input fuzzing, 50–59*. IEEE.
93. Cheng, L., et al. (2019). *Optimizing seed inputs in fuzzing with machine learning* (pp. 244–245). IEEE.
94. Becker, S., Abdelnur, H., State, R., & Engel, T. (2010). *An autonomic testing framework for ipv6 configuration protocols* (pp. 65–76). Springer.
95. Fang, K., & Yan, G. (2018). *Emulation-instrumented fuzz testing of 4g/lte android mobile devices guided by reinforcement learning* (pp. 20–40). Springer.
96. Drozd, W., & Wagner, M. D. (2018). Fuzzergym: A competitive framework for fuzzing and learning. *arXiv preprint arXiv:1807.07490.*
97. Li, X., Liu, X., Chen, L., Prajapati, R., & Wu, D. (2022). *Fuzzboost: Reinforcement compiler fuzzing* (pp. 359–375). Springer.
98. Ouyang, Y., Wu, Z., Mu, Q., & Wang, Q. (2014). *Leveraging markov chain and optimal mutation strategy for smart fuzzing*. IET.
99. Wang, Y., Wu, Z., Wei, Q., & Wang, Q. (2019). Neufuzz: Efficient fuzzing with deep neural network. *IEEE Access, 7*, 36340–36352.
100. Meade, T., Zhang, S., & Jin, Y. (2016). *Netlist reverse engineering for high-level functionality reconstruction* (pp. 655–660).
101. Meade, T., et al. (2016). *Gate-level netlist reverse engineering for hardware security: Control logic register identification* (pp. 1334–1337).
102. Brunner, M., Baehr, J., & Sigl, G. (2019). *Improving on state register identification in sequential hardware reverse engineering* (pp. 151–160).

103. Geist, J., et al. (2020). *RELIC-FUN: Logic identification through functional signal comparisons* (pp. 1–6).
104. Botero, U. J., et al. (2021). Hardware trust and assurance through reverse engineering: A tutorial and outlook from image analysis and machine learning perspectives. *JETC, 17*, 1–53.
105. Gassend, B., et al. (2002). *Silicon physical random functions* (pp. 148–160).
106. Guajardo, J., et al. (2007). *FPGA intrinsic PUFs and their use for IP protection* (pp. 63–80). Springer.
107. Chen, B., et al. (2017). *A robust SRAM-PUF key generation scheme based on polar codes* (pp. 1–6). IEEE.
108. Lim, D., et al. (2005). Extracting secret keys from integrated circuits. *Transactions on Very Large Scale Integration (VLSI) Systems, 13*, 1200–1205.
109. Majzoobi, M., et al. (2008). *Testing techniques for hardware security* (pp. 1–10). IEEE.
110. Chen, Q., et al. (2011). *The bistable ring PUF: A new architecture for strong physical unclonable functions* (pp. 134–141). IEEE.
111. Sahoo, D. P., et al. (2017). A multiplexer-based arbiter PUF composition with enhanced reliability and security. *Transactions on Computers, 67*, 403–417.
112. Sahoo, D. P., et al. (2015). A case of lightweight PUF constructions: Cryptanalysis and machine learning attacks. *TCAD, 34*, 1334–1343.
113. Ganji, F., Tajik, S., & Seifert, J.-P. (2015). *Why attackers win: On the learnability of xor arbiter pufs* (pp. 22–39). Springer.
114. Ganji, F., et al. (2016). *Strong machine learning attack against pufs with no mathematical model* (pp. 391–411). Springer.
115. Ganji, F., et al. (2016). Pac learning of arbiter pufs. *Journal of Cryptographic Engineering, 6*, 249–258.
116. Dumoulin, V., Rao, W., & Devroye, N. (2023). Active learning for fast and slow modeling attacks on arbiter pufs. *arXiv preprint arXiv:2308.13645.*
117. Nguyen, P. H., et al. (2016). Security analysis of arbiter PUF and its lightweight compositions under predictability test. *TODAES, 22*, 1–28.
118. Nguyen, P. H., et al. (2018). The interpose PUF: Secure PUF design against state-of- the-art machine learning attacks. *Cryptology ePrint Archive.*
119. Ruhrmair, U., et al. (2010). *Modeling attacks on physical unclonable functions* (pp. 237–249).
120. Tobisch, J., & Becker, G. T. (2015). *On the scaling of machine learning attacks on PUFs with application to noise bifurcation* (pp. 17–31). Springer.
121. Schuster, D., & Hesselbarth, R. (2014). *Evaluation of bistable ring PUFs using single layer neural networks* (pp. 101–109). Springer.
122. Awano, H., & Sato, T. (2018). *Ising-PUF: A machine learning attack resistant PUF featuring lattice like arrangement of arbiter-PUFs* (pp. 1447–1452). IEEE.
123. Zalivaka, S. S., Ivaniuk, A. A., & Chang, C.-H. (2017). *Low-cost fortification of arbiter PUF against modeling attack* (pp. 1–4). IEEE.
124. Ma, Q., et al. (2018). *A machine learning attack resistant multi-PUF design on FPGA, 97–104.* IEEE.
125. Zhang, J., et al. (2021). CT PUF: Configurable tristate PUF against machine learning attacks for iot security. *IEEE IoT Journal, 9*, 14452–14462.
126. Millwood, O., et al. (2023). Puf-phenotype: A robust and noise-resilient approach to aid group-based authentication with dram-pufs using machine learning. *IEEE Transactions on Information Forensics and Security, 18*, 2451.
127. Blocklove, J., Garg, S., Karri, R., & Pearce, H. (2023). *Chip-chat: Challenges and opportunities in conversational hardware design* (pp. 1–6). IEEE.
128. Thakur, S., et al. (2023). Autochip: Automating hdl generation using llm feedback. *arXiv preprint arXiv:2311.04887.*

Chapter 14
Hardware Accelerators for Artificial Intelligence

S. M. Mojahidul Ahsan, Tamzidul Hoque, Md Sakib Hasan, Mrittika Chowdhury, and Anurag Dhungel

In this chapter, we explore the specialized hardware accelerators designed to enhance artificial intelligence (AI) applications, focusing on their necessity, development, and impact on the AI field. It covers the transition from traditional computing systems to advanced AI-specific hardware, addressing the growing demands of AI algorithms and the inefficiencies of conventional architectures. The discussion extends to various types of accelerators, including GPUs, FPGAs, and ASICs, and their roles in optimizing AI workloads. Additionally, it touches on the challenges and considerations in designing and implementing these accelerators, along with future prospects in the evolution of AI hardware. This comprehensive overview aims to provide readers with a clear understanding of the current state and future trends in AI hardware development, ensuring it is accessible to both experts and beginners in the field.

S. M. M. Ahsan
Electrical Engineering and Computer Science, University of Kansas, Lawrence, KS, USA
e-mail: smmojahidul.ahsan@ku.edu

T. Hoque (✉)
Electrical Engineering and Computer Science, University of Kansas, Lawrence, KS, USA
e-mail: hoque@ku.edu

M. S. Hasan · M. Chowdhury · A. Dhungel
Electrical and Computer Engineering, University of Mississippi, Oxford, MS, USA
e-mail: mhasan5@olemiss.edu; adhungel@go.olemiss.edu

A. Iranmanesh, H. Sayadi (eds.), *AI-Enabled Electronic Circuit and System Design*,
https://doi.org/10.1007/978-3-031-71436-8_14

14.1 Introduction to Hardware Accelerators for AI

14.1.1 Overview of AI Advancements and Impacts

Artificial intelligence (AI) and deep learning have emerged as powerful forces driving innovation across a vast spectrum of industries and applications. From revolutionizing the way we interact with technology to tackling complex scientific challenges, these advancements are shaping the future of our world. Figure 14.1 shows the transformation of AI over the past decade marked by a series of groundbreaking innovations, beginning with the introduction of AlexNet in 2012, which revolutionized the field of deep learning. This journey has seen the development of increasingly sophisticated algorithms, culminating in the latest generation of large language models (LLMs). These advancements have significantly expanded AI capabilities, enabling more complex and nuanced understanding and generation of human language, and transforming a wide range of applications across industries.

In the current landscape, several key factors stand out as crucial for the ongoing advancement of AI.

- *Availability of data*: The explosion of data across various fields has provided fertile ground for the training of deep learning models. From social media posts and medical records to satellite imagery and financial transactions, the abundance of data has been instrumental in fueling progress in AI.
- *Increased computational power*: Advancements in hardware, such as graphics processing units (GPUs) and specialized AI accelerator chips, have greatly enhanced the building and training of increasingly sophisticated neural networks. This increase in processing capability enables faster training durations and the handling of larger datasets, continuously expanding the horizons of AI potential.
- *Deep learning architectures*: Deep neural networks, including convolutional neural networks (CNNs) and recurrent neural networks (RNNs), have excelled in various applications such as image classification, natural language understanding, and language translation. Their success in these domains has significantly

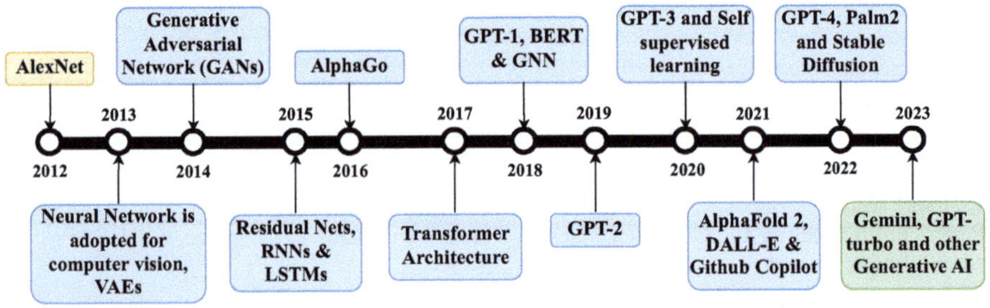

Fig. 14.1 The evolution of artificial intelligence over the past decade, marking significant milestones from the introduction of AlexNet in 2012, which revolutionized neural networks, to the development of sophisticated generative pretrained transformer (GPT) models

advanced the capabilities of AI in processing and interpreting visual and textual data.

- *Generative AI*: This subfield of AI focuses on creating new data, such as text, images, code, and music. Generative models like generative pretrained transformers (GPTs) can generate human-quality content that is often indistinguishable from the real thing. This has opened up exciting possibilities for creative applications, personalized content generation, and data augmentation.

14.1.2 AI Hardware Accelerators: Overcoming Traditional Limits

The world of artificial intelligence is rapidly evolving, demanding ever-increasing computational power to handle complex algorithms and massive datasets. Traditional computing architectures, such as the Von Neumann architecture, have long served as the backbone of information processing. However, these architectures are increasingly falling short when it comes to the specific demands of modern AI, particularly in the realm of neural network processing.

- *Bottlenecks of traditional CPUs*

  Traditional central processing units (CPUs) with Von Neumann architecture face significant challenges with deep learning, primarily due to inefficiencies in managing the high volume of multiply-accumulate (MAC) operations essential for neural network computation. This issue is compounded by the limited parallel processing capabilities, narrower vector processing units, and lower memory bandwidth of modern CPU architectures creating bottlenecks in data handling. Additionally, the complex cache hierarchies introduce latency, and the general-purpose design of CPUs, including their instruction sets and control flows, leads to underutilization in the predictable, repetitive tasks of neural network processing [11]. These factors, combined with the energy-intensive nature of neural network computations, underscore the inadequacy of traditional computing systems for the parallel, data-heavy demands of AI workloads, especially as the depth and complexity of neural networks continue to grow.
- *GPUs: A stepping stone, but not the solution*

  GPUs, with their thousands of cores, excel at parallel processing, allowing them to perform a large number of simple computations simultaneously, which is a common requirement in AI and deep learning algorithms. Moreover, GPUs incorporate cutting-edge memory solutions, such as high bandwidth memory (HBM) and GDDR6X, to support the intensive data transfer required for these computations. This high bandwidth enables GPUs to efficiently perform parallel processing tasks, such as matrix multiplications, and accumulate operations found in deep learning algorithms, thereby significantly speeding up AI model training and inference processes. Despite their superiority over CPUs in AI applications, GPUs are increasingly overshadowed by their energy-intensive nature as AI algorithms grow in complexity and data requirements. This substantial power consumption stems from their adherence to the Von Neumann architecture, which leads to notable limitations in power efficiency, memory bandwidth, and latency. These constraints not only escalate operational costs but also impede the scalability and sustainability of AI advancements [12]. Consequently, while GPUs represent a significant advancement over CPUs, they are not the ultimate solution for the evolving landscape of AI, particularly in an era where energy efficiency and area benchmarks are paramount.

- *Need for specialized hardware accelerators*: *Tailored for AI efficiency*

  Recognizing the shortcomings of traditional architectures, researchers have turned to the development of specialized hardware accelerators for AI. These accelerators are intended to address the unique computational requirements of AI in their design, focusing on the following key aspects:

  - *High parallelism*: AI accelerators typically employ massive parallelism, utilizing hundreds or even thousands of processing cores to handle the massive volume of MAC operations simultaneously. This dramatically increases efficiency and performance compared to the conventional computing methods.
  - *Near-memory processing*: To address the bottleneck of data movement between external memory and the processing unit, AI accelerators employ near-memory or in-memory computing. This approach significantly reduces latency and energy consumption by minimizing the distance data travels, and processing data closer to where it is stored.
  - *High memory bandwidth*: Essential for supporting the high-throughput demands of AI workloads, specialized accelerators are equipped with high memory bandwidth. This feature enables the rapid transfer of data between memory and processing units, crucial for feeding the computational appetite of AI algorithms and preventing processing cores from becoming data-starved.
  - *Energy efficiency*: At the heart of AI accelerator design, achieving energy efficiency is paramount. This is accomplished through a multilayered optimization strategy, spanning from refined parallel algorithms and circuit-level enhancements in memory and computational units to cutting-edge device-level breakthroughs, such as the integration of energy-saving nonvolatile memory technologies for synaptic functions.
  - *Reduced instruction set computing (RISC)*: Unlike general-purpose CPU architectures, AI accelerators often utilize RISC architectures. These architectures are designed for simpler, dedicated instructions, resulting in faster execution and reduced energy consumption. Moreover, AI accelerators may use simplified or specialized instruction sets that are optimized for computations most frequently performed in AI algorithms.
  - *Reduced-precision arithmetic*: AI applications often exhibit inherent resilience to noise and imperfection. Leveraging this characteristic, AI accelerators frequently employ reduced-precision arithmetic, sacrificing some accuracy for significant gains in computational efficiency and memory

bandwidth. The specialized AI hardware is capable of using this technique for efficient computing.

The shift toward specialized hardware for AI acceleration is driven by the limitations of traditional computing architectures in keeping pace with the demands of advanced AI and neural network algorithms. This move is critical for harnessing the full potential of AI, enabling more efficient and powerful processing capabilities tailored for the complexity and depth of modern AI computations. Such dedicated hardware is essential for the next wave of AI innovations, providing the foundation for more sophisticated, real-time AI applications and significantly advancing the field.

14.2 AI Algorithms and Their Hardware Implementation

14.2.1 Overview of Key AI Algorithms

AI algorithms, serving as the cornerstone of contemporary technological breakthroughs, are designed to tackle a diverse array of tasks ranging from intricate data analysis to the autonomous operation of systems. These sophisticated algorithms demand extensive computational power to function effectively, highlighting the pivotal role of computational resources in enabling their full potential. The synergy between advanced AI algorithms and the computational infrastructure paves the way for groundbreaking developments in AI, fostering the creation of more efficient, powerful, and adaptable applications across various fields. The interplay between these sophisticated algorithms and their hardware implementation is crucial in advancing the field of AI, leading to more efficient, powerful, and versatile applications. A classification of modern key AI algorithms is shown in Fig. 14.2. Let us dig deeper into the functionalities and applications of key AI algorithms, categorized into three major areas:

1. *Supervised learning*
   - *Conventional machine learning*:
     - *Linear regression*: This algorithm learns a linear relationship between features and a continuous target variable. It operates by fitting a linear equation to the data points, optimizing the line to minimize the sum of the squared differences between the predicted and actual values. This straightforward yet effective method is widely used in regression tasks such as predicting housing prices based on factors like size and location.
     - *Decision trees*: This algorithm classifies data points by asking a series of yes/no questions based on features. Each question splits the data into smaller subsets until reaching a final classification. Decision trees offer interpretable results and are efficient for handling large datasets.

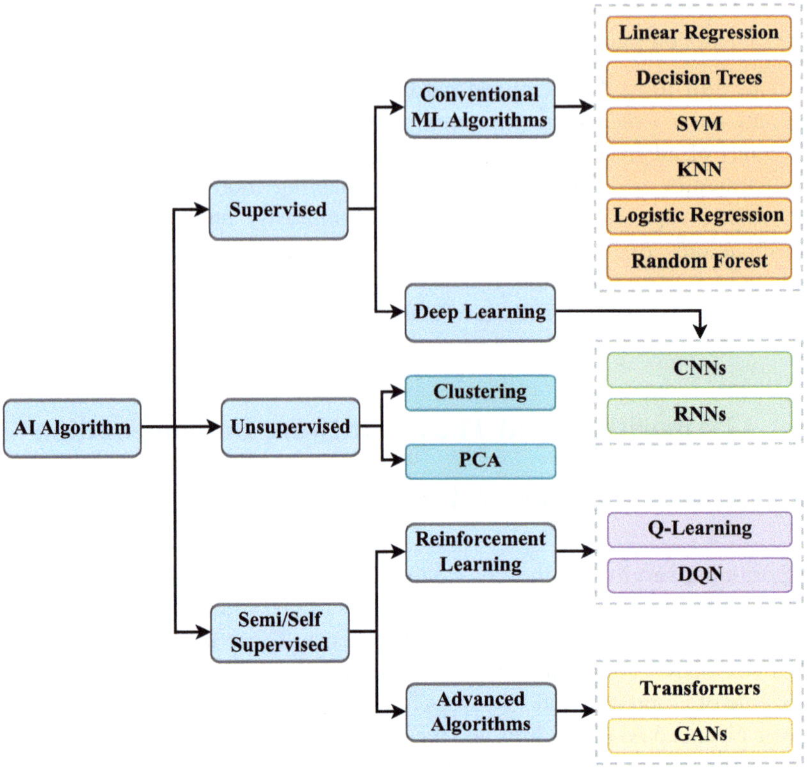

Fig. 14.2 A classification of key AI algorithms

- *Support vector machine (SVM)*: SVM is a powerful classification algorithm that finds the best hyperplane separating different classes in the feature space. It maximizes the margin between data points of different categories, offering robustness in classification tasks.
- *K-nearest neighbors (KNN)*: The K-nearest neighbors (KNN) algorithm classifies a data point based on the majority vote of its "k-" nearest neighbors in the feature space. Versatile and straightforward, KNN is used for both classification and regression tasks by analyzing the distances between data points to find the most similar neighbors. Its simplicity and effectiveness make it popular in various machine-learning applications.
- *Logistic regression*: Logistic regression is used for binary classification. It estimates the probability that a given input point belongs to a certain class. This is done by applying a logistic function to a linear equation, providing a probabilistic foundation for classification.
- *Random forest*: Random forest is an ensemble learning technique that builds multiple decision trees and combines their results for improved

accuracy and stability. It is effective for both classification and regression tasks, reducing overfitting and enhancing predictive performance.

- *Deep Learning*:
  - *Convolutional neural networks (CNNs)*: These networks excel at image recognition and computer vision tasks. They learn to extract features from images by applying convolution filters over the input data. CNNs are widely used in applications like face recognition, object detection, and medical image analysis.
  - *Recurrent neural networks (RNNs)*: These networks are developed to process sequential data like text and speech. They have internal loops that allow them to remember information from previous inputs, making them ideal for tasks like language translation and machine translation.

2. Unsupervised learning
   - *Clustering*: Clustering algorithms group similar data points based on their characteristics, identifying patterns and structures in unlabeled data. This technique is useful for tasks such as segmenting customers by purchase history. Common clustering methods include K-means and hierarchical clustering, which organize data into meaningful groups to uncover insights and trends.
   - *Principal component analysis (PCA)*: This algorithm reduces the dimensionality of data by identifying the most important features. By projecting the data onto a lower-dimensional space, PCA helps visualize data and improve the performance of other algorithms.
3. Semi–/self-supervised learning
   - *Reinforcement learning*:
     - *Q-learning*: This algorithm learns an optimal policy for an agent in a given environment. It estimates the expected reward for taking a specific action in each state and chooses the action with the highest expected reward. Q-learning has been successfully applied in robotics and game playing.
     - *Deep Q-networks (DQN)*: This algorithm utilizes Q-learning with a neural network to handle complex state spaces. DQN has achieved superhuman performance in various Atari games, demonstrating its potential for solving complex decision-making problems.
   - *Advanced algorithms*
     - *Transformers*: Transformers are advanced neural networks that utilize the attention mechanism to evaluate relationships within an input sequence. This capability makes them highly effective for natural language understanding and generation tasks. They are integral to large language models like BERT and GPT-3, which excel in applications ranging from text translation to conversational agents.

– *Generative adversarial networks (GANs)*: GANs feature a generator that creates new data and a discriminator that evaluates its authenticity. This adversarial process enhances the performance of both networks, leading to the creation of highly realistic outputs such as images, music, and text. By constantly challenging each other, the generator produces increasingly convincing data, while the discriminator becomes more adept at identifying fakes, resulting in the progressive improvement of the system's capabilities.

These algorithms' computational complexity and requirements are a driving force behind the evolution of specialized hardware. The hardware not only needs to handle vast amounts of data efficiently but also provides the speed necessary for real-time processing and learning.

14.2.2 Case Studies of Hardware Accelerator for AI

Artificial intelligence (AI) algorithms are increasingly being deployed in various domains, such as computer vision, natural language processing, robotics, and healthcare. However, the computational complexity and resource requirements of these algorithms pose significant challenges for their efficient and scalable implementation on hardware platforms. In this section, we review some case studies of specific hardware specially designed for handling fast and energy-efficient AI computation, covering different types of AI algorithms and hardware platforms. Selected hardware packages and boards are demonstrated in Fig. 14.3.

- *Tensor processing unit (TPU)*: One example of a state-of-the-art AI algorithm implementation on a hardware accelerator is the Google TPU (tensor processing unit), a custom ASIC (application-specific integrated circuit) specifically designed to accelerate deep neural networks (DNNs). The TPU uses a systolic array architecture, facilitating high-throughput matrix multiplication, essential for DNN operations. It also supports reduced-precision arithmetic and data compression techniques, enhancing both the performance and energy efficiency of DNN inference. The TPU has been deployed in powering various Google services, including Google Search, Google Photos, Google Translate, and AlphaGo. A detailed architectural explanation of the TPU is provided in Sect. 14.3.
- *IBM TrueNorth and NorthPole*: Another example of a neuromorphic chip specifically designed for faster and energy-efficient AI computation is the IBM TrueNorth chip. This neuromorphic processor emulates the structure and function of biological neurons and synapses. TrueNorth [6] contains 4096 cores, each with 256 neurons and a 256 $\times$ 256 binary synapse array. It operates asynchronously, generating and propagating spikes among neurons. TrueNorth supports neural network models like CNNs, RNNs, and SNNs, and is applied in image recognition, video analysis, and anomaly detection. A recent advancement is IBM's NorthPole chip [7]. It features a neural-inspired architecture for

Fig. 14.3 Modern hardware accelerators: (**a**) TPU v2 was unveiled at Google I/O in May 2017 [1], (**b**) NVIDIA Tesla P100: The World's First AI Supercomputing Data Center GPU [2], (**c**) Nvidia H100 Hopper chip [3], (**d**) Intel Loihi 2 Neuromorphic Chip [4], (**e**) ALINX AX7Z020: SoC FPGA Development Board [5]

enhanced performance, energy, and area efficiency compared to comparable architectures. NorthPole integrates computation with memory, reducing off-chip memory requirements. It is a digital system allowing for tailored bit precision, optimizing power usage. This chip, using a 12-nanometer process, achieves significant improvements in energy, space, and time metrics over prevalent architectures, even those using more advanced technology processes.

- *Intel Loihi series*: The Intel Loihi chip, a neuromorphic processor, mirrors the behavior and learning processes of biological neurons and synapses. It features 128 cores, each with 1024 neurons and 128 K synapses, programmable for various tasks. The chip is adept at adapting to new inputs and environments, thanks to its support for spike-timing-dependent plasticity and online learning. Loihi is compatible with multiple neural network architectures like CNNs, RNNs, and SNNs. Its applications include gesture recognition, path planning, reinforcement learning, and optimization. The Intel Loihi 2 neuromorphic chip is a significant evolution from its predecessor, offering up to 10 times the performance. It supports more advanced and flexible neuromorphic computing architectures, allowing for a wide range of applications. Loihi 2 introduces improved features such as generalized event-based messaging, enhanced neuron model programmability, and advanced learning capabilities. These upgrades enable the chip to support more complex algorithms and applications, positioning it at the forefront of neuromorphic computing research and development.

- *Graphics processing unit (GPU)*: In addition to specialized hardware, advanced GPUs are crucial in AI advancements, particularly in data centers. NVIDIA's P, H, and V series GPUs, for example, excel in accelerating DNNs, especially during training phases. Their parallel architecture allows for numerous simultaneous floating-point operations, ideal for DNN processes. GPUs also feature high memory bandwidth and performant interconnects, handling vast data quantities efficiently. They can be combined with CPUs or FPGAs, creating heterogeneous systems that leverage the strengths of various hardware components.
- *Field-programmable gate arrays (FPGAs)*: FPGAs are increasingly recognized for their critical role in accelerating deep learning algorithms. Unlike traditional hardware, FPGAs offer a unique blend of programmability and performance, making them highly adaptable to the evolving needs of AI computations. Their ability to be reconfigured for specific tasks allows for optimized processing of complex neural networks, enhancing both efficiency and speed. This adaptability, coupled with their capacity for parallel processing, positions FPGAs as a pivotal technology in advancing the field of AI, particularly in areas requiring real-time processing and energy-efficient solutions.
- *Memory-centric computation*: In-memory computing, leveraging traditional memory technologies such as SRAM and DRAM, represents a promising approach to addressing the von Neumann bottleneck. This paradigm integrates computation directly within the memory units, reducing data transfer latency and energy consumption. For instance, SRAM-based in-memory computing designs enable fast and energy-efficient operations by conducting logical and arithmetic functions directly within the memory arrays [28]. Similarly, DRAM-based architectures utilize the inherent parallelism and large storage capacities of DRAM to accelerate AI workloads [29]. These memory-centric approaches can significantly enhance the performance of AI applications by minimizing the data movement between the CPU and memory, thus overcoming one of the critical limitations of conventional von Neumann architecture.
- *Emerging NVM-based process-in-memory*: Emerging memory technologies like RRAM, PCM, and MRAM enable a new paradigm called process-in-memory (PIM). This architecture integrates memory and processing elements within the same chip, enabling data to be processed directly within the memory, and eliminating the need for data movement between memory and processing units. This drastically reduces energy consumption and improves performance, making PIM a key technology for future AI hardware.

  - *Resistive random-access memory (RRAM)*: RRAM utilizes the resistance change of materials to store data, offering high density and low power consumption. This makes it promising for resource-constrained edge AI applications. For example, RRAM-based accelerators have been developed for image recognition tasks on mobile devices, achieving significant performance improvements compared to traditional CPUs and GPUs [8].
  - *Phase change memory (PCM)*: PCM stores data by reversibly switching a material's phase between crystalline and amorphous *states. This allows for*

high performance, low power consumption, and nonvolatility, making it suitable for various AI applications. PCM-based accelerators have demonstrated promising results in tasks like natural language processing and deep learning, offering faster processing and reduced energy consumption compared to traditional memory technologies [9].

- *Magnetoresistive random-access memory (MRAM)*: MRAM uses magnetic fields to store data, offering low latency, energy efficiency, and nonvolatility [10]. This makes it ideal for applications requiring fast and reliable data access, such as real-time AI applications. MRAM-based accelerators are currently developing, with potential applications in areas like autonomous vehicles and robotics, where fast and accurate decision-making is critical.
- *Ferroelectric field-effect transistor (FeFET)*: FeFET leverages the polarization properties of ferroelectric materials to store data, combining nonvolatility with high speed and low energy consumption [10]. This unique combination makes FeFET particularly appealing for AI applications that demand efficient data processing and storage capabilities. FeFET-based accelerators are emerging as a potent solution for AI tasks, offering significant advantages in terms of density, power efficiency, and speed. Early implementations of FeFET technology in AI hardware have shown promising results.

These emerging devices create new research opportunities for PIM technology. Tools like NVSim [25], NeuroSIM [26], and NVMexplorer [27] can be utilized to investigate these devices for various innovative applications.

To summarize, a variety of creative solutions are being employed to address the computational needs of AI applications in different areas. It covers a range of innovations, from Google's TPU with its systolic array design that optimizes deep neural network tasks to IBM's brain-inspired neuromorphic chips. Each example illustrates a distinct method for boosting both efficiency and energy conservation. Intel's Loihi chips and high-end GPUs merge neuromorphic concepts with parallel computation, whereas FPGAs provide versatility and adaptiveness. Furthermore, the advent of NVM-based technologies marks the beginning of a new phase in memory-centric computing designs, setting the stage for substantial progress in AI hardware. The next section will provide a side-by-side comparison of these technologies, evaluating their respective strengths and weaknesses.

14.2.3 Comparative Analysis of Different Hardware Solutions for AI

The rapid evolution of artificial intelligence (AI) has spurred an equally significant demand for specialized hardware solutions capable of handling its increasingly sophisticated algorithms. While traditional CPUs have played a crucial role in AI development, they are often outmatched by the computational demands of modern machine-learning (ML) and neural network (NN) algorithms. This has led to the rise of alternative hardware platforms specifically designed to accelerate AI tasks. Here, we will explore these diverse hardware solutions, providing insight into their comparative strengths and weaknesses.

A. GPUs

- *Strengths*: Initially designed for rendering graphics in gaming and visual applications, GPUs have evolved significantly over the past 30 years to become crucial for AI applications. The high data throughput and massive parallelism of GPUs, with their hundreds of cores, allow them to excel in parallel calculations such as matrix multiplications. This makes them ideal for deep learning, big data analytics, and genomic sequencing. In particular, GPUs are highly effective in training AI models, where the ability to process large, similar datasets simultaneously is essential. Extensively used in neural networks and accelerated AI operations, GPUs provide the computational power necessary for handling large volumes of identical or unstructured data. Today, their advanced capabilities make them a staple in data centers and cloud applications, continuing to play a significant role in the AI revolution.
- *Weaknesses*: GPUs, while powerful, face challenges in multi-GPU setups due to complexity and scalability issues. Programming for these systems requires specialized knowledge, and communication overhead between GPUs can hinder performance. Memory management is critical, as large datasets may not fit in a single GPU. Additionally, the high cost and increased power consumption are significant considerations. Compatibility with AI frameworks and diminishing performance gains with added GPUs further complicate their use. Maintenance and limited availability of cloud services also pose challenges.

B. FPGAs (field-programmable gate arrays)

- *Strengths*: FPGAs, positioned between CPUs and GPUs, offer a unique blend of benefits for AI acceleration. Unlike CPUs with limited parallelism and GPUs that may lack power efficiency, FPGAs provide reconfigurable hardware tailored to specific AI tasks. They enable custom hardware accelerators that can be programmed to align perfectly with an AI algorithm's requirements, resulting in enhanced performance and energy efficiency. FPGAs excel due to their parallel processing capabilities, ideal for AI operations like matrix multiplications. Additionally, they allow for bespoke hardware customization, offloading intensive computations for faster execution. Crucially, FPGAs are well suited for low-latency inference in real-time applications, offering immediate predictions. Their ability to be fine-tuned for specific tasks also makes them highly energy-efficient, particularly beneficial for edge computing and IoT applications.
- *Weaknesses*: Despite their versatility, FPGAs pose challenges in programming complexity, requiring expertise in hardware description languages like VHDL or Verilog, which can be a barrier for some AI practitioners. They

also face resource constraints, including limited logic gates, memory, and DSP blocks, making optimization of AI algorithms within these confines a complex task. Moreover, scalability with FPGAs in large clusters or cloud environments can be more intricate compared to GPUs. FPGAs are highly efficient for specific algorithms or tasks, but adapting them to different AI models may necessitate substantial reconfiguration, limiting their flexibility in certain scenarios.

C. Neuromorphic integrated circuits (ICs)

- *Strengths*: Neuromorphic ICs, inspired by the human brain, excel in efficiency and speed for specific AI tasks. Their design, mimicking neurons and synapses, allows for lower power consumption and faster data processing, particularly in pattern recognition and sensory data interpretation. This makes them ideal for edge computing applications in AI.
- *Weaknesses*: The complexity of mimicking biological structures means neuromorphic ICs are still in the early development stages. They may not yet match the versatility or raw power of more established platforms like GPUs for a broad range of AI tasks. Their specialized nature also poses a challenge for programming and integration into existing technology stacks.

D. Application-specific integrated circuits (ASICs)

- *Strengths*: AI ASICs are custom hardware designed and optimized for specific AI computations, allowing for the creation of innovative architectures tailored to particular models or applications. This specialization in task execution leads to exceptional energy efficiency and a smaller footprint compared to other accelerators. Their design focus on specific functions also results in higher speed and performance, making ASICs highly effective for high-performance, energy-sensitive applications within AI. The combination of tailored architecture, power efficiency, and compact size positions ASICs as a superior choice for specialized AI computations.
- *Weaknesses*: The biggest drawback of ASICs is their lack of flexibility. Their custom design locks them into performing a single function, making them unsuitable for adapting to evolving AI algorithms and changing requirements. Additionally, their high development cost can make them less attractive for smaller projects.

E. Emerging devices

- *Strengths*: Emerging devices in AI hardware applications are advancing with novel architectures and memory technologies. Process-in-memory and near-memory computing, for instance, are changing how data is handled, bringing computation closer to storage, thereby improving speed and efficiency. Innovations in nonvolatile memory technologies like ReRAM, PCM, and MRAM are also pivotal, as they enable matrix-vector multiplication operations directly within the device, a critical operation in AI algorithms. Additionally, new devices like FeFET (ferroelectric field effect transistor) are emerging, promising advancements in AI hardware design due to their unique properties. These technologies and architectures collectively represent a significant leap forward in AI hardware, offering improvements in speed, energy efficiency, and computing power.

- *Weaknesses*: Emerging device-based AI hardware accelerators, despite their potential, face challenges like variation, aging, and device variabilities, leading to inconsistent behavior. Process compatibility issues also present hurdles in integrating these technologies into existing manufacturing systems. However, engineers and physicists are actively working to optimize these devices for AI hardware design. They are focused on addressing these variabilities and compatibility issues to harness the full potential of these advanced technologies for efficient and reliable AI computation.

The diverse landscape of hardware solutions for AI presents developers and researchers with a range of options. Understanding the strengths and weaknesses of each platform and carefully considering application requirements is key to selecting the optimal hardware for successful AI implementation. As the field of AI continues to evolve, we can expect further innovations in hardware design, pushing the boundaries of performance and efficiency to new heights.

14.3 AI Hardware Accelerator Architectures

In this section, we will discuss some of the significant advancements in chronological order, exploring the innovative landscape of AI accelerator architectures developed by research communities. These architectures have significantly contributed to the evolution of specialized AI processing hardware. Tailored specifically for the demands of AI and machine-learning algorithms, they represent a departure from traditional computing paradigms, focusing instead on optimizing computational efficiency, reducing power consumption, and accelerating AI workloads. From the integration of novel memory technologies to the adoption of specialized computational models like neuromorphic computing and processing-in-memory (PIM), these architectures showcase the forefront of hardware design tailored to the unique requirements of AI applications. This chronological examination highlights the progressive advancements and groundbreaking contributions that have shaped the current state and future directions of AI accelerator architectures.

14.3.1 NeuFlow Architecture

The NeuFlow architecture, introduced in 2011 [14], is a dynamically reconfigurable dataflow processor tailored for vision-related tasks and implemented on FPGA. It features a scalable hardware structure and employs a dataflow compiler called Lu-

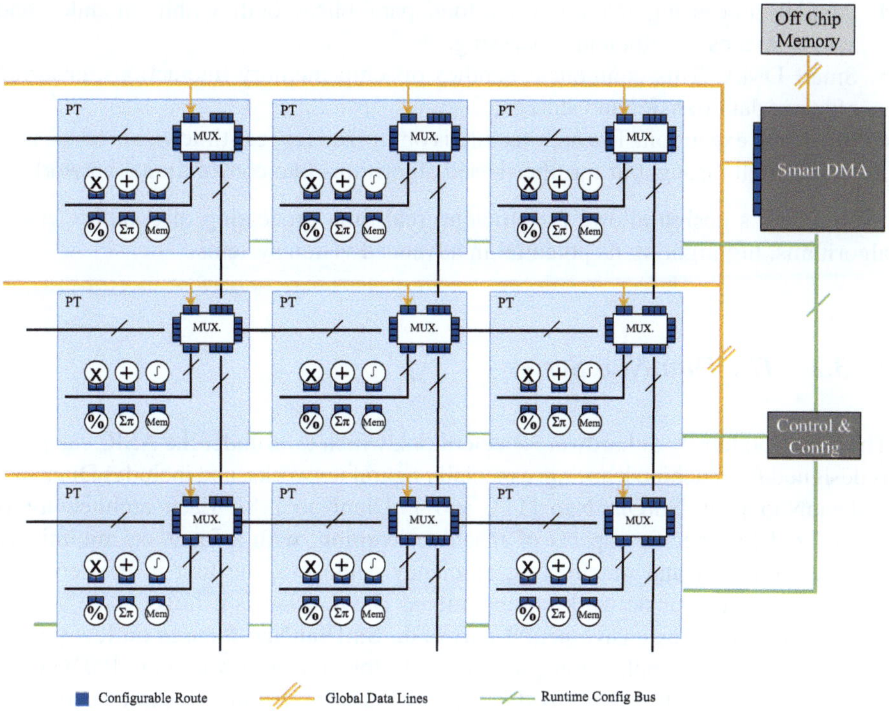

Fig. 14.4 Dynamic dataflow architecture for NeuFlow. (The figure is recreated from [14])

aFlow to convert high-level algorithmic flow-graphs into machine code. NeuFlow is engineered to handle real-time object detection, classification, and localization in intricate scenes, providing notable improvements in processing speed and energy efficiency. The architecture uses a 2D grid of processing tiles (PTs), each with processing operators and routing multiplexers, and a Smart DMA for efficient off-chip memory interfacing. It achieves high throughput by exploiting parallelism within modules and across images, with a focus on filter-based algorithms like convolutional networks. The NeuFlow system demonstrates a new approach to dataflow computing in vision systems, combining high performance with low power consumption. An illustration of this architecture is shown in Fig. 14.4. The NeuFlow architecture operates as follows:

1. High-level programming: NeuFlow begins with a high-level programming interface, where algorithms are represented as flow-graphs.
2. LuaFlow compiler: These flow-graphs are translated into machine code by the LuaFlow compiler.
3. Processing tiles (PTs): The machine code is executed on a 2D grid of PTs, each consisting of processing elements and routing multiplexers.

4. Parallel processing: NeuFlow exploits parallelism both within modules and across images for efficient processing.
5. Smart DMA: This component handles off-chip memory interactions, ensuring efficient data transfer and storage.
6. Real-time execution: The architecture is optimized for real-time vision tasks, providing high throughput for filter-based algorithms like convolutional networks.

NeuFlow's design allows for efficient, real-time processing of complex vision algorithms, highlighting its potential in advanced vision systems.

14.3.2 The DianNao Series

The DianNao family of hardware accelerators, which falls under the ASIC category, is designed for machine learning, especially neural networks, and includes DianNao, DaDianNao [16], ShiDianNao [17], and PuDianNao [18]. Each architecture is specialized for specific aspects of machine learning, with a focus on minimizing memory transfers and maximizing efficiency. DianNao, the first in the series, is optimized for neural network computations. DaDianNao is a larger-scale system for handling more extensive neural networks. ShiDianNao focuses on low-power, embedded system applications, particularly for image processing. PuDianNao extends the capabilities to a broader range of machine-learning techniques beyond neural networks. These architectures demonstrate significant improvements in speed and energy efficiency over traditional computing systems, notably in handling neural network computations. The baseline DianNao architecture is centered around its neural functional units (NFUs), which are crucial for performing arithmetic operations integral to neural networks. These NFUs are designed to efficiently carry out multiplications and additions, operations that are fundamental in neural network computations. To support this processing, DianNao incorporates specific data buffers: input, weight, and output buffers. These buffers play a key role in reducing the dependency on external memory access, thereby enhancing overall computational efficiency. The data in DianNao flows through these NFUs in a pipelined manner. This pipeline approach allows for the simultaneous processing of multiple data points, significantly speeding up the neural network computations and ensuring a more efficient flow of data through the system.

- *DaDianNao* builds on the DianNao design but scales it up with multiple interconnected tiles. This expansion allows it to handle larger neural networks more efficiently. A distinctive feature of DaDianNao is the integration of a large on-chip embedded DRAM (eDRAM). This eDRAM stores neural network parameters, significantly reducing data movement and enhancing computational speed. By having each tile work in parallel, DaDianNao achieves a higher processing capacity, making it well suited for more complex neural network tasks. The architecture is depicted in Fig. 14.5.

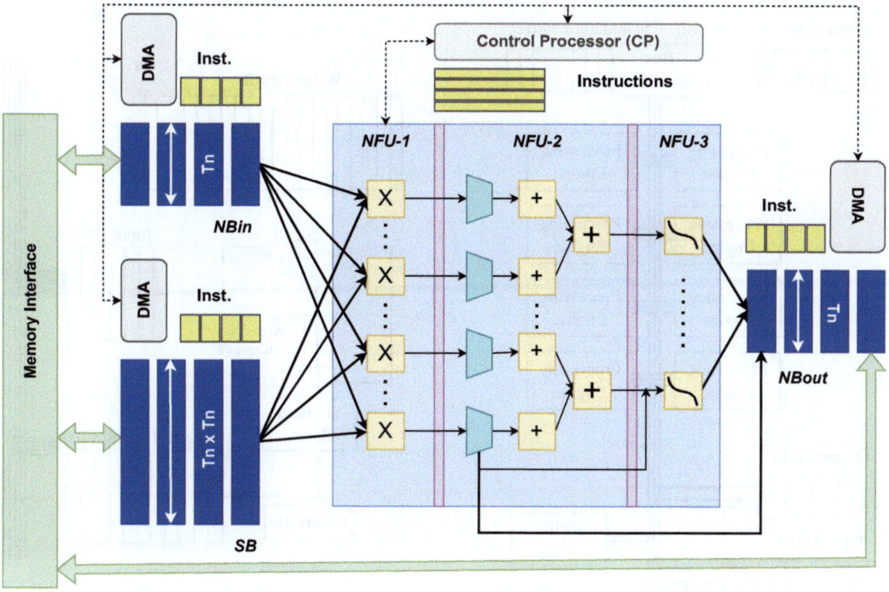

Fig. 14.5 DaDianNao accelerator architecture. (The figure is recreated from [16])

- *ShiDianNao* takes a different direction, focusing on low-power applications, particularly in image processing. Its architecture is simplified compared to DianNao, aiming to reduce energy consumption. This makes ShiDianNao ideal for embedded systems where power efficiency is crucial. The design targets high-throughput vision processing tasks, optimizing the architecture for embedded applications that require efficient image-processing capabilities.
- *PuDianNao* extends the capabilities of the DianNao series to support a broader range of machine-learning algorithms, not just neural networks. It includes specialized hardware units dedicated to different machine-learning tasks, offering a more versatile approach to machine-learning computations. This general-purpose orientation makes PuDianNao a flexible solution, capable of handling various types of machine-learning algorithms beyond the scope of traditional neural network processing.

14.3.3 The Neural Processing Unit (NPU)

The neural processing unit (NPU) falls under the category of a digital ASIC. NPU operates by transforming select segments of general-purpose programs into neural network (NN) models [13]. This process, known as the parrot transformation, involves identifying program segments that can tolerate approximation without

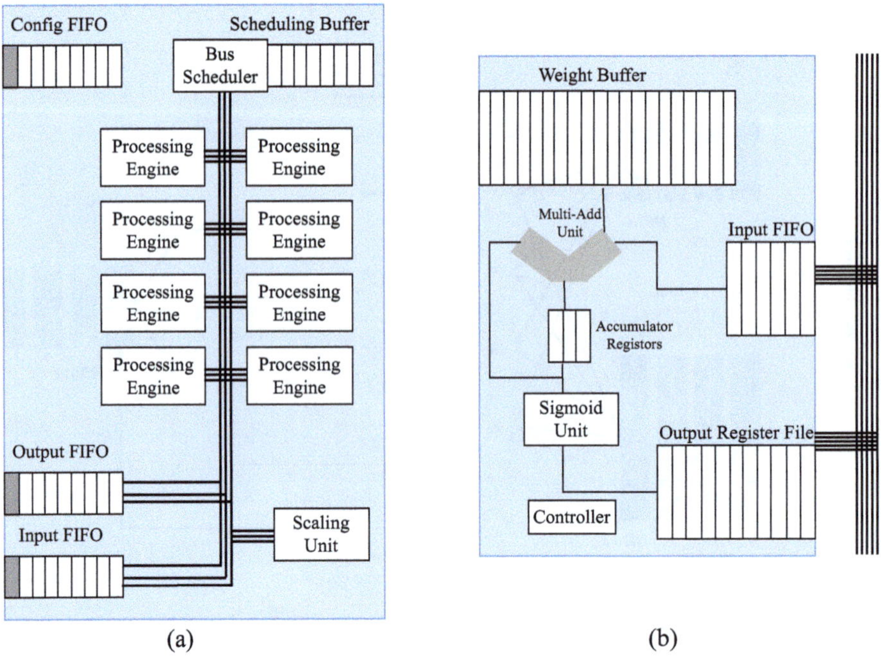

Fig. 14.6 The neural processing unit (NPU) architecture (**a**) 8-processing engine (PE) NPU, (**b**) single-processing engine (PE). (The figure is recreated from [13])

significantly impacting overall accuracy. Once transformed, these segments are executed on the NPU, which is specially designed to efficiently process NN models. The integration of NPUs into traditional computing systems involves a coprocessor model, where NPUs work alongside standard CPUs. The CPU handles regular precise computations, while the NPU accelerates approximate computations. This dual-processor approach allows for a balance between accuracy and computational efficiency. The architecture of the NPU, particularly in the operational framework depicted in Fig. 14.6, showcasesthe detailed workflow of how program segments are processed. This includes dataflow, control mechanisms, and the interaction between the NPU and the main CPU. The advantage of this setup lies in its ability to significantly improve performance and reduce energy consumption for suitable tasks, with the trade-off being a manageable decrease in computational accuracy. The neural processing unit (NPU) architecture is shown in Fig. 14.6. This specialized processing unit for AI works in the following steps at the architectural level:

1. Identification of approximable code segments: The CPU identifies code segments within a general-purpose program that are suitable for approximation.

2. Parrot transformation: These identified segments are transformed into neural network models using the parrot transformation. This involves mapping the program's logic and data flow into a neural network structure.
3. Execution on NPU: The transformed neural network model is then executed on the NPU. The NPU is specialized for efficient neural network computations, particularly those involving operations like matrix multiplications and activations.
4. Interaction with CPU: The NPU operates as a coprocessor to the CPU. It handles the approximate computations, while the CPU continues to process tasks requiring high accuracy.
5. Output integration: The results from the NPU are integrated back into the main program flow on the CPU, ensuring that the overall program objectives are met despite the approximation in certain segments.
6. Performance and energy efficiency: The NPU provides enhanced performance and energy efficiency for the approximate computing tasks, with a tolerable reduction in accuracy.

This architecture leverages the strengths of both neural network processing and traditional CPU computations to achieve an optimal balance of speed, energy efficiency, and accuracy.

14.3.4 RENO Architecture

RENO is a highly efficient, reconfigurable accelerator designed for neuromorphic computing [15], which falls under the neuromorphic IC category among different hardware solutions for AI. It leverages memristor-based crossbar arrays for enhanced mixed-signal computation, speeding up artificial neural network (ANN) executions. RENO's architecture consists of hierarchically arranged memristor-based crossbar (MBC) arrays that support various ANN topologies through a mixed-signal interconnection network (M-Net). The design demonstrates significant performance speedups and energy savings compared to general-purpose processors and other digital neural processing units. It balances efficiency, fault tolerance, and flexibility in neural network computations. The architecture is depicted in Fig. 14.7 and different features of this architecture are listed below:

1. The Core of RENO: At the heart of RENO lies its memristor-based crossbar (MBC) arrays. These arrays are fundamental to RENO's ability to process artificial neural networks (ANNs) efficiently. Unlike traditional computing systems, RENO's MBC arrays enable simultaneous memory storage and computation, significantly enhancing processing speeds and reducing energy consumption.
2. Hierarchical architecture: RENO's architecture is hierarchically structured, allowing for flexible and efficient handling of various ANN topologies. This hierarchical design not only streamlines computation but also provides a framework for scaling and integrating larger and more complex neural networks.

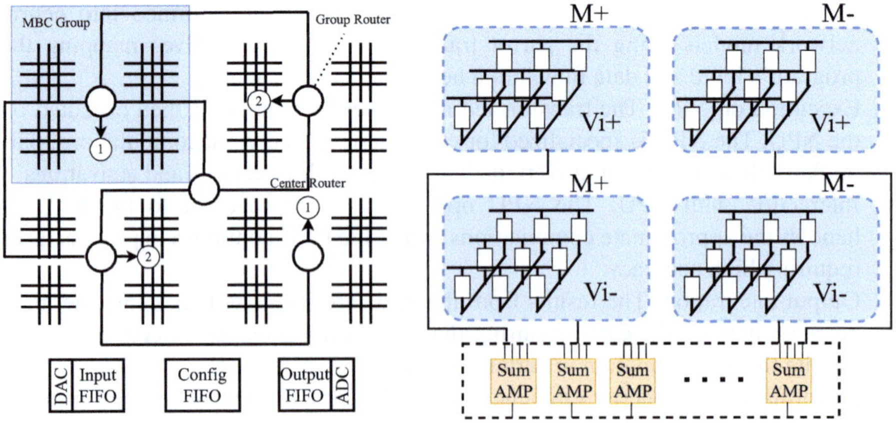

Fig. 14.7 The RENO architecture. (The figure is recreated from [15])

3. Mixed-signal interconnection network (M-Net): A pivotal component of RENO is its mixed-signal interconnection network, known as M-Net. M-Net facilitates communication between different layers and elements of the neural network, ensuring seamless dataflow and processing across the entire architecture.
4. Performance and efficiency: When compared to general-purpose processors and digital neural processing units, RENO demonstrates a substantial leap in performance. It achieves remarkable speedups in ANN executions, coupled with significant energy savings. This efficiency is a result of its mixed-signal computation approach and the inherent advantages of memristor technology.
5. Balancing flexibility and fault tolerance: RENO's design adeptly balances efficiency with flexibility and fault tolerance. It can adapt to various neural network structures, offering a reconfigurable platform for diverse computing needs. Additionally, RENO incorporates mechanisms to handle potential faults in memristor arrays, ensuring reliable operation.

RENO stands as a testament to the innovative advancements in neuromorphic computing. With its unique architecture and superior performance, RENO paves the way for new possibilities in AI and machine-learning applications. As technology evolves, RENO's impact on neuromorphic computing will undoubtedly be a subject of keen interest and continued research.

14.3.5 Neurocube Architecture

Neurocube, introduced in 2016 redefines neuromorphic computing with its unique digital architecture integrated with high-density 3D memory [20]. This design is specifically aimed at enhancing neural network computations, leveraging the depth

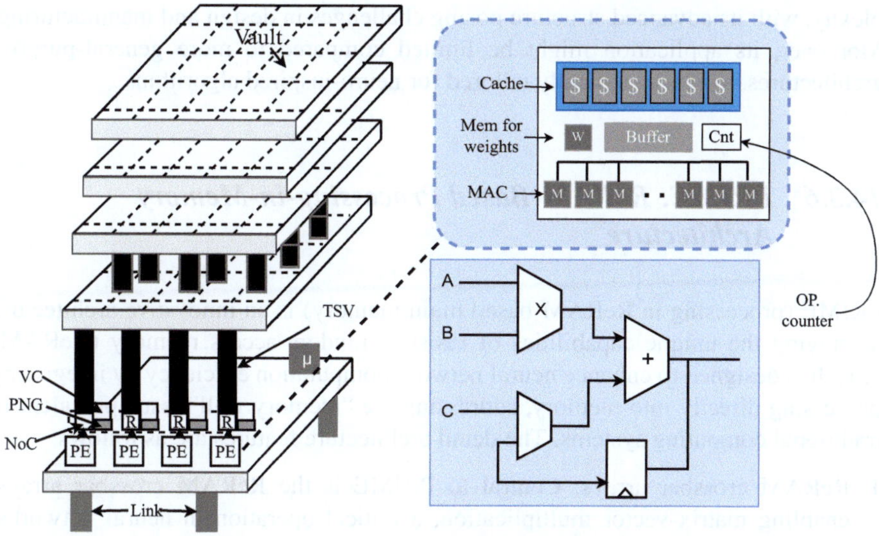

Fig. 14.8 Neurocube architecture (left) and processing elements (PEs) organization (right). (The figure is recreated from [20])

of 3D memory to achieve superior computational performance and power efficiency. The architecture is depicted in Fig. 14.8 has the following features:

1. In-memory neuromorphic processing: At its core, Neurocube combines computing layers within a 3D memory stack. This allows for direct processing within memory layers, reducing the data movement significantly.
2. Memory-Centric Neural Computing (MCNC): Neurocube utilizes the MCNC approach, harnessing memory-centric techniques for efficient data-driven neural algorithms. This leads to a substantial increase in both energy and computational efficiency.
3. Programmable neurosequence generator: Key to its flexibility, Neurocube employs memory-based state machines, enabling it to program and adapt to a wide range of neural network algorithms.
4. Distinctive features: Unlike conventional architectures that distinctly separate memory and computation, Neurocube's integration of a compute layer within a 3D memory stack sets it apart, resulting in higher operational efficiency and reduced latency.

Neurocube offers several advantages, including enhanced computational efficiency, where it excels in neural computations by providing rapid processing speeds with lowered energy consumption. It also boasts scalability and flexibility, matching GPU-like programmability while offering more efficient power and computational handling. Additionally, Neurocube's advanced memory integration, through the incorporation of 3D high-density memory, significantly boosts memory bandwidth and minimizes latency. However, Neurocube has limitations such as design complexity, with its advanced structure posing challenges in design and manufacturing. Moreover, its application might be limited compared to more general-purpose architectures, as it is specifically tailored for neuro-inspired algorithms.

14.3.6 PRIME: ReRAM-Based Processing-in-Memory Architecture

PRIME (processing in ReRAM-based main memory) is an innovative architecture leveraging the unique capabilities of resistive random-access memory (ReRAM) [24]. It is designed to enhance neural network computation efficiency by integrating processing directly into memory, addressing the "memory wall" issue prevalent in traditional computing systems. The detail architecture features are as follows:

1. ReRAM crossbar arrays: Central to PRIME is the ReRAM crossbar arrays, enabling matrix-vector multiplication, a critical operation in neural networks. ReRAM-based crossbar arrays perform matrix-vector multiplication differently than CMOS-based systems. Each crossbar junction represents a synaptic weight, and the array inherently executes MAC operations through Ohm's law and Kirchhoff's law, making the process highly parallel and energy-efficient.
2. Processing-in-memory (PIM): The architecture adopts a PIM approach, combining storage and computation in memory arrays to reduce data movement. This integration allows for simultaneous data storage and processing, significantly reducing the need for data to be moved between separate memory and processing units, which is a major source of energy consumption and latency in traditional architectures.
3. Configurable arrays: PRIME features configurable ReRAM arrays, which can switch between functioning as neural network accelerators or as standard memory storage. This flexibility enables the system to adapt to various computational demands efficiently, providing the advantage of using the same hardware for different purposes—either for high-performance neural network computations or for regular memory storage tasks.
4. Software/hardware interface: PRIME includes a sophisticated software/hardware interface that ensures seamless integration and implementation of diverse neural network models. This interface acts as a bridge between the computational hardware and the software algorithms, enabling the effective execution of various neural network architectures and paradigms on the PRIME platform. This design facilitates versatility and ease of adaptation to different neural network requirements and models (Fig. 14.9).

PRIME offers several advantages due to its innovative design. PRIME enhances efficiency in neural network computations through its ReRAM-based processing-in-memory (PIM) architecture, enabling simultaneous processing and storage of data within the memory arrays. This design reduces the need for time-consuming

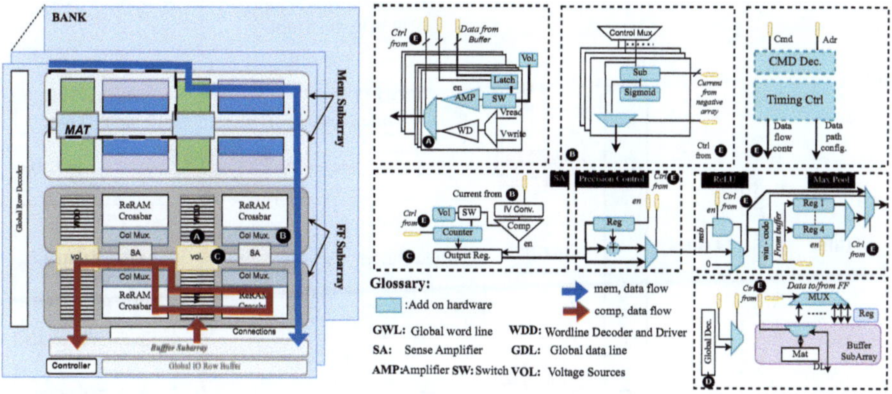

Fig. 14.9 The PRIME framework illustrated. On the left, the memory bank layout is shown, with bold blue and red lines indicating pathways for standard memory operations and computational tasks. On the right, enhancements specific to PRIME are highlighted: (**a**) a wordline driver with multiple voltage levels; (**b**) a column multiplexer with analog subtraction and sigmoid functions; (**c**) a versatile sense amplifier with counters for multi-level signals, including ReLU functions and 4-to-1 max pooling; (**d**) the linkage between flip-flop (FF) and buffer subarrays; (**e**) the PRIME management unit. (The figure is recreated from [24])

and energy-intensive data transfers between separate processing and storage units. Furthermore, the PIM approach inherently minimizes data movement, significantly cutting down on energy consumption and improving operational efficiency by processing data directly within the memory arrays. PRIME's configurable ReRAM arrays provide remarkable flexibility, supporting various neural network architectures and accommodating a wide range of computational needs, enhancing the architecture's utility across different application scenarios. However, PRIME also has several limitations, primarily due to its reliance on ReRAM technology. The integration of processing and memory in ReRAM arrays increases the complexity of design and fabrication, posing challenges in integrating ReRAM technology at a large scale while ensuring consistent performance. Additionally, ReRAM is not as standardized as CMOS technology and may face reliability issues such as retention, cycle-to-cycle, and device-to-device variations, which can affect the performance and predictability of PRIME. The architecture is specifically designed for neural network computations, limiting its effectiveness for other types of computations and restricting its utility in more general-purpose computing scenarios. Finally, ReRAM technologies are known for variability and stability challenges, including retention time, endurance, and variability in electrical characteristics, which can lead to performance inconsistencies and impact the overall efficiency and reliability of the PRIME architecture.

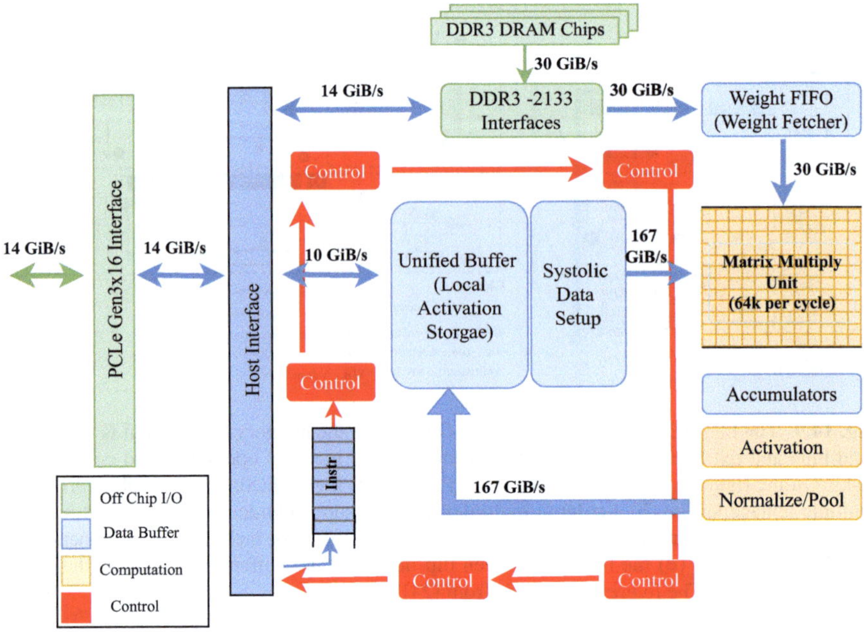

Fig. 14.10 Tensor processing unit (TPU) architecture block diagram. (The figure is recreated from [19])

14.3.7 Tensor Processing Unit (TPU)

The tensor processing unit (TPU) is an AI ASIC developed by Google, tailored for machine-learning tasks [19]. TPUs are designed to accelerate neural network computations, primarily focusing on inference operations rather than training. TPUs address the need for specialized hardware to handle the computational demands of large-scale neural networks efficiently. They offer significant speedups and energy efficiency over general-purpose CPUs and GPUs in machine-learning applications. The details of TPU architecture are depicted in Fig. 14.10. TPU has the followingarchitectural features:

1. Matrix multiply unit: The TPU's core is a matrix multiply unit, which performs high-throughput 8-bit multiply-and-add operations, essential for NN computations.
2. Memory and data management: It includes large on-chip memories to reduce external memory access. A notable feature is the unified buffer, which stores intermediate data, and a weight FIFO for staging weights.
3. Execution model: TPUs follow a deterministic execution model, differing from the time-varying optimizations in CPUs and GPUs. This model helps meet the stringent latency requirements of NN applications.

4. Systolic data flow: The architecture utilizes a systolic array approach for dataflow, enhancing energy efficiency and reducing the read/write operations from the unified buffer.
5. Software compatibility: TPU is designed to be compatible with existing machine-learning frameworks, such as TensorFlow, ensuring easy integration into existing workflows.
6. Host integration: Operates as a coprocessor connected via the PCIe I/O bus, allowing seamless integration into existing server architectures.
7. Performance efficiency: TPUs are optimized for high-throughput per watt, significantly outperforming contemporary CPUs and GPUs in efficiency for machine-learning tasks.

TPUs offer several key advantages for neural network inference tasks. They are highly efficient and deliver exceptional throughput, significantly outperforming traditional CPUs and GPUs in these applications. TPUs also consume less power compared to CPUs and GPUs, making them more energy-efficient. Additionally, TPUs are designed to handle low-precision computations effectively, which are common in neural network operations, further enhancing their performance in these specific tasks. However, TPUs also have certain limitations. They are less flexible for tasks outside of neural network inference, making them less versatile compared to general-purpose processors. TPUs also rely heavily on specific software ecosystems, such as TensorFlow, which can limit their usability with other software platforms. Moreover, TPUs come with higher upfront costs and complexity compared to general-purpose processors, which can be a barrier to adoption for some users.

14.3.8 Eyeriss v2 Architecture

Eyeriss v2, shown in Fig. 14.11, is a successor to the Eyeriss architecture [22] implemented as an ASIC. It is tailored for efficient processing of compact and sparse deep neural networks (DNNs) and was published in 2019 by a research group at MIT [23]. It addresses the challenges posed by varying layer shapes and sizes and the demand for energy-efficient hardware in mobile devices. The architecture details are given below:

1. Hierarchical mesh network (HM-NoC): A key feature of Eyeriss v2, HM-NoC adapts to different bandwidth and data reuse requirements. It efficiently handles the dataflow across the processor, enhancing throughput and energy efficiency.
2. Processing element (PE) design: The architecture includes PEs tailored for processing sparse DNNs. It utilizes compressed sparse column (CSC) format for data storage and movement, reducing storage costs and improving energy efficiency.

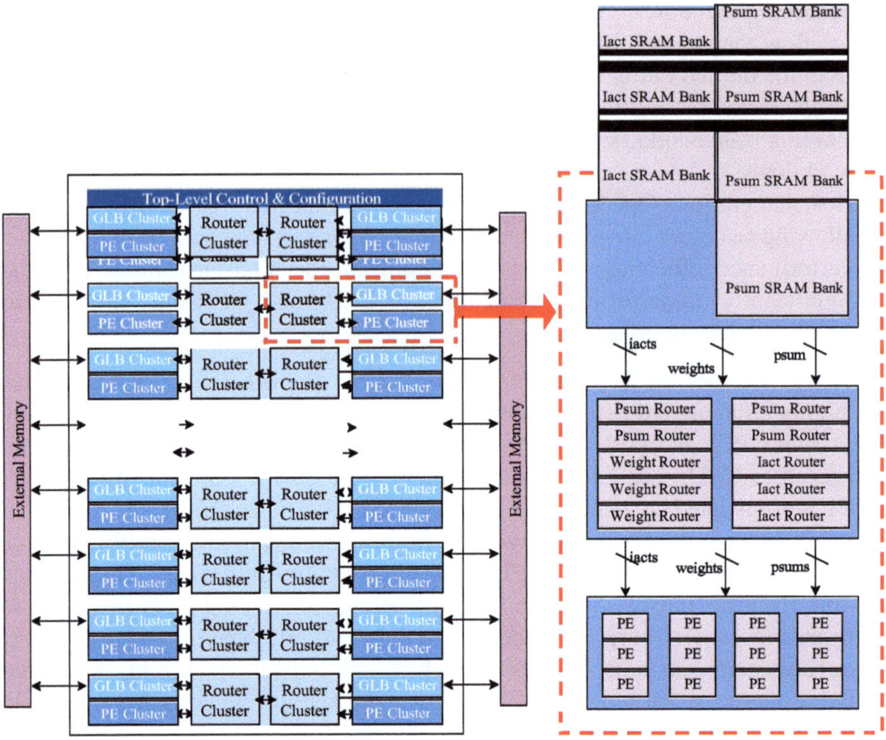

Fig. 14.11 Eyeriss v2 top-level architecture. (The figure is recreated from [23])

3. Row-stationary (RS) dataflow: Eyeriss v2 adopts RS dataflow, optimizing data reuse and minimizing data movement. This approach is particularly effective for the compact DNN layers found in modern mobile-oriented networks.
4. Run-length coding (RLC) in Eyeriss v2: RLC, a compression technique, is not explicitly mentioned in Eyeriss v2's architecture. However, the use of CSC for data compression can be seen as a related approach to handling sparse data efficiently. This methodology plays a crucial role in optimizing data storage and transfers in sparse DNNs, directly impacting the overall performance and energy efficiency of Eyeriss v2.

Eyeriss v2 offers several advantages in deep neural networks (DNNs), particularly for mobile devices. It excels in energy-efficient processing, a crucial feature for battery-powered mobile applications. The architecture is specifically designed to handle the complexities and challenges associated with compact and sparse DNNs, which are becoming increasingly common in mobile technology. Additionally, Eyeriss v2's hierarchical mesh network-on-chip (HM-NoC) provides flexible data handling, ensuring high throughput and efficiency across various DNN structures. However, Eyeriss v2 also presents certain limitations. The advanced architecture, particularly the HM-NoC, introduces significant design and implementation com-

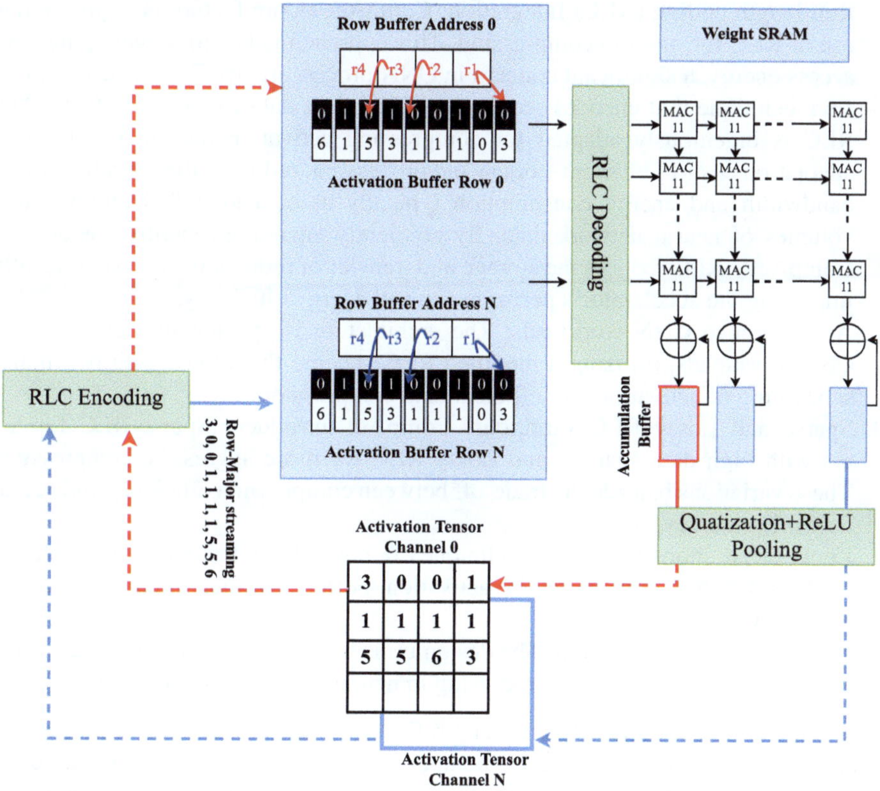

Fig. 14.12 Illustration of the CompAct architecture. The RLC encoding and decoding blocks ensure that activations are always stored on-chip in a compressed form. (The figure is recreated from [21])

plexities. Furthermore, being tailored for compact and sparse DNNs, Eyeriss v2 may have limited applicability for other types of neural networks or general-purpose computing tasks.

14.3.9 RLC Compressed Form Architecture (CompAct)

CompAct, implemented as ASIC, represents a breakthrough in CNN accelerator architecture, focusing on efficient activation compression to reduce power consumption [21]. It is a systolic array-based accelerator, particularly tailored for convolutional neural networks (CNNs), with an emphasis on minimizing the energy costs associated with high-frequency memory accesses. The architecture is illustrated in Fig. 14.12. The RLCcompressed form architecture has the following features:

1. Run-length coding (RLC) Integration: CompAct's core feature is its innovative use of RLC for on-chip compression. This approach is key to reducing memory access energy, a significant concern in CNN processing. RLC is a data compression technique that encodes sequences of identical data elements. In CompAct, RLC is ingeniously adapted for compressing activation data in CNNs. This adaptation ensures efficient storage and processing, reducing the overall memory bandwidth and energy consumption typically associated with handling large volumes of neural network data. By efficiently encoding repetitive sequences, CompAct's RLC reduces the storage and transfer of redundant data, significantly impacting the accelerator's performance and energy efficiency.
2. Adaptation to CNN workloads: The architecture is specifically optimized for CNN workloads, utilizing a modified RLC scheme that aligns with row-major scheduling of activations for efficient data compression.
3. Sparse and Lossy RLC variations: CompAct introduces Sparse-RLC for lay ers with high data sparsity and Lossy RLC for more aggressive compression. These variations balance the trade-off between compression efficiency and neural network accuracy.
4. Look-ahead snoozing (LAS) technique: A novel LAS technique is employed to decrease leakage energy in activation buffers, further enhancing energy efficiency.
5. Systolic array-based design: The design utilizes a systolic array structure, which is advantageous for parallel processing in neural network computations.

CompAct's integration of Run-Length Coding (RLC) for on-chip compression addresses the specific challenge of memory access energy in convolutional neural networks (CNNs). This architecture stands out due to its ability to switch between Sparse and Lossy RLC, offering a customizable approach to balance compression and accuracy. CompAct significantly enhances energy efficiency by reducing the activation buffer energy, leading to lower total chip energy consumption. Its tailored RLC scheme effectively minimizes the memory footprint, enhancing performance without sacrificing accuracy. Furthermore, the architecture's adaptability to different CNN workloads makes it versatile and efficient in various application scenarios. However, the sophisticated integration of RLC and its variations introduces complexity in design and implementation. Additionally, especially with Lossy RLC, there is a potential trade-off between compression efficiency and the accuracy of neural network outputs. In summary, CompAct is a sophisticated and efficient accelerator architecture for CNNs, combining innovative compression techniques and energy-saving strategies to address the specific challenges of neural network computations.

14.4 Design Considerations and Optimization Technique

14.4.1 Design Considerations

Designing an AI accelerator involves various considerations, shaped by the unique demands of neural network computations and the specific strengths of different architectures:

- *Efficiency in data movement*: Efficiency in data movement is a pivotal consideration in designing neural network accelerators, as seen in the architectures of PRIME, Neurocube, Eyeriss v2, and TPU. These architectures demonstrate innovative approaches to minimize data transfers, a significant bottleneck in performance and energy consumption. For instance, PRIME's processing-in-memory architecture, leveraging ReRAM crossbar arrays, integrates computation within memory units, significantly reducing data movement. Similarly, the TPU's systolic array design enables efficient dataflow through its matrix multiply units, minimizing the need for frequent memory access. In practical terms, AI accelerator design should focus on minimizing memory access. This involves optimizing data paths and storage within the architecture to ensure that data is processed as close to its storage location as possible, thereby reducing the energy and time costs of data movement. The goal is to create a balance between processing power and data accessibility, tailoring the architecture to handle the specific demands of neural network computations efficiently.
- *Adaptability to network sparsity*: Adaptability to network sparsity is a critical design consideration for neural network accelerators, as sparsity in neural networks can significantly impact computational efficiency and energy consumption. Efficient processing of sparse neural networks requires architectures that can dynamically adapt to varying levels of sparsity. This involves employing techniques such as sparse matrix operations, which can skip over zero-valued elements to reduce unnecessary computations, and incorporating dynamically configurable units that activate only for relevant computations. For instance, the architecture of Eyeriss v2, designed to handle compact and sparse DNNs, exemplifies this approach. It adopts strategies that optimize for sparsity, such as compressed data formats and selective processing, to improve energy efficiency and performance. Designers can learn from these architectures to create systems that are more adept at handling the unique challenges posed by sparse neural networks.
- *Energy consumption*: Energy efficiency is paramount in neural network accelerators, especially for applications in embedded systems. Beyond quantization and dynamic voltage and frequency scaling (DVFS), other techniques like low-power memory design are crucial. This involves using energy-efficient memory types, such as SRAM or nonvolatile memory, which consume less power than traditional DRAM. Clock gating and power gating techniques can also be employed to reduce power usage in idle circuitry. Additionally, optimizing the

memory hierarchy to reduce access to higher power-consuming memory units, and employing approximate computing methods where high precision is not required, can further lower energy consumption. These strategies not only reduce the power usage during intensive computations but also minimize the overall energy footprint of the hardware, making the accelerators more suitable for energy-constrained environments.

- *Scalability*: Scalability in neural network accelerators is critical to adapt to the diverse requirements of different neural network models. This extends beyond simple modular design to encompass a range of techniques aimed at ensuring the accelerator can handle networks of varying sizes and complexities. Adaptive precision, where the computational precision is adjusted based on the requirement of the task, is one approach. This not only aids in resource management but also in energy efficiency. Scalability also involves employing advanced interconnect architectures that can efficiently manage the increased dataflow in larger networks. These architectures should support both vertical scaling, where the capabilities of an individual unit are enhanced, and horizontal scaling, where more units are added to increase throughput. Another aspect is the use of reconfigurable computing elements, like FPGAs, which allow the accelerator to adapt to different network topologies. Additionally, the integration of heterogeneous computing elements, combining CPUs, GPUs, and custom ASICs, can provide the flexibility to optimize performance for various types of neural network computations. Lastly, scalability must also consider the software stack, ensuring that the accelerator's hardware is fully utilized through efficient software frameworks and libraries that can adapt to the changing hardware configurations.
- *Hardware-software synergy*: Optimal performance in neural network accelerators hinges on the deep integration between hardware and software. This goes beyond using specialized hardware instructions or adapting hardware to specific algorithms. It involves developing software that can dynamically interact with hardware features, including real-time reconfigurability and adaptability to various computational loads. Advanced machine-learning techniques, such as hardware-aware neural architecture search (NAS), play a pivotal role in this co-optimization process, allowing for the simultaneous design of both hardware and software to maximize performance and efficiency. Additionally, the synergy should consider the use of compiler optimizations that can translate high-level neural network models into efficient hardware-specific instructions, bridging the gap between algorithmic design and hardware execution. This also includes the development of robust APIs and SDKs that provide an abstraction layer, enabling developers to efficiently utilize hardware resources without needing deep hardware expertise. Furthermore, embracing emerging technologies like in-memory computing and neuromorphic computing in the hardware design can open new avenues for software to exploit these novel architectures, leading to breakthroughs in processing speed and energy efficiency. This comprehensive approach ensures that the hardware is not only capable of running current

neural network models efficiently but is also future-proofed against the evolving landscape of machine-learning algorithms.
- *Processing parallelism*: Handling the vast parallel computations in neural networks demands architectures capable of concurrent processing. This extends beyond mere multiple-processing elements working in unison. One example is the use of systolic arrays, as seen in Google's TPU, where data flows across a grid of processors, enabling highly efficient matrix operations. Another approach is the employment of SIMD (single instruction, multiple data) architectures, commonly found in GPUs, which are adept at handling vectorized operations. Additionally, techniques like pipelining can be leveraged, where different stages of a computation are overlapped, and data parallelism, where the same operation is performed on different data sets in parallel. More advanced techniques include spatiotemporal parallelism, which combines both spatial and temporal aspects to maximize resource utilization. This approach can be particularly effective in recurrent neural network (RNN) computations, where both spatial and temporal data dependencies are prevalent. Moreover, emerging technologies like neuromorphic computing, which mimics the parallel processing capabilities of the human brain, present new frontiers in parallel processing architecture. These technologies offer the potential for massively parallel processing capabilities, far exceeding traditional computing architectures.

In summary, AI accelerator design must be evaluated for the specific needs of their applications to choose the most appropriate form of design trade-offs. By understanding and leveraging these various architectural features, designers can significantly enhance the computational efficiency of AI accelerators.

14.4.2 Optimization Techniques

The explosive growth of deep learning applications has fueled the demand for efficient hardware accelerators. Traditional processors struggle with the immense computational demands of deep learning algorithms, leading to high energy consumption and long latency. Hardware accelerators offer a promising solution by exploiting the inherent parallelism and data-intensive nature of these algorithms. However, designing efficient hardware accelerators for deep learning requires careful consideration of various optimization techniques. This section delves into these techniques, exploring their methodologies and impact on accelerator performance.

14.4.2.1 Architectural Optimization Techniques

1. *Dataflow optimization*:
    - Data reuse: Deep learning models often exhibit data reuse patterns, where the same data is processed multiple times. Optimizing dataflow by exploiting

these patterns can significantly reduce memory bandwidth requirements and improve performance. Techniques such as data reuse buffers and loop transformations can effectively manage data movement and minimize redundant transfers.

- Pipelining: Pipelining breaks down complex computations into smaller, independent stages that can be executed concurrently. This allows for efficient utilization of hardware resources and improves throughput. By carefully designing the pipeline stages and balancing their workload, accelerators can achieve high performance with minimal latency.
- Memory hierarchy design: Optimizing the memory hierarchy plays a crucial role in ensuring efficient data access. A well-designed memory hierarchy utilizes different memory types with varying access speeds and capacities to effectively manage the trade-off between performance and cost. Techniques like caching and prefetching can further improve data access latency and reduce memory bandwidth consumption.

2. *Computational optimization*

- Quantization: Reducing the precision of weights and activations from 32-bit floating-point to lower precision formats like 8-bit integers can significantly improve performance and reduce memory footprint. Quantization techniques like fixed-point arithmetic and custom quantization schemes can achieve significant performance gains while maintaining acceptable accuracy.
- Pruning: Pruning removes redundant or insignificant connections from the neural network, leading to a smaller model with minimal accuracy loss. This can significantly reduce computational complexity and memory requirements, which is particularly beneficial for resource-constrained environments.
- Approximate computing: Approximate computing techniques trade-off some accuracy for significant gains in performance and energy efficiency. Techniques like stochastic computing and voltage scaling can be utilized to approximate computations in a controlled manner, leading to faster and more energy-efficient accelerators.

3. *Hardware Codesign*

- Algorithmic-hardware codesign: This approach involves jointly optimizing the neural network architecture and the hardware design to achieve the best performance and efficiency. Techniques like exploiting sparsity in the network weights and designing custom computing units for specific operations can lead to significant performance gains.
- Near-memory computing: By placing processing units closer to the memory, data movement overhead can be significantly reduced, leading to improved energy efficiency and performance. Technologies like memristors and in-memory computing offer promising avenues for near-memory processing in deep learning accelerators.

14.4.2.2 Emerging Technologies

Neuromorphic computing: Neuromorphic chips mimic the architecture and function of the human brain, offering a potential path to highly efficient deep- learning hardware. While still in its early stages, neuromorphic computing holds immense promise for future generations of deep learning accelerators.

– Quantum computing: Quantum computers leverage the principles of quantum mechanics to perform computations that are impossible for classical computers. While still under development, quantum computing has the potential to revolutionize deep learning by enabling significantly faster and more efficient training and inference.

14.4.2.3 Impact of Optimization Techniques

Optimizing hardware accelerators for deep learning using the techniques mentioned above leads to several significant benefits:

- Increased performance: Optimization techniques can substantially improve the performance of deep learning accelerators, leading to faster inference and training times. This allows for running complex models on resource-constrained devices and real-time applications.
- Reduced energy consumption: Optimization techniques can significantly reduce the energy consumption of deep learning accelerators, making them more sustainable and environmentally friendly. This is particularly crucial for battery-powered devices and large-scale data centers.
- Improved cost-effectiveness: By reducing hardware complexity and resource requirements, optimization techniques can lead to more cost-effective deep learning accelerators. This makes them more accessible to a wider range of users and applications.
- Enhanced accuracy: Although some optimization techniques may involve sacrificing some accuracy, others can even lead to improved accuracy by better utilizing hardware resources and exploiting specific properties of deep learning models.

The field of hardware accelerator design for deep learning is rapidly evolving, with new optimization techniques constantly emerging. By leveraging these techniques effectively, engineers can design efficient accelerators that deliver high performance, low energy consumption, and improved cost-effectiveness. As deep learning applications continue to grow in complexity and demand, the development and optimization of hardware accelerators will play a critical role in enabling their widespread adoption across diverse domains.

14.5 Applications and Future

Hardware accelerators are used in various applications where deep learning models play a critical role. These applications include the following.

14.5.1 Revolutionizing Industries

The diverse applications of hardware accelerators for deep learning models have revolutionized countless industries, but their complex computational needs pose significant challenges. Hardware accelerators, specialized hardware designed to efficiently execute these models, are emerging as the key to unlocking their full potential. By dramatically reducing training times and inference speeds, they enable faster innovation and deployment across diverse fields.

1. *Computer Vision*
   - Image and video recognition: From facial recognition in security systems to automatic image tagging on social media platforms, hardware accelerators power the real-time analysis of vast amounts of visual data.
   - Object Detection: Self-driving cars rely on hardware accelerators for real-time object detection on the road, ensuring safe navigation. Similarly, retail stores utilize these tools for inventory management and theft prevention.
   - Autonomous Vehicles: The development of autonomous vehicles hinges on the efficient processing of sensor data. Hardware accelerators enable real-time obstacle detection and path planning, making self-driving cars a reality.
2. *Natural Language Processing*
   - Machine Translation: Breaking down language barriers is becoming easier with hardware accelerators. These accelerators power real-time translation tools, facilitating communication across cultures and languages.
   - Sentiment analysis: Businesses rely on sentiment analysis to understand customer opinions and improve their products and services. Hardware accelerators accelerate this process, allowing for faster and more accurate insights.
   - Chatbot development: Chatbots are taking over customer service and online interactions. Hardware accelerators enable natural and engaging conversations by powering the complex algorithms behind these virtual assistants.
3. *Speech Recognition*
   - Voice assistants: From Siri to Alexa, voice assistants are changing the way we interact with technology. Hardware accelerators power these tools, enabling natural language understanding and accurate response generation.

- Smart speakers: Smart speakers are revolutionizing the way we listen to music and access information. Hardware accelerators make these devices responsive and intelligent, providing users with a seamless experience.
- Transcription services: Transcribing audio and video content is becoming easier and more efficient with hardware acceleration. This technology is transforming workflows in various industries, including education, media, and law.

4. *Robotics*

- Navigation: Robots rely on accurate navigation to operate in complex environments. Hardware accelerators enable them to perceive their surroundings and plan safe and efficient routes.
- Object manipulation: Robotic arms are becoming increasingly sophisticated, performing delicate tasks in manufacturing and healthcare. Hardware accelerators ensure the precise and controlled movement of these robots.
- Decision-making: Robots are being tasked with making complex decisions in autonomous systems. Hardware accelerators allow them to process information quickly and accurately, leading to better decision-making capabilities.

5. *Healthcare*

- Medical image analysis: Hardware accelerators enable the rapid analysis of medical images, such as X-rays and MRIs. This leads to faster diagnosis, more accurate treatment plans, and improved patient outcomes.
- Drug discovery: The search for new drugs is a complex and time-consuming process. Hardware accelerators can analyze vast amounts of data, leading to faster development of life-saving medications.
- Personalized medicine: Tailoring healthcare to individual patients is becoming a reality with hardware-accelerated deep learning. This approach leads to more effective and personalized treatment plans.

6. *Finance*

- Fraud detection: The financial industry relies on robust fraud detection systems to protect against financial crimes. Hardware accelerators enable real-time analysis of financial transactions, preventing fraudulent activities.
- Risk assessment: Evaluating financial risk is crucial for investors and lenders. Hardware accelerators provide faster and more accurate risk assessments, leading to better financial decisions.
- Algorithmic trading: High-frequency trading relies on split-second decision-making. Hardware accelerators enable faster execution of trades, providing a competitive edge in the financial markets.

7. *Internet of Things (IoT)*

- Edge computing: Processing data directly at the edge of the network is crucial for many IoT applications. Hardware accelerators enable real-time analysis of sensor data, leading to faster response times and improved decision-making.

- Sensor data analysis: The vast amount of data generated by IoT devices needs to be processed efficiently. Hardware accelerators enable real-time analysis of this data, unlocking valuable insights and enabling smarter devices.
- Anomaly detection: Identifying unusual patterns in sensor data is critical for various applications, such as predictive maintenance and fault detection. Hardware accelerators power these anomaly detection algorithms, ensuring timely identification and response to potential problems.

14.5.2 The Future of Hardware Accelerator

The field of hardware acceleration for deep learning is experiencing a thrilling transformation, with innovative technologies and applications emerging rapidly. This exciting landscape is driven by several key trends:

1. *Orchestrating heterogeneity*: Instead of relying on a single dominant technology, the future lies in leveraging the strengths of diverse hardware approaches. This involves combining different types of accelerators, such as GPUs, TPUs, FPGAs, and ASICs, to achieve optimal performance and energy efficiency for specific tasks.
2. *Tailoring for unmatched performance*: The future moves away from a one-size-fits-all approach toward domain-specific optimization. This involves designing accelerators specifically for certain deep learning models or applications through hardware-specific algorithms, model compression techniques like pruning and quantization, and hardware-aware training.
3. *Mimicking the brain's efficiency*: Neuromorphic computing, inspired by the brain's unparalleled processing capabilities, aims to create hardware systems that mimic its architecture. Utilizing components like memristors, these systems promise ultra-low-power consumption, high parallelism, and adaptability, opening doors for future hardware acceleration advancements.
4. *Democratizing access*: To broaden the accessibility of hardware acceleration, seamless integration with software tools and frameworks is crucial. This includes developing high-level programming interfaces, automated optimization tools, and open-source platforms to empower a wider range of developers.
5. *Collaborative innovation*: Shaping the future of hardware acceleration necessitates a collaborative effort from diverse stakeholders. Hardware vendors must continuously refine their technologies to address evolving needs, while software developers create user-friendly tools and frameworks to unlock hardware acceleration's full potential. Researchers, meanwhile, push performance boundaries by exploring new frontiers in hardware architecture and algorithms. Finally, end-users provide invaluable feedback and drive demand for innovative solutions across various industries. By fostering this collaborative spirit and embracing emerging trends, we can unlock a future of unprecedented innovation and progress in hardware acceleration for deep learning. This will pave the way for

new possibilities across diverse domains, propelling artificial intelligence to new heights.

14.6 Conclusion

In this chapter, a comprehensive exploration is undertaken into the evolution and significance of specialized hardware designed to optimize artificial intelligence (AI) processes. It discusses the transformative shift from conventional computing paradigms to the adoption of AI-specific hardware such as GPUs, FPGAs, and ASICs, emphasizing their pivotal roles in augmenting the efficiency and performance of AI tasks. This chapter also encapsulates the challenges encountered in the development of these accelerators, their influence on advancing AI capabilities, and the anticipation of future innovations in this domain. By weaving together technical insights, practical examples, and forward-looking perspectives, this chapter aims to furnish a nuanced understanding of the current state and emerging trends in AI hardware development, catering to a diverse audience ranging from industry practitioners to academic researchers.

References

1. Sapunov, G. *TPU v2 was unveiled at Google I/O in May 2017*.https://moocaholic.medium.com/hardware-for-deep-learning-part-4-asic-96a542fe6a813948
2. NVIDIA. *NVIDIA Tesla P100: The World's First AI Supercomputing Data Center GPU*. https://www.nvidia.com/en-us/data-center/tesla-p100/
3. Shankland, S. *Nvidia H100 Hopper chip*. https://www.cnet.com/pictures/see-nvidias-h100-hopper-chip-up-close/4/
4. Intel. *Intel Advances Neuromorphic with Loihi 2, New Lava Software Framework and New Partners*. https://www.intel.com/content/www/us/en/newsroom/news/intel-unveils-neuromorphic-loihi-2-lava-software.htmlgs.4dc193
5. *ALINX AX7Z020: SoC FPGA Development Board*. https://www.xilinx.com/products/boards-and-kits/1-1bu8447.html
6. Akopyan, F., et al. (2015). TrueNorth: Design and tool flow of a 65 mW 1 million neu- ron programmable neurosynaptic chip. *IEEE Transactions on Computer-Aided Design of Integrated Circuits and Systems, 34*(10), 1537–1557. https://doi.org/10.1109/TCAD.2015.2474396
7. Modha, D. S., et al. (2023). Neural inference at the frontier of energy, space, and time. *Science, 382*, 329–335. https://doi.org/10.1126/science.adh1174
8. Zahoor, F., Hussin, F. A., Isyaku, U. B., et al. (2023). Resistive random access memory: Introduction to device mechanism, materials and application to neuromorphic computing. *Discover Nano, 18*, 36. https://doi.org/10.1186/s11671-023-03775-y
9. Le Gallo, M., & Sebastian, A. (2020). An overview of phase-change memory device physics. *Journal of Physics D: Applied Physics, 53*, 213002. https://doi.org/10.1088/1361-6463/ab7794
10. Chang, T.-C., Chang, K.-C., Tsai, T.-M., Chu, T.-J., & Sze, S. M. (2016). Resistance random access memory. *Materials Today, 19*(5), 254–264., ISSN 1369-7021. https://doi.org/10.1016/j.mattod.2015.11.009
11. Capra, M., Bussolino, B., Marchisio, A., Shafique, M., Masera, G., & Martina, M. (2020). An updated survey of efficient hardware architectures for accelerating deep convolutional neural net- works. *Future Internet, 12*, 113. https://doi.org/10.3390/fi12070113

12. Nurvitadhi, E., Sheffield, D., Sim, J., Mishra, A., Venkatesh, G., & Marr, D. (2016). Accelerating Binarized Neural Networks: Comparison of FPGA, CPU, GPU, and ASIC. In *2016 International Conference on Field-Programmable Technology (FPT)* (pp. 77–84). https://doi.org/10.1109/FPT.2016.7929192. keywords: Neurons;Random access memory;Biological neural networks;Field programmable gate arrays;Graphics processing units;Hardware;System-on-chip;Deep learning;binarized neural networks;FPGA;CPU;GPU;ASIC;data analytics;hardware accelerator.
13. Esmaeilzadeh, H., Sampson, A., Ceze, L., & Burger, D. (2012). Neural acceleration for general-purpose approximate programs. In *2012 45th Annual IEEE/ACM international symposium on microarchitecture* (pp. 449–460). https://doi.org/10.1109/MI-CRO.2012.48. keywords: Approximate Computing;Neural Networks;Accelerator;Neural Pro- cessing Unit;NPU.
14. Farabet, C., Martini, B., Corda, B., Akselrod, P., Culurciello, E., & Le-Cun, Y. (2011). NeuFlow: A runtime reconfigurable dataflow processor for vision. In *CVPR 2011 workshops* (pp. 109–116). https://doi.org/10.1109/CVPRW.2011.5981829. keywords: Tiles;Computer architecture;Runtime;Field pro- grammable gate arrays;Hardware;Convolvers;Feature extraction.
15. Liu, X., et al. (2015). RENO: A high-efficient reconfigurable neuromorphic computing accelerator design. In *2015 52nd ACM/EDAC/IEEE design automation conference (DAC)* (pp. 1–6). https://doi.org/10.1145/2744769.2744900. keywords: Active appearance model;Artificial neural networks;Training;Arrays;Routing;Accuracy;Memristors.
16. Chen, T., Zidong, D., Sun, N., Wang, J., Wu, C., Chen, Y., & Temam, O. (2014). DianNao: A small-footprint high-throughput accelerator for ubiqui-tous machine-learning. *ACM SIGARCH Computer Architecture News, 42*(1), 269–284. https://doi.org/10.1145/2654822.2541967
17. Du, Z., et al. (2015). ShiDianNao: Shifting vision processing closer to the sensor. In *2015 ACM/IEEE 42nd Annual international symposium on computer architecture (ISCA)* (pp. 92–104). https://doi.org/10.1145/2749469.2750389. keywords: Kernel;Neural networks;Sensors;Energy efficiency;Smart phones;Filtering;Neurons.
18. Liu, D., Chen, T., Liu, S., Zhou, J., Zhou, S., Te-man, O., Feng, X., Zhou, X., & Chen, Y. (2015). PuDianNao: A polyvalent machine learning accelerator. *SIGPLAN Not, 50*(4), 369–381. https://doi.org/10.1145/2775054.2694358
19. Jouppi, N. P., Young, C., Patil, N., Patterson, D., Agrawal, G., Bajwa, R., Bates, S., Bhatia, S., Boden, N., Borchers, A., Boyle, R., Cantin, P., Chao, C., Clark, C., Coriell, J., Daley, M., Dau, M., Dean, J., Gelb, B., & Yoon, D. H. (2017). In-datacenter performance analysis of a tensor processing unit. In *Proceedings of the 44th annual international symposium on computer architecture. ISCA '17: The 44th annual international symposium on computer architecture.* ACM. https://doi.org/10.1145/3079856.3080246
20. Kim, D., Kung, J., Chai, S., Yalamanchili, S., & Mukhopadhyay, S. (2016). Neurocube: A programmable digital neuromorphic architecture with high-density 3D memory. In *2016 ACM/IEEE 43rd annual international symposium on computer architecture (ISCA)* (pp. 380–392). https://doi.org/10.1109/ISCA.2016.41. keywords: Neurons;Random access memory;Computer architecture;Three-dimensional displays;Biological neural net- works;Artificial neural networks;Neural nets;Neurocomputers;Neuromorphic computing.
21. Zhang, J. (J.), Raj, P., Zarar, S., Ambardekar, A., & Garg, S. (2019). CompAct: On-chip compression of activations for low power systolic array based CNN acceleration. *ACM Transactions on Embedded Computing Systems (TECS), 18*(5s) Article 47, 1–24. https://doi.org/10.1145/3358178
22. Chen, Y. -H., Krishna, T., Emer, J. S., & Sze, V. (2017). Eyeriss: An energy-efficient reconfigurable accelerator for deep convolutional neural networks. *IEEE Journal of Solid- State Circuits, 52*(1), 127–138. https://doi.org/10.1109/JSSC.2016.2616357. keywords: Shape;Random access memory;Computer architecture;Throughput;Clocks;Neural networks;Hardware;Convolutional neural networks (CNNs);dataflow processing;deep learning;energy-efficient accelerators;spatial architecture.
23. Chen, Y. -H., Yang, T. -J., Emer, J., & Sze, V. (2019). Eyeriss v2: A flexible accelerator for emerging deep neural networks on mobile devices. *IEEE Journal on Emerging and Selected Topics in Circuits and Systems, 9*(2), 292–308. https://doi.org/10.1109/JET-

CAS.2019.2910232. keywords: Hardware;Shape;Arrays;Parallel processing;Mobile handsets;Bandwidth;Deep neural network accelerators;deep learning;energy-efficient accelerators;dataflow processing;spatial architecture.
24. Chi, P., et al. (2016). PRIME: A Novel Processing-in-Memory Architecture for Neural Net- work Computation in ReRAM-Based Main Memory. In *2016 ACM/IEEE 43rd Annual In- ternational Symposium on Computer Architecture (ISCA)* (pp. 27–39). https://doi.org/10.1109/ISCA.2016.13. keywords: Artificial neural networks;Random access memory;Microprocessors;Acceleration;Biological neural networks;Memory management;processing in memory;neural network;resistive random access memory.
25. Dong, X., Xu, C., Xie, Y., & Jouppi, N. P. (2012). NVSim: A circuit-level performance, energy, and area model for emerging nonvolatile memory. In *IEEE transactions on computer-aided design of integrated circuits and systems* (Vol. 31, pp. 994–1007). https://doi.org/10.1109/TCAD.2012.2185930. keywords: Nonvolatile memory;Arrays;Phase change random access memory;Wires;Distributed databases;Integrated circuit modeling;Analytical circuit model;MRAM;NAND Flash;nonvolatile memory;phase-change random-access memory (PCRAM);resistive random-access memory (ReRAM);spin-torque-transfer memory (STT-RAM).
26. Peng, X., Huang, S., Luo, Y., Sun, X., & Yu, S. (2019). DNN+NeuroSim: An end-to-end benchmark- ing framework for compute-in-memory accelerators with versatile device technologies. In *2019 IEEE International Electron Devices Meeting (IEDM)* (pp. 32.5.1–32.5.4). https://doi.org/10.1109/IEDM19573.2019.8993491
27. Pentecost, L., et al. (2022). NVMExplorer: A framework for cross-stack comparisons of embedded non-volatile memories. In *IEEE international symposium on high-performance computer architecture (HPCA)*. IEEE. https://doi.org/10.1109/hpca53966.2022.00073
28. Agrawal, A. (2021). *Compute-in-memory primitives for energy- efficient machine learning* [Thesis, Purdue University Graduate School]. https://doi.org/10.25394/PGS.15048825.v1.
29. Seshadri, V., et al. (2017). Ambit: In-Memory accelerator for bulk bitwise operations using commodity DRAM technology. In *2017 50th annual IEEE/ACM international symposium on microarchitecture (MICRO)* (pp. 273–287).

Chapter 15
Implementation of Associative Learning Using Cognitive-Inspired Robotic System

Tianze Liu, Noah Zins, Yan Zhang, and Hongyu An

15.1 Introduction

Deep neural networks (DNNs) have revolutionized artificial intelligence (AI), achieving remarkable success across various cognitive tasks through extensive training on large datasets [1]. These networks refine their predictions by minimizing errors via backpropagation, necessitating vast computational resources and substantial power consumption [2–6]. However, this dependence on large datasets and high power consumption presents significant challenges for autonomous operations, especially in environments with stringent Size, Weight, and Power (SWaP) constraints, such as planetary robotics [1, 2]. Planetary rovers, for example, must operate autonomously with limited energy and without human intervention [2].

To overcome these limitations, our research focuses on enhancing the autonomous capabilities of intelligent robots by emulating the associative learning processes observed in animals, mainly through the use of neuromorphic systems. However, applying neuromorphic systems to robotics presents its own set of challenges. These include the design of efficient learning algorithms, integrating neuromorphic hardware with robotic systems, and adapting animal learning behaviors to the robot's environment. Inspired by the brain's structure and function, Neuromorphic systems offer a more energy-efficient approach to artificial intelligence. Associative learning, a natural self-learning mechanism in animals, enables them to adapt to their environment by forming connections between

T. Liu · N. Zins · H. An (✉)
Department of Electrical and Computer Engineering, Michigan Technological University, Houghton, MI, USA
e-mail: tianzel@mtu.edu; nwzins@mtu.edu; hongyua@mtu.edu

Y. Zhang (✉)
Department of Biological Sciences, Michigan Technological University, Houghton, MI, USA
e-mail: yzhang49@mtu.edu

A. Iranmanesh, H. Sayadi (eds.), *AI-Enabled Electronic Circuit and System Design*,
https://doi.org/10.1007/978-3-031-71436-8_15

concurrent events. This capability can be adapted for robots, allowing them to link information and experiences and adjust their behavior in dynamic environments like Mars.

Our study is structured into two main parts. The first part explores the theoretical aspects of associative learning, providing a comprehensive understanding of this natural self-learning mechanism in animals. The second part applies these principles to robotics, employing a large-scale associative learning system. This system, powered by Intel's Loihi neuromorphic chip, offers a more energy-efficient approach to artificial intelligence. It allows a mobile robot to learn associations between light and vibration signals as conditional and unconditional stimuli, mimicking the fear conditioning process observed in animals. We aim to demonstrate a significant advancement in autonomous robotics by implementing this system in a real-world scenario.

This research marks a significant milestone as the first real-world application of a cognitive-inspired associative learning paradigm using a mobile robot. By integrating the Loihi chip, we enhance signal processing speed and energy efficiency and herald a new era for autonomous robotics. In our experiment, the robot learns to associate light stimuli with vibration stimuli, mimicking fear conditioning processes observed in animals. The synaptic weights are adjusted using Hebbian learning principles, enabling the robot to adapt its responses based on the learned associations.

The key contributions of this chapter include the following:

1. Successfully demonstrate a novel associative learning paradigm in a real-world setting with a mobile robot.
2. Implementing a neural assembly designed for decision-making and applying it to the associative learning process within the robot.

Our findings demonstrate the vast potential of neuromorphic systems to replicate animal learning behaviors in robots and mark a significant leap forward in the field of autonomous robotics. This research paves the way for robots to adapt and learn from their environment, opening up new possibilities for their applications.

15.2 Background

By examining and mimicking the biological processes that facilitate associative learning in animals, we have developed an innovative self-learning paradigm for neuromorphic systems. This section explores recent advancements in neuromorphic computing and hardware and analyzes the mechanisms that support associative learning, from cellular functions to behavioral responses.

15.2.1 Neuromorphic Computing

Our research has pioneered a transformative advancement in neuromorphic systems, introducing a self-learning paradigm that emulates the biological processes underlying animal associative learning. This section delves into our significant strides in neuromorphic computing and hardware, dissecting the mechanisms that enable associative learning from cellular activities to behavioral responses. By doing so, we offer profound insights into the future potential of neuromorphic systems.

Neuromorphic systems, drawing inspiration from the astounding efficiency of the human brain, can potentially revolutionize artificial intelligence (AI) [7–12]. The human brain, a testament to nature's ingenuity, can execute complex tasks with a mere 20 watts of power [13]. This efficiency contrasts with artificial neural networks (ANNs), which demand extensive training and large datasets. The brain's adaptability, a product of its parallel processing capabilities, high connectivity, flexible network topology, integrated data memory and computation, and spike-based information representation, underscores the transformative potential of neuromorphic systems in AI and computing, instilling hope and inspiration for the future.

The human brain comprises billions of neurons and trillions of synapses, creating a highly interconnected three-dimensional network. Each neuron can communicate with over ten thousand other neurons simultaneously. Neurons function as signal processors, integrating incoming spiking signals and transmitting new spike sequences to other neurons through synapses. The strength of these signals is modulated by synaptic connection strength, which can be dynamically adjusted through synaptic plasticity [13–15]. When presynaptic and postsynaptic neurons fire together, their connection is strengthened, a principle known as Hebbian learning [16–19].

Neuromorphic systems, with their groundbreaking architecture, aim to overcome the inefficiencies of the von Neumann architecture, which separates computing and memory units, necessitating frequent data transfers between them. In contrast, neuromorphic systems use sparse, event-driven computation, activating only the required computing resources in response to specific events. By emulating brain architecture, including neurons, synapses, and learning methods, neuromorphic computing offers a more efficient future for computers and AI [20].

These systems harness specialized chips with artificial neurons to operate spiking neural networks (SNNs), encoding information through spike sequences akin to those in biological nervous systems. Intel's Loihi chips serve as a prime illustration of this technology [21, 22]. Unlike traditional GPUs and CPUs based on the von Neumann architecture, Loihi chips are purpose-built for neuromorphic computing and asynchronous SNNs. Unveiled in 2017, the first-generation Loihi-1 chips boast 130,000 electronic neurons and 130 million synapses across 128 neuromorphic cores, all housed within a compact 60 mm^2 chip size made possible by Intel's advanced 14 nm process [21, 22]. They employ digital leaky integrate-and-fire neurons with mesh-configured communication and configurable synapses that

Fig. 15.1 Kapoho Bay with two onboard Loihi chips

Table 15.1 Introduction to Loihi and Loihi 2 chips

Feature	Loihi 1	Loihi 2
Technology	Intel 14 nm	Intel 4 (7 nm)
Die Area	60 mm^2	31 mm^2
Max # Neurons/Chip	128,000	1 million
Max # Synapses/Chip	128 million	120 million
Neuron Model	Generalized digital LIF	Fully programmable

support weight-sharing and compression. Synaptic plasticity in Loihi-1 chips can be fine-tuned using various biologically plausible learning rules, including Hebbian learning, spike-timing-dependent plasticity (STDP), and reward-modulated rules [21, 22]. Neurons fire when accumulated spikes reach a threshold within a specific time frame, transmitting signals to connected neurons.

Loihi-1 chips are seamlessly integrated with several neuromorphic platforms, which provide interfaces to combine the chip with other computer systems or Field-Programmable Gate Array (FPGA) devices. A shining example is the Kapoho Bay platform, which vividly demonstrates the versatility and collaborative potential of Loihi-1 chips, as depicted in Fig. 15.1. This cooperative aspect invites you, our esteemed audience, to be part of the transformative journey of neuromorphic systems.

Nahuku is a 32-chip Loihi board with a standard FPGA Mezzanine Card (FMC) connector, facilitating communication with the Arria FPGA development board. Pohoiki Spring is a large-scale Loihi system equipped with 100 million neurons and designed to function as a server for remote access. The second generation of Loihi chips, Loihi-2, was introduced in late 2021 [23]. Fabricated using Intel's four processes, previously called 7 nm technology, Loihi-2 has a reduced chip area of 31 mm^2, compared to the 60 mm^2 of the first-generation Loihi chips. Loihi-2 stands out with its unique feature of offering fully programmable neuron models, a significant improvement over the fixed neuron models in the earlier generation. These neuron behaviors can be customized with microcode instructions that support basic bitwise and mathematical operations. Designed for neuromorphic computing and edge devices, Loihi-2 provides enhanced computational and energy efficiency through parallel processing. A comparison between the two generations of Loihi chips is summarized in Table 15.1.

Loihi 1 utilizes a Leaky Integrate and Fire (LIF) model to implement neurons. The LIF model is favored because it captures neural dynamics' essential information processing functions while computationally simple enough for efficient evaluation. The behavior of LIF neurons is described by the following equations [24]:

$$C_m \frac{dV_m}{dt} = G_L (E_L - V_m) + A * I_{\text{app}}, \text{ if } V_m > V_{\text{th}} \text{ then } V_m = V_{\text{reset}}, \tag{15.1}$$

$$\tau_{RC} = C_m / G_L, \tag{15.2}$$

where C_m is the membrane capacitance, G_L is the leak conductance, E_L is the leak potential, V_m is the membrane potential, A is the input signal gain, I_{app} is the input current, and τ_{RC} is the membrane RC time constant. The Integrate and Fire (IF) model is a simplified version of the Leaky Integrate and Fire (LIF) model. Suppose neurons function similarly to LIF neurons but do not have a decaying membrane potential. This simplification is achieved by setting the equation's membrane resistance-capacitance (RC) time constant to infinity.

15.2.2 Associative Learning

Associative learning in rodents, particularly rats, has been extensively studied using the T-maze paradigm, a classical experimental setup designed to investigate the cognitive processes underlying spatial navigation and memory. In this maze, rats are typically exposed to a simple T-shaped structure with distinct arms and a choice point. The arms of the maze present different spatial cues or stimuli, and the rodent learns to associate specific outcomes or rewards with particular arm choices.

During the initial phases of T-maze experiments, rodents often display exploratory behaviors, navigating through the maze without a clear preference for either arm. However, as the learning trials progress, a remarkable shift occurs. Through repeated exposure to the maze, rats begin to form associations between specific stimuli or cues in the arms and the outcomes associated with those choices. This associative learning process is notably demonstrated when the rats consistently choose one arm over the other, guided by anticipating a positive result, such as a food reward.

Furthermore, T-maze studies have revealed that rats are adaptable to modifying their choices based on changes in the spatial cues or the introduction of new stimuli. For instance, if the location of the reward is switched from one arm to the other, rats exhibit a learning curve as they adjust their navigational preferences to align with the updated reward contingencies. This flexibility in adapting to changing conditions underscores the dynamic nature of associative learning in T-maze experiments.

The neural mechanisms underlying T-maze associative learning involve intricate interactions within the rat's brain, particularly in regions associated with memory, spatial navigation, and reward processing. The T-maze paradigm continues to serve as a valuable tool for unraveling the complexities of associative learning in rodents, shedding light on the fundamental principles governing cognitive processes and behavior in mammalian brains.

15.3 Fear Conditioning with Cognitive-Inspired Robots

This section outlines the methodology and design process for simulating fear conditioning in rats using mobile robotics. Instead of the traditional electric shock as the unconditional stimulus (US) in fear conditioning experiments, we used a vibration platform to deliver vibrations. Similarly, a mounted light source replaced the buzzer tone typically used as the conditional stimulus (CS). To ensure reliable results, the Leaky Integrate and Fire (LIF) neurons were efficiently implemented in Nengo, a neuromorphic simulator developed by Applied Brain Research [25]. For several experiments, Intel's Loihi chip served as the backend for Nengo, enhancing data processing efficiency. The Robot Operating System (ROS) framework managed the cognitive-inspired robot, miming certain aspects of human cognition, ensuring smooth data transfer to and from the Nengo program [26].

15.3.1 Simulation and Preliminary Testing

The central focus of our experiment is to modify signal pathways using Hebbian learning, a crucial aspect of our research. To demonstrate essential associative learning using Nengo, we constructed a primary network. This network includes two programmatically controlled input nodes representing the CS and US. As illustrated in Fig. 15.2, each input node is connected to a specific LIF neuron that activates in response to the stimuli. These neurons are labeled as CS neurons and US neurons. A third LIF neuron, the response neuron, is also established to indicate the network's response to the stimuli.

The values used for the LIF parameters from Eqs. (15.1 and 15.2) are given for the three LIF neurons in Table 15.2.

For all LIF neurons in Nengo, the firing threshold is fixed at 1 V, and input gain is modified instead. The initial value for τ_{RC} Nengo's standard implementation of

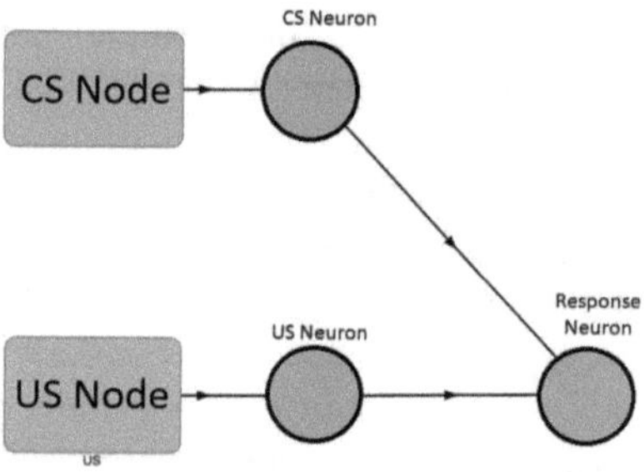

Fig. 15.2 Neural network for demonstrating Hebbian learning

Table 15.2 LIF neuron parameters for simple associative learning

Neuron types	τ_{RC}	A	V_{reset} (V)	V_{th} (V)
US neuron	0.02	1.0	0.01	1.0
CS neuron	0.02	1.0	0.01	1.0
Response neuron	0.02	1.0	0.01	1.0

spiking LIF neurons is 0.02, which was not modified as the neurons exhibited the desired behavior. V_{reset} was set at 0.01 V so the neurons fire when they receive the default programmable input stimulus without modifying the default gain of 1.0. The US neuron's output spikes are relayed to the response neuron via an unmodifiable synaptic connection modeled by the lowpass filter with the impulse response:

$$h(t) = \frac{1}{\tau} e^{\frac{-t}{\tau}} \tag{15.3}$$

The synaptic time constant, τ, is measured in seconds. We used a τ value of 0.005 seconds for all synapses, the default model in Nengo, which effectively serves our purpose. Synaptic connections have "weights," scalar values that influence the output signal, acting as a gain for the synapse output current. Inhibitory connections, which decrease the postsynaptic neuron's potential, are created by assigning a negative value to the synapse weight. The Hebbian learning rule, a fundamental principle in neuroscience, states that synapses between simultaneously active neurons will be strengthened. This principle, detailed in Eq. (15.4) of our neural network model, guides the synapses to strengthen when neurons fire together.

$$\Delta w = \eta r_i r_j, \tag{15.4}$$

where w is the synaptic weight, η is the learning rate constant, r_i is the firing rate of the presynaptic neuron, and r_j is the firing rate of the postsynaptic neuron. According to this equation, synaptic strength increases in proportion to the firing rates of both neurons when they fire simultaneously. Initially, the synaptic weight is set to 0.0001, allowing the US input signal to activate the response neuron, while the CS input signal alone does not elicit a response. The learning rate, empirically set at 2×10^{-5}, balances learning speed and network stability. A higher learning rate could cause instability, while a lower rate might lead to slow convergence. Each training cycle involves presenting the US and CS for two seconds each, with a one-second overlap, followed by one second without any stimulus, as shown in Fig. 15.3. This timing ensures sufficient exposure to the stimuli without overwhelming the network.

Our research initially demonstrates the network's response to both stimuli, followed by training cycles designed to increase the synaptic weight. After sufficient repetitions, the synapse strength rises to the point where the presentation of the CS alone triggers the response neuron, driven solely by the output of the CS neuron. This result highlights the potential of our approach, successfully demonstrating associative learning in the network.

After successfully validating the associative learning proof of concept in Nengo, we embarked on a pivotal phase of the mobile robotics experiment. Implementing a vibration platform as the unconditional stimulus (US) was significant. This was due to the inherent challenges of rapid collisions causing instability within the Gazebo simulation environment. We meticulously recorded precise vibration data to overcome this and seamlessly integrated it into the simulation. The vibration table was calibrated to generate 15 Hz vibrations with an amplitude of 1.2 mm, the

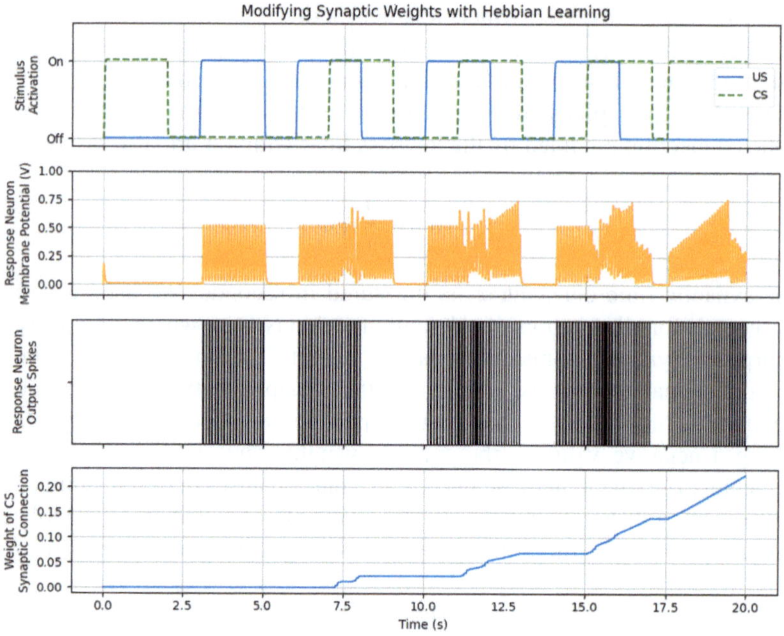

Fig. 15.3 Hebbian learning causing synaptic weight modification

lowest available setting, to ensure minimal impact on the cognitive-inspired robot. The robot's onboard IMU was instrumental in measuring these vibrations.

With Clearpath's ROS software operating on the robot, we received the IMU data and published the robot's acceleration to an IMU topic, making this data accessible to other ROS nodes. This setup allowed us to record and replay real-time vibration data and simulate real-world conditions. Nengo was executed within a ROS node that subscribed to the IMU topic, establishing a direct communication channel with Nengo. The raw acceleration data during vibration, as shown in Fig. 15.4, underwent minimal preprocessing before being integrated into the network.

This configuration was pivotal in enabling a realistic simulation of the robot's response to the unconditional stimulus. It ensured that the data utilized in Nengo accurately mirrored the conditions the robot would encounter in the real world. This meticulous approach to data handling and simulation setup was not just important but crucial for the success of the associative learning experiments. It provided a robust foundation for testing and validation, enhancing the credibility of our results.

The z-axis had an average magnitude of 9.81 m/s^2 due to Earth's gravity, which was subtracted to bring the resting acceleration for all three axes to zero. Initially, the resultant acceleration, as shown in Eq. (15.5), was used to evaluate the vibration state of the robot. However, the z-axis acceleration exhibited significant deviations from its resting value compared to the other axes, suggesting that z-axis acceleration alone was sufficient to indicate vibration. In a meticulous process, the preprocessing equation was then simplified as follows:

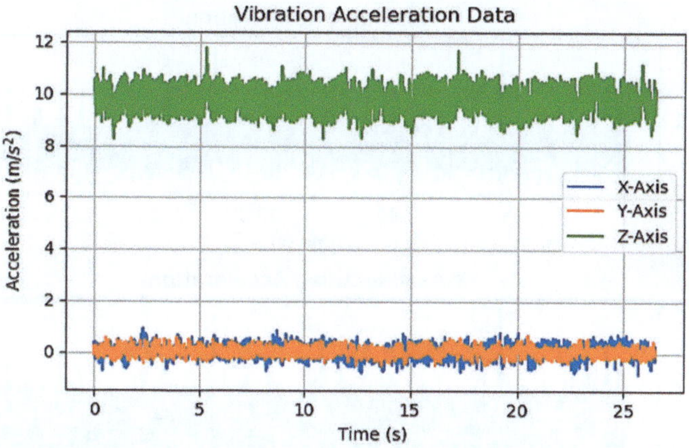

Fig. 15.4 Acceleration data from vibration table at 15 Hz

$$a_{res} = \sqrt{{a_x}^2 + {a_y}^2 + (a_z - 9.8)^2} \tag{15.5}$$

However, as shown in Fig. 15.4, the z-axis acceleration deviates significantly more from its resting value than the other two axes. This observation indicates that measuring the z-axis acceleration alone is sufficient to detect vibrations. Consequently, the preprocessing equation was simplified to the following:

$$a_r = \mid a_z - 9.8 \mid \tag{15.6}$$

This new equation is ideal as preprocessing should be minimized to take full advantage of the neuromorphic system's potential for efficiency. A comparison of a_{res} and a_r during vibration is shown in Fig. 15.5.

The rectified z-axis acceleration, which showed trends and magnitudes similar to the resultant acceleration, was suitable for vibration preprocessing. Therefore, the US input node was adjusted to receive IMU data and use the rectified z-axis acceleration as its output. This node functioned as a spike generator with a firing rate proportional to the input value, albeit with some modifications from the original design. To minimize preprocessing, raw values from the node remained unscaled and unnormalized. Instead, we carefully tuned the US neuron's input synapse and LIF parameters, now designated as the vibration detection neuron, to achieve the desired response to vibration signals. The parameters chosen for the simulation experiment, outlined in Table 15.3, were meticulously calibrated, reflecting the precision of our research methodology.

The default τ_{RC} constant of 0.02 s is maintained, as it is sufficient for the desired functionality. The other two parameters, gain (A) and bias (V_{reset}) are empirically calculated and optimized based on the experimental setup to ensure they produce the desired responses. These parameters were derived for the vibration detection neuron

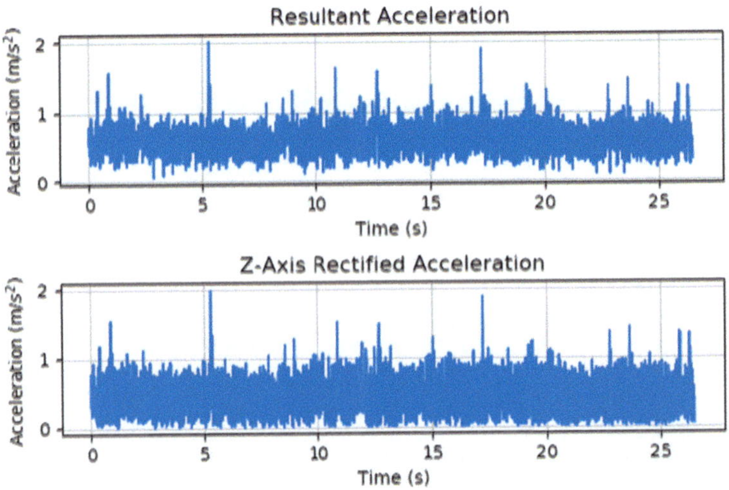

Fig. 15.5 Comparison of resultant acceleration, a_{res}, and rectified z-axis acceleration, a_r

Table 15.3 LIF neuron parameters for associative learning in mobile robotics simulation

Neuron types	τ_{RC}	A	V_{reset} (V)	V_{th} (V)
(US) Vibration neuron	0.02	1.3	0.6	1.0
(CS) Brightness neuron	0.02	0.9	0.15	1.0
(Response) Movement neuron	0.02	1.0	0.01	1.0

to ensure the neuron continuously sends output spikes only when the vibration platform is enabled, as shown in Fig. 15.6.

Moreover, the filter time constant of the synapse connecting the input to the vibration neuron is increased to 0.2 s. This modification helps to ensure that the vibration neuron does not fire in response to minor, abrupt movements of the cognitive-inspired robot, such as halting.

Our experiment involved using a simulated stereo camera to measure the brightness of a light, a departure from the traditional use of a buzzer tone in rat experiments. While Gazebo does not directly support the ZED 2 stereo camera on our cognitive-inspired robot, it does provide functionality for the Bumblebee2 camera, which serves our needs. We focused on the right camera of the Bumblebee2, as a single camera was sufficient for our experiment and met our requirements. The Gazebo environment, as shown in Fig. 15.7, was meticulously designed to closely mimic our real-world experimental setup, complete with a light panel and a placeholder for the vibration platform.

The light panel has a circular light centered in the right camera frame when the cognitive-inspired robot is on the vibration platform, as shown in Fig. 15.8.

Integrating image data with Nengo was a complex task, surpassing the challenges of processing acceleration data. It necessitated the creation of a custom "Nengo process" and additional support structures. These Nengo processes, versatile in their

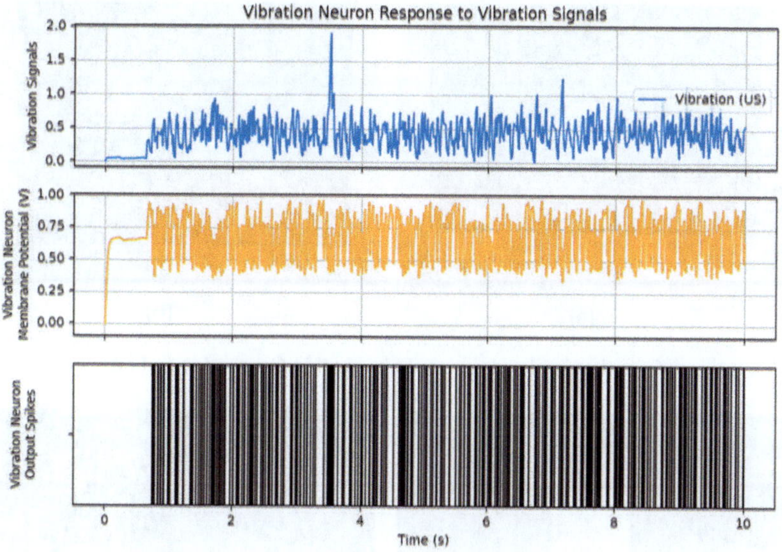

Fig. 15.6 Vibration neuron response to acceleration

Fig. 15.7 Gazebo environment setup for associative learning experiment

functions, were utilized to generate node outputs and simulate dynamical systems for neuron groups. Specifically, we developed a Nengo process, dubbed the "camera process," to manage image data for our experiment. This process retrieves images from a dedicated ROS service, reduces their resolution, and converts pixel values from 0 to 255 to a continuous range of −1 to 1, rendering them suitable as stimulus inputs. The ROS service, subscribing to the image data topic, stores the most recent frame, acting as a single-sample buffer to prevent data overload.

Initial tests indicated that the framerate of the image data was inadequate, necessitating a shift to a compressed image stream. This shift entailed the development of additional software to interface with and decode the images, resulting in specific modifications to Nengo and the ROS service. These modifications were crucial in ensuring the successful implementation of the camera process and establishing a steady stream of images. The CS node was subsequently altered to utilize the

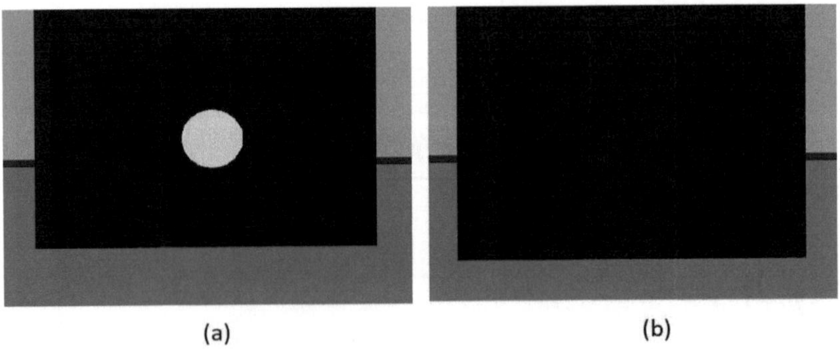

Fig. 15.8 Images from the simulated camera light panel in Gazebo (on and off)

Fig. 15.9 Resized images for brightness detection in simulation (on and off)

camera process, enabling the output of rescaled pixel-intensity data to the network. This marked a significant milestone in our experiment, showcasing the successful integration of image data with Nengo. As shown in Fig. 15.9, the image is rescaled to 5×3 pixels. Hence, one center pixel contains the approximate average value of the entire light, and the camera process' extraneous computations are minimized.

The output of the CS node is derived from the intensity of the central pixel, which is then connected to the brightness neuron. Other outputs from the CS node remain unused. The Leaky Integrate and Fire (LIF) parameters for the brightness neuron, illustrated in Fig. 15.10, were meticulously and empirically optimized to ensure the desired functionality, much like the adjustments made for the vibration neuron. These parameters enable the brightness neuron to fire continuously when the light panel is active and to cease firing immediately once the light is turned off, as shown in Fig. 15.10.

Finally, the response neuron was adapted into a movement neuron by creating a "movement output node" that issues movement commands to the cognitive-inspired robot, mimicking the fear response seen in rats. To enhance stability during Nengo's execution and avoid issues such as division by zero, a slight bias was added to V_{reset} the movement neuron, ensuring smooth integration with the movement node.

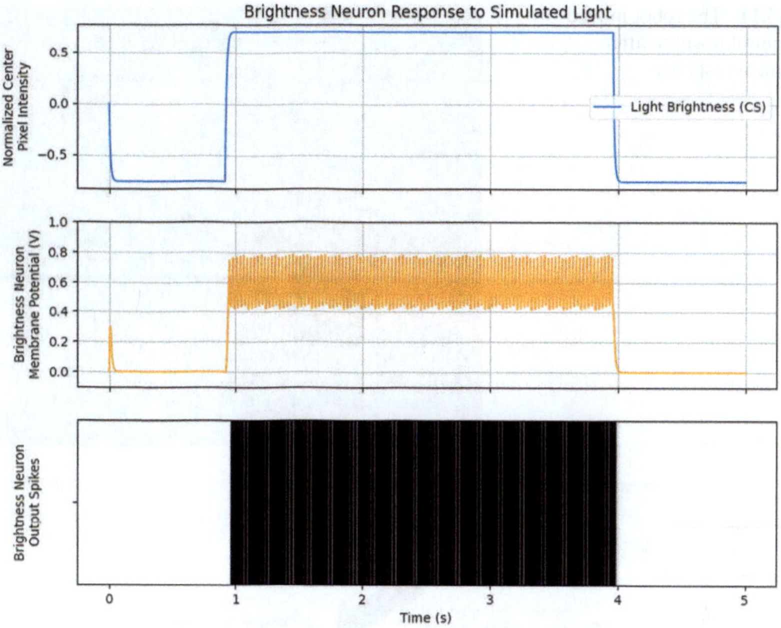

Fig. 15.10 Brightness neuron activating with light stimulus in simulation

The movement node is crucial to the system, facilitating communication with the cognitive-inspired robot through a ROS movement topic. Nengo publishes the movement data, which is then received and processed by the Jackal software. Clearpath's software on the robot interprets these commands, controls the motor drivers, and uses wheel encoders and other sensors to calculate odometry. When the movement neuron fires, the movement node directs the robot backward at a speed of 0.3 m/s. This movement command is maintained for approximately one second after the last spike, after which the robot returns to a neutral position, as depicted in Fig. 15.11.

Figure 15.12 shows that the associative learning network and three new functioning subsystems are ready to be tested in the simulation experiment.

Implementing real-time modifications in Nengo is a significant milestone in our research. This innovation underscores the crucial role of multithreaded programming in aligning Nengo's simulation time steps with real-time operations and successfully bridges the gap between Nengo and ROS/Gazebo. Traditionally, Nengo runs time steps as quickly as possible, while ROS and Gazebo aim for real-time execution, often resulting in Nengo receiving time-stretched, 'slow-motion' data. Our approach has successfully overcome this challenge, allowing Nengo networks to operate with a cumulative difference of less than one timestep from perfect real-time. This achievement significantly boosts performance and eliminates the need for further modifications.

Fig. 15.11 The robot moved to a neutral position after movement response

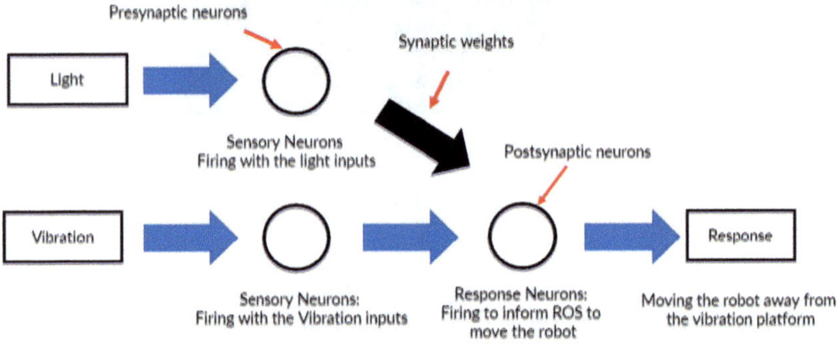

Fig. 15.12 System for associative learning with mobile robotics experiment

This advancement, made possible by our real-time modifications, has been instrumental in transitioning from simulation to real-world experiments and scaling up the network size. The impact of this transition is significant, as it opens up new possibilities for our research. Although NengoGUI offers real-time simulation execution, our custom techniques have proven far more accurate and consistent in timing. This accuracy is due to our custom real-time Nengo, which synchronizes the average time step execution with real-time rather than focusing on each step. NengoGUI's higher resource consumption and computational limitations make it less suitable for larger, more complex networks. Our custom real-time solution addresses these limitations, highlighting its critical value in the field and paving the way for exciting future research.

The real-time modifications enabled the successful simulation of the experiment within the Gazebo environment. The experiment followed a systematic process: the light was initially activated to show the absence of response to the CS, then the vibration table was 'activated' by replaying vibration data, demonstrating the US triggering the movement response. We then conducted training cycles with overlapping periods of the two stimuli. Finally, the light was reactivated to trigger the movement response without the vibration stimulus, showcasing successful associative learning through the Hebbian modification of synaptic weight. Throughout the process, the learning rate was consistently maintained at 2×10^{-5}, demonstrating the stability and reliability of our real-time modifications.

15.3.1.1 Experimental Validation

One of the most exciting achievements of our research was successfully transitioning the associative learning experiment from a simulated environment to a real-world application. This milestone opens up numerous possibilities for practical applications. This success was made possible through a series of modifications. Significantly, the vibration input node and vibration neuron had already been optimized for real-world vibration signals, eliminating the need for further adjustments. Moreover, the existing movement response neuron and output node proved adequate for the real-world setup. The new experimental configuration, shown in Fig. 15.13, primarily differed in its approach to brightness perception, showcasing the adaptability and scalability of our method. This setup was effectively tailored to accommodate specific real-world conditions, demonstrating its robustness and practicality. For example, adapting to [specific real-world conditions] revealed [explain the implications of these adaptations], underscoring the flexibility of our approach. This ability to seamlessly adjust the experimental setup to real-world conditions highlights the strength of our research methodology and its potential for practical applications in dynamic environments.

The cognitive-inspired robot was positioned on a testing platform constructed from nine wooden panels, each measuring 23 inches by 23 inches and standing eight inches tall. The center panel, highlighted in red in Fig. 15.13, served as a vibration platform with a vibration table underneath, delivering 15 Hz vibration signals that the IMU monitored. Adding a new background, light panel, and environmental lighting resulted in a completely different set of images for the updated camera, as depicted in Fig. 15.14.

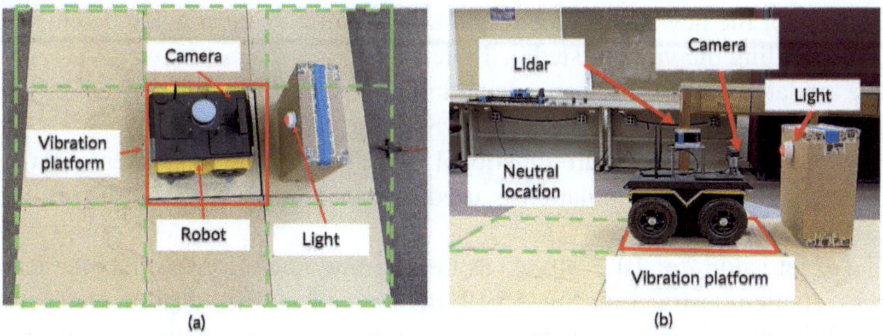

Fig. 15.13 Experimental setup for real-world associative learning experiment

Fig. 15.14 Images from the ZED 2 camera showing the light panel on and off

Fig. 15.15 Example images of the light panel from the real-world experiment

Table 15.4 LIF neuron parameters for associative learning in real-world robotics experiment

Neuron types	τ_{RC}	A	V_{reset} (V)	V_{th} (V)
(US) Vibration neuron	0.02	1.3	0.6	1.0
(CS) Brightness neuron	0.02	0.3	−1.0	1.0
(Response) Movement neuron	0.02	1.0	0.01	1.0

The new camera requires a revised Nengo camera process and ROS service to input image data into the network, but the 5×3 image center pixel output setup remains the same (Fig. 15.15).

Because of the new environment, the LIF parameters must again be empirically derived, yielding the values listed in Table 15.4.

The updated brightness neuron parameters reproduce the desired brightness detection function of the neuron, as shown in Fig. 15.16.

Now that all subsystems are functional in the real-world experimental setup, the same process from the simulation is used to train the network, as shown in Figs. 15.17 and 15.18.

The exact value 2×10^{-5} is used as the learning rate η for the Hebbian learning process. As demonstrated in Fig. 15.19, the strength of the synaptic connection in the conditional pathway increases as the stimuli are presented simultaneously.

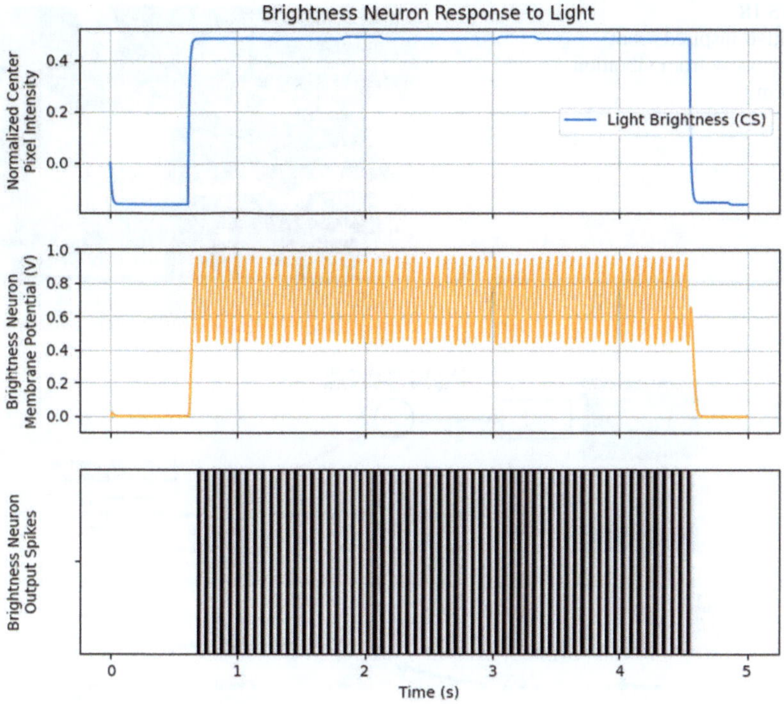

Fig. 15.16 Brightness neuron firing in response to the activated light panel

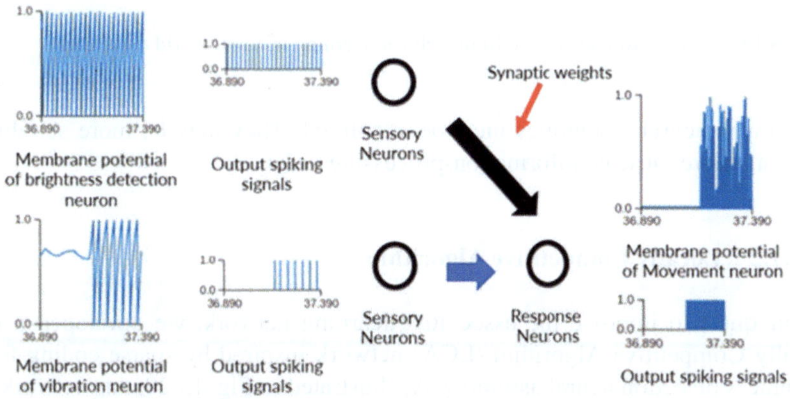

Fig. 15.17 Membrane potentials and spiking outputs of brightness detection, vibration, and movement neurons

The initial findings, illustrated in Fig. 15.19, demonstrated effective associative learning in a mobile robotics adaptation of the fear conditioning experiment. Although the system leverages biological principles to achieve associative learning,

Fig. 15.18 Cognitive-inspired robot moving away from vibration platform

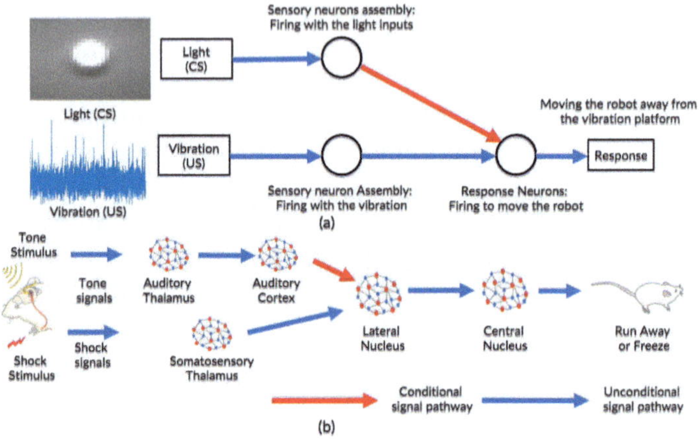

Fig. 15.19 Synaptic weight change from Hebbian learning in a real-world experiment

the current neural assemblies must be simplified. They may be more suitable for handling more intricate information-processing tasks.

15.3.1.2 Locally Competitive Algorithm

In our quest to improve the associative learning network, we developed a novel Locally Competitive Algorithm (LCA) network inspired by sparse coding for the brightness detection neural assembly. As illustrated in Fig. 15.20, this network used the same camera process to connect with the image stream. However, the images were resized to 24 × 48 pixels to simulate a more realistic image processing scenario. This successful implementation of the LCA network highlighted our research's innovative approach and effectiveness, paving the way for handling more complex information-processing tasks.

To minimize redundant neural computations, we focused on using only the center region of the image containing the light as input to the LCA network. Usually, in convolutional LCA, input patches overlap, leading to connections between feature

Fig. 15.20 Higher resolution images of the light panel on (**a**) and off (**b**)

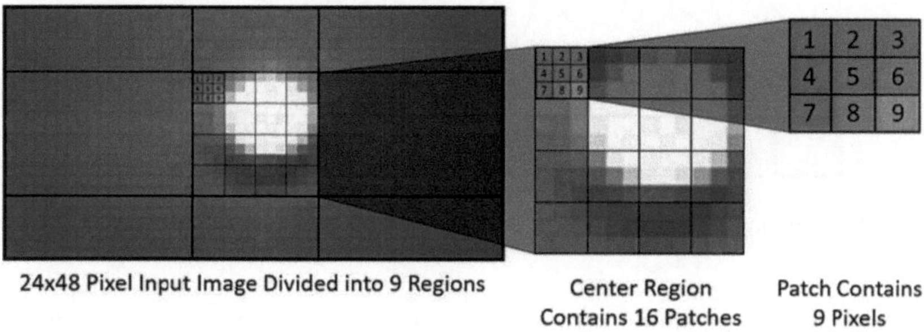

Fig. 15.21 Image division layout showing regions and patch structure

neurons of these patches. We selected a convolutional stride of three to prevent this overlap and isolate each LCA network. This strategy aimed to preserve the necessary functionality while simplifying the implementation and reducing computational requirements. Each of the 16 patches in the center region was fed into individual LCA optimization networks, as depicted in Fig. 15.21.

To efficiently manage neural activity computations, we focused solely on the center region containing the light for input to the LCA network. Unlike traditional convolutional LCA, which uses overlapping input patches and creates connections between feature neurons, we used a convolutional stride of three to avoid overlaps and isolate each LCA network. This method simplifies the implementation of the LCA network and reduces computational resource requirements while maintaining the desired functionality. As illustrated in Fig. 15.21, each of the 16 center region patches is processed by individual LCA optimization networks (Fig. 15.22).

This configuration is termed single layer because it comprises one layer of "feature neurons" that solve the LCA optimization function in conjunction with the synapses connecting these neurons to the inputs.

$$\dot{u} = \frac{1}{\tau}\left(\Phi^T x - u - \left(\Phi^T \Phi - I\right), a = T_\lambda(u)\right. \tag{15.7}$$

$$T_\lambda(u) = 0 \text{ if } u \le \lambda, \quad \text{else } \; T_\lambda(u) = u - \lambda \tag{15.8}$$

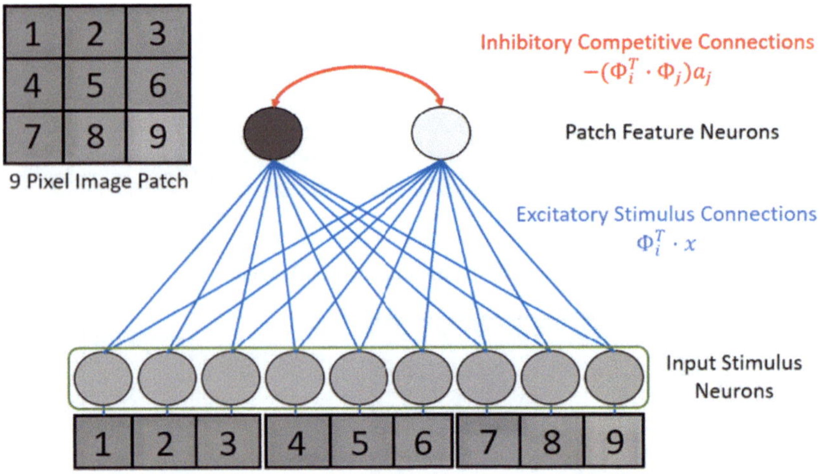

Fig. 15.22 Single-layer LCA network for brightness detection in one image patch

The input x is transmitted from the output of the camera process to the LCA layer, which then solves for the "sparse code." This layer adjusts the firing rates of each feature neuron to converge to the coefficients a_i. Each feature neuron receives input associated with a dictionary atom or feature Φ_i, as the weights of the synapses connecting it to the input are elements of the vector Φ_i.

For this experiment, a dark feature was manually created for the dictionary by selecting vector elements that are all negative, reflecting pixel intensities ranging from -1 to 1. Similarly, a light feature was created using positive vector elements. All feature vectors Φ_i must have a unit norm to ensure the magnitude of the feature vectors does not impact the sparsity penalty in Eq. (15.3). Ideally, the dictionary Φ should be overcomplete, meaning the dimension should be more significant than x's. However, this experiment uses only two features—light and dark. Hence, the neural assembly for brightness detection can only be described as "inspired by" sparse coding. While the under-complete dictionary does not fully solve the sparse coding problem and may not converge to a solution, the underlying competitive mechanisms are considered sufficient for achieving the desired functionality.

It could be overcomplete to enhance the dictionary by increasing the number of features and, consequently, the number of feature neurons, exceeding the input size. Training with spiking neurons could also generate learned dictionary features using a modified LCA implementation [27]. These additional feature neurons could remain solely positive and negative, representing variations of light and dark features with minimal theoretical changes. Alternatively, more complex features could be introduced to reduce unnecessary neural activity and improve the network's accuracy and efficiency.

Alternatively, developing features like gradients or other essential image elements could minimize the unwanted activity of the light feature neurons by offering better representations of the less-defined patches of the image. Additionally, expanding the input to accept RGB pixel values instead of scalars would enable more selective dictionary features, theoretically reducing unwanted activity in feature neurons when presented with dissimilar inputs. The brightness perception network incorporates a second layer following the LCA layer, which integrates the output spikes from the corresponding patch feature neurons in each region, as illustrated in Fig. 15.23.

A proper LCA implementation necessitates that the LCA layer (feature neurons) use the Integrate and Fire neuron model. Conveniently, the existing brightness neuron from Table 15.4 fulfills the required function of the Layer 2 neuron in Fig. 15.23 without any modifications. Table 15.5 summarizes the parameters for the neurons in the network.

Layer 1's V_{reset} parameter was empirically fine-tuned to achieve the desired functionality. The synapse model was adjusted by eliminating the filter and multiplying the weights by the spikes to implement spiking LCA. The synaptic weights connecting to the feature neurons were set to match the values of the correspond-

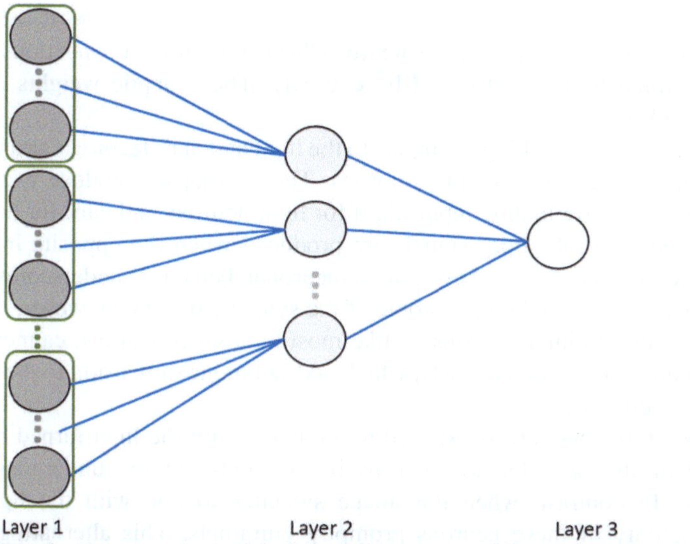

Fig. 15.23 Partition of brightness perception network used for the images' center region (note that only the neurons associated with the light feature are depicted)

Table 15.5 LCA neuron parameters for light detection

Neuron types	τ_{RC}	A	V_{reset} (V)	V_{th} (V)
LCA neuron (Layer 1)	0.02	1.0	$-\lambda = 0.85$	1.0
Light detector neuron (Layer 2)	0.02	0.3	−1.0	1.0

Table 15.6 Synaptic weights for LCA neurons

Presynaptic connection	Postsynaptic connection	Weight
Input stimulus	Light feature neuron	0.111
Input stimulus	Dark feature neuron	−0.111
Dark feature neuron	Light feature neuron	−1.0
Light feature neuron	Dark feature neuron	−1.0

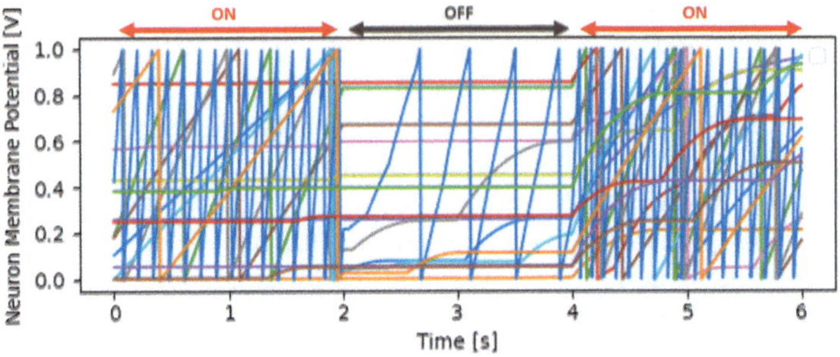

Fig. 15.24 Membrane potentials of light feature neurons of the center region

ing dictionary atom. Feature neurons have inhibitory connections with weights determined by the matrix $-\left(\Phi_i^T \bullet \Phi_j\right) a_j$. The synaptic weights are detailed in Table 15.6.

The synapses linking the inputs to the light and dark feature neurons are precisely designed to generate rapid responses. These synapses produce excitatory signals when the corresponding input (light for light neurons and dark for dark neurons) is detected, and inhibitory signals are produced when the opposite input is present. This quick reaction, a key trait of neuronal behavior, underscores the system's efficiency and is a direct outcome of the synaptic design. However, it is fascinating to note that Loihi 1 neurons, unlike most biological neurons, cannot be excitatory and inhibitory simultaneously, which adds a unique dimension to the functioning of neural networks.

When the network is exposed to an image with the light turned on, the activity level of the light feature neurons in the center region undergoes a significant surge. In contrast, when the image switches to one with the light turned off, the activity of these neurons promptly plummets. This alternating pattern, with each image displayed for two seconds, creates a unique sequence of neuronal activity. As depicted in Fig. 15.24, this sequence vividly showcases the network's dynamic response to changing stimuli, a phenomenon that never fails to inspire awe, highlighting its efficiency and adaptability.

The output spikes from the light feature neurons drive the activity of the second-layer brightness detection neuron, which is shown in Fig. 15.25. The neuron's output reliably spikes when the light is on and not off.

Integrating the brightness detection LCA network with the associative learning network was smooth, as both were implemented using Nengo. The existing network was modified by replacing the brightness neuron with the LCA network, which

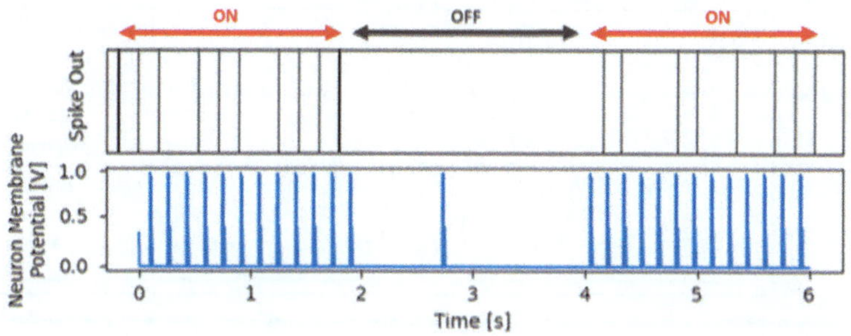

Fig. 15.25 Membrane potential and spike output of layer two brightness detection neuron corresponding to the center region

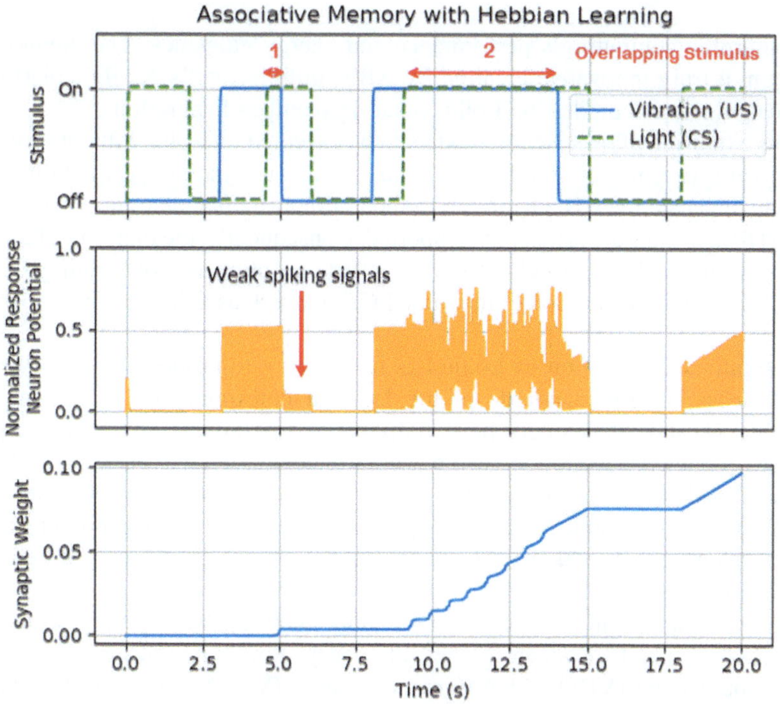

Fig. 15.26 Weight change from Hebbian learning in an experiment with LCA

now serves as the output from the camera process. No further adjustments were necessary because the layer two neuron, which connects to the learning synapse or conditional pathway, remains the same as the previous brightness neuron. With this updated configuration, the associative learning experiment was repeated, yielding significant results, as shown in Fig. 15.26.

The next stage of our experiment involves deploying it onto Intel's Loihi neuromorphic hardware, a step that enhances real-world applicability and leverages

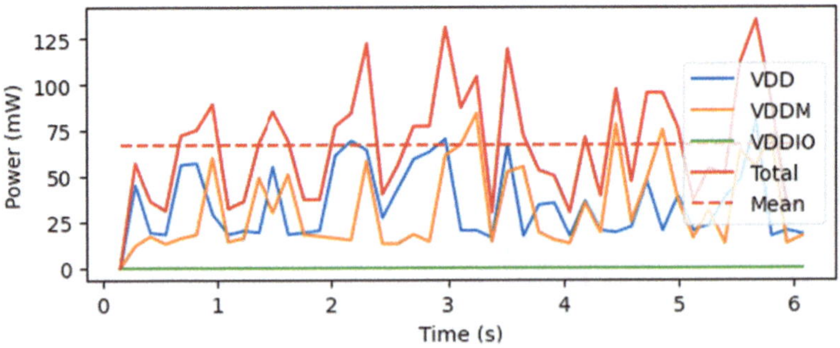

Fig. 15.27 Power consumption of LCA network running on Loihi

neuromorphic computing's performance and energy efficiency. The Nengo Loihi program, a truly innovative creation by ABR, utilizes Intel's NxSDK platform to run Nengo network models on Loihi, ensuring seamless integration.

This extension allows Nengo models to be executed on Loihi without modifying the network neurons and modules, a testament to our system's adaptability. However, the extension needs adaptation to incorporate Hebbian learning because only the STDP learning rule is natively supported. Consequently, the majority of the LCA network runs on Loihi. Intel's NxSDK software, optimized for running LCA on Loihi, facilitates the creation of the same LCA network used in previous experiment stages.

A significant limitation of Loihi 1 is its inability to produce inhibitory output spikes necessary for standard spiking LCA, a constraint not present in Loihi 2. While NxSDK developers have proposed a workaround for Loihi 1, which involves [detailed description of the workaround], its effectiveness is still being evaluated. Additional architectural adjustments required for Loihi include using a 4-bit resolution for synaptic weights, as opposed to the 24-bit floating-point resolution used by Nengo, and reducing the simulation time step from 1 ms to 20 ms when running on Loihi.

Power consumption is meticulously monitored during the execution of the LCA network on Loihi to evaluate energy efficiency. Power measurements are taken for compute logic (VDD), SRAM memory units (VDDM), and the IO interface (VDDIO), with the results shown in Fig. 15.27.

Power measurements are made throughout the experiment, with values averaged and reported every eight steps. These measurements are crucial as they provide insights into the system's energy efficiency, a key consideration for real-world applications. Figure 15.27 shows that VDDIO's power consumption is negligible compared to the other two measurements. VDD and VDDM have approximately equal average power consumptions of 30 and 29 mW, respectively. This indicates that the system operates within acceptable power consumption limits, making it a viable option for energy-efficient neuromorphic computing applications.

Table 15.7 Comparison of scale and association capability with other state-of-the-art works

	Neuron	Synapse	Dataset	Learning scheme	Biology scenarios
[28]	6	3	N/A	Simulation	N/A
[29]	3	1	N/A	Simulation	N/A
[30]	5	6	N/A	Simulation	N/A
[31]	3	1	N/A	Simulation	N/A
[32]	3	1	N/A	Simulation	N/A
[33]	3	2	N/A	Simulation	N/A
[34]	3	2	N/A	Simulation	Cellular Association in Aplysia
[35]	20	100	Pretrained with datasets	Simulation	N/A
This work	1419	1420	No dataset for pretraining	Experiment	Fear conditioning of rats

In the experiments, the synaptic weights between the brightness detection neuron and the movement neuron are modified during the training process. As a result, after associative memory learning, the mobile robot will move away from the vibration platform under the stimulus of light, even with no vibration signal presented, demonstrating successful online learning in real time. Compared to other state-of-the-art associative memory works listed in Table 15.7, we reproduce the classic fear conditioning experiments of rats using a mobile robot and the Loihi chip rather than simply simulation. In addition, the scales of our neural networks outperform other works.

15.4 Conclusion

This study introduces a novel self-learning paradigm based on associative learning, specifically fear conditioning, using a mobile robot equipped with a neuromorphic system (Loihi chip) for online learning. The robot learns to respond to light signals through repeated simultaneous exposure to light (conditional stimulus) and vibration (unconditional stimulus), mimicking the fear conditioning process observed in rats. The simplicity and computational efficiency of Leaky Integrate and Fire (LIF) neurons are utilized for detecting lights and vibrations, with Hebbian learning facilitating signal pathway modifications.

Future research avenues include improving the Locally Competitive Algorithm (LCA) implementation for broader applicability, applying sparse coding and LCA to natural language processing, and streamlining vibration detection. Enhancing the system's performance could involve incorporating Hebbian learning into Nengo Loihi and choosing the appropriate development platform (Intel's Lava) and camera (DVS). These considerations are crucial for optimizing outcomes in neuromorphic computing, robotics, and AI research.

Acknowledgment This work was supported by the Robust Intelligence program in Directorate for Computer and Information Science and Engineering (CISE) of National Science Foundation under Award Number 2245712.

References

1. Goodfellow, I., Yoshua, B., & Aaron, C. (2016). Deep learning. *Deep Learning*, 785. https://doi.org/10.1016/B978-0-12-391420-0.09987-X
2. Devlin, J., Chang, M.-W., Lee, K., & Toutanova, K. (2018). Bert: Pre-training of deep bidirectional transformers for language understanding. *arXiv preprint arXiv*, 1810.04805.
3. Goodfellow, I., Bengio, Y., Courville, A., & Bengio, Y. (2016). *Deep learning*. MIT Press.
4. Sun, C., Shrivastava, A., Singh, S., & Gupta, A. (2017, 2017). Revisiting unreasonable effectiveness of data in deep learning era. *Proceedings of the IEEE International Conference on Computer Vision*, 843–852. https://doi.org/10.1109/ICCV.2017.97
5. Sengupta, S., et al. (2020). A review of deep learning with special emphasis on architectures, applications and recent trends. *Knowledge-Based Systems, 194*, 105596.
6. An, H., Al-Mamun, M. S., Orlowski, M. K., Liu, L., & Yi, Y. (2020). Robust deep reservoir computing through reliable Memristor with improved heat dissipation capability. *IEEE Transactions on Computer-Aided Design of Integrated Circuits and Systems*.
7. Zins, N., Zhang, Y., Yu, C., & An, H. (2023). Neuromorphic computing: A path to artificial intelligence through emulating human brains. In *Frontiers of Quality Electronic Design (QED)* (pp. 259–296). Springer.
8. Roy, K., Jaiswal, A., & Panda, P. (2019). Towards spike-based machine intelligence with neuromorphic computing. *Nature, 575*(7784), 607–617.
9. Mead, C. (1990). Neuromorphic electronic systems. *Proceedings of the IEEE, 78*(10), 1629–1636.
10. An, H. (2020). *Powering next-generation artificial intelligence by designing three-dimensional high-performance neuromorphic computing system with memristors*. Virginia Tech.
11. Bai, K., & Yi, Y. (2019). Opening the "Black Box" of silicon chip design in neuromorphic computing. In *Bio-inspired technology*. IntechOpen.
12. Bai, K., & Yi, Y. (2018). DFR: An energy-efficient analog delay feedback reservoir computing system for brain-inspired computing. *ACM Journal on Emerging Technologies in Computing Systems (JETC), 14*(4), 45.
13. Kandel, E. R., Schwartz, J. H., Jessell, T. M., Siegelbaum, S. A., & Hudspeth, A. (2000). *Principles of neural science*. McGraw-Hill.
14. Baird, E., Srinivasan, M. V., Zhang, S., & Cowling, A. (2005). Visual control of flight speed in honeybees. *Journal of Experimental Biology, 208*(20), 3895–3905.
15. Kern, R., Boeddeker, N., Dittmar, L., & Egelhaaf, M. (2012). Blowfly flight characteristics are shaped by environmental features and controlled by optic flow information. *Journal of Experimental Biology, 215*(14), 2501–2514.
16. Kempter, R., Gerstner, W., & Van Hemmen, J. L. (1999). Hebbian learning and spiking neurons. *Physical Review E, 59*, 4498–4514. https://doi.org/10.1103/PhysRevE.59.4498
17. Levy, N., Horn, D., Meilijson, I., & Ruppin, E. (2000). Distributed synchrony of spiking neurons in a Hebbian cell assembly. *Advances in Neural Information Processing Systems 12, 14*, 129–135.
18. Van Rossum, M. C., Bi, G. Q., & Turrigiano, G. G. (2000). Stable Hebbian learning from spike timing-dependent plasticity. *Journal of Neuroscience, 20*(23), 8812–8821.
19. Caporale, N., & Dan, Y. (2008). Spike timing–dependent plasticity: A Hebbian learning rule. *Annual Review of Neuroscience, 31*, 25–46.
20. Bai, K., An, Q., & Yi, Y. (2019). Deep-DFR: A memristive deep delayed feedback reservoir computing system with hybrid neural network topology. In *Proceedings of the 56th annual design automation conference 2019* (p. 54). ACM.

21. Davies, M., et al. (2021). Advancing neuromorphic computing with Loihi: A survey of results and outlook. *Proceedings of the IEEE.*
22. Davies, M., et al. (2018). Loihi: A neuromorphic manycore processor with on-chip learning. *IEEE Micro, 38*(1), 82–99.
23. Orchard, G., et al. (2021). Efficient neuromorphic signal processing with Loihi 2. In *2021 IEEE workshop on signal processing systems (SiPS)* (pp. 254–259). IEEE.
24. Miller, P. (2018). *An introductory course in computational neuroscience*. MIT Press.
25. Bekolay, T., et al. (2014). Nengo: A Python tool for building large-scale functional brain models. *Frontiers in Neuroinformatics, 7*, 48.
26. DiLuoffo, V., Michalson, W. R., & Sunar, B. (2018). Robot Operating System 2: The need for a holistic security approach to robotic architectures. *International Journal of Advanced Robotic Systems, 15*(3), 1729881418770011.
27. Parpart, G. G., et al. (2022). Dictionary learning with accumulator neurons. In *ICONS*.
28. Yang, J., Wang, L., Wang, Y., & Guo, T. (2017). A novel memristive Hopfield neural network with application in associative memory. *Neurocomputing, 227*, 142–148. https://doi.org/10.1016/j.neucom.2016.07.065
29. Liu, X., Zeng, Z., & Wen, S. (2016). Implementation of memristive neural network with full-function Pavlov associative memory. *IEEE Transactions on Circuits and Systems I: Regular Papers, 63*(9), 1454–1463.
30. Hu, X., Duan, S., Chen, G., & Chen, L. (2017). Modeling affections with memristor-based associative memory neural networks. *Neurocomputing, 223*, 129–137. https://doi.org/10.1016/j.neucom.2016.10.028
31. Moon, K., et al. (2014). Hardware implementation of associative memory characteristics with analogue-type resistive-switching device. *Nanotechnology, 25*(49), 495204. https://doi.org/10.1088/0957-4484/25/49/495204
32. Ziegler, M., et al. (2012). An electronic version of Pavlov's dog. *Advanced Functional Materials, 22*(13), 2744–2749. https://doi.org/10.1002/adfm.201200244
33. Pershin, Y. V., & Di Ventra, M. (2010). Experimental demonstration of associative memory with memristive neural networks. *Neural Networks, 23*(7), 881–886.
34. An, H., Zhou, Z., & Yi, Y. (2017). Memristor-based 3D neuromorphic computing system and its application to associative memory learning. *In 2017 IEEE 17th International Conference on Nanotechnology, NANO 2017* (pp. 555–560). New York: IEEE. https://doi.org/10.1109/NANO.2017.8117459
35. An, H., An, Q., & Yi, Y. (2019). Realizing behavior level associative memory learning through three-dimensional memristor-based neuromorphic circuits. In *IEEE Transactions on Emerging Topics in Computational Intelligence*. New York: IEEE.

Chapter 16
Enabling Memory-Augmented Neural Networks for Efficient Edge Applications

Elias Khatami Zenozian, Mehdi Kamal, Ali Afzali-Kusha, and Massoud Pedram

16.1 Introduction

Deep learning-based networks have achieved remarkable success in machine learning, demonstrating its effectiveness in numerous application domains, including computer vision, natural language processing, and big data analysis [1–3]. The precision of the systems based on deep learning relies on substantial computational resources as well as memory capacities in both the training and inference stages in these systems. More specifically, the use of deep neural networks (DNNs) involves computationally expensive training of the deep model, where millions of parameters are determined through a repeated parameter adjustment and fine-tuning process. The computations require significant memory capacities too. In the inference phase, the model computations for obtaining the output(s) based on the inputs should also be carried out. Again, the computation cost will be very high mainly due to, for example, high dimensionality of the input data (e.g., a high-resolution image or a long text) and significant numbers of tensor computations that need to be performed [4, 5]. In the inference phase, the calculations related to the model evaluation should be repeated. To improve the accuracy, increasingly more complex network architectures may need to be employed, which exacerbates the problem. These all make using DNNs in a hardware-implemented system a challenging task.

One class of recurrent neural network (RNN) models is the memory-augmented neural network (MANN), which builds upon the foundation of recurrent neural

E. Khatami Zenozian · A. Afzali-Kusha (✉)
School of Electrical and Computer Engineering, University of Tehran, Tehran, Iran
e-mail: eliaskhatami@ut.ac.ir; afzali@ut.ac.ir

M. Kamal · M. Pedram
Department of Electrical and Computer Engineering, University of Southern California, Los Angeles, CA, USA
e-mail: mehdi.kamal@usc.edu; pedram@usc.edu

A. Iranmanesh, H. Sayadi (eds.), *AI-Enabled Electronic Circuit and System Design*,
https://doi.org/10.1007/978-3-031-71436-8_16

networks (RNNs). Unlike traditional neural networks, MANNs utilize external memory structures to augment their internal memory capabilities rather than relying solely on the standard internal memory mechanism. MANNs demonstrate remarkable efficiency in handling long-term dependent data, exemplified by developments from Facebook [6, 7], neural Turing machines (NTMs) [8], and differentiable neural computers (DNCs) [9]. Equipped with a differentiable associative memory, MANNs utilize a scratchpad or working memory to store previous context and input sequences (e.g., sentences of a story), enhancing the models' learning and reasoning capabilities [10]. The compelling reasoning capability of MANNs has rendered them appealing for various application domains, including, but not limited to, simple dialogue systems, machine translation, video summarization, document reading, and question-answering (QA) tasks [7], [11–16]. It should be noted that the architectures and capabilities of MANN network models differ from those of transformers, such as BERT [17] and GPT [18], which are also used in large language models. More specifically, MANNs incorporate associate memory modules, enabling them to store and retrieve information from previous inputs. This allows MANNs to perform tasks involving memory access, complex reasoning, and decision-making efficiently [11–16]. On the other hand, the transformers are neural network architectures that excel in sequence processing tasks, particularly in natural language understanding and generation [18–22].

Depending on the application, DNNs may be implemented on hardware architectures with different computation/memory resources. Among them, in many applications, edge computing [23, 24], which has limited resources, is an important implementation platform. In edge computing, data processing and analysis occur on or near the "edge" of the network, typically at or near the location where the (model input) data is obtained. Fortunately, MANNs have valuable capabilities for applications relying on edges. They include contextual decision-making [23], personalized recommendations [24, 25], real-time anomaly detection [26, 27], natural language understanding [13, 16], and adaptation to user preferences [24, 28] which are required in various applications. By leveraging memory modules and memorizing the tasks, MANNs empower edge devices to make intelligent decisions autonomously, provide tailored recommendations locally, detect anomalies in real-time, process natural language locally, and adapt to user preferences [29, 30]. The features provided by edge computing reduce reliance on cloud services, ensure privacy, improve response times, and enable efficient operation in resource-constrained environments.

Deploying MANNs on edge devices faces the key challenges of limited computational power, memory, and energy resources. This necessitates efficient resource management to make the realization of MANNs feasible [31]. In addition, efficient memory management is crucial for optimizing the performance of DNN models having large numbers of parameters [32]. Real-time responsiveness, which demands inference time reduction, is crucial. This may be achieved by invoking more efficient algorithms for designing MANNs and deploying them on edge devices. Data privacy and security are the other challenges requiring robust mechanisms to protect sensitive information stored in the memory [33]. Additionally, MANNs should adapt their model and memory requirement due to the dynamic range of the edge environments available in these applications. Overcoming these challenges through optimization techniques, resource management, secure protocols, and adaptive mechanisms are mandatory for the effective deployment of MANNs in applications using edge devices.

Various methods for enabling the deployment of neural networks on edge devices with limited resources have been suggested [34–37]. Pruning and reducing the bit-width are two strategies to decrease the size of neural networks (NNs), resulting in improved speed and energy efficiency. Pruning is a technique that can significantly reduce the energy consumption of neural networks [38–40]. By pruning (selectively removing) unnecessary connections or parameters in a neural network, the size and computational requirements of the NN model are reduced. This technique provides acceptable model accuracy with lowered computational requirements giving rise to energy reduction in inference phases.

For the case of the second technique (i.e., low-precision computing), by reducing the bit-width of neural networks, the network memory footprint and computational requirements are reduced [41–44]. More specifically, the reduction corresponds to the precision of network parameters and activation function operations where the fixed-point or lower-precision representations may be invoked instead of higher-precision floating-point (FP) numbers. This reduction in bit-width is directly translated to lower power consumption during both memory accesses and arithmetic computations [45]. Naturally, this yields lower energy consumption. As another advantage, the reduced bit-width enables more efficient storage and transfer of network parameters using lower bandwidth and further energy consumption. Reducing the bit-width, however, introduces quantization errors which reduce the model accuracy. Invoking both techniques of pruning and reduced bit-width will make the accuracy degradation more compared to the case of using one of the techniques alone. The accuracy drop may be over the acceptable threshold forcing us to explore other alternatives. One of these alternatives is the use of another number system which could provide higher accuracy for reduced bit-widths.

The *Posit* number system is a representation of numbers that has been proposed to address the limitations of traditional floating-point systems [46]. Its potential advantages include improved precision, range, and representation uniformity [47]. In the Posit system, numbers are represented using a regime, exponent, and fraction, allowing for higher dynamic range and precision. Unlike the floating-point system, Posit does not employ "gradual underflow," a mechanism that gradually increases the precision of numbers (by considering subnormal values) as they approach zero [46]. Instead, it utilizes tapered precision, a novel approach in which the precision dynamically adjusts based on the magnitude of the number being represented. This distinction underscores the unique design and advantages of the Posit number representation scheme compared to traditional floating-point systems [46]. Owing to this, the preciseness of Posit for small values (i.e., in the range of $[-1, 1]$) is higher than that of floating-point. This makes Posit suitable for applications having small values requiring high-accuracies, such as scientific computation [48] and machine learning [49]. Thus, the ongoing research aims to explore its practical implementation and integration into various computing domains.

Based on the above explanations, this chapter deals with the techniques to enhance performance and decrease the energy consumption of the MANNs to make them suitable for resource-constraint edge devices. More specifically, two pruning techniques of [50] proposed for MANNs are explored. The techniques focus on decreasing both the number of parameters and the computational requirements during the inference phase. Furthermore, the use of the Posit number system in representing the MANN model and performing the necessary computations are described. Posit is invoked to facilitate the utilization of low-bit-width representation in MANNs. The efficiencies of the techniques in improving the performance of the MANNs are analyzed by performing comparative studies. The analysis should provide practical insights for the optimal implementation of MANNs.

16.2 Background

Edge computing which is performed using devices with constrained computational resources may require real-time processing capabilities. The processing may be performed on (deep) neural network models which is the driving force for efficient implementations of the models. In particular, devices such as smartphones, IoT sensors, and embedded systems leverage artificial intelligence based on DNNs to perform tasks locally, reducing latency and safeguarding privacy [51].

Moreover, natural language processing is a pivotal application area of neural networks on edge devices, particularly for voice assistants and text-based communication. In these areas, to enhance user experience and accessibility, recurrent neural networks (RNNs) and transformers are utilized for tasks such as speech recognition, language translation, and chatbots [52]. Additionally, for industrial applications, edge devices equipped with neural networks may analyze sensor data in real-time to identify anomalies and defects to predict required maintenance and monitor quality for quality control purposes [53]. In the following subsections, first, we briefly discuss different optimization techniques aimed at diminishing the computational power and memory requirements associated with deep neural network models in the inference phase. Then, we describe the use of Posit number system [46] which may be used as another technique for reducing the aforementioned requirements for the neural network implementations.

16.2.1 Challenges and Optimization Techniques of Deploying Neural Networks on Edge Devices

In spite of increase in the computational power of edge devices, such as smartphones, IoT sensors, and embedded systems [54], they still have limited memory, processing capabilities, and energy resources. More specifically, the challenges of deploying neural networks on edge devices include the following:

1. *Limited computational power*: Edge devices are equipped with low-power CPUs, and, in some cases, specialized accelerators like graphics processing units (GPUs) or neural processing units (NPUs). These resources are significantly less powerful than cloud servers, making efficient resource utilization essential.
2. *Limited memory capacity*: Edge devices often have constrained memory capacities. Storing and processing large neural network models with millions of parameters is impractical without memory optimization.
3. *Limited available energy*: Many edge devices are battery-powered. Excessive computations can quickly drain device available (battery) energy.
4. *Limited real-time processing capability*: Numerous edge applications, such as object detection and autonomous navigation, require real-time or near-real-time processing capabilities where high inference latencies are not tolerable [33].

To address the challenges, several optimization techniques have been developed for deploying deep neural networks efficiently on these devices for inference tasks [55]. The techniques, which aim at reducing the computational load, model size, and energy consumption while maintaining high accuracy, are categorized into the following:

1. *Quantization*: Quantization, an efficient optimization technique which may enhance the performance of neural networks on edge devices at the price of lower output NN model accuracy induced through reduced accuracies of the model parameters and activations. Here, the number of bits for the representations is decreased. The reduction typically involves the transformation of high-precision numerical representations, often 32-bit floating-point numbers (FP32), into lower-bit fixed-point numbers [56]. The reduced computational requirements of the technique provide a reduction in the memory requirement, improved inference speed, and reduced energy consumption [44].

   In the quantization technique, two approaches, each addressing a different part of the model, have been widely used [57]:

   - *Weight quantization*: In this approach, model weights, which have an instrumental role in the precision of a neural network output, are quantized. A common practice involves reducing the number of bits for the representation of the weight values to 8 (or even lower). This significantly lowers the memory requirement for storing the model size and the corresponding computations enhancing the performance in the inference phase. This approach is typically implemented using techniques such as post-training quantization (PTQ) and quantization-aware training (QAT) [58].
   - *Activation quantization*: Similar to the weight quantization approach, the activation quantization is based on truncating the number of bits required for the representation of the activation values. It can be executed symmetrically (allocating an equal number of bits for positive and negative values, centered around zero) or asymmetrically, depending on the specific requirements of the edge device [59]. In the activation quantization, to take full advantage of the

benefits, one should ensure that the entire network components are compatible with the lower-precision numerical representation [60].

2. *Pruning*: This strategic optimization process can significantly reduce the model size and computational requirements without intolerably degrading the accuracy [61, 62]. Several pruning methods, which offer varying degrees of granularity and flexibility, have been frequently employed:
   (a) *Magnitude-based pruning*: This method is based on setting the value to zero for the weights smaller than a predefined threshold [63]. This way, a considerable number of parameters may be omitted reducing the number of computations.
   (b) *Structured pruning*: In this method, where specific architectural components can be safely removed to optimize the model, the optimization process takes a broader approach by targeting entire neurons or channels for removal from the network [64].
   (c) *Iterative pruning*: As an iterative approach, the pruning process is performed in multiple stages where each stage is followed by a retraining phase providing some recovery from induced lost accuracy [65]. The gradual accuracy decrease resulted from each pruning iteration provides selecting a better optimum point in the trade-off (pareto-optimal) space of computational cost/memory requirement and model performance (accuracy) given the constraints of the edge devices.
3. *Knowledge distillation*: Knowledge distillation is a process where a smaller, more lightweight model (student) learns from a larger, more complex model (teacher). The teacher model guides the student in the learning process to be trained more efficiently [66]. The knowledge distillation process primarily occurs on cloud hardware before the distilled model is transferred to the edge device. This ensures that the computationally intensive tasks involved in the distillation are handled by the cloud more substantial resources. Once distilled, the lightweight student model is then deployed in the edge device to perform efficient real-time inference without the burden of the distillation process. More specifically, the transfer learning technique enables deploying smaller neural networks on edge devices, with the larger model residing in the cloud [67]. Moreover, knowledge distillation synergizes with transfer learning, another significant methodology for edge devices. Transfer learning allows these devices to leverage pre-trained models, avoiding the need to train from scratch. Instead, edge devices can fine-tune models for specific tasks, saving computational resources and time. This strategy is particularly well suited for edge inference, where the goal is to make real-time decisions with limited computational resources. Integrating knowledge distillation and transfer learning empowers edge devices to benefit from the expertise of larger models in the cloud, enabling efficient deployment and utilization of neural networks in diverse applications on the edge [68].

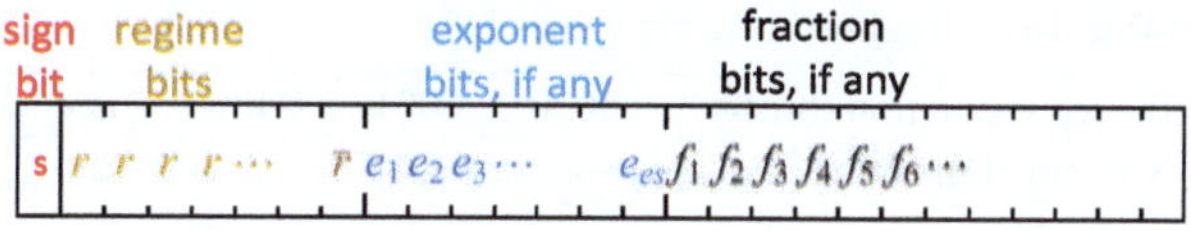

Fig. 16.1 General Posit format for finite and non-zero values [46]

16.2.2 Posit Number System

The Posit numbers may replace standard floating-point numbers in the computations required for the deep neural networks [46]. This system offers several advantages over the floating-point one, including a larger dynamic range, higher precision, better compactness, simpler hardware, and exceptionally straightforward exception handling [46]. Operations with Posit numbers never experience overflow to infinity or underflow to zero and use NaN (not-a-number) to represent exceptions instead of a bit pattern [46]. Posit processing unit consumes less hardware resources compared to the floating-point one, leading to reduced energy consumption and a smaller silicon footprint [46]. Supported by a single chip, Posit operations per second (POPS) can significantly surpass floating-point operations per second (FLOPS) using similar hardware resources [46].

The structure of an n-bit Posit format which consists of four components of sign, regime, exponent (with es exponent bits), and fraction is shown in Fig. 16.1 [46]. The components are explained in the following [46]:

- *Sign bit* (s): For positive numbers, the sign bit is zero, while for negative numbers, it is one. It should be noted that if the number is negative, the concatenation of the regime, exponent, and fraction are in two's complement format.
- *Regime bits* (r): Binary strings that begin with a sequence of all 0's or all 1's are considered as the start of a regime. The regime ends when the next bit differs from the preceding one or when the end of the string is reached. Let us assume m is the number of consecutive identical bits within a regime. If there are m 0's, its corresponding numerical value (denoted as k) is considered as $-m$. If they are m 1's, we consider k as $m - 1$. The regime also indicates the scaling factor useed^k, where useed equals to $2^{2^{es}}$ (the number of exponent bits can be from 0 to es, where es is the exponent field (e) size). For example, consider the 4-bit binary string shown in Table 16.1 (in a bit string, x means don't care).
- *Exponent bits* (e): These bits are considered as an unsigned integer. There is no bias, as is in the case of floating-point numbers, and the scaling factor is 2^e. The number of remaining bits on the right side of the regime determines the possible range, which can vary from 0 to es exponent bits. This is a compressed method to express the precision of conic numbers, where the numbers close to absolute 1 have higher precision compared to very large or very small numbers, which are less common in computations.
- *Fraction bits* (f): If there are bits remaining after the regime and exponent bits, these bits denote the fraction part, like the fraction part in the representation of floating-point numbers. Similarly, there is an implicit bit of 1. Unlike the floating-

Table 16.1 Regime bit string values [46]

Bit representation (binary)	0000	0001	001x	01xx	10xx	110x	1110	1111
Numerical meaning, k	−4	−3	−2	−1	0	1	2	3

point numbers, there are no subnormal numbers with a hidden bit of 0 in Posit numbers.

Based on the provided explanations, the corresponding real number (x) which is represented by the Posit format could be obtained from

$$x = \begin{cases} 0 & p = 0 \\ \pm\infty & p = -2^{n-1} \\ \text{sign}(p) \times \text{useed}^k \times 2^e \times \left(1 + \frac{f}{2^h}\right) & \text{otherwise} \end{cases}$$

where p is the bit string and h is the number of bits allocated to the fraction. Table 16.2 illustrates the representable numbers in 4-bit Posit, FP, and integer (INT) number systems which reveals the dynamic range and the granularity for each numbering system for this example. In addition, it shows the corresponding positive and negative numbers. In the Table 16.3, Also, the dynamic ranges of these number systems for 8, 16, and 32 bits are compared.

As mentioned before, compared to FP, Posit provides higher precision in representing small numbers around zero. Table 16.4 shows the precision of different number systems for different range. For a better evaluation of the systems, in the case of the MANN application, we also provide the precision of representing the weights in the MANN. The average relative errors (relative errors $= \left| \frac{x-q}{x} \right|$, where x is the true or exact value and q is the approximate or estimated value) in the cases of the 8- and 16-bit Posit (i.e., Posit8 and Posit16) and floating-point (i.e., FP8 and FP16) number systems for the representation of a set of 1 M data, uniformly distributed in the ranges of (0,1] and (0,0.1], as well as for the representation of the parameters of the MANN network it was trained under bAbI dataset [69]. The range of the weight values of MANN network is [−2,2], with the mean (variance) of 0.0016 (0.0199). Thus, most of the values are in the range of [−0.1,0.1]. The weight range of the majority of deep neural networks typically falls within the range of 1 to −1. This implies Posit could provide higher precision in representing the weights compared to the FP [70–72]. The reported figures show the higher accuracy of the Posit system compared to those of the floating-point system.

16.2.3 Quantization, Posit, and MANNs

As mentioned above, the quantization technique has been used as a key technique for reducing the requirements for the implementation of deep NN models at the price

Table 16.2 Representable numbers in 4-bit Posit, FP, and INT number systems

4-bit pattern representation	Value for Posit4 (es = 2)	Value for Posit4 (es = 1)	Value for FP4 (Exponent size = 3)	Value for FP4 (Exponent size = 2)	Value for 4bit-fixed point (Signed integer)
0000	0	0	+0.0	+0.0	+0
0001	+1/256	+1/16	+0.25	+0.25	+1
0010	+1/16	+1/4	+0.5	+1.0	+2
0011	+1/4	+1/2	+1.0	+1.5	+3
0100	+1	+1	+2.0	+2.0	+4
0101	+4	+2	+4.0	+3.0	+5
0110	+16	+4	+8.0	+∞	+6
0111	+256	+16	+∞	NaN	+7
1000	±∞	±∞	−0.0	−0.0	−8
1001	−1/256	−1/16	−0.25	−0.25	−7
1010	−1/16	−1/4	−0.5	−1.0	−6
1011	−1/4	−1/2	−1.0	−1.5	−5
1100	−1	−1	−2.0	−2.0	−4
1101	−4	−2	−4.0	−3.0	−3
1110	−16	−4	−8.0	−∞	−2
1111	−256	−16	−∞	NaN	−1

Table 16.3 Dynamic ranges of various data types

Number of bits	Number system	Maxpos	Minpos
8 bits	FP8 (Exponent size = 3 bits)	248	9.75e-4
	FP8 (Exponent size = 4 bits)	15.5	3.90e-3
	Posit8 (es = 1)	**4096**	**2.44e-4**
	Posit8 (es = 2)	**1.68e7**	**5.96e-8**
	Fixed point (unsigned integer)	127	1
16 bits	FP16 (Exponent size = 5 bits)	65,535	5.97e-8
	Posit16 (es = 1)	**2.68e8**	**3.27e-9**
	Posit16 (es = 2)	**7.20e16**	**1.39e-17**
	Fixed point (unsigned integer)	2.14e9	1
32 bits	FP32 (Exponent size = 8 bits)	3.40e38	1.40e-45
	Posit32 (es = 3)	**2.2e72**	**6.3e-73**
	Fixed point (unsigned integer)	2.14e9	1

Table 16.4 Relative errors of Posit and FP number systems for representing different sets of data

Data	Number system	Mean relative error
(0,1] Step = 10^{-6}	FP16 (Exponent size = 5 bits)	1.94e-4
	Posit16 (es = 1)	**6.32e-5**
	FP8 (Exponent size = 3 bits)	8.72e-2
	FP8 (Exponent size = 4 bits)	6.8e-2
	Posit8 (es = 1)	**1.67e-2**
(0,0.1] Step = 10^{-7}	FP16 (Exponent size = 5 bits)	2.19e-4
	Posit16 (es = 1)	**1.83e-4**
	FP8 (Exponent size = 3 bits)	8.98e-2
	FP8 (Exponent size = 4 bits)	7.12e-2
	Posit8 (es = 1)	**5.48e-2**
MANN Weights	FP16 (Exponent size = 5 bits)	2.28e-4
	Posit16 (es = 1)	**1.55e-4**
	FP8 (Exponent size = 3 bits)	5.18e-2
	FP8 (Exponent size = 4 bits)	5.01e-2
	Posit8 (es = 1)	**3.93e-2**

model accuracy. Some of the previous works focused on the quantization strategies for deep neural networks (DNNs) include BinaryConnect [73], logarithmic binary representation [74], Dorefa-Net [41], 16-bit fixed point with Flexpoint scaling [44], and FP16 training with PyTorch [75]. In the case of MANNs, considerably lower research activities have been performed on low-bit numerical systems compared to the DNN models.

In addition, the use of Posit number system for deep NNs (DNNs) has been considered by researchers. The works include frameworks such as Deep Positron [71], Cheetah [42], PositNN [76], logarithmic-Posit arithmetic [77], and Deep Pensive [70]. The frameworks make use of Posit for multiplication and accumulation demonstrating improved accuracy compared to that of the traditional fixed point and floating-point representations. While the Deep Positron focuses on FPGA-based inference acceleration, the Cheetah framework explores different Posit formats for training and inference with the focus on hardware complexity optimization. PositNN employs a combination of 16-bit, 8-bit, and 10-bit Posits for training CNNs, demonstrating comparable accuracy to 16-bit floating-point. Additionally, logarithmic-Posit arithmetic exhibits promising results in terms of negligible accuracy loss and higher energy efficiency. Deep Pensive, a comprehensive framework, utilizes Posits for all the computational steps in CNNs and demonstrates the potential of 16-bit Posits to achieve the same accuracy as 32-bit floating-point. It, however, needed a 32-bit accumulator for certain functions. In addition to these studies, exploring the use of Posit numbers within the domain of memory-augmented neural networks (MANNs) should be investigated to determine the level of its efficiency improvement for the MANN implementation on edge devices.

16.3 Memory-Augmented Neural Networks

For several application domains, machine learning models with an easy way to access and integrate long-term memory in the inference and training phases are required. As an example, consider the process of answering questions after reading a book or text. This can be performed using a language modeler like a RNN (recurrent neural network), which predicts the next set of words from a series of words. Figure 16.2 illustrates the general structure of long short-term memory (LSTM), a specific type of RNN. RNNs, however, have limited memory having difficulties even with simple tasks, such as copying a sequence from the input to the output of the network without distortion or loss of information [78, 79]. This challenge is also present in other domains, such as visual and auditory tasks that require long-term memory to effectively answer questions about movies or podcasts. MANNs are emerging network architectures appropriate for this need. The primary concept involves integrating effective learning strategies established in the machine learning literature for inference with a memory component capable of both read and write operations [80]. The models learn how to work effectively with the memory component to realize efficient machine learning models capable of integrating long-term memory.

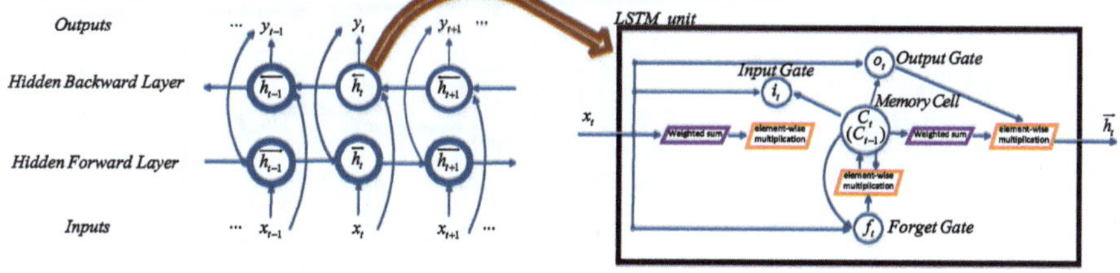

Fig. 16.2 The general structure of a bidirectional LSTM [81]

16.3.1 MANN Structure

In the process of question answering (QA), the MANN model initially receives sets of sentences that provide a narration or description of an event, and stores them in its augmented memory. Subsequently, a question related to the information in the story is presented to the network. To answer this question effectively, the network engages in several iterations of an attention-based inference method. Attention-based inference is a method where the network dynamically focuses on different parts of the input sequence during each iteration. This is achieved by assigning importance weights to different elements of the input sequence, allowing the network to selectively attend to relevant information while ignoring irrelevant or less important parts. By iteratively adjusting these attention weights, the network can refine its understanding of the input sequence over multiple steps, ultimately leading to a more accurate and context-aware answer to the question at hand [82]. Within each inference layer with attention, the network assesses the connections and correlations between each article and the question sentence. These sequential layers of inference with attention are commonly referred to as "steps." The information acquired through these steps is passed into a fully connected (FC) layer to generate the (accurate) final response.

The general datapath of the MANN, which consists of four primary data processing components, is illustrated in Fig. 16.3. The components include the following [80]:

1. *Input feature map* (**I**), which transforms the input vectors into the internal feature representations.
2. *Generalization* (**G**), which updates the existing memory based on the new input (in end-to-end networks, this memory is not updated).
3. *Output feature map* (**O**), which computes a new output based on the new input states and actual memory.

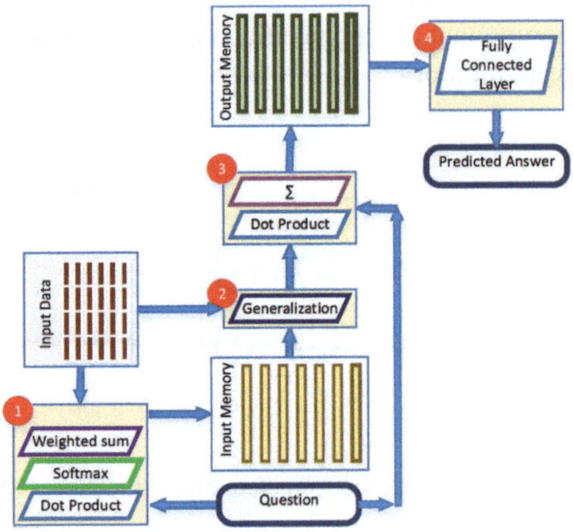

Fig. 16.3 The general structure of memory-augmented neural networks [80]

4. *Response* (**R**) converts the output to the desired response form.

For the inference equations of **I**, **O**, and **R** in these networks, used to calculate the output, see [6].

Despite its success in a wide range of QA tasks, the use of MANN is limited due to the requirement of knowing the actual positions of the supporting statements [83]. This means that, in addition to the final output, the model must have access to the location of each relevant clue in the memory during training. This necessitates additional annotations, such as explicit indications of where each supporting statement is located within the input sequence. However, this requirement sometimes imposes unrealistic demands on certain datasets, particularly those lacking such annotations. As a result, it becomes challenging for datasets without these annotations to effectively utilize MANNs for question-answering tasks. To address this limitation, a modified MANN architecture called MEMNN has been introduced in [7]. This architecture eliminates the need for additional annotations by incorporating mechanisms to automatically identify and attend to relevant clues in the input sequence, thereby overcoming the obstacle posed by the lack of explicit positional annotations. One of the advantages of this architecture is that it can be trained end-to-end, training entire system, from input to output, as a single, integrated model. This means requiring less supervised information during the training [7]. Also, it does not need *Generalization* (G) data processing component.

The inference equations of **I**, **O**, and **R** in these networks used to calculate the output are given in the following [7]:

I: $p_i = \text{Softmax}(u^T . m_i)$ where $m_i = Ax_i$ (x_i is the embedding vectors of the input sentences) and $u = Bq$ (q is the input query).
O: $o = \sum_i p_i . c_i$ where $c_i = Cx_i$.
R: $\hat{a} = \text{Softmax}\,(W\,(o + u))$.

In the aforementioned relationships, A and C represent embedding tensors comprising learnable parameters whose values are determined during the training. These tensors have dimensions of $V \times d$, where d stands for the embedding dimension, and V represents the total number of words in the dictionary for embedding [50]. Note that the embedding process involves converting input tokens (e.g., input words) into high-dimensional vectors through learned embeddings, enabling the model to capture semantic relationships and contextual information of the inputs.

In these network structures, two memory units, denoted as m_i and c_i, are employed. These memory units store embedded representations of sentences using a bag-of-words (BoW) approach [84]. If we consider the number of sentences in the input story as n_s, the size of these memory units equals $d \times n_s$. Furthermore, the word count in a given story is denoted as n_w. Consequently, the entire network includes a tensor with the dimensions of $n_s \times n_w$. It is worth noting that if the number of sentences or words falls short of n_s or n_w, padding with zeros is utilized. Each sentence in the story is preserved within a tensor with the dimensions of $n_s \times n_w$. The elements of the tensor are obtained by mapping each word in the story to its corresponding entry in the dictionary. For instance, in a story comprising

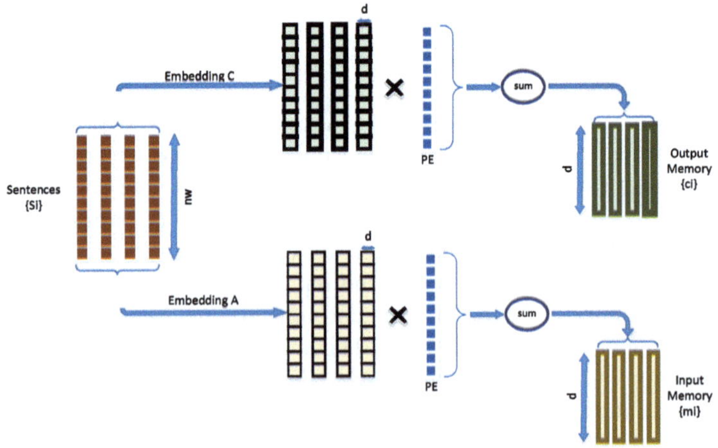

Fig. 16.4 Story sentence embedding operation in MANNs [50]

a sentence like "The cat sat on the table," each word of it is associated with an index in the dictionary. If "The" corresponds to index 1, "cat" to index 2, and so forth, then the tensor representation of the story would map each word to its respective index. In this case the sentence is represented as [1, 2, 3, 4, 5, 6, 0, 0, 0, 0], where empty slots are padded with zeros to maintain consistent tensor dimensions.

As illustrated in Fig. 16.4, the embedding process involves treating the numbers (the indices assigned to each word within the dictionary) within each sentence as the row indices of the embedding matrices of A or C. The rows of the matrices, effectively map each word to a vector of size $1 \times d$. Consequently, each sentence with the dimensions of $n_{\mathrm{w}} \times 1$ is transformed into a tensor of size $n_{\mathrm{w}} \times d$. A temporary *encoding* tensor named *PE* (positional encoding) is utilized to maintain the word order in these networks. *PE* is designed to have dimensions matching those of the sentence. The sentence tensor is then element-wise multiplied by *PE*, where each element of PE represents the positional encoding information for the corresponding word in the sentence. This multiplication integrates the positional encoding information directly into the embedding of the words, ensuring that the network can effectively capture the sequential order of the words in the input sequence. The values within each column of a tensor are summed together through this multiplication. Subsequently, a transpose operation is performed to yield a tensor with the dimensions of $d \times 1$. The resultant tensors are then stored in a memory, denoted as c_i or m_i, forming a tensor with the dimensions of $d \times n_{\mathrm{s}}$.

In the case of the question sentence embedding, it should be noted that this sentence also comprises of n_{w} words. This embedding is performed using another tensor known as B, which has the same dimensions as $V \times d$. Similar to the previous process, the question sentence undergoes embedding, resulting in the creation of a column vector with the size of $d \times 1$ symbolized as u (Fig. 16.5).

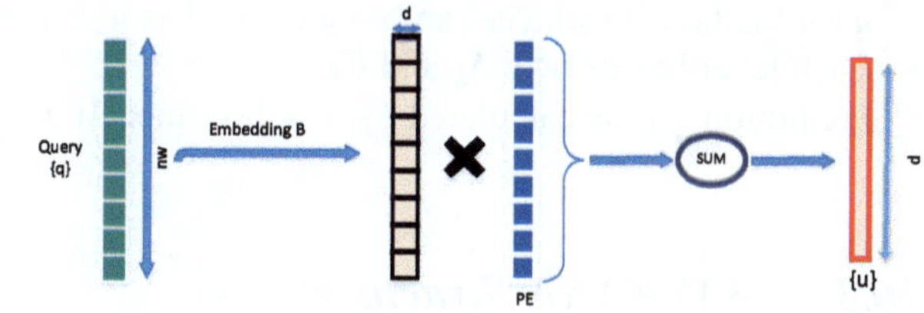

Fig. 16.5 Question sentence embedding operation in MANNs [50]

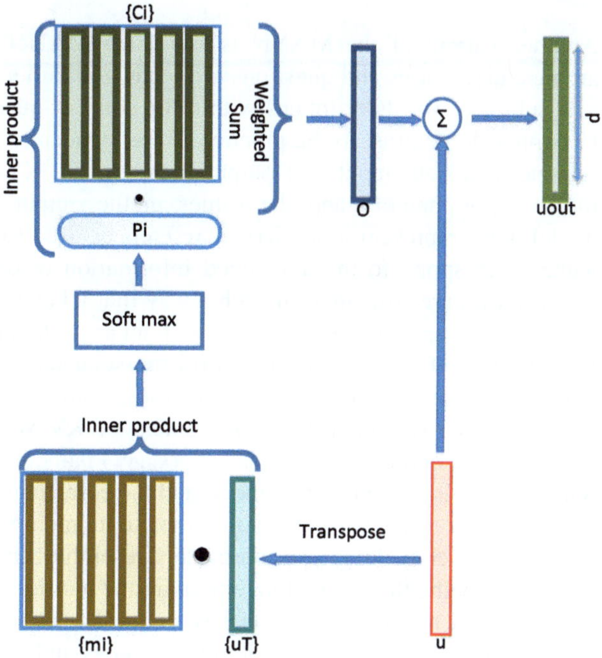

Fig. 16.6 Operations which are performed in each layer of the MANN network [50]

Following the embedding procedures, the inference steps, as previously discussed, are executed. A visual representation of these steps for a single layer is illustrated in Fig. 16.6. In the concluding stage of the output generation, referred to as the *R* step mentioned earlier, the network incorporates a linear fully connected (FC) layer. In this step, a final weight matrix *W*, with the dimensions of $V \times d$, is used to perform multiplication by the output feature vector u (u_{out}) obtained from $u_{\text{out}} = o + u$. This operation results in the production of the final answer (output of the network) in the form of a vector with dimensions of $1 \times V$, representing the probabilities or scores associated with each element in the output vocabulary. This vector is often subjected to a *Softmax* function to obtain a probability distribution, providing the model's prediction for the answer based on the input and memory information.

In MEMNN networks, where k steps or layers are used in the inference stage for generating the output, a better accuracy may be achieved through increasing the layers in the following way [80]:

- u for the $(k+1)$st layer can be calculated as $u_{k+1} = u_k + o_k$.
- Each layer has its own A_k and C_k.
- Prediction can be calculated by $\hat{a} = \text{Softmax}\,(W.u_{k+1})$.

16.3.2 KV-MANN Structure

Another variant of the MANN is key-value MANN (KV-MANN) [13]. In this architecture, stories and questions are organized into n_s windows (a window refers to a subset of words or tokens within the story or question [13]), each containing n_w words. In contrast to the practice of the standard networks in storing complete sentences in both input and output memory, the KV-MANN model stores keys in the input memory and the values in the output memory. For this network model, keys represent identifiers or references used to retrieve information, while values correspond to the associated information or data. Within these networks, the windows are structured in such a way that a keyword from the story sentences and questions appears consecutively without any inherent semantic connection [13]. For instance, if we were to transform the sentence "Alex went to school" while considering the two-word window, it would result in expressions like (Alex went), (went to), and (to school). In the context of KV-MANNs, the input and output memories are defined as (k_1, v_1), ..., (k_n, v_n) pairs. The key window is embedded within the cells of the input memory, and the corresponding values of these keys are stored in their respective cells within the output memory.

The input embedding process closely resembles that of MANNs, with the key distinction being that we no longer require the multiplication of the encoding tensor (*PE*). In this network, both the key and value are embedded through a shared matrix denoted as A. By embedding both keys and values using the same matrix, the network can effectively capture and utilize the relationships between them, facilitating more robust and efficient memory retrieval and processing. Additionally, sharing the embedding matrix reduces the number of parameters in the model and promotes parameter sharing, which can improve generalization and mitigate overfitting. The embedding is performed only once at the onset of the operation, and the same cells are employed across all layers. While the inference stage is the same as that of MANNs, the tensor product of the outputs (u) is calculated differently. In this scenario, after the addition of the vectors o and u, the resulting tensor is multiplied by the tensor R_k, having the dimensions of $d \times d$. A KV-MANN structure, consisting of three sequential steps, is shown in Fig. 16.7.

16.3.3 Comparative Analysis and Challenges of MANN

In the rapidly evolving landscape of neural networks, comparing memory-augmented neural networks (MANNs) with traditional neural networks (e.g., RNN)

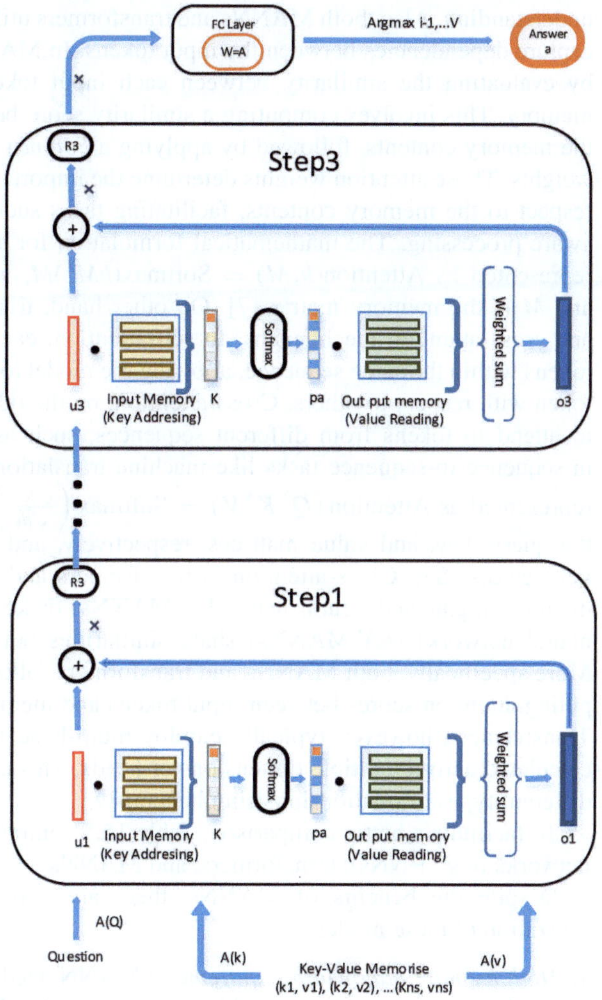

Fig. 16.7 Steps taken in the KV-MANN network with three layers or steps [50]

and state-of-the-art networks like transformers is crucial to understand the unique advantages MANNs offer. RNNs, once a staple in sequential data tasks, are now considered outdated and are no longer the main research focus for applications such as question answering due to their limitations in maintaining long-term dependencies. MANNs, however, address these limitations by incorporating an external memory component, enabling them to store and retrieve information over extended periods effectively. This capability is particularly advantageous for complex reasoning and tasks requiring long-term memory, where traditional RNNs fall short.

Meanwhile, the architecture of MANNs shares similarities with transformers. Both MANNs and transformers excel in processing sequential data and capturing long-range dependencies, making them suitable for tasks requiring contextual understanding. Also, both MANNs and transformers utilize attention mechanisms to capture dependencies between the input tokens. In MANNs, attention is calculated by evaluating the similarity between each input token and the contents of the memory. This involves computing a similarity score between the input tokens and the memory contents, followed by applying a *Softmax* function to obtain attention weights. These attention weights determine the importance of each input token with respect to the memory contents, facilitating tasks such as reasoning and context-aware processing. The mathematical formulation for attention in MANNs can be represented as $\text{Attention}(h, M) = \text{Softmax}(hM^T)M$, where h is the current state and M is the memory matrix [7]. On other hand, transformers use self-attention and cross-attention mechanisms. In self-attention, each token attends to all other tokens within the same sequence, allowing the model to weigh the relevance of each token with respect to others. Cross-attention, on the other hand, enables the model to attend to tokens from different sequences, such as encoder-decoder attention in sequence-to-sequence tasks like machine translation. Mathematically, it can be represented as $\text{Attention}\,(Q, K, V) = \text{Softmax}\left(\frac{QK^T}{\sqrt{d_k}}\right)V$, where Q, K, and V are the query, key, and value matrices, respectively, and d_k is the dimension of the key vectors [85]. Cross-attention in transformers and the attention mechanism in memory-augmented neural networks (MANNs) or key-value memory-augmented neural networks (KV-MANNs) share similarities but also have key differences. More specifically, both MANNs and transformers follow a similar process of computing attention scores between input tokens and memory/memory-like structures. Transformers, however, typically employ multi-head attention, where attention is calculated across multiple dimensions or heads. This enables the model to capture different types of relationships simultaneously.

To facilitate a better comparison, Table 16.5 contrasts the features of traditional networks (e.g., RNNs), transformers, and MANNs.

Despite the benefits of MANNs, there are two main obstacles to the full utilization of these models:

1. *High memory bandwidth requirement*: MANN needs a large amount of memory bandwidth. During the embedding operations, MANN searches for the embedding tensor to convert input sentences into internal state vectors. While larger embedding (i.e., each token is represented by a vector with more dimensions) is useful for solving complex problems [86], it requires higher memory. In addition to the embedding operations, the inference operations cause high memory traffic. The traffic, for example, consists of the input/output memory accesses which increase quickly with the growth of network size, a large amount of intermediate data between each layer inserted by the network, including inner products, *Softmax* function, and weighted sums.
2. *Computational complexity*: MANNs require multiple matrix multiplication operations, including inner product, weighted sum, and those pertaining to the FC layer. Also, the *Softmax* function uses powers that require many integer multiplications. In MANNs, the computation amount increases with the size of the input data with the time complexity of $O(n^a)$ where $a \geq 2.375$ [87].

Table 16.5 Features of traditional networks (e.g., RNN), transformers, and MANNs

Feature	Traditional networks (e.g., RNN)	Transformers (e.g., BERT, GPT)	Memory-augmented neural networks (MANNs)
Memory handling	Limited memory capacity	Improved memory handling	Enhanced memory capabilities with associative memory
Long-term dependencies	Struggles with long-term dependencies	Better with attention mechanisms	Excels with long-term dependencies due to memory augmentation
Computational complexity	Moderate to high	High	Can be high, but optimized with memory management techniques
Latency	Lower	Higher due to complex architectures	Potentially lower with optimizations (e.g., A^2P-MANN, Full Posit)
Energy consumption	Lower	Higher	Optimized for lower consumption with efficient memory use
Application fields	Basic tasks, image recognition, etc.	Natural language processing, complex reasoning	Natural language processing, suitable for tasks requiring contextual memory (e.g., QA, dialogue systems)
Scalability	Limited	Better scalability	Scalable with efficient memory use and management techniques
Flexibility	Less flexible	More flexible	Highly flexible due to memory augmentation and dynamic learning
Implementation complexity	Lower	High	Higher, but offers significant performance benefits
Training efficiency	Moderate	Requires extensive data and computation	Efficient with memory-augmented approaches, reduces redundant computations

The high memory bandwidth and computational requirements, cause high energy/power consumption and increase the latency in the inference phase. To keep both within bounds, the use of optimization techniques is inevitable. In [87], an innovative hardware accelerator leveraging three key methods to enhance efficiency for MANNs was proposed. First, a column-based algorithm was introduced to address the substantial memory bandwidth usage issue by reducing the number of generated intermediate values. The algorithm, also, defers the *Softmax* division operation until the end of the computations. The second method, known as "zero-skipping," streamlines weighted sum computation during inference by excluding probability vectors below a predefined threshold (i.e., θ_{zs}), thereby reducing the computational workload considerably. Additionally, to optimize memory bandwidth requirement a cache memory was used to embed internal state vectors of the inputs. Also, a MANN hardware accelerator to optimize the computation load was introduced in [9]. The hardware accelerator operates on a dataflow architecture (DFA), enabling direct data flow between processing elements to minimize memory access and reduce energy consumption. Initially, the method analyzes the statistical distribution of words in the vocabulary to inform the network's final output. Subsequently, a confidence threshold is established for each logit. During execution, upon reaching the final FC layer, these logits are computed sequentially, with the first logit exceeding its confidence threshold considered as the network's answer, thus terminating further computations.

16.4 MANN Model Pruning

In Sect. 16.2, three different pruning methods of deep neural networks, namely, magnitude-based pruning, structured pruning, and iterative pruning based on three fundamental principles were described. In this part, we discuss in detail the work of A^2P-MANN which is a dynamic solution designed to enhance the computation reduction of MANNs by decreasing the necessary number of inference steps [50]. As stated previously, the increase in the number of consecutive layers or inference steps improves the average accuracy in MANNs. In many applications, two, three, or even more steps (hops) are typically employed. Nevertheless, the investigations have revealed that a large number of questions do not require more than one step to identify the correct answer, rendering the remaining computations redundant. The solution of [50] is based on the identification of the *easy questions* resulting in a reduced number of steps, thereby, decreasing the computational workload and energy consumption during the inference phase. In addition, as mentioned before, the final answer to the questions is derived through an FC layer in the last stage of the network. The observations in [50] suggested that a substantial portion of the weights in this final layer exerts minimal influence on the ultimate outcome of the MANN. Thus, pruning this layer could be carried out without a substantial impact on the network's ultimate accuracy. Such pruning reduces the network's execution time and the memory required for weight storage. In the subsequent sections, we explain the implementation of A^2P-MANN in more details.

16.4.1 A^2P-MANN Method

The two approaches aimed at diminishing the computational workload of MANNs are as follows:

1. Integration of a multilayer perceptron network called input classifier network (ICN) at the output of the initial step to dynamically adjust the number of required steps.
2. Pruning the final fully connected (FC) layers.

The ICN network, which consists of two compact linear layers, classifies the feature vector produced at the output of the initial step, denoted as u^2, into either of the two categories of *easy* or *hard*. The visual representation of this structure

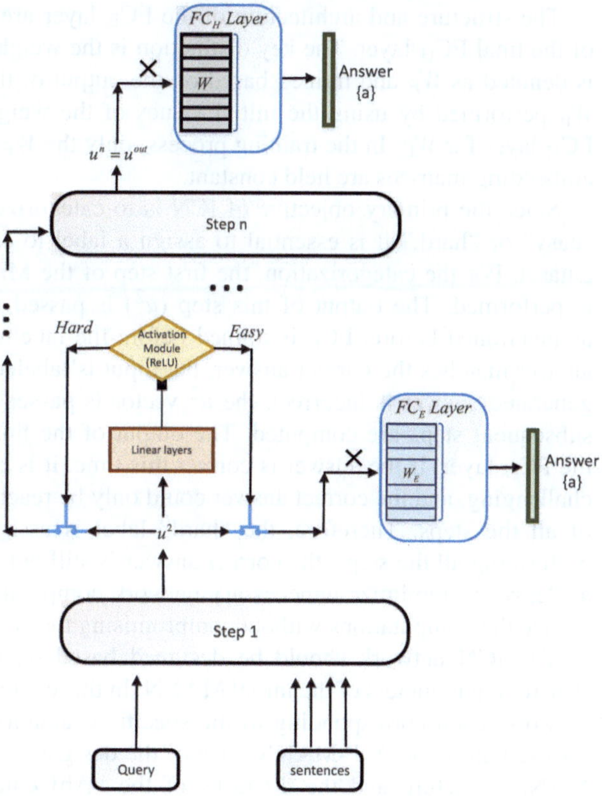

Fig. 16.8 General structure of the proposed A^2P-MANN method [50]

is provided in Fig. 16.8. If the ICN network categorizes the input query as *easy*, the output of that step will be used to generate the answer. If the ICN network determines the input query to be *hard*, however, the computations for the subsequent steps are executed to obtain the final answer. The final layer of ICN has two neurons for "hard" and "easy" and a *Softmax* function is used to transform the outputs into probabilistic logits with values between zero and one.

When the ICN classifies the vector u^2 as "easy," the initial step directly computes the final answer using vector u^2. In essence, this vector is fed into a new FC layer (FC_E layer in Fig. 16.8) with the same dimensions as those of the final FC layer (FC_H layer in Fig. 16.8). The output is obtained by performing a matrix multiplication with the weight matrix of this new FC layer. In other words, for each input query, the initial step computes u^2 which is provided to the ICN. If the ICN output indicates the "easy" category, the u^2 vector is fed into the FC_E layer to obtain the answer to the question. In this scenario, only one step is utilized to generate the output. If the ICN output, however, indicates the "hard" category, the u^2 vector is passed to the second step. Finally, the last hop output is provided to the FC_H layer to reach the answer to the question.

The structure and architecture of the FC_E layer are almost identical to the that of the final FC_H layer. The key distinction is the weight matrix of this layer which is denoted as W_E and trained based on the output of the first step. The training of W_E performed by using the initial values of the weight matrix W from the final FC_H layer for W_E. In the training process, only the W_E matrix is updated while the embedding matrices are held constant.

Since the primary objective of ICN is to categorize the input queries as either "easy" or "hard," it is essential to assign a label to each input of the considered dataset. For the categorization, the first step of the MANN process for each input is performed. The output of this step (u^2) is passed to the FC_E layer. Note that as mentioned before, FC_E is trained before the labeling process. If the generated answer matches the correct answer, the input is labeled as "easy." If, however, the generated answer is incorrect, the u^2 vector is passed to the second step, and the subsequent steps are computed. The output of the final step (u^{out}) is provided to the FC_H layer. If the answer is correct this time, it is concluded that the input was challenging, and the correct answer could only be reached through the computation of all the steps. Therefore, the "hard" label is assigned to this query. If, after performing all the steps, the correct answer is still not reached, the input is labeled as "easy" to minimize unnecessary network computations. This approach aims to reduce the computations without compromising the accuracy.

The ICN network should be designed based on the desired dataset and the structural parameters of the initial MANN. In this context, two variations of the ICN network, each corresponding to the specific evaluation dataset, may be designed for evaluating the A^2P-MANN. One of the designs is tailored for the foundational MANN structure and the 20 tasks of the bAbI dataset. This dataset has been considered as the primary benchmark for MANNs in different references (see, e.g., [7, 87–89]). The second ICN structure is intended for the WikiMovies dataset, which is primarily used to evaluate KV-MANNs. The results obtained for these two variations revealed that employing small linear layers to implement the ICN leads to an acceptable level of accuracy [50]. More specifically, a two-layer linear ICN network can achieve approximately 86% accuracy for the MANN network with the bAbI dataset and 84% accuracy for the KV-MANN network with the WikiMovies dataset. The first layer of these ICN networks transfers the input vector u^2 with a dimension of d to a hidden layer with fewer neurons. The dimension d was set to 40 for the base MANN and 500 for the KV-MANN. The numbers of neurons in the hidden layer of the ICN designed for the base MANN and the KV-MANN, were also selected as 32 and 200, respectively. Here, the ReLU function was adopted as the activation function for the neurons in the hidden layer where its output is transferred to the output layer with two neurons corresponding to "hard" and "easy" classes.

Since the accuracy of ICN may not reach 100%, it is inevitable to have false positive inputs which refer to "hard" inputs that are mistakenly identified as "easy" by the ICN. Such misclassifications can lead to a significant decrease in A^2P-MANN's accuracy. Using an activation module at the output of ICN, similar to what suggested in reference [7], can reduce these false positive classifications improving the accuracy. This suggested module makes use of a confidence value (in other words, a probability value) of z, and a confidence threshold to determine whether the ICN network is sufficiently confident in classifying the input as "easy" or not. When the output of the first step is fed to ICN, the ICN output is then provided to this module. It is known that the ICN output contains two probability values, or logits, for the "easy" and "hard" classes. In this module, if the maximum logit corresponds to the "easy" class (indicating that the input is labeled as "easy" by the ICN), the maximum logit value is compared with the confidence threshold. If it is greater than the threshold, it is implied that the ICN is sufficiently confident about this input "easy" classification. If, however, the maximum logit value is lower than the confidence threshold, it signifies that the ICN lacks sufficient confidence in labeling the input as "easy." This module, therefore, does not trust the ICN classifications and passes the u^2 vector to the subsequent steps so that all those stages are executed.

The approach has the drawback of increasing the number of false negative inputs, which are the "easy" inputs mistakenly identified as "hard." This diminishes the advantage of reduced computations. Fortunately, such errors where the network follows its original (not shorten) path, do not impact the final accuracy of MANN.

The assessment of the proposed method indicates that more than 65% and 85% of the words in the vocabulary are never the correct answers to any of the questions in the bAbI and WikiMovies datasets, respectively [50]. In other words, these words have never been labeled correctly during the training phase. In fact, the MANN network has learned not to select the correct final answer from these words. This phenomenon is directly related to the fact that only a limited number of words in the vocabulary have been used as answers for the queries relevant to the dataset. Most of the words in the vocabulary consist of prepositions, question words, articles, etc., which logically do not serve as answers to the questions. It is, therefore, suggested that the rows corresponding to these words, termed "unused rows," be removed from the weight matrices of the FC_H and FC_E layers, namely, matrices W and W_E, respectively, without degrading the final accuracy.

The second proposed pruning method is based on removing "unimportant rows" from the weight matrices of the FC layers. The decrease in the accuracy caused by this method is also minimal yet controllable. To find unimportant rows, a threshold θ_p is proposed in A^2P-MANN technique where all the values in each row are compared with this threshold. Rows with a large number (denoted as N_p) of weights with absolute values less than the threshold (symbolized as θ_p) are considered as "unimportant rows."

By employing these methods, a significant portion of the computations in the output layer is eliminated, and even the possibility of parallelizing the computations at this layer is provided. In some cases, depending on the values considered for θ_p and N_p, which themselves are functions of the maximum tolerable drop in the accuracy for the target application, the two methods may suggest the removal of similar rows to a large extent. This implies that they may pinpoint to the same rows for the pruning process. In Fig. 16.9, a color-coded representation of the weight matrices and their pruned values are depicted. The embedding dimension is depicted on the horizontal axis, while the vocabulary words are represented on the vertical axis. As is observed, a significant portion of the image is purple, indicating many

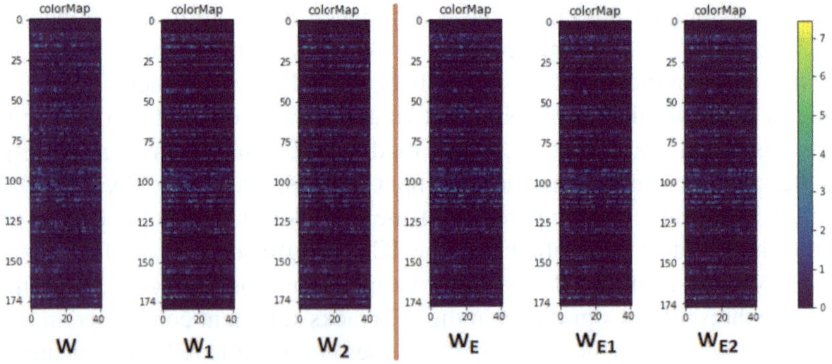

Fig. 16.9 The values of W and W_{E} and their pruning by the method of (1) removing unused rows and (2) removing unimportant rows

values are below the pruning threshold (almost close to the zero) in these matrices and only a limited number of rows are above the threshold (bright and yellow colors). This pictorially demonstrates the ability of the pruning methods to remove a large of rows for these dataset without causing a substantial decrease in accuracy as the results presented in [50] indicates. It should be noted that the same values for θ_{p} and N_{p} have been used here for both W and W_{E} matrices, which are used for $\mathrm{FC_H}$ and $\mathrm{FC_E}$ layers, respectively.

To properly implement the pruning operation, the indices of the rows intended for deletion, first, is identified using the first or the second method. The row indices not identified are stored in a list which form a new matrix containing only the important rows. Now, when the network is used in the inference phase, upon reaching the FC layer, instead of using the original matrix, the pruned new matrix is used. When the maximum argument (*argmax*) operation, which identifies the index of the maximum value in a list of values, is executed at the conclusion of the network, the resulting index is appended to the list of significant indices to determine the accurate and definitive index, representing the final answer to the question.

It should be noted that, even though "unused rows" and "unimportant rows" in the final matrix multiplication in the FC layers are eliminated, these rows and their corresponding words in the embedding matrices are essential for the inference computations and should not be removed. During the FC layer computations, the network effectively calculates the similarity between the feature vector u^1 or u^{out} and each of the rows. Thus, the word with the highest similarity or alignment will be assigned the highest probability logits. On the other hand, the words corresponding to the rows proposed to be deleted are not among the correct tags in the dataset. During the training phase, therefore, these rows in the matrices W and W_{E} reach the smallest absolute values to have the least alignment with the u^1 or u^{out} vector and, as a result, assign the lowest probability logits to themselves. This is why their absolute values become so small that they can be eliminated from the final layer. During the inference steps, the network measures the alignment and similarity of each sentence with the question sentence. For this operation, all the words and their corresponding rows in the embedding matrices are essential and should not be removed to prevent the final accuracy degradation.

Matrix pruning in the A^2P-MANN method typically results in unstructured sparsity where individual elements of a matrix are set to zero without any particular pattern or structure. To obtain gain on reducing the inference time from the pruning, the employed hardware platform should support sparse matrix multiplication. Currently, Google's tensor processing units (TPUs) and NVIDIA's ampere GPUs support sparse multiplication [4, 90]. Also, in the literature, a variety of architectures for efficient sparse matrix multiplication which may be employed in the costume realization of the technique have been suggested (see, e.g., [91–94]). If needed, algorithmic modifications to adapt A^2P-MANN pruning techniques may be invoked when considering constraints of the existing hardware [92]. This may involve exploring hybrid approaches that combine structured and unstructured sparsity to maximize the utilization of available hardware resources [93]. Also, fine-tuning and adaptation strategies can be employed to optimize the pruning process based on the specific characteristics of the underlying hardware architecture [94], thereby mitigating the impact of unstructured sparsity on overall system performance improvement [95].

The scalability imposed due to the computational complexity of memory-augmented neural networks (MANNs) may be related to the required floating-point operations (FLOPs). The computational complexities (denoted by CC) of *single hop* for pre-embedded ($CC_{H,E}$) and interactive ($CC_{H,I}$) applications are derived from [50]

$$CC_{H,E} = \begin{cases} n_s \times (4 \times d + 12) - 1, \text{conventional MANN} \\ n_s \times (4 \times d + 12) - 1 + 2 \times d^2, \text{KV} - \text{MANN} \end{cases}$$

$$CC_{H,I} = n_s \times [(4 \times n_w + 2) \times d + 12] - 1, \text{conventional MANN}$$

where d represents the embedding dimension, n_s is the number of sentences in the input story, and n_w is the word count in the given story. Here, each individual addition and multiplication (division and exponentials) operation is counted as 1 (4) FLOP [96]. By employing $CC_{H,E}$ and $CC_{H,I}$, the computational complexity of an m-hop (including the final fully connected (FC) layer) in the cases of pre-embedded (CC_E) and interactive (CC_I) applications are determined from [50]

$$CC_E = \begin{cases} (2 \times n_w - 1) \times d + m \times CC_{H,E} + V \times (2 \times d - 1), \text{conventional MANN} \\ (n_w - 1) \times d + m \times CC_{H,E} + V \times (2 \times d - 1), \text{KV} - \text{MANN} \end{cases}$$

$$CC_I = (2 \times n_w - 1) \times d + m \times CC_{H,I} + V \times (2 \times d - 1), \text{conventional MANN}$$

where V denotes the total number of words in the dictionary. As the dataset size increases, the number of FLOPs increases, leading to increased computational volume. Based on the details provided for the proposed A^2P-MANN, the computational complexity reduction (CR or computation reduction in FLOPs) of the models for pre-embedded and interactive applications can be obtained from [50]

$$\mathrm{CR_E} = \zeta E \times (m-1) \times \mathrm{CC_{H,E}} + P_\mathrm{R} \times [V \times (2 \times d - 1)] \\ - [2 \times L1 \times (d+2) + 25]$$

$$\mathrm{CR_I} = \zeta E \times 2 \times \mathrm{CC_{H,I}} + P_\mathrm{R} \times [V \times (2 \times d - 1)] - [2 \times L1 \times (d+2) + 25]$$

where ζE represents the probability of an input query being classified as an easy input by the input classifier network (ICN), P_R is the ratio of pruned rows to the total number of rows in the final FC layers, and $L1$ indicates the number of ICN hidden layer neurons. The last terms in the formulas denote the overhead of the considered ICN network, which should be adjusted when employing an ICN with more hidden layers [50].

16.5 Posit MANN

In DNNs, the number system can be considered as a design choice affecting efficiency of the network. In addition, reducing the bit-width, known as low-precision computing, has been considered as an approach to reduce energy consumption by reducing the memory footprint and computational requirements (see, e.g., [97–99]). Different number systems, including logarithmic systems [100, 101], fixed-point, continuous fractions [102, 103], floating-point numbers [104], and Posit [46] have been introduced considered for the use in DDNs. While floating-point computations are prevalent in modern computers, they may face intolerable accuracy degradation for low-bit representations [105]. To avoid the degradation, higher-bit representation which requires increased resource consumption should be invoked. As another alternative, one may utilize another number system which provides a higher accuracy for the same number of bit representation as that of the floating-point system. The Posit number system has demonstrated its potential as a promising alternative to the floating-point system [46]. In this section, we deal with the use of the Posit number system for the implementation of the MANN.

Because of the absence of support for the Posit number system on current computing platforms such as CPU, GPU, and TPU, two approaches for the implementation of the MANN with Posit number (Posit-based MANN) were adopted. In the first one, a fake quantization [60] approach to implement the Posit-based MANN was used. In this approach, a quantization unit is introduced at the output of each computing unit within the MANN. In addition, in this approach, called as the *partial Posit* MANN, the input data undergo quantization before being fed into the MANN framework. The quantization unit (denoted as Posit_Q) is responsible for converting its input from the FP32 format to the desired Posit format (e.g., 16-bit Posit with an

```
function Posit_Q (FP32 x):
  # Step 1: Transfer sign bit
  Posit_sign = x[31]
  # Step 2: Calculate 'e - bias' and determine the first '1' location
  exponent = x[30:23]
  bias = binary(127)
  first_1_location = exponent - bias
  # Determine 'k' and 'e' using the relationship 'the first 1 location = K * 2^es + e'
  es = 1  # Value of es
  k = first_1_location div (2 ** es)
  e_Posit = first_1_location mod (2 ** es)
  # Repeat bits based on 'k' and 'm' for find regim bits string
  m = abs(k)
  Posit_regime = "1" * (m + 1) if k > 0 else "0" * m
  Posit_regime += "0" if m == 1 else "1"
  # Represent 'e' in es bits for the exponent
  Posit_exponent = e_Posit
  # Step 3: Allocate remaining bits to the Posit fraction part
  Posit_fraction = x[22:22-(16-(m-2))]
  # Construct the 16-bit Posit representation
  Posit16 O = {Posit_sign, Posit_regime, Posit_exponent, Posit_fraction}
return O
```

Fig. 16.10 The pseudo-code of the propose converting function

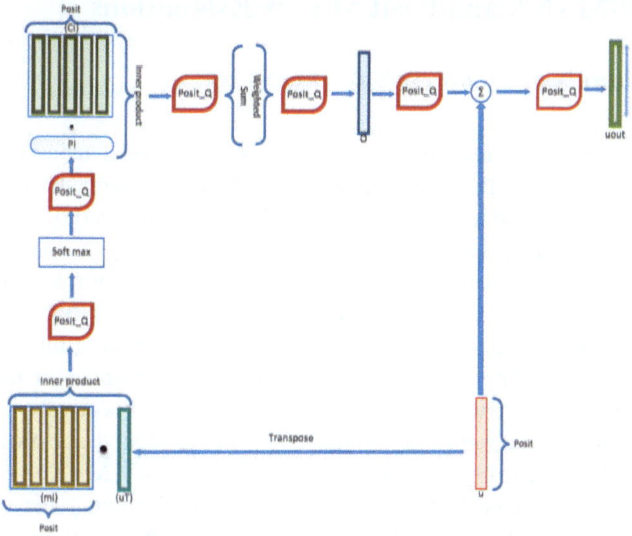

Fig. 16.11 The operations performed in each layer of the MANN network using the Partially posit MANN

exponent size (es) of 1). To convert FP32 to Posit, a suggested efficient conversion mechanism is invoked. The pseudo-code for converting FP32 to Posit16 with es = 1 is presented in Fig. 16.10 and the Posit-based implementation of a MANN layer where a Posit_Q unit is placed after each computing unit is depicted in Fig. 16.11.

In the second approach, the MANN structure is instantiated utilizing the Posit data type, the Posit number system with 16-bit integer is modeled at the software level by defining it as a new data type where all the Posit arithmetic operations are emulated based on this data type. This approach is called the *full Posit* MANN.

Given the common use of MAC operations in machine learning applications, a Posit MAC operation with a 32-bit quire format [46] for the internal accumulation is introduced. A quire in the Posit arithmetic is an extended precision accumulator designed to allow the exact accumulation of the products without rounding errors. The quire typically has significantly more bits than the Posit itself, enabling it to hold intermediate results of multiple operations, especially multiplications, in the fixed-point format without the risk of precision loss. More specifically, in this operation, the multiplication is performed using the Posit number system and the accumulation is performed within the integer domain whose corresponding data type is referred to as quire [46]. In addition to the inference phase, both approaches discussed above are applied for the backpropagation step computations, including loss calculation, weight updating, as well as derivation operation. This means that the Posit MANNs are trained using the Posit number system.

16.6 A^{2}P-MANN and Posit MANN Evaluations

16.6.1 Models and Datasets of the Study

The evaluation was carried out on the MANN base model using 20 tasks from the bAbI dataset as well as on the KV-MANN model with 13 tasks from the WikiMovies dataset. The test data for the bAbI dataset consisted of 20,000 samples, each containing a story (with number of sentences of n_s), a question sentence, and an answer word label. Each task consisted of 1000 samples, resulting in a total of 20,000 samples (20 tasks). The considered base MANN model is a three-layer model that employs Positional encoding (PE matrix multiplication). The parameters d (embedding dimension), n_s (number of sentences or memory size), and d (vocabulary size) were set to 40, 50, and 174, respectively. The test data for the WikiMovies dataset comprised 10,000 movie-related questions that could be categorized into 13 different types of tasks. In the KV-MANN model, before the inference phase, facts (Wikipedia documents about movies) were stored in the added memory in the form of a key-value structure. The parameters n_s (memory size or number of document windows), d, and v were set to 50, 500, and 238,000, respectively, for this dataset. Also, when using KV-MANN for the bAbI dataset, the parameters n_s, d, and v were set to 40, 50, and 174, respectively.

In this study, the effectiveness of the A^{2}P-MANN approach is evaluated using both MANN with three steps (for the bAbl dataset) and KV-MANN with two steps (for the WikiMovies dataset). Additionally, the partial Posit implementation on both MANN and KV-MANN with three steps (for bAbl datasets) was studied while the fully Posit implementation was exclusively assessed using MANN with three steps (for the bAbl datasets). For the A^{2}P-MANN approach, in the case of both considered MANN architectures, ICNs were trained concurrently on all the tasks, and a weighted cross-entropy loss function was employed during the training.

Additionally, the Adam optimizer [106] was used with a learning rate of 0.01 (0.001) for the training of ICNs, in the case of MANN/bAbI (KV-MANN/WikiMovies) benchmarks. For the partial Posit implementation, the basic MANN network was trained using the SGD optimizer [107] with an initial learning rate of 0.01, gradually being decreased to 0.005 after 25 epochs. Also, the batch size during each training epoch was set to 32 [7]. For the KV-MANN network in most tasks, the SGD optimizer with a learning rate of 0.01 was employed. For the full Posit implementation, the MANN network was trained using the SGD optimizer with a constant learning rate of 0.01, and the batch size of 32 during each training epoch [7].

16.6.2 A^2P-MANN Evaluation

A^2P-MANN introduced two pruning methods for removing unused or unimportant rows for the weight matrices of the final FC layers. For the bAbI/MANN (WikiMovies/KV-MANN) benchmark, the values of 0.1 (0.05) and 13 (240) for the parameters θ_p and N_p, respectively, were used. The values provided negligible loss of accuracy. Furthermore, more than 90% of the selected unimportant rows were among the unused rows. It should be noted that increasing θ_p or decreasing N_p leads to the removal of more rows degrading the final accuracy.

The ICN network was configured by three scenarios for distinguishing between the easy and hard inputs. They were (1) no confidence threshold: Represented by NC, (2) global confidence threshold, and (3) PerTask confidence threshold [50]. Figure 16.12 illustrates the percentage of easy input detection for various tasks under each of these three scenarios. The last three bars in both graphs represent the average results for all the tasks in each scenario. As is evident from the figure, for the MANN (KV-MANN) network on the bAbI (WikiMovies) dataset, ~75% (70%) of the inputs are classified as easy in the NC, ~67% (63%) in the Global, and ~63% (59%) in the PerTask scenarios.

The final accuracy loss of A2P-MANN compared to the conventional MANN and KV-MANN under two main scenarios of PerTask and global are shown in Fig. 16.13. The figure shows that, in the case of the bAbI tasks, the worst-case accuracy reduction (loss) compared to conventional MANN is 4.6% for task 14. In some cases (e.g., tasks 11 and 13), however, the final accuracy improves by as much as 2.7% compared to the case of conventional MANN. In the case of WikiMovies tasks, the worst-case corresponds to the "*movie to director*" task with a 3.4% reduction in the accuracy, while for some tasks (e.g., "*movie to imdbrating*" and "*movie to imdbvotes*"), an accuracy improvement of about 1.1% is achieved. As mentioned earlier, the selected parameters for pruning methods (as well as the zero skipping method) have no adverse impact on the final accuracy. This means that the reduction in the accuracy is a result of ICN performance and FC_E layer efficiency. It is worth noting that considering separate confidence thresholds for tasks (PerTask scenario) in the A^2P-MANN method results in better final accuracy compared to that of the

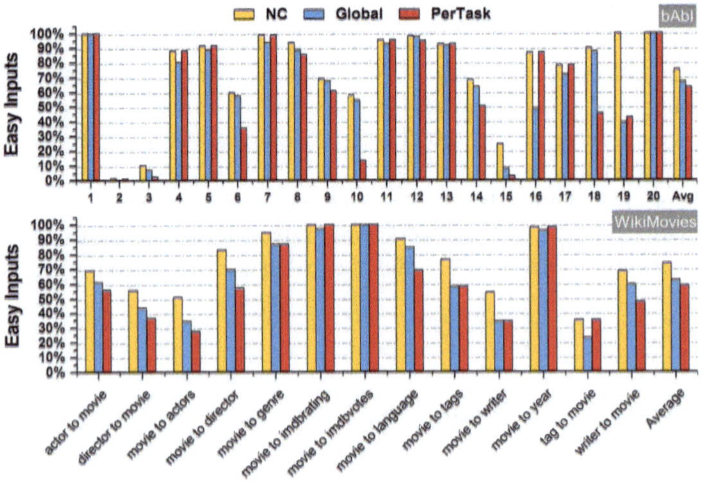

Fig. 16.12 Percentage of easy entries at the ICN output for all three considered confidence threshold scenarios [50]

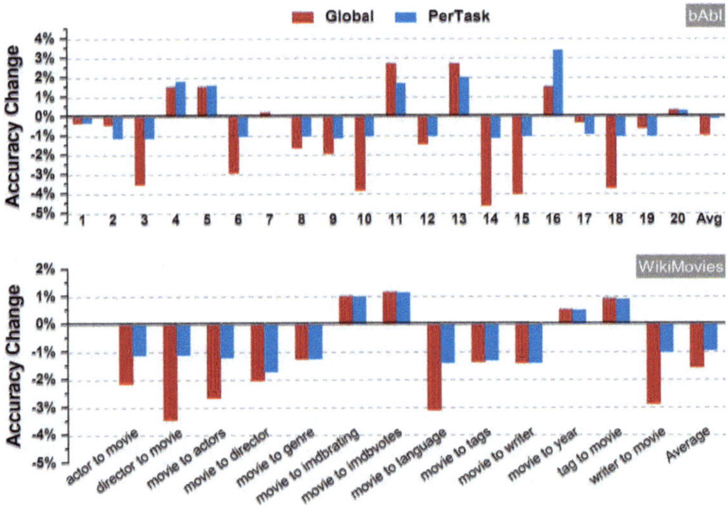

Fig. 16.13 The accuracy degradation of the A^2P-MANN method for the two benchmarks compared to the basic MANN primary networks (upper diagram) and KV-MANN (lower diagram) in the two scenarios presented for the confidence threshold [50]

global scenario. In the PerTask scenario, we aimed to limit the accuracy drop for each task to 1%. On average, the PerTask (global) scenario led to a 0.1% (1%) accuracy decrease in A^2P-MANN compared to the base MANN network for the bAbI dataset. Similarly, for WikiMovies tasks, on average, the A^2P-MANN method in the PerTask (global) scenario resulted in an accuracy drop of approximately 0.9% (1.5%) compared to the initial KV-MANN network.

To assess the performance of A^{2}P-MANN in reducing the required computations, the reduction in computations or, in other words, FLOPs with and without the zero skipping methods for pre-embedded and interactive application types are considered [50]. In interactive applications, the user can provide both the data (relevant story or document) and the question that needs to be answered to the network, whereas in pre-embedded applications, the user typically only inputs the question sentence into the embedded network. Using the parameter θ of 0.01 [108], a reduction of 81% of the computations involving weight summation is achieved. Here, the accuracy has not been degraded. It is essential to note that the zero skipping method results in a 40% (23%) decrease in the computations for interactional (pre-embedded) applications in the case of the bAbI dataset [50]. Furthermore, for pre-embedded applications with WikiMovies tasks, this method leads to an approximately 21% reduction in the computations. On the other hand, the A^{2}P-MANN inference method alone, in both global and PerTask scenarios, provide average reductions of 45.6% (44.8%) and 44.2% (42.5%) in the computations for bAbI tasks in interactional (pre-embedded) applications compared to those of the base MANN. Similarly, for the WikiMovies tasks, on average, the A^{2}P-MANN method provides around a 60% (59%) reduction in the computations for the pre-embedded applications in the Global (PerTask) scenario.

When the zero skipping method is concurrently used with the A^{2}P-MANN method for the bAbI tasks, we generally have about a 68% (59%) and 66% (57.5%) reductions in computations for the interactional applications in the global and PerTask scenarios, respectively. Additionally, for the WikiMovies tasks, in the combined scenario of A^{2}P-MANN and zero skipping, provide a computational reduction of approximately 75% (74%) in the global (PerTask) scenario.

The normalized latencies (NL $= \frac{\text{Latency}_{\text{current scenario}}}{\text{Latency}_{\text{reference scenario}}}$) of A^{2}P-MANN with respect to the conventional MANN and KV-MANN on CPU (Intel Core i7-4790) and GPU (Nvidia 1080ti with 12 GB memory) platforms are illustrated that the proposed A^{2}P-MANN inference compared to conventional MANN/bAbI on the CPU, achieves an average latency reduction of 40% (43%) for pre-embedded (interactive) application types. Similarly, on the GPU, an average latency reduction of 33% (41%) is attained [50]. Note that the A^{2}P-MANN approach, which is designed to optimize the memory usage and reduce the computational complexity, could be indeed applied to other types of memory-augmented neural networks (MANNs) beyond just MEMNN and KV-MANN which were considered in the chapter. It originates from sufficiently general underlying principles and techniques of the approach. Examples of other MANNs include neural Turing machines (NTMs) [8], differentiable neural computers (DNCs) [9], and sparse access memory networks [83] which share the fundamental principle of enhancing neural networks with an external memory component despite having unique mechanisms for the memory access and management. The first reason is that the core idea of improving memory access efficiency, which relies on differentiable read/write operations to manage memory accesses, is applicable across different MANN architectures. Second, the improvements proposed in A^{2}P-MANN, such as zero-skipping and pruning for memory operations, are modular integrable into the memory mechanisms of other

MANNs with minimal modifications. Lastly, the scalability of the techniques used in A^{2}P-MANN provides the opportunity for its use across a wide range of different MANN-based models to optimize their computation, latency, and memory usage features.

16.6.3 Partial Posit MANN Evaluation

In this subsection, the 16-bit (Posit16) and 8-bit (Posit8) Partial Posit MANNs are compared to the 32-bit (FP32) and 16-bit (FP16) floating-point implementations of the MANN and KV-MANN. The *es* parameter value in the Posit implementation was 1. The accuracies of the MANN and KVMAMM under the considered number systems are shown in Figs. 16.14 and 16.15, respectively. As the results reveal, in almost all the cases, the accuracies of the Poist16 (Posit8) were similar to the FP32 (FP16). In few cases also Poist16 (Posit8) led to higher accuracies. On average, the accuracy of the MANN (KV-MANN) in the cases of the FP32, Poist16, FP16 and Posit8 were 89.6% (72.0%), 90.4% (73.1%), 86.3% (66.2%) and 88.2% (67.8%), respectively. The superiority of the Posit, as explained before in Sect. 16.2, is mainly due to its higher precision in representing small numbers around the zero (see Table 16.4).

16.6.4 Full Posit MANN Evaluation

Now, the performance of the 16-bit Full Posit (Posit16) MANN utilizing an *es* parameter value of 1 is compared with that of the 32-bit floating-point (FP32)

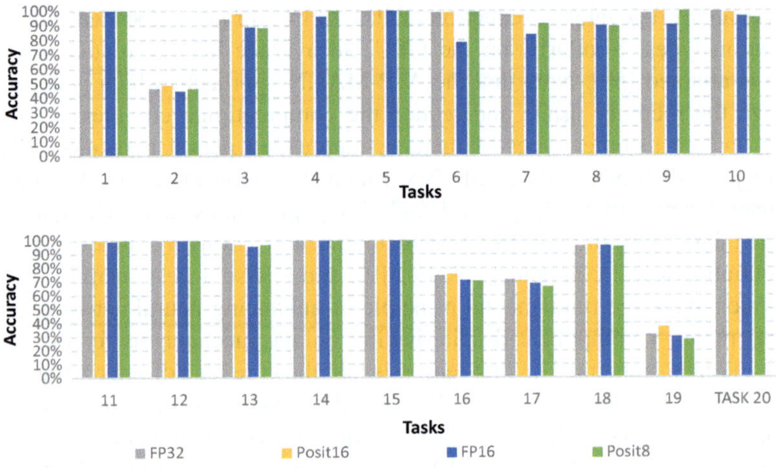

Fig. 16.14 MANN accuracy under different number systems after 60 training epochs

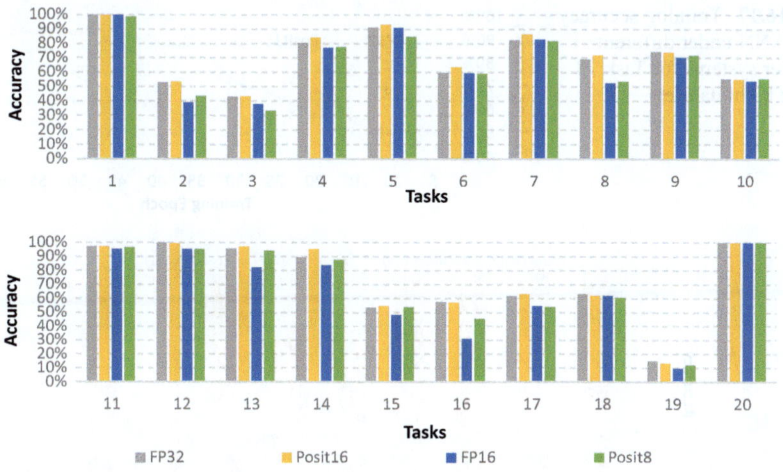

Fig. 16.15 KV-MANN accuracy under different number systems after 200 training epochs

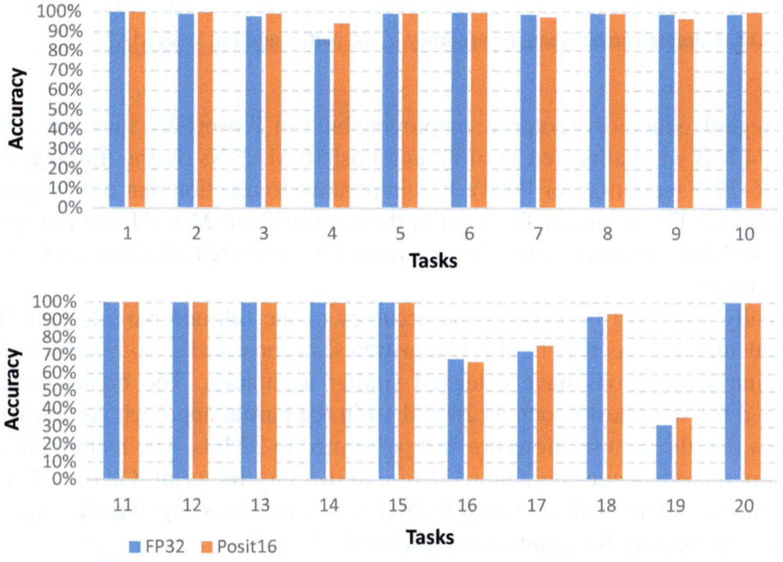

Fig. 16.16 Training Accuracy of the full Posit MANN after 60 epochs

implementation. The accuracy results for various tasks in MANN are depicted in Fig. 16.16. The outcomes reveal that, in nearly all instances, Posit16 exhibits accuracies comparable to that of FP32. Additionally, Posit16 for some tasks provides higher accuracies. On average, the accuracies of MANN for FP32 and Posit16 are 92.03% and 92.82%, respectively. Also, in Fig. 16.17, the training accuracies of the Posit16 and FP32 under different epochs for Task 18 of the bAbI dataset are compared.

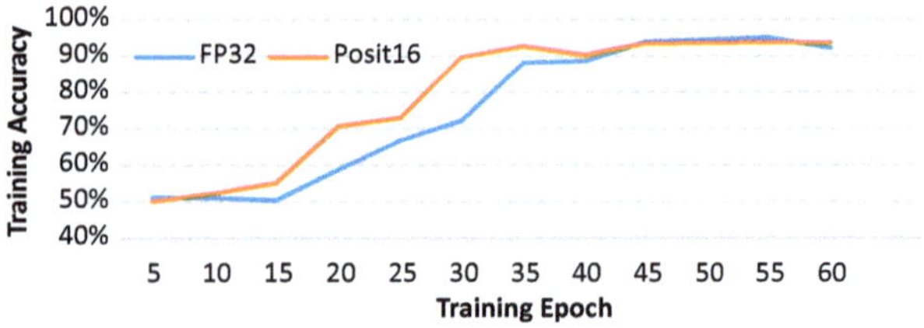

Fig. 16.17 Training accuracy of MANN under different number systems for Task 18 of the bAbI dataset

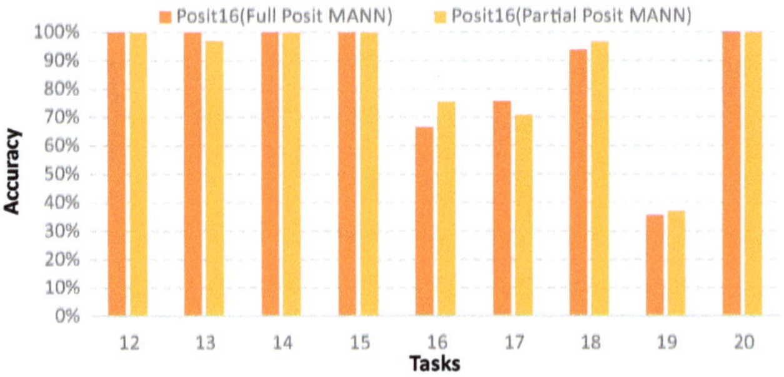

Fig. 16.18 Comparison between the accuracy of the partial and full Posit MANN

In regard to memory footprint, employing built-in TensorFlow functions, 248 MB and 174 MB were allocated to FP32 and Posit16 MANNs during training, respectively (~30% reduction for Posit). It is imperative to mention that utilizing axillary variables (in FP or integer formats) in the case of Posit MANN and the quires in the Posit MAC implementation has decreased the expected ideal memory footprint reduction (50%).

Finally, Fig. 16.18 compares the accuracy of the full and partial Posit MANN implementations. As this figure demonstrates, in most tasks, both partial and full Posit implementations have achieved similar accuracies. The higher accuracies obtained for some tasks such as 28 and 16 in the partial approach are attributed to the fact that the computations are still performed in 32 bits. The method, however, does not provide a considerable reduction in memory footprint. On the other hand, by using the full method, comparable (higher in some cases, e.g., task 17) accuracies with lower memory footprints were achieved.

The normalized latencies (average of 100 runs) of the Full Posit MANN (NL $= \frac{\text{Latency}_{\text{Posit16}}}{\text{Latency}_{\text{FP32}}}$) on CPU (Intel Core i7-8550U) platform are illustrated in Fig. 16.19. The results show that an average latency reduction of 32.8% is achieved by the proposed Full Posit MANN Evaluation.

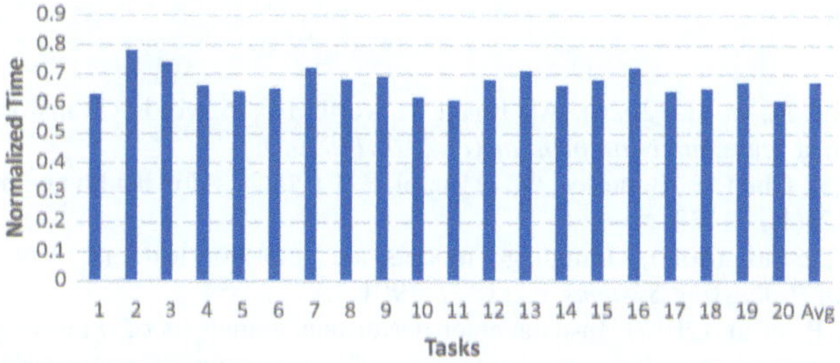

Fig. 16.19 The normalized latency of the Full Posit MANN with respect to the MANN/bAbI on the CPU platform

16.7 Conclusion

After a review of the significance of optimizing deep neural networks for edge devices, we delved into the methods and anticipated challenges in this domain. Our exploration underscored the critical role of memory-augmented neural networks in edge applications and entailed an examination of the essential structures and prerequisites necessary for their optimization. Thus, this chapter introduced two approaches for improving the efficiency of the memory-augmented neural network (MANN) structure. These approaches were based on decreasing the computational and memory requirements these networks providing opportunities for their efficient deployments on edge devices. The first approach, referred to as A^2P-MANN, focused on reducing the number of steps involved in analyzing the input data (query) as well as pruning the weights in the fully connected (FC) layers. The second approach invoked the Posit number system to considerably degrade accuracy of the network in the case of using lower bit-widths for the number representation. The studies demonstrated that A^2P-MANN can achieve a 42% reduction in the computations compared to the original MANN implementation while employing Posit for implementing the original MANN, provided a lower memory usage of ~30% without loss of accuracy. Although no hardware implementation has been performed, the proposed A^2P-MANN approaches could be implemented on FPGA based on the MANN implementation design discussed in [87] by simply adding the ICN (a small NN). In addition, many hardware implementations for Posit arithmetic units have been introduced in the literature (see, e.g., [109, 110]), which may be utilized in place of the binary arithmetic unit in the design suggested in [87], to implement the Posit-MANN.

References

1. Dosovitskiy, A., et al. (2020). An image is worth 16x16 words: Transformers for image recognition at scale. *arXiv preprint arXiv:2010.11929.*
2. Rajpurkar, P., Chen, E., Banerjee, O., & Topol, E. J. (2022). AI in health and medicine. *Nature Medicine, 28*(1), 31–38.
3. Brown, T., et al. (2020). Language models are few-shot learners. *Advances in Neural Information Processing Systems, 33*, 1877–1901.
4. Jouppi, N. P., et al. (2017). In-datacenter performance analysis of a tensor processing unit. In *Proceedings of the 44th annual international symposium on computer architecture* (pp. 1–12). ACM/IEEE.
5. Xu, B., Wang, N., Chen, T., & Li, M. (2015). Empirical evaluation of rectified activations in convolutional network. *arXiv preprint arXiv:1505.00853.*
6. Weston, J., Chopra, S., & Bordes, A. (2014). Memory networks. *arXiv preprint arXiv:1410.3916.*
7. Sukhbaatar, S., Weston, J., & Fergus, R. (2015). End-to-end memory networks. *Advances in Neural Information Processing Systems, 28*, 2440–2448.
8. Graves, A., Wayne, G., & Danihelka, I. (2014). Neural turing machines. *arXiv preprint arXiv:1410.5401.*
9. Graves, A., et al. (2016). Hybrid computing using a neural network with dynamic external memory. *Nature, 538*(7626), 471–476.
10. LeCun, Y. (2019). 1.1 deep learning hardware: Past, present, and future. In *2019 IEEE international solid-state circuits conference-(ISSCC)* (pp. 12–19). IEEE.
11. Collier, M., & Beel, J. (2019). Memory-augmented neural networks for machine translation. *arXiv preprint arXiv:1909.08314.*
12. Feng, L., Li, Z., Kuang, Z., & Zhang, W. (2018). Extractive video summarizer with memory augmented neural networks. In *Proceedings of the 26th ACM international conference on multimedia* (pp. 976–983). ACM.
13. Miller, A., Fisch, A., Dodge, J., Karimi, A.-H., Bordes, A., & Weston, J. (2016). Key-value memory networks for directly reading documents. *arXiv preprint arXiv:1606.03126.*
14. Dodge, J., et al. (2015). Evaluating prerequisite qualities for learning end-to-end dialog systems. *arXiv preprint arXiv:1511.06931.*
15. Hill, F., Bordes, A., Chopra, S., & Weston, J. (2015). The goldilocks principle: Reading children's books with explicit memory representations. *arXiv preprint arXiv:1511.02301.*
16. Weston, J. E. (2016). Dialog-based language learning. *Advances in Neural Information Processing Systems, 29*, 829–837.
17. Devlin, J., Chang, M.-W., Lee, K., & Toutanova, K. (2018). BERT: Pre-training of deep bidirectional transformers for language understanding. *arXiv preprint arXiv:1810.04805.*
18. Topal, M. O., Bas, A., & van Heerden, I. (2021). Exploring transformers in natural language generation: GPT, BERT, and XLNet. *arXiv preprint arXiv:2102.08036.*
19. Wolf, T., et al. (2020). Transformers: State-of-the-art natural language processing. In *Proceedings of the 2020 conference on empirical methods in natural language processing: System demonstrations* (pp. 38–45). Association for Computational Linguistics.
20. Wolf, T., et al. (2019). Huggingface's transformers: State-of-the-art natural language processing. *arXiv preprint arXiv:1910.03771.*
21. Dong, L., et al. (2019). Unified language model pre-training for natural language understanding and generation. *Advances in Neural Information Processing Systems, 32*, 13042–13054.
22. Liu, Y., Shen, G., Tao, G., An, S., Ma, S., & Zhang, X. (2022). PICCOLO: Exposing complex backdoors in NLP transformer models. In *2022 IEEE symposium on security and privacy (SP)* (pp. 2025–2042). IEEE.
23. Khadka, S., Chung, J. J., & Tumer, K. (2019). Neuroevolution of a modular memory-augmented neural network for deep memory problems. *Evolutionary Computation, 27*(4), 639–664.

24. De Divitiis, L., Becattini, F., Baecchi, C., & Del Bimbo, A. (2021). Garment recommendation with memory augmented neural networks. In *International conference on pattern recognition* (pp. 282–295). Springer.
25. Mi, F., & Faltings, B. (2020). Memory augmented neural model for incremental session-based recommendation. *arXiv preprint arXiv:2005.01573.*
26. Wang, T., Xu, X., Shen, F., & Yang, Y. (2021). A cognitive memory-augmented network for visual anomaly detection. *IEEE/CAA Journal of Automatica Sinica, 8*(7), 1296–1307.
27. Gao, H., Qiu, B., Barroso, R. J. D., Hussain, W., Xu, Y., & Wang, X. (2022). TSMAE: A novel anomaly detection approach for internet of things time series data using memory-augmented autoencoder. *IEEE Transactions on Network Science and Engineering, 10*, 2978–2990.
28. Karunaratne, G., et al. (2021). Robust high-dimensional memory-augmented neural networks. *Nature Communications, 12*(1), 2468.
29. Mao, R., et al. (2022). Experimentally realized memristive memory augmented neural network. *arXiv preprint arXiv:2204.07429.*
30. Hung, J.-M., Li, X., Wu, J., & Chang, M.-F. (2020). Challenges and trends indeveloping nonvolatile memory-enabled computing chips for intelligent edge devices. *IEEE Transactions on Electron Devices, 67*(4), 1444–1453.
31. Mach, P., & Becvar, Z. (2017). Mobile edge computing: A survey on architecture and computation offloading. *IEEE Communications Surveys & Tutorials, 19*(3), 1628–1656.
32. Jiang, C., et al. (2020). Energy aware edge computing: A survey. *Computer Communications, 151*, 556–580.
33. Yao, X., Farha, F., Li, R., Psychoula, I., Chen, L., & Ning, H. (2021). Security and privacy issues of physical objects in the IoT: Challenges and opportunities. *Digital Communications and Networks, 7*(3), 373–384.
34. Kum, S., Kim, Y., & Moon, J. (2019). Deploying deep neural network on edge-cloud environment. In *2019 International conference on information and communication technology convergence (ICTC)* (pp. 242–244). IEEE.
35. Hadidi, R., Cao, J., Xie, Y., Asgari, B., Krishna, T., & Kim, H. (2019). Characterizing the deployment of deep neural networks on commercial edge devices. In *2019 IEEE international symposium on workload characterization (IISWC)* (pp. 35–48). IEEE.
36. Zhou, L., Samavatian, M. H., Bacha, A., Majumdar, S., & Teodorescu, R. (2019). Adaptive parallel execution of deep neural networks on heterogeneous edge devices. In *Proceedings of the 4th ACM/IEEE symposium on edge computing* (pp. 195–208). ACM/IEEE.
37. Isakov, M., Gadepally, V., Gettings, K. M., & Kinsy, M. A. (2019). Survey of attacks and defenses on edge-deployed neural networks. In *2019 IEEE high performance extreme computing conference (HPEC)* (pp. 1–8). IEEE.
38. Gobieski, G., Lucia, B., & Beckmann, N. (2019). Intelligence beyond the edge: Inference on intermittent embedded systems. In *Proceedings of the twenty-fourth international conference on architectural support for programming languages and operating systems* (pp. 199–213). ACM.
39. Reagen, B., et al. (2018). Ares: A framework for quantifying the resilience of deep neural networks. In *Proceedings of the 55th annual design automation conference* (pp. 1–6). ACM/IEEE.
40. Han, S., Pool, J., Tran, J., & Dally, W. (2015). Learning both weights and connections for efficient neural network. *Advances in Neural Information Processing Systems, 28*, 1135–1143.
41. Zhou, S., Wu, Y., Ni, Z., Zhou, X., Wen, H., & Zou, Y. (2016). DoReFa-Net: Training low bitwidth convolutional neural networks with low bitwidth gradients. *arXiv preprint arXiv:1606.06160.*
42. Langroudi, H. F., Carmichael, Z., Pastuch, D., & Kudithipudi, D. (2019). Cheetah: Mixed low-precision hardware & software co-design framework for DNNs on the edge. *arXiv preprint arXiv:1908.02386.*
43. Mellempudi, N., Srinivasan, S., Das, D., & Kaul, B. (2019). Mixed precision training with 8-bit floating point. *arXiv preprint arXiv:1905.12334.*

44. Köster, U., et al. (2017). Flexpoint: An adaptive numerical format for efficient training of deep neural networks. *Advances in Neural Information Processing Systems, 30*, 1742–1752.
45. Li, J., Chen, G., Jin, M., Mao, W., & Lu, H. (2024). AE-Qdrop: Towards accurate and efficient low-bit post-training quantization for a convolutional neural network. *Electronics, 13*(3), 644.
46. Gustafson, J. L., & Yonemoto, I. T. (2017). Beating floating point at its own game: Posit arithmetic. *Supercomputing Frontiers and Innovations, 4*(2), 71–86.
47. Gustafson, J. (2017). Posit arithmetic. In *Mathematica Notebook describing the posit number system*. John L. Gustafson.
48. Mallasén, D., Del Barrio, A. A., & Prieto-Matias, M. (2023). Big-PERCIVAL: Exploring the native use of 64-bit posit arithmetic in scientific computing. *arXiv preprint arXiv:2305.06946*.
49. Azizi, S., Nazemi, M., Kamal, M., & Pedram, M. (2023). Low-precision mixed-computation models for inference on edge. *arXiv preprint arXiv:2312.02210*.
50. Ahmadzadeh, M., Kamal, M., Afzali-Kusha, A., & Pedram, M. (2023). A^2P-MANN: Adaptive attention inference hops pruned memory-augmented neural networks. *IEEE Transactions on Neural Networks and Learning Systems, 34*, 8284–8296.
51. Ahmed, S., Hossain, M. F., Kaiser, M. S., Noor, M. B. T., Mahmud, M., & Chakraborty, C. (2021). Artificial intelligence and machine learning for ensuring security in smart cities. In *Data-driven mining, learning and analytics for secured smart cities: Trends and advances* (pp. 23–47). Springer.
52. Reddy, V. M., Vaishnavi, T., & Kumar, K. P. (2023). Speech-to-text and text-to-speech recognition using deep learning. In *2023 2nd international conference on edge computing and applications (ICECAA)* (pp. 657–666). IEEE.
53. Yu, W., Liu, Y., Dillon, T., & Rahayu, W. (2022). Edge computing-assisted IoT framework with an autoencoder for fault detection in manufacturing predictive maintenance. *IEEE Transactions on Industrial Informatics, 19*(4), 5701–5710.
54. Chang, Z., Liu, S., Xiong, X., Cai, Z., & Tu, G. (2021). A survey of recent advances in edge-computing-powered artificial intelligence of things. *IEEE Internet of Things Journal, 8*(18), 13849–13875.
55. Li, E., Zeng, L., Zhou, Z., & Chen, X. (2019). Edge AI: On-demand accelerating deep neural network inference via edge computing. *IEEE Transactions on Wireless Communications, 19*(1), 447–457.
56. Lin, D., Talathi, S., & Annapureddy, S. (2016). Fixed point quantization of deep convolutional networks. In *International conference on machine learning* (pp. 2849–2858). PMLR.
57. Yang, J., et al. (2019). Quantization networks. In *Proceedings of the IEEE/CVF conference on computer vision and pattern recognition* (pp. 7308–7316). IEEE.
58. Jiang, C. (2021). Efficient quantization techniques for deep neural networks. In *2021 International conference on signal processing and machine learning (CONF-SPML)* (pp. 271–277). IEEE.
59. Dobronets, B. S., & Popova, O. A. (2012). Numerical probabilistic analysis under aleatory and epistemic uncertainty. In *Book of abstracts*. Citeseer.
60. Gholami, A., Kim, S., Dong, Z., Yao, Z., Mahoney, M. W., & Keutzer, K. (2022). A survey of quantization methods for efficient neural network inference. In *Low-power computer vision* (pp. 291–326). Chapman and Hall/CRC.
61. Molchanov, P., Tyree, S., Karras, T., Aila, T., & Kautz, J. (2016). Pruning convolutional neural networks for resource efficient inference. *arXiv preprint arXiv:1611.06440*.
62. Blalock, D., Gonzalez Ortiz, J. J., Frankle, J., & Guttag, J. (2020). What is the state of neural network pruning? *Proceedings of Machine Learning and Systems, 2*, 129–146.
63. Li, G., Qian, C., Jiang, C., Lu, X., & Tang, K. (2018). Optimization based layer-wise magnitude-based pruning for DNN compression. In *Proceedings of the 27th international joint conference on artificial intelligence* (pp. 2383–2389). AAAI Press.
64. Anwar, S., Hwang, K., & Sung, W. (2017). Structured pruning of deep convolutional neural networks. *ACM Journal on Emerging Technologies in Computing Systems (JETC), 13*(3), 1–18.

65. Gordienko, Y., Kochura, Y., Taran, V., Gordienko, N., Bugaiov, A., & Stirenko, S. (2019). Adaptive iterative pruning for accelerating deep neural networks. In *2019 XIth international scientific and practical conference on electronics and information technologies (ELIT)* (pp. 173–178). IEEE.
66. Tanghatari, E., Kamal, M., Afzali-Kusha, A., & Pedram, M. (2023). Federated learning by employing knowledge distillation on edge devices with limited hardware resources. *Neurocomputing, 531*, 87–99.
67. Fang, W., Xue, F., Ding, Y., Xiong, N., & Leung, V. C. (2020). EdgeKE: An on-demand deep learning IoT system for cognitive big data on industrial edge devices. *IEEE Transactions on Industrial Informatics, 17*(9), 6144–6152.
68. Shuvo, M. M. H., Islam, S. K., Cheng, J., & Morshed, B. I. (2022). Efficient acceleration of deep learning inference on resource-constrained edge devices: A review. *Proceedings of the IEEE, 111*(1), 42–91.
69. Weston, J., et al. (2015). Towards ai-complete question answering: A set of prerequisite toy tasks. *arXiv preprint arXiv:1502.05698.*
70. Murillo, R., Del Barrio, A. A., & Botella, G. (2020). Deep PeNSieve: A deep learning framework based on the posit number system. *Digital Signal Processing, 102*, 102762.
71. Carmichael, Z., Langroudi, H. F., Khazanov, C., Lillie, J., Gustafson, J. L., & Kudithipudi, D. (2019). Deep positron: A deep neural network using the posit number system. In *2019 Design, automation & test in Europe conference & exhibition (DATE)* (pp. 1421–1426). IEEE.
72. Lu, J., Fang, C., Xu, M., Lin, J., & Wang, Z. (2020). Evaluations on deep neural networks training using posit number system. *IEEE Transactions on Computers, 70*(2), 174–187.
73. Courbariaux, M., Bengio, Y., & David, J.-P. (2015). BinaryConnect: Training deep neural networks with binary weights during propagations. *Advances in Neural Information Processing Systems, 28*, 3123–3131.
74. Miyashita, D., Lee, E. H., & Murmann, B. (2016). Convolutional neural networks using logarithmic data representation. *arXiv preprint arXiv:1603.01025.*
75. Micikevicius, P., et al. (2017). Mixed precision training. *arXiv preprint arXiv:1710.03740.*
76. Langroudi, H. F., Carmichael, Z., Gustafson, J. L., & Kudithipudi, D. (2019). Positnn framework: Tapered precision deep learning inference for the edge. In *2019 IEEE space computing conference (SCC)* (pp. 53–59). IEEE.
77. Johnson, J. (2018). Rethinking floating point for deep learning. *arXiv preprint arXiv:1811.01721.*
78. Bengio, Y., Simard, P., & Frasconi, P. (1994). Learning long-term dependencies with gradient descent is difficult. *IEEE Transactions on Neural Networks, 5*(2), 157–166.
79. Hochreiter, S., & Schmidhuber, J. (1997). Long short-term memory. *Neural Computation, 9*(8), 1735–1780.
80. Feng, H., & Shi, F. (2021). Deep learning in natural language processing, edited by Li Deng and Yang Liu. Singapore: Springer, 2018. ISBN 9789811052088. XVII+ 329 pages. *Natural Language Engineering, 27*(3), 373–375.
81. Graves, A., Mohamed, A.-r., & Hinton, G. (2013). Speech recognition with deep recurrent neural networks. In *2013 IEEE international conference on acoustics, speech and signal processing* (pp. 6645–6649). IEEE.
82. Li, P., Yu, H., Zhang, W., Xu, G., & Sun, X. (2020). SA-NLI: A supervised attention based framework for natural language inference. *Neurocomputing, 407*, 72–82.
83. Rae, J., et al. (2016). Scaling memory-augmented neural networks with sparse reads and writes. *Advances in Neural Information Processing Systems, 29*, 3628–3636.
84. Baeza-Yates, R., & Ribeiro-Neto, B. (1999). *Modern information retrieval.* ACM Press.
85. Ashish, V. (2017). Attention is all you need. *Advances in Neural Information Processing Systems, 30*, I.
86. Bordes, A., Usunier, N., Chopra, S., & Weston, J. (2015). Large-scale simple question answering with memory networks. *arXiv preprint arXiv:1506.02075.*

87. Jang, H., Kim, J., Jo, J.-E., Lee, J., & Kim, J. (2019). MnnFast: A fast and scalable system architecture for memory-augmented neural networks. In *Proceedings of the 46th international symposium on computer architecture* (pp. 250–263). ACM/IEEE.
88. Stevens, J. R., Ranjan, A., Das, D., Kaul, B., & Raghunathan, A. (2019). Manna: An accelerator for memory-augmented neural networks. In *Proceedings of the 52nd Annual IEEE/ACM international symposium on microarchitecture* (pp. 794–806). ACM/IEEE.
89. Park, S., Jang, J., Kim, S., & Yoon, S. (2019). Energy-efficient inference accelerator for memory-augmented neural networks on an FPGA. In *2019 Design, automation & test in Europe conference & exhibition (DATE)* (pp. 1587–1590). IEEE.
90. Krashinsky, R., Giroux, O., Jones, S., Stam, N., & Ramaswamy, S. (2020). *NVIDIA Ampere architecture in-depth*. NVIDIA blog: https://devblogs.nvidia.com/nvidia-ampere-architecture-in-depth.
91. Han, S., et al. (2016). EIE: Efficient inference engine on compressed deep neural network. *ACM SIGARCH Computer Architecture News, 44*(3), 243–254.
92. Wen, W., Wu, C., Wang, Y., Chen, Y., & Li, H. (2016). Learning structured sparsity in deep neural networks. *Advances in Neural Information Processing Systems, 29*, 2074–2082.
93. Narang, S., Elsen, E., Diamos, G., & Sengupta, S. (2017). Exploring sparsity in recurrent neural networks. *arXiv preprint arXiv:1704.05119*.
94. Han, S., Mao, H., & Dally, W. J. (2015). Deep compression: Compressing deep neural networks with pruning, trained quantization and huffman coding. *arXiv preprint arXiv:1510.00149*.
95. Mao, H., et al. (2017). Exploring the regularity of sparse structure in convolutional neural networks. *arXiv preprint arXiv:1705.08922*.
96. Thant, H. A., San, K. M., Tun, K. M. L., Naing, T. T., & Thein, N. (2005). Mobile agents based load balancing method for parallel applications. In *6th Asia-Pacific symposium on information and telecommunication technologies* (pp. 77–82). IEEE.
97. Jain, S., Venkataramani, S., Srinivasan, V., Choi, J., Chuang, P., & Chang, L. (2018). Compensated-DNN: Energy efficient low-precision deep neural networks by compensating quantization errors. In *Proceedings of the 55th annual design automation conference* (pp. 1–6). ACM/IEEE.
98. Latotzke, C., & Gemmeke, T. (2021). Efficiency versus accuracy: A review of design techniques for DNN hardware accelerators. *IEEE Access, 9*, 9785–9799.
99. Yin, S., Srivastava, G., Venkataramanaiah, S. K., Chakrabarti, C., Berisha, V., & Seo, J.-s. (2017). Minimizing area and energy of deep learning hardware design using collective low precision and structured compression. In *2017 51st Asilomar conference on signals, systems, and computers* (pp. 1907–1911). IEEE.
100. Kingsbury, N. G., & Rayner, P. J. (1971). Digital filtering using logarithmic arithmetic. *Electronics Letters, 2*(7), 56–58.
101. Swartzlander, E. E., & Alexopoulos, A. G. (1975). The sign/logarithm number system. *IEEE Transactions on Computers, 100*(12), 1238–1242.
102. Vuillemin, J. (1988). Exact real computer arithmetic with continued fractions. In *Proceedings of the 1988 ACM conference on LISP and functional programming* (pp. 14–27). ACM.
103. Kornerup, P., & Matula, D. W. (1985). Finite precision lexicographic continued fraction number systems. In *1985 IEEE 7th symposium on computer arithmetic (ARITH)* (pp. 207–213). IEEE.
104. Kornerup, & Matula. (1983). Finite precision rational arithmetic: An arithmetic unit. *IEEE Transactions on Computers, 100*(4), 378–388.
105. Carmichael, Z., Langroudi, H. F., Khazanov, C., Lillie, J., Gustafson, J. L., & Kudithipudi, D. (2019). Performance-efficiency trade-off of low-precision numerical formats in deep neural networks. In *Proceedings of the conference for next generation arithmetic 2019* (pp. 1–9). ACM.
106. Zhang, Z. (2018). Improved Adam optimizer for deep neural networks. In *2018 IEEE/ACM 26th international symposium on quality of service (IWQoS)* (pp. 1–2). IEEE.

107. Keskar, N. S., & Socher, R. (2017). Improving generalization performance by switching from adam to SGD. *arXiv preprint arXiv:1712.07628*.
108. Mikolov, T., Chen, K., Corrado, G., & Dean, J. (2013). Efficient estimation of word representations in vector space. *arXiv preprint arXiv:1301.3781*.
109. Jugade, C., Ingole, D., Sonawane, D. N., Kvasnica, M., & Gustafson, J. (2022). A memory efficient FPGA implementation of offset-free explicit model predictive controller. *IEEE Transactions on Control Systems Technology, 30*(6), 2646–2657.
110. Nambi, S., Ullah, S., Sahoo, S. S., Lohana, A., Merchant, F., & Kumar, A. (2021). ExPAN(N)D: Exploring posits for efficient artificial neural network design in FPGA-based systems. *IEEE Access, 9*, 103691–103708.

Index

A
Advantage Actor Critic (A2C), 442–443, 450, 453, 457–462
AI-enabled cybersecurity, 343–382
AI-Enabled Static Timing Analysis (STA), 109–151
AI Hardware Accelerators, 499–501, 510–524
Analog-Mixed Signal Systems (AMS), 44
Application-Specific Integrated Circuits (ASICs), 2, 3, 6, 104, 114, 117, 226, 228, 250, 312, 420, 422, 504, 505, 509, 512, 519–521, 523, 526, 532, 533
Artificial intelligence (AI), 2, 50, 84, 226, 233, 257, 310, 343, 388, 429, 497, 537, 568
Associative learning, 537–562

B
Behavioral malware detection, 349, 428, 429

C
Convolutional neural networks (CNNs), 170, 173, 195, 196, 200, 282, 310, 350, 356–357, 359, 376, 404, 415, 503, 523

D
Data knowledge, 84, 89, 92, 93, 103
Data mining, 89, 101, 429
Deep learning models, 104, 191, 215, 360, 376, 377, 379, 388, 405, 498, 527, 530, 532
Deep neural networks (DNNs), 15, 86, 114, 117, 375–377, 393–394, 419, 498, 504, 521, 522, 537, 565, 568, 569, 571, 572, 574, 584, 599
Digital design flow, 3, 109–150
Digital Twin, 42, 50–56, 74, 78
Direct memory access (DMA), 388, 394, 406, 409–411, 419, 511–514
Dueling Deep Q-Networks (DQN), 251, 441, 442, 454, 457–461, 502, 503

E
Electronic design automation (EDA), 1, 2, 6, 7, 11, 16, 19, 25, 33, 42, 56, 64, 109, 116–119, 128, 134, 140, 149, 156, 157, 206, 209

F
Field-programmable gate arrays (FPGAs), 6, 46, 114, 117, 119, 225–252, 266, 312, 328–329, 336, 403, 412–414, 416, 420–422, 456, 483, 488, 497, 505–508, 510, 526, 532, 533, 540, 574, 599
Fuzzing, 479–484, 491

A. Iranmanesh, H. Sayadi (eds.), *AI-Enabled Electronic Circuit and System Design*,
https://doi.org/10.1007/978-3-031-71436-8

G

Graphics processing units (GPUs), 68, 167, 203, 228, 266, 483, 497–500, 506–509, 520, 521, 526, 527, 532, 533, 539, 569, 589
Graph learning, 155–185
Graph neural networks (GNN), 23, 111, 157, 204, 234, 325, 470, 497

H

Hardware accelerator, 42, 62, 497–533, 583
Hardware-assisted security, 427–462
Hardware Performance Counters (HPCs), 348, 367–370, 372–379, 381, 429, 431–433, 445–447, 449, 453–455, 462
Hardware Trojan Detection, 327–328, 476–478, 484, 486

I

In-Memory Computing, 500, 506
Interconnect timing, 157, 167, 170, 172–176, 181, 184, 185

L

Logic locking, 469, 470, 472–475, 479

M

Machine learning (ML)
  in cybersecurity, 386
  in hardware security, 403
  models, 14, 83, 132, 163, 191, 233, 265, 309, 350, 394, 436, 575
Machine learning-based malware detection, 382, 446
Malware detection techniques, 347–349, 365, 366, 427–430
Malware Feature Extraction, 349, 370
Memory, 14, 62, 83, 205, 255, 317, 350, 387, 429, 480, 499, 539, 565
Model extraction attacks, 419
Multi-armed bandits (MAB), 439–441

N

Neural networks (NNs), 8, 42, 86, 111, 156, 191, 233, 269, 309, 350, 389, 442, 471, 539, 566
Neuromorphic computing, 14, 302, 505, 510, 516, 526, 529, 532, 539–541, 560

P

Performance estimation, 25, 27, 28, 30, 31, 238
Physical unclonable functions (PUFs), 330–336, 394, 401, 486–491
Pin-to-pin delay estimation, 139, 140
Pin-to-pin delay prediction, 140
Placement, 5, 46, 110, 156, 189, 230, 267, 325, 351
Power efficiency in AI processing, 94
Power-efficient, 83, 87
Power, performance, and area (PPA), 109, 212, 222
Power, performance, and area (PPA) optimization, 212
Processing-in-memory (PIM), 506, 507, 510, 518, 519

R

RC network, 156, 159–161
Register transfer level (RTL), 3, 41, 109, 229, 230
Reinforcement learning (RL), 7, 116, 192, 241, 274, 349, 389, 427, 476, 502
Reverse engineering, 310, 323–330, 336, 337, 412, 416, 474, 476, 479, 484, 485, 491
Reverse engineering of ICs and FPGAs, 323
Robotics, 1, 9, 271, 273, 283, 284, 295, 296, 298–299, 437, 444, 503, 504, 507, 531, 537–562
RTL generation, 109–110, 112, 120, 127, 128, 145, 146, 148–150

S

Side-Channel Analysis (SCA), 309–311, 315, 317, 318, 320–323, 336, 467, 475
Signoff, 155–185
Spiking neural networks (SNN), 504, 505, 539
Static and dynamic malware detection, 348
Static Timing Analysis (STA), 110, 112, 124, 125
SysMLv2, 42–48, 50, 53, 69, 71, 77, 78
System modeling language (SysML), 41, 43

T

Temporal Difference Learning (Q-Learning), 441
Tensor Processing Units (TPU), 504, 505, 507, 520–521, 525, 527, 589, 590
Through silicon via (TSV), 208, 258–260, 262, 263, 269, 274, 276, 278, 281
Timing path analysis, 112

Transfer learning in cybersecurity, 367
Trojan, 323, 327, 328, 345, 368, 369, 445, 468, 475–479, 484–486
Trustworthy hardware design, 469, 474
TSV defects, 275, 279

V
Variational analysis, 12, 16
Videos, 83, 227, 504, 566
VLSI, 1, 8, 9, 16, 30–32, 189–190, 192, 193, 197–200, 202–204, 214, 220, 230, 469

Z
Zero-day malware detection, 343–382, 449, 455, 460

MIX
Papier aus verantwortungsvollen Quellen
Paper from responsible sources
FSC® C105338

If you have any concerns about our products,
you can contact us on
ProductSafety@springernature.com

In case Publisher is established outside the EU,
the EU authorized representative is:
Springer Nature Customer Service Center GmbH
Europaplatz 3, 69115 Heidelberg, Germany

Printed by Libri Plureos GmbH
in Hamburg, Germany